Jose R. Pérez-Castiñeira
Chemistry and Biochemistry of Food

Also of Interest

Food Analysis.
Using Ion Chromatography
Edward Muntean, 2022
ISBN 978-3-11-064438-8, e-ISBN (PDF) 978-3-11-064440-1

Essential Oils.
Sources, Production and Applications
Edited by: Rajendra Chandra Padalia,Dakeshwar Kumar Verma, Charu
Arora and Pramod Kumar Mahish, 2023
ISBN 978-3-11-079159-4, e-ISBN (PDF) 978-3-11-079160-0

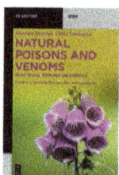

Natural Poisons and Venoms.
Plant Toxins: Terpenes and Steroids
Eberhard Teuscher and Ulrike Lindequist, 2023
ISBN 978-3-11-072472-1, e-ISBN (PDF) 978-3-11-072473-8

Chemical Technicians.
Good Laboratory Practice and Laboratory Information Management Systems
Mohamed Elzagheid, 2023
ISBN 978-3-11-119110-2, e-ISBN (PDF) 978-3-11-119149-2

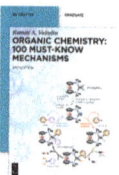

Organic Chemistry: 100 Must-Know Mechanisms
Roman Valiulin, 2023
ISBN 978-3-11-078682-8, e-ISBN (PDF) 978-3-11-078683-5

Jose R. Pérez-Castiñeira

Chemistry and Biochemistry of Food

2nd, Revised and Extended Edition

DE GRUYTER

Author
Prof. Dr. Jose R. Pérez-Castiñeira
Instituto de Bioquímica Vegetal y Fotosíntesis
Universidad de Sevilla
CSIC
Avda. Américo Vespucio 49
41092 Sevilla
Spain
jroman@us.es

ISBN 978-3-11-110834-6
e-ISBN (PDF) 978-3-11-111187-2
e-ISBN (EPUB) 978-3-11-111330-2

Library of Congress Control Number: 2023947623

Bibliographic information published by the Deutsche Nationalbibliothek
The Deutsche Nationalbibliothek lists this publication in the Deutsche Nationalbibliografie;
detailed bibliographic data are available on the Internet at http://dnb.dnb.de.

A Isabel, Manuel y Santiago
Gracias por todo. . .

Preface

Three years after the publication of the first edition of *Chemistry and Biochemistry of Food*, this second edition comes to light. I have updated some of the contents and introduced new ones, as well as checked figures and tables and corrected some minor mistakes. The new contents are basically related to biochemical and physiological aspects of foods and nutrition. There is a reason for this: although I graduated in Chemistry many years ago, Biochemistry (that is, the chemical processes that occur within living beings) always exerted a special fascination on me. For this reason I decided to dedicate my scientific career to this scientific discipline. Today, I can say that I made the right decision, as I have thoroughly enjoyed Biochemistry for nearly 40 years. Moreover, the continuous advances in the field make it even more fascinating with each passing day.

Several things have changed in my professional life since 2020; thus, after more than 20 years teaching the subject "Chemistry and biochemistry of food" at the Faculty of Chemistry of the University of Sevilla, I decided to allow younger professors to take over the wonderful task of introducing the molecular bases of nutrition to the students. I have also changed my research line and, after several decades studying the molecular aspects of proton translocation across biological membranes, I decided to join a research group involved in the study of the evolution of photoperiodic signaling and how it affects plant development. This has been a real challenge that has allowed me to enter the amazing world of intrinsically disordered proteins (IDPs). In any case, my interest in every aspect of food (bio)chemistry is as strong as ever.

Obviously, a number of people must be thanked for their contribution to this second edition of *Chemistry and Biochemistry of Food*: first of all, my wife, Isabel, and my two sons, Manuel and Santiago, who have supported and encouraged me throughout the whole process of elaborating both editions of the book. This has been extremely important to me and I deeply appreciate it.

I want to reiterate my acknowledgement to the colleagues that have contributed again to the book by writing several chapters: Dr. José María Vega Piqueres, Emeritus Professor of Biochemistry at the University of Sevilla; Dr. Javier Vigara Fernández, Associate Professor of Biochemistry at the University of Huelva; Dr. María Montaña Durán-Barrantes, Associate Professor of Chemical Engineering at the University of Sevilla; and Dr. Victoria Valls-Bellés, Associate Professor of Physiology at the University Jaume I of Castellón de la Plana.

I also thank again Dr. José Manuel Martínez Rivas, researcher at the Instituto de la Grasa of the Spanish Research Council (CSIC); Professor José María Ortega, from the Department of Plant Biochemistry and Molecular Biology of the University of Sevilla (USE); as well as Professors Inmaculada Robina and Carmen Ortiz Mellet, from the Department of Organic Chemistry, also of the USE, for their corrections and comments to the first versions of Chapters 4 (Lipids), 2 (Water), and 3 (Carbohydrates), respectively.

https://doi.org/10.1515/9783111111872-202

My dear friends and colleagues Drs. Aurelio Serrano and Federico Valverde have indirectly contributed to this book by being always ready to discuss new and fresh ideas about every aspect of science (and much more...). Thanks to the members of the Department of Plant Biochemistry and Molecular Biology at the USE and the Instituto de Bioquímica Vegetal y Fotosíntesis (IBVF). The USE and the Spanish Research Council (CSIC) have provided online access to an exhaustive list of scientific journals and books; this has definitely facilitated my work and is gratefully acknowledged.

This book is also dedicated to the students who have attended my lectures during all these years and to those who have had their Final Years' Projects, Master and/or Doctoral Theses supervised by me. I have always thought that you learn while you teach.

I will always be indebted to Professors José M. Vega Piqueres, Ernesto Carmona, David K. Apps, Ramón Serrano, and Manuel Losada for their contribution to my scientific and personal formation.

Thanks a lot to De Gruyter for their interest in continuing with the publication of the book, and specially to my editor, Jessika Kischke, for being such a pleasure to work with her.

Finally, I want to express my gratitude to the rest of my family: my parents (the best in the world), Delia, Sean, and Chris.

Sevilla, Alcaidesa, Isla Cristina

Contents

1 Food, nutrition, and diet

1.1 Basic concepts

1.1.1 Feeding, nutrition, food, and nutrients

Living organisms are open systems in thermodynamics terms, that is, they need to interchange matter and energy with the surrounding environment. In heterotrophic organisms like humans, this task is accomplished by taking up certain molecules from the exterior, using them as substrates for energy-releasing chemical reactions, utilizing this energy to perform the so-called biological functions, and releasing the final products of those reactions into the environment.

The molecules we obtain from outside of our bodies are termed **nutrients**, and they are usually the components of complex chemical systems known as **foods**. **Nutrition** is the process by which our bodies are able to obtain and assimilate the nutrients contained in a given food. Let us take milk as an example: this is the basic food for newborn mammals; however, milk is not a pure substance, but it is composed of many different nutrients that are essential to survive during the first months of our lives. Nutrients include proteins, carbohydrates, lipids, minerals, and vitamins; however, other non-nutritional chemical substances usually occur along with nutrients in foods. Some of these substances are contaminants, that is, they appear in the food accidentally and may pose a threat to the consumer (Chapter 11), but others are real components and some of them have been reported to interfere with the assimilation of nutrients. The latter are known as antinutrients and include chemicals such as lectins, oxalates, goitrogens, phytoestrogens, phytates, and tannins [1]. Some of these compounds will be discussed in subsequent chapters.

Feeding (or eating) is the process of finding food, introducing it into our mouths, masticating it (if necessary), and swallowing it; therefore, it is a voluntary, conscious, and educable action. Nutrition implies a myriad of chemobiological mechanisms integrated in the gastrointestinal tract and controlled in a nonvoluntary fashion by the autonomic nervous system.

1.1.2 Diet and dietetics

Diet can be defined as the type and range of food that a person or an animal regularly eats [2]. With very few exceptions (like table sugar, composed almost exclusively of sucrose), the different types of food contain more than one nutrient; however, no food contains all the necessary nutrients in the required concentrations to sustain a healthy life. **Dietetics** is a branch of science that applies the current knowledge of physiology, biochemistry, epidemiology, and other subjects to the study of nutrition.

https://doi.org/10.1515/9783111111872-001

The main objective of dietetics is the optimization of the human diet, which is not an easy task, as each individual may have his/her own particularities. Qualified dietitians perform a thorough study of their patients in order to prescribe an optimal diet for them considering genetic, metabolic, and even social and personal aspects.

1.1.3 Biological roles of nutrients

Nutrients play three major functions within our bodies:
- **Energetic**, by supplying the energy necessary to support all biological processes.
- **Structural**, by providing the basic units of the large molecules (biomolecules) that shape our bodies.
- **Regulatory**, by participating in the control of the chemical reactions that sustain life.

Nutrients can be classified into five basic types according to their chemical properties and/or biological roles: carbohydrates, lipids, proteins, vitamins, and minerals. A sixth nutrient with special properties has to be added to this list: water. Not all nutrients perform the same functions and some of them may play different ones (Table 1.1).

Table 1.1: Biological functions of nutrients.

Nutrient	Function		
	Energetic	Structural	Regulatory
Carbohydrates	+	+	+
Lipids	+	+	+
Proteins	+	+	+
Vitamins	–	–	+
Minerals	–	+	+
Water	–	+	+

Carbohydrates (Chapter 3) are generally regarded as energetic nutrients; however, they play a wide array of biological roles. Thus, they are an important part of the extracellular matrix. They are involved in cellular recognition and can also attach to polypeptides, thereby regulating the biological activity of the resulting proteins. No wonder that the study of the important biological functions of the carbohydrates, glycobiology, is a very active field in natural sciences nowadays [3].

Lipids (Chapter 4) are regarded by many as the biomolecules with the highest content in energy per unit of mass; in fact, our bodies use some members of this family of compounds as an efficient storage and source of energy. However, lipids are a heterogeneous group of molecules that can play other nonenergetic biological roles.

Thus, phospholipids are essential to build the cellular membranes, whereas hormones like testosterone or progesterone are also lipids.

Proteins (Chapter 5) are rightly considered as "the molecules of life" with permission of the nucleic acids (whose role in nutrition is a matter of some debate, as we shall see in Chapter 5). Proteins perform virtually every biological function; however, strictly speaking, the basic units that form proteins, **amino acids**, are the actual nutrients we need. There are only 21 proteinogenic amino acids, 9 of them being considered essential nutrients for healthy humans; however, other amino acids may be regarded as "conditionally essential," because the body cannot synthesize them in sufficient quantities under certain circumstances [4].

Vitamins (Chapter 6) are a heterogeneous family of organic molecules whose functions are mainly regulatory. They are also essential nutrients that can be classified according to their solubility in water. In many cases, the absence of a given vitamin in our diet may produce a specific disease or disorder (hypovitaminosis). Nowadays, due to the availability of dietary supplements, some disorders associated with an excess of some vitamins in the diet (hypervitaminosis) have also been reported.

Minerals (Chapter 7) are inorganic substances that usually appear on the Earth's surface as salts, although there are exceptions like selenium that can be incorporated by humans as selenocystein or selenomethionine, two amino acids. We usually obtain minerals from other living organisms, although common salt and drinking water are also sources of minerals. These nutrients play a wide variety of functions like bone formation (calcium and phosphate), oxygen transport in the blood (iron), or modulation of enzymatic activities (calcium, magnesium, zinc, etc.). Minerals are also essential nutrients.

Water (Chapter 2) is a very special nutrient because although about 70% of human body weight is water we do not directly obtain energy from this molecule. Water participates directly or indirectly in the vast majority of biochemical reactions because its physicochemical properties are of paramount importance in the chemistry of life. Moreover, the three-dimensional structures of complex biological systems like cellular membranes and proteins are also determined by the interactions that biomolecules such as phospholipids, cholesterol (Chapter 4), and polypeptides (Chapter 5) have with water molecules.

1.2 Factors that influence diet

Dietary choices are influenced by a large number of interconnected factors. Although a profound analysis of these factors is beyond the scope of this book, a nonexhaustive list of some of them with brief comments of each is shown below:

– *Availability of food.* The major factor that influences an individual's diet is the range of foods available. For our ancestors, this basically depended on the geography and the climate, as primitive societies presumably collected the foods from their surroundings. The expansion of humans also implied expansion of food (crops and animals), that is, food globalization is not a modern trend but has been going on for tens of thousands of years [5]. During the last decades, this process has intensified and, along with the scientific and technological advances (see below), it has widened the range of available foods, thereby changing dietary choices around the world. In developing countries, feeding a growing human population is still a major issue of concern; however, the availability of food significantly increased worldwide during the second half of the twentieth century [6].

– *Tradition, culture, and religion.* This factor is related to the previous one because, as stated above, human communities tend to eat what they have at hand. One example is the so-called Mediterranean diet, based on olive oil, grains, vegetables, and fish, that is, the foods available in the Middle East for millennia [7]. This diet is also deeply related to religious tradition [7] and has been subjected to important changes due to the contributions introduced along the history by the different empires that have dwelt on the Mediterranean shores (Romans, Arabs, Spanish, Ottomans, among others) [8]. Religions have also played an important role by forbidding some foods like pork for Jews and Muslims.

– *Technological advances.* Humans are among the most successful species on the biosphere: we have been able to adapt to virtually all sorts of ecosystems. One of the main reasons for this success is the development of technologies like agriculture, livestock farming, fishing, or cooking that has allowed us to obtain nutrients from many different sources. Advances in transportation have also been a major issue along the history, as illustrated by the changes in European diet that occurred after the discoveries by the Portuguese and Spaniards since the beginning of the fifteenth century [9].

– *Socioeconomic status.* Regardless of the availability of food, it is also clear that the socioeconomic status is a major determinant of dietary choice. This has been thoroughly studied in developed countries and, although the social mechanisms by which this actually happens are complex, people of lower socioeconomic status tend to have energy-dense diets that have been associated with higher disease risk and mortality rates [10].

– *Advertising.* In the European Union, food and drink industry is the main manufacturing industry with a turnover of 1,098 billion euros in 2015, employing more than 4 million workers [11]. This means that food industry is a major actor in worldwide economy and also that competition is fierce not only within Europe but also between European companies and those from countries like the USA and China and emerging economies like Brazil, India, or South Africa. As a consequence, food companies try to influence consumer behavior in order to increase their profits. To this end, they spend millions of euros in advertising and lobbying, a topic that has been studied in detail elsewhere [12].

– *Social and cultural factors*. The perception of the ideal human body has evolved along our history [13]; however, during the second half of the twentieth century, the new trend of the "thin ideal" has been imposed for females, whereas a muscular body is expected for males in order to be "attractive," overweight being associated with negative features [14]. These trends, profusely broadcast by mass media, may be in conflict with the instincts and biological mechanisms developed during our history, when food shortage was probably the norm [13]. In current developed societies, where food is widely accessible, many people, influenced by trends, think of themselves as being overweight and attempt to lose weight on some form of diet not usually prescribed by qualified professionals. These are known as "fad diets" [15]. In Europe and the USA, the diet industry is estimated to have an annual turnover in excess of $150 billion. The problems associated with dieting are very complex, and there are strong evidences suggesting that dieting may be promoting exactly the opposite of what it is intended to achieve [16]. Other popular terms nowadays are "functional foods" and "superfoods," which refer to foods and food products that are supposed to provide health benefits; however, this must be taken with caution because of the difficulty to obtain scientific evidence supporting these claims [17].

1.3 Nutritional requirements and basic recommendations

Since the discovery that some diseases are linked to a lack or shortage of certain nutrients, like scurvy and vitamin C (Chapter 3), there has been an interest by scientists and authorities to propose guidelines for healthy diets [18]. During the last decades, an enormous quantity of information has been supplied by scientists, and nowadays there are hundreds of scientific journals that regularly publish a myriad of articles about different aspects of food and nutrition. As a consequence, many public and private institutions are dedicated to evaluate and update this information and provide technical advice to health authorities worldwide. In the European Union, the European Food Safety Agency (EFSA) provides scientific advice and communication on many issues associated with food and nutrition. It was funded in 2002 following a series of food crises in the late 1990s. The EFSA operates independently of the European legislative and executive institutions (Commission, Council, Parliament), as well as member states, and produces scientific opinions and advice that form the basis for European policies and legislation [19]. The remit of EFSA covers:
– Food and feed safety
– Nutrition
– Animal health and welfare
– Plant protection
– Plant health
– Impact of the food chain on the biodiversity of plant and animal habitats

The available information about foods and nutrition is usually too technical for the standard consumer and, sometimes, even for health professionals. To solve this problem, health authorities publish the so-called food-based dietary guidelines, aimed at helping both professionals and the public to have a healthy lifestyle. According to the Food and Agriculture Organization of the United Nations (FAO),

> food-based dietary guidelines (also known as dietary guidelines) are intended to establish a basis for public food and nutrition, health and agricultural policies and nutrition education programmes to foster healthy eating habits and lifestyles. They provide advice on foods, food groups and dietary patterns to provide the required nutrients to the general public to promote overall health and prevent chronic diseases [20].

Dietary guidelines elaborated by the authorities of many countries are available at the FAO web page (http://www.fao.org/nutrition/education/food-dietary-guidelines/en/). A significant example is the *Dietary Guidelines for Americans*, jointly published every 5 years by the US Department of Agriculture and Department of Health and Human Services. A summary that includes five major guidelines for the general public can be found at the beginning of this document. These guidelines, along with a few key recommendations, are the bases of a healthy diet not only for Americans but also for people worldwide [21]:

The Guidelines:

1. Follow a healthy eating pattern across the life span. All food and beverage choices matter. Choose a healthy eating pattern at an appropriate calorie level to help achieve and maintain a healthy body weight, support nutrient adequacy, and reduce the risk of chronic disease.
2. Focus on variety, nutrient density, and amount. To meet nutrient needs within calorie limits, choose a variety of nutrient-dense foods across and within all food groups in recommended amounts.
3. Limit calories from added sugars and saturated fats and reduce sodium intake. Consume an eating pattern low in added sugars, saturated fats, and sodium. Cut back on foods and beverages higher in these components to amounts that fit within healthy eating patterns.
4. Shift to healthier food and beverage choices. Choose nutrient-dense foods and beverages across and within all food groups in place of less healthy choices. Consider cultural and personal preferences to make these shifts easier to accomplish and maintain.
5. Support healthy eating patterns for all. Everyone has a role in helping to create and support healthy eating patterns in multiple settings nationwide, from home to school to work to communities.

Several "Key Recommendations" provide further guidance on how individuals can follow these five guidelines by describing what a "healthy eating pattern" is [22]:

Consume a healthy eating pattern that accounts for all foods and beverages within an appropriate calorie level.

A healthy eating pattern includes:
- A variety of vegetables from all of the subgroups – dark green, red and orange, legumes (beans and peas), starchy, and other
- Fruits, especially whole fruits
- Grains, at least half of which are whole grains
- Fat-free or low-fat dairy, including milk, yogurt, cheese, and/or fortified soy beverages
- A variety of protein foods, including seafood, lean meats and poultry, eggs, legumes (beans and peas), and nuts, seeds, and soy products
- Oils

A healthy eating pattern limits:
- Saturated fats and trans fats (Chapter 4), added sugars, and sodium
- Consume less than 10% of calories per day from added sugars
- Consume less than 10% of calories per day from saturated fats
- Consume less than 2,300 mg per day of sodium
- If alcohol is consumed (Appendix 4), it should be done in moderation – up to one drink per day for women and up to two drinks per day for men – and only by adults of legal drinking age

In tandem with the recommendations above, Americans of all ages – children, adolescents, adults, and older adults – should meet the Physical Activity Guidelines for Americans[1] to help promote health and reduce the risk of chronic disease.

In an effort to make dietary guidelines and recommendations easily available to the general public, many health authorities publish visual representations such as pyramids and plates that inform on the recommended relative contributions of different food groups to the diet [21, 23]. Some examples are shown in Figure 1.1.

Recently, healthy diets have been defined by a panel of several international institutions (including FAO and WHO) as follows:

They
1. start early in life with early initiation of breastfeeding, exclusive breastfeeding until six months of age, and continued breastfeeding until two years of age and beyond combined with appropriate complementary feeding;
2. are based on a great variety of unprocessed or minimally processed foods, balanced across food groups, while restricting highly processed food and drink products;
3. include wholegrains, legumes, nuts and an abundance and variety of fruits and vegetables;
4. can include moderate amounts of eggs, dairy, poultry and fish, and small amounts of red meat;

---

1 Available in [22].

Spain   **Belgium**

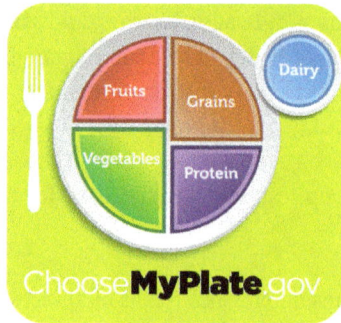

USA

Figure 1.1: Visual representations of official food-based dietary guidelines from Spain, Belgium (Flemish), and the USA. These guidelines and others from different countries are available at https://www.fao.org/nutrition/education/food-dietary-guidelines/en/.

5. include safe and clean drinking water as the fluid of choice;
6. are adequate (i.e. reaching but not exceeding needs) in energy and nutrients for growth and development and meet the needs for an active and healthy life across the life cycle;
7. are consistent with WHO guidelines for reducing the risk of diet-related NCDs and ensuring health and well-being for the general population; and
8. contain minimal levels of, or if possible no, pathogens, toxins or other agents that can cause food-borne disease. According to WHO the characteristics of a healthy diet include the following: less than 30% of total energy intake comes from fats, with a shift in fat consumption away from saturated fats towards unsaturated fats and the elimination of industrial trans-fats; less than 10% of total energy intake comes from free sugars (preferably less than 5%); at least 400 g of fruits and vegetables are consumed per day; and not more than 5 g of salt (which should be iodized) is consumed per day.

Very importantly, it is recognized that healthy diets may vary according to individuals [24].

1.4 Sustainability

Food production exerts a considerable impact of the planet's resources, while many people are not adequately nourished. Data on the environmental impact on current food (especially crop and livestock) production and consumption are overwhelming [23]:
– It contributes to 20–30% of anthropogenic greenhouse gas emissions.
– It is the leading cause of deforestation, land use change, and biodiversity loss.
– It accounts for 70% of all human water use, being a major source of water pollution.
– Unsustainable fishing practices deplete stocks of species we consume and also cause wider disruption to the marine environment.

The impacts of climatic and environmental change are already making food production more difficult and unpredictable in many regions of the world. In order to address the multiple social, health, and environmental challenges caused by food systems, it is necessary to move toward diets that are both healthy and respectful of environmental limits.

In September 2015, Heads of States and High Representatives of all members of the United Nations announced the 2030 Agenda for Sustainable Development, whose main objective is "eradicating poverty in all its forms and dimensions." The Agenda established 17 Sustainable Development Goals (SDGs) and 169 associated targets. The 2nd SDG is "End hunger, achieve food security and improved nutrition and promote sustainable agriculture," whereas the 13th one is "Take urgent action to combat climate change and its impacts" [25]. Therefore, providing the world population with healthy diets and sustainable food systems is not only a major challenge but also a global objective. In line with the SDGs, Willet and colleagues from the so-called EAT-Lancet Commission have made a proposal for healthy diets and sustainable food production that can be globally applicable. The universal healthy reference diet proposed by these authors "largely consists of vegetables, fruits, whole grains, legumes, nuts, and unsaturated oils, includes a low to moderate amount of seafood and poultry, and includes no or a low quantity of red meat, processed meat, added sugar, refined grains, and starchy vegetables" [26].

1.5 Personalized dietary patterns

Nowadays there is a general consensus about the association of overweight and obesity with conditions and diseases like type II diabetes, cardiovascular diseases, and certain types of cancer (Chapter 14). Although obesity is considered a complex pathology that results from the interaction of environmental and genetic factors, dietary habits play a significant role in the increasing incidence of this disease; however, interventions aimed at altering these habits in order to improve public health have had limited impact so far. Some authors argue that personalization of interventions may be more effective in changing eating behaviors than general guidelines and recommendations. Personalized nutrition uses information on individual characteristics to develop targeted nutritional advice that assists people to achieve a dietary change that can result in health benefits. The concept of "precision nutrition" further suggests that this kind of interventions can be achieved by understanding the complex relationships among an individual, his/her food consumption, and his/her phenotype (including health) [27]. Some relatively new scientific disciplines are involved in the development of personalized/precision nutrition. These include:

- Nutrigenetics, the study of the effects of genetic variation on dietary response [28]
- Nutrigenomics, the study on the evolutionary aspects of diet and the role of nutrients in gene expression [28]
- Exposome science, concerned with the collection of environmental factors (stress, physical activity, and diet) to which an individual is exposed and which may affect health [27]

These disciplines are based on a series of high-throughput technologies, generically known as "omics," that allow the identification, characterization, and quantification of all biomolecules that are involved in the structure, function, and/or dynamics of a given biological system [29]. In the case of nutrition, the main "omics" involved are:

- Epigenomics, the study of the whole set of epigenetic changes that modify the expression and function of the genetic material of an organism (Chapter 14).
- Metabolomics, the large-scale study of small molecules, commonly known as metabolites, within cells, tissues, or organisms. Collectively, these small molecules (within a mass range of 50–1,500 Da) and their interactions within a biological system are known as the "metabolome" [30].
- Microbiomics, the study of the microbiome, the totality of microbes in specific environments, such as the human gut [27].

According to Ordovas and colleagues [27], there are two key questions with respect to personalized nutrition:

- Can this approach produce greater, more appropriate, and sustained changes in behavior than conventional approaches?
- Do these changes result in better health and well-being?

Some limited information suggests that the answer to the first of these questions is yes, but the second question remains unanswered, because no study has been carried out at a large scale, in an appropriate population group and over a sufficiently long time. These authors reached the conclusion that "much research and regulation is required before personalized nutrition can deliver the expected benefits" [27].

An interesting summary of some of the concepts and views described in this chapter can be found in [31].

References

[1]   Petroski W, Minich DM. Is there such a thing as "Anti-Nutrients"? A narrative review of perceived problematic plant compounds. Nutrients 2020;12:2929. https://doi.org/10.3390/nu12102929

[2]   Diet definition and meaning | Collins English Dictionary n.d. https://www.collinsdictionary.com/dictionary/english/diet (accessed March 31, 2018).

[3]   Varki A, Sharon N. Historical background and overview. In: Varki A, Cummings RD, Esko JD, Freeze HH, Stanley P, Bertozzi CR, et al., editors. Cold Spring. 2nd ed. Harbor (NY): Cold Spring Harbor Laboratory Press; 2009.

[4]   Lopez MJ, Mohiuddin SS. Biochemistry, essential amino acids. StatPearls, Treasure Island (FL): StatPearls Publishing; 2023.

[5]   Henn BM, Cavalli-Sforza LL, Feldman MW. The great human expansion. Proc Natl Acad Sci 2012;109:17758–64. https://doi.org/10.1073/pnas.1212380109

[6]   The State of Food and Agriculture 2000 n.d. http://www.fao.org/docrep/x4400e/x4400e09.htm (accessed March 31, 2018).

[7]   Berry EM, Arnoni Y, Aviram M. The Middle Eastern and biblical origins of the Mediterranean diet. Public Health Nutr 2011;14:2288–95. https://doi.org/10.1017/S1368980011002539

[8]   Altomare R, Cacciabaudo F, Damiano G, Palumbo V, Gioviale M, Bellavia M, et al. The Mediterranean diet: A history of health. Iran J Public Health 2013;42:449–57.

[9]   Nunn N, Qian N. The Columbian exchange: A history of disease, food, and ideas. J Econ Perspect 2010;24:163–88. https://doi.org/10.1257/jep.24.2.163

[10] Darmon N, Drewnowski A. Does social class predict diet quality? Am J Clin Nutr 2008;87:1107–17. https://doi.org/10.1093/ajcn/87.5.1107

[11] Data and Trends Report 2017 n.d. http://www.fooddrinkeurope.eu/uploads/publications_documents/DataandTrends_Report_2017.pdf (accessed March 31, 2018).

[12] Nestle M. Food politics: How the food industry influences nutrition and health. University of California Press; 2013.

[13] Eknoyan GA History of obesity, or how what was good became ugly and then bad. Adv Chronic Kidney Dis 2006;13:421–7. https://doi.org/10.1053/j.ackd.2006.07.002

[14] Grogan S. Body image: Understanding body dissatisfaction in men, women and children. Taylor & Francis; 2016.

[15] Tahreem A, Rakha A, Rabail R, Nazir A, Socol CT, Maerescu CM, et al. Fad diets: Facts and fiction. Front Nutr 2022;9:960922. https://doi.org/10.3389/fnut.2022.960922

[16] Dulloo AG, Montani J-P. Pathways from dieting to weight regain, to obesity and to the metabolic syndrome: An overview. Obes Rev 2015;16:1–6. https://doi.org/10.1111/obr.12250

[17] Galanakis CM, Aldawoud TMS, Rizou M, Rowan NJ, Ibrahim SA. Food ingredients and active compounds against the coronavirus disease (COVID-19) pandemic: A comprehensive review. Foods 2020;9:1701. https://doi.org/10.3390/foods9111701

[18] Mozaffarian D, Ludwig DS. Dietary guidelines in the 21st century – a time for food. JAMA 2010;304:681–2. https://doi.org/10.1001/jama.2010.1116

[19] European Food Safety Authority | Trusted science for safe food. Eur Food Saf Auth n.d. http://www.efsa.europa.eu/en (accessed December 22, 2019).

[20] Home. Food Agric Organ U N n.d. http://www.fao.org/nutrition/education/food-dietary-guidelines/home/en/ (accessed December 22, 2019).

[21] US Department of Health and Human Services. Dietary guidelines for Americans 2015–2020. Skyhorse Publishing Inc.; 2017.

[22] Current Guidelines – health.gov n.d. https://health.gov/paguidelines/second-edition/ (accessed December 22, 2019).

[23] Fischer CG, Garnett T, Food and agriculture organization of the United Nations, University of Oxford, food climate research network. Plates, pyramids, and planets: developments in national healthy and sustainable dietary guidelines : a state of play assessment. 2016.

[24] Contribution of terrestrial animal source food to healthy diets for improved nutrition and health outcomes. FAO; 2023. https://doi.org/10.4060/cc3912en

[25] Agenda for Sustainable Development. https://sdgs.un.org/sites/default/files/publications/21252030%20Agenda%20for%20Sustainable%20Development%20web.pdf (accessed May 18, 2023).

[26] Willett W, Rockström J, Loken B, Springmann M, Lang T, Vermeulen S, et al. Food in the anthropocene: The EAT–Lancet Commission on healthy diets from sustainable food systems. The Lancet 2019;393:447–92. https://doi.org/10.1016/S0140-6736(18)31788-4

[27] Ordovas JM, Ferguson LR, Tai ES, Mathers JC. Personalised nutrition and health. BMJ 2018:bmj.k2173. https://doi.org/10.1136/bmj.k2173

[28] Simopoulos AP. Nutrigenetics/Nutrigenomics. Annu Rev Public Health 2010;31:53–68. https://doi.org/10.1146/annurev.publhealth.031809.130844

[29] Vailati-Riboni M, Palombo V, Loor JJ. What are omics sciences? In: Ametaj BN, editor. Periparturient Dis. Dairy Cows Syst. Biol. Approach. Cham: Springer International Publishing; 2017, p. 1–7. https://doi.org/10.1007/978-3-319-43033-1_1

[30] What is metabolomics? EMBL-EBI Train Online 2013. https://www.ebi.ac.uk/training/online/course/introduction-metabolomics/what-metabolomics (accessed December 23, 2019).

[31] Li KJ, Burton-Pimentel KJ, Brouwer-Brolsma EM, Vergères G, Feskens EJM. How can new personalized nutrition tools improve health? Front Young Minds 2022;10:738922. https://doi.org/10.3389/frym.2022.738922

2 Water

2.1 Introduction

In the summer of the year 2018, subglacial liquid water was reported in Mars [1]. This finding prompted a worldwide expectation about the possibility of the existence of life in our neighboring planet because, instinctively and rightly, we humans link water in its liquid form to life.

Water is a stable molecule from the chemical point of view under the conditions prevailing on the Earth; hence, its occurrence is not only in the atmosphere and the crust but also in the mantle [2]. Liquid water presents a large number of physico-chemical properties that make it essential for what we call "life": it acts as a building material, solvent, reaction medium, reactant, carrier for nutrients and waste products, thermoregulator, lubricant, and shock absorber [3]. Consequently, the chemical reactions of living organisms are directly or indirectly determined by water even when it is not explicitly mentioned.

One of the most important roles of water in the biosphere is as an electron donor in oxygenic photosynthesis, a biochemical process devised by the ancestors of cyanobacteria. When these organisms developed the capacity to use water to capture the energy from the Sun and transform it into biologically useful chemical energy, they changed our planet's history from that point on. In oxygenic photosynthesis, water molecules are split using the energy from sunlight and molecular oxygen (O_2) is released as a by-product. This was a spectacular breakthrough in Earth's history that led to the so-called great oxidation event, which very likely implied the annihilation of many organisms; however, the availability of oxygen triggered the evolution of aerobic respiration and novel biosynthetic pathways, facilitating complex multicellularity and the biodiversity we observe today [4]. Oxygenic photosynthesis (and, thus, water) is considered the largest source of O_2 in the atmosphere [5].

Another point worth of consideration is that although water is known as "the universal solvent," it does not dissolve many molecules. Liquid water has a very stable dynamic three-dimensional structure in which the hydrogen bond plays an essential role [6]. Some molecules (hydrophilic or "water-loving") integrate into this structure establishing an energetically stable system, whereas others (hydrophobic or "water-hating") disrupt the water structure and, consequently, they are "rejected," that is, not dissolved. The most important consequence for life of this property of liquid water is that many biomolecules adapt their structures to the constraints imposed by the surrounding water molecules in order to fit within them, thereby yielding stable biomolecular complexes. This is how water, a "simple" chemical substance, fundamentally determines the structure (and thus the function) of proteins, biological membranes, or chromatin, composed by thousands of atoms [6, 7].

https://doi.org/10.1515/9783111111872-002

2.2 Water as a nutrient

Water is the most abundant chemical component of the body, comprising 50–60% of an adult body weight with a range from 45% to 75% according to some sources and from 30% to 70% according to others [8, 9]. Among the factors that can alter the ratio of total body water to weight are sex, age, level of fatness/leanness, climate, salt intake, and certain diseases [8]. Fatness is one of the major factors that determine the proportion of water in the body, actually, the percentage of water content with respect to the fat free mass is less variable than with respect to total body weight, this value being around 73% [10]. Water is not evenly distributed within our body: in adults, about two-thirds of total water is in the intracellular space, whereas one-third is extracellular [3]. The volume and composition of the intracellular and extracellular fluids (ECFs) must be maintained within certain limits compatible with life and this is accomplished by a set of complex homeostatic physiological and biochemical processes.

Water is a very special nutrient: although it is very abundant in our body it is often ignored in reports of dietary surveys and nutrition, the attention being more focused on the minerals and other substances dissolved in it. However, we need water and cannot synthesize it by ourselves (at least, in the quantities we need); therefore, it is an essential nutrient. Then, why has it been regarded as "neglected, unappreciated, and under researched" [9]? Probably, this is due to a number of peculiarities of this molecule, among of which are: (1) the way our metabolism utilizes water differs from other nutrients; (2) it is an "inorganic" molecule, composed of one atom of oxygen and two atoms of hydrogen, that is, it does not contain carbon (C), the element considered to be characteristic of life; and (3) it does not require digestion to be incorporated into our bodies.

2.3 Water balance in the body: inputs, outputs, and regulation

Intake and loss of water must be more or less equal over a 24-h period. Water intake is normally composed of beverages, food, and metabolically produced water (see below); conversely, water is continually lost from the body through respiration, urine formation, insensible loss (evaporation from lungs and skin), sweat, and feces. Direct measurements of total amount of water gained and lost are not possible, which means that significant variations of inputs and outputs of water are observed in different studies, as illustrated in Table 2.1, which show data taken from two different sources [3, 11]. The assessment of water balance is further complicated by the fact that these data can be subjected to important changes due to factors such as physical activity, climate, and individual physiology [9].

At the molecular level, water is transported across biological membranes by three different mechanisms: (1) by diffusion through the lipid matrix; (2) through some membrane-embedded channels or solute transporters; and (3) through specific

Table 2.1: Water balance in sedentary adults living in temperate climate.

Water inputs (mL/day)			Water outputs (mL/day)		
Beverages	1,575[a]	>1,000[b]	Respiration and sweat	750[a]	900[b]
Foods	675[a]	1,000[b]	Feces	100[a]	200[b]
Metabolic water	300[a]	400[b]	Urine	1,600[a]	600–2,400[b]
Total	2,550[a]	2,400[b]	Total	2,550[a]	1,700–3,500

Data taken from references [3][a] and [11][b].

water channels known as aquaporins. A common feature to these mechanisms is that water movement is driven by gradients of its own chemical potential [12, 13]. The chemical potential of water in our bodies is essentially determined by the concentration of solutes or electrolytes in the biological fluids, usually measured as the number of moles of osmotically active species (osmoles) per kilogram of water ("osmolality") or per liter of solution ("osmolarity"). Osmolarity and osmolality values do not differ much in dilute solutions; therefore, they are often used interchangeably [3]. The most important contributors to osmolarity in the body fluids are electrolytes like sodium (Na^+), chloride (Cl^-), potassium (K^+), calcium (Ca^{2+}), magnesium (Mg^{2+}), sulfate (SO_4^{2-}), or phosphate (PO_4^{3-}) and solutes such as glucose or urea [14, 15].

Total body water is precisely maintained under moderate conditions (within 0.2% of total weight over a 24-h period) by a complex system involving the hypothalamus, the neurohypophysis, and the kidneys. Essentially, it is a feedback mechanism in which some membrane proteins (osmoreceptors) located on the hypothalamus sense increases in plasma osmolality, thus triggering secretion of vasopressin (also known as antidiuretic hormone) to the bloodstream and thirst. Vasopressin increases reabsorption of water in the kidneys, a process that, along with drinking, should restore osmolarity to normal levels [3, 16]. By means of this physiological process, osmolarity of plasma is tightly controlled and rarely varies by more than 2% around a set point of 0.280–0.290 "osmoles/L," which increases somewhat with age [17]. In clinical studies, plasma osmolality can be directly measured (by the degree of freezing point depression, for example) or predicted by using equations that consider plasma concentrations of sodium, potassium, glucose, and urea [14].

Not only maintaining the overall water balance in the body is essential for survival but its distribution between the different compartments within living organisms must also be carefully regulated, as water determines the absorption and distribution of all nutrients [11]. Cells spend a significant amount of energy in active transport of solutes and electrolytes across their membranes in order to maintain an appropriate composition inside and outside, thereby adjusting the distribution of water and, consequently, their own volume [16, 18] (see Chapter 13 for more details on transport of water and electrolytes).

2.4 Water requirements

Human water requirements depend on a number of factors such as age, metabolism, climate, physical activity, or diet; however, due to the precise mechanisms of water balance control, normal hydration is compatible with a wide range of fluid intake in healthy adults. Inputs are composed of three major sources: drinking water and beverages (70–80% of total intake), food (20–30%), and some water produced from the oxidation of macronutrients (<2%). These percentages are not fixed and depend on the type of beverages and food we ingest, which may contain variable contents of water: from 85% to >90% in the former and from 40% to >80% in the latter [3].

Normal hydration status is "the presumed condition of healthy individuals who maintain water balance," according to the study of the European Food Safety Authority published in 2010 about dietary reference values for total water intakes [17]. In this report, a number of values for adequate intake (AI) of water were recommended:
– 100–190 mL/kg per day during the first half of the first year of life (based on water intake from human milk in exclusively breast-fed infants)
– 800–1,000 mL/day for the age period 6–12 months
– 1,100–1,200 mL/day during the second year of life (value is defined by interpolation, as intake data are not available)
– 1,300 mL/day for boys and girls 2–3 years of age
– 1,600 mL/day for boys and girls 4–8 years of age
– 2,100 mL/day for boys 9–13 years of age; 1,900 mL/day for girls 9–13 years of age
– Adolescents of 14 years and older are considered as adults with respect to adequate water intake and the adult values apply
– Adults: females would have to be 2.0 L/day and for males 2.5 L/day; same AIs are recommended for the elderly as for adults, because both renal concentrating capacity and thirst are decreasing with age
– No data on habitual water intake in pregnant women were found and the same water intake as in nonpregnant women plus an increase in proportion to the increase in energy intake (300 mL/day) was proposed
– For lactating women, 700 mL/day above the AIs of nonlactating women of the same age

The EFSA Panel also included several important considerations concerning these values:
– Recommended AIs apply only under conditions of moderate environmental temperature and physical activity levels.
– Water losses incurred under extreme conditions of external temperature and physical exercise, which can be up to about 8,000 mL/day, have to be replaced with appropriate amounts. In such instances, concomitant losses of electrolytes have to be replaced adequately to avoid hypo-osmolar disturbances.

- Too high intake of water which cannot be compensated by the excretion of very dilute urine (maximum urine volumes of about 1 L/h in adults) can lead to hyponatremic, hypo-osmolar water intoxication with cerebral edema.
- Maximum daily amount of water that can be tolerated has not been defined.
- The reference values for total water intake include water from drinking water, beverages of all kinds, and from food moisture.

The EFSA Report has certain limitations due to the uncertainties in the methods used to assess water intake from food and beverages in different European countries. This means that a consistent and validated methodology to reduce potential sources of errors in the elaboration of these reports is very much needed [19].

The process of losing body water that may eventually lead to a deficit is called dehydration, which can be classified as isotonic, hypertonic, or hypotonic:

- Isotonic dehydration occurs when both water and solutes are lost from the ECF, for example, through vomiting, diarrhea, or inadequate intake.
- Hypertonic dehydration occurs when water loss exceeds salt loss, for example, through inadequate water intake, excessive sweating, osmotic diuresis, or diuretic drugs.
- Hypotonic dehydration occurs when more sodium than water is lost, for example, in some instances of high sweat or gastrointestinal fluid losses or when fluid and electrolyte deficits are treated with water replacement only.

Increasing dehydration has been linked to reductions in exercise performance, thermoregulation, and appetite (fluid losses of more than 1%); severe performance decrements, difficulties in concentration, headaches, irritability, and sleepiness as well as increase in body temperature and in respiratory rates (fluid deficits of 4%), and, ultimately, death (fluid deficits of 8%). Dehydration has also been linked to impairment of cognitive performance and motor function, such as fatigue, mood, target shooting, discrimination, choice reaction time, visual-motor tracking, short-term and long-term memory, and attention. Populations at particular risk of dehydration include the very young and the elderly; therefore, assessment of the hydration status is of paramount importance for health. Among the techniques that can be used to accomplish this goal are: measurement of body weight, tracer techniques, bioelectrical impedance, determination of plasma osmolarity, and measures of some urine indices [3, 17].

2.5 Water resources and supply

Water is not only necessary for direct human consumption but also for food production, including agriculture, livestock breeding, and processing, activities that require water of varying quality. The guidelines for drinking water quality published by the World Health Organization [20] state that "water is essential to sustain life, and a satisfactory (adequate, safe and accessible) supply must be available to all [. . .]. Safe drinking-water [. . .] does not represent any significant risk to health over a lifetime of consumption, including different sensitivities that may occur between life stages."

Water supply is one of the major challenges for humankind because the number of people who do not have access to water without any potential health hazards is estimated to be around 3.5–4 billion, nearly five times the current United Nations estimate [21]. This will continue to be a major problem in the near future, as the world's population is projected to increase by slightly more than 1 billion people over the next 13 years, reaching 8.6 billion in 2030, and to increase further to 9.8 billion in 2050 and 11.2 billion by 2100 according to the United Nations [22]. Actually the so-called water crisis is listed among the most important risks by the World Economic Forum in its last report issued on 2018 [23].

One of the most important bottlenecks for safe water supply is the shortage of freshwater sources in our planet. Although more than 70% of Earth's surface is covered by water, about 97% is saline and is not easily usable for human use and consumption, only around 2.5% is fresh water. A mere 1.2% of the latter is on the surface, the rest being held in glaciers and ice caps (65.7%), as well as in groundwater (30%). To complicate matters, only a low percentage of surface water is in lakes (21%) or rivers (0.5%), the rest being as permafrost (69%; Table 2.2) [24]. Moreover, global distribution is not the complete picture because freshwater sources are unevenly distributed and sometimes they are not close to populated or agricultural areas; therefore, sustainable water technologies must be developed and made available worldwide to ensure a proper supply for most people [25, 26].

Table 2.2: Global use of rainwater.

Blue water (groundwater, rivers, and lakes)		Green water (available for uptake by plants)	
Total rainwater 110,000 km³/year			
Total	40,000 km³	**Total**	70,000 km³
Withdrawn for human usage	4,000 km³		
Agriculture (irrigation, livestock breeding and washing, aquaculture)	2,760 km³*	Agriculture (rain-fed crops)	6,400 km³
Industry (including food and beverages industry)	760 km³*	Evaporated (from forests, uncultivated vegetation, and wetlands)	Rest
Domestic (including water for drinking and cooking)	480 km³*		

Elaborated with data taken from references [24, 29–31].
*Average global values. A significant variability occurs worldwide. See AQUASTAT website for further details.

In socioeconomic and environmental terms, water resources can be classified as renewable and nonrenewable. Groundwater and surface water are considered renewable whereas deep aquifers, which do not have a significant replenishment rate on the human time scale, are nonrenewable water resources [27]. Unfortunately, not all surface water or groundwater is accessible for use; actually, exploitable water resources are significantly smaller than the natural resources and depend on factors such as the capability of storing floodwater behind dams or extracting groundwater, among others [28].

2.6 The importance of water in food production

Water is not only an essential nutrient for all living organisms but it is also crucial in many aspects of food production and processing: irrigation, livestock watering, aquaculture (initial production) or as an ingredient, transport medium, and hygiene aid (processing). Both water and food may be vehicles for the transmission of infectious agents or toxic chemicals and people can be exposed to these agents in three different ways:
- Through contaminated water incorporated into foods
- Through foods irrigated with or harvested from contaminated water
- Through foods that have come into contact with contaminated water during processing

Therefore, a proper supply of quality water is vital to food security and production [32]. In the European Union, the quality of the water used in food manufacturing must meet the basic standards governing the quality of drinking water [33] amended in 2003, 2009, and 2015. In December 2020, the EU adopted a recast of the Drinking Water Directive that entered into force in January 2021 (*Official Journal of the European Union*, L 435, December 23, 2020, p. 0001; available at https://eur-lex.europa.eu/eli/dir/2020/2184/oj). This Directive updates water quality standards, tackling pollutants of concern, such as endocrine disruptors and microplastics, and applies to all water, either in its original state or after treatment, intended for drinking, cooking, food preparation, or other domestic purposes in both public and private premises.

Atmospheric precipitation on the continents and islands is the origin of the renewable water resources. It has been estimated that an average of 110,000 km^3 of rain falls over the continents annually. More than 70,000 km^3 infiltrates the soil and is available for uptake by plants. This fraction is known as "green water." Around 40,000 km^3 ("blue water") flows in groundwater and surface water (rivers and lakes). Green water can be either productive, if taken up by crops or natural vegetation, or nonproductive, if evaporated from soil and open water; actually, only 6,400 km^3 of rainwater is estimated to fall on cultivated areas, the rest being evaporated from forest, natural (uncultivated) vegetation, and wetlands. As far as blue water is concerned, only 10% of it is withdrawn for human use (around 4,000 km^3): agriculture uses an average 69%, including irrigation, livestock breeding, and washing and aquaculture; 20% is for industrial uses, which includes the food and beverage industry, and the rest (around 12%) is for domestic use, including drinking and cooking water (Table 2.2).

Rain-fed agriculture, which depends on "green water," covers 80% of the world's cultivated land, and is responsible for about 60% of crop production, whereas irrigated agriculture (that uses "blue water" although sometimes relies on green water in addition [27]) covers about 20% of cultivated land and accounts for 40% of global food production.

The food processing and manufacturing industry uses water for a number of processes such as transport, steam generation, cooling, and cleaning and also as an ingredient [34]. This means an additional 5–10% of freshwater consumption with respect to freshwater withdrawal for irrigation in a country like the USA or 30% of all the water used for manufacturing in Australia [35]. In the European Union, the food and beverage industry is the largest manufacturing sector in terms of turnover, value-added, and employment, and it has been reported to be responsible for 1.8% of Europe's total water use [36]. However, water supply is not the only problem, it is also very important to understand how demand, sources of pollution, water reuse, and contamination of food through water affect food safety. Pressures on water demand are expected to increase in the next decades creating a vicious cycle: increasing exploitation of water resources leads to more water pollution that further reduces the availability of clean water. Sustainable alternatives need to be implemented in all areas of human activity, but especially during food production and processing [26, 32].

Another point of interest is the possibility to recycle water for subsequent uses. This is feasible in water for industrial and domestic uses, where most of the water returns to rivers after use; however, in agriculture, a large part of water is released in the atmosphere by evapotranspiration because this is the physiological mechanism by which plants take up nutrients from the soil [29].

Two very important concepts, "water footprint" and "virtual water," have been introduced to correctly assess the problems facing water supply in the next future:
- The water footprint is "a measure of human's appropriation of freshwater resources [. . .] measured in terms of water volumes consumed (evaporated or incorporated into a product) or polluted per unit of time" [37]. The interesting feature of this concept is that it is calculated considering not only blue and green water consumption but also the wastewater production ("gray water").
- The "virtual water" content of an agricultural or industrial product is the overall water used in the production process of that product. Virtual water allows the calculation of "hidden water" trade flows between countries; for example, a water-scarce country imports products that require a lot of water in their production and export products or services that require less water. This implies net import of virtual water.

The water footprint of a certain good is calculated considering its virtual water content [37, 38]. Table 2.3 shows the water footprint of a number of food products.

Table 2.3: Water footprint of some selected food products from vegetable and animal origin.

Product	Water footprint (m^3/ton or L/kg)
Sugar crops	197
Vegetables	322
Starchy roots	387
Fruits	962
Cereals	1,644
Oil crops	2,364
Legumes	4,055
Nuts	9,063
Milk	1,020
Eggs	3,265
Chicken meat	4,325
Butter	5,553
Pig meat	5,988
Sheep/goat meat	8,763
Beef	15,415

Source: [39].

2.7 The importance of water in the properties of foods

The water content of a food plays a major role in its properties; however, rather than the total moisture content, the parameter "water activity" (a_w) has been shown to correlate more closely with microbial, chemical, and physical properties of foods. Water activity is a thermodynamic concept, intrinsically linked to the concept of (water) chemical potential, and is defined as the ratio of the fugacity of water (f_w) in a system and the fugacity ($f°_w$) of pure liquid water at a given temperature. Fugacity is a measure of the escaping tendency of a particular component within a system and can be used to calculate the deviation from the ideality in a gas phase. It can be replaced by the total or partial vapor pressure in the equilibrium with an error less than 0.1% in many systems under normal conditions. Therefore, a working definition of a_w may be the ratio of the equilibrium partial vapor pressure of water in the system (p_w) to the equilibrium partial vapor pressure ($p°_w$) of pure liquid water at the same temperature and total pressure [40]. Although there may be inherent problems with applying the concept of a_w, which is an equilibrium thermodynamic concept, it may help to understand what happens in systems like foods, that are not usually in strict equilibrium [41].

Water activity is considered one of the most important parameters in food preservation and processing. William J. Scott was the first to show in 1953 that there was a significant correlation between a_w and microbial growth and toxin production in foods. Subsequent studies during the 1960s and 1970s shed information about this parameter and its influence on the chemical, enzymatic, and physical stability of foods that affected the characteristics (color, flavor, texture, etc.) and acceptability of raw and processed food products [42]. Consequently, several food preservation techniques rely on controlling (normally decreasing) a_w in order to reduce the rates of microbial growth and chemical reactions by using food stability maps as a function of water activity [41].

Water activity is related to the moisture content of a food in complex way: an increase in a_w is usually accompanied by an increase in water content in a nonlinear fashion. The relationship between water activity and moisture content at a given temperature is called the moisture (or water) sorption isotherm, which illustrates the steady-state amount of water held by the food solids as a function of a_w (or percentage of relative humidity) at constant temperature. Water vapor sorption by foods depends on many factors, including chemical composition, physical–chemical state of ingredients, and physical structure. This makes sorption isotherms valuable tools for food scientists because they can be used to predict which reactions will decrease stability at a given moisture or which ingredient can be used to change the a_w in order to increase the stability. They also help to predict moisture gain or loss in a package with known moisture permeability [41].

Glass transition is also an important water-related concept in food technology that has been reported to play an important role in the properties of dehydrated and frozen foods. It is a relaxation process occurring in food solids that can be defined as a transformation of a supercooled liquid to a highly viscous, solid-like glass. Unlike water activity, a property of water molecules, glass transition is a property of amorphous food components (such as carbohydrates and proteins) which is affected by the extent of water plasticization (softening) of the solids [43].

Glass transition and water activity are two completely different parameters, but they are complementary and both can be used to explain food deterioration or stability. The a_w is an important parameter affecting microbial growth, whereas glass transition is a property related to changes in food structure and microstructure, crystallization, rates of diffusion-controlled reactions, and, possibly, stabilization of microbial cells and spores [43].

A more profound analysis on the physicochemical properties of water and its role in the chemistry of foods can be found elsewhere [44].

References

[1]   Orosei R, Lauro SE, Pettinelli E, Cicchetti A, Coradini M, Cosciotti B, et al. Radar evidence of subglacial liquid water on Mars. Science 2018. https://doi.org/10.1126/science.aar7268.
[2]   Bertaux JL, Carr, M., Des Marais, D.J., Gaidos, E. Conversations on the habitability of worlds: The importance of volatiles. Space Sci Rev 2007;129:123–65. https://doi.org/10.1007/s11214-007-9193-3.

[3]   Jéquier E, Constant F. Water as an essential nutrient: The physiological basis of hydration. Eur J Clin Nutr 2010;64:115–23. https://doi.org/10.1038/ejcn.2009.111.

[4]   Fischer WW, Hemp J, Johnson JE. Evolution of oxygenic photosynthesis. Annu Rev Earth Planet Sci 2016;44:647–83. https://doi.org/10.1146/annurev-earth-060313-054810.

[5]   Hamilton TL, Bryant DA, Macalady JL. The role of biology in planetary evolution: Cyanobacterial primary production in low-oxygen Proterozoic oceans. Environ Microbiol 2016;18:325–40. https://doi.org/10.1111/1462-2920.13118.

[6]   Brini E, Fennell CJ, Fernandez-Serra M, Hribar-Lee B, Lukšič M, Dill KA. How water's properties are encoded in its molecular structure and energies. Chem Rev 2017;117:12385–414. https://doi.org/10.1021/acs.chemrev.7b00259.

[7]   Sharp KA. Water: Structure and properties. Encycl Life Sci., vol. 10, John Wiley & Sons; 2001. https://doi.org/10.1038/npg.els.

[8]   Chumlea WC, Guo SS, Zeller CM, Reo NV, Siervogel RM. Total body water data for white adults 18 to 64 years of age: The Fels Longitudinal Study. Kidney Int 1999;56:244–52. https://doi.org/10.1046/j.1523-1755.1999.00532.x.

[9]   Rush EC. Water: Neglected, unappreciated and under researched. Eur J Clin Nutr 2013;67:492–5. https://doi.org/10.1038/ejcn.2013.11.

[10]   Wang Z, Deurenberg P, Wang W, Pietrobelli A, Baumgartner RN, Heymsfield SB. Hydration of fat-free body mass: Review and critique of a classic body-composition constant. Am J Clin Nutr 1999;69:833–41. https://doi.org/10.1093/ajcn/69.5.833.

[11]   Bahn A. Water, electrolytes, and acid-base balance. Essent Hum Nutr. 5th Ed. J. Mann & A. S. Truswell, Oxford University Press; 2017, p. 110–30.

[12]   Disalvo EA, Pinto OA, Martini MF, Bouchet AM, Hollmann A, Frías MA. Functional role of water in membranes updated: A tribute to Träuble. Biochim Biophys Acta BBA – Biomembr 2015;1848:1552–62. https://doi.org/10.1016/j.bbamem.2015.03.031.

[13]   Madeira A, Moura TF, Soveral G. Detecting aquaporin function and regulation. Front Chem 2016;4:3. https://doi.org/10.3389/fchem.2016.00003.

[14]   Hooper L, Abdelhamid A, Ali A, Bunn DK, Jennings A, John WG, et al. Diagnostic accuracy of calculated serum osmolarity to predict dehydration in older people: Adding value to pathology laboratory reports. BMJ Open 2015;5. https://doi.org/10.1136/bmjopen-2015-008846.

[15]   Rundgren A, Svensen CH. Fluid balance, regulatory mechanisms and electrolytes. Fluid Ther Surg Patient, CRC Press; 2018, p. 1–28.

[16]   Knepper MA, Kwon T-H, Nielsen S. Molecular physiology of water balance. N Engl J Med 2015;373:196. https://doi.org/10.1056/NEJMc1505505.

[17]   EFSA. Scientific Opinion on dietary reference values for water. EFSA J 2010;8. https://doi.org/10.2903/j.efsa.2010.1459.

[18]   Danziger J, Zeidel ML. Osmotic homeostasis. Clin J Am Soc Nephrol CJASN 2015;10:852–62. https://doi.org/10.2215/CJN.10741013.

[19]   Gandy J, Bellego LL, König J, Piekarz A, Tavoularis G, Tennant DR. Recording of fluid, beverage and water intakes at the population level in Europe. Br J Nutr 2016;116:677–82. https://doi.org/10.1017/S0007114516002336.

[20]   World Health Organization. Guidelines for drinking-water quality. 4th ed. Geneva: World Health Organization; 2011.

[21]   Tortajada C, Biswas AK. Water as a human right. Int J Water Resour Dev 2017;33:509–11. https://doi.org/10.1080/07900627.2017.1321237.

[22]   United Nations, Department of Economic and Social Affairs, Population Division. World Population Prospects: The 2017 Revision, Key Findings and Advance Tables. vol. Working Paper No. ESA/P/WP/248. 2017.

[23]   World Economic Forum. Global risks 2018. 13th ed. Geneva: World Economic Forum; 2018.

[24] Shiklomanov IA, Rodda JC, editors. World water resources at the beginning of the twenty-first century. Cambridge, UK ; New York: Cambridge University Press; 2003.

[25] UNESCO. Water for people – water for life. The United Nations World Water Development Report. UNESCO and Berghahn Books. Paris: 2003.

[26] WWAP (United Nations World Water Assessment Programme)/UN-Water. The United Nations World Water Development Report. 2018. Nature-Based Solutions for Water. Paris: UNESCO; 2018.

[27] Mancosu N, Snyder LR, Kyriakakis G, Spano D. Water scarcity and future challenges for food production. Water 2015;7. https://doi.org/10.3390/w7030975.

[28] FAO F. Agriculture organisation of the United Nations, 2003: Review of world water resources by country. Water Rep 2003;23.

[29] United Nations WWAP (United, UN-Water. Water in a Changing World. vol. 1. Earthscan; 2009.

[30] AQUASTAT database n.d. http://www.fao.org/nr/water/aquastat/data/query/index.html?lang=en (accessed December 24, 2019).

[31] SIWI I, IUCN I. Let it reign: The new water paradigm for global food security. Final Rep CSD-13 Stockholm International Water Institute 2005.

[32] Kirby RM, Bartram J, Carr R. Water in food production and processing: Quantity and quality concerns. Food Control 2003;14:283–99. https://doi.org/10.1016/S0956-7135(02)00090-7.

[33] EC. Council Directive 98/83/EC of 3 November 1998 on the quality of water intended for human consumption; Off J Eur Communities 1998;L 330: 32–54.

[34] Badui S. Química De Los Alimentos. Pearson Educación de México, SA de CV; 2006.

[35] Meneses YE, Stratton J, Flores RA. Water reconditioning and reuse in the food processing industry: Current situation and challenges. Trends Food Sci Technol 2017;61:72–9. https://doi.org/10.1016/j.tifs.2016.12.008.

[36] Valta K, Kosanovic T, Malamis D, Moustakas K, Loizidou M. Water consumption and wastewater generation and treatment in the Food and Beverage Industry. Desalination Water Treat 2013;53:12.

[37] Mekonnen M, Hoekstra AY. National water footprint accounts: the green, blue and grey water footprint of production and consumption. Delft, The Netherlands: UNESCO-IHE Institute for Water Education; 2011.

[38] Mekonnen M, Hoekstra A. The green, blue and grey water footprint of farm animals and animal products. 2010.

[39] Mekonnen MM, Hoekstra AY. A global assessment of the water footprint of farm animal products. Ecosystems 2012;15:401–15. https://doi.org/10.1007/s10021-011-9517-8.

[40] Reid DS. Water activity: Fundamentals and relationships. Water Act Foods Fundam Appl, Blackwell Publishing, Oxford; 2007; p. 15–28.

[41] Labuza TP, Altunakar B. Water activity prediction and moisture sorption isotherms. Water Act Foods Fundam Appl 2007;109:154.

[42] Chirife J, Fontana Jr AJ. Introduction: Historical highlights of water activity research. Water Act Foods Fundam Appl 2007:3–13.

[43] Roos YH. Water activity and glass transition. Water Act Foods Fundam Appl, Blackwell Publishing, Ames, IA; 2007; p. 29–45.

[44] Fennema's food chemistry. 5th Ed. CRC Press; 2017. https://doi.org/10.1201/9781315372914.

3 Carbohydrates

3.1 Definition, terminology, and classification

Carbohydrates are a family of biomolecules composed, in principle, of carbon (C), hydrogen (H), and oxygen (O) atoms whose basic general formula is $C_n(H_2O)_m$. The fact that many carbohydrates have two atoms of hydrogen per atom of oxygen in their molecules is responsible for the somewhat misleading name that these molecules have, as "hydrate" means "containing water." Carbohydrates have no water in their chemical composition, as we shall see in this chapter, although they are usually hydrated to different degrees both in vivo and in vitro.

Carbohydrates are usually constituted by an indeterminate number of basic units linked forming polymers. The connections among these units may occur by means of different linkage types, thereby allowing many structural variations [1].

There are several important terms related to carbohydrates:

- **Monosaccharides**: They are the basic nonhydrolyzable units of carbohydrates. Monosaccharides can be chemically altered yielding derivatives that may also form polymers. These alterations may involve the addition of elements such as nitrogen (N), sulfur (S), or phosphorus (P).
- **Oligosaccharides**. Molecules composed of 2–12 (20 for some authors) linked units (residues) of monosaccharides by a specific type of chemical bond known as **glycosidic bond**. Oligosaccharides are denoted according to the number of monosaccharide residues they have: **disaccharides** (two units), **trisaccharides** (three units), **tetrasaccharides** (four units), and so on.
- **Sugar:** Many monosaccharides and disaccharides are sweet, hence their trivial name **sugars**, although table sugar is only composed of sucrose, a disaccharide. Carbohydrates are also known as **glycids** (from the Greek *glykys, glykeros*: sweet) or **saccharides** (from the Latin *saccharum*: sugar).
- **Polysaccharides**: Polymers composed of more than 12 (or 20) residues of monosaccharides. Polysaccharides are named as **homo-** or **heteropolysaccharides** depending on whether they are composed of a single type or more than one type of monosaccharide, respectively. Oligo- and polysaccharides can be linear or branched and they may undergo further chemical modifications (see below).
- **Glycan:** Although this term is considered a synonym of polysaccharide by the IUPAC [2], biochemists use it for those oligo- or polysaccharides that bind to other biomolecules.
- **Glycoconjugates:** Biochemical compounds consisting of a glycan moiety linked to proteins or lipids. The process by which a glycan is attached to another biomolecule is known as glycosylation when it is catalyzed by an enzyme, or glycation when it is just a chemical nonmediated process. There are several types of glycoconjugates:

https://doi.org/10.1515/9783111111872-003

- **Glycoproteins/glycopeptides**: Proteins/peptides (Chapter 5) containing oligosaccharide chains covalently attached to amino acid side chains.
- **Peptidoglycans**: Generally composed of linear heteropolysaccharides covalently linked with short peptides.
- **Glycolipids**: Lipids (usually 1,2-diacylglycerols, see Chapter 4) with a monosaccharide or oligosaccharide linked.
- **Lipopolysaccharides**: Complex glycolipids that occur in the outer membrane of Gram-negative bacteria. They induce immune response in many species, including humans.

The large number of different basic units (monosaccharides and derivatives) that can be linked by several types of chemical bonds along with the possibility to form glycoconjugates allow carbohydrates to constitute or participate in an ample variety of biological compounds. Some members of this family, such as the extracellular polysaccharides of plants, fungi, and arthropods (cellulose and chitin), are the most abundant biomolecules on the Earth. Glycoconjugates are also very common: more than 50% of all human proteins are estimated to be glycosylated. Therefore, carbohydrates are of paramount importance for life, both qualitatively and quantitatively, and they are involved in a wide array of biological functions [3]:

- Energy supply
- Formation of biological structures
- Morphogenesis
- Protein folding
- Transcriptional regulation
- Information exchange between cells
- Cellular recognition

The structural and functional diversity of carbohydrates has led to the development of a relatively new branch of biochemistry and molecular biology, glycobiology, which can be defined as "the study of the structure, biosynthesis, biology and evolution of saccharides [. . .] that are widely distributed in nature and of the proteins that recognize them" [4]. Glycobiology aims at integrating traditional carbohydrate chemistry and biochemistry with the current knowledge of molecular and cellular biology. Other interesting terms are glycomics, the study of the glycome, that is, all the glycans that a cell or tissue produces under specified conditions; and glycoproteomics, whose goal is the identification and quantitation of the glycans that interact with proteins. The last two terms describe those aspects of glycobiology that require a systems-level analysis to be appropriately studied [5].

This chapter focuses mainly on the importance of the energetic function of carbohydrates and the composition and biological roles of the members of this family most commonly involved in human nutrition.

3.2 Monosaccharides and oligosaccharides

3.2.1 Monosaccharides

Monosaccharides are the only members of the carbohydrates that respond to the molecular formula $(CH_2O)_n$, that is, in the general formula shown at the beginning of this chapter $[C_n(H_2O)_m]$, they are the particular case in which "n" is equal to "m." Chemically, monosaccharides are polyhydroxy aldehydes (**aldoses**) or polyhydroxy ketones (**ketoses**), typically consisting of a chain of hydroxymethylene (–CHOH) units (with the only exception of dihydroxyacetone), which terminates at one end with a hydroxymethyl group (–CH$_2$OH) and at the other with an aldehyde group (–CHO, aldoses) or an α-hydroxy ketone group (HOCH$_2$–CO–, ketoses) [6]. Their chemical formulae are often depicted in the two-dimensional Fischer projection (Figures 3.1–3.3), in which the carbon atoms are aligned vertically and numbered from top to bottom following the rules of organic chemistry nomenclature (the carbon of the carbonyl group is C-1 in aldoses and C-2 in ketoses). This projection was originally proposed by Emil Fischer in the late nineteenth century for monosaccharides and, although problematic for many organic molecules, it remains acceptable nowadays for carbohydrates and their derivatives [7].

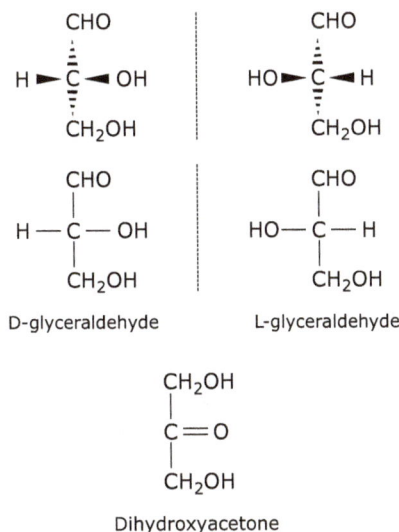

Figure 3.1: Chemical structures of glyceraldehyde and dihydroxyacetone.

Glyceraldehyde is the simplest aldose, and dihydroxyacetone is the simplest ketose (Figure 3.1). Addition of successive hydroxymethylene (–CHOH) units to these molecules after the carbonyl group can be used to formulate the aldose and ketose families, respectively (Figure 3.2). From the biological point of view, the most prominent monosaccharides are those containing between three and seven carbon atoms, but, in

(A)

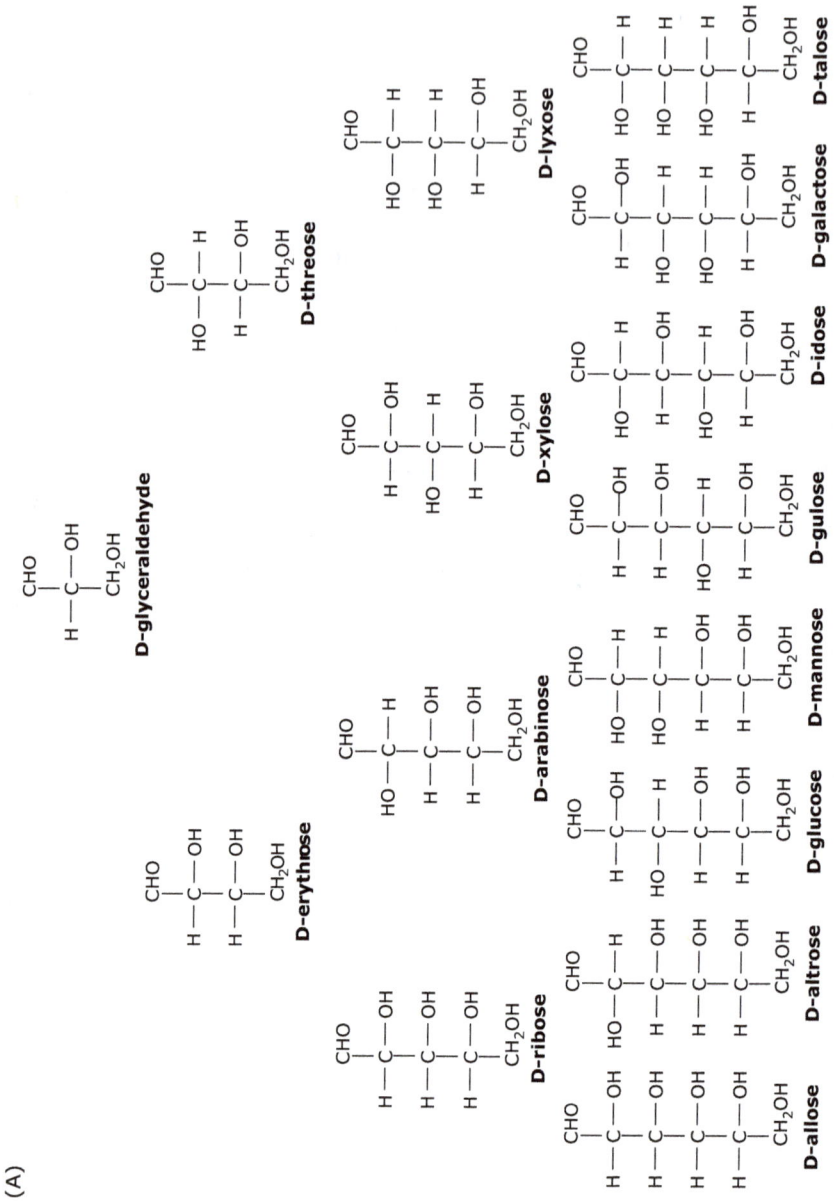

Figure 3.2: The D-aldose (A) and D-ketose (B) families of monosaccharides.

(B)

CH₂OH
|
C=O
|
CH₂OH

Dihydroxyacetone

CH₂OH
|
C=O
|
H — C — OH
|
CH₂OH

D-erythrulose

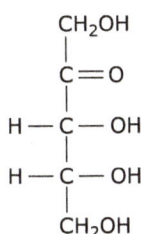

D-ribulose

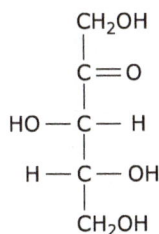

D-xylulose

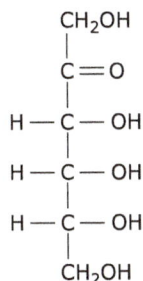

D-psicose

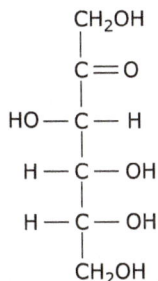

D-fructose

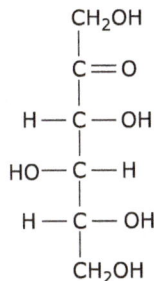

D-sorbose

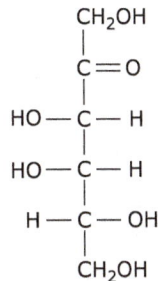

D-tagatose

Figure 3.2 (continued)

nutritional terms, the most important and abundant are those that have six [(CH₂O)₆ or $C_6H_{12}O_6$], the so-called hexoses (Table 3.1). Hexoses containing an aldehyde group are known as aldohexoses and those containing an α-hydroxy ketone group, keto-hexoses or hexuloses.

With the only exception of dihydroxyacetone, all monosaccharides present a number of chiral carbons (linked to four different atoms or atom groups) equal to that of hydromethylene units: aldoses of n carbon atoms contain $n-2$ chiral centers, whereas ketoses contain $n-3$. Molecules that differ only in the configuration of one or more chiral carbons are known as **stereoisomers**. The number of possible stereo-isomers for aldoses and ketoses are 2^{n-2} and 2^{n-3}, respectively [6]; for example, aldohexoses have 4 stereogenic centers (6–2), that is, 16 possible isomeric forms, whereas ketohexoses have 3 stereogenic centers (6–3), hence, 8 isomeric forms (Figure 3.2). Those stereoisomers that are nonsuperimposable mirror images of each other are known as **enantiomers** (or enantioisomers), and the rest are called **diastereomers** (or diastereoisomers). Enantiomers have the same physicochemical properties, with the only exception of their optical properties, as they rotate the plane of polarized light in opposite directions. **Epimers** are stereoisomers that differ in the configuration of a single stereogenic center.

The overall configuration of monosaccharides is determined by that of the stereogenic center furthest from the carbonyl group in a Fischer projection: if the hydroxyl group is on the right, the overall configuration is D, otherwise the configuration is L. Most naturally occurring sugars are D (Figure 3.2 and Table 3.1). The classical D–L terminology for isomerism preceding the trivial names of monosaccharides (and amino acids, as we shall see in Chapter 4) is preferred by biochemists over the Cahn–Ingold–Prelog rules [8]. The latter are more systematic, but rather cumbersome for these compounds; thus, the IUPAC systematic name for D-glucose would be (2R,3S,4R,5R)-2,3,4,5,6-pentahydroxyhexanal.

Table 3.1: Some monosaccharides of nutritional interest.

	Sources
Pentoses	
L-Arabinose	Plant exudates, hemicelluloses, pectines (dietary fiber)
D-Xylose	Hemicelluloses (dietary fiber)
2-Deoxy-D-ribose	DNA
D-Ribose	RNA
Aldohexoses	
L-Fucose (6-deoxy-L-galactose)	Human milk, plant exudates, and mucous membranes
D-Galactose	Milk (as lactose), oligo- and polysaccharides, cerebrosides
D-Glucose	Frequent and ubiquitous (sugar, milk, starch, cellulose, etc.)
D-Mannose	Homo- or heteropolysaccharides, glycoconjugates
Ketohexoses (or hexuloses)	
D-Fructose	Fruits, honey, saccharose

The most abundant in human nutrition are italicized.

3.2.2 Cyclic forms of monosaccharides

Monosaccharides are readily soluble in water due to the presence of hydroxyl groups that form hydrogen bonds with the surrounding water molecules. In aqueous solutions, monosaccharides exist as an equilibrium mixture of acyclic and cyclic forms. In the latter, one of the hydroxyl groups reacts with the carbonyl group producing cyclic hemiacetals (Figure 3.3). Five- and six-membered rings are structurally favored, the former being named furanoses s and the latter pyranoses. The formation of cyclic forms in monosaccharides implies the occurrence of a new stereogenic center, known as anomeric carbon, and, consequently, two more stereoisomers named anomers, designated as α or β. Figure 3.3 shows the anomers resulting in the cyclation of D-glucose in aqueous solution drawn in Fischer and Haworth projections along with their respective chair conformations.

For D monosaccharides, the conversion of a Fischer projection into a Haworth projection proceeds as follows:
- The cyclic structure is depicted as a planar ring system with oxygen atom at the top right-hand corner in pyranoses (like glucose) and at the top center in furanoses.
- The numbering of the ring carbons increases in a clockwise direction.
- Groups (–H or –OH) directed to the right in the Fischer structure are given a downward orientation.
- Groups directed to the left in the Fischer structure are given an upward orientation.
- Terminal –CH$_2$OH groups are given an upward orientation.

For L sugars, all the steps are the same, but the terminal group is projected downward [6]. The planar Haworth structures are distorted representations of the actual molecules, but they have the advantage of allowing a quick evaluation of stereochemistry around the monosaccharide ring. In 1949, Richard Reeves remodeled the Haworth projection by applying the ring conformations of cyclohexane to describe the structures of the main conformations of pyranoses in solution, leaving the downward or upward orientation of ring substituents unaltered. These conformations were given the names *boat* (B), *chair* (C), *envelope* (E), *half-chair* (H), and *skew* (S). There are 2 C, 6 B and S, and 12 H and E conformations, making a total of 38 canonical configurations for pyranoses. Most aldohexoses prefer the *chair* conformation [9]. In contrast to pyranoses, furanoses can interconvert between multiple ring conformations (including pyranose rings in chair conformation), complicating the interpretation of structure–function relationships for these molecules [10].

Formation and hydrolysis of hemiacetal groups of monosaccharides in aqueous solution result in an equilibrium between cyclic and open-chain forms and even, to a lesser degree, in a mixture of pyranose and furanose forms. When a crystalline sample of a pure anomer is dissolved in water at neutral pH, there is a change in the optical rotation of the solution as the equilibrium between the different anomeric forms is established. This effect is known as mutarotation [9].

Anomerism is extremely important in the biochemistry of carbohydrates because structurally, functionally, and nutritionally different oligo- and polysaccharides are obtained by linking different anomers of the same monosaccharide.

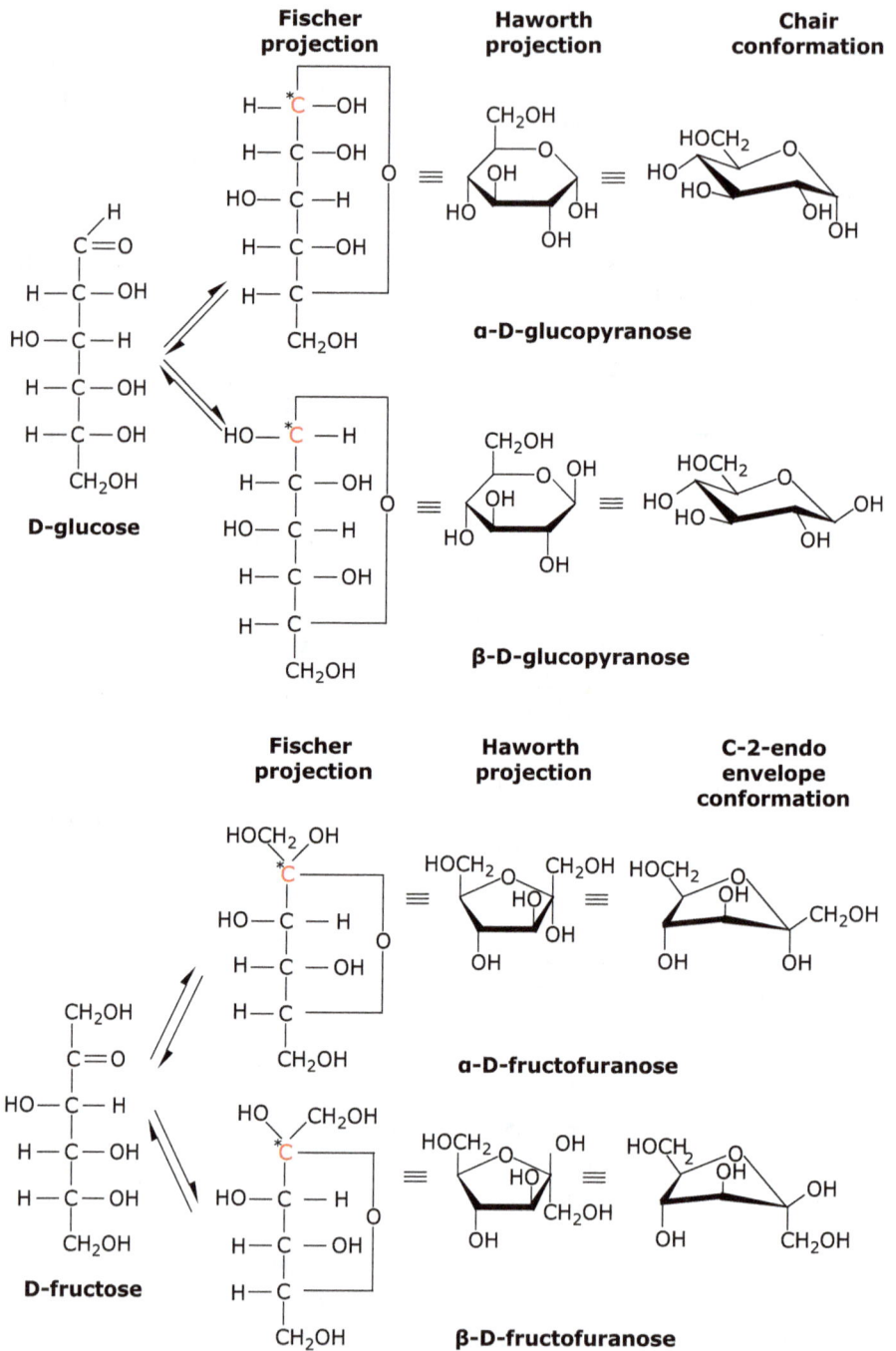

Figure 3.3: Formation of the cyclic forms of D-glucose (an aldose) and D-fructose (a ketose). The anomeric carbon is colored in red and marked with an asterisk.

3.2.3 Disaccharides and oligosaccharides

A condensation reaction may occur between the anomeric center of one monosaccharide and a hydroxyl group of a second monosaccharide producing a disaccharide and a water molecule (Figure 3.4). The resulting linkage is known as glycosidic bond which can be described in chemical terms as the reaction between a hemiacetal group and an alcohol group to form an acetal [6]. Disaccharides are the simplest oligo-

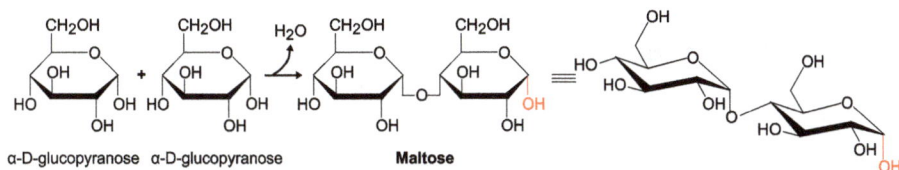

a-D-glucopyranose   a-D-glucopyranose   Maltose

Figure 3.4: Formation of maltose, a disaccharide, from two molecules of α-D-glucopyranose. In aqueous solution, the monosaccharide whose anomeric carbon is not involved in the glycosidic bond (shown in red) alternates between the acyclic and cyclic (α and β) forms.

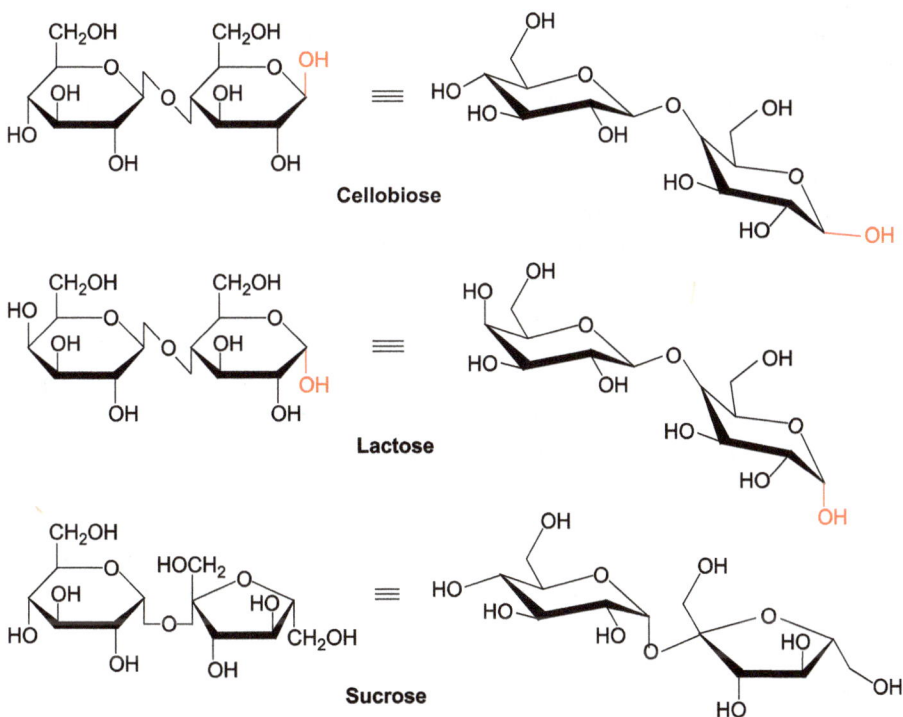

Cellobiose

Lactose

Sucrose

Figure 3.5: Disaccharides of nutritional interest. In aqueous solution, the monosaccharide whose anomeric carbon is not involved in the glycosidic bond (shown in red) alternates between the acyclic and cyclic (α and β) forms.

saccharides: a new monosaccharidic unit can be linked by means of another glycosidic bond to a disaccharide having a free anomeric hydroxyl group, thus yielding a trisaccharide. This process can be repeated, in theory unlimitedly, to obtain oligo- and polysaccharides.

Four disaccharides can be highlighted for their abundance and biological roles: maltose, cellobiose, lactose, and sucrose (Figures 3.4 and 3.5):

– **Maltose** is obtained when a glycosidic bond is formed between the anomeric carbon of a molecule of α-D-glucopyranose and the C-4 of another D-glucopyranose (α or β). This bond is usually denoted as α(1→4) and it is the most abundant glycosidic bond in starch and glycogen, polysaccharides with an important energy store function in many living organisms as we shall see later in this chapter. The major source of maltose is the enzyme-mediated hydrolysis of starch. The α(1→4) glycosidic bond is readily hydrolyzed during the digestion by specific enzymes (maltase-glucoamylase (MGA) in the small intestine); consequently, both maltose and starch are good sources of glucose for humans.

– **Cellobiose** is obtained when a glycosidic bond is formed between the anomeric carbon of a molecule of β-D-glucopyranose and the C-4 of another D-glucopyranose (α or β). This bond, denoted as β(1→4), is present in cellulose, the most abundant biomolecule on the biosphere that performs an important structural function in plants and some bacteria. Humans do not have genes encoding enzymes capable of hydrolyzing this bond, a situation that has important consequences for our diet (see below).

– **Lactose** is the principal sugar of mammals' milk. In human milk, lactose concentration varies between 6.4 and 7.6 g/dL [11]. This disaccharide results from the formation of a glycosidic bond between the anomeric carbon of a β-D-galactopyranose and the C-4 of a D-glucopyranose [β(1→4) glycosidic bond]. It is hydrolyzed by lactase (also known as lactose galactohydrolase and lactase-phlorizin hydrolase, EC 3.2.1.108), a membrane-associated glycoprotein located at the brush border of enterocytes, the absorptive cells located at the small intestine epithelium. Insufficient levels of lactase activity cause the accumulation of lactose in the bowels, triggering a number of symptoms known as **lactose intolerance (LI)**, one of the most common digestive problems in humans [12] (see below).

– **Sucrose** results from the linking of the anomeric carbons of a α-D-glucopyranose (C-1) and a β-D-fructofuranose (C-2) with a glycosidic bond . It is a major product of plant photosynthesis, playing important biological roles in growth, development, and storage among others [13]. It is also the most common, almost universal, sweetener, being industrially obtained from sugarcane in tropical and subtropical countries, and from sugar beet in temperate zones. Although competition is increasing from alternative and synthetic sweeteners, worldwide production of sugar in 2017 was around 180 million metric tons, and it is expected to rise steadily up to more than 200 million metric tons in the next decade [14]. The glycosidic bond of sucrose is hydrolyzed by the enzyme sucrase-isomaltase (SI, EC 3.2.1.48), expressed in the human intestinal brush border, and also by extracellular invertases (or β-D-fructofuranosidases, EC 3.2.1.26) present in the microbial flora of the oral cavity [15]. Mutations in the gene coding for the human SI are the cause of congenital SI deficiency, a rare disease that has been linked to irritable bowel syndrome [16].

Maltose, cellobiose, and lactose have two anomeric carbons (one per participating monosaccharide), one of them being involved in the glycosidic bond. In aqueous solutions, hydrolysis of the other hemiacetal group allows the presence of cyclic structures with different configurations as well as an acyclic structure bearing an aldehyde group. As a consequence, these disaccharides exhibit mutarotation and participate in oxidation and reduction reactions through the carbonyl group, being considered reducing sugars. Sucrose, on the contrary, does not show these properties because both anomeric carbons become "fixed" due to their participation in the glycosidic bond. Consequently, sucrose is a nonreducing sugar.

3.2.4 Derivatives of monosaccharides

Although monosaccharides are in principle composed of carbon, hydrogen, and oxygen atoms linked according to the basic formula $(CH_2O)_n$, a high number of derivatives of these substances occur in living organisms (Figure 3.6). Among these are:

- **Aminosugars:** These are monosaccharides in which a hydroxyl (–OH) group is replaced with an amine group (–NH_2), most commonly at the C-2 of aldoses, yielding 2-amino-2-deoxy sugars.
- **Deoxy sugars:** These are monosaccharides in which a hydroxyl group is replaced with a hydrogen atom. The most important member of these derivatives is 2-deoxy-D-ribose, a component of the deoxyribonucleic acid (DNA).
- **Acidic sugars:** The aldehyde group (–CHO) and/or the hydroxymethyl group (–CH_2OH) at the other end can be oxidized to a carboxyl group (–COOH). Three types of acidic sugars can be obtained depending on the location of the carboxyl group: **aldonic** (at the C-1), **uronic** (at the last carbon), and **aldaric** (two –COOH groups at both ends of the carbon backbone) acids.
- **Polyalcohols (alditols):** These are polyhydroxylated derivatives resulting from the replacement of a carbonyl group with a CHOH group.
- **Esterified sugars:** An acid can react with one of the hydroxyl groups of the monosaccharide molecule producing an ester derivative.
- **Glycosides:** These are monosaccharides (or oligosaccharides) bound to a noncarbohydrate moiety (known as **aglycone**) by glycosidic bonds. There are several types of glycosides depending on the type of bond established between the sugar and the aglycone:
  - *O***-Glycosides** are usually formed when the –OH of the hemiacetal group of a monosaccharide (or an oligosaccharide) reacts with an aliphatic or aromatic

alcohol through a nucleophilic substitution reaction that results in a bond between the two moieties through an oxygen atom and a loss of a water molecule. This type of link is present in *O*-linked glycoproteins. Other interesting examples of *O*-glucosides are steviol glycosides (Chapter 9), a group of highly sweet diterpene glycosides discovered in plant species such as *Stevia rebaudiana* and others that are becoming increasingly popular as sweeteners [17]. "Cardiac glucosides" are also a family of *O*-glycosides in which the aglycones are sterol derivatives. They are found in plants like *Digitalis purpurea* and include molecules (digitonin, ouabain, etc.) with strong pharmacologic activities in conditions such as congestive heart failure or cancer [18].

–   *N*-**Glycosides** are generated when the sugar component is attached to the aglycon, through a nitrogen atom, establishing as a result of a C–N–C linkage. Nucleosides, essential components of nucleic acids, are *N*-glycosides of ribose or 2-deoxyribose as the sugar and a purine or a pyrimidine as the aglycon. The C–N–C linkage is also present in *N*-linked glycoproteins.

–   In *C*-**glycosides**, the anomeric center of the sugar moiety is linked to the aglycone by means of a carbon–carbon bond. *C*-Glycosides occur in microbes, plants, and insects, where they serve a diverse range of functions. In plants, the most abundant *C*-glycosylated products are the flavonoids, a large group of polyphenolic compounds. From a dietary perspective, these compounds have been suggested to have both positive and negative biological activities [19].

–   In *S*-**glycosides**, the sugar moiety is attached to a sulfur atom of the aglycone. Glucosinolates are an example of *S*-β-glucosides present primarily in the species of the order Brassicales (that include mustard, cabbage, or horseradish) that release toxic chemicals like nitriles, isothiocyanates, epithionitriles, and thiocyanates in reactions catalyzed by thioglucosidases. This mechanism is believed to be part of the plant's defense against insects and pathogens [20].

3.3 Vitamin C

3.3.1 Properties and biosynthesis

Vitamin C, or L-ascorbic acid, is a monosaccharide derivative first identified in the 1920s by Albert von Szent Györgyi as a chemical compound that could prevent and cure scurvy. A decade earlier, Casimir Funk (1884–1967) had attributed scurvy, pellagra, rickets, and beriberi to nutritional deficiencies of certain factors that he called "vitamins." Vitamin C was the name proposed for the unidentified anti-scurvy factor [21, 22]. Scurvy is a potentially life-threatening condition, already described by the Egyptians in the Papyrus of Ebers (1550 BCE), where, in addition to diagnosis, treat-

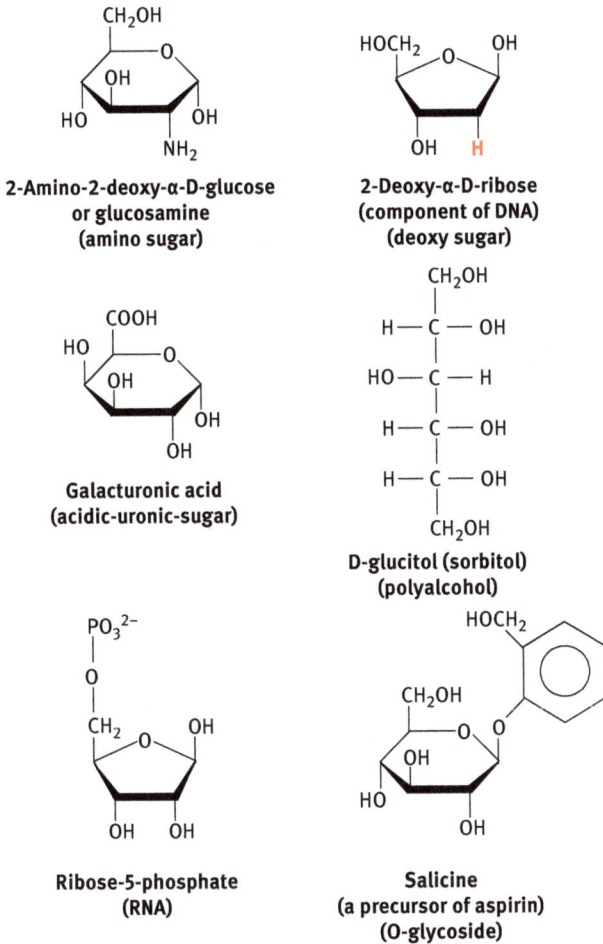

Figure 3.6: Some derivatives of monosaccharides.

ment by eating onions and vegetables was also recommended; however, the first formal description of scurvy is attributed to Hippocrates [21].

Vitamin C is vitally important for many species, and several biosynthetic pathways for this molecule (and analogues) have been characterized in different organisms. Plants and most animals synthesize L-ascorbate from glucose; in the latter, this synthesis takes place in the kidneys (primitive fish, amphibians, and reptiles) or in the liver (mammals). Surprisingly, some fishes, birds, and mammals (including guinea pigs, some bats, gorillas, chimps, orangutans, and humans) have lost the capability to produce it over the course of evolution. Human cells cannot perform the crucial last step of their vitamin C biosynthetic route: the conversion of L-gulono-1,4-lactone into L-ascorbic acid, catalyzed by the enzyme L-gulonolactone oxidase (EC 1.1.3.8). The gene that codes for this protein is actually present in humans, although as a nonfunctional pseudogene due to the accumulation of several mutations. Several hypotheses have been proposed to explain this intriguing situation, which results in the necessity to introduce L-ascorbate in human diet [22, 23].

3.3.2 Biological roles

Vitamin C is a water-soluble reducing agent that can undergo two consecutive oxidations resulting in the formation of ascorbate radicals and dehydroascorbic acid (Figure 3.7). All known physiological and biochemical functions of ascorbate are due to its action as an electron donor [22, 24]:

- Millimolar levels of L-ascorbate have been found in tissues and organs subjected to oxidizing conditions such as eyes (solar radiation) and lungs (high concentrations of oxygen and the so-called reactive oxygen species), where it supposedly serves to protect cells and tissues from oxidative damage.
- L-Ascorbate has been reported to interact synergistically with vitamin E, a lipid-soluble antioxidant, in low-density lipoproteins and lipid membrane oxidation (Chapter 8).
- Vitamin C is an essential cofactor for enzymes like dopamine β-hydroxylase (DBH), peptidyl glycine α-amidating monooxygenase (PAM), and many members of the Fe^{2+}-2-oxoglutarate-dependent (2-OGDD) family of dioxygenases. In all these enzymes, vitamin C acts by maintaining the copper (DBH and PAM) or iron (2-OGDDs dioxygenases) cations, essential for full activity, in their reduced forms, Cu^+ and Fe^{2+}, respectively. Malfunctioning of DBH, involved in dopamine and adrenaline biosynthesis, has been associated with the neurological dysfunction and lassitude symptoms of scurvy.

Among the 2-OGDD dioxygenases, procollagen-proline 3-dioxygenase (EC 1.14.11.7), procollagen-proline 4-dioxygenase (EC 1.14.11.2), and procollagen-lysine 5-dioxygenase (EC 1.14.11.4) are responsible for the insertion of the hydroxyl groups onto prolyl and lysyl residues of procollagen. This chemical alteration is essential for the conversion of procollagen into collagen. Nonfunctional collagen accounts for the symptoms of scurvy associated with problems in the formation and maintenance of bones and blood vessels, as the different types of collagen are major constituents of the extracellular matrix (ECM) of these tissues and others like teeth, tendons, ligaments, and skin.

Collagens have also been involved in the formation of a physical barrier against invasion and metastasis of cancer cells. In the late 1970s, Cameron, Pauling, and Leibovitz proposed that megadoses of ascorbate could inhibit cancer cell invasion among other beneficial effects [25]. More experimental evidence is needed in order to ascertain if high levels of ascorbate can influence cancer growth by mechanisms related to collagen synthesis (see below).

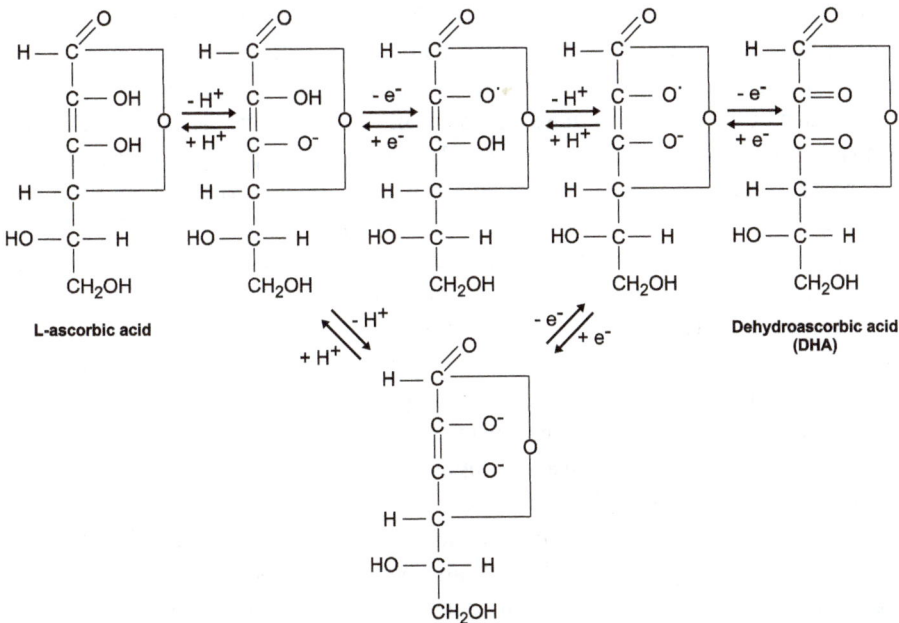

Figure 3.7: The redox chemistry of L-ascorbic acid (vitamin C) (adapted from [24]).

Another member of the 2-OGDD dioxygenases, the hypoxia-inducible transcription factor (HIF)-proline dioxygenase (EC 1.14.11.2), regulates the HIF, which controls the expression of more than 60 gene products. The HIF-proline deoxygenase also requires ascorbate to be active; therefore, the availability of vitamin C may have a profound effect on cell functions such as angiogenesis, cell survival, glucose uptake, glycolysis, and iron homeostasis. Some histone demethylases are also 2-OGDD dioxygenases and require ascorbate for full catalytic activity; therefore, vitamin C might also influence the epigenetic (Figure 14.6 of Chapter 14) status of some cells and tissues [26, 27].

3.3.3 Role of L-ascorbate in cancer prevention and treatment

The anticancer role of vitamin C has been discussed and studied for decades. The Nobel Laureate Linus Pauling was one of his most prominent advocates, proposing megadoses of L-ascorbate (g/day) to prevent several types of cancer and/or enhance treatments [25]. Some mechanisms of protection by vitamin C against cancer at molecular level have been proposed. Thus, the control of 2-OGDDs by L-ascorbate has been reported to have the potential to impact tumor growth at all stages of tumor progression [26]. However, despite the growing interest in L-ascorbate as a cancer treatment, there remains a great deal of controversy over its clinical use, because studies performed during many years have yielded conflicting results. It has been recently reported that a major reason for the controversies about the importance of vitamin C in human health is the interpretation and quality of the available clinical data, as the complexity of the chemistry and pharmacology of vitamin C is rarely taken into account when designing these studies. Examples of these controversies are the placebo-controlled trials performed in the 1980s, which failed to show any benefit from oral ascorbate. Today, we know that oral administration cannot produce significant increases in plasmatic ascorbate concentrations, tightly controlled by the absorption in the guts and excretion in the kidneys [28]. On the other hand, the antioxidant effects of L-ascorbate may hamper the activity of some chemotherapeutic agents [29].

Paradoxically for a molecule with a reputation of being an antioxidant, millimolar concentrations of ascorbate result in the generation of considerable amounts of hydrogen peroxide (H_2O_2), a strong oxidizing agent, in vitro. This reaction is believed to occur also in vivo and is enhanced by catalytic iron (Fenton reaction, which produces the highly reactive OH radical) or copper cations present in variable concentrations in blood plasma and cells. Intravenous administration of ascorbate can be used to raise plasma concentration up to the millimolar range and is currently being investigated as a cancer therapy. The hypothesis behind this treatment is that it will increase H_2O_2 levels, thereby killing some types of cancer cells, which seem to be more sensitive to oxidative stress than their normal counterparts [23, 25].

Despite all the controversies, the fact that vitamin C is required to maintain full function of such a wide array of biochemical mechanisms indicates that an appropriate intake is important to optimize metabolism and enjoy a healthy life.

3.3.4 Occurrence, stability, and dietary reference values for vitamin C

Although vitamin C can be theoretically obtained from many foods, including fresh meat, it is a very labile compound and more than 90% of intake is estimated to come from plant sources, normally eaten raw, like fruits and leafy vegetables (apples, bananas, oranges, broccoli, etc.) [30–32]. In any case, the precise and accurate measurement of ascorbate content from biological samples is a real challenge for analytical chemists. Problems arise from the lability of ascorbate and some of its products as well as from the fact that these compounds seem to be in chemical and biological equilibrium in their natural sources. Moreover, L-ascorbate gets oxidized with storage and type of processing and preparation; thus, frying in oil seems to destroy most of it, whereas steaming and boiling are apparently less damaging to the vitamin C content, which can also change depending on growth conditions and season. Another significant source of L-ascorbate are some processed foods, where it is used as an additive to protect against oxidation [32]. More detailed information about the vitamin C (and other nutrients) content in foods can be obtained from public sources like the Ciqual Table published online by the French Agency for Food, Environmental and Occupational Health & Safety (available at https://ciqual.anses.fr/).

The intestinal absorption of orally ingested vitamin C occurs through the sodium-dependent vitamin C transporter SVCT1, which is also present in the epithelium of the proximal renal tubuli, where it governs the active reabsorption of ascorbate in the kidneys. Another isoform of this transporter, SVCT2, is responsible for the distribution of vitamin C from plasma to tissues and is located in most cell types [33]. On the other hand, dehydroascorbic acid, the oxidized form of vitamin C, can also enter cells through glucose transporters, where it will be reduced back to ascorbate [27] (see Chapter 13 for more information about transport across biological membranes). The distribution of vitamin C is highly differential between organs, and a whole range of mechanisms has evolved to ensure that oxidized vitamin C is recycled back to the biologically active reduced form in many cells. Consequently, the overall vitamin C levels within the human body result from a balance between ingestion, absorption, renal reuptake, and recycling of dehydroascorbic acid to ascorbate. In practical terms, a reliable information on the overall vitamin C status can only be obtained by analyzing blood samples from a fasted individual; furthermore, the particular chemical properties of L-ascorbate (see above) require a careful handling of samples and sound methodologies [28].

In a document issued in 2013, the Panel on Dietetic Products, Nutrition and Allergies of the European Food Safety Authority established a series of average requirements (ARs) and population reference intake (PRI) for different groups. For healthy adults, the ARs were determined from the quantity of vitamin C that balances metabolic losses and allows the maintenance of an adequate body pool indicated by a plasma ascorbate concentration of around 50 μmol/L during fasting. The values proposed were [34]:

- Men: AR of 90 mg/day of vitamin C and PRI of 110 mg/day
- Women: AR of 80 mg/day and PRI of 95 mg/day (extrapolated from those for men on the basis of differences in reference body weight)
- For infants aged 7–11 months, PRI of 20 mg/day
- For children and adolescents, the ARs for vitamin C were extrapolated from the ARs for adults taking into account differences in reference body weight, and PRIs were derived, ranging from 20 mg/day for 1- to 3-year-old children, to 100 and 90 mg/day for boys and girls aged 15–17 years, respectively
- For pregnant and lactating women, vitamin C intake of 10 and 60 mg/day in addition to the PRI of nonpregnant nonlactating women is proposed
- Several health outcomes possibly associated with vitamin C intake were also considered but data were found to be insufficient to establish dietary reference values

It must be underlined that there are significant discrepancies in the criteria used by different health authorities to establish dietary recommendations for daily vitamin C intake; actually, figures may vary by more than fivefold. These differences seem to arise from (a) the interpretation of the scientific data underlying the recommendations, (b) the use of obsolete data in some cases, or (c) the omission by some international authorities of aspects that modify vitamin C requirements, such as gender, age, pregnancy, lactation, smoking, or body weight. Therefore, some authors have encouraged the authorities to reassess their dietary recommendations in accordance with the latest scientific information available for this essential nutrient [35].

3.4 Polysaccharides: fiber

Polysaccharides are polymers of more than 12 units of monosaccharides (or derivatives) linked by glycosidic bonds established between the anomeric carbon of one sugar and any of the hydroxyl groups of another mono- or oligosaccharide, thus allowing the synthesis of branched products. In addition, each anomeric carbon is a stereogenic center and, therefore, every glycosidic linkage can be constructed having either the α- or β-configuration. These properties, in combination with the possibility to choose from many different basic units, give rise to an enormous diversity of structures and chemicobiological properties among polysaccharides.

As mentioned at the beginning of this chapter, polysaccharides are classified as homo- or heteropolysaccharides depending on whether they are composed of a single type or more than one type of basic units, respectively. Three members of this family can be underlined for their importance in human nutrition and metabolism: starch, glycogen, and cellulose.

3.4.1 Starch

Starch is a homopolysaccharide composed of α-D-glucose units, which constitutes an important storage of energy and carbon for plants and other photosynthetic organisms like algae and cyanobacteria. It accounts for a significant fraction of foods like cereals grains (that contain from 60% to 80% of dry weight of this carbohydrate), legumes (25–50%), tubers (60–90%), and some green or immature fruits (as much as 70%).

Starch also has uses in the industries of food (control of consistency and texture of sauces and soups), textiles (laundry sizing), cosmetics (absorbents, thickeners, and film-forming agents), paper (enhancing paper strength and printing properties), pharmaceutical (drug fillers), and packaging (binders) [36].

Starch is composed of two different polymers: amylose and amylopectin (Figure 3.8).

– **Amylose** has been traditionally defined as a linear polymer of α-D-glucose units linked by α(1→4) glycosidic bonds, like in maltose, but it is well established that some molecules are slightly branched by α(1→6) linkages. Amylose chains vary in molecular weight from a few thousands to a million. The α(1→4)-glycosidic bonds determine the helical structure of the resulting chains.

– **Amylopectin** is also composed of polymers of α-D-glucose linked by α(1→4) bonds but, in this case, they present branching points every 24 to 30 residues. At these points, the free anomeric carbon of one chain establishes a bond with the C-6 of a α-D-glucose unit (α(1→6) glycosidic bonds) of another. Side chains can also be branched themselves, thus establishing a highly branched structure that may have a molecular weight as high as 200 million (Figure 3.8) [37].

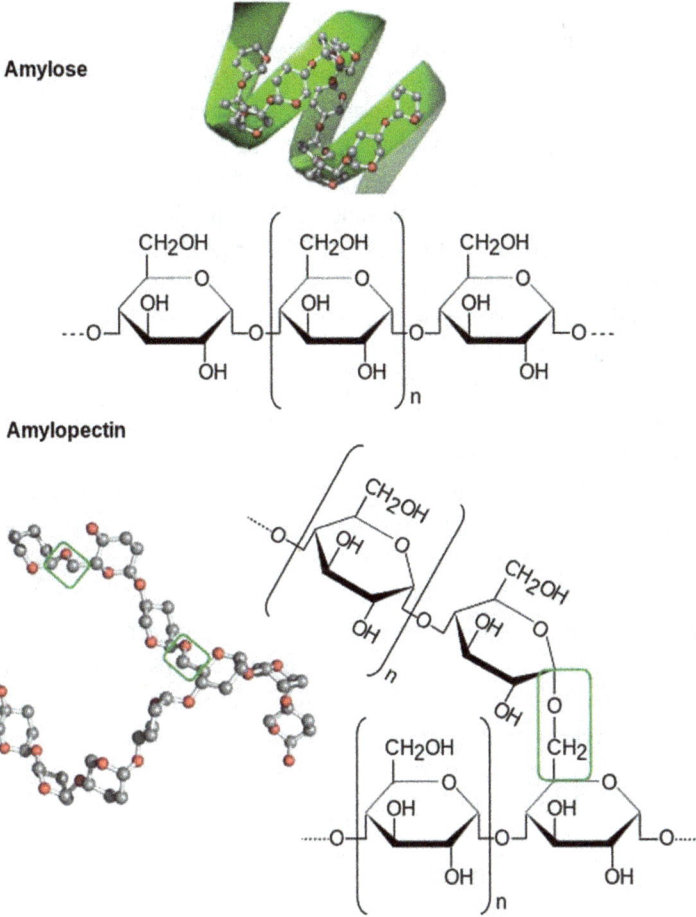

Figure 3.8: Starch structure. Two fractions can be distinguished in starch: amylose and amylopectin. The latter presents a much higher degree of branching. The coordinate files for amylose and amylopectin were obtained from "A Database of Polysaccharide 3D structures (https://polysac3db.cermav.cnrs.fr/home.html)" and the figures were drawn using the program PyMOL (Schrödinger, LLC, New York, NY, USA) after removing –OH and –CH₂OH groups for clarity. α(1→6) glycosidic bonds of amylopectin are highlighted with green rectangles.

Amylopectin usually accounts for 70–90% of total starch (with amylose comprising between 10% and 30%) depending on the botanical species and even on growth conditions for a given cultivar. Moreover, starch is biosynthesized as semicrystalline granules with varying polymorphic types and degrees of crystallinity, where amylose and amylopectin occur as single or double helical conformations due to the structural characteristics conferred by the α(1→4) glycosidic bonds [37]. This molecular structure is reflected in features like the granule size, the amylose/amylopectin ratio, or the arrangement of starch components in the granule, all of them having an important role in their digestibility and, hence, in their nutritional value [38].

Starch is one of the main sources of energy and the major source of carbohydrates in human diet. During the process of digestion, it is gradually hydrolyzed to smaller components until the basic unit, D-glucose, by the consecutive action of several enzymes in the small intestine. α-Amylase (1,4-α-D-glucan glucohydrolase, EC 3.2.1.1), also found in salivary and pancreatic secretions, hydrolyzes internal α(1→4) linkages, by-passing the α(1→6) linkages of amylopectin. This produces a mixture of linear and branched oligosaccharides that have to be further processed to glucose to be absorbed. This task is accomplished by two enzymes located at the intestinal mucosa: MGA (EC 3.2.1.20 and 3.2.1.3), which digests only linear regions, and SI (EC 3.2.1.48 and 3.2.1.10), which also digests branch points [39].

Traditionally, three fractions have been distinguished according to the rate and efficiency of starch digestion: rapidly digestible, slowly digestible, and resistant starch (RDS, SDS, and RS, respectively). RS is the fraction that escapes digestion and enters the large intestine, where it seems to have beneficial effects through fermentation by symbiotic bacteria to short-chain fatty acids (SCFAs, small organic monocarboxylic acids with a chain length of up to six carbons atoms, such as acetic, propionic, butyric, lactate, and formic acids) [40]. The classification in RDS, SDS, and RS is probably too simplistic, as digestion of starch is a complex process that depends on its intrinsic properties, the presence of other accompanying nutrients, and the individual that consumes it, as mastication, gastric processing, in vivo enzyme concentrations, residence time, and hormonal responses are also relevant factors [41]. Starch digestion is an active field of research as the degree and rate of glucose release from starchy foods are believed to play an important role in eating disorders like obesity and type II diabetes.

3.4.2 Glycogen

Glycogen is another homopolysaccharide of α-D-glucose that serves as an osmotically neutral storage of glucose in cells, being present in organisms from bacteria and archaea to humans. Its structural properties are similar to those of amylopectin, that is, the basic units are linked by α(1→4) bonds with branching points. Glycogen presents a higher level of ramification than amylopectin, the α(1→6) glycosidic bonds occurring approximately every 12 residues instead of every 24–30 residues as in the case of the latter. Precise three-dimensional structures of glycogen are difficult to determine. A well-accepted model for glycogen structure proposes the occurrence of inner B-chains, which would normally contain two branch points and outer unbranched A-chains (Figure 3.9). This model predicts glycogen particles of around 44 nm of diameter; however, some studies done in skeletal muscle glycogen have indicated that the average diameter is closer to 25 nm [42]. Dietary glycogen is hydrolyzed by the same set of salivary and intestinal enzymes as amylopectin.

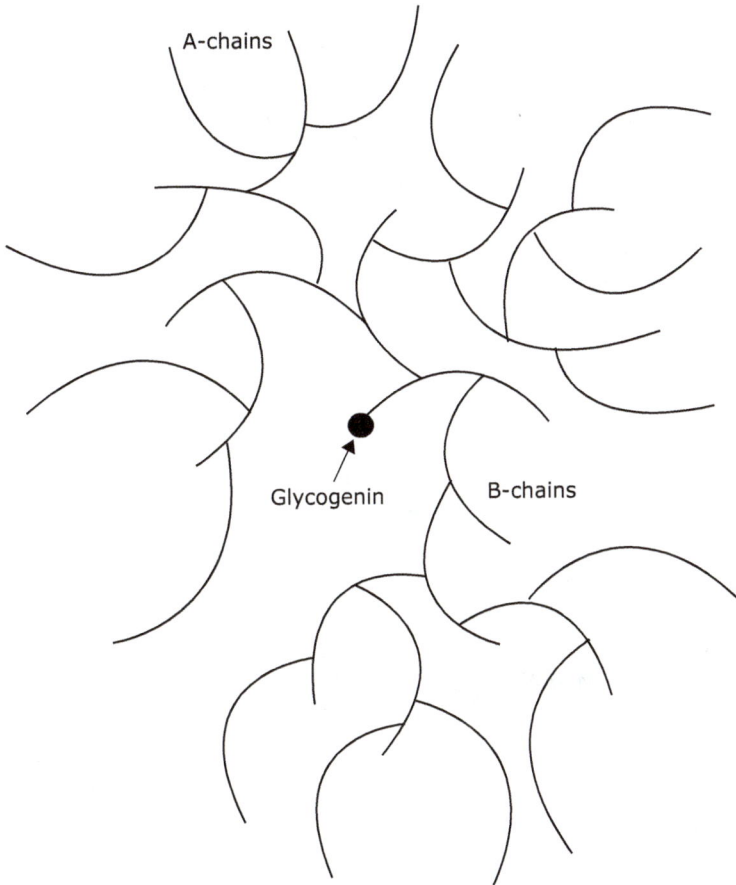

Figure 3.9: Model of a basic structural unit of glycogen granules that shows unbranched (A) and branched (B) chains. The average chain length is 13 residues and branch points occur every 12 residues on average. A full-size glycogen molecule in this model would comprise a total of ~ 55,000 glucose residues, a molecular mass of ~ 10^7 kDa, and a diameter of ~ 44 nm. Glycogenin is a protein that glycosylates itself to form an oligosaccharide primer chain [42].

3.4.3 Cellulose

Cellulose is also a homopolysaccharide of D-glucose; however, in contrast to starch and glycogen, it has no ramifications and its monosaccharide units are linked by glycosidic β(1→4) bonds. The latter are responsible for the enormous differences in structural and functional properties of cellulose compared to those of starch and glycogen, as the most stable conformation for cellulose implies 180° rotation of each glucopyranose relative to its neighbors. This yields straight extended chains with molecular weights of over 1,000,000 that can align next to each other establishing numerous hydrogen bonds (Figure 3.10) [43]. Cellulose is the most abundant biomolecule on the biosphere, comprising over 50% of all the carbon vegetation, and plays an important structural role in photosynthetic organisms, such as higher plants, eukaryotic algae, and cyanobacteria, being the major component of their cell walls [44].

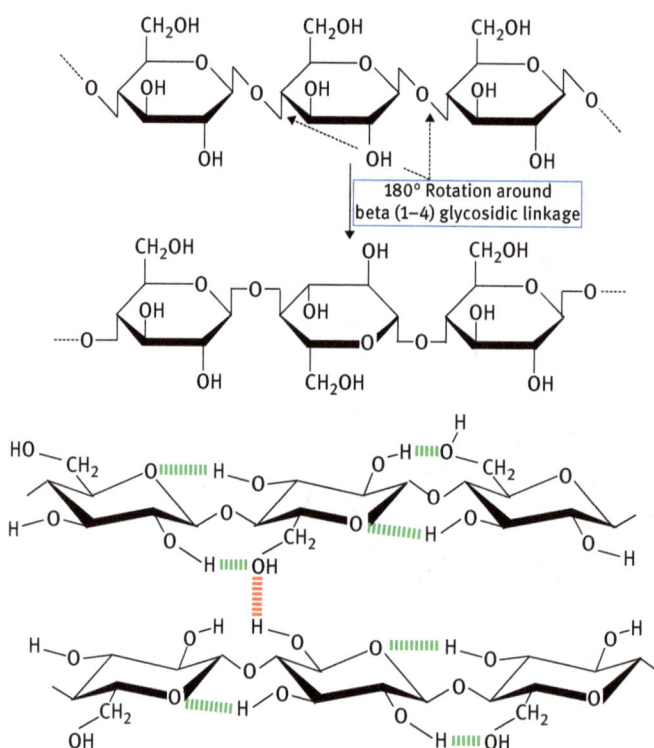

Figure 3.10: Structural characteristics of cellulose. β-D-Glucose residues rotate around the glycosidic link, thereby optimizing interactions by intrachain (in green) and interchain (in red) hydrogen bonds.

The glycosidic β(1→4) bond can only be cleaved in vivo by cellulases, a family of hydrolytic enzymes present in microorganisms such as actinomycetes, bacteria, and fungi [45]. Some of these microorganisms live in the stomach of ruminants allowing them to digest this polypeptide, thus using it as a source of glucose. In humans, the only cellulase activity is associated with some symbiotic bacteria of the colon [46], which means that cellulose can pass unaltered through most of our intestinal tract due to its mechanochemical resistance to pH changes and hydrolytic attack. Fermentation of polysaccharides by intestinal microorganisms can produce SCFAs that we subsequently absorb and metabolize, providing a small fraction (<10%) of our dietary energy [47]. Thus, although cellulose does not play a major energetic role in human nutrition, it is essential for the correct functioning of the digestive system as the major component of the dietary fiber (DF).

3.4.4 Other important polysaccharides in human nutrition: dietary fiber

DF can be defined as dietary components resistant to degradation by mammalian enzymes. DF is composed of lignin, the so-called nonstarch polysaccharides (NSPs), and other compounds such as nondigestible oligosaccharides and RS [44, 48].

Lignin is the generic term for a large group of aromatic polymers that are deposited predominantly in the walls of secondarily thickened plant cells, making them rigid and impervious and protecting them from microbial degradation. The main building blocks of lignin are the hydroxycinnamyl alcohols (or monolignols) coniferyl alcohol and sinapyl alcohol, with typically minor amounts of *p*-coumaryl alcohol (Figure 3.11) [49].

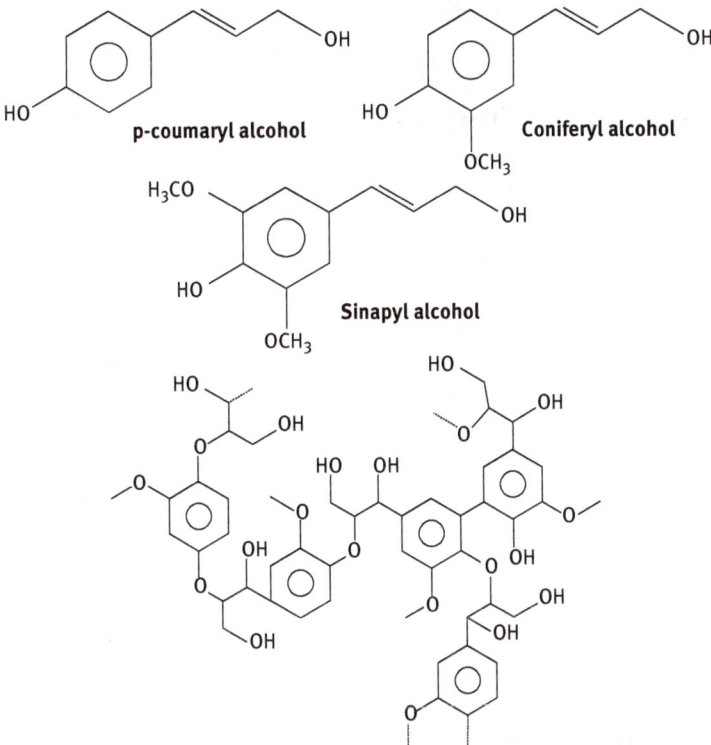

Figure 3.11: Structures of lignin and its basic units (monolignols) (retrieved and adapted from [50]).

NSPs are mixtures of polysaccharides that differ according to the sequence and composition of monosaccharides, the types of linkages present, the number of monosaccharides with acidic groups (e.g., uronic acids), and the presence of branches (Figure 3.12). Major components of NSPs are [44]:

- **Cellulose**
- **Hemicelluloses** are composed of a mixture of polysaccharides that bind to cellulose microfibrils by hydrogen bonds in natural sources of fiber. They include xyloglucans (backbones of 1→4-linked β-D-glucose residues with xylose-containing side chains linked by α- or β(1→6) glycosidic bonds), **xylans** (backbones of β(1→4)-linked D-xylose residues to which differently structured side chains can be attached), **galactomannans** (backbones of β(1→4)-linked D-mannose residues, which can bear α(1→6)-linked D-galactose residues), and **galactoglucomannans** (galactomannans with some β(1→4)-linked D-glucoses in the backbone). Hemicellulose compositions differ depending on their origins [51].
- **Pectins** are complex polysaccharides largely made up of the D-galacturonate connected by α(1→4) linkages. Many of these residues can be methylated at their carboxylic groups and up to 20% of them can be replaced by other monosaccharides [52].
- **β-Glucans** are glucose-containing polysaccharides whose structures vary according to different sources; thus, mushroom β-glucans have short β(1→6)-linked branches to a β(1→3) backbone, whereas β-glucans of oats and barley are linear β (1→4) linkages separating shorter chain of β(1→3) structures [53].
- **Pentosans**, also known as **arabinoxylans**, vary significantly in different food sources. In cereals, they consist of a backbone of β(1→4)-linked D-xylose residues to which α-L-arabinose units are linked as side branches [54].
- Other polysaccharide-containing compounds that are not components of cell walls are also classified as dietary like gums and mucilages, composed of complex polysaccharides (some of them sulfated or carboxylated) from various sources like the endosperm of plant seeds (guar gum), plant exudates (tragacanth), tree or shrub exudates (gum Arabic and karaya gum), sea weed extracts (agar), or bacteria (xanthan gum) (see below). The chemical compositions of gums and mucilages vary depending on the sources and also include proteins and minerals [55].
- **Fructans** are reserve polysaccharides in certain plants. The two main kinds of fructans are inulin and levans. Inulin is a low-molecular-weight polysaccharide containing D-fructose residues (normally, between 20 and 40) linked by β(2→1) glycosidic bonds with a D-glucose residue linked to the last fructose residue also with a β(2→1) bond, thus producing a nonreducing end. Levans, found mainly in grasses, have a similar composition to that of inulins, although they have higher molecular weights and some degree of branching [56].

Two types of DF can be distinguished based on their behavior in water: insoluble DF (cellulose, hemicellulose, and lignin) forms nonviscous fibers very slowly (or not) fermented by bacteria in the colon, whereas soluble DF (pectin, gums, and mucilages)

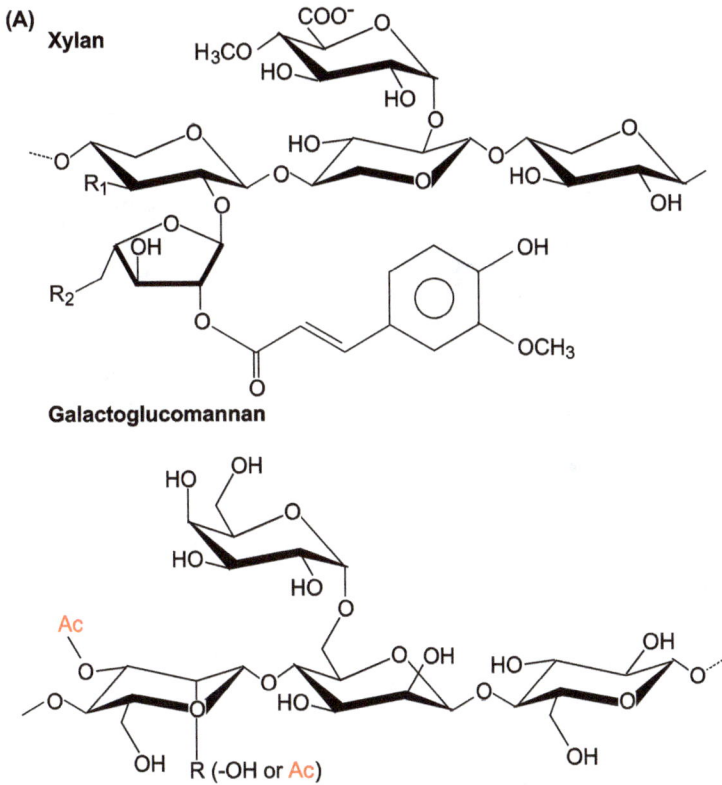

Figure 3.12: Nonstarch polysaccharides (NSPs). (A) Basic structural units of xylan and galactoglucomannan (Ac: acetyl group, in red), two components of hemicelluloses (adapted from reference [57]). (B) Basic structural units of pectin (taken from [58]), β-glucan (adapted from [59], with the β-(1→3) linkage is colored in red), and inulin (adapted from [60]).

forms colloidal suspensions when mixed with water (hydrocolloids) and is fermented by the colonic microbiota (Table 3.2) [48]. Many beneficial effects have been linked to ingestion of DF [44]:

– Provides bulk to gut contents due to its water-binding capacity and allows easy passage through the human intestine preventing constipation.
– Meals rich in DF are calorically less dense and lower in fat and they are processed slowly in the human body, promoting earlier satiety.
– NSPs contained in DF are fermented into SCFAs by the microflora residing in the colon. SCFAs have been reported to have a number of health-promoting effects such as stimulating growth of certain colonic bacteria (prebiotic effect) and colon epithelial cells (increasing the absorptive capacity of the epithelium), and lowering the pH of the colon. The latter inhibits growth of pathogenic organism, increases mineral absorption (especially Ca^{2+}, Mg^{2+}, and $Fe^{2+/3+}$ cations), maintains

(B)

Pectin backbone

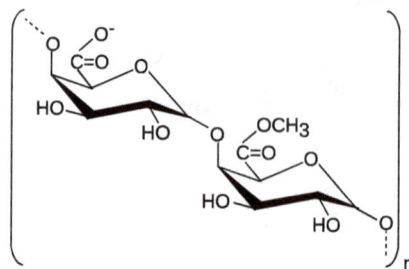

β-glucan

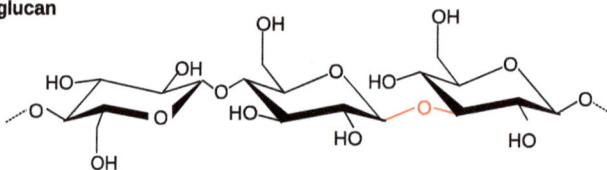

Inulin

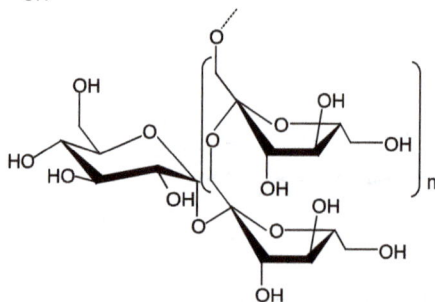

Figure 3.12 (continued)

normal bowel structure and function, prevents or alleviates diarrhea, and stimulates blood flow and fluid and electrolyte uptake in the colon.
– Restricts the production of harmful by-products of protein degradation associated
with various types of ulcerative colitis and cancer.
– High NSP- and polysaccharide-containing foods are generally a source of phytoestrogens that can provide protection against breast cancer.
– Dietary intake of NSPs has been reported to reduce the risk of dietary problems
such as obesity, coronary disease, diabetes, inflammatory bowel diseases like di-
verticulitis and ulcerative colitis, and colon cancer.

Table 3.2: Classification of dietary fiber components based on chemical and physical properties.

Dietary fiber component	Chemical classification	Physical classification
Lignin	Noncarbohydrate	Insoluble, nonfermentable, nonviscous
Cellulose	NSP:[1] cellulose	Insoluble, nonfermentable, nonviscous
Hemicellulose	NSP:[1] noncellulose	(In)soluble, polysaccharide (non)fermentable, (non)viscous
Pectin	NSP:[1] noncellulose	Soluble, polysaccharide fermentable, (non)viscous
Gum	NSP:[1] noncellulose	Soluble, polysaccharide fermentable, (non)viscous
Fructo-oligosaccharides	Nondigestible	Soluble, oligosaccharide fermentable, (non)viscous
Galacto-oligosaccharides	Nondigestible	Soluble, oligosaccharide fermentable, (non)viscous
Resistant starch	Resistant starch	Soluble fermentable, (non)viscous

Data obtained from [42].
[1]NSP, nonstarch polysaccharide.

3.5 Properties and importance of carbohydrates in the food industry

Carbohydrates are widely used in food industry as additives; however, their functional attributes depend on their chemical composition:

– Sucrose (table sugar) is one of the most commonly added sugars in food, its most notable function being its sweet taste, which improves the palatability of food and contributes to the flavor profile by interacting with other ingredients to enhance or lessen certain flavors [61]. In addition, due to their physicochemical properties, sucrose and other simple sugars (mono- and disaccharides) bind water, increase viscosity, and alter texture of food products, provide bulk, serve as energy for fermentation, provide glaze and sparkle, serve as a precursors for flavor and color development and act as food preservatives [61, 62]. Apart from these general properties, the functionality of a given mono- or disaccharide within a specific food system must be evaluated when formulating a product, as individual molecules may provide different characteristics when used as food additives [62].

– Starch is used as a source for sweeteners and in native or modified forms [63]. It may be directly used in the food industry as baking flour, being one of the major components responsible for the structure and properties of the final products; however, most native starches are limited in their direct application because they are unstable when subjected to changes in temperature, pH, and shear forces. During heat treatment, for example, starch granules are gelatinized and lose their structural organization; these disaggregated starch molecules form a gel and then retrograde gradually into semicrystalline aggregates on cooling [64]. Several physical, chemical, and enzymatic methods have been developed to produce modified starches with different physicochemical properties in order to make them more suitable for the food industry. Modified starches are used as modifiers of texture, viscosity, adhesion, moisture retention, gel formation, and films [63, 65].

- Cellulose and derivatives also have important applications in food industry [56]:
  - Microcrystalline cellulose, prepared by treating cellulose with hydrochloric acid, is used as a bulking agent and as a fat replacer in emulsion-based food products.
  - Carboxymethylcellulose, an anionic derivative prepared by treating cellulose with alkali and monochloroacetic acid, is used in the food industry as a thickener, stabilizer, and suspending agent.
  - Methylcellulose, prepared by treating alkali cellulose with methyl chloride and derivatives like hydroxypropylmethylcellulose are used as emulsifiers and stabilizers in low-oil and/or no-oil salad dressings.
- Pectins are used as gelling agents in various foods like dairy, bakery, and fruit products, whereas gums and mucilages have broad applications in the food industry as emulsifiers and stabilizers and flavor encapsulation agent, bulking agent, to prevent sugar crystallization and DF supplement [56].
- Several polysaccharides of algal and bacterial origins are also used in the food industry (Figure 3.13):
  - Alginate, a polyuronate that comprises β-linked D-mannuronate and L-guluronate (in varying lengths and arrangements), is a major cell wall component of brown macroalgae (seaweed) [52]. Alginates tend to form gels upon contact with divalent ions, a capacity that has been used in modern cooking [56].
  - Carrageenan (CGN) belongs to a group of viscosifying polysaccharides that are extracted from certain species of red seaweeds in the family Rhodophyceae. It is composed of a linear backbone of galactose sugars that have varying amounts of sulfate attached. CGN is used primarily as a gelling, thickening, and stabilizing agent in dairy products [66].
  - Agar, obtained from red-purple algae of the Rhodophyceae class, is a linear polysaccharide built up of the repeating disaccharide unit of β(1→3)-linked-D-galactose and α(1→4)-linked 3,6-anhydro-L-galactose residues. In contrast to CGNs, agar is only lightly sulfated and may contain methyl groups. It is considered the best gelling agent and used the baking industry because of its heat-resistant characteristics [56].
  - Xanthan gum, isolated from the bacterium *Xanthomonas campestris*, is a non-gelling gum with a basic structure similar to that of cellulose, widely used in the food industry, mainly in dairy and baked products [56].

- Partial hydrolysis of starch using specific enzymes or dilute acids results in oligo-saccharides with different degrees of polymerization that may vary from 2 to 20 monosaccharide units. The differences in the composition of the final products are often expressed by the dextrose equivalent (DE), which corresponds to the grams of reducing sugars, expressed as glucose, per 100 g of dry matter. Dried glucose syrups are, by definition, dried starch hydrolysis products with a DE greater than 20, whereas maltodextrins are defined as dried starch hydrolysis products with a DE equal to or lower than 20, but higher than 3 [67]. Cyclic poly-mers, known as cyclodextrins, are also produced during the hydrolysis of starch. Maltodextrins have been widely employed in the food industry as fat replacers and bulking agents, encapsulation of flavor, and colorants. Syrups are used as sweeteners, stabilizers of the fat-water emulsion, and inhibitors of oxidative deg-radation. Cyclodextrins have been used to stabilize aroma compounds during ex-trusion and thermal processing of foods [68].

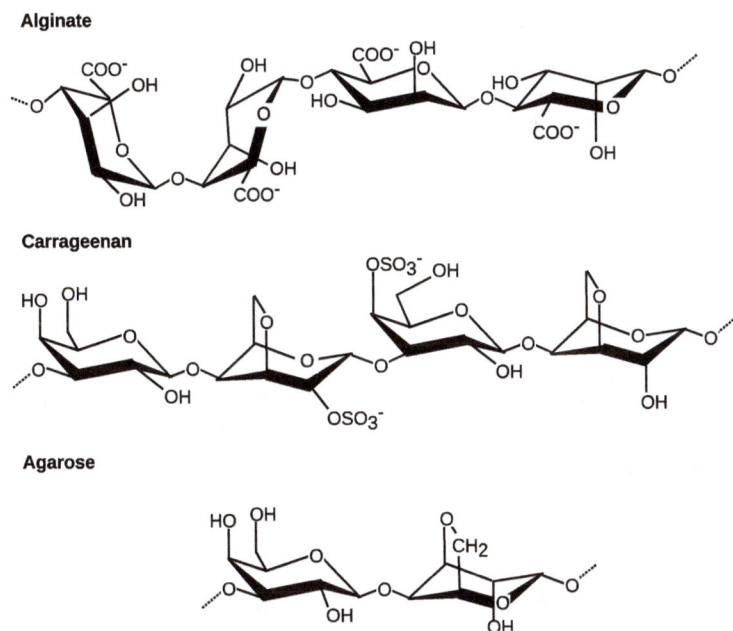

Figure 3.13: Basic structural units of some polysaccharides of bacterial and algal origin used in the food industry. Alginate, carrageenan, and agarose structures were taken from references [58, 69, 70], respectively. Agarose comprises about 70% of agar, the rest being agaropectin, a non-gelling fraction, which consists of β(1→3)-linked D-galactose units, and some of which are sulfated [70].

3.6 Browning reactions of sugars

Browning is one of the most important chemical processes that take place during food processing and storage, affecting the quality of food in either positive or negative ways. Two types of browning can be distinguished: enzymatic and nonenzymatic [71].

Enzymatic browning consists in the oxidation of phenolic compounds to quinones, which subsequently polymerize to form a series of colored molecules known as melanins. The first step in these reactions is catalyzed by polyphenol oxidases (PPOs), copper-containing enzymes that occur in most plants and many other organisms. They have catechol oxidase activity (oxidation of o-diphenols to their corresponding o-quinones, EC 1.10.3.1) and many also have the ability to hydroxylate monophenols to o-diphenols (tyrosinase activity, EC 1.14.18.1). In the literature, the designation "PPO" sometimes includes laccases, which can also oxidize p-diphenols (EC 1.10.3.2) [71, 72]. Enzymatic browning may affect food (especially fruits, vegetables, and seafood) in either positive or negative ways. On the one hand, they may contribute to the overall acceptability of foods such as tea, coffee, and dried fruits like raisins; on the other hand, it is considered a devastating process for many tropical and subtropical fruits and vegetables. Enzymatic browning can affect color, flavor, and nutritional value of these foods. More detailed information about this important process and the strategies followed to control its undesired effects can be found elsewhere [71].

Several types of nonenzymatic browning reactions have been characterized to date [71]:

– **Maillard reaction**. It is a complex set of chemical reactions named after the French physicist and chemist Louis Camille Maillard (1878–1936), who initially described it. Maillard reaction products (MRPs) may have desired and undesired effects during cooking and processing of foods. Among the factors that affect this reaction are: temperature (maybe the most important), pH, and the chemical composition of the food [71]. On the other hand, MRPs can be beneficial or toxic to the consumers; consequently, understanding and controlling the Maillard reaction is of paramount importance for food safety. Furthermore, it has been proposed that some products of the Maillard reaction, collectively known as "advanced glycation end products," can be formed in vivo and are implicated in metabolic diseases and aging, by altering the structure and function of biomolecules like proteins and nucleotides [73]. Three stages have been proposed to occur in the chemical mechanism of this process (Figure 3.14) [71, 74]:
    – Condensation of free amino groups from amino acids, peptides, or proteins (Chapter 5) with the carbonyl group of a reducing sugar, followed by the formation of a Schiff's base and Amadori rearrangement, producing 1-amino-1-deoxy-2-ketoses (Amadori products). The latter are precursors of several compounds important for the formation of characteristic flavors, aromas, and brown polymers.

- Breakdown of Amadori compounds, dehydration, and fragmentation of sugar molecules and degradation of amino acids, and formation of products such as pyruvaldehyde and diacetyl (butane-2,3-dione).
- Aldol condensation and formation of melanoidins, brown-colored heterocyclic nitrogenous compounds. Three theories on melanoidin formation have been proposed: (1) by polymerization of low-molecular-weight products of Maillard reaction, such as furans and pyrroles, (2) by cross-linking of these products with proteins, and (3) by polymerization of sugar degradation products, formed in the early stages of the Maillard reaction [75].

(A)

(B)

(C)

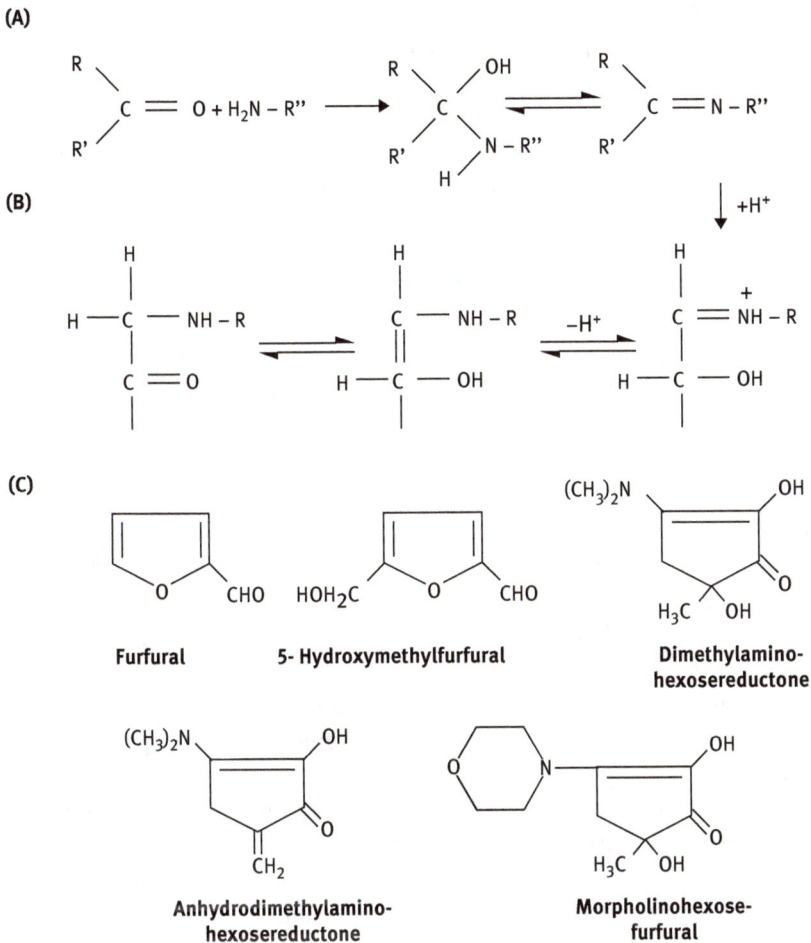

Figure 3.14: The Maillard reaction. (A) Schiff base formation. (B) Amadori rearrangement. (C) Some premelanoidins, products of the Maillard reaction considered to be precursors of melanoidins. Chemical formulae retrieved and adapted from: Deshpande, S.S. (2002). Handbook of Food Toxicology (1st ed.). CRC Press. https://doi.org/10.1201/9780203908969

- **Caramelization.** This process occurs when sugars are heated over their melting point at acidic or alkaline pH and many of the products formed are similar to those resulting from the Maillard reaction. The chemical mechanism of caramelization includes formation of isomeric carbohydrates, dehydration, and oxidation reactions producing a complex mixture of volatile and nonvolatile compounds. Like the Maillard reaction, caramelization also depends on temperature (and duration of heating), pH, and food composition. It is a widely used process in cooking and food processing to alter coloring and flavoring [71].
- **Degradation of ascorbic acid** (AA). This process has been associated with browning of citrus products for a long time. AA is a very labile compound that can be degraded in multiple ways: in the presence or absence of oxygen and enzymatically or nonenzymatically. As a consequence, the exact route of AA degradation in foods is highly variable and dependent upon the particular system, being influenced by factors like temperature, pH, and the presence of oxygen, enzymes, amino acids, and certain cations among others [71].

3.7 Importance of carbohydrates in the diet

The World Health Organization (WHO) recommended in its guidelines published in 1998 that 55% of total dietary energy should be sourced from carbohydrates [76]; however, scientific evidence published after this report made it necessary to clarify a number of issues concerning the role that carbohydrates play in human health:
- Carbohydrates are important for glucose homeostasis, oxidative metabolism, and gastrointestinal function; however, as we have seen previously, it is a term that includes a large variety of biochemical compounds (sugars, starch, RS, NSPs, etc.) that exert different physiological effects. The physical and chemical properties of the different carbohydrates affect the rates at which they are hydrolyzed and absorbed in the human gastrointestinal tract, with impacts on digestive physiology, gut hormone signaling, and postprandial metabolism [77].
- Different sources of carbohydrates cannot be considered equally healthy for human nutrition; thus, free sugars (mono- and disaccharides that required little or no digestion) are considered cariogenic substances that lead to an unhealthy diet, weight gain, and increased risk of medical conditions like type II diabetes and cardiovascular disease [78]. For these reasons, the WHO updated its guidelines in subsequent reports [78–80], where it advised people to consume less than 10% of energy as free sugars and suggested that additional health benefits

may be obtained if free sugar intake is reduced to 5% or less of total energy. By contrast, sources of carbohydrates like whole grains and DF are considered healthy. These sources also include components (RSs and micronutrients such as magnesium, folate, vitamin B6, and vitamin E), which are considered beneficial for health [81].

– Food preparation, cooking, and preservation as well as industrial processing may alter our carbohydrate sources with beneficial or detrimental effects [81].

– The glycemic response to different carbohydrate-containing foods, measured by the glycemic index (GI), is considered important in human nutrition by some authors. GI refers to the effect in blood glucose levels produced by a given food, relative to the effect of an equal amount of a substance defined as having a GI of 100 (usually white bread or glucose). Foods classified as high GI include refined grain products, white bread, and potato, whereas low GI foods include whole-grain products, legumes, and fruits [77]. However, both the quantity and quality of carbohydrates influence the glycemic response, and the GI only provides a measure of the latter; therefore, the concept of glycemic load (GL) was introduced to account for the overall glycemic effect of a portion of food (Table 3.3). The GL of a typical serving of food is the product of the amount of available carbohydrate in that serving and the GI of the food. The higher the GL, the greater the expected elevation in blood glucose and in the insulinogenic effect of the food. The long-term consumption of a diet with a relatively high GL (adjusted for total energy) is associated with an increased risk of type II diabetes and coronary heart disease [82]. However, recent studies suggest that it is unlikely that the GI of a food or diet is directly linked to disease risk or health outcomes [83].

– There is evidence about the beneficial effects of DF (discussed above) and certain synthetic and natural oligosaccharides; however, more research is needed to obtain information on the optimal dose of these compounds for human health as well as their possible involvement in gut physiology, gut microbiota, and immunomodulation [77, 84].

During the last decade, some international health authorities, apart from WHO, have issued their own dietary recommendations for carbohydrates. These recommendations and the differences among them have been reviewed elsewhere [85]. Besides this, an interesting study on the role of carbohydrates in health and disease has been recently published by Clemente-Suárez and colleagues [86].

Table 3.3: Glycemic index and glycemic load of selected foods.

Food	Glycemic index	Serve (g)	Carbohydrates per serve (g)	Glycemic load
White bread, wheat flour	71	30	14	10
Wholemeal (whole wheat) bread	72	30	12	9
Potato, type not specified, boiled	66	150	19	13
Potato, type not specified, boiled in salted water	76	150	34	26
Basmati, white rice, boiled 12 min	52	150	28	15
Basmati, white rice, organic, boiled 10 min	57	150	40	23
Sweet corn, boiled	60	80	18	11
Apple, raw	40	120	16	6
Oranges, raw	40	120	11	4
Orange juice	46	250	26	12
Rockmelon/cantaloupe, raw	70	120	6	4
Watermelon, raw	72	120	6	4
Pineapple, raw	66	120	10	6

Elaborated with data obtained from the official website for the glycemic index and international GI database based in the Boden Institute of Obesity, Nutrition, Exercise and Eating Disorders and Charles Perkins Centre at the University of Sydney (http://www.glycemicindex.com/foodsearch.php).

3.8 Metabolism of carbohydrates: glycolysis and gluconeogenesis

Metabolism can be defined as the sum of all the chemical transformations that take place in a living cell or organism. The vast majority of the metabolic reactions are catalyzed by enzymes, macromolecules (usually proteins) characterized by their high efficiency and specificity both for a reaction and a substrate (although the so-called moonlighting enzymes can mediate more than one reaction and/or perform other functions in vivo [87]). Metabolism has four main objectives [43]:
–   Obtain chemical energy
–   Convert nutrient molecules into cell's own molecules
–   Polymerize monomeric precursors into macromolecules (proteins, nucleic acids, and polysaccharides)
–   Synthesize and degrade biomolecules required for the cellular functions

Glucose is probably the most prominent member of the carbohydrate family of compounds in metabolic terms, occupying a central position in the metabolism of many living organisms. On the one hand, it is an excellent fuel and, on the other, it supplies a wide array of metabolic intermediates for biochemical reactions. In our cells, glucose has four major fates: synthesis of glycogen, synthesis of the ECM, oxidation to pyruvate to provide ATP and metabolic intermediates, and oxidation to yield NADH or NADPH (for reductive biosynthetic processes and/or protection from oxidative stress – Chapter 8) and ribose-5'-phosphate (for nucleic acid synthesis reactions) [43].

3.8.1 Glycolysis

Glycolysis is a series of metabolic reactions (pathway) that converts a molecule of glucose into two molecules of pyruvate. It takes place in the cytosol of all cells of the body and occurs in the absence (anaerobic conditions) or in the presence of oxygen (aerobic conditions).

Glycolysis is a central metabolic pathway, many of its intermediates providing branch points to other pathways such as the synthesis of amino acids or lipids. It consists of 10 chemical reactions that are traditionally classified into 2 consecutive phases:

- **Preparatory phase:** Energy in the form of ATP (two molecules per molecule of glucose) is invested in splitting a molecule of glucose (six carbons) into one molecule of glyceraldehyde-3-phosphate (G3P, three carbons) and another of dihydroxyacetone-phosphate (DHAP, three carbons).
- **Pay-off phase:** G3P is oxidized to pyruvate in five consecutive steps. The molecule of DHAP also enters this part of the pathway after being transformed to G3P via an isomerization reaction, thus yielding a maximum of two molecules of pyruvate per glucose. In this phase, two molecules of NADH, a powerful biological reducing agent, and four molecules of ATP are produced per molecule of glucose that enters the pathway.

Figure 3.15 shows a general scheme of glycolysis with the different enzymes involved in the catalysis of each step. Overall, two molecules of pyruvate, NADH and ATP, are obtained in this degradative (catabolic) pathway, whose global chemical equation can be written as follows:

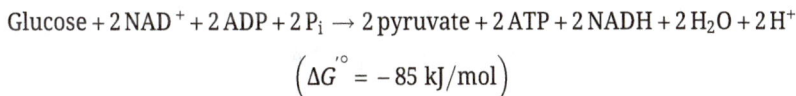

$$\text{Glucose} + 2\,\text{NAD}^+ + 2\,\text{ADP} + 2\,P_i \rightarrow 2\,\text{pyruvate} + 2\,\text{ATP} + 2\,\text{NADH} + 2\,H_2O + 2\,H^+$$

$$\left(\Delta G^{'o} = -85\ \text{kJ/mol}\right)$$

The whole process of glycolysis is thermodynamically favorable, with a negative value of its free energy (endergonic) under standard conditions (1 M concentrations for all chemical compounds involved, pH 7 and 25 °C) as well as under cellular conditions. Seven of the ten reactions of glycolysis are considered to be reversible and three irreversible ($\Delta G^{'\circ} \ll 0$) in vivo.

The final product of glycolysis, pyruvate, is a useful molecule for the cell, having different fates depending on the metabolic status (Figure 3.16):

- It can enter the mitochondria where it will be further oxidized to CO_2 and H_2O providing more reducing power (NADH and $FADH_2$) and ATP.
- It may be reduced to L-lactate by NADH, the latter being oxidized back to NAD^+. The conversion of D-glucose to L-lactate is known as **lactic fermentation**, and it takes place in skeletal muscular cells during vigorous contraction due to insufficient supply of oxygen. Oxidation of NADH is important to avoid its accumulation, which would alter the cytosolic NAD^+/NADH ratio and, thus, many cellular biochemical processes [88]. Some cells like erythrocytes, which lack mitochondria, and many types of tumor cells obtain much of the ATP they need from this pathway under aerobic conditions. In the case of tumor cells, this process is known as the "Warburg effect" and it is being thoroughly studied nowadays in the hope that its understanding may lead to better treatments for some cancers [89]. Lactic fermentation is also performed by some bacteria, being the basis of dairy products like yoghourt.
- Pyruvate may also provide the carbon skeleton for the amino acid alanine or fatty acid synthesis (via acetyl-CoA) [43].

Other monosaccharides like galactose of fructose can be degraded through the glycolytic pathway with a similar yield as glucose.

3.8.2 Gluconeogenesis

Gluconeogenesis is the synthesis of glucose from pyruvate. Seven of the enzymatic reactions of this pathway are the reverse of glycolytic reactions, whereas the three irreversible steps of the latter are bypassed by a separate set of gluconeogenic enzymes (Figure 3.15). Both pathways are, therefore, irreversible and take place in the cytosol. Unlike glycolysis, gluconeogenesis occurs only in the liver, renal cortex, and enterocytes. The global reaction for this anabolic (biosynthetic) pathway can be written as follows:

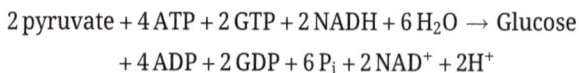

$$2\,\text{pyruvate} + 4\,\text{ATP} + 2\,\text{GTP} + 2\,\text{NADH} + 6\,H_2O \rightarrow \text{Glucose}$$
$$+ 4\,\text{ADP} + 2\,\text{GDP} + 6\,P_i + 2\,NAD^+ + 2H^+$$

Gluconeogenesis is energetically expensive for the cell, as it requires spending the equivalent to six molecules of ATP (4 ATP + 2 GTP) and two molecules of NADH per

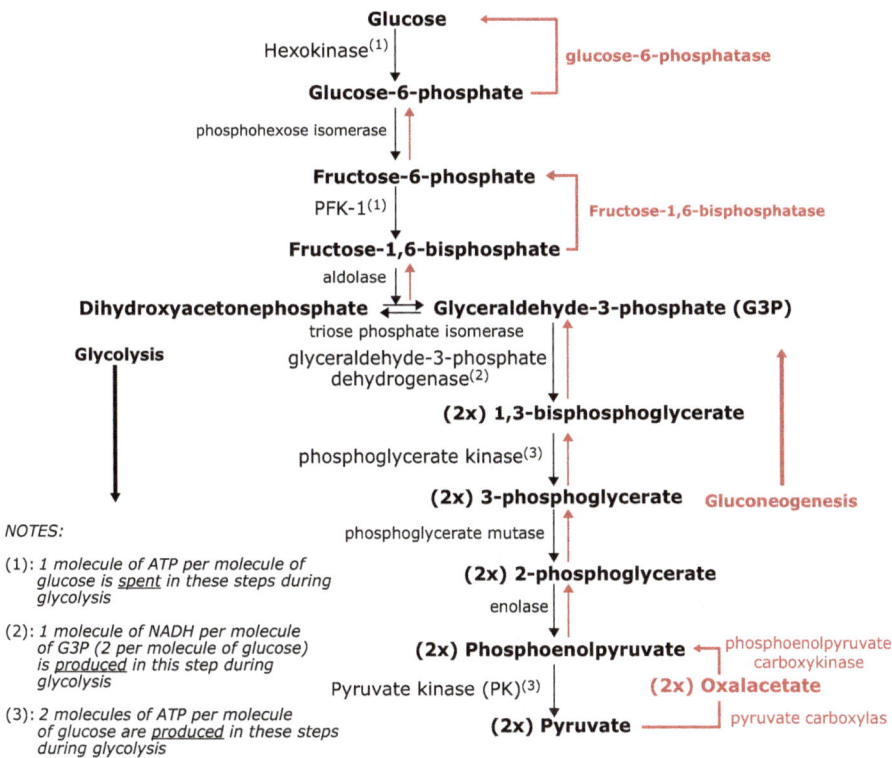

Figure 3.15: Schematic representation of glycolysis versus gluconeogenesis. Glycolysis is a central metabolic pathway in most cells; however, gluconeogenesis occurs only in the liver, renal cortex, and enterocytes. Black arrows and red arrows indicate glycolytic and gluconeogenic steps, respectively (adapted from "File:WP534 de.png" by original = Kristina Hanspers, Alexander Pico, Martijn van Iersel, Susan Coort, Thomas Kelder, Jildau Bouwman, Kdahlquist, Nick Fidelman deutsch = Ghilt und Maxxl2; licensed under CC BY 3.0; available at: https://commons.wikimedia.org/wiki/File:WP534_de.png).

molecule of glucose synthesized; therefore, glycolysis would only provide one-third of the energy needed to obtain glucose from pyruvate. This means that a tight reciprocal regulation of both pathways is needed in order to avoid a futile cycle, that is, a situation in which energy is spent uselessly. In any case, gluconeogenesis is a very important pathway, present in many organisms, because it allows them to obtain glucose from certain amino acids via oxalacetate, an intermediate of the citric acid cycle. This is particularly crucial in mammals because they cannot use fatty acids to synthesize glucose. Gluconeogenesis is the main pathway used by the liver to maintain glucose levels in blood, when supply from the diet or from the reserves (glycogen, see below) is not sufficient. It must be underlined that tissues like brain, erythrocytes, or renal medulla among others depend on glucose to obtain their energy [43].

Glucose

(Glycolysis)

COO^-

$C = O$

CH_3

Pyruvate

Fermentation   Transamination

COO^-

$HO \;-\; C \;-\; H$

CH_3

L-lactate

Transport into
mitochondria

L-alanine
(an amino acid)

Acetyl-CoA

Fatty acids
(via Krebs' cycle)

Krebs' cycle
+
Oxidative phosphorylation

Energy (ATP) +
reducing power (NADH)
+ final products (CO_2, H_2O)

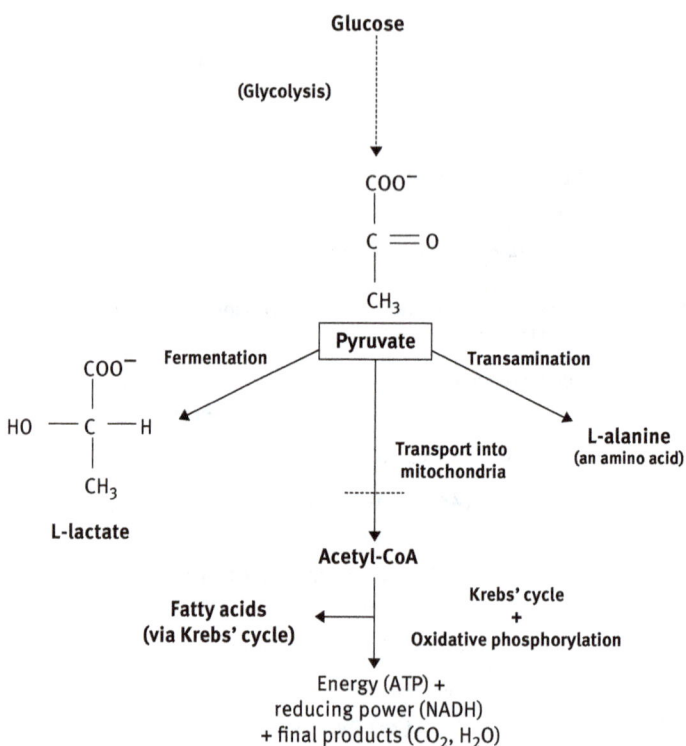

Figure 3.16: Different fates of pyruvate within human cells.

3.8.3 Regulation of glycolysis and gluconeogenesis

Thousands of chemical reactions that involve the synthesis and degradation of a myriad of substances take place within any living cell. These reactions, collectively known as metabolism, must be coordinated in space and time and adapt to the different circumstances that an organism may encounter. Biochemical reactions and metabolic fluxes, that is, the rates of substrate conversions, must be perfectly regulated to provide energy and sustain what we call "life." Regulation is a central concept in biochemistry and draws much attention in this field of research; however, many of its aspects remain to be established. Several interrelated levels of regulation have been described in organisms and cells: control of enzyme synthesis (which involves DNA transcription to yield mRNA, followed by splicing, stabilization, and translation of the latter), covalent modification of polypeptides, protein–protein interaction and allosteric regulation (see below), among others [90]. Some examples of regulation of the glycolytic pathway are:

- **In vivo role of key enzymes.** The three irreversible steps of glycolysis are relatively slow and crucial for the regulation of the whole pathway. Common features of these reactions are that they are kept far from equilibrium in vivo and that the enzymes involved (hexokinase, phosphofructokinase-1 (PFK-1), and pyruvate kinase) are allosteric [43]. The concept of allostery was introduced and developed in the 1960s by Monod, Wyman, and Changeaux to explain the behavior of certain proteins whose kinetic properties can be finely modulated by some substances, known as "effectors," which are not necessarily analogues of their substrates [91]. Many metabolites (participating either in glycolysis or in other pathways) are positive or negative effectors of the mentioned enzymes in vivo, thereby controlling both the levels of metabolic intermediates and the flux of glycolysis in coordination with other metabolic routes. The concept of allostery has evolved along the decades, and nowadays it is considered a fundamental concept in life sciences, known sometimes as "the second secret of life" (the genetic code being "the first secret") [92].
- **Reciprocal regulation of glycolysis and gluconeogenesis.** In cells like hepatocytes, both glycolysis and gluconeogenesis can take place and, thus, the possibility of a futile cycle, a situation that must be controlled in order to optimize cellular bioenergetics. Several mechanisms of reciprocal regulation have been elucidated:
  - When pyruvate is translocated to the mitochondria, it can be converted either to acetyl-CoA, in a reaction catalyzed by the enzymic complex pyruvate dehydrogenase (PDH) or to oxaloacetate (OAA) by pyruvate carboxylase (PC, EC 6.4.1.1). OAA can be converted to phosphoenolpyruvate, thus initiating gluconeogenesis. Acetyl-CoA is a negative effector of PDH and a positive effector of PC; thus, its concentration determines the rates of glycolysis and gluconeogenesis. High concentrations of this metabolite inhibit PDH, slowing glycolysis, and activate PC, speeding up gluconeogenesis. Low concentrations of acetyl-CoA have opposite effects.
  - PFK-1 catalyzes the phosphorylation of fructose-6-phosphate with ATP to yield fructose-1,6-bisphosphate and ADP. This reaction is irreversible in vivo, and the reverse gluconeogenetic step, catalyzed by fructose-1,6-bisphosphatase (FBPase-1), is slightly different: fructose-1,6-bisphosphate is hydrolyzed giving fructose-6-phosphate and inorganic phosphate as products. Both PFK-1 and FBPase are allosteric enzymes whose respective effectors are metabolites related to the energetic status of the cell. ATP and citrate are negative effectors of PFK; therefore, when the concentration of either of them is high (the cell has plenty of energy), the enzyme is inhibited and glycolysis, an energy-yielding pathway, is slowed. High concentrations of ADP or AMP, indicating a low energy status in the cell, activate PFK speeding up glycolysis to provide energy (ATP) and reducing power (NADH). AMP simultaneously slows down gluconeogenesis by inhibiting FBPase.

–   PFK-1 and FBPase-1 can be reciprocally regulated by hormones (insulin and glucagon) in a rather complicated fashion that involves allosteric control and covalent modifications (phosphorylation). Hormone-mediated regulation of other enzymes of the glycolytic pathway, like pyruvate kinase, also involves phosphorylation/dephosphorylation.

3.8.4 Glycogen synthesis and degradation

Glycogen synthesis is the main fate of excess glucose in many cells, although it is especially important in skeletal muscle cells and hepatocytes. Glycogen is an efficient way to store energy in a rapidly mobilizable form and it also minimizes the osmotic effects that a high concentration of glucose might exert in the cytosol. In animal cells, glycogen is usually found forming granules that contain the carbohydrate and a number of proteins related to its metabolism in an organelle-like structure known as "glycosome." Glycosomes are located in the cytosol either free or associated with actin filaments and/or sarcoplasmic reticulum of skeletal muscle cell. The subcellular distribution of glycosomes, their structure, as well as the regulation of glycogen metabolism may vary among different cell types [93].

The major steps in **glycogen synthesis from glucose** are:

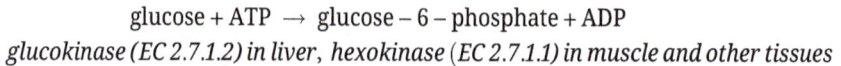

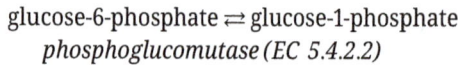

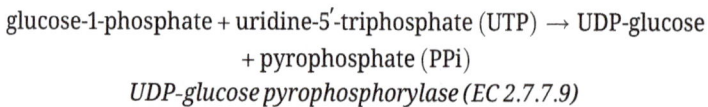

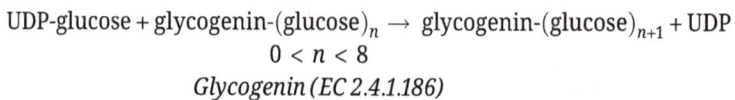

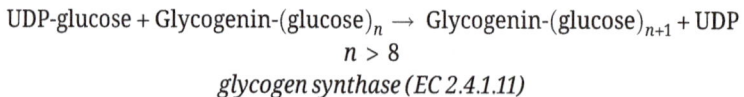

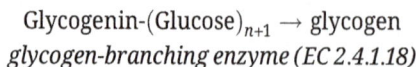

$$\text{glucose} + \text{ATP} \rightarrow \text{glucose} - 6 - \text{phosphate} + \text{ADP}$$
glucokinase (EC 2.7.1.2) in liver, hexokinase (EC 2.7.1.1) in muscle and other tissues

$$\text{glucose-6-phosphate} \rightleftharpoons \text{glucose-1-phosphate}$$
phosphoglucomutase (EC 5.4.2.2)

$$\text{glucose-1-phosphate} + \text{uridine-5}'\text{-triphosphate (UTP)} \rightarrow \text{UDP-glucose} + \text{pyrophosphate (PPi)}$$
UDP-glucose pyrophosphorylase (EC 2.7.7.9)

$$\text{UDP-glucose} + \text{glycogenin-(glucose)}_n \rightarrow \text{glycogenin-(glucose)}_{n+1} + \text{UDP}$$
$$0 < n < 8$$
Glycogenin (EC 2.4.1.186)

$$\text{UDP-glucose} + \text{Glycogenin-(glucose)}_n \rightarrow \text{Glycogenin-(glucose)}_{n+1} + \text{UDP}$$
$$n > 8$$
glycogen synthase (EC 2.4.1.11)

$$\text{Glycogenin-(Glucose)}_{n+1} \rightarrow \text{glycogen}$$
glycogen-branching enzyme (EC 2.4.1.18)

Glycogen synthesis cannot be initiated by the reaction between UDP-glucose and a single molecule of glucose, instead, a protein known as glycogenin takes a glucose residue from a molecule of UDP-glucose in a reaction catalyzed by itself, a process that is repeated until an oligosaccharide of eight residues is bound to glycogenin. Then, glycogen synthase adds successive glucose residues using UDP-glucose as donor. A branching enzyme gives shape to the final structure of glycogen (see Figure 3.9).

Glucose is released from glycogen by phosphorolysis mediated by the enzyme glycogen-phosphorylase (EC. 2.4.1.1) that transfers an α(1,4)-linked glucose residue from the nonreducing end of a glycogen chain to inorganic phosphate yielding glucose-1-phosphate. The latter can be converted to glucose-6-phosphate by the phosphoglucomutase, thus entering the glycolytic pathway (Figure 3.17). A debranching enzyme is needed to hydrolyze α(1,6)-linked glucose residue.

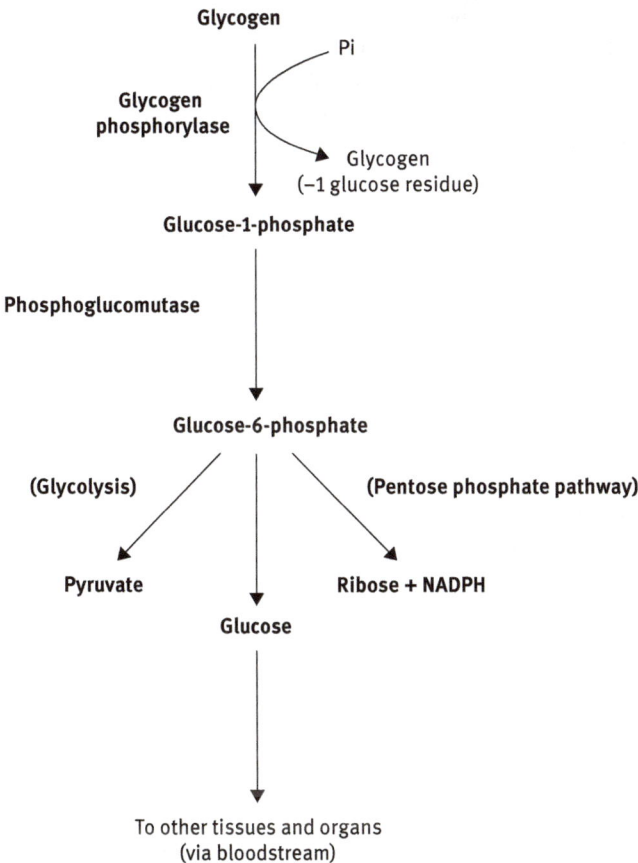

Figure 3.17: Glycogen degradative metabolism and possible fates of glucose (adapted from "File Glc6P fate.svg" by Zlir'a; available at: https://commons.wikimedia.org/wiki/File:Glc6P_fate.svg).

Glycogen synthesis and degradation is a good example of hormonal reciprocal regulation. Adrenaline (aka epinephrine) and glucagon inhibit glycogen synthesis and increase degradation, by hormone-mediated phosphorylation, one of the most important types of covalent modification. Insulin, released in blood under conditions of hyperglycemia, stimulates the synthesis of glycogen by a similar mechanism that involves phosphatases and kinases [43, 94].

3.8.5 Pentose phosphate pathway of glucose oxidation

The pentose phosphate pathway (PPP) provides an alternative to glycolysis in all cells. About 10–20% of glucose oxidation has been reported to occur via PPP in most cells, whereas glycolysis is responsible for the remaining 80–90%. Both pathways are cytosolic and do not require oxygen but, unlike glycolysis, PPP does not generate ATP, its two main functions being the synthesis of ribose-5'-phosphate, required for nucleotide and nucleic acid synthesis, and the production of NADPH, a reducing agent in many biosynthetic pathways and a protectant against oxidative damage [95]. Two phases can be distinguished in PPP (Figure 3.18):

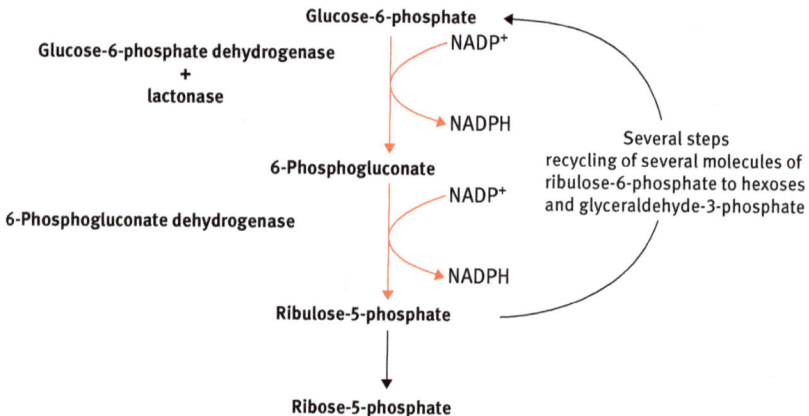

Figure 3.18: Schematic diagram of the pentose phosphate pathway (PPP). Red lines indicate steps of the oxidative phase.

- The **oxidative phase**, which consists of three irreversible reactions that yield NADPH and pentose phosphate.
- The **nonoxidative phase**, a set of reversible reactions that reconvert pentose phosphates into glucose-6-phosphate, and two glycolytic intermediates: fructose-6-phosphate and glyceraldehyde-3-phosphate.

Glucose-6-phosphate dehydrogenase (G6PDH) catalyzes the first reaction of PPP, which involves oxidation of glucose-6-phosphate and generation of the first molecule of NADPH. This is the rate-limiting step of the pathway; therefore, G6PDH is tightly controlled by extracellular stimuli and signaling pathways that regulate its expression and modulate its activity via post-translational mechanisms. Moreover, G6PDH is also subjected to allosteric regulation by their own catalytic products (the $NADP^+$/NADPH ratio is one of the main modulators) and other metabolites [96].

PPP has raised much attention due to its importance for tumor cells and the discovery of several heritable disorders related to genes encoding enzymes of this pathway. Actually, reduced activity or deficiency of G6PDH is the most common heritable human enzyme defect, present in more than 400 million people. The most frequent clinical manifestations are neonatal jaundice and acute hemolytic anemia, often triggered by oxidant agents, fava beans (favism), and certain drugs. In red blood cells, PPP is the exclusive source of NADPH, which is required for the replenishment of reduced glutathione (GSH), essential for detoxification of hydrogen peroxide, oxygen radicals, as well as for the maintenance of hemoglobin and other blood proteins in their reduced state [95, 96].

Vicia faba is a plant of the Leguminosae family that produces beans (broad or fava beans) containing high concentrations of two β-glucosides: vicine and convicine. These compounds undergo hydrolysis by glucosidases that are present both in the beans and in the gastrointestinal tract releasing the respective aglycones: divicine (2,6-diamino-4,5-dihydroxypyrimidine) and isouramil (6-amino-2,4,5-trihydroxypyrimidine). Divicine and isouramil are highly reactive redox compounds with antifungal and pesticide activity, which probably are beneficial for the beans. The erythrocytes of people with low levels of G6PDH cannot produce enough NADPH to inactivate the oxygen radicals produced by these substances, being seriously damaged (hemolysis) by oxidative attack after ingestion of fava beans. Although most patients of G6PD deficiency remain asymptomatic throughout their life, others may show clinical manifestations: in newborns it is recognized as a serious risk factor for the development of neonatal hyperbilirubinemia and kernicterus (symptoms include lethargy, extreme sleepiness, and poor muscle tone), and in adults, symptoms may include pallor, jaundice, fatigue, splenomegaly, and dark urine [97]. Favism is common in Southern Europe, Middle East, and Southeast Asia [98].

3.9 Regulation of blood glucose

Blood transports oxygen and nutrients to the cells and removes waste products. It also carries hormonal signals, thus playing a major role in the integration of metabolism at the organismal level. Nearly 50% of blood volume is occupied by three types of cells: erythrocytes, leukocytes, and platelets, the rest being plasma, composed of water (90%) and solutes. The latter consist of a large variety of proteins, nutrients, metabolites, inorganic ions, and hormones dissolved or suspended in the liquid milieu. The overall composition has to be maintained within certain limits in a dynamic steady state essential for life. This is accomplished by the combined function of multiple tissues and organs like liver, lungs, kidneys, and others [43].

Glucose is one of the most important metabolites in human blood, and its levels must be maintained in a range between 3.5 and 5.5 mmol/L (0.6–1 g/L) before meals. This concentration depends on the rate of glucose entering the circulation balanced by the rate of glucose removal. Glucose may enter the bloodstream via intestinal absorption after food ingestion or from the glycogenolysis and gluconeogenesis that occur in the liver (contribution of renal gluconeogenesis is significant only during periods of extreme starvation). Overall, glucose concentration in blood is regulated by a complex mechanism that basically involves three hormones: insulin, glucagon, and adrenaline, although others are also implicated (see also Chapter 14 for control of energy balance by these hormones) [99]:

- **Insulin** and **amylin** are polypeptidic hormones co-secreted by pancreatic β-cells in a molar ratio of approximately 50:1, the ratio falling to ~ 20:1 in the peripheral circulation, due to insulin removal by the different insulin-sensitive tissues. Insulin is the only known hormone that can lower blood glucose concentration [100], the latter being the most potent stimulus of its secretion, and exerts its actions through binding to specific receptors present mainly in cells of the adipose tissue, liver, and muscle. Insulin signals cells to increase their uptake of glucose, promotes glycogenesis and glycolysis in liver and muscle, and inhibits glucagon secretion from pancreatic cells. It is also important to underline that insulin influences other important aspects of metabolism; thus, it increases fatty acid and triacylglycerol synthesis in liver and adipocytes, respectively, and regulates protein synthesis and degradation in tissues like skeletal muscle [101] (Figure 14.4). Amylin complements the effects of insulin.
- **Glucagon** is a polypeptidic hormone produced by pancreatic α-cells with opposite effects to those of insulin (Figure 14.5). Glucagon secretion increases when plasma glucose falls below the normal range. It regulates glucose production in the liver by promoting gluconeogenesis and glycogenolysis and induces ketogenesis (Chapter 4). Glucagon also stimulates fatty acid mobilization from the adipose tissue.
- **Adrenaline** (aka epinephrine), produced from the amino acid tyrosine by adrenal medulla cells, prepares the body for a fight-or-flight response acting mainly on muscle, adipose tissue, and liver. Many of its effects are similar to those exerted by glucagon; actually, adrenaline increases glucagon secretion and decreases insulin secretion (Table 3.4).
- **Glucagon-like peptide** (GLP-1) and **glucose-dependent insulinotropic peptide** (aka gastric inhibitory polypeptide, GIP). Ingestion of food causes a potent release of insulin ("incretin effect"), indicating that signals from the gut are important in the hormonal regulation of glucose disappearance. The most important incretin

hormones for glucose homeostasis are GIP and GLP-1, synthesized by enteroendo-crine K-cells, predominantly found in the duodenum, and L-cells, present mainly in the ileum and colon. GIP stimulates insulin secretion and regulates fat metabo-lism, but does not inhibit glucagon secretion or gastric emptying, whereas GLP-1 has a wide range of glucoregulatory effects and regulates gastric emptying and gut motility.

– **Cortisol** (sometimes called "the stress hormone") is a biological steroid derived from cholesterol (Chapter 4) produced by the cortical cells of the adrenal gland, which increases gluconeogenesis in the liver. Like all steroid hormones, cortisol diffuses across the plasma membrane of the target cells and travels to the nucleus where it changes the level of expressions of certain genes after binding to specific nuclear receptors [43].

– **Ghrelin** is a polypeptide derived from a 117 amino acid precursor (preproghrelin) produced by X/A-like cells within gastric oxyntic glands of the stomach. Preprogh-relin is subsequently cleaved into a small signal peptide, ghrelin and obestatin. Ghrelin exerts a wide range of physiological effects including modulation of insu-lin secretion, albeit in a rather complex way [102].

Table 3.4: Metabolic effects of adrenaline.

Increase	Decrease
Glycogen breakdown in muscle and liver	Glycogen synthesis in muscle and liver
Gluconeogenesis in liver	Insulin secretion
Glycolysis in muscle	
Fatty acid mobilization in adipose tissue	
Glucagon secretion	

Adapted from [43].

3.10 Diabetes: types, metabolic changes, and dietary recommendations

Diabetes is a Latin word derived from ancient Greek that can be translated as "excess of urine" and refers to a group of diseases characterized by this symptom. Two major types of diabetes can be distinguished:

– **Diabetes insipidus (DI).** Patients with this pathology excrete abnormally large volumes of dilute urine. Three fundamentally different defects can lead to DI: (1) pituitary DI, due to inadequate production and secretion of antidiuretic hormone (ADH, aka vasopressin); (2) primary polydipsia, due to suppression of ADH secre-tion by excessive fluid intake; and (3) nephrogenic DI due to renal insensitivity to the antidiuretic effect of ADH [103].

- **Diabetes mellitus (DM)** refers to a heterogeneous group of disorders character-ized by hyperglycemia due to a deficit in insulin production, action, or both. One of its main symptoms is the presence of high concentrations of glucose in the urine, which causes frequent urination due to osmotic effects. There are two major forms of DM: type I DM (previously known as insulin-dependent, juvenile, or childhood-onset diabetes) and type II DM (formerly called non-insulin-dependent or adult-onset diabetes) [104]. Other forms of the disease that involve mutations in one (monogenic diabetes) or several (polygenic diabetes) genes have also been identi-fied. These rarer types of DM include neonatal diabetes, maturity onset diabetes of the young, DM caused by mutations in mitochondrial DNA, and others that are cur-rently under study [105]. According to a recent report by the WHO, "Globally, an estimated 422 million adults were living with diabetes in 2014, compared to 108 million in 1980. The global prevalence (age-standardized) of diabetes has nearly doubled since 1980, rising from 4.7% to 8.5% in the adult population [. . .]. Over the past decade, diabetes prevalence has risen faster in low- and middle-income coun-tries than in high-income countries." The WHO also estimates that diabetes caused 1.5 million deaths in 2012 and higher than optimal blood glucose caused an addi-tional 2.2 million deaths, by increasing the risks of cardiovascular and other dis-eases. Forty-three percent of these 3.7 million deaths occur before the age of 70 years [104]. It must be borne in mind that both type I and type II diabetes are diseases with considerable variations in their clinical presentation and progression; actually, some individuals cannot be clearly classified as having one type of diabe-tes or the other at the time of diagnosis [106]. Another type of DM is **gestational diabetes** caused by a failure of the β-cells to cope with the decrease in sensitivity of the gestating mother's cells to insulin, a natural mechanism aimed at diverting glucose toward the growing fetus. Hormones such as prolactin, lactogen, and kiss-peptin, produced by the placenta, seem to play a major role in the necessary adap-tations of the β-cells during pregnancy; however, the mechanisms involved are not fully understood [107].

3.10.1 Type I diabetes mellitus (T1DM)

Type 1 diabetes mellitus (T1DM) is an autoimmune disease, that is, cells of the body's immune system cause destruction of pancreatic β-cells, leading to a deficiency of insu-lin production. It is a relatively rare form of the disease affecting 5–10% of diabetics that is usually diagnosed in childhood and is not associated with excess body weight [108]. Typical symptoms of this disease are: polyuria (excessive urination), polydipsia (excessive thirst that leads to excessive drinking), weight loss, ketonemia and ketonu-ria (high concentrations of ketone bodies in blood and urine, respectively), and acido-sis (lowering of blood pH). The last three symptoms are due to production of ketone bodies by the liver (Chapter 4), which cannot sense the high levels of blood glucose due to the absence of insulin. This stimulates the liver to deplete citric acid intermediates to produce glucose by gluconeogenesis and to use acetyl-CoA to produce ketone bodies, an alternative fuel to glucose for the brain. Treatment of T1DM involves insulin injections and a careful control of the balance between dietary intake and insulin dose [43]; however, even with the most advanced treatments available today, only about 21% of people with T1DM are able to achieve near-normal blood glucose levels. This means that much research is still necessary to find an optimal control and, eventually, a cure for this disease. The last scientific evidence has shown that T1DM may be more complex than previously thought; thus, it has been reported that pancreatic β-cells are not only the victims of an immune attack but they probably contribute to this abnormal behavior of the immune system. Consequently, it has been suggested that T1DM "should no longer be considered only a disease of the immune system, but also a disease of the β-cell" [109].

3.10.2 Type II diabetes mellitus (T2DM)

Type 2 diabetes mellitus (T2DM) is associated with both impaired insulin release and impaired insulin action (insulin resistance), thus resulting in an ineffective use of this hormone by the body. It is a progressive disease that starts with insulin resistance and evolves to diabetes as β-cells gradually fail. Although elevated blood lipid levels have been previously reported to be responsible for β-cell failure (lipotoxicity), current evidence suggests that it has rather increased blood glucose levels, the main factor that causes the disease [100]. T2DM accounts for the vast majority of people with diabetes around the world (around 90% of diagnosed cases) and its symptoms may be similar to those of T1DM, but are often less marked or absent. As a result, the disease may go undiagnosed for several years, until complications have already arisen. Although it is usually considered as an adult disease, many cases are already being reported in children. T2DM is a complex disease caused by an interplay between genetic and environmental factors, such as ethnicity, family history of diabetes, and previous gestational diabetes, which can combine with age, overweight and obesity, unhealthy diet, physical inactivity (Chapter 14), and smoking to increase risk [104]. Identification of the genetic and epigenetic factors implicated in T2DM is a major challenge for the scientific community [110, 111].

The hallmark of T2DM is the gradual development of insulin resistance, that is, muscle and liver show lower sensitivity to insulin. In the initial stages of the disease, pancreatic β-cells produce enough hormone to overcome this situation but they eventually fail. There are intermediate conditions between normal blood glucose levels and T2DM. This transient situation, known as prediabetes, can be defined by the presence of impaired fasting glucose (IFG) and/or impaired glucose tolerance (IGT), or glycated hemoglobin (HbA1c or simply A1C). IFG is defined by an elevated fasting plasma glucose (FPG) concentration and IGT by an elevated 2 h plasma glucose concentration after a 75 g glucose load on the oral glucose tolerance test in the presence of an FPG concentration of <7 mmol/L. The WHO and the American Diabetes Association (ADA) use different cutoff values for IFG (WHO: 6.1–6.9 mmol/L; ADA: 5.6–6.9 mmol/L) but the same values for IGT (7.8–11.0 mmol/L). IFG has been reported to be the result of hepatic insulin resistance, whereas IGT seems to be the result of insulin resistance in skeletal muscle; in any case, pancreatic β-cell dysfunction seems to be common to both. Prediabetes has also been identified by mildly elevated glycated hemoglobin (HbA1c), although IFG and IGT remain the current recommendations for the identification and diagnosis of this disorder. People with IGT or IFG are at increased risk of heart attacks and strokes [112, 113]. Transition from prediabetes to T2DM is not inevitable; therefore, understanding the former may be crucial to reducing the global T2DM epidemic [104].

The risk of developing T2DM is also related to the so-called metabolic syndrome (MS) or X syndrome, an accumulation of several disorders that are diagnosed when at least three of the following conditions are met [114]:

- Elevated waist circumference: ≥102 cm in males, ≥88 cm in females
- Elevated triglycerides (Chapter 4): ≥150 mg/dL (1.7 mmol/L)
- Reduced high-density lipoproteins (Chapter 4): <40 mg/dL (1.0 mmol/L) in males, <50 mg/dL (1.3 mmol/L) in females
- Elevated blood pressure: Systolic ≥ 130 mm Hg and/or diastolic ≥ 85 mm Hg
- Elevated fasting glucose: ≥100 mg/dL

Individuals with MS often show symptoms like abnormal clotting and chronic inflammation and have an elevated risk of developing other pathologies like atherosclerotic cardiovascular disease and vascular and neurological complications.

Although the pathogenesis of T2DM is not completely understood, overweight is one of the most important environmental factors associated with this disease; actually, it has been reported that 80% of the individuals with T2DM are obese. However, many obese individuals do not develop T2DM, which demonstrates that obesity is not the primary cause of the disease and suggests that genetic factors are also involved [43, 100].

Untreated T2DM may result in microvascular complications (retinopathy, neuropathy, or nephropathy) that can lead to limb amputations, blindness, and kidney failure. It also doubles the risk of macrovascular complications compared to the general population. The global prevalence of diabetes has been predicted to reach more than 7.5% of the world's total population in 2030, paralleling the aging and body mass index (BMI), thus confirming the relationship between obesity and diabetes. Although medication that increases insulin sensitivity or production is available to treat T2DM, there is a consensus among the scientific community that prevention, which basically implies a combination of healthy diet, weight loss, and physical exercise, is the best approach to tackle this epidemic [115].

3.11 Other problems related to the metabolism of carbohydrates

DM (especially T2DM that accounts for the vast majority of diagnosed cases) is considered one of the four major noncommunicable diseases (NCDs, or chronic diseases) that directly or indirectly kill millions of people worldwide every year. However, there are many other diseases or disorders related to carbohydrate metabolism that deserve to be mentioned:

– **Galactosemia**. This term encompasses three distinct enzymatic deficiencies in the pathway that allows galactose to enter glycolysis (Leloir pathway). The most prevalent and severe is classic galactosemia or type I galactosemia, caused by deficient activity of galactose-1-phosphate uridylyltransferase (EC 2.7.7.12), an enzyme required by galactose to enter the glycolytic pathway. It is an autosomal recessive disorder caused by mutations in the GALT gene, whose prevalence is 1:16,000–60,000 live births. Dietary restriction of galactose has been the therapeutic basis for classic galactosemia; however, this treatment is insufficient to prevent severe long-term complications, such as cognitive, social, and reproductive impairments. The exact pathogenic mechanisms underlying this disease are still not fully understood [116, 117].

– **Lactose intolerance (LI)**. This term refers to symptoms related to the consumption of lactose-containing foods: gas, bloat, abdominal cramps, and pain, sometimes associated with diarrhea, nausea, and/or vomiting. The most common cause of LI is the genetically determined adult onset lactose maldigestion due to loss of intestinal lactase. In these patients, consumption of lactose quantities that overwhelm residual lactase is fermented by intestinal bacteria producing gases (hydrogen, carbon dioxide, and methane) and SCFAs. These metabolic products, and their rate of production, are thought to contribute to symptoms of LI. Avoidance of lactose-containing (usually dairy) products is the most usual treatment for LI; however, symptoms could be modified by other food intolerances and the presence of functional gastrointestinal disorders. Therefore, a careful study must be made before prescribing a lactose-free diet [118].

– **Disorders of fructose metabolism**. Three inborn errors related to fructose metabolism have been identified to date: essential fructosuria, hereditary fructose intolerance, and FBPase deficiency [119].

– **Glycogen storage diseases (GSDs)**. GSDs are a group of rare inherited metabolic disorders that result from a defect in any one of several enzymes required for either glycogen synthesis or glycogen degradation. They can be divided into those with hepatic involvement, which present as hypoglycemia, and those which are associated with neuromuscular disease and weakness. The severity of the GSDs range from those that are fatal in infancy if untreated to mild disorders with a normal life span [120, 121].

– **Dental caries** is an oral disease that results in the removal of inorganic ions (basically, calcium and phosphate) from the enamel of teeth (demineralization). It is commonly accepted that dental caries is a result of the carbohydrate metabolism of the community of microorganisms (microbiota) that cover the surface of human oral cavity. Bacteria, like non-mutans streptococci and actinomyces, usually found in healthy individuals, produce organic acids by fermentation of monosaccharides (lactic, acetic, and others) that demineralize enamel at a rate that can be compensated by the natural remineralization. However, frequent intake of carbohydrates like sucrose or starch that can be readily hydrolyzed in the oral cavity producing fermentable monosaccharides like glucose and fructose increases the rate of organic acid production. This process induces changes in the oral microbiota by increasing the presence of acidogenic and acidophilic bacteria (such as *Streptococcus mutans*), which ultimately shift the demineralization/remineralization balance toward net mineral loss [122].

References

[1]   Dwek RA. Glycobiology: Toward understanding the function of sugars. Chem Rev 1996;96:683–720.

[2]   Nič M, Jirát J, Košata B, Jenkins A, McNaught A, editors. IUPAC Compend. Chem. Terminol. 2.1.0, Research Triangle Park, NC: IUPAC; 2009. https://doi.org/10.1351/goldbook.G02645.

[3]   Springer S, Gagneux P. Glycan evolution in response to collaboration, conflict, and constraint. J Biol Chem 2013:jbc. R112. 424523.

[4]   Varki A, Kornfeld S. Historical background and overview. In: Varki A, Cummings RD, Esko JD, Stanley P, Hart GW, Aebi M, et al., editors. Essent. Glycobiol. 3rd edition. Cold Spring Harbor (NY): Cold Spring Harbor Laboratory Press; 2015.

[5]   Rudd P, Karlsson NG, Khoo K-H, Packer NH. Glycomics and glycoproteomics. In: Varki A, Cummings RD, Esko JD, Stanley P, Hart GW, Aebi M, et al., editors. Essent. Glycobiol. 3rd edition. Cold Spring Harbor (NY): Cold Spring Harbor Laboratory Press; 2015.

[6]   Seeberger PH. Monosaccharide diversity. In: Varki A, Cummings RD, Esko JD, Stanley P, Hart GW, Aebi M, et al., editors. Essent. Glycobiol. 3rd edition. Cold Spring Harbor (NY): Cold Spring Harbor Laboratory Press; 2015.

[7]   Brecher J. (IUPAC Recommendations 2006). Pure Appl Chem 2006:74.

[8]   Moss GP. Basic terminology of stereochemistry (IUPAC Recommendations 1996). Pure Appl Chem 1996;68:2193–222.

[9]   Schombs M, Gervay-Hague J. Glycochemistry. Glycochemical Synth. Wiley-Blackwell; 2016, p. 1–34. https://doi.org/10.1002/9781119006435.ch1.

[10]   Wang X, Woods RJ. Insights into Furanose Solution Conformations: Beyond the Two-State Model. J Biomol NMR 2016;64:291–305. https://doi.org/10.1007/s10858-016-0028-y.

[11]   Ballard O, Morrow AL. Human Milk Composition: Nutrients and Bioactive Factors. Pediatr Clin North Am 2013;60:49–74. https://doi.org/10.1016/j.pcl.2012.10.002.

[12] Amiri M, Diekmann L, von Köckritz-Blickwede M, Naim HY. The Diverse Forms of Lactose Intolerance and the Putative Linkage to Several Cancers. Nutrients 2015;7:7209–30. https://doi.org/10.3390/nu7095332.

[13] Jiang S-Y, Chi Y-H, Wang J-Z, Zhou J-X, Cheng Y-S, Zhang B-L, et al. Sucrose metabolism gene families and their biological functions. Sci Rep 2015; 5. https://doi.org/10.1038/srep17583.

[14] Michèle P. OECD-FAO agricultural outlook 2018–2027, 2018:10.

[15] Markopoulos AK. Oral Microbial Flora. Handb Atten Deficit Hyperact Disord ADHD Interdiscip Perspect 2016:53.

[16] Garcia-Etxebarria K, Zheng T, Bonfiglio F, Bujanda L, Dlugosz A, Lindberg G, et al. Increased Prevalence of Rare Sucrase-isomaltase Pathogenic Variants in Irritable Bowel Syndrome Patients. Clin Pract J Am Gastroenterol Assoc 2018;16:1673–6. https://doi.org/10.1016/j.cgh.2018.01.047.

[17] Ceunen S, Geuns JMC. Steviol Glycosides: Chemical Diversity, Metabolism, and Function. J Nat Prod 2013;76:1201–28. https://doi.org/10.1021/np400203b.

[18] Newman RA, Yang P, Pawlus AD, Block KI. Cardiac glycosides as novel cancer therapeutic agents. Mol Interv 2008;8:36.

[19] Brazier-Hicks M, Evans KM, Gershater MC, Puschmann H, Steel PG, Edwards R. The C-Glycosylation of Flavonoids in Cereals♦. J Biol Chem 2009;284:17926–34. https://doi.org/10.1074/jbc.M109.009258.

[20] Rask L, Andréasson E, Ekbom B, Eriksson S, Pontoppidan B, Meijer J. Myrosinase: Gene family evolution and herbivore defense in Brassicaceae. Plant Mol Biol 2000;42:93–113.

[21] Magiorkinis E, Beloukas A, Diamantis A. Scurvy: Past, present and future. Eur J Intern Med 2011;22:147–52.

[22] De Tullio MC. The mystery of vitamin C. Nat Educ 2010;3:48.

[23] C. De Tullio M. The Function of Ascorbic Acid Through Occam's Razor: What We Know, What We Presume and What We Hope For. Ascorbic Acid – Biochem. Funct. Work. Title, IntechOpen; 2023. https://doi.org/10.5772/intechopen.109434.

[24] Du J, Cullen JJ, Buettner GR. Ascorbic acid: Chemistry, biology and the treatment of cancer. Biochim Biophys Acta BBA-Rev Cancer 2012;1826:443–57.

[25] Cameron E, Pauling L, Leibovitz B. Ascorbic acid and cancer: A review. Cancer Res 1979;39:663–81.

[26] Kuiper C, Vissers MCM. Ascorbate as a Co-Factor for Fe- and 2-Oxoglutarate Dependent Dioxygenases: Physiological Activity in Tumor Growth and Progression. Front Oncol 2014;4. https://doi.org/10.3389/fonc.2014.00359.

[27] Camarena V, Wang G. The epigenetic role of vitamin C in health and disease. Cell Mol Life Sci 2016;73:1645–58. https://doi.org/10.1007/s00018-016-2145-x.

[28] Lykkesfeldt J. On the effect of vitamin C intake on human health: How to (mis)interprete the clinical evidence. Redox Biol 2020;34:101532. https://doi.org/10.1016/j.redox.2020.101532.

[29] Vernieri C, Nichetti F, Raimondi A, Pusceddu S, Platania M, Berrino F, et al. Diet and supplements in cancer prevention and treatment: Clinical evidences and future perspectives. Crit Rev Oncol Hematol 2018;123:57–73. https://doi.org/10.1016/j.critrevonc.2018.01.002.

[30] Gallie DR. Increasing Vitamin C Content in Plant Foods to Improve Their Nutritional Value – Successes and Challenges. Nutrients 2013;5:3424–46. https://doi.org/10.3390/nu5093424.

[31] Oroian M, Escriche I. Antioxidants: Characterization, natural sources, extraction and analysis. Food Res Int 2015;74:10–36. https://doi.org/10.1016/j.foodres.2015.04.018.

[32] Frikke-Schmidt H, Tveden-Nyborg P, Lykkesfeldt J. Chapter 9. Vitamin C–Ascorbic acid. Vitam. Prev. Hum. Dis. De Gruyter; 2011.

[33] Tsukaguchi H, Tokui T, Mackenzie B, Berger UV, Chen X-Z, Wang Y, et al. A family of mammalian Na+-dependent L-ascorbic acid transporters. Nature 1999;399:70–5. https://doi.org/10.1038/19986.

[34] Scientific Opinion on Dietary Reference Values for vitamin C. EFSA J 2013;11:3418. https://doi.org/10.2903/j.efsa.2013.3418.

[35] Carr AC, Lykkesfeldt J. Discrepancies in global vitamin C recommendations: A review of RDA criteria and underlying health perspectives. Crit Rev Food Sci Nutr 2021;61:742–55. https://doi.org/10.1080/10408398.2020.1744513.

[36] Santana ÁL, Meireles MAA. New starches are the trend for industry applications: A review. Food Public Health 2014;4:229–41.

[37] Buléon A, Colonna P, Planchot V, Ball S. Starch granules: Structure and biosynthesis. Int J Biol Macromol 1998;23:85–112.

[38] Bello-Perez LA, Flores-Silva PC, Agama-Acevedo E, Tovar J. Starch digestibility: Past, present, and future. J Sci Food Agric 2018. https://doi.org/10.1002/jsfa.8955.

[39] Nichols BL, Avery S, Sen P, Swallow DM, Hahn D, Sterchi E. The maltase-glucoamylase gene: Common ancestry to sucrase-isomaltase with complementary starch digestion activities. Proc Natl Acad Sci 2003;100:1432–7.

[40] Silva YP, Bernardi A, Frozza RL. The Role of Short-Chain Fatty Acids From Gut Microbiota in Gut-Brain Communication. Front Endocrinol 2020;11:25. https://doi.org/10.3389/fendo.2020.00025.

[41] Dhital S, Warren FJ, Butterworth PJ, Ellis PR, Gidley MJ. Mechanisms of starch digestion by α-amylase – Structural basis for kinetic properties. Crit Rev Food Sci Nutr 2017;57:875–92.

[42] Roach PJ, Depaoli-Roach AA, Hurley TD, Tagliabracci VS. Glycogen and its metabolism: Some new developments and old themes. Biochem J 2012;441:763–87. https://doi.org/10.1042/BJ20111416.

[43] Nelson DL, Cox MM, Lehninger AL. Lehninger principles of biochemistry. New York: W.H. Freeman; 2013.

[44] Kumar V, Sinha AK, Makkar HPS, de Boeck G, Becker K. Dietary roles of non-starch polysaccharides in human nutrition: A review. Crit Rev Food Sci Nutr 2012;52:899–935. https://doi.org/10.1080/10408398.2010.512671.

[45] Sharma A, Tewari R, Rana SS, Soni R, Soni SK. Cellulases: Classification, Methods of Determination and Industrial Applications. Appl Biochem Biotechnol 2016;179:1346–80. https://doi.org/10.1007/s12010-016-2070-3.

[46] Robert C, Bernalier-Donadille A. The cellulolytic microflora of the human colon: Evidence of microcrystalline cellulose-degrading bacteria in methane-excreting subjects. FEMS Microbiol Ecol 2003;46:81–9. https://doi.org/10.1016/S0168-6496(03)00207-1.

[47] McNeil NI. The contribution of the large intestine to energy supplies in man. Am J Clin Nutr 1984;39:338–42. https://doi.org/10.1093/ajcn/39.2.338.

[48] Vanhauwaert E, Matthys C, Verdonck L, De Preter V. Low-residue and low-fiber diets in gastrointestinal disease management, Volume 6; 2015. https://doi.org/10.3945/an.115.009688.

[49] Vanholme R, Demedts B, Morreel K, Ralph J, Boerjan W. Lignin Biosynthesis and Structure. Plant Physiol 2010;153:895. https://doi.org/10.1104/pp.110.155119.

[50] Lignin. n.d. http://polymerdatabase.com/polymer%20classes/Lignin%20type.html (accessed December 29, 2019).

[51] Kubicek CP. Fungi and lignocellulosic biomass. Somerset, United States: John Wiley & Sons, Incorporated; 2012.

[52] Hobbs JK, Lee SM, Robb M, Hof F, Barr C, Abe KT, et al. KdgF, the missing link in the microbial metabolism of uronate sugars from pectin and alginate. Proc Natl Acad Sci U S A 2016;113:6188–93. https://doi.org/10.1073/pnas.1524214113.

[53] Jayachandran M, Chen J, Chung SSM, Xu B. A critical review on the impacts of β-glucans on gut microbiota and human health. J Nutr Biochem 2018;61:101–10. https://doi.org/10.1016/j.jnutbio.2018.06.010.

[54]   Revanappa SB, Nandini CD, Salimath PV. Structural characterisation of pentosans from hemicellulose B of wheat varieties with varying chapati-making quality. Food Chem 2010;119:27–33. https://doi.org/10.1016/j.foodchem.2009.04.064.

[55]   Mirhosseini H, Amid BT. A review study on chemical composition and molecular structure of newly plant gum exudates and seed gums. Food Res Int 2012;46:387–98. https://doi.org/10.1016/j.foodres.2011.11.017.

[56]   Izydorczyk M, Cui S, Wang Q. Polysaccharide gums: Structures, functional properties, and applications. In: Cui, S.W., editor. Food Carbohydr. Chem. Phys. Prop. Appl. Boca Raton: FL: CRC Press; 2005.

[57]   Ciric M. Metasecretome Phage Display: A New Approach for Mining Surface and Secreted Proteins from Microbial Communities : a Thesis Presented in Partial Fulfillment of the Requirements for the Degree of Doctor of Philosophy in Biochemistry at Massey University, Palmerston North, New Zealand. Massey University; 2014.

[58]   Hamman JH. Chitosan Based Polyelectrolyte Complexes as Potential Carrier Materials in Drug Delivery Systems. Mar Drugs 2010;8:1305–22. https://doi.org/10.3390/md8041305.

[59]   Beta-glucan n.d. https://water.lsbu.ac.uk/water/glucan.html (accessed April 16, 2023).

[60]   Shoaib M, Shehzad A, Omar M, Rakha A, Raza H, Sharif HR, et al. Inulin: Properties, health benefits and food applications. Carbohydr Polym 2016;147:444–54. https://doi.org/10.1016/j.carbpol.2016.04.020.

[61]   Goldfein KR, Slavin JL. Why sugar is added to food: Food science 101. Compr Rev Food Sci Food Saf 2015;14:644–56.

[62]   Clemens R, M. Jones J, Kern M, Lee S-Y, J. Mayhew E, L. Slavin J, et al. Functionality of Sugars in Foods and Health. Compr Rev Food Sci Food Saf 2016;15. https://doi.org/10.1111/1541-4337.12194.

[63]   Waterschoot J, Gomand SV, Fierens E, Delcour JA. Production, structure, physicochemical and functional properties of maize, cassava, wheat, potato and rice starches. Starch – Stärke 2015;67:14–29. https://doi.org/10.1002/star.201300238.

[64]   Copeland L, Blazek J, Salman H, Tang MC. Form and functionality of starch. Food Hydrocoll 2009;23:1527–34.

[65]   Alcázar-Alay SC, Meireles MAA. Physicochemical properties, modifications and applications of starches from different botanical sources. Food Sci Technol 2015;35:215–36.

[66]   McKim JM. Food additive carrageenan: Part I: A critical review of carrageenan in vitro studies, potential pitfalls, and implications for human health and safety. Crit Rev Toxicol 2014;44:211–43.

[67]   Hofman DL, van Buul VJ, Brouns FJPH. Nutrition, Health, and Regulatory Aspects of Digestible Maltodextrins. Crit Rev Food Sci Nutr 2016;56:2091–100. https://doi.org/10.1080/10408398.2014.940415.

[68]   Xie SX, Liu Q, Cui SW. Starch modification and applications. In: Cui, S.W. editor. Food Carbohydr. Chem. Phys. Prop. Appl. Boca Raton: FL: CRC Press; 2005.

[69]   Carrageenan. n.d. https://water.lsbu.ac.uk/water/carrageenan.html (accessed April 16, 2023).

[70]   Syamdidi, Irianto H, Irianto G. Agar-abundant marine carbohydrate from seaweeds in Indonesia: Principles and applications; 2016, p. 255–62. https://doi.org/10.1201/9781315371399-19.

[71]   Corzo-Martinez M, Corzo N, Villamiel M, Del Castillo MD. Browning reactions. Food Biochem Food Process 2012;56.

[72]   Sullivan ML. Beyond brown: Polyphenol oxidases as enzymes of plant specialized metabolism. Front Plant Sci 2015; 5. https://doi.org/10.3389/fpls.2014.00783.

[73]   Chaudhuri J, Bains Y, Guha S, Kahn A, Hall D, Bose N, et al. The role of advanced glycation end products in aging and metabolic diseases: Bridging association and causality. Cell Metab 2018;28:337–52.

[74] Tamanna N, Mahmood N. Food Processing and Maillard Reaction Products: Effect on Human Health and Nutrition. Int J Food Sci 2015; 2015. https://doi.org/10.1155/2015/526762.

[75] Moreira ASP, Nunes FM, Domingues MR, Coimbra MA. Coffee melanoidins: Structures, mechanisms of formation and potential health impacts. Food Funct 2012;3:903–15. https://doi.org/10.1039/C2FO30048F.

[76] Carbohydrates in human nutrition. Report of a Joint FAO/WHO Expert Consultation. FAO Food Nutr Pap 1998;66:1–140.

[77] Lovegrove A, Edwards CH, De Noni I, Patel H, El SN, Grassby T, et al. Role of polysaccharides in food, digestion, and health. Crit Rev Food Sci Nutr 2017;57:237–53.

[78] Guideline: Sugars intake for adults and children. Geneva: World Health Organization; 2015.

[79] World Health Organization. Global action plan for the prevention and control of noncommunicable diseases 2013–2020. Geneva: World Health Organization; 2013.

[80] World Health Organization. Global status report on noncommunicable diseases 2014. Geneva: World Health Organization; 2014.

[81] Green H. Dietary carbohydrates: A food processing perspective. Nutr Bull 2015;40:77–82.

[82] Foster-Powell K, Holt SH, Brand-Miller JC. International table of glycemic index and glycemic load values. Am J Clin Nutr 2002;76:5–56.

[83] Vega-López S, Venn B, Slavin J. Relevance of the glycemic index and glycemic load for body weight, diabetes, and cardiovascular disease. Nutrients 2018;10:1361.

[84] Giese EC, Barbosa AM, Dekker RF. Pathways to bioactive oligosaccharides: Biological functions and potential applications. Handb Carbohydr Polym Dev Prop Appl 2010:279–309.

[85] Buyken AE, Mela DJ, Dussort P, Johnson IT, Macdonald IA, Stowell JD, et al. Dietary carbohydrates: A review of international recommendations and the methods used to derive them. Eur J Clin Nutr 2018;72:1625–43. https://doi.org/10.1038/s41430-017-0035-4.

[86] Clemente-Suárez VJ, Mielgo-Ayuso J, Martín-Rodríguez A, Ramos-Campo DJ, Redondo-Flórez L, Tornero-Aguilera JF. The Burden of Carbohydrates in Health and Disease. Nutrients 2022;14:3809. https://doi.org/10.3390/nu14183809.

[87] Huberts D, van der Klei IJ. Moonlighting proteins: An intriguing mode of multitasking. Biochim Biophys Acta 2010;1803:520–5. https://doi.org/10.1016/j.bbamcr.2010.01.022.

[88] Sun F, Dai C, Xie J, Hu X. Biochemical Issues in Estimation of Cytosolic Free NAD/NADH Ratio. PLoS One 2012;7. https://doi.org/10.1371/journal.pone.0034525.

[89] Vander Heiden MG, Cantley LC, Thompson CB. Understanding the Warburg effect: The metabolic requirements of cell proliferation. Science 2009;324:1029–33. https://doi.org/10.1126/science.1160809.

[90] Metallo CM, Heiden MGV. Understanding metabolic regulation and its influence on cell physiology. Mol Cell 2013;49:388–98. https://doi.org/10.1016/j.molcel.2013.01.018.

[91] Monod J, Wyman J, Changeux JP. On the nature of allosteric transitions: A plausible model. J Mol Biol 1965;12:88–118.

[92] Liu J, Nussinov R. Allostery: An Overview of Its History, Concepts, Methods, and Applications. PLOS Comput Biol 2016;12:e1004966. https://doi.org/10.1371/journal.pcbi.1004966.

[93] Prats C, Graham TE, Shearer J. The dynamic life of the glycogen granule. J Biol Chem 2018:jbc.R117.802843. https://doi.org/10.1074/jbc.R117.802843.

[94] Berg JM, Tymoczko JL, Stryer L. Biochemistry. Basingstoke: W.H. Freeman; 2012.

[95] Wamelink MMC, Struys EA, Jakobs C. The biochemistry, metabolism and inherited defects of the pentose phosphate pathway: A review. J Inherit Metab Dis 2008;31:703–17. https://doi.org/10.1007/s10545-008-1015-6.

[96] Patra KC, Hay N. The pentose phosphate pathway and cancer. Trends Biochem Sci 2014;39:347–54. https://doi.org/10.1016/j.tibs.2014.06.005.

[97] Richardson SR, O'Malley GF. Glucose 6 phosphate dehydrogenase deficiency. StatPearls, Treasure Island (FL): StatPearls Publishing; 2023.

[98] Luzzatto L, Arese P. Favism and Glucose-6-Phosphate Dehydrogenase Deficiency. N Engl J Med 2018;378:1068–9. https://doi.org/10.1056/NEJMc1801271.

[99] Aronoff SL, Berkowitz K, Shreiner B, Want L. Glucose Metabolism and Regulation: Beyond Insulin and Glucagon. Diabetes Spectr 2004;17:183–90.

[100] Haythorne E, Ashcroft FM. Metabolic regulation of insulin secretion in health and disease. The Biochemist 2021;43:4–8. https://doi.org/10.1042/bio_2021_116.

[101] James HA, O'Neill BT, Nair KS. Insulin Regulation of Proteostasis and Clinical Implications. Cell Metab 2017;26:310–23. https://doi.org/10.1016/j.cmet.2017.06.010.

[102] Poher A-L, Tschöp MH, Müller TD. Ghrelin regulation of glucose metabolism. Peptides 2018;100:236–42.

[103] Robertson GL. Diabetes insipidus: Differential diagnosis and management. Best Pract Res Clin Endocrinol Metab 2016;30:205–18. https://doi.org/10.1016/j.beem.2016.02.007.

[104] Roglic G, World Health Organization, editors. Global report on diabetes. Geneva, Switzerland: World Health Organization; 2016.

[105] Misra S, Owen KR. Genetics of Monogenic Diabetes: Present Clinical Challenges. Curr Diab Rep 2018;18. https://doi.org/10.1007/s11892-018-1111-4.

[106] ElSayed NA, Aleppo G, Aroda VR, Bannuru RR, Brown FM, Bruemmer D, et al. 2. Classification and Diagnosis of Diabetes: *Standards of Care in Diabetes – 2023*. Diabetes Care 2023;46:S19–40. https://doi.org/10.2337/dc23-S002.

[107] Smith LIF, Bowe JE. The pancreas and the placenta: Understanding gestational diabetes and why some islets fail to cope with pregnancy. The Biochemist 2021;43:42–6. https://doi.org/10.1042/bio_2021_115.

[108] Barr AJ. The biochemical basis of disease. Essays Biochem 2018;62:619–42.

[109] Weaver SA, Felton JL, Evans-Molina C. A parallax view of type 1 diabetes. The Biochemist 2021;43:22–6. https://doi.org/10.1042/bio_2021_112.

[110] Prasad RB, Groop L. Genetics of type 2 diabetes – pitfalls and possibilities. Genes 2015;6:87–123.

[111] Jackson M, Marks L, May GH, Wilson JB. The genetic basis of disease. Essays Biochem 2018;62:643–723.

[112] Nathan DM, Davidson MB, DeFronzo RA, Heine RJ, Henry RR, Pratley R, et al. Impaired fasting glucose and impaired glucose tolerance: Implications for care. Diabetes Care 2007;30:753–9.

[113] Yip WCY, Sequeira IR, Plank LD, Poppitt SD. Prevalence of Pre-Diabetes across Ethnicities: A Review of Impaired Fasting Glucose (IFG) and Impaired Glucose Tolerance (IGT) for Classification of Dysglycaemia. Nutrients 2017;9. https://doi.org/10.3390/nu9111273.

[114] Grundy SM. Metabolic syndrome update. Trends Cardiovasc Med 2016;26:364–73.

[115] Riobó Serván P. Obesity and diabetes. Nutr Hosp 2013;28 Suppl 5:138–43. https://doi.org/10.3305/nh.2013.28.sup5.6929.

[116] Coelho AI, Rubio-Gozalbo ME, Vicente JB, Rivera I. Sweet and sour: An update on classic galactosemia. J Inherit Metab Dis 2017;40:325–42. https://doi.org/10.1007/s10545-017-0029-3.

[117] Pasquali M, Yu C, Coffee B. Laboratory diagnosis of galactosemia: A technical standard and guideline of the American College of Medical Genetics and Genomics (ACMG). Genet Med 2018;20:3–11. https://doi.org/10.1038/gim.2017.172.

[118] Szilagyi A, Ishayek N. Lactose Intolerance, Dairy Avoidance, and Treatment Options. Nutrients 2018;10. https://doi.org/10.3390/nu10121994.

[119] Steinmann B, Santer R. Disorders of fructose metabolism. In: Saudubray J-M, van den Berghe G, Walter JH, editors. Inborn Metab. Dis., Berlin, Heidelberg: Springer Berlin Heidelberg; 2012, p. 157–65. https://doi.org/10.1007/978-3-642-15720-2_9.

[120] Chen MA, Weinstein DA. Glycogen storage diseases: Diagnosis, treatment and outcome. Transl Sci Rare Dis 2016;1:45–72.

[121] Adeva-Andany MM, González-Lucán M, Donapetry-García C, Fernández-Fernández C, Ameneiros-Rodríguez E. Glycogen metabolism in humans. BBA Clin 2016;5:85–100.

[122] Takahashi N. Oral Microbiome Metabolism: From "Who Are They?" to "What Are They Doing?" J Dent Res 2015;94:1628–37. https://doi.org/10.1177/0022034515606045.

4 Lipids

4.1 Definition and classification

Lipids are a heterogeneous family of biomolecules whose common characteristic is a low or null solubility in water and high solubility in nonpolar organic solvents such as chloroform or hexane. This definition includes a wide variety of biological compounds that can be classified according to different criteria, such as behavior upon hydrolyzation or chemical structure and composition [1]. According to the first criterion, lipids are classified as

- Hydrolyzable (saponifiable) lipids: Those that yield fatty acids (FAs) when subjected to hydrolysis in the presence of strong acids (soaps in the presence of strong bases such as NaOH or KOH).
- Nonhydrolyzable (unsaponifiable) lipids: They do not produce FAs/soaps upon hydrolysis.

Considering chemical structure and composition, food chemists distinguish three classes of lipids:

- Simple lipids: Esters of FAs and biological alcohols like glycerol (propane-1,2,3-triol). This group is of paramount importance due to their abundance in foods, as they include fats and oils.
- Compound, complex, or conjugated lipids: Lipids linked or associated to other nonlipid molecules producing substances with amphiphilic properties. These include phospholipids, glycolipids, and lipoproteins.
- Derived lipids: Hydrophobic substances not included in the above groups, such as free FAs, carotenoids, lipophilic vitamins, steroids, or pigments and volatile scents from plants.

Similarly to the rest of biomolecules, the backbones of lipids are composed mainly of carbon, hydrogen, and oxygen although they may contain additional elements such as phosphorus, sulfur, and/or nitrogen.

Lipids are molecules of hydrophobic nature implicated in important biological functions:

- Energetic: The most energy-dense biomolecules are lipids, containing up to 9 kilocalories (kcal, approximately 37.6 kilojoules or kJ) per gram; therefore, they are the molecules of choice for efficient energy storage in vertebrates and other species [1].
- Structural: Phospholipids, glycolipids, and sterols play an essential role in the structure and function of biological membranes, the complex structure that mediates the interchange of mass, energy, and information between the cell and its environment [2].

https://doi.org/10.1515/9783111111872-004

- Insulation and protection: Some lipids are important components of animal skin and feathers as well as plant cuticles [3].
- Cell signaling and regulation: Many intra- and extracellular chemical messengers are lipids [4].
- Antioxidants: Vitamin E is the most important lipophilic protectant against oxidative damage of biological membranes. Some carotenoids have also been reported to be antioxidants (Chapter 8).
- Electron carriers like coenzyme Q10 in the mitochondrial respiratory chains.
- Enzymatic co-factors: Some hydrophobic molecules, like vitamin K, are required by some enzymes to exert their biological functions [5].

The importance of the different roles played by lipids in vivo led biochemists to launch the LIPIDS MAPS ("LIPID Metabolites and Pathways Strategies") initiative in 2002 that set off the development of lipidomics. Lipidomics aims at the identification, characterization, and quantification of the lipidome, that is, the entire spectrum of lipids and derived metabolites in a biological system [4, 6]. In 2005, an updated definition as well as a comprehensive classification of these molecules was introduced based on their biochemical properties. Lipids were defined as "hydrophobic or amphipathic small molecules that may originate entirely or in part by carbanion-based condensations of thioesters and/or by carbocation-based condensation of isoprene units" [7]. Additionally, the term simple lipids was proposed for those that yield up to two types of products upon hydrolysis (e.g., FAs, sterols, and acylglycerols), and complex lipids for those yielding three or more products (e.g., glycerophoshopilids and glycosphingolipids). Naturally occurring lipids were classified into eight categories that cover eukaryotic and prokaryotic sources (Table 4.1 and Figure 4.1), each category being subdivided into classes and subclasses. The LIPID MAPS initiative also proposed an appropriate nomenclature as well as guidelines for drawing the structures of these biomolecules [7, 8]. Another commonly accepted definition describes simple lipids as those that do not have FAs in their molecules (terpenes, hydrocarbons, steroids, and prostaglandins) and complex lipids as those that contain FAs (acylglycerides, phospholipids, glycolipids, sphingolipids, and waxes).

In this chapter, we shall focus on the chemical and biochemical aspects of a relatively small number of lipid molecules of major nutritional interest, namely, FAs, triacylglycerols (TAGs), carotenoids, and steroids. Other important lipids such as lipophilic (fat-soluble) vitamins will be discussed in Chapter 6.

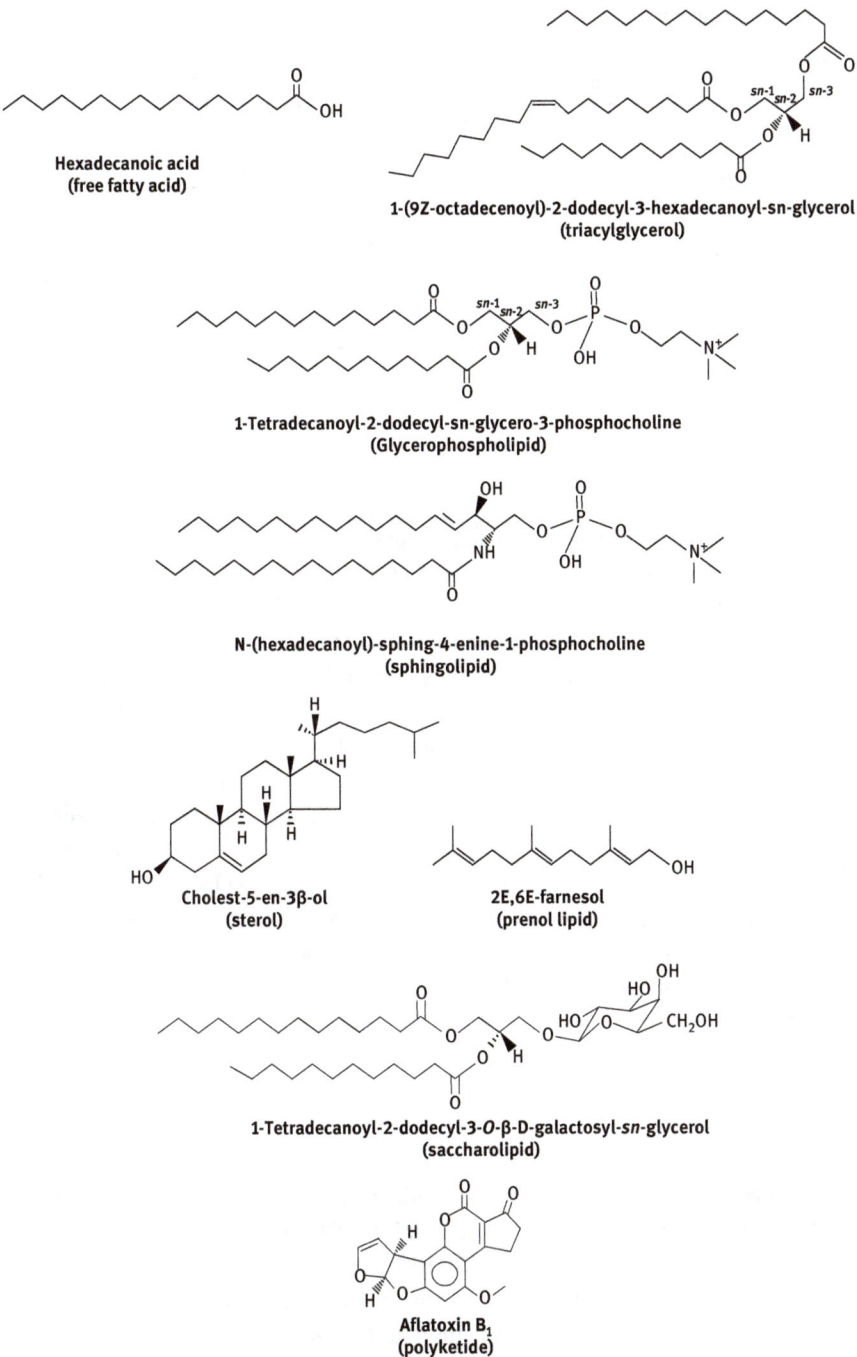

Hexadecanoic acid
(free fatty acid)

1-(9Z-octadecenoyl)-2-dodecyl-3-hexadecanoyl-sn-glycerol
(triacylglycerol)

1-Tetradecanoyl-2-dodecyl-sn-glycero-3-phosphocholine
(Glycerophospholipid)

N-(hexadecanoyl)-sphing-4-enine-1-phosphocholine
(sphingolipid)

Cholest-5-en-3β-ol
(sterol)

2E,6E-farnesol
(prenol lipid)

1-Tetradecanoyl-2-dodecyl-3-O-β-D-galactosyl-sn-glycerol
(saccharolipid)

Aflatoxin B₁
(polyketide)

Figure 4.1: Chemical structures of the lipid compounds are mentioned in Table 4.1. All chemical formulae were retrieved and adapted from https://www.lipidmaps.org/data/structure/LMSDSearch.php.

Table 4.1: Comprehensive lipid classification for lipidomics.

Category	Abbreviation	Example
Fatty acids	FA	Hexadecanoic acid
Glycerolipids	GL	1-Hexadecanoyl-2-(9Z-octadecenoyl)-sn-glycerol
Glycerophospholipids	GP	1-Hexadecanoyl-2-(9Z-octadecenoyl)-sn-glycero-3-phosphocholine
Sphingolipids	SP	N-(Hexadecanoyl)-sphing-4-enine-1-phosphocholine
Sterol lipids	ST	Cholest-5-en-3β-ol
Prenol lipids	PR	2E,6E-Farnesol
Saccharolipids	SL	1-Tetradecanoyl-2-dodecyl-3-O-β-D-galactosyl-sn-glycerol
Polyketides	PK	Aflatoxin B_1

Source: [7, 8].

4.2 Fatty acids

FAs do not usually occur as free molecules in foods, but as components of other bio-molecules. In their most basic form, FAs are repeating series of methylene groups ($-CH_2-$) with a methyl group (CH_3-) at one end and a carboxylic group ($-COOH$) at the other, resulting in linear hydrocarbon chains that range from 4 to 36 carbon atoms in living organisms, although the most common have between 12 and 24 [9]. Due to the mechanism of FA biosynthesis, they usually have an even number of carbon atoms in their molecules and can also be dehydrogenated in vivo to introduce carbon–carbon double bonds ($>C=C<$). These bonds are also known as unsaturations; consequently, FAs are saturated or unsaturated depending on the absence or presence of double bonds in their backbones, respectively. Although many variations of the structure described above can be found in living cells [7, 8], "basic" linear FAs participate in molecules that constitute the major part of the lipid fraction in foods and contain most of their metabolically usable energy. For these reasons, they are important contributors to the physical, chemical, and nutritional properties of foods.

4.2.1 Nomenclature of fatty acids

Table 4.2 shows the names, structures, and symbols of some FAs of biological interest. Systematic names follow the International Union of Pure and Applied Chemistry-International Union of Biochemistry (IUPAC-IUB) recommendations; however, trivial names are widely used by biochemists and food chemists. A numerical notation indicating the total number of carbon atoms and double bonds (and their location within the carbon chain) is also very useful for the most common naturally occurring FAs.

Table 4.2: Names and symbols of biochemically important fatty acids.

Numerical symbol	Structure	Systematic name[a]	Trivial name[a]
4:0	CH_3-$[CH_2]_2$-COOH	Butanoic	Butyric
6:0	CH_3-$[CH_2]_4$-COOH	Hexanoic	Caproic
8:0	CH_3-$[CH_2]_6$-COOH	Octanoic	Caprylic
10:0	CH_3-$[CH_2]_8$-COOH	Decanoic	Capric
12:0	CH_3-$[CH_2]_{10}$-COOH	Dodecanoic	Lauric
14:0	CH_3-$[CH_2]_{12}$-COOH	Tetradecanoic	Myristic
16:0	CH_3-$[CH_2]_{14}$-COOH	Hexadecanoic	Palmitic
16:1	CH_3-$[CH_2]_5$CH=CH$[CH_2]_7$-COOH	9-Hexadecenoic	Palmitoleic
18:0	CH_3-$[CH_2]_{16}$-COOH	Octadecanoic	Stearic
18:1(9)	CH_3-$[CH_2]_7$CH=CH$[CH_2]_7$-COOH	cis-9-Octadecenoic[b]	Oleic
18:1(11)[c]	CH_3-$[CH_2]_5$CH=CH$[CH_2]_9$-COOH	11-Octadecenoic	Vaccenic
18:2(9,12)[c]	CH_3-$[CH_2]_3(CH_2$ CH=CH$)_2[CH_2]_7$-COOH	cis,cis-9,12-Octadecadienoic[b]	Linoleic
18:3(9,12,15)[c]	CH_3-(CH$_2$CH=CH$)_3[CH_2]_7$-COOH	9,12,15-Octadecatrienoic	(9,12,15)-Linolenic
18:3(6,9,12)[c]	CH_3-$[CH_2]_3(CH_2$CH=CH$)_3[CH_2]_4$-COOH	6,9,12-Octadecatrienoic-	(6,9,12)-Linolenic
18:3(9,11,13)[c]	CH_3-$[CH_2]_3(CH=CH)_3[CH_2]_7$-COOH	9,11,13-Octadecatrienoic	Eleostearic
20:0	CH_3-$[CH_2]_{18}$-COOH	Icosanoic	Arachidic
20:2(8,11)[c]	CH_3-$[CH_2]_6(CH_2$CH=CH$)_2[CH_2]_6$-COOH	8,11-(E)Icosadienoic	
20:3(5,8,11)[c]	CH_3-$[CH_2]_6(CH_2$CH=CH$)_3[CH_2]_3$-COOH	5,8,11-(E)Icosatrienoic	
20:4(5,8,11,14)[c]	CH_3-$[CH_2]_3(CH_2$CH=CH$)_4[CH_2]_3$-COOH	5,8,11,14-(E)Icosatetraenoic	Arachidonic
22:0	CH_3-$[CH_2]_{20}$-COOH	Docosanoic	Behenic
22:6(4,7,10,13,16,19)[c]	CH_3-(CH$_2$CH=CH$)_6[CH_2]_2$-COOH	all-cis-4,7,10,13,16,19-Docosahexaenoic acid	Cervonic acid (DHA)
24:0	CH_3-$[CH_2]_{22}$-COOH	Tetracosanoic	Lignoceric

(continued)

Table 4.2 (continued)

Numerical symbol	Structure	Systematic name[a]	Trivial name[a]
24:1	$CH_3-[CH_2]_7-CH=CH[CH_2]_{13}-COOH$	*cis*-15-Tetracosenoic	Nervonic
26:0	$CH_3-[CH_2]_{24}-COOH$	Hexacosanoic	Cerotic
28:0	$CH_3-[CH_2]_{26}-COOH$	Octacosanoic	Montanic

Source: [10].

[a]Endings in "-oic/ic" are substituted by "-ate" and "-yl/-oyl" in fatty acid salts (e.g., sodium palmitate) and radicals (e.g., palmitoyl-CoA), respectively.

[b]*E/Z* notation is preferred over *cis/trans*, but the latter is still widely used in biochemistry and food chemistry books.

[c]Double bonds can also be specified in numerical symbols by adding the Greek letter "Δ" with the position numbers as superscripts (e.g., 20:4 $\Delta^{5,8,11,14}$ for arachidonic acid).

Unsaturations determine the structure of FAs (Figure 4.2) as well as other physical and chemical properties. According to the number and position of double bonds, FAs can be classified as

- Saturated fatty acids (SFAs)): They do not contain double bonds.
- Monounsaturated fatty acids (MUFAs): They contain only one double bond.
- Polyunsaturated fatty acids (PUFAs): They contain more than one double bond.

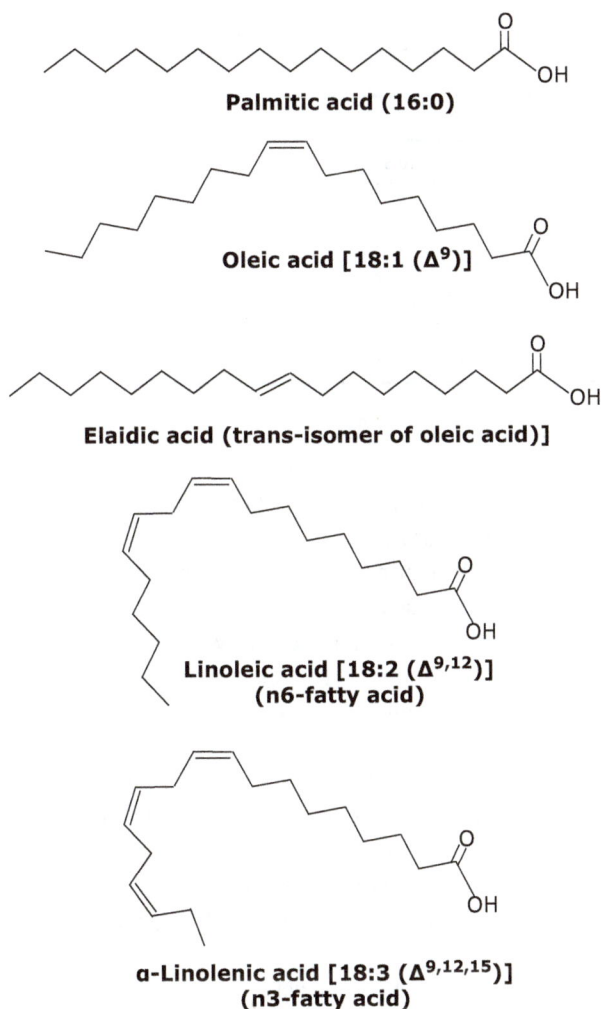

Palmitic acid (16:0)

Oleic acid [18:1 (Δ^9)]

Elaidic acid (trans-isomer of oleic acid)]

Linoleic acid [18:2 ($\Delta^{9,12}$)]
(n6-fatty acid)

α-Linolenic acid [18:3 ($\Delta^{9,12,15}$)]
(n3-fatty acid)

Figure 4.2: Chemical structures of some representative fatty acids. All chemical formulae were retrieved and adapted from https://www.lipidmaps.org/data/structure/LMSDSearch.php.

As a general rule, the longer the chain and the fewer double bonds, the lower is the solubility in water and the higher are the melting points. SFAs have highly flexible structures and form stable aggregates in which molecules are fully extended and pack tightly in nearly crystalline arrays. These aggregates are responsible for the high melting points of this type of FA. In unsaturated FAs (MUFAs and PUFAs), the double bonds force kinks in the hydrocarbon chain and packing is less tight, thus lowering the melting points with respect to the homologous SFAs [9]. As FAs are major components of glycerolipids, they are responsible for most of the physico-chemical properties of these compounds (Table 4.3).

Table 4.3: Melting points of selected fatty acids and their corresponding triacylglycerol (TAG).

Numerical symbol	Trivial name	Melting point (°C)[a]	TAG melting point (°C)
12:0	Lauric	43.29 (0.34)	46.29 (0.04)
14:0	Myristic	53.47 (0.31)	57.35 (0.19)
16:0	Palmitic	62.20 (0.19)	65.45 (0.27)
18:0	Stearic	69.29 (0.19)	72.67 (0.32)
18:1(9)	Oleic	12.82 (0.15)	3.98 (0.59)
18:2(9,12)	Linoleic	−7.15 (0.69)	−12.70 (0.35)
22:0	Behenic	79.54 (0.34)	82.50 (0.08)
24:1(15)	Nervonic	43.23 (0.11)	41.43 (0.09)

Source: [11].
[a]Numbers between parentheses are standard deviations.

In natural FAs, substituents linked to the carbon atoms implicated in double bonds usually occur in cis-configuration; however, processes like hydrogenation, meat irradiation, and food frying can lead to the formation of FAs that may contain one or more double bonds in trans-configuration, the so-called trans-fatty acids (TFAs). TFAs have been reported to have a negative impact on human health when present in the diet, a frequent situation in industrialized countries [12]. Dietary recommendations for FAs usually consider SFAs, MUFAs, PUFAs, and TFAs as separated groups with specific nutritional properties (see below).

In the standard IUPAC-IUB recommendations for FA nomenclature, carbon atoms are numbered starting from the principal functional group, that is, C-1 is the carbon atom of the carboxylic group (–COOH); the rest are numbered successively (C-2, C-3, C-4, etc.) as we move away from C-1. In this nomenclature, the Greek letter "delta" (Δ) is used to specify the position of a double bond. Therefore, delta-12 (Δ^{12}) and delta-15 (Δ^{15}) FAs are those that have double bonds at positions 12 and 15, respectively, starting from the carboxylic group (Figure 4.2) In a different nomenclature, the carbon located at the farthest position from COOH, that is, that of the methyl group, is generically known as "omega" (ω) carbon. In this nomenclature, omega-3 ($\omega 3$ or n-3) and omega-6 ($\omega 6$ or n-6) are those that have double bonds at positions 3 and 6, respectively, starting from the methyl group (Figure 4.2). n-3 and n-6 FAs cannot be synthesized by humans; however, they must be included in the diet because they are precursors of an important group of biomolecules, eicosanoids, or icosanoids (Section 4.10), consequently, they are known as "essential fatty acids." Vegetable and fish oils are important dietary sources of these types of FAs.

An important subset of FAs from the physiological point of view is short-chain fatty acid (SCFAs). SCFAs are small organic monocarboxylic acids with a chain length of up to six carbons atoms. Acetate and propionate, with chain lengths of two and three carbon atoms, respectively, are normally included in this group although they are not technically "fatty acids." SCFAs are the main metabolites produced in the colon by bacterial fermentation of dietary fibers and resistant starch (Chapter 3) and they have been reported to play roles in gastrointestinal (GI) physiology, immune function, host metabolism, and even in the development and homeostasis of the central nervous system [13].

4.3 Triacylglycerides: edible fats and oils

The terms fat and oil are used in normal speech for solid and liquid lipids, respectively, at room temperature (20–25 °C). All foods contain lipids at varying concentrations [14]:
–   Vegetables and fruits contain an average 0.3% (expressed as percentage of wet mass) with some exceptions like avocado, whose edible portion contains about 20% lipids
–   Muscle tissue of lean beef, fish, white poultry, and shell fish: 2%
–   Cow's milk: 3.7%
–   Grains: 2–4%
–   Fatty pork: 30%
–   Egg yolk: 32%
–   Fillets of fatty fish (tuna, salmon, etc.): 35%
–   Oil bearing nuts and seeds: from 20% in soybeans to 65% in walnuts

Vegetable oils are the main lipid sources in the human diet (Table 4.4), especially those obtained from palm, soybean, rapeseed, and sunflower seed, that account for more than 75% of world production; however, animal fats like butter, lard, tallow, and fish oil still provide a significant part of global fat production and consumption (around 13%) [15]. Other minor sources include nuts, cereals, and legumes [16].

Dietary fats and oils are mainly constituted by esters of FAs and glycerol that can be subdivided into TAGs, the most abundant, 1,2- or 1,3-di(acyl)glycerols, and 1- or 2-mono(acyl)glycerols, according to the number and position of acyl groups. In the case of TAGs, when the two primary hydroxyl groups of glycerol are esterified with different FAs, the resulting molecule displays chiral asymmetry. Although the conventional D/L or R/S systems can be used to designate the enantiomers, the most convenient way to describe natural TAGs is by the so-called "stereospecific numbering" (*sn*) system: glycerol is depicted in a Fischer projection (Chapter 3) with the secondary hy-

Table 4.4: Production of major vegetable oils worldwide in 2020.

Source	Production in million metric tons
Palm	75.88
Soybean	58.62
Rapeseed	25.09
Sunflower seed	20.61
Palm kernel	8.00
Peanut	4.61
Cottonseed	4.25
Coconut	2.61
Olive	3.37

Source: https://ourworldindata.org/grapher/vegetable-oil-produc
tion (with data obtained from the Food and Agriculture
Organization of the United Nations). Retrieved on 6/5/2023.

droxyl group to the left of C-2, which is numbered as *sn*-2, the carbon atoms above and below are then denoted as *sn*-1 and *sn*-3, respectively (Figure 4.3). The digestibility of TAGs and, thus, the absorption and the metabolic effects of their constituent FAs are determined by the type and position of the latter [17].

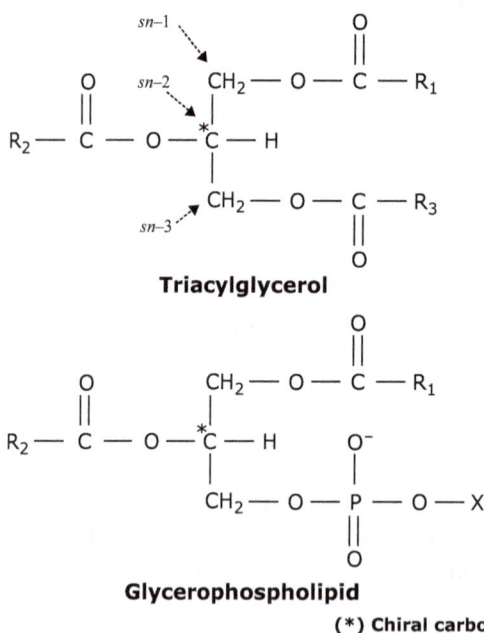

Figure 4.3: Stereospecificity in triacylglycerols and glycerophospholipids. X represents certain polar molecules (see main text for further details).

In diacylglycerides, the remaining hydroxyl group can be esterified by phosphoric acid, which may form a diester (phosphodiester linkage) with polar molecules like 2-aminoethanol, choline, glycerol, inositol, or serine. These compounds are known as phosphoglycerides or glycerophospholipids that belong to the family of phospholipids. The *sn* system can also be used with glycerophospholipids. Other minor lipid components of food that contain FAs are waxes, glycolipids, and sphingolipids (Figure 4.4) [9, 18].

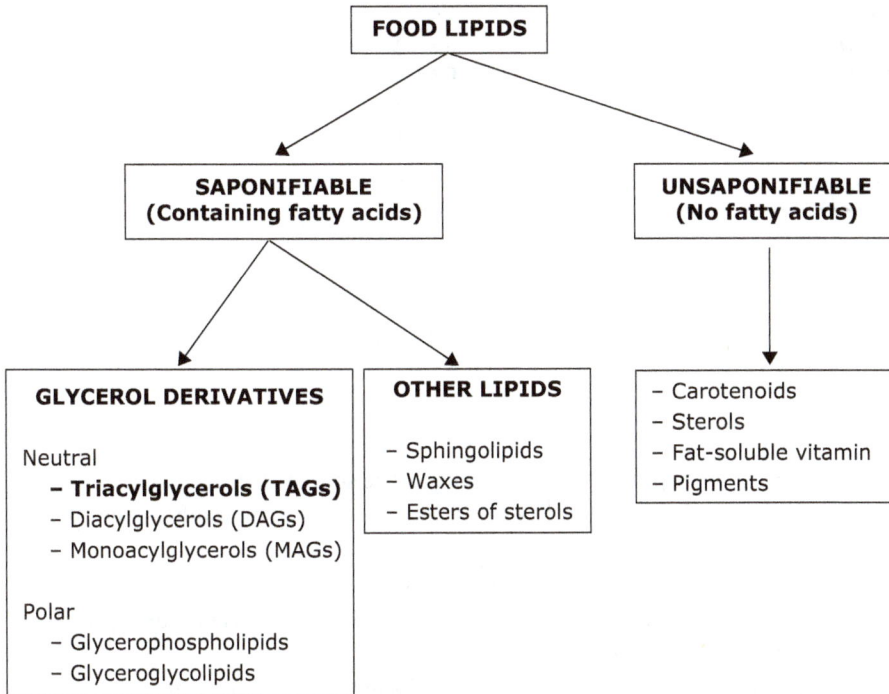

Figure 4.4: Classification of the main lipid molecules found in food. TAGs (in bold letters) are the most abundant, containing up to 97% of the energy associated to lipids.

TAGs are hydrophobic nonpolar molecules, as all the hydroxyl groups of glycerol are bound in ester linkages, and together with DAGs and MAGs are considered neutral lipids. In contrast, phospolipids, glycolipids, and sphingolipids are known as polar lipids because they present a highly polar or charged group usually attached to the third carbon. Polar lipids are important components of biological membranes, whereas TAGs provide energy and insulation in vertebrates and plant oilseeds and oil fruits [9]. TAGs are by far the greatest source of dietary FA and contain up to 97% of the energy associated to lipids [19]. For this reason the terms "triacylglycerides" and "fats" are often used as synonyms. TAGs are also important contributors to organoleptic properties of foods (Chapter 9).

A minor fraction devoid of FAs, that is, composed of unsaponifiable lipids, occur in natural fats and oils (Table 4.5). This fraction contains molecules with important biological properties such as fat-soluble vitamins, sterols, and carotenoids (Sections 4.8 and 4.12).

Table 4.5: Content of unsaponifiable fraction in various oils.

Fat/oil	Unsaponifiable (wt%)
Soya	0.6–1.2
Sunflower	0.3–1.2
Cocoa	0.2–0.3
Peanut	0.2–4.4
Olive	0.4–1.1
Palm	0.3–0.9
Rapeseed	0.7–1.1

Sources: [20, 21].

4.4 Production of oils and fats for human consumption

4.4.1 Animal sources

There are three major procedures depending on the source [22]:
- Carcass fats are produced by the rendering process which usually implies the following steps:
  - comminution of material by mechanical means;
  - heating, either directly (dry rendering) or with steam (wet rendering);
  - separation of fat by centrifugation.
- Fish oils are usually obtained by wet rendering followed by pressure or extraction with food grade alcohols (ethanol or isopropanol).
- Milk fat is produced by centrifugation after heating raw milk, which usually contains between 3% and 4% of fat. Centrifugation produces skimmed milk, containing 0.05% fat, and cream, containing 30–40%. Cream is subsequently pasteurized and churned.

4.4.2 Vegetable sources

There are major differences in the production procedure of vegetable oils/fats depending on if the material is obtained from fruits or from seeds.
- The major fruit oils produced nowadays are palm oil, avocado oil, and olive oil. The latter is a minor oil at global level (see Table 4.4), but it is the most important

oil in Mediterranean countries, especially in Spain, Italy, and Greece. Actually, 70% of the global olive production takes place in the European Union, generating a value of 7,000 million euros each year [23].

- Palm oil: It is obtained from the mesocarp of fruits of the tropical plant *Elaeis guineensis* by breaking the fruits in a steam-jacketed vessel kept at 90–100 °C and fitted with beater arms. The resulting material is fed to a screw press that produces a liquid stream and a press cake. Previously, palm bunches are cut from the palm trees, sterilized, and subjected to a stripping procedure to release the fruits [22]. Palm kernels can be processed as seeds for further oil production, but shows a different FA composition with respect to the palm oil.

- Olive oil: The fruit of the olive tree (*Olea europaea*) is a drupe composed of three principal tissues: endocarp, mesocarp, and epi- or exocarp. The endocarp is the fruit pit, a hard, woody structure, which surrounds the seed in the fruit center. The mesocarp or fruit pulp is the major and edible part of the olive fruit and the main site for oil metabolism and storage. Oil deposition takes place in the cytoplasm as large oil droplets. The epicarp, exocarp, or fruit skin constitutes the external tissue [24]. Preliminary procedures like harvesting and transport of olives may significantly affect oil quality and yield. Oil production involves three main steps: crushing of fruits, malaxation or kneading, and separation. During crushing, olives are ground into a fine paste in order to open the mesocarp cells and release the oil. Traditionally, granite millstones were used for this purpose, but nowadays crushing is performed with metal hammer mills equipped with a single or double sieve. The malaxation/kneading process is essential to obtain a continuous oily phase from the emulsion (oil droplets dispersed in aqueous medium) liberated during fruit milling. Addition of co-adjuvants like micronized talc improves the paste structure and reduces emulsions. An efficient malaxation is essential to obtain optimal oil yields and contributes to the composition and quality of the final product. The separation step can be done either by pressure or by centrifugation. Separation by pressure is a traditional discontinuous process that produces a solid and a liquid phases, the latter containing a mixture of oil and vegetation wastewater (known as "alpechín" in Spain) that can be separated by vertical centrifugation. Alpechín is highly polluting and must be correctly processed and disposed of. In the separation by centrifugation, two systems that use specially designed centrifuges coexist. The three-way (or three phases) system has one outlet for pomace ("orujo" in Spanish) and one for each of the liquid phases (oil and alpechín). In this system, the olive paste is loaded with water (around 50% of the fruit dry weight) thereby increasing the amount of alpechín generated. In the two-way system, there is one outlet for oil and another for alpechín plus pomace (a mixture called "alperujo" in Spanish), no addition of water is usually needed [25, 26]. After the centrifugation step, oil still has between 1 and 3% wastewater, whereas wastewater from the three-way system has oil content in the range of 0.5–2%. Vertical centrifuges are normally used to separate the oil from the aqueous phase in both cases [25]. Part of the oil released after milling and kneading may be collected before the separation by selective filtration or percolation. This method is based on the differences in surface tension between the liquid (oily and aqueous) phases of the olive paste and a stainless steel surface. Olive oil production has increased during the last decades due to its organoleptic properties and alleged health benefits (see below). This has brought about environmental challenges that are being tackled by the scientific community with the goal of not only minimizing polluting wastes but also obtaining by-products of added value [27].

- Seed oil: The general procedure for large-scale production of oil seeds (soybean, rapeseed, and sunflower seed, among others) involve the following steps [22, 28]:
  - Cleaning and drying of seeds to less than 10% moisture: Seeds may be stored for prolonged time periods under conditions (aeration, prevention of infestation by insects and rodents, etc.) that minimizes mold infection, mycotoxin contamination, and biological degradative processes.
  - Dehulling
  - Grinding
  - Cooking with steam
  - Extraction in a screw or hydraulic press, which separates oil from the rest of the material (cake).
  - Flaking of the pressed cake for later extraction of residual fat with solvents such as "food grade" hexane.
  - Oil can be directly extracted with solvents from products with low oil content, such as soybean (around 20% oil content, in contrast to rapeseed or sunflower seed that may contain more than 40%).

4.5 Oil refining

Crude oils and fats normally need to be purified before they can be used for human consumption. The purification procedure is known as "refining" in Europe (in the USA, the term "refining" is limited to the removal of free FAs). Refining produces edible oil with characteristics such as bland flavor and odor, clear appearance, light color, stability to oxidation, and suitability for frying. A notable exception is olive oil, whose refining is considered detrimental since its attractive organoleptic, nutritional, and technological properties are diminished in the process. Virgin olive oils (Section 4.8) are more expensive than refined olive oils because of their pleasant, rather delicate flavor and aroma, presence of bioactive compounds, high oxidative stability, and limited production volume.

The steps that lead to a fully refined oil are [22]:

- Removal of gums (phospholipids): This is usually achieved by extraction with water or with acidic (phosphoric or citric) solutions. Gum removal is essential for subsequent processing and more than one degumming steps can be performed.
- Removal of free FAs: Two methods are used:
  - Chemical: Treatment with a slight excess of sodium hydroxide solution, followed by the washing out of soaps and hydrated phospholipids.
  - Physical ("steam refining"): Free FAs are volatilized by a steam stripping process at reduced pressure and elevated temperature.

- Removal of absorbable compounds ("bleaching"): This procedure does more than just decrease the oil absorbance by removing coloring components; thus, natural or acid-activated clay minerals also catalyzes the decomposition of peroxides and removes all kinds of polar compounds. Other adsorbents, such as activated carbon, can remove polycyclic aromatic hydrocarbons.
- Removal of waxes: Because of their high melting point, waxes tend to crystallize at refrigerator temperature; therefore, they are removed by a process of fractional crystallization called "winterization" in Europe.
- Deodorization is essentially a steam distillation process carried out at low pressures (2–6 mbar) and elevated temperatures (180–220 °C) that eliminates volatile components, mainly aldehydes and ketones, with low threshold values for detection by taste or smell [29].

Although some of these procedures have negative connotations, undesirable changes in oil composition and losses of components such as tocopherols and sterols can be minimized using well-controlled conditions. By contrast, refining has some advantages:

- Some impurities, including oxidized components, trace metals, and coloring materials, are partially removed during the removal of gums and further reduced during bleaching.
- Steps like neutralization, deodorization, or the use of active carbon for bleaching also contribute significantly to the removal of contaminants like aflatoxins, organophosphorus pesticides, and polycyclic aromatic hydrocarbons.

4.6 Deterioration of oils and fats

The major processes of lipid deterioration include hydrolysis (lipolysis) and oxidation. Both processes may or may not be catalyzed by enzymes.

4.6.1 Lipolysis

As described above, TAGs and phospholipids are the most abundant molecules of lipid nature in food sources. In these molecules, FAs form ester links with hydroxyl groups of glycerol or sphingosine that can be hydrolyzed by different ways:
– Chemically in a spontaneous reaction that depends on the temperature and the presence of water. This situation is likely to occur in food processing methods that involve heating, like frying.
– Biochemically catalyzed by endogenous lipases and phospholipases present in many food sources like milk, oilseeds, cereals, fruits, and vegetables.
– Biologically, mediated by lipases and phospholipases of microorganisms (bacteria or fungi) that contaminate the food.

The result of lipolysis is the presence of free FAs, which may have detrimental effects in the food flavor, the so-called hydrolytic rancidity [1]. Both free and linked FAs may undergo oxidation.

4.6.2 Lipid oxidation

Lipid oxidation is the most common deteriorative process in food processing and storage. It normally occurs in the presence of molecular oxygen (O_2) at conjugated double bonds (>C=C-C=C<) or nonconjugated bonds linked by a methylene group (>C=C-CH$_2$-C=C<, also known as isolenic double bonds). The former are present in vitamin A, carotenoids, and xantophylls, and the latter in PUFAs like linoleic and araquidonic acids (Figure 4.5). Consequently, these are the lipid molecules more prone to deleterious chemical reactions with oxygen [1, 30]. Steroids from animals (cholesterol and derivatives) and plants (collectively known as phytosterols) can also be subjected to oxidation at the double bond present in their tetracyclic cyclopenta[α]-phenanthrene structure (Figure 4.12) [31].
Traditionally, lipid oxidation has been reported to proceed in three stages [33]:
– Initiation: Removal of a hydrogen atom, usually from a FA, forming an alkyl radical

$$RH \rightarrow R^{\cdot} + H^{\cdot}$$

– Propagation: Reaction of the alkyl radical with oxygen, yielding peroxyl radicals that may abstract a hydrogen atom from another unsaturated FA, thus forming a hydroperoxide (ROOH)

$$R^{\cdot} + O_2 \rightarrow ROO^{\cdot}$$
$$ROO^{\cdot} + RH \rightarrow ROOH + R^{\cdot}$$

$$R_1 - CH = CH - CH_2 - CH = CH - R_2$$

R_1: $CH_3- (CH_2)_4$
R_2: $(CH_2)_7- COOH$ H˙

$$R_1 - CH = CH - \overset{\cdot}{C}H - CH = CH - R_2$$

$$R_1 - \overset{\cdot}{C}H - CH = CH - CH = CH - R_2 \rightleftharpoons R_1 - CH = CH - CH = \overset{\cdot}{C}H - CH - R_2$$

O_2 O_2

$$R_1 - \underset{\underset{\overset{|}{O}}{|}{\overset{|}{O}}}{CH} - CH = CH - CH = CH - R_2 \qquad R_1 - CH = CH - CH = CH - \underset{\underset{\overset{|}{O}}{|}{\overset{|}{O}}}{CH} - R_2$$

R_3H   R_3H

R_3˙   R_3˙

$$R_1 - \underset{\underset{\overset{|}{OH}}{|}{\overset{|}{O}}}{CH} - CH = CH - CH = CH - R_2 \qquad R_1 - CH = CH - CH = CH - \underset{\underset{\overset{|}{OH}}{|}{\overset{|}{O}}}{CH} - R_2$$

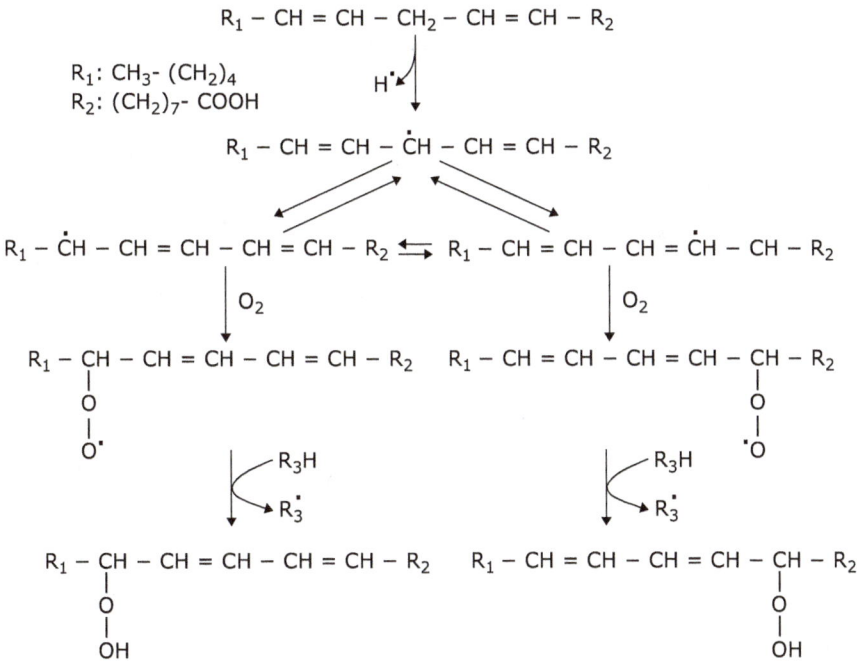

Figure 4.5: Mechanism of linoleic acid oxidation (adapted from references [32, 33]).

Peroxyl radicals and hydroperoxides can also be generated by direct reaction of FA with oxygen (see below)

–   Termination: Two radicals react forming a nonradical species

$$R˙ + R˙ \rightarrow RR$$

$$ROO˙ + R˙ \rightarrow ROOR$$

$$ROO˙ + ROO˙ \rightarrow ROOR + O_2$$

The ground state of oxygen, the most abundant in the atmosphere, is triplet oxygen ($^3O_2$), whose reactivity toward biomolecules is extremely low due to spin restrictions [30]. On the contrary, the first excited state of the molecule of oxygen, singlet oxygen ($^1O_2$), can readily react with electron-rich double bonds producing peroxyl radicals (ROO˙). Some molecules, known as photosynthesizers (cholorophylls, porphyrins, ribo-flavin, myoglobin, and synthetic colorants), can absorb energy from light and transfer it to $^3O_2$ producing $^1O_2$. Photosynthesizers may also react directly with unsaturated FAs at low oxygen concentrations [30, 34].

Alternatively, lipid oxidation can start with the reduction of ground-state triplet ox-ygen by other molecules present in food, thus producing reactive oxygen species (ROS), such as superoxide anion (O^{2-}), and hydroxyl (OH˙), peroxyl (ROO˙), alkoxy (RO˙), and peroxide (ROO˙) radicals [30]. Formation of ROS will be studied in Chapter 8.

The reaction of oxygen with unsaturated FAs to form hydroperoxides may also be catalyzed by enzymes like lipoxygenases (LOXs), very abundant in plants. The function of these enzymes in vivo is to catalyze the first step in the synthesis of important metabolites called oxylipins from PUFAs, some of them with food-related applications like aroma production in olive oil [35]. However, LOX activity has also been reported to have negative implications for foods [36].

Lipid hydroperoxides are frequently subjected to fragmentation either by enzymes like hydroperoxide lyase or by nonenzymatic mechanisms, the latter triggered by light, heat, or transition metals. A wide variety of short-chain oxidation products like the so-called reactive carbonyl species (RCS) can be generated as the result of these reactions. RCS present a highly complex chemistry that includes polymerization and cyclation reactions as well as interactions with other biomolecules like amino acids and proteins. This second phase of lipid peroxidation plays an important role in plant metabolism, although it also alters organoleptic properties and nutritional value of foods with health-promoting or detrimental implications [1, 33, 37].

A number of factors have been shown to influence lipid oxidation [1, 30]:
– The lipid composition of the food
– The presence of antioxidants or pro-oxidants
– The presence of cations of transition metals like copper and iron
– Temperature
– Food irradiation

Lipid oxidation can occur during manufacturing, storage, and processing of foods from all natural sources; therefore, methods for the determination of oxidation has been developed. Comprehensive reviews of these analytical methods can be found elsewhere [38, 39].

4.7 Importance in the diet: fat as energy source

In a report issued in 2010, the World Health Organization (WHO) established a series of recommendations and dietary guidelines for the total fats and FAs in human nutrition, based on classifying FAs according to the number of double bonds. The main recommendations were [16]:
– Energy intake (calories) should be in balance with energy expenditure.
– Minimum fat consumption should be 15% of the total energy intake (E) for most individuals and 20% E for women in reproductive age and undernourished adults to provide essential FAs and fat soluble vitamins. The maximum total fat intake must be 30–35% E for most individuals.
– It is recommended that saturated fats (SFAs) should be replaced by PUFAs (n-3 and n-6 FA), and in any case, SFAs must provide less than 10% of total energy intake. This is based on convincing evidence about the beneficial effects on PUFAs

in cardiovascular health; moreover, studies with animal models and humans have provided some evidence of anti-inflammatory efficacy of n-3-PUFA, especially in rheumatoid arthritis. So far, no conclusive evidence has been found between consumption of SFAs and cancer.

–   The acceptable range for total PUFAs consumption ranges between 6% and 11% E. It must be underlined that PUFAs basically refers to linoleic (18:2 $\Delta^{9,12}$) and α-linolenic (18:3 $\Delta^{9,12,15}$) acids, an n-6 FA and an n-3 FA, respectively. Both are recognized as essential. It is also considered that individual PUFAs may have different properties.

–   The estimated average requirement for linoleic acid is 2% E and an adequate intake of 2–3% E is proposed, whereas the total n-3 FA intake can range between 0.5% and 2% E. Other long-chain n-3 FAs are also admitted as part of a healthy diet. No specific recommendation for an n-6/n-3 FA ratio was proposed.

–   Previous recommendations that TFAs must be less than 1% E may need to be revised, as there is convincing evidence that these compounds increase the risk of coronary heart disease, sudden cardiac death, metabolic syndrome, and type II diabetes. The determination of intake for MUFAs is calculated by difference in % E, so that

$$\text{MUFAs} = \text{Total fat intake} - \text{SFAs} - \text{PUFAs} - \text{TFAs}$$

This is based on evidence supporting the beneficial effects of replacing SFAs and carbohydrates with MUFAs (basically oleic acid) for cardiovascular health.

–   Intake of long-chain PUFAs like arachidonic (20:4 $\Delta^{5,8,11,14}$) and docosahexanoic (22:6 $\Delta^{4,7,10,13,16,19}$) acids seems to be important for optimal neural function; however, dietary recommendations for these acids in order to treat or prevent disorders, like depression, Alzheimer's disease, schizophrenia, or Huntington's disease, need to be established.

The WHO admits that there are limitations in these general recommendations, as individual FAs may have unique biological properties and health effects. In any case, in October 2018, a report with updated dietary guidelines confirmed these recommendations for FAs, albeit suggesting total avoidance of industrially produced TFAs [40]. Some considerations can be added to the WHO guidelines:

–   FA profiles of lipid sources must be adequate for their intended use; thus, cooking oils must contain a higher proportion of MUFAs, which are more stable under high temperature, while margarines and spreads have to be rich in SFAs in order to be solid at room temperature. Oils intended for direct consumption, such as salad oils, can contain more MUFAs [41].

–   Increasing evidence suggests that not only the FA composition of a given fat is important for its nutritional properties (Table 4.6) but also their stereospecific positioning. TAGs are mainly hydrolyzed by pancreatic lipases yielding *sn*2-monoacylglycerols that are efficiently absorbed and free FAs (Section 4.9). Free

MUFAs, PUFAs, and medium-chain FAs (up to 12 carbons) are better absorbed than free long-chain SFAs; however, the absorption coefficient of all types of FAs significantly increase when they esterify *sn*-2 positions in dietary TAGs. The stereospecificity of TAGs is a characteristic hallmark of the different lipid sources (Table 4.7); therefore, the metabolic effect of FAs may be different depending on the lipid source that is being consumed [42, 43]:

- In cocoa butter and palm and soybean oils, SFAs (palmitic and stearic acids) are mainly located on the *sn*-1 and/or *sn*-3 positions and unsaturated FAs (oleic and linoleic acids) on the *sn*-2 position.
- In peanut and olive oils, unsaturated FAs are equally distributed among the three stereospecific positions.
- Animal fats have SFAs mainly on the *sn*-2 position and unsaturated FAs on the *sn*-1 and/or *sn*-3 positions. In particular, palmitic acid preferentially occupies the *sn*-2 position in bovine milk fat and pork fat (lard), whereas oleic acid is mostly present at *sn*-1 and/or *sn*-3 positions. Exceptionally, in beef tallow, palmitic is mainly located in *sn*-1 and/or *sn*-3 positions.

A summary of dietary recommendations for fat intake as described by several food- and health-related organizations can be found at https://knowledge4policy.ec.europa. eu/health-promotion-knowledge-gateway/dietary-fats-table-4_en

The stereospecificity shown by TAGs in animal fats might explain why SFAs from these sources are more efficiently absorbed than those from plant origin, whereas plant oils are good suppliers of MUFAs and PUFAs [45].

Food industries use a process known as interesterification (IE) in the production of margarines, cooking fats, and shortenings as an alternative to partial hydrogenization (that produces TFAs). IE involves the rearranging of acyl groups in TAGs, generally at random, changing properties like melting profiles and crystallization behavior without changing FA composition; however, IE has been reported to alter the digestibility and thus the absorption properties of TAG components. Research is currently being conducted not only to study the effects on IE on the metabolic effects of dietary lipids but also to design fats with healthier nutritional profiles [45, 46]. In clinical applications, IE is already being used to provide products well-absorbed by infants or patients with fat malabsorption disorders [42]. For these applications, a more specific enzyme-based interesterification is needed, significantly increasing the cost of the process [46].

Table 4.6: Fatty acid composition of oils from several crops.

	8:0	10:0	C12:0	14:0	16:0	18:0	18:1	18:2	18:3	20:1
Palm			ND–0.5	0.5–2.0	39.3–47.5	3.5–6.0	36.3–44.0	9.0–12.0	ND–0.5	ND–0.4
Soybean			ND–0.1	ND–0.2	8.0–13.5	2.5–5.4	17.0–30.0	48.0–59.0	4.5–11.0	0–0.5
Canola				ND–0.2	2.5–7.0	0.8–3.0	51.0–70.0	15.0–30.0	5.0–14.0	0.1–4.3
Sunflower			ND–0.1	ND–0.2	5.0–7.6	2.7–6.5	14.0–39.4	48.3–74.0	0–0.3	0–0.3
Sunflower (HO)				ND–0.1	2.6–5.0	2.9–6.2	70.0–90.7	2.1–20.0	ND–3.0	0.1–0.5
Olive (CODEX)					7.5–20.0		55.0–83.0	3.5–21.0	Max 1.0	Max 0.4
Corn			ND–0.3	ND–0.3	8.6–14.0	ND–3.3	20.0–42.0	34.0–65.6	0–1.2	0.2–0.6
Grape seed				ND–0.3	5.5–11.0	3.0–6.5	12.0–28.0	58.0–78.0	0–1.0	0–0.3
Cocoa butter					22.6–30.4	30.2–36.0	29.2–36.4	1.3–4.0	ND–0.5	
Coconut	4.6–10	5.0–8.0	45.1–53.2	16.8–21.0	7.5–10.2	2.0–4.0	5.0–10.0	1.0–2.5	ND–0.2	ND–0.2
Palm kernel	2.4–6.2	2.6–5.0	45.0–55.0	14.0–18.0	6.5–10.0	1.0–3.0	12.0–19.0	1.0–3.5	ND–0.2	ND–0.2

Source: [44].

Table 4.7: Distribution of FAs in TAG molecules of chosen plant oils and animal fats.

Fat/oil source	Position	Fatty acid					
		16:0	18:0	18:1	18:2	18:3	C20–C24
Palm	sn-1	60	3	27	9	–	–
	sn-2	13	<0.5%	68	18	–	–
	sn-3	72	8	14	3	–	–
Soybean	sn-1	14	6	23	48	9	–
	sn-2	1	<0.5%	22	70	7	–
	sn-3	13	6	28	45	9	–
Rapeseed (high erucic)	sn-1	4	2	23	11	–	53
	sn-2	1		37	36	6	6
	sn-3	4	3	17	4	20	70
Sunflower	sn-1	11	3	17	70	–	–
	sn-2	1	1	22	76	–	–
	sn-3	10	9	28	54	–	–
Peanut	sn-1	14	5	59	19	–	4
	sn-2	2	<0.5%	59	39	–	1
	sn-3	11	5	57	10	–	15
Olive	sn-1	13	3	72	10	1	–
	sn-2	1	–	83	14	1	–
	sn-3	17	4	74	5	–	–
Maize (corn)	sn-1	1	3	28	50	1	
	sn-2	18	<0.5%	27	70	1	
	sn-3	2	3	31	52	1	
Cacao butter	sn-1	34	50	12	1	1	1
	sn-2	2	2	87	9	–	–
	sn-3	37	53	9	<0.5%	–	2
Cattle	sn-1	41	6	17	20	4	1
	sn-2	17	6	9	41	5	1
	sn-3	22	6	24	37	5	1
Pig	sn-1	10	2	30	51	6	–
	sn-2	72	5	2	13	3	–
	sn-3	<0.5%	2	7	70	18	–
Chicken	sn-1	47	7	8	31	5	1
	sn-2	13	5	6	55	19	1
	sn-3	31	7	3	49	8	1
Cow milk	sn-1	34	2	10	30	2	–
	sn-2	32	4	10	19	4	–
	sn-3	32	4	10	19	4	–

Source: [17, 20].

4.8 Olive oil: the healthy oil

Observational studies have confirmed the benefits of the so-called Mediterranean diet (Chapter 15) on the prevention of cardiovascular disease (CVD), the major cause of death in industrialized countries, although clinical trial evidence is more limited [47]. The Mediterranean diet is based on olive oil as the major source of dietary lipids; therefore, this oil is regarded by many authors as "the healthy oil."

4.8.1 Types of olive oil

According to the International Olive Council (IOC), virgin olive oils are "obtained from the fruit of the olive tree (*Olea europaea* L.) solely by mechanical or other physical means under conditions, particularly thermal conditions, that do not lead to alterations in the oil, and which have not undergone any treatment other than washing, decantation, centrifugation and filtration" [48]. The main definitions and designations related to olive oil established by the IOC are:

-   Extra virgin olive oil: free acidity, expressed as oleic acid, of not more than 0.8 g per 100 g.
-   Virgin olive oil: free acidity, expressed as oleic acid, of not more than 2 g per 100 g.
-   Ordinary virgin olive oil: virgin which has a free acidity, expressed as oleic acid, of not more than 3.3 g per 100 g.
-   Lampante virgin olive oil, not fit for consumption as it is, has a free acidity, expressed as oleic acid, of more than 3.3 g per 100 g. It must be refined for human consumption.
-   Refined olive oil is the olive oil obtained from virgin olive oils by refining methods which do not alter the native glyceridic structure. It has a free acidity, expressed as oleic acid, of not more than 0.3 g per 100 g.
-   Olive oil is the denomination of a blend of refined olive oil and virgin olive oils fit for consumption. It has a free acidity, expressed as oleic acid, of not more than 1 g per 100 g.
-   Olive pomace oil is the oil obtained by treating olive pomace with solvents or other physical treatments.

The IOC also fixes other analytical parameters for the different types of olive oil [49].

4.8.2 The unsaponifiable fraction of olive oil

Until recently, most of the protective effects of olive oil were associated almost exclusively to the high proportion of unsaturated FAs; however, some constituents of the unsaponifiable fraction (UF) also called "minor components," have been reported to exert beneficial effects. These so-called bioactive compounds can be found and obtained not only from the unsaponifiable fraction of olive oil but also from wastes and by-products of the olive industry and include chemical species such as [27, 50]:

- Tocopherols, especially α-tocopherol (vitamin E), considered as potent lipophilic antioxidants that exert a protective effect on membrane lipids against oxidation.
- Phytosterols, like β-sitosterol, that have a recognized effect on lowering cholesterol concentrations by interfering its absorption in the intestinal tract.
- Triterpenoids like squalene that has been suggested to have anticarcinogenic effects.
- β-Carotene, a provitamin A carotenoid, and lutein are considered beneficial (Section 4.12).
- Flavonoids (Chapter 8)
- Secoiridoids
- Phenolic acid and derivatives, including polyphenols that have been reported to be involved in biological functions such as cell signaling and redox balance as well as in the inhibition of toxic amyloid aggregation [51].

The unsaponifiable fraction is also responsible for the color of olive oil, an essential attribute for the acceptability by the consumers. The characteristic color of an olive oil depends on the concentration and the ratio of different pigments associated to the UF like chlorophylls, pheophytins (chlorophylls without the coordinated magnesium), and carotenoids.

4.9 Fat and adipose tissue

According to the WHO recommendations, around 30% of energy intake must come from lipids, although in many industrialized countries this percentage rises up to 40%. We have previously seen that most of this energy is associated to TAGs that also act as providers of essential FAs and carriers of important nutrients such as fat-soluble vitamins. In mammals, the lipids used for energy production and synthesis of other molecules have two main origins: diet and adipose tissue (AT).

4.9.1 Lipids from the diet: digestion and absorption

The hydrophobicity of lipids influences the whole mechanism of digestion and adsorption of these biomolecules, as all the fluids in our GI tract (saliva, gastric juice, etc.) are aqueous. In order to minimize contact with surrounding water molecules, lipids organize in microscopic spherical particles, that are converted in the duodenum to small micelles (emulsification) by bile salts, a family of cholesterol-derived natural detergents, so that they can be efficiently digested by intestinal enzymes. TAGs, phospholipids, and cholesterol esters, the predominant dietary lipids, are hydrolyzed by lipases, phospholipases, and cholesterol esterases, releasing free FAs (FFAs), *sn2*-monoacylglycerols, lysophospholipids, and free cholesterol, that are incorporated into micelles with bile salts and taken up by enterocytes, the cells of the intestinal mucosa [52].

Enterocytes reconvert FAs to TAGs and esterify cholesterol and lysophospholipids with FAs. These compounds are combined with apolipoproteins to form aggregates known as chylomicrons (Section 4.14) that are released into the lymphatic system and, then, into the bloodstream. Chylomicrons are responsible for transporting TAGs and sterols to the AT and muscle, where they are used for storage and energy production, respectively. The remnants of chylomicrons travel to the liver where they are taken up by receptor-mediated endocytosis [53].

The liver can use FAs and glycerol to produce energy and derived metabolites, but it is also the main responsible for supplying appropriate levels of nutrients to the rest of organs and tissues and for the removal of waste and toxic products. To accomplish this goal, the hepatocytes, the cells that comprise more than 70% of the liver mass, must have an enormous metabolic versatility that allow them to degrade and synthesize a wide variety of biomolecules and metabolites.

Hepatocytes pack excess or newly synthesized lipids into another type of lipoproteins, very low density lipoproteins (VLDL), and release them into the bloodstream in order to be delivered to the rest of the tissues (Section 4.14).

4.9.2 Lipids from the adipose tissue: mobilization of TAGs

AT was traditionally considered as a passive store of energy and thermal insulator. However, AT is in fact an active endocrine tissue with an essential role in regulating body mass and optimizing the energetic status of the organism (Chapter 14).

The main cells of AT at metabolic levels are adipocytes that take up most of the TAGs carried by and VLDL. The mechanism of uptake implies extracellular hydrolysis of TAGs by the enzyme lipoprotein lipase (EC 3.1.1.34), followed by the entry of FAs through the plasma membrane of the adipocyte either by diffusion or by a protein-mediated mechanism in the case of long-chain FA (LCFA, more than 14 carbon atoms) [54]. Glycerol uptake is mediated by aquaglyceroporins [55]. TAGs are resynthesized inside the adipocyte where they form a spherical hydrophobic droplet in the cytosol. This is an optimal situation for energy store not only in terms of the energy density but also in terms of space as, unlike glycogen particles (Chapter 3), the hydrophobicity of these droplets prevents swelling by interaction with water molecules, thus minimizing their volume within the cell. Lipid droplets have a general structure in which a core of sterol esters and TAGs is surrounded by a monolayer of phospholipids [9].

Under conditions of starvation or emergency, stored TAGs are hydrolyzed by lipases activated by hormones like glucagon and adrenaline. The resulting glycerol and

FA molecules are released in the bloodstream and delivered to the kidney and other organs and tissues, where they will be used to provide energy. This process is known as mobilization of TAGs. Due to their low solubility in aqueous media, mobilized FA must bind to the specialized protein serum albumin in order to be transported.

4.10 Degradation and biosynthesis of fatty acids in animals

FAs that reach the liver may come from two different sources, as explained above: via chylomicron remnants or by mobilization of fats. Once inside the hepatocyte, FAs have different destinations depending on the necessities of the cell and/or the rest of the organism:

- Degradation to obtain energy and/or metabolites that can be used for synthesis of other biomolecules.
- Synthesis of phospholipids and TAGs that can be used by liver cells or packed as VLDL and exported to the bloodstream.

4.10.1 Fatty acid degradation

Three pathways of FA oxidation have been identified in eukaryotes: α, ω, and β. ω-Oxidation takes place in microsomes, whereas α- and β-oxidation can occur in peroxisomes and mitochondria [56]. Mitochondrial β-oxidation is the primary catabolic pathway in the liver for most FAs and occurs in three stages:

1. Activation of FA in the cytosol: This step is common to all types of oxidation and ensures that FAs do not diffuse out of the cell:

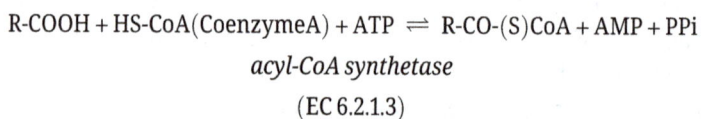

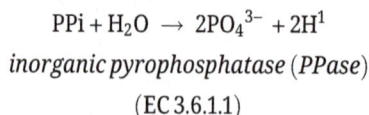

$$\text{R-COOH} + \text{HS-CoA(CoenzymeA)} + \text{ATP} \rightleftharpoons \text{R-CO-(S)CoA} + \text{AMP} + \text{PPi}$$

acyl-CoA synthetase

(EC 6.2.1.3)

$$\text{PPi} + \text{H}_2\text{O} \rightarrow 2\text{PO}_4^{3-} + 2\text{H}^1$$

inorganic pyrophosphatase (PPase)

(EC 3.6.1.1)

The second reaction is necessary to pull the first reaction toward the products, according to Le-Chatelier's principle, thereby increasing the yield of acyl-CoA (R-CO-(S)CoA). In bioenergetic terms, this is equivalent to spending two molecules of ATP. Coenzyme A is a ubiquitous acyl carrier synthesized from ATP, panthotenic acid (vitamin B5), and cystein, the latter providing a thiol group that forms the thioesther bond with the carboxylic group (Chapter 7).

2. Transport to the interior of the mitochondria in three steps with the carnitine shuttle (Figure 4.6). This step is not required by medium-chain FAs, such as octanoate, that can freely enter mitochondria:

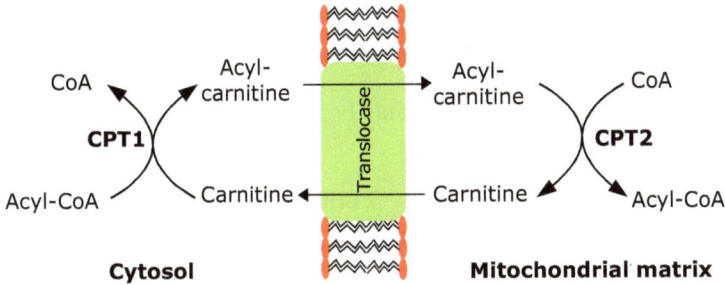

Figure 4.6: Transport of fatty acids into the mitochondria via the carnitine shuttle. CPT1 and CPT2: carnitine palmitoyltransferase 1 and 2, respectively (adapted from https://fr.wikipedia.org/wiki/Carnitine#/media/Fichier:Acyl-CoA_from_cytosol_to_the_mitochondrial_matrix.svg).

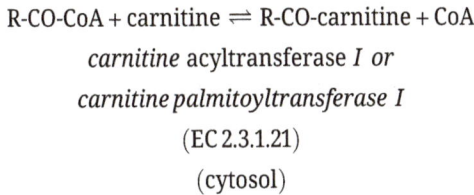

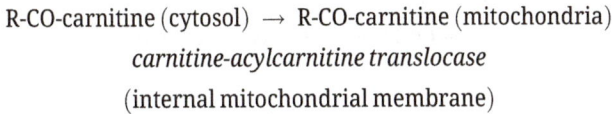

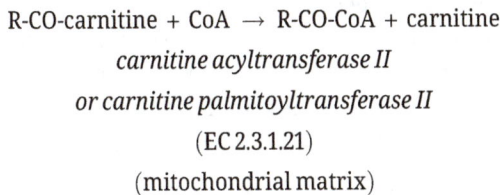

$$\text{R-CO-CoA} + \text{carnitine} \rightleftharpoons \text{R-CO-carnitine} + \text{CoA}$$

carnitine acyltransferase *I or*

carnitine palmitoyltransferase I

(EC 2.3.1.21)

(cytosol)

$$\text{R-CO-carnitine (cytosol)} \rightarrow \text{R-CO-carnitine (mitochondria)}$$

carnitine-acylcarnitine translocase

(internal mitochondrial membrane)

$$\text{R-CO-carnitine} + \text{CoA} \rightarrow \text{R-CO-CoA} + \text{carnitine}$$

carnitine acyltransferase II

or carnitine palmitoyltransferase II

(EC 2.3.1.21)

(mitochondrial matrix)

3. β-Oxidation: An oxidative cycle that involves shortening of acyl-CoAs into fragments of two carbon atoms that form molecules of acetyl-CoA. Each cycle has four steps:

(a) Dehydrogenation of the bond that links C-2 (carbon α) and C-3 (carbon β), yielding an unsaturated *trans*-enoyl-CoA and a molecule of $FADH_2$:

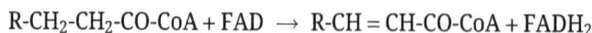

$$R\text{-}CH_2\text{-}CH_2\text{-}CO\text{-}CoA + FAD \rightarrow R\text{-}CH = CH\text{-}CO\text{-}CoA + FADH_2$$

acyl-CoA dehydrogenase

(ECs 1.3.8.1, 1.3.8.7, 1.3.8.8, and 1.3.8.9)

(b) Addition of a molecule of water to the double bond, mediated by an enoyl-CoA hidratase, generating a β-hydroxyacyl-CoA:

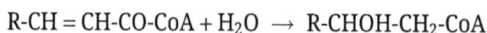

$$R\text{-}CH = CH\text{-}CO\text{-}CoA + H_2O \rightarrow R\text{-}CHOH\text{-}CH_2\text{-}CoA$$

enoyl-CoA hydratase

(EC 4.2.1.17)

(c) Dehydrogenation of the β-hydroxyacyl-CoA, yielding a β-ketoacyl-CoA and NADH:

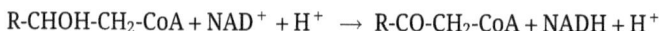

$$R\text{-}CHOH\text{-}CH_2\text{-}CoA + NAD^+ + H^+ \rightarrow R\text{-}CO\text{-}CH_2\text{-}CoA + NADH + H^+$$

β-hydroxyacyl-CoA dehydrogenase

(EC 1.1.1.35)

(d) Reaction of the β-ketoacyl-CoA with another molecule of CoA to yield acetyl-CoA and an acyl-CoA shortened by two carbon atoms with respect to the initial acyl-CoA:

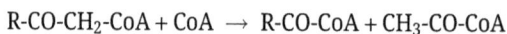

$$R\text{-}CO\text{-}CH_2\text{-}CoA + CoA \rightarrow R\text{-}CO\text{-}CoA + CH_3\text{-}CO\text{-}CoA$$

acyl-CoA acetyltransferase(thiolase)

(EC 2.3.1.16)

The process of β-oxidation is repeated until the initial acyl-CoA of n carbon atoms is converted into $n/2$ molecules of acetyl-CoA (CH_3-CO-CoA). The latter can enter the citric acid cycle to yield more reducing power ($FADH_2$ and NADH) and ATP [2, 9], or it may be used to synthesize ketone bodies (Section 4.10.4).

The overall balance for a typical FA like palmitic acid (16:0) is

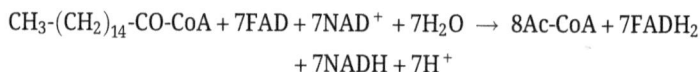

$$CH_3\text{-}(CH_2)_{14}\text{-}CO\text{-}CoA + 7FAD + 7NAD^+ + 7H_2O \rightarrow 8Ac\text{-}CoA + 7FADH_2$$
$$+ 7NADH + 7H^+$$

A maximum of 106 molecules of ATP can be obtained from each molecule of palmitic acid provided that all the acetyl-CoA molecules enter the citric acid cycle and all the molecules of $FADH_2$ and NADH are oxidized in the respiratory chain [9].

Unsaturated FAs require additional steps to connect with this cycle, whereas FAs with an odd number of carbon atoms are also subjected to β-oxidation. In this case, propionyl-CoA (CH_3-CH_2-CO-CoA), instead of acetyl-CoA, is generated in the last cycle.

Propionyl-CoA undergoes a series of chemical steps in order to enter the citric acid cycle as succinyl-CoA [9].

4.10.2 Biosynthesis of fatty acids

FA biosynthesis mainly occurs in the cytosol of mammalian cells although its primary substrates are molecules of acetyl-CoA located inside the mitochondria that must be transported to the cytosol. Since mitochondrial membranes are impermeable to acetyl-CoA, the exit occurs through the so-called citrate shuttle. In the first step of the citric acid cycle, acetyl-CoA is condensed with oxalacetate (OAA) to form citrate in a step catalyzed by the mitochondrial enzyme citrate synthase. Under certain conditions (Section 4.10.3), citrate may leave the mitochondria toward the cytosol through a citrate transporter located at the mitochondrial inner membrane. In the cytosol, citrate is converted back into acetyl-CoA and OAA by the cytosolic enzyme citrate lyase (Figure 4.7).

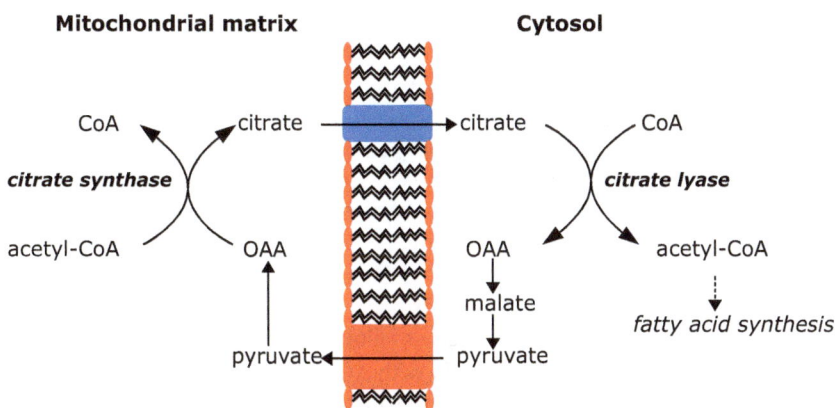

Figure 4.7: Export of acetyl-CoA from the mitochondria to the cytosol via the citrate shuttle (adapted from https://commons.wikimedia.org/wiki/File:Citrat-Shuttle.svg#filelinks).

Cytosolic acetyl-CoA must be carboxylated to malonyl-CoA before being used as substrate for the synthesis of FAs. This reaction is catalyzed by acetyl-CoA carboxylase (EC 6.4.1.2), a key enzyme in the regulation of FA biosynthesis:

$$CH_3\text{-}CO\text{-}CoA + HCO_3^- + ATP \longrightarrow {}^-OOC\text{-}CH_2\text{-}CO\text{-}CoA + ADP + PO_4^{3-} (Pi)$$

Seven molecules of malonyl-CoA molecules are condensed to produce palmitate in a series of cyclic enzymatic reactions, mediated by the fatty acid synthase (FAS, EC 2.3.1.85), that resembles a reverse FA β-oxidation, although there are significant differences between them. In mammals, fatty acid synthase is a homodimer composed of two identical polypeptides, each of them containing seven domains with distinct functions or enzymatic activities (Figure 4.8): acyl-carrier protein (ACP), malonyl/acetyl-CoA-ACP transferase (MAT); β-ketoacyl-ACP synthase (KS), β-ketoacyl-ACP reductase (KR), β-hydroxyacyl-ACP dehydratase (DH), enoyl-ACP reductase (ER), and thioesterase (TE).

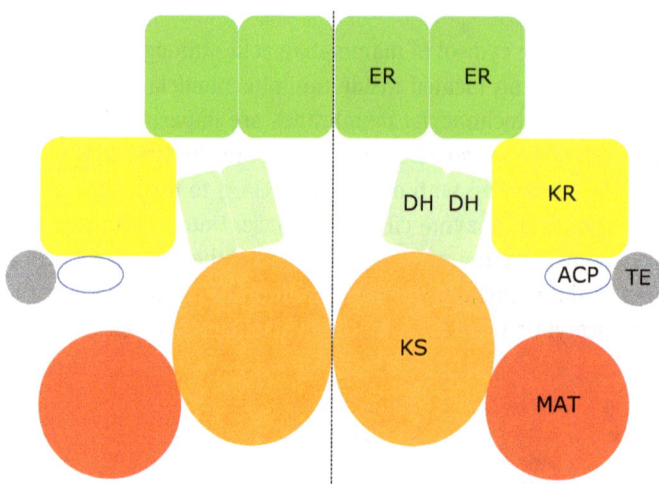

Figure 4.8: Schematic diagram of the mammalian fatty acid synthase.
This protein is a homodimer composed of two identical polypeptides (separated by a vertical dashed line), each of them containing seven domains with distinct functions or enzymatic activities (shown in different colors): acyl-carrier protein (ACP), malonyl/acetyl-CoA-ACP transferase (MAT); β-ketoacyl-ACP synthase (KS), β-ketoacyl-ACP reductase (KR), β-hydroxyacyl-ACP dehydratase (DH), enoyl-ACP reductase (ER), and thioesterase (TE) (adapted from [57]).

The steps of FA synthesis can be summarized as follows:
–   Initially, a molecule of acetyl-CoA reacts with ACP releasing CoA and leaving an acetyl group linked to ACP that is rapidly transferred to the KS domain; then, a similar process occurs between a malonyl-CoA and ACP. Both reactions are mediated by the MAT domain:

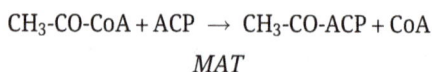

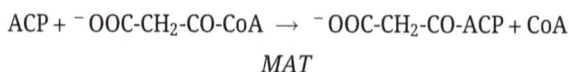

$$CH_3\text{-}CO\text{-}CoA + ACP \rightarrow CH_3\text{-}CO\text{-}ACP + CoA$$
$$MAT$$

$$(CH_3\text{-}CO\text{-}ACP + KS \rightarrow CH_3\text{-}CO\text{-}KS + ACP)$$

$$ACP + {}^-OOC\text{-}CH_2\text{-}CO\text{-}CoA \rightarrow {}^-OOC\text{-}CH_2\text{-}CO\text{-}ACP + CoA$$
$$MAT$$

–   Condensation:

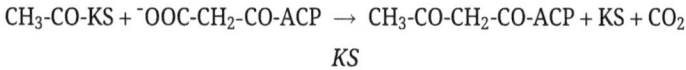

$$CH_3\text{-}CO\text{-}KS + {}^-OOC\text{-}CH_2\text{-}CO\text{-}ACP \rightarrow CH_3\text{-}CO\text{-}CH_2\text{-}CO\text{-}ACP + KS + CO_2$$
$$KS$$

–   Reduction of the carbonyl group:

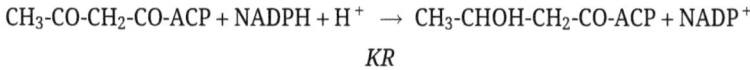

$$CH_3\text{-}CO\text{-}CH_2\text{-}CO\text{-}ACP + NADPH + H^+ \rightarrow CH_3\text{-}CHOH\text{-}CH_2\text{-}CO\text{-}ACP + NADP^+$$
$$KR$$

–   Dehydration:

$$CH_3\text{-}CHOH\text{-}CH_2\text{-}CO\text{-}ACP \rightarrow CH_3\text{-}CH = CH\text{-}CO\text{-}ACP + H_2O$$
$$DH$$

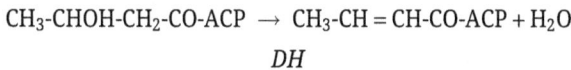

–   Reduction of the double bond:

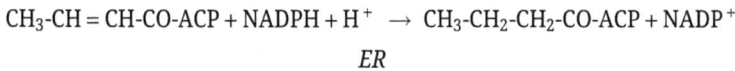

$$CH_3\text{-}CH = CH\text{-}CO\text{-}ACP + NADPH + H^+ \rightarrow CH_3\text{-}CH_2\text{-}CH_2\text{-}CO\text{-}ACP + NADP^+$$
$$ER$$

The result of these reactions is the formation of a butyryl residue (four carbon atoms) that is transferred from ACP to the KS domain, then, ACP binds a new malonyl group thus starting a new cycle. This process is repeated to continue the elongation in fragments of two carbons to finally produce palmitoyl-ACP. Palmitic acid (16:0) is released from the ACP domain of the FAS complex in a reaction mediated by the thioesterase (TE) domain.

The global balance for palmitic acid synthesis is

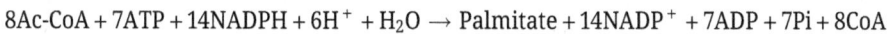

$$8Ac\text{-}CoA + 7ATP + 14NADPH + 6H^+ + H_2O \rightarrow Palmitate + 14NADP^+ + 7ADP + 7Pi + 8CoA$$

The synthesis of FAs with a greater number of carbons requires elongation systems present in the smooth endoplasmic reticulum (ER) and mitochondria [58]. The introduction of double bonds is mediated by fatty acyl-CoA desaturases, like stearoyl-CoA Δ^9-desaturase (SCD, EC 1.14.19.1), that requires O_2, NAD(P)H, and other enzymes (cytochrome b5 and cytochrome b5 reductase) to produce oleyl (18:1 Δ^9)-CoA and palmitoleyl (16:1 Δ^9)-CoA from stearoyl (18:0)-CoA and palmitoyl (16:0)-CoA, respectively. Two isoforms of SCD (SCD1 and SCD5) have been identified in humans, and it has been suggested that SCD1 may play a prominent role in the development of obesity-related chronic metabolic diseases, including nonalcoholic fatty liver disease, insulin resistance, and hyperlipidemia [59]. Desaturases from humans and other species cannot introduce additional double bonds between C-10 and the methyl-terminal end, and this limitation has a major impact in our diet.

4.10.3 Regulation of the metabolism of fatty acids

– Regulation of β-oxidation: The major limiting step of β-oxidation is the conversion of acyl-CoA to acyl-carnitine, a reaction catalyzed by the cytosolic enzyme carnitil-acyl transferase I (carnitine palmitoyltransferase or CPT1). CPT1 is subjected to a tight regulation that includes allosteric inhibition by malonyl-CoA and control at transcriptional level. An increase in the cytosolic levels of malonyl-CoA inhibits transport of FAs to the mitochondria and, thus, their degradation. This mechanism links the regulation of β-oxidation to FA biosynthesis.
– Regulation of FA synthesis: In general terms, FA synthesis is regulated by the energy status of the cell through the enzyme citrate synthase: when mitochondrial ATP levels are low, citrate follows the citric acid cycle; on the contrary, high levels of ATP stop the citric acid cycle and citrate is transferred to the cytoplasm. Acetyl-CoA carboxylase, the key enzyme in the regulation of FA synthesis, is allosterically activated by citrate thus increasing the levels of malonyl-CoA, the main substrate of fatty acid synthase. At the same time, malonyl-CoA inhibits CPT1, thereby blocking the access of FA to the mitochondria, as explained above. Acetyl-CoA carboxylase can also be regulated by phosphorylation mediated by the enzyme AMP-activated protein kinase (AMPK). AMPK acts in response to low cellular energy (high [AMP]/[ATP] ratio) [56].

Besides the short-term regulatory mechanisms just described, the metabolism of FAs is also regulated at longer term by hormones (glucagon, adrenalin, and insulin), transcription factors like PPARα and epigenetic mechanisms [56, 60] (Chapter 14). Although the mechanisms are complex and not completely understood at molecular level, the rationale behind all of them is to optimize the energy resources of the organism in every circumstance. Thus, high levels of glucose in blood, a signal of nutrient abundance, trigger insulin secretion by pancreatic β-cells. Insulin favors energy storage by increasing the synthesis of FAs, TAGs, phospholipids, and VLDL in the liver and the uptake of TAGs in AT and muscle. Glucagon, which is secreted into the bloodstream when glucose levels are low, exerts the opposite effect to insulin: increases FA degradation to obtain energy and produce ketone bodies, an alternative fuel to glucose. Adrenaline, the emergency hormone, has similar effects to those of glucagon, that is, mobilize all the energy resources, in this case to prepare the body for a "fight-or-flight" response.

4.10.4 Metabolism of the ketone bodies

During fasting, untreated type I diabetes mellitus or a high-fat diet with a low or null content in carbohydrates, mitochondrial oxalacetate is derived to the production of glucose by gluconeogenesis, thus blocking the citric acid cycle. On the other hand, acetyl-CoA is synthesized in the mitochondria from FAs that come either from the diet or

TAGs mobilization (fasting and diabetes). Under these conditions, acetyl-CoA is diverted to the synthesis of a group of metabolites, collectively known as ketone bodies (KB), that include β-hydroxybutyrate, acetoacetate, and acetone. KB synthesis, also called ketogenesis, is a biochemical process that primarily takes place in the mitochondria of hepatocytes, although kidney epithelia, astrocytes, and enterocytes are also capable of producing KB to a lesser extent. The substrate for ketogenesis is acetyl-CoA, the FA β-oxidation product, and the process is as follows [61]:

–   Condensation of two molecules of acetyl-CoA to yield acetoacetyl-CoA:

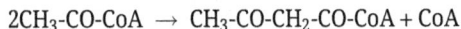

$$2CH_3\text{-}CO\text{-}CoA \rightarrow CH_3\text{-}CO\text{-}CH_2\text{-}CO\text{-}CoA + CoA$$

acetoacetyl-CoA thiolase

(ACAT1, EC 2.3.1.9)

–   A third acetyl-CoA molecule reacts with acetoacetyl-CoA to produce 3-hydroxy-3-methylglytaryl-CoA (HMG-CoA):

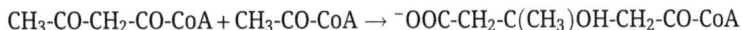

$$CH_3\text{-}CO\text{-}CH_2\text{-}CO\text{-}CoA + CH_3\text{-}CO\text{-}CoA \rightarrow {}^-OOC\text{-}CH_2\text{-}C(CH_3)OH\text{-}CH_2\text{-}CO\text{-}CoA$$

mitochondrial HMG-CoA synthetase

EC 2.3.3.10

–   HMG-CoA is transformed into acetoacetate and acetyl-CoA:

$$\text{-}OC\text{-}CH_2\text{-}C(CH_3)OH\text{-}CH_2\text{-}CO\text{-}CoA \rightarrow CH_3\text{-}CO\text{-}CH_2\text{-}COO^- + CH_3\text{-}CO\text{-}CoA$$

HMG-CoA lyase

(HMGCL, EC 4.1.3.4)

Acetoacetate has two alternative destinations:
–   Reduction to β-hydroxybutyrate, the most abundant ketone body in the bloodstream:

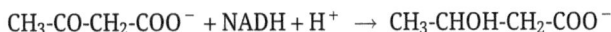

$$CH_3\text{-}CO\text{-}CH_2\text{-}COO^- + NADH + H^+ \rightarrow CH_3\text{-}CHOH\text{-}CH_2\text{-}COO^-$$

NADH-dependent β-hydroxybutyrate dehydrogenase

(BDH, EC 1.1.1.30)

–   Decarboxylation into volatile acetone, the simplest ketone body. This reaction has been suggested to be mediated by the enzyme acetoacetate decarboxylase; however, the gene coding for this protein has not been identified in humans, although the activity was reported to be present in human serum [62]. Other authors suggest that this reaction occurs spontaneously, at least in certain tissues like lungs, where acetone can be released by exhalation [61]:

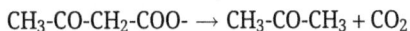

$$CH_3\text{-}CO\text{-}CH_2\text{-}COO\text{-} \rightarrow CH_3\text{-}CO\text{-}CH_3 + CO_2$$

?

Ketone bodies can be used as fuels by almost all cells, including neurons (but excluding hepatocytes), in a process called ketolysis, which is basically the opposite to ketogenesis [61]. An excess of ketone bodies can cause serious medical problems (acidosis or ketosis), very common in periods of severe starvation or uncontrolled diabetes. The presence of acetone in the air exhaled by untreated type I diabetic patients is a symptom of this life-threatening condition.

Diets aimed at inducing ketogenesis, the so-called ketogenic dietss, are extremely popular in countries with a high prevalence of overweight and obesity. The idea behind these diets is that keeping dietary carbohydrates very low, with varying levels of protein and fat, will reduce insulin secretion and induce the use of stored fat to produce energy. Eventually, this metabolic change would end up in weight loss and an improvement in glucose levels and lipid profiles in blood. The classical ketogenic diet recommends 1 gram of protein per kilogram of body weight, 10–15 g of carbohydrates per day, and the remaining calories from fat. Ketogenic diets were initially used to treat diseases like epilepsy and type I diabetes, however, since the 1970s they have been increasingly used as a means to reduce weight. Scientific evidence shows significant results on weight loss in the short term, although the efficacies of ketogenic diets in the long term are comparable to those of other hypocaloric diets; moreover, long-term safety and effects on other health issues are not clear and require further studies [63].

4.11 Essential fatty acids: metabolism of arachidonate

Desaturases from humans, other animal species, and yeast cannot introduce double bonds at carbon atoms located at positions 3 and 6 from the methyl-terminal end to produce the so-called n-3 and n-6 FAs. For example, humans can synthesize oleic acid (18:1 Δ^9) from acetyl-CoA, which may be obtained not only from other FAs, but also from glucose or some aminoacids; however, PUFAs like linoleic (18:2 $\Delta^{9,12}$, LA) or α-linolenic (18:1 $\Delta^{9,12,15}$, ALA) acids cannot be subsequently obtained from oleic acid.

LA and ALA are precursors of other n-3 and n-6 FAs like arachidonic acid (ARA, 20:4 $\Delta^{5,8,11,14}$, n-6), eicosapentaenoic acid (EPA, 20:5 $\Delta^{5,8,11,14,17}$, n-3), and docosahexaenoic acid (DHA, 22:6 $\Delta^{4,7,10,13,16,19}$, n-3), that play key regulatory roles in the human body [64]. For this reason, LA and ALA are considered essential FAs and must be included in the diet providing between 6 and 11% of our total energy intake (see above). Plants and other photosynthetic organisms (as well as bacteria) have a set of desaturases that allow them to synthesize both linoleic and α-linolenic acid; therefore, vegetable oils are good sources of essential FAs. Other important sources of LA, ALA, ARA, EPA, and DHA are fatty fishes like tunas, sardines, or salmons that obtain these molecules in the diet from marine algae and accumulate them at high concentrations in their tissues.

The synthesis of arachidonic acid derivatives, known as eicosanoids for having 20 carbon atoms in their molecules, is particularly important for human health. Eicosanoid production occurs in virtually all tissues and usually starts by releasing arachidonic acid from cell membrane phospholipids in a process catalyzed mainly by enzymes of the phospholipase A2 (PLA2) superfamily. Free arachidonic acid can be processed in several ways, resulting in the generation of many derivatives with different functions (Figure 4.9) [65].

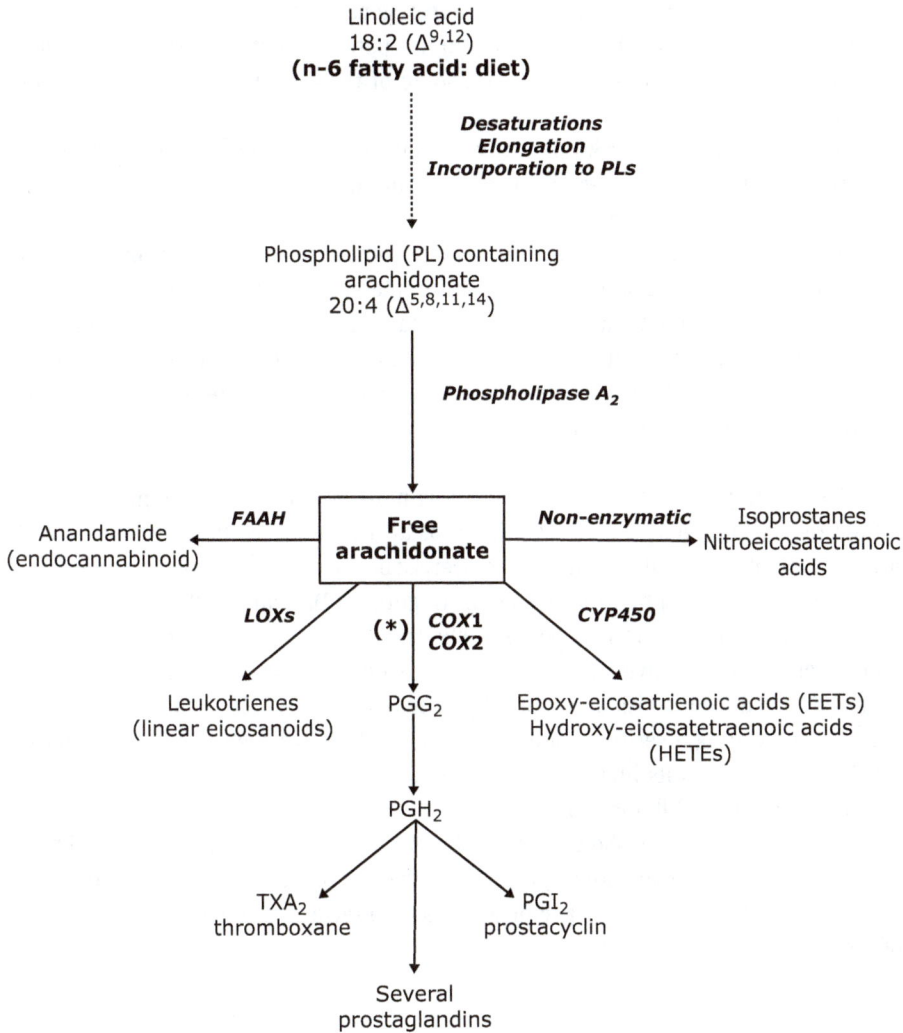

Linoleic acid
18:2 ($\Delta^{9,12}$)
(n-6 fatty acid: diet)

Desaturations
Elongation
Incorporation to PLs

Phospholipid (PL) containing
arachidonate
20:4 ($\Delta^{5,8,11,14}$)

Phospholipase A$_2$

| FAAH | **Free arachidonate** | *Non-enzymatic* | Isoprostanes Nitroeicosatetranoic acids |

Anandamide
(endocannabinoid)

LOXs   (*) COX1 COX2   *CYP450*

Leukotrienes
(linear eicosanoids)   PGG$_2$ Epoxy-eicosatrienoic acids (EETs)
Hydroxy-eicosatetraenoic acids
(HETEs)

PGH$_2$

TXA$_2$
thromboxane   PGI$_2$
prostacyclin

Several
prostaglandins

(*) Inhibited by NSAIDs

Figure 4.9: Metabolism of arachidonic acid.

- Cycloxygenase (COX) pathway: Arachidonic acid is subjected to two consecutive reactions: dioxygenation to yield prostaglandin G2 (PGG$_2$) and a subsequent peroxidation that converts PGG$_2$ to prostaglandin H$_2$ (PGH$_2$). These two reactions are catalyzed by the enzyme prostaglandin G/H synthase. Two different isoforms of this protein encoded by the genes COX1 and COX2 have been identified in humans. Prostaglandin PGH_2 is subsequently transformed in different eicosanoids like thromboxane A_2 (TXA_2), prostacyclin (PGI_2), and several prostaglandins (PGs). These molecules have been reportedly implicated in inflammation, vasoconstriction, platelet aggregation, GI mucosa protection, renal function, neurodegeneration, and cancer.

- LOX pathway: Four LOX are involved in this pathway that produces a set of linear eicosanoids, known as leukotrienes, some of them showing anti-inflammatory properties.
- Cytochrome P450 (CYP450) pathway, mainly restricted to the liver, produces derivatives that regulate vessel constriction and dilation, among other effects, and inhibit COX2 expression.
- The FA amide hydrolase (FAAH) pathway synthesizes an endocannabinoid, anandamide, from arachidonic acid. Anandamide supports tissue regeneration and cellular proliferation via interaction with cannabinoid type 1 (CB1) receptors.
- Nonenzymatic processing of arachidonic acid: ROS and reactive nitrogen species may oxidize the double bonds of arachidonic acid, thus generating isoprostanes and nitroeicosatetraenoic acids.

A series of compounds collectively known as nonsteroidal anti-inflammatory drugs inhibit the cyclooxygenase activity of the prostaglandin G/H synthase. Among NSAIDs are popular pharmaceuticals drugs like acetylsalicylic acid (aspirin), ibuprofen, naproxen, and acetaminophen (paracetamol) (Figure 4.10), many of them sold without prescription in many countries. NSAIDs relieve pain, reduce fever and inflammation, and act as antithrombolytic agents; moreover, some of them, like aspirin, have been reported to reduce the risk and mortality of many cancers [66]. Aspirin acts by irreversibly acetylating a serine residue at position 529 in prostaglandin G/H synthase I (COX1); this explains its higher antithrombotic action with respect to other NSAIDs that are reversible inhibitors [67].

All NSAIDs have been shown to have important side effects related to the GI tract, cardiovascular system, and kidneys. The involvement of eicosanoids in human health and the development of more specific NSAIDs with fewer side effects are active fields of research nowadays [66, 68, 69].

4.12 Nonhydrolyzable lipids: importance in the diet

Most natural sources of fats and oils contain a nonhydrolyzable (nonsaponifiable or unsaponifiable) fraction (UF) that comprises between 0.2 and 2% of total lipids (Table 4.5). This fraction includes a variety of compounds that include hydrocarbons, triterpenic alcohols, steroids, phenolics, tocopherols, carotenoids, pigments, and volatile com-

**Acetylsalicylic acid
(Aspirin)**

Ibuprofen

**Acetaminophen
(Paracetamol)**

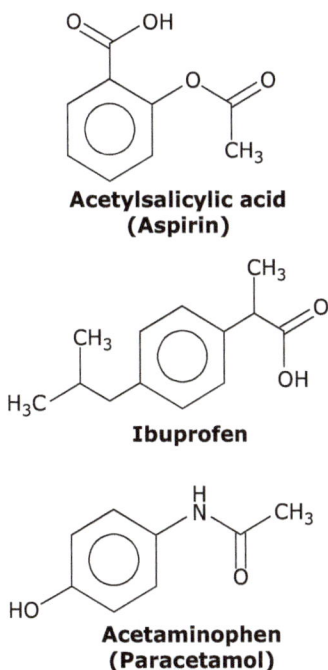

Figure 4.10: Chemical structures of some common nonsteroidal anti-inflammatory drugs (NSAIDs).

pounds [20, 21, 27]. UF seems to be responsible for some of the beneficial effects of certain oils, especially olive oil, whose UF has been shown to have anti-inflammatory and antioxidant effects, among others [21, 27].

Due to their importance for human diet, some aspects of the chemistry and biochemistry of carotenoids and sterols will be discussed. Other lipid compounds with important properties for human health will be discussed in Chapters 6 and 8.

4.12.1 Carotenoids

Carotenoids are a type of tetraterpenoids, hydrocarbons of biological origin derived from eight molecules of isoprene (2-methyl-1,3-butadiene, $CH_2=C(CH_3)CH=CH_2$) that, consequently, have 40 carbon skeletons. More specifically, IUPAC defines carotenoids as "tetraterpenoids (C_{40}), formally derived from the acyclic parent Ψ,Ψ-carotene (lycopene, Figure 4.11) by hydrogenation, dehydrogenation, cyclization, oxidation, or combination of these processes" [70]. In food, the isoprenoid units of carotenoids are normally joined so that the sequence is reversed at the center and can be cyclized at one or both ends. A distinctive characteristic of carotenoids is their extensive conjugated double-bond system, which is responsible for the yellow, orange, or red color they present. They normally occur as *all-trans* isomeric forms in natural sources, although *cis*-isomers are being increasingly reported. Depending on their composition, two major types of

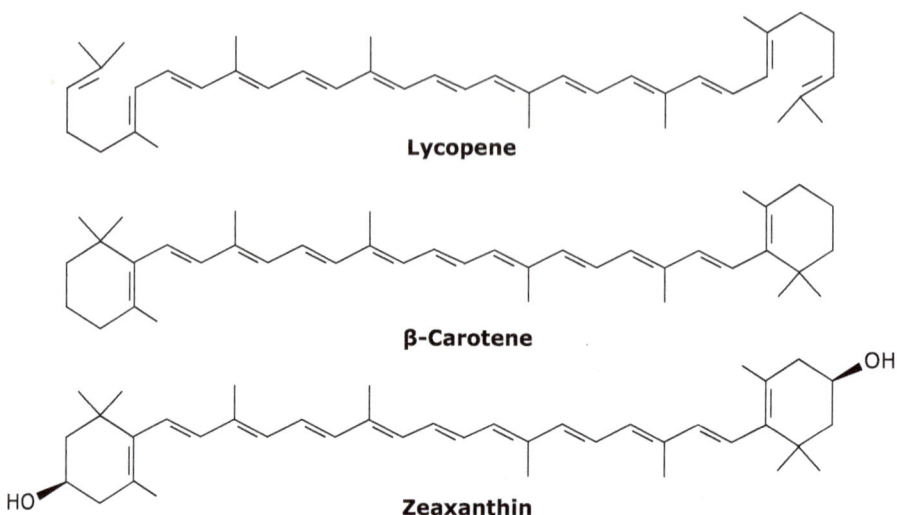

Lycopene

β-Carotene

Zeaxanthin

Figure 4.11: Chemical structures of some carotenoids. Chemical formulae were retrieved and adapted from https://www.lipidmaps.org/data/structure/LMSDSearch.php.

carotenoids occur in food: carotenes, made up of only carbon and hydrogen, and xanthophylls, which also contain oxygen [71].

Carotenoids are synthesized by photosynthetic organisms where they accomplish two major functions [72]:

– Absorption of photons in the blue–green region of the solar spectrum to transfer the energy to (bacterio-) chlorophylls. This capacity expands the range of light that can drive photosynthesis thus enhancing its efficiency.

– Protection of photosynthetic organisms from the harmful effects of excess exposure to light.

Carotenoids also occur in animals although they cannot synthesize them de novo, therefore, they obtain these nutrients from the diet and accumulate them unchanged or slightly modified [71]. More than 30 different carotenoids are regularly identified in human serum and breast milk; however, none of these molecules are considered essential nutrients because they have not been implicated in vital metabolic pathways or linked to specific deficiencies or diseases. For this reason, no formal dietary recommendation has been established by the European Food Safety Authority [73].

Dietary carotenoids can be classified according to their capacity to generate vitamin A (Chapter 6) in vivo: provitamin A carotenoids, that include α- and β-cryptoxanthine, as well as non-provitamin A carotenoids, that include lycopene, lutein, and zeaxanthin.

The biological activities and the efficacy of these compounds depend on their content in food sources, dietary intake, and bioavailability. Many epidemiological studies suggest that the plasma β-carotene concentration to obtain healthy effects is 0.4 μmol/

l, which can be achieved with intakes of 2–4 mg per day. Consumptions of food rich in β-carotene have also been associated with lower risks of chronic diseases and protection against oxidative stress.

β-Carotene is also an important source of vitamin A, whose dietary recommendation is 800 μg/day. In most Europeans countries, food authorities assume that 4.8 mg of β-carotene are needed to meet vitamin A requirements (a conversion factor of 6), whereas the US authorities established a conversion factor of 12 for β-carotene and 24 for other provitamin A carotenes. On the other hand, excess of vitamin A and derivatives have been reported to have teratogenic effects, whereas β-carotene supplementation seems to increase lung cancer risk in smokers (see Chapter 6 for further details).

Proposed intake recommendations for non-provitamin A carotenoids are 10–20 mg/day for lutein and 5.7–15 mg/day for lycopene, as they seem to be safer for human consumption and have been reported to be active in immunomodulation. In any case, more scientific evidence is needed in order to establish formal dietary recommendations for carotenoids and their possible use as supplements with beneficial effects [73].

4.12.2 Steroids

Steroids are naturally occurring compounds based on the cyclopenta[α]-phenanthrene carbon skeleton, partially or completely hydrogenated; they usually have methyl groups at C-10 and C-13, and often an alkyl group at C-17. Bond scissions, ring expansions, and/or ring contractions of the skeleton may occur thus producing a wide variety of related compounds. One of the most important variations among steroids are sterols, that contain a hydroxy group in the C3 position [70] (Figure 4.12).

Steroids are synthesized in vivo by the polymerization of six molecules of isoprene that initially produce a hydrocarbon of 30 carbons known as squalene; therefore, they are triterpenoids or derivatives. Squalene is cycled and subjected to a series of about 20 reactions whose final product is cholesterol in animal cells (see below), whereas slightly different sterols like stigmasterol and ergosterol are produced in plants and fungi, respectively [9]. Plant sterols are collectively named phytosterols and are being used as food supplements to reduce cholesterol uptake.

Cholesterol is an important component of mammalian cellular membranes where it is required for proper permeability, fluidity, organelle identity, and protein function. It is also the precursor of bile acids, steroid hormones, and vitamin D [74]; however, cholesterol is not an essential nutrient because human cells can synthesize it de novo from other nutrients like monosaccharides, FAs, or aminoacids.

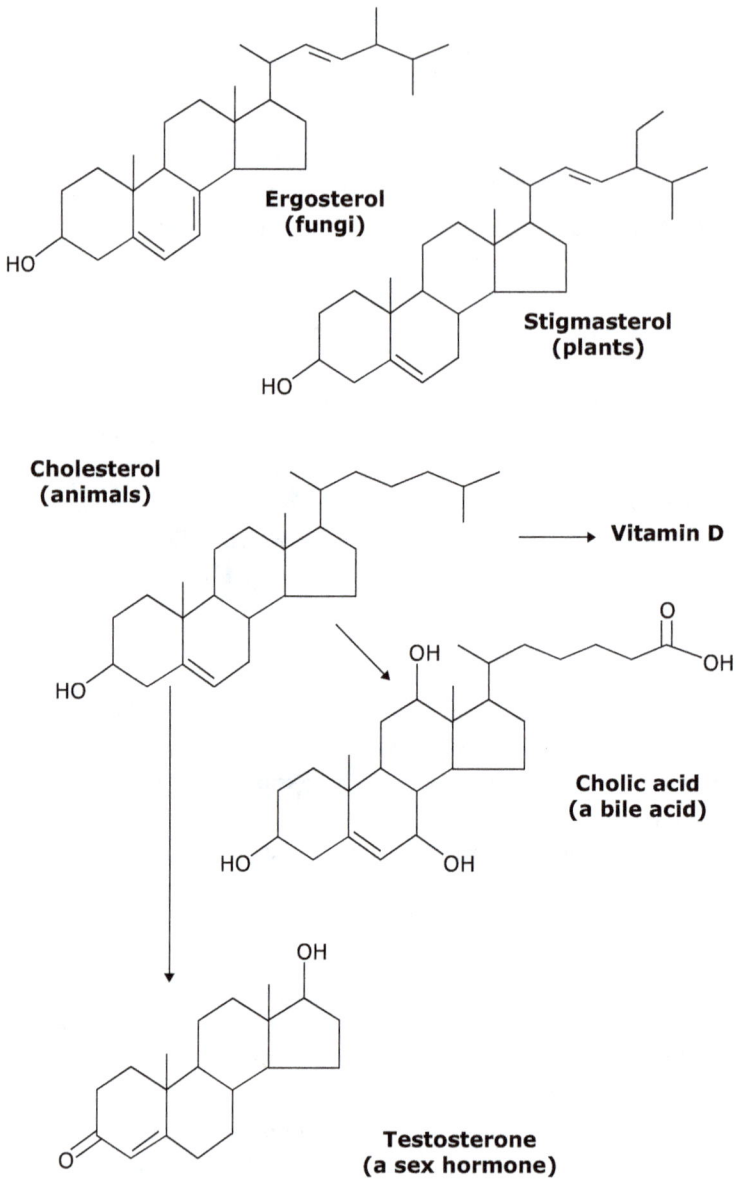

Figure 4.12: Chemical structures of sterols from fungi, plants, and animals and derivatives of cholesterol. Chemical formulae were retrieved and adapted from https://www.lipidmaps.org/data/structure/ LMSDSearch.php.

4.13 Metabolism of cholesterol

In humans, cholesterol may be taken up in the GI tract or synthesized de novo, mainly in liver and the small intestine [75]. Absorbed cholesterol may come directly from the diet (exogenous) or from the bile (endogenous). Exogenous cholesterol has been estimated in approximately 5 mg/day/kg body weight in a typical Western diet, whereas endogenous cholesterol accounts for 1,000 mg more. Enterocytes take up about 50% of the total cholesterol that travels through the small intestine by endocytosis, esterify the absorbed molecules with FAs, and incorporate them, along with TAGs, into chylomicrons, that are exported to the bloodstream [76]. Most of the cholesterol absorbed by the enterocytes reaches the hepatocytes as chylomicrons remnants.

In the hepatocytes, cholesterol has several destinations:
- Liver pools
- Bile production
- Incorporation into newly synthesized lipoproteins (see below)

Hepatocytes can also synthesize cholesterol from acetyl-CoA if needed. For simplicity, the process of cholesterol synthesis is normally considered to occur in four stages [9]:
1. Synthesis of mevalonate from acetyl-CoA
2. Synthesis of 3-isopentenyl pyrophosphate (IPP), also known as activated isoprene
3. Condensation of six isoprene units to form squalene
4. Cyclation of squalene producing a tetracyclic product that is subsequently converted into cholesterol

The complete pathway can be found elsewhere [77].

Cholesterol (and some intermediates of its synthesis pathway) not only has a major role in the structure and function of biological membranes but it is also the precursor of a number of important compounds that include steroid hormones, bile acids (Figure 4.12), and vitamin D (Chapter 6).

The last step of stage 1, that involves the reduction of 3-hydroxy-3-methylglutaryl-CoA to mevalonate, is a major rate-limiting step of the pathway; therefore, the enzyme that mediates this step, known as HMG-CoA reductase (HMGR, EC: 1.1.1.34), is considered to be extremely important for the regulation of the route. HMGR is an ER-resident integral membrane protein subjected to a complex regulation at several levels (transcriptional, translational, posttranslational, and degradation) to ensure that sterol synthesis meets cellular requirements [78]. These mechanisms of regulation can be modulated by hormones like insulin and glucagon and by intra- and extracellular levels of cholesterol [2, 9, 79].

Statins, a series of molecules of fungal origin, have been used as successful cholesterol lowering medications, thus contributing to the prevention of CVD. These molecules are strong inhibitors of HMGR, illustrating the important role of this enzyme in the homeostasis of cholesterol in the body. In any case, overall control of cholesterol synthesis seems to be extremely complex and many other regulatory mechanisms affecting more enzymes of this pathway have already been described [79–81]. The advances in the understanding of cholesterol metabolism have allowed the development of many cholesterol-lowering therapies beyond the statins. The success of these therapies are beneficial not only to CVD but also to other health conditions such as certain types of cancer, virus infections or Alzheimer's, Parkinson's, and Huntington's diseases [79].

4.14 Cholesterol and lipid transport in the bloodstream: apolipoproteins

Cholesterol and triglycerides are insoluble in water and must be associated with proteins in order to be transported in the bloodstream. These complexes, collectively known as lipoproteins, usually have a central hydrophobic core of nonpolar lipids (primarily cholesterol esters and TAGs) surrounded by a hydrophilic membrane consisting of phospolipids, free cholesterol, and apolipoproteins (Figure 4.13). Plasma lipoproteins are divided into seven classes based on size, lipid composition, and apolipoproteins: chylomicrons, chylomicron remnants, VLDL, IDL, low-density lipoprotein (LDL), high-density lipoprotein (HDL), and Lp (a) (Table 4.8):

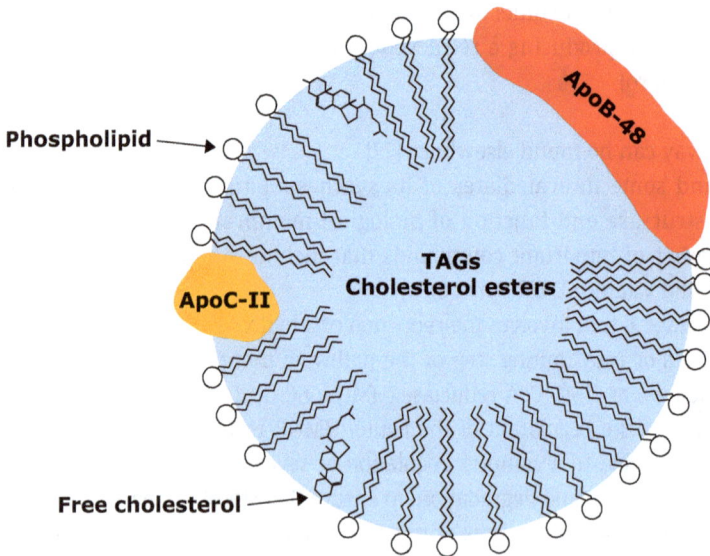

Figure 4.13: Structure of a chylomicron (adapted from [9, 82]).

- Chylomicrons: They are involved in the transport of dietary TGAs and cholesterol to peripheral tissues (mainly AT and muscles) and liver. These particles contain several apolipoproteins (Table 4.6), Apo B-48 being the core structural one. The size of chylomicrons varies depending on the amount of fat ingested. The removal of TGAs from by peripheral tissues results in smaller cholesterol-enriched particles called chylomicron remnants that are basically taken up by the liver and are considered pro-atherogenic (see below).
- Very low density lipoproteins (VLDL): They are produced by the liver and are enriched in TAGs. Apo B-100 is their core structural protein. They transport TGAs to AT, fundamentally, and muscle, and their size also vary depending on the quantity of TGAs carried in the particle.
- IDL (intermediate density lipoproteins or VLDL remnants). They result from the removal of triglycerides from VLDL by muscle and AT; therefore, they are enriched in cholesterol. They contain apolipoprotein B-100 and E. IDL particles are pro-atherogenic.
- Low density lipoproteins (LDL): These particles derive from VLDL and IDL particles and they are even further enriched in cholesterol. LDL carries the majority of the cholesterol that is in the circulation. The predominant apolipoprotein is Apo B-100. They carry cholesterol to extrahepatic tissue where there is a specific receptor for that protein. LDL consists of a spectrum of particles varying in size and density. Small dense LDL particles are considered to be more pro-atherogenic than large LDL particles.
- High density lipoproteins (HDL): These particles play an important role in reverse cholesterol transport from peripheral tissues to the liver. They increase their cholesterol content by interacting with other lipoproteins to take it to the liver and be eliminated by the bile. They are considered antiatherogenic. HDL particles also have antioxidant, anti-inflammatory, antithrombotic, and antiapoptotic properties. HDL particles are enriched in cholesterol and phospholipids, Apo A-I being the core structural protein. HDL particles are very heterogeneous and can be classified based on density, size, charge, or protein composition.
- Lipoprotein (a): (Lp (a)) is an LDL particle that has apolipoprotein (a) attached to Apo B-100 via a disulfide bond. This particle is pro-atherogenic. Its physiologic function is unknown.

4.15 Lipids and the mechanism of atherogenesis

Atherogenesis is the development of atheromatous ("gruel-looking") plaques, also known as atheroma, in the inner lining of the arteries; whereas the term atherosclerosis, sometimes used as a synonym, refers to the process of hardening ("sclerosis") that these plaques undergo in the evolution of this chronic disease. Atherosclerosis is considered the underlying cause of heart attack, stroke, and peripheral vascular disease, major causes of death in industrial countries; however, its pathophysiology is a com-

Table 4.8: Properties and composition of the main lipoproteins.

Lipoprotein	Density (g/ml)	Size (nm)	Major lipids	Major apoproteins
Chylomicrons	<0.930	75–1200	Triglycerides	Apo B-48, ApoC-I, -II, -III, ApoE, ApoA-I, -II, -IV,-V
Chylomicron remnants	0.930–1.006	30–80	Triglycerides Cholesterol	Apo B-48, Apo E
VLDL	0.930–1.006	30–80	Triglycerides	Apo A-V, Apo B-100, Apo E, Apo C-I, -II, -III
IDL	1.006–1.019	25–35	Triglycerides Cholesterol	Apo B-100, Apo E, Apo C-II, -III
LDL	1.019–1.063	18–25	Cholesterol	Apo B-100
HDL	1.063–1.210	5–12	Cholesterol Phospholipids	Apo A-I, -II, -IV,-V, Apo C-I, -II, -III, Apo E
Lp (a)	1.055–1.085	~30	Cholesterol	Apo B-100, Apo (a)

Source: [83, 84].

plex and not fully understood process. The current model of atheroma formation involves the following stages (Figure 4.14) [85]:

1. Recruitment of pro-inflammatory monocytes to the tunica intima layer of the medium and large size arteries.
2. In the intima, monocytes differentiate into macrophages that take up modified lipid species such as oxidized or glycated low-density lipoprotein (oxLDL and glyLDL), leading to the formation of the so-called foam cells (lipid-laden macrophages).
3. Foam cells secrete pro-inflammatory mediators, ROS, and tissue factor procoagulants that amplify local inflammation and promote thrombotic complications. T cells also enter the intima, thus contributing to this process.
4. Smooth muscle cells from the tunica media (the middle layer of the artery wall) are recruited into the intima where they proliferate and produce extracellular matrix molecules like collagen and elastin, thus forming a fibrous cap that covers the plaque.
5. Cellular debris and extracellular lipids derived from inefficient clearance of dead cells form the so-called necrotic core of the plaque.
6. Plaques may produce flow-limiting stenoses or provoke thrombi that can interrupt blood flow.

Elevated concentration of cholesterol in blood serum, especially that associated to LDL (LDL-cholesterol or LDL-C), has been linked to increase risk of CVD, type 2 diabetes, liver disease progression, and several types of cancer. One of the most important proofs of the association between LDL-C and CVD is familial hypercholesterolemia

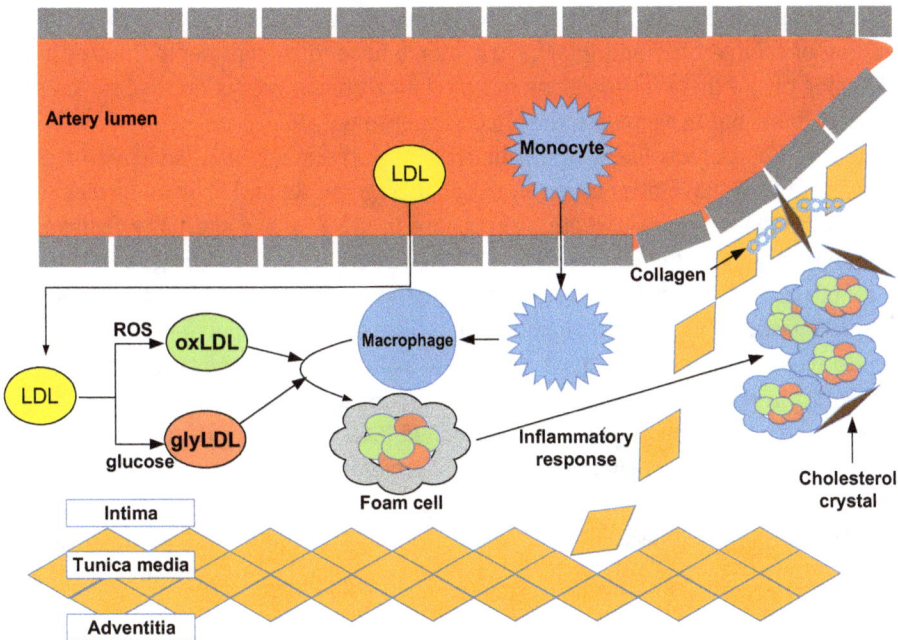

Figure 4.14: Cartoon showing the current model of the development of atheromatous plaques (adapted from references [79, 85]).

(FH), a disease characterized by severely elevated LDL-C levels that is inherited in an autosomal dominant manner. Patients with FH present atherosclerotic plaque deposition and increased risk for CVD, especially coronary artery disease, which usually manifest as angina and myocardial infarction. FH is associated with pathogenic mutations in one of the following genes: LDLR, APOB, or PCSK9. LDLR codes for the LDL receptor (LDL-R) located at the plasma membrane of several cell types, APOB codes for protein Apo B-100, responsible for the interaction between LDL and LDL-R, and PCSK9 codes for a protein that regulates the number of LDL receptors in the cell surface [86].

Although Dietary Guidelines for Americans recommended that individuals should eat as little cholesterol as possible in order to maintain correct levels of the different lipoproteins [87], dietary cholesterol is only one of several factors that influence serum levels. The FA composition of dietary fats has also been reported to affect the levels and composition of the different lipoproteins present in the bloodstream. In this sense, The Seven Countries study prompted the "Diet Heart" hypothesis that high intakes of SFA and cholesterol and low intakes of PUFA increase the level of total cholesterol and ultimately result in the development of coronary heart disease [88]. In any case, few studies have been able to demonstrate consistent associations between CVDs and any specific dietary lipids, although the evidence of lower CHD risk when

PUFAs replace SFAs and TFAs in the diet is convincing. Others factors such as the ingestion of soluble fiber and total caloric intake have to be considered. Interestingly, changing one of these factors alone has a relatively small impact on total serum cholesterol levels, but in aggregate, the effect is substantial [76].

Other lipoproteins like triglyceride-rich VLDL remnants, IDL, chylomicron remnants, and Lp(a) have been reported to be atherogenic. Remarkably, all of them contain Apo B-100 or Apo B-48 (both proteins are products of the alternative splicing of the same gene) and the involvement of these two proteins in atherogenesis is also a topic of active research. Other important questions currently under study are the role in cardiovascular events of lipid oxidation products carried by the lipoproteins, or the protective role that HDL (or Apo A-I, its major structural component) appears to have. Currently, new therapeutic agents directly targeting apolipoproteins have been developed and their effectiveness is being evaluated [84]. More thorough coverage of atherogenesis and the involvement of lipids/lipoproteins can be found elsewhere [79, 83, 84, 89].

Although many aspects of CVD remain to be established, it is advisable to sustain a healthy life in order to minimize the risk. This implies no smoking, physical exercise, balanced caloric intake, and consumption of MUFAs, PUFAs, and dietary fiber, always under the guidance of an expert in nutrition.

References

[1]   Badui S. Química De Los Alimentos. Pearson Educación de México, SA de CV; 2006.
[2]   Berg JMark, Tymoczko JL, Stryer Lubert. Biochemistry. Basingstoke: W.H. Freeman; 2012.
[3]   Jia Y, Gan Y, He C, Chen Z, Zhou C. The mechanism of skin lipids influencing skin status. J Dermatol Sci 2018;89:112–9. https://doi.org/10.1016/j.jdermsci.2017.11.006.
[4]   Dennis EA. Liberating chiral lipid mediators, inflammatory enzymes, and lipid maps from biological grease. J Biol Chem 2016;291:24431–48. https://doi.org/10.1074/jbc.X116.723791.
[5]   Stafford DW. The vitamin K cycle. J Thromb Haemost JTH 2005;3:1873–8. https://doi.org/10.1111/j.1538-7836.2005.01419.x.
[6]   Spener F, Lagarde M, Géloên A, Record M. Editorial: What is lipidomics? Eur J Lipid Sci Technol 2003;105:481–2. https://doi.org/10.1002/ejlt.200390101.
[7]   Fahy E, Subramaniam S, Brown HA, Glass CK, Merrill AH, Murphy RC, et al. A comprehensive classification system for lipids. J Lipid Res 2005;46:839–61. https://doi.org/10.1194/jlr.E400004-JLR200.
[8]   Fahy E, Subramaniam S, Murphy RC, Nishijima M, Raetz CRH, Shimizu T, et al. Update of the lipid maps comprehensive classification system for lipids. J Lipid Res 2009;50 Suppl:S9–14. https://doi.org/10.1194/jlr.R800095-JLR200.
[9]   Nelson DL, Cox MM, Lehninger AL. Lehninger principles of biochemistry. New York: W.H. Freeman; 2013.
[10]   The nomenclature of lipids (Recommendations 1976) IUPAC-IUB Commission on Biochemical Nomenclature. Biochem J 1978;171:21–35.

[11] Knothe G, Dunn RO. A comprehensive evaluation of the melting points of fatty acids and esters determined by differential scanning calorimetry. J Am Oil Chem Soc 2009;86:843–56. https://doi.org/10.1007/s11746-009-1423-2.

[12] Martin CA, Milinsk MC, Visentainer JV, Matsushita M, de-Souza NE. Trans fatty acid-forming processes in foods: A review. An Acad Bras Ciênc 2007;79:343–50. https://doi.org/10.1590/S0001-37652007000200015.

[13] Silva YP, Bernardi A, Frozza RL. The role of short-chain fatty acids from gut microbiota in gut-brain communication. Front Endocrinol 2020;11:25. https://doi.org/10.3389/fendo.2020.00025.

[14] Wang D, Lin H, Kan J, Liu L, Zeng X, Shen grong. Food Chemistry. Hauppauge. United States: Nova Science Publishers, Incorporated; 2012.

[15] Palm Oil Fact Slide n.d. http://www.mpoc.org.my/Palm_Oil_Fact_Slides.aspx# (accessed February 20, 2019).

[16] Food and agriculture organization of the United Nations, editor. Fats and fatty acids in human nutrition: report of an expert consultation: 10–14 November 2008, Geneva. Rome: Food and Agriculture Organization of the United Nations; 2010.

[17] Triacylglycerols Part 1, triglycerides, stereospecific distributions, animals, plants, microorganisms,-structure, occurrence n.d. http://www.lipidhome.co.uk/lipids/simple/tag1/index.htm (accessed March 25, 2019).

[18] Book G. Compendium of chemical terminology. Int Union Pure Appl Chem 2014;528.

[19] Meynier A, Genot C. Molecular and structural organization of lipids in foods: Their fate during digestion and impact in nutrition. EDP Sciences; 2017.

[20] Belitz H-D, Grosch W, Schieberle P. Food chemistry. 4th rev. and extended edition. Berlin: Springer; 2009.

[21] Cardeno A, Sanchez-Hidalgo M, Aparicio-Soto M, Alarcón-de-la-Lastra C. Unsaponifiable fraction from extra virgin olive oil inhibits the inflammatory response in LPS-activated murine macrophages. Food Chem 2014;147:117–23. https://doi.org/10.1016/j.foodchem.2013.09.117.

[22] Dijkstra AJ, Segers JC. Production and refining of oils and fats. Lipid Handb. CD-ROM Third Ed. 3rd edition. Boca Raton: CRC Press; 2007, p. 150–269.

[23] Jimenez-Lopez C, Carpena M, Lourenço-Lopes C, Gallardo-Gomez M, Lorenzo JM, Barba FJ, et al. Bioactive compounds and quality of extra virgin olive oil. Foods 2020;9:1014. https://doi.org/10.3390/foods9081014.

[24] Rapoport HF, Fabbri A, Sebastiani L. Olive biology. In: Rugini E, Baldoni L, Muleo R, Sebastiani L, editors. Olive tree genome. Cham: Springer International Publishing; 2016, p. 13–25. https://doi.org/10.1007/978-3-319-48887-5_2.

[25] Uceda M, Beltrán G, Jiménez A. Olive oil extraction and quality. Grasas Aceites 2006;57:25–31.

[26] Maqueda JE. Estudio analítico comparado entre el aceite de acebuchina y el aceite de oliva virgen. Universidad de Sevilla, 2005.

[27] Otero P, Garcia-Oliveira P, Carpena M, Barral-Martinez M, Chamorro F, Echave J, et al. Applications of by-products from the olive oil processing: Revalorization strategies based on target molecules and green extraction technologies. Trends Food Sci Technol 2021;116:1084–104. https://doi.org/10.1016/j.tifs.2021.09.007.

[28] Processing and refining edible oils n.d. http://www.fao.org/3/v4700e/V4700E0a.htm (accessed March 24, 2019).

[29] Deodorization n.d. https://lipidlibrary.aocs.org/edible-oil-processing/deodorization (accessed May 6, 2023).

[30] Johnson DR, Decker EA. The role of oxygen in lipid oxidation reactions: A review. Annu Rev Food Sci Technol 2015;6:171–90.

[31] García-Llatas G, Rodríguez-Estrada MT. Current and new insights on phytosterol oxides in plant sterol-enriched food. Chem Phys Lipids 2011;164:607–24. https://doi.org/10.1016/j.chemphyslip.2011.06.005.

[32] Kubow S. Routes of formation and toxic consequences of lipid oxidation products in foods. Free Radic Biol Med 1992;12:63–81. https://doi.org/10.1016/0891-5849(92)90059-P.

[33] Do Q, Lee DD, Dinh AN, Seguin RP, Zhang R, Xu L. Development and application of a peroxyl radical clock approach for measuring both hydrogen-atom transfer and peroxyl radical addition rate constants. J Org Chem 2021;86:153–68. https://doi.org/10.1021/acs.joc.0c01920.

[34] Min DB, Boff JM. Chemistry and reaction of singlet oxygen in foods. Compr Rev Food Sci Food Saf 2002;1:58–72.

[35] Sánchez-Ortiz A, Pérez AG, Sanz C. Synthesis of aroma compounds of virgin olive oil: Significance of the cleavage of polyunsaturated fatty acid hydroperoxides during the oil extraction process. Food Res Int 2013;54:1972–8. https://doi.org/10.1016/j.foodres.2013.03.045.

[36] Baysal T, Demirdöven A. Lipoxygenase in fruits and vegetables: A review. Enzyme Microb Technol 2007;40:491–6.

[37] de Dios Alché J. A concise appraisal of lipid oxidation and lipoxidation in higher plants. Redox Biol 2019:101136.

[38] Barriuso B, Astiasarán I, Ansorena D. A review of analytical methods measuring lipid oxidation status in foods: A challenging task. Eur Food Res Technol 2013;236:1–15. https://doi.org/10.1007/s00217-012-1866-9.

[39] Abeyrathne EDNS, Nam K, Ahn DU. Analytical methods for lipid oxidation and antioxidant capacity in food systems. Antioxidants 2021;10:1587. https://doi.org/10.3390/antiox10101587.

[40] Healthy diet n.d. https://www.who.int/news-room/fact-sheets/detail/healthy-diet (accessed November 9, 2019).

[41] Dyer JM, Stymne S, Green AG, Carlsson AS. High-value oils from plants. Plant J 2008;54:640–55. https://doi.org/10.1111/j.1365-313X.2008.03430.x.

[42] Alfieri A, Imperlini E, Nigro E, Vitucci D, Orrù S, Daniele A, et al. Effects of plant oil interesterified triacylglycerols on lipemia and human health. Int J Mol Sci 2018;19:104. https://doi.org/10.3390/ijms19010104.

[43] Karupaiah T, Sundram K. Effects of stereospecific positioning of fatty acids in triacylglycerol structures in native and randomized fats: A review of their nutritional implications. Nutr Metab 2007;4:16. https://doi.org/10.1186/1743-7075-4-16.

[44] Massimo C, Lucio T, Jesus MA, Giovanni L, Caramia GM. Extra virgin olive oil and oleic acid. Nutr Clin Diet Hosp 2009;29:12–24.

[45] Berry SE, Bruce JH, Steenson S, Stanner S, Buttriss JL, Spiro A, et al. Interesterified fats: What are they and why are they used? A briefing report from the Roundtable on Interesterified Fats in Foods. Nutr Bull 2019;44:363–80. https://doi.org/10.1111/nbu.12397.

[46] Damodaran S, Parkin K, Fennema OR, editors. Fennema's food chemistry. 4th edition. Boca Raton: CRC Press/Taylor & Francis; 2008.

[47] Rees K, Takeda A, Martin N, Ellis L, Wijesekara D, Vepa A, et al. Mediterranean-style diet for the primary and secondary prevention of cardiovascular disease. Cochrane Database Syst Rev 2019;3: CD009825. https://doi.org/10.1002/14651858.CD009825.pub3.

[48] Designations and definitions of olive oils – International Olive Council n.d. http://www.internationaloliveoil.org/estaticos/view/83-designations-and-definitions-of-olive-oils (accessed March 27, 2019).

[49] Standards – International Olive Council n.d. http://www.internationaloliveoil.org/estaticos/view/222-standards (accessed March 26, 2019).

[50] Mapelli-Brahm P, Hernanz-Vila D, Stinco CM, Heredia FJ, Meléndez-Martínez AJ. Isoprenoids composition and colour to differentiate virgin olive oils from a specific mill. LWT 2018;89:18–23. https://doi.org/10.1016/j.lwt.2017.10.021.

[51] Rigacci S, Stefani M. Nutraceutical properties of olive oil polyphenols. An itinerary from cultured cells through animal models to humans. Int J Mol Sci 2016;17. https://doi.org/10.3390/ijms17060843.

[52] Wang TY, Liu M, Portincasa P, Wang DQ-H. New insights into the molecular mechanism of intestinal fatty acid absorption. Eur J Clin Invest 2013;43:1203–23. https://doi.org/10.1111/eci.12161.

[53] Zanoni P, Velagapudi S, Yalcinkaya M, Rohrer L, von Eckardstein A. Endocytosis of lipoproteins. Atherosclerosis 2018;275:273–95. https://doi.org/10.1016/j.atherosclerosis.2018.06.881.

[54] Wu Q, Ortegon AM, Tsang B, Doege H, Feingold KR, Stahl A. FATP1 is an insulin-sensitive fatty acid transporter involved in diet-induced obesity. Mol Cell Biol 2006;26:3455–67. https://doi.org/10.1128/MCB.26.9.3455-3467.2006.

[55] Rodríguez A, Catalán V, Gómez-Ambrosi J, Frühbeck G. Aquaglyceroporins serve as metabolic gateways in adiposity and insulin resistance control. Cell Cycle 2011;10:1548–56. https://doi.org/10.4161/cc.10.10.15672.

[56] Mashek DG. Hepatic fatty acid trafficking: Multiple forks in the road. Adv Nutr 2013;4:697–710.

[57] Maier T, Leibundgut M, Ban N. The crystal structure of a mammalian fatty acid synthase. Science 2008;321:1315–22.

[58] Jump DB. Mammalian fatty acid elongases. Methods Mol Biol Clifton NJ 2009;579:375–89. https://doi.org/10.1007/978-1-60761-322-0_19.

[59] ALJohani AM, Syed DN, Ntambi JM. Insights into Stearoyl-CoA Desaturase-1 regulation of systemic metabolism. Trends Endocrinol Metab 2017;28:831–42. https://doi.org/10.1016/j.tem.2017.10.003.

[60] Mittelstraß K, Waldenberger M. DNA methylation in human lipid metabolism and related diseases. Curr Opin Lipidol 2018;29:116–24. https://doi.org/10.1097/MOL.0000000000000491.

[61] Grabacka M, Pierzchalska M, Dean M, Reiss K. Regulation of ketone body metabolism and the role of PPARα. Int J Mol Sci 2016;17:2093. https://doi.org/10.3390/ijms17122093.

[62] van Stekelenburg GJ, Koorevaar G. Evidence for the existence of mammalian acetoacetate decarboxylase: With special reference to human blood serum. Clin Chim Acta Int J Clin Chem 1972;39:191–9.

[63] McGaugh E, Barthel B. A review of ketogenic diet and lifestyle. Mol Med 2022;119:84–8.

[64] Saini RK, Keum Y-S. Omega-3 and omega-6 polyunsaturated fatty acids: Dietary sources, metabolism, and significance – A review. Life Sci 2018;203:255–67. https://doi.org/10.1016/j.lfs.2018.04.049.

[65] Sonnweber T, Pizzini A, Nairz M, Weiss G, Tancevski I. Arachidonic acid metabolites in cardiovascular and metabolic diseases. Int J Mol Sci 2018; 19. https://doi.org/10.3390/ijms19113285.

[66] Danielak A, Wallace JL, Brzozowski T, Magierowski M. Gaseous mediators as a key molecular targets for the development of gastrointestinal-safe anti-inflammatory pharmacology. Front Pharmacol 2021;12:657457. https://doi.org/10.3389/fphar.2021.657457.

[67] Funk CD, Funk LB, Kennedy ME, Pong AS, Fitzgerald GA. Human platelet/erythroleukemia cell prostaglandin G/H synthase: cDNA cloning, expression, and gene chromosomal assignment. FASEB J Off Publ Fed Am Soc Exp Biol 1991;5:2304–12.

[68] Kodela R, Chattopadhyay M, Velázquez-Martínez CA, Kashfi K. NOSH-aspirin (NBS-1120), a novel nitric oxide- and hydrogen sulfide-releasing hybrid has enhanced chemo-preventive properties compared to aspirin, is gastrointestinal safe with all the classic therapeutic indications. Biochem Pharmacol 2015;98:564–72. https://doi.org/10.1016/j.bcp.2015.09.014.

[69] Bruno A, Tacconelli S, Patrignani P. Variability in the response to non-steroidal anti-inflammatory drugs: Mechanisms and perspectives. Basic Clin Pharmacol Toxicol 2014;114:56–63.

[70] IUPAC Gold Book n.d. http://goldbook.iupac.org/ (accessed March 7, 2019).

[71] Rodriguez-Amaya DB, OMNI (Project). A guide to carotenoid analysis in foods. Washington, D.C.: ILSI Press; 2001.

[72] Hashimoto H, Uragami C, Cogdell RJ. Carotenoids and photosynthesis. Subcell Biochem 2016;79:111–39. https://doi.org/10.1007/978-3-319-39126-7_4.

[73] Toti E, Chen C-YO, Palmery M, Villaño Valencia D, Peluso I. Non-provitamin A and provitamin A carotenoids as immunomodulators: Recommended dietary allowance, therapeutic index, or personalized nutrition? Oxid Med Cell Longev 2018;2018:1–20. https://doi.org/10.1155/2018/4637861.

[74] Espenshade PJ, Hughes AL. Regulation of sterol synthesis in eukaryotes. Annu Rev Genet 2007;41:401–27. https://doi.org/10.1146/annurev.genet.41.110306.130315.

[75] Lopez AM, Chuang J-C, Turley SD. Measurement of rates of cholesterol and fatty acid synthesis in vivo using tritiated water. In: Gelissen IC, Brown AJ, editors. Cholest. Homeost. Methods Protoc. New York, NY: Springer New York; 2017, p. 241–56. https://doi.org/10.1007/978-1-4939-6875-6_18.

[76] Grundy SM. Does dietary cholesterol matter? Curr Atheroscler Rep 2016;18:68. https://doi.org/10.1007/s11883-016-0615-0.

[77] Cerqueira NM, Oliveira EF, Gesto DS, Santos-Martins D, Moreira C, Moorthy HN, et al. Cholesterol biosynthesis: A mechanistic overview. Biochemistry 2016;55:5483–506.

[78] Burg JS, Espenshade PJ. Regulation of HMG-CoA reductase in mammals and yeast. Prog Lipid Res 2011;50:403–10. https://doi.org/10.1016/j.plipres.2011.07.002.

[79] Duan Y, Gong K, Xu S, Zhang F, Meng X, Han J. Regulation of cholesterol homeostasis in health and diseases: From mechanisms to targeted therapeutics. Signal Transduct Target Ther 2022;7:265. https://doi.org/10.1038/s41392-022-01125-5.

[80] Sirtori CR. The pharmacology of statins. Pharmacol Res 2014;88:3–11. https://doi.org/10.1016/j.phrs.2014.03.002.

[81] Sharpe LJ, Brown AJ. Controlling cholesterol synthesis beyond 3-Hydroxy-3-methylglutaryl-CoA reductase (HMGCR). J Biol Chem 2013;288:18707–15. https://doi.org/10.1074/jbc.R113.479808.

[82] Plasma Lipoproteins, HDL, LDL, VLDL, apoproteins, cholesterol, triacylglycerols : composition, structure and biochemistry n.d. https://www.lipidhome.co.uk/lipids/simple/lipoprot/index.htm (accessed January 1, 2020).

[83] Feingold KR, Grunfeld C. Introduction to lipids and lipoproteins. In: Feingold KR, Anawalt B, Boyce A, Chrousos G, Dungan K, Grossman A, et al., editors. Endotext. South Dartmouth (MA): MDText.com, Inc.; 2000.

[84] Mehta A, Shapiro MD. Apolipoproteins in vascular biology and atherosclerotic disease. Nat Rev Cardiol 2022;19:168–79. https://doi.org/10.1038/s41569-021-00613-5.

[85] Libby P, Ridker PM, Hansson GK. Progress and challenges in translating the biology of atherosclerosis. Nature 2011;473:317–25. https://doi.org/10.1038/nature10146.

[86] Youngblom E, Pariani M, Knowles JW. Familial hypercholesterolemia. In: Adam MP, Ardinger HH, Pagon RA, Wallace SE, Bean LJ, Stephens K, et al., editors. GeneReviews®, Seattle (WA): University of Washington, Seattle; 1993.

[87] Xu Z, McClure ST, Appel LJ. Dietary cholesterol intake and sources among U.S adults: Results from National Health and Nutrition Examination Surveys (NHANES), 2001–2014. Nutrients 2018;10:771. https://doi.org/10.3390/nu10060771.

[88] Seven Countries Study | The first study to relate diet with cardiovascular disease. – The Seven Countries Study (SCS for short) is the first major study to look at dietary components and patterns and lifestyle as risk factors for cardiovascular disease, over multiple countries and extended periods of time. Seven Ctries Study First Study Relate Diet Cardiovasc Dis n.d. https://www.sevencoun triesstudy.com/ (accessed March 28, 2019).

[89] Linton MF, Yancey PG, Davies SS, Jerome WG, Linton EF, Song WL, et al. The role of lipids and lipoproteins in atherosclerosis. In: Feingold KR, Anawalt B, Boyce A, Chrousos G, Dungan K, Grossman A, et al., editors. Endotext. South Dartmouth (MA): MDText.com, Inc.; 2000.

5 Proteins

5.1 Introduction

Proteins are biomolecules whose basic structures are conformed by unbranched chains of 21 different amino acids linked by the so-called peptide bond. This establishes a backbone of carbon and nitrogen atoms, the latter becoming a basic element of proteins in addition to carbon, hydrogen, and oxygen, as well as sulfur, present in two amino acids.

Unlike lipids and carbohydrates, proteins are the direct expression of the genetic information contained in nucleic acids, according to the "central dogma of molecular biology" proposed by Francis Crick in 1958 [1]. The main idea in the central dogma is that the information required for the proper functioning of living systems is contained in the deoxyribonucleic acid (DNA), composed of sequences of four nucleotides (Section 5.11). This information is transcribed to a similar molecule, the messenger ribonucleic acid (RNA) or mRNA and, then, translated to a sequence of amino acids to synthesize proteins. Very importantly, DNA is able to make copies of itself, the so-called replication so that the information can be transferred to the next generation of cells (Figure 5.1). Although some changes have been introduced in the central dogma to account for new discoveries like reverse transcription, the central Crick's idea remains. The process of protein synthesis, known as translation, is based on the instructions contained in mRNA, by following the so-called genetic code, in which each

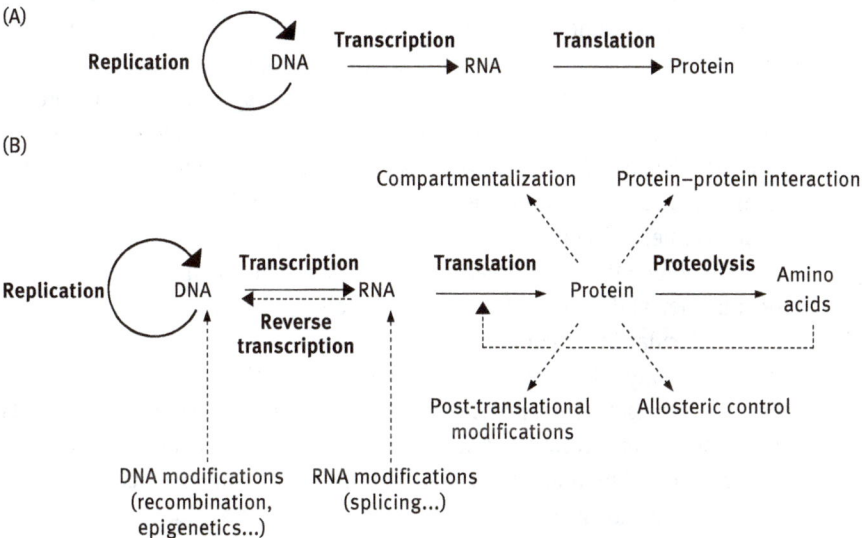

Figure 5.1: The central dogma of molecular biology: (A) the original Francis Crick's proposal and (B) an updated version.

https://doi.org/10.1515/9783111111872-005

amino acid of the protein is specifically encoded by a group of three nucleotides or codons. The genetic code is degenerate (or redundant) because each amino acid is encoded by more than one codon, with the only exceptions of methionine and tryptophan. It is also almost universal, some exceptions being found in mitochondria and some bacteria and protozoa [2].

Proteins are extremely complex macromolecules, exhibiting a virtually endless variety of sizes and structures. Classically, the function of a given protein was thought to be determined by its three-dimensional structure, which, in turn, is ultimately driven by its sequence of amino acids. This structure–function paradigm has been challenged during the last two decades by the identification of proteins that perform functions in the cell without attaining a stable three-dimensional structure either in the whole polypeptide ("intrinsically disordered proteins" or IDPs) or in certain regions (proteins with intrinsically disordered regions or IDRs). IDPs and proteins with IDRs have been reported to be one-third of the total proteins in eukaryotes, and their properties and involvement in cellular processes are being thoroughly studied [3].

Other basic terms and concepts related to proteins are:

– Proteinogenic amino acids: These are a group of α-amino acids and their general chemical structure is shown in Figure 5.2. "R" represents the 21 different chemical groups, or side chains, present in naturally occurring canonical amino acids, known as proteinogenic to distinguish them from others that occur in living organisms but do not participate in proteins.

In all proteinogenic amino acids, with the only exception of glycine (see below), the carbon atom adjacent to the carboxylic group (carbon α) has four different substituents and, therefore, is a chiral center. This implies the occurrence of two optical isomers, traditionally termed as L and D, that can be depicted in Fischer projections. All amino acids found in natural proteins are L enantiomers. Biochemists usually represent amino acids with the carboxylic group deprotonated and the amino group protonated (dipolar ion or "zwitterion"), as this is how they appear in solution at pH 7 (Figure 5.2). Proteinogenic amino acids are usually known by their trivial names and can be classified according to their chemical properties that are basically determined by their respective R groups (Figure 5.3) [4]:

– Glycine, in which R is a hydrogen atom, is a class on its own because it has no optical isomer and its side chain is very small, not contributing significantly to intramolecular interactions.

– Aliphatic R groups: They are nonpolar and hydrophobic and include alanine, valine, leucine, isoleucine, proline, and methionine. The last two amino acids are special: proline, because its side chain is cyclic and produces a secondary amino group, and methionine, which contains sulfur. Valine, leucine, and isoleucine are collectively known as branched-chain amino acids (BCAAs).

– Aromatic R groups: The amino acids included in this group are phenylalanine, tyrosine, and tryptophan.

- Polar, uncharged R groups: These include serine, threonine, asparagine, glutamine, and cysteine. Serine and threonine have –OH groups in their side chains, whereas asparagine and glutamine have amide groups. Cysteine is the other amino acid that contains sulfur and can interact with another cysteine, thus producing the so-called disulfide bond (Figure 5.3).
- Positively charged R groups (in solution at pH 7): Lysine, histidine, and arginine are included in this group.
- R groups that contain a carboxylic group: Two amino acids, aspartic acid and glutamic acid, contain a –COOH group, which is deprotonated in solution at pH 7, thus producing a negatively charged side chain.
- Selenocysteine ($R = -CH_2SeH$) has been recently included as the 21st naturally occurring amino acid, after being demonstrated that it is incorporated into proteins during translation, that is, it is encoded by the genetic code, albeit in a rather special way. More than 20 proteins containing selenocysteine have been identified to date in humans, where they seem to be implicated in important functions such as protection against oxidative damage [5].

- Peptide bond: The carboxyl group bound to the carbon α of one amino acid can interact with the amino group of another, thus forming a substituted amide linkage, known as "peptide bond," which yields a dimer or dipeptide and a molecule of water (Figure 5.4). This process can be repeated producing tripeptides, tetrapeptides, and so on, with no limits, at least in theory, with respect to the type and number of amino acids that can be linked by successive peptide bonds. This explains the amazing variety of proteins that occur in the biosphere. The peptide bond has a partial double-bond character that keeps six atoms in a single plane, thus introducing a constraint in the backbones of polypeptides with important consequences for the final protein structure [4].
- Oligopeptides: Chains of up to 10 amino acids linked by peptide bonds ("residues").
- Polypeptides: Chains of more than 10 amino acid residues with molecular weights below 10,000 (a generally accepted, albeit not restrictive, number).
- Protein: This term was proposed by Berzelius in 1838 that derives from the Greek word "proteios," meaning "of the first rank or position" [6]. It is normally used for polypeptides of more than 10,000 of molecular weight [4, 7]; however, the concept "protein" implies a function and biologically active proteins often require more than one polypeptide and/or nonpeptidic chemical species, the latter known as cofactors. **Cofactors** are called **prosthetic groups** if they are covalently bound to the polypeptide(s) or **coenzymes** if their binding to the polypeptide is mediated by weak interactions (hydrogen bonds, van der Waals forces, etc.). The polypeptidic portion of a protein is termed **apoprotein**, whereas the complete functional complex that includes polypeptide(s) and cofactor(s) is known as **holoprotein**. Among the compounds that can act as cofactors are carbohydrates (glycoproteins), lipids (lipoproteins), RNA (ribonucleoproteins), organic molecules (like vi-

tamins), organometallic complexes (such as the heme group), and metallic cations (metalloproteins).

– Proteome: This represents the entire set of proteins present in a cell, a tissue, or an organism at a particular time under certain conditions. This term was proposed by Wilkins and colleagues and defined as "the entire PROTein complement expressed by a genOME, or by a cell or tissue type" [8].

– Proteomics: The application of technologies for the identification and quantification of the proteome [9].

– Deep learning methods: Methods like RoseTTAFold [10] and AlphaFold [11] have been able to achieve structure prediction accuracies for proteins far beyond those obtained with classical models. These methods have already had an impact on areas like protein interaction and assembly modeling, protein design, and small-molecule drug discovery. The potential for deep learning methods in biology and medicine is expected to be enormous [12].

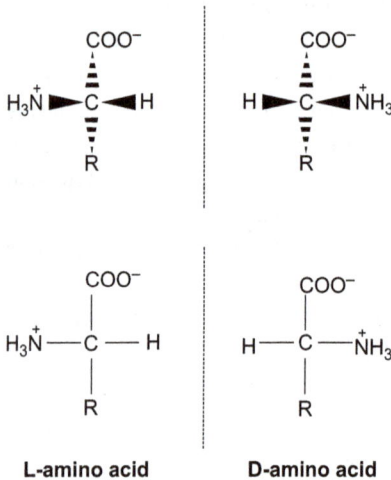

L-amino acid   **D-amino acid**   Figure 5.2: Fischer projection of α-amino acids.

5.2 Functions of proteins

Dietary proteins undergo a complex process of digestion in the gastrointestinal tract (GIT) that basically involves enzyme-mediated hydrolysis (proteolysis) of most of the peptidic bonds. Proteolysis begins in the mouth and continues along the GIT, so that, eventually, only free amino acids and small peptides are taken up by enterocytes in the small intestine (Chapters 12 and 13). Consequently, food proteins are the source of the building blocks necessary to make our own proteins following the instructions contained in our DNA.

$$
\begin{array}{c}
COO^- \\
| \\
H_3\overset{+}{N}-C-H \\
| \\
H \\
\text{Glycine}
\end{array}
\qquad
\begin{array}{c}
COO^- \\
| \\
H_3\overset{+}{N}-C-H \\
| \\
CH_3 \\
\text{Alanine}
\end{array}
$$

$$
\begin{array}{c}
COO^- \\
| \\
H_3\overset{+}{N}-C-H \\
| \\
CH \\
\diagup\ \diagdown \\
H_3C \quad CH_3 \\
\text{Valine}
\end{array}
\qquad
\begin{array}{c}
COO^- \\
| \\
C-H \\
\diagup \quad \diagdown \\
H_2\overset{+}{N} \qquad CH_2 \\
| \qquad\quad | \\
H_2C - CH_2 \\
\text{Proline}
\end{array}
$$

$$
\begin{array}{c}
COO^- \\
| \\
H_3\overset{+}{N}-C-H \\
| \\
CH_2 \\
| \\
CH \\
\diagup\ \diagdown \\
H_3C \quad CH_3 \\
\text{Leucine}
\end{array}
\qquad
\begin{array}{c}
COO^- \\
| \\
H_3\overset{+}{N}-C-H \\
| \\
HC-CH_3 \\
| \\
CH_2 \\
| \\
CH_3 \\
\text{Isoleucine}
\end{array}
$$

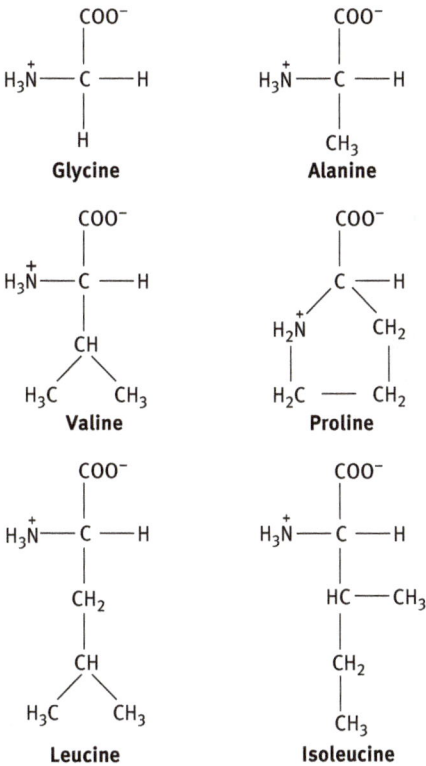

Figure 5.3: Chemical formulae in Fischer projection of proteinogenic amino acids.

Amino acids themselves are also precursors of a number of important metabolites such as creatine and glutathione, hormones like adrenaline, and neurotransmitters such as gamma-aminobutyric acid, histamine, or serotonin. Moreover, the carbon skeleton of amino acids can be used to obtain metabolically useful energy and metabolites like glucose. All these properties make amino acids very versatile nutrients, although the use of the carbon skeleton implies the necessity of eliminating the nitrogen, with important metabolic consequences (see below).

Proteins are involved in virtually all biological functions:

- Enzymatic: The vast majority of biochemical reactions are catalyzed by specific extremely efficient biological catalysts known as enzymes. With the only exception of ribozymes, composed of RNA [13], all known enzymes are proteins.
- Transport: This term may refer to two different processes, both mediated by proteins:
  - Distribution of substances like lipids, fatty acids, or oxygen via bloodstream.
  - Movement of solutes across biological membranes.
- Muscular contraction: Actin and myosin.
- Immune system: Antibodies.
- Hormones, like insulin, glucagon, or leptin.

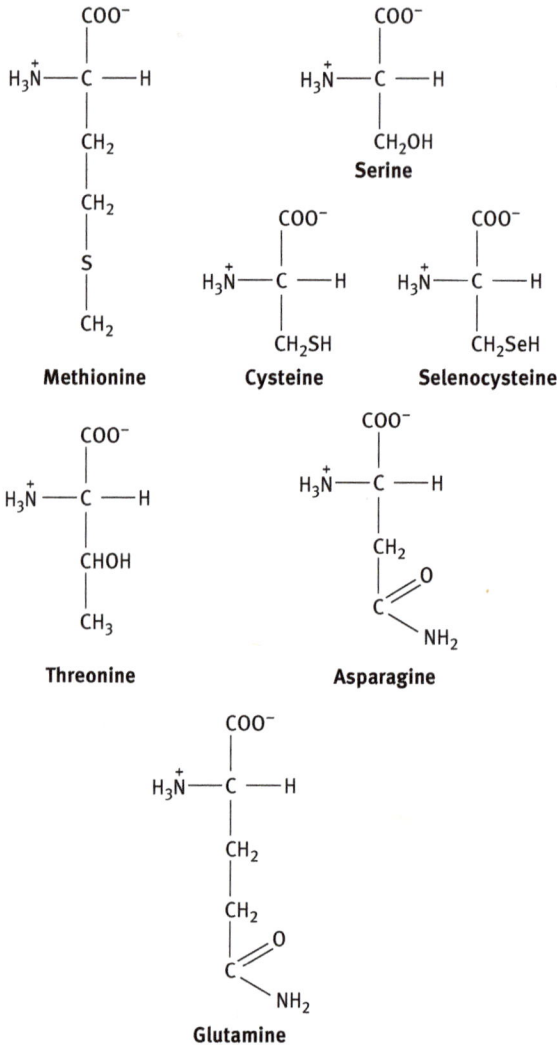

Methionine

Serine

Cysteine

Selenocysteine

Threonine

Asparagine

Glutamine

Figure 5.3 (continued)

- Structural: Keratins (hair, nails, and horns in mammals, and feathers in birds), collagen (bones and teeth), fibroin (silk), and elastin (ligament, tendons, etc.).
- Storage of energy and amino acids, like milk caseins.
- DNA packaging: Histones.
- Control of transcription and translation.

Around 20,000 different proteins are estimated to be present in humans if we accept the hypothesis of "one gene = one protein," that is, if only the sequences of the human genome that are potentially translatable are accounted for; however, it is estimated

Lysine

Histidine

Arginine

Aspartate

Glutamate

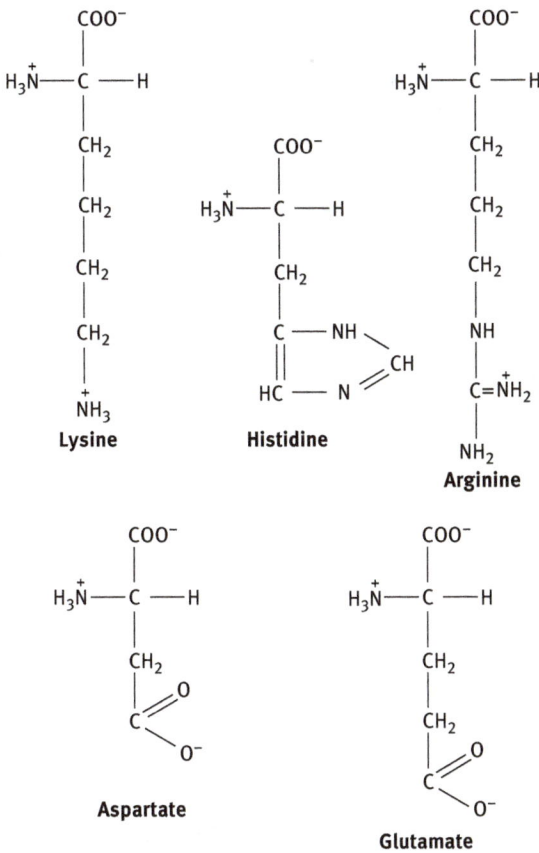

Figure 5.3 (continued)

that about 100 different proteins can potentially be produced from a single gene by processes like alternative splicing, single amino acid polymorphisms, or posttranslational modifications (PTMs) [14]. PTMs are also responsible for the occurrence of certain variations in amino acids such as hydroxyproline and hydroxylysine.

5.3 Requirements of proteins in human nutrition

Table 5.1 summarizes the dietary requirements proposed by the World Health Organization (WHO) for the different groups of population in a report of the Joint WHO/Food and Agriculture Organization of the United Nations (FAO)/UNU Expert Consultation on Protein and Amino Acid Requirements in Human Nutrition, held in 2002 and published in 2007. In this report, dietary requirement is defined as "the amount of protein or its constituent amino acids, or both, that must be supplied in the diet in

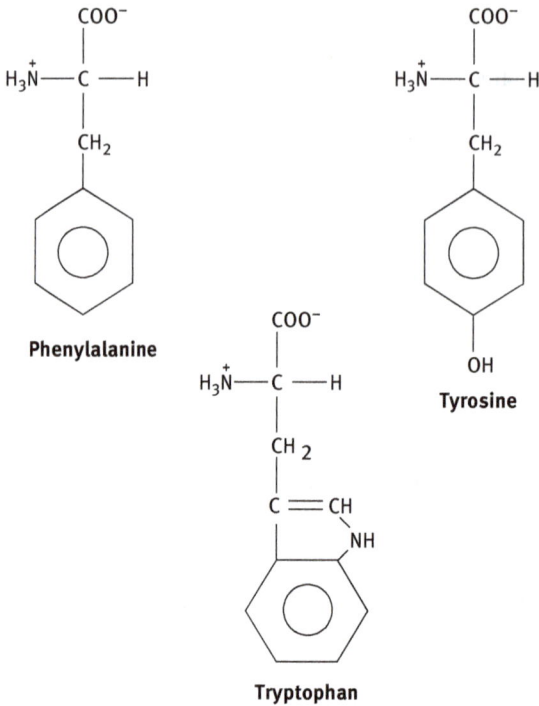

Phenylalanine

Tyrosine

Tryptophan

Figure 5.3 (continued)

order to satisfy the metabolic demand and achieve nitrogen equilibrium" [15]. Calculation of protein requirements is extremely difficult, as there are a number of factors that have to be considered:

–   Metabolic demand, that is, the flux of amino acids needed to sustain biochemical processes like protein synthesis, conversion to other metabolites, or oxidation of carbon skeletons to obtain energy. The metabolic demand of amino acids is influenced by factors like genetics, sex, age, lifestyle, or environmental conditions, and by special circumstances such as growth, pregnancy, or lactation.

–   Efficiency of protein use, which is associated with the digestibility of foods' protein, the rate of absorption of amino acids, and the availability of the latter at cellular level.

–   Obligatory nitrogen losses. Even when dietary intake of nitrogen is zero, there is a continuous loss of nitrogen from the body via urine, feces, sweat, skin, hair, and other secretions. There must be a minimum protein intake to compensate for this loss and enable nitrogen balance.

–   Energy intake. All stages of the metabolism of proteins and amino acids require energy; therefore, there is an interdependence between total energy intake, metabolic demand, and efficiency of protein use. It is generally acknowledged that rec-

(A)

(B)

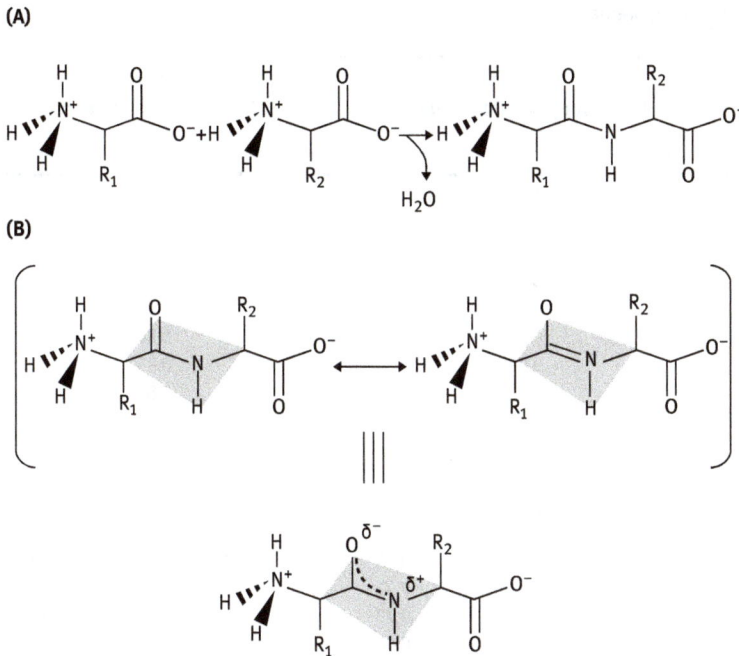

Figure 5.4: Formation and planar structure of the peptide bond.

ommendations for protein intake must be linked to total energy intake; however, the complexity of the mechanisms involved, not completely understood, complicates the task of proposing adequate protein-to-energy ratios in the diet.

- Micronutrients. Metabolism (including those pathways that involve amino acids) is critically dependent on an adequate status of nonenergetic nutrients like vitamins and minerals (Chapters 6 and 7).
- Protein quality. There are several amino acids that are considered indispensable, or essential, for humans and must be provided by the diet. The rest of the proteogenic amino acids can be synthesized *de novo* by the body in adequate amounts, at least in principle. Dietary protein must guarantee an adequate supply of essential amino acids (EAAs) to satisfy human needs under every circumstance.

Links to dietary recommendations for protein intake published by different authorities can be found at https://knowledge4policy.ec.europa.eu/health-promotion-knowledge-gateway/dietary-protein-intake-adults-3_en.

Table 5.1: Dietary protein requirements by humans of all age groups.

Age (years)	AR for protein (g/kg body weight per day)		PRI for protein (g/kg body weight per day)	
	Male	Female	Male	Female
0.5	1.12		1.31	
1	0.95		1.14	
1.5	0.85		1.03	
2	0.79		0.97	
3	0.73		0.90	
4	0.69		0.86	
5	0.69		0.85	
6	0.72		0.89	
7	0.74		0.91	
8	0.75		0.92	
9	0.75		0.92	
10	0.75		0.91	
11	0.75	0.73	0.91	0.90
12	0.74	0.72	0.90	0.89
13	0.73	0.71	0.90	0.88
14	0.72	0.70	0.89	0.87
15	0.72	0.69	0.88	0.85
16	0.71	0.68	0.87	0.84
17	0.70	0.67	0.86	0.83
≥ 18	0.66		0.83	
	Pregnancy			
First trimester	+0.52 g/day (a)		+1 g/day (b)	
Second trimester	+7.2 g/day (a)		+9 g/day (b)	
Third trimester	+23 g/day (a)		+28 g/day (b)	
	Lactation			
	+15 g/day (a)		+19 g/day (b)	
	+10 g/day (a)		+13 g/day (b)	

Source: [15, 16]. Average requirement (AR) and population reference intake (PRI) are defined in Appendix 1.
(a) In addition to the AR for protein of nonpregnant, nonlactating women.
(b) In addition to the PRI for protein of nonpregnant, nonlactating women.

5.3.1 Importance of protein in the diet: essential amino acids and evaluation of protein quality

The quality of a given protein is very much related to its content in EAAs and its digestibility. The terms "indispensable" (or essential) and "dispensable" amino acids were originally defined both in dietary terms and also in relation to their role in supporting protein synthesis and growth. From a strict nutritional point of view, EAAs are defined as those amino acids that cannot be synthesized *de novo* (net synthesis from simpler precursors) or that are inadequately synthesized by the body relative to its needs. EAAs must be provided from the diet to meet optimal requirements, whereas non-EAAs (NEAAs) are those that can be synthesized *de novo* in adequate amounts by the body. A third group of amino acids are frequently termed "conditionally essential" because there are limitations to the rate at which they can be produced. These limitations can be due to several reasons: (a) the synthesis can depend on the provision of another amino acid, (b) the synthesis may be limited to certain tissues, or (c) the metabolic demand of the amino acid cannot be met under some circumstances [17, 18].

Nine amino acids have been traditionally considered essential: histidine, isoleucine, leucine, lysine, methionine, phenylalanine, threonine, tryptophan, and valine (Table 5.2). A tenth amino acid, arginine, is indispensable for young mammals. In adults, although it is not required in the diet to maintain nitrogen balance, long-term arginine deficiency can result in metabolic, neurological, or reproductive dysfunction [18].

Table 5.2: Essential amino acid (EAA) profile of human milk.

Amino acid (mg/g of total protein)	
Histidine	21
Isoleucine	55
Leucine	96
Lysine	69
Methionine + cysteine	33
Phenylalanine + tyrosine	94
Threonine	44
Tryptophan	17
Valine	55

Source: [19].

Classically, three kinds of experimental approaches have been used to evaluate the nutritional quality of proteins: biological, chemical, and enzymatic/microbial methods [20].

- **Biological methods** are based on weight gain or nitrogen retention in test animals (rats and sometimes humans) when fed with a protein-containing diet using a protein-free diet as a control and guaranteeing an adequate supply of energy. Parameters like **protein efficiency ratio (PER)** and **net protein ratio (NPR)** can be calculated with the experimental data thus obtained:

$$\text{Protein efficiency ratio (PER)} = \frac{\text{weight (g) gained}}{\text{gram protein consumed}}$$

$$\text{Net protein ratio (NPR)} = \frac{(\text{weight gain}) - (\text{weight loss of protein} - \text{free group})}{\text{protein ingested}}$$

PER and NPR measure the capacity of the protein source to support maintenance and growth. Corrections have to be introduced to adapt the results to humans when data are obtained with laboratory animals. Other biological methods rely on measuring nitrogen balance. This is based on the fact that metabolically active proteins in the human body, as well as most dietary proteins, have a nitrogen content of 16% (w/w); therefore, the amount of protein in a purified sample can be calculated by measuring its N content and multiplying the resulting value by 6.25 [21]. Among the values that can be calculated using nitrogen balance data are:

-   **True (fecal) digestibility (TD)** gives information on the percentage of nitrogen intake absorbed by the body:

$$TD = \frac{(\text{N ingested}) - (\text{N fecal} - \text{N endogenous fecal})}{\text{N ingested}} \times 100$$

-   **Biological value (BV)** quantifies the percentage of nitrogen retained by the body:

$$BV = \frac{(\text{N ingested}) - (\text{N fecal} - \text{N endogenous fecal}) - (\text{N urine} - \text{N endogenous urine})}{\text{N ingested} - (\text{N fecal} - \text{N endogenous fecal})} \times 100$$

-   **Net protein utilization (NPU)** is obtained from the product of TD and BV:

$$NPU = \frac{(\text{N ingested}) - (\text{N fecal} - \text{N endogenous fecal}) - (\text{N urine} - \text{N endogenous urine})}{\text{N ingested}} \times 100$$

The values of endogenous fecal and urine nitrogen are obtained from test animals fed with a protein-free diet.

-   **Chemical methods** are based on determining the content of amino acids in a given protein and comparing with the EAA pattern of an ideal reference protein. The amino acid composition of human milk should be the basis of the scoring pattern to evaluate protein quality in foods for infants under 1 year of age, whereas the ideal pattern of EAAs for all age groups except infants has been estimated and tabulated in Table 5.3. Each EAA in a test protein is given an **amino acid chemical score (AAS)**, defined as

$$AAS = \frac{\text{mg amino acid per gram of test protein}}{\text{mg same amino acid per gram of reference protein}} \times 100$$

The EAA with the lowest score is the limiting amino acid in the test protein and provides the overall chemical score for the latter. Lysine, threonine, tryptophan, and sulfur amino acids are often the limiting amino acids in food proteins. The amino acid chemical scores corrected by protein digestibility (PDCAAS) have been

Table 5.3: Recommended amino acid scoring patterns for children, adolescents, and adults.

Age group	His	Ile	Leu	Lys	Met + Cys	Phe + Tyr	Thr	Trp	Val
Up to 6 months					(values in Table 5.2)				
From 6 months to 3 years	20	32	66	57	27	52	31	8.5	4.3
Older than 3 years	16	30	61	48	23	41	25	6.6	40

Source: [19].

widely used for evaluating protein quality. PDCAAS is calculated by multiplying the AAS of the limiting amino acids by true protein digestibility (TD) [15].

- **Enzymatic and microbial methods**. In enzymatic methods, test proteins are digested with enzymes like pepsin, pancreatin, or trypsin under standard assay conditions, in order to provide information on innate digestibility of proteins and changes in protein quality induced by processing. In the second method, bacteria like *Clostridium perfringens* or different types of streptococci, and the protozoan *Tetrahymena pyriformis* have been used to determine the nutritional value of proteins.

The methods and parameters described above have been traditionally used to evaluate and compare different protein sources (Table 5.4); however, their suitability for evaluating the quality of dietary protein has been hotly debated for decades.

Table 5.4: Protein quality parameters of different protein sources.

Protein source	Biological value (BV)	Net protein utilization (NPU)	PDCAAS*	DIAAS*
Wheat	56–68	53–65	51	45 (Lys)
Cooked rice	–	–	62	60
Soy protein	74	61	100	–
Beef	80	73	92	–
Milk	91	82	100	114
Casein	77	76–82	100	–
Whey protein	104	92	100	–
Egg	100	94	100	113

Source: [22].
***PDCAAS:** amino acid chemical scores corrected by protein digestibility (values in percentage). **DIAAS**, digestible indispensable amino acid score.

Some of the criticisms made, especially against the calculation of true fecal digestibility (TD), are:

-   TD does not account for individual amino acids.
-   TD values are not reliable to evaluate the nitrogen absorbed by the human body due to the metabolic activity of gut microbiota.
-   TD values are usually determined using test animals.
-   PDCAAS values may be overestimated by the limited bioavailability of specific forms of amino acids.

In a report of a FAO expert consultation published in 2013, a number of recommendations and suggestions were proposed in an attempt to meet these criticisms [19]:

-   PDCAAS was recommended to be replaced by a new protein quality measure, DIAAS (digestible indispensable amino acid score), defined as

$$DIASS = \frac{mg\ digestible\ EAA\ per\ g\ test\ protein}{mg\ same\ EAA\ per\ g\ of\ reference\ protein} \times 100$$

The main difference between DIAAS and PDCAAS is that true ileal (determined at the end of the small intestine) amino acid digestibility for dietary EAAs is used rather than a single fecal crude protein digestibility value. The determination of digestion and absorption of amino acids at ileal level is technically, economically, and ethically difficult in humans, as it requires the use of invasive procedures. Therefore, test animals like rats and pigs are routinely used to collect experimental data that are extrapolated to humans [23].

-   Dietary amino acids must be treated as individual nutrients; therefore, data in food tables should be given on an individual amino acid bases.
-   The use of methods based on stable radioactive isotopes is encouraged. Two methods are mentioned in the report:
    -   Net postprandial protein utilization (NPPU) uses [^{15}N]-labeled proteins to measure the metabolic fate of dietary nitrogen after its consumption in humans. Calculation of NPPU incorporates the determination of true ileal amino acid digestibility.
    -   The indicator amino acid oxidation (IAAO) method is based on the principle that excess amino acids are not stored and must be partitioned between incorporation into proteins or oxidation. When an EAA is deficient for protein synthesis, all other amino acids are in excess and will be oxidized. The rate of oxidation can be followed by using a [^{13}C]-labeled indicator amino acid and measuring radioactivity in breath and urine samples. Oxidation of the indicator amino acid will decrease (reflecting incorporation into proteins) with increasing intake of the limiting amino acid, until no further change is observed. This point indicates the estimated average requirement of the limiting amino acid [24].

– The use of high-throughput studies of complex metabolic profiles (metabolomics) in plasma and urine samples also offers promising perspectives for the
evaluation of dietary protein quality in humans.

Regardless of these recommendations, some authors point out that experimental approaches based on nitrogen balance and isotope studies are short-term experiments
and do not take into consideration functional needs of amino acids in the long term
[25]. Recent advances in biochemistry and molecular and cellular biology have shown
that amino acids are not only building blocks of proteins or precursors of important
metabolites, but they are also involved in regulation of gene expression and protein
function, thereby affecting traits like growth, reproduction, physiology, health, and
longevity. Interestingly, this involvement is not only valid for EAAs but also for those
regarded as nonessential [26–28]. In conclusion, much is yet to be known about the
role of dietary proteins and amino acids in human health and a multidisciplinary approach will be required to further study this important issue.

5.4 Protein sources and composition

Table 5.5 shows that plant-based protein dominates protein supply globally; however,
important differences in the pattern of protein intake are observed when considered
data by continent. In North America, Europe, and Oceania, the main protein sources
are from animal origin, whereas in Africa nearly 77% of protein intake comes from
plant origin. Therefore, there is a clear pattern in which high-income countries tend
to consume protein from animal origin, whereas low-income countries rely on plants
for protein supply.

In the last 50 years, the daily intake of protein increased from to 39 to 52 g per
capita from 1961 to 2011 in high-income countries, mainly from animal origin, although this value has remained almost constant since the 1980s and FAO projections
for 2030 and 2050 show a modest increase. In the same period of time, the daily per
capita availability of protein from animal products rose from 9 to 20 g (an increase of
116%) in low- and middle-income countries, and this value is projected to reach 22 g
by 2030 and 25 g by 2050 [29]. Guaranteeing an adequate protein intake in the context
of a healthy diet for the world population will be an enormous challenge in the near
future; moreover, these objectives must be accomplished by fulfilling the Sustainable
Development Goals established by the United Nations [30].

Table 5.5: (A) Protein sources in human diet: global data. (B) Differences in protein supply pattern among continents.

(A)

		Protein supply	
		g/capita/day	Percentage
Plant sources		**50.8**	**60.1**
	Cereals – excluding beer	32,5	38.4
	Pulses	4.3	5.1
	Vegetables	5.0	5.9
	Fruits – excluding wine	1.2	1.4
	Starchy roots	2.4	2.8
	Oil crops	3.4	4.0
	Tree nuts	0.4	0.5
Animal sources		**33.5**	**39.6**
	Meat	14.5	17.2
	Eggs	3.2	3.8
	Milk – excluding butter	8.9	10.5
	Fish, seafood	5.6	6.6
	Offals	1.2	1.4

(B)

	Protein supply (g/capita/day)	
Continent	**Plant sources**	**Animal sources**
Africa	51.5	14.2
Asia	53.8	29.7
Central America	45.5	40.5
South America	38.7	49.4
Northern America	43.0	74.5
Europe	44.5	61.1
Oceania	37.4	61.1

Data corresponding to 2020. Source: FAOSTAT [31].

5.4.1 Plant-based protein

Two of the most important sources of plant-based protein worldwide are cereals and pulses (Table 5.6).

5.4.1.1 Cereal proteins

All cereals are similar in overall composition, being low in protein and high in carbohydrates (Table 5.7). These chemical components are not uniformly distributed in the

Table 5.6: Average protein content in g/100 g and % of food energy from protein of some plant-derived raw foods (edible parts only).

Plant-derived foods	Protein content (g/100 g)	Energy from protein (%)
Rice (white)	6.1	7
Root vegetables (average, light colored)	1.3	11
Leaf vegetables (average, dark green)	1.4	43
Fruits[a]	0.2–1.2	2–6
Nuts (mix, without salt)	16.5	11
Bread (multigrain, whole grain)	11.2	16
Legumes (bean, brown and white, average)	22.2	30
Seed blend[b]	22.7	16

Source: [32].
[a]Apple, honeydew melon, pear, orange, kiwi, grape, banana, prune.
[b]Sunflower, pumpkin, pine nut seed.

grain and different cultivars of a given type of cereal exhibit variability in composition, which also depends on culture conditions.

Wheat proteins have been traditionally classified into four solubility classes called Osborne fractions that were subsequently extended to other cereals [33]:
- Albumins, soluble in water.
- Globulins, soluble in salt solutions, but insoluble in water.
- Prolamins, soluble in 70–90% alcohol. This fraction is called gliadin in wheat, secalin in rye, hordein in barley, avenin in oats, and zein in maize.
- Glutelins, soluble in alcohol only in the presence of reducing agents.

Table 5.7: Proximate composition of cereal grains in percentage of dry weight.

Cereal	Protein[a]	Fat[a]	Ash	Fiber[a]	Carbohydrate[b]
Brown rice	7.3	2.2	1.4	0.8	64.3
Sorghum	8.3	3.9	2.6	4.1	62.9
Rye	8.7	1.5	1.8	2.2	71.8
Oats	9.3	5.9	2.3	2.3	62.9
Maize	9.8	4.9	1.4	2.0	63.6
Wheat	10.6	1.9	1.4	1.0	69.7
Barley	11.0	3.4	1.9	3.7	55.8
Pearl millet	11.5	4.7	1.5	1.5	63.4

Source: [34].
[a]Crude.
[b]Available.

The distribution of these fractions varies among different cereals (Table 5.8). Albumins and globulins (20–25% of grain proteins) mainly comprise metabolic and protective proteins such as enzymes and enzyme inhibitors, whereas prolamins and glutelins (75–80% of grain proteins) serve as storage proteins and produce the so-called gluten.

From a biochemical point of view, gluten can be defined as a protein network formed from covalent (disulfide bond) and noncovalent interactions between gliadins (a family of proteins of the prolamin fraction) and glutenins (present in the glutelin fraction) of wheat, rye, and barley. The network development occurs during the mixing of flour and water, associated with mechanical energy, thus producing a cohesive dough with viscoelastic properties necessary for the preparation of bakery products and pasta [20, 35]. The baking quality of flours from different cereals (and varieties) depends on the capacity to form gluten, which, in turn, depends on the composition of glutelins. Some cereals, like rice, even though they have a high glutelin fraction, cannot form gluten.

Table 5.8: Analytical characterization (Osborne fractions) of flours from different cereals.

g/100 g of flour	Wheat	Rye	Barley	Oats
Water	13.23 ± 0.17	11.30 ± 0.09	12.85 ± 0.09	11.8 ± 0.16
Crude protein	11.28 ± 0.08	7.13 ± 0.09	7.66 ± 0.10	8.07 ± 0.04
Albumins/globulins	1.22 ± 0.01	1.84 ± 0.09	1.24 ± 0.03	2.37 ± 0.04
Prolamins	5.94 ± 0.07	2.53 ± 0.03	3.13 ± 0.06	1.29 ± 0.03
Glutelins	2.98 ± 0.04	0.55 ± 0.01	1.10 ± 0.02	1.01 ± 0.05
Gluten[a]	8.92 ± 0.11	3.08 ± 0.04	4.23 ± 0.08	1.29 ± 0.03[b]

Source: [36].
Values are given as mean ± standard deviation (n = 3).
[a]Sum of prolamin and glutelin fractions.
[b]Only the oat prolamin fraction is considered as oat gluten

Gluten proteins from wheat, rye, barley, and, in rare cases, oat, are also responsible for triggering hypersensitivity reactions such as celiac disease (CD), non-celiac gluten sensitivity, and wheat allergy (see below). It has been hypothesized that maize prolamins could also be harmful for a very limited subgroup of patients suffering from CD [37].

Bioactive peptides from cereals (and legumes) have been reported to exert a wide range of physiological effects (antioxidant, anti-inflammatory, cholesterol-lowering, satiety, antidiabetic, and others) in vitro and in animal models. The content of these peptides depends on cultivars, environmental conditions, and agronomical practices. Studies regarding peptide bioavailability in humans and how processes such as digestion can modify peptide activity are needed to deepen into this interesting field [38, 39].

Cereal proteins contain low levels of some EAAs like lysine (Table 5.9); however, this can be compensated either by eating extra quantities of the cereal or, more reasonably, by mixing with other proteins of plant origin like pulses. Therefore, it is possible to follow a vegetarian diet with a good protein quality.

Table 5.9: Recommended daily allowances (RDA, for a 70 kg man) and their composition in essential amino acids (EAAs), of sample vegetal foods.

	Wheat	Maize	Rice	Soybeans	Beans	Peas	Potato	Quinoa	RDA
Grams of protein in 100 g of edible food part	11	8.7	6.7	38.9	10.2	5.5	2.1 g	19.6	
Essential amino acid (mg)									
Lysine	239	258	257	3,047	714	348	92	1,025	**2,100**
Histidine	228	251	165	1,170	303	85	28	478	**700**
Threonine	310	334	246	1,843	428	310	59	849	**1,050**
Cysteine + methionine	454	307	257	1,183	238	95	51	565	**1,050**
Valine	452	472	438	2,176	616	226	99	961	**1,820**
Isoleucine	403	350	306	2,222	556	201	68	808	**1,400**
Leucine	741	1,028	590	3,689	885	342	96	1,399	**2,730**
Phenylalanine + tyrosine	855	761	588	3,970	963	345	132	1,542	**1,750**
Tryptophan	116	61	84	618	113	54	/	726	**280**
Total EAAs (mg)	**3,798**	**3,822**	**2,931**	**19,918**	**4,816**	**2,006**	**624**	**8,353**	**12,880**

Source: [40].

5.4.1.2 Pulses

Pulses are defined by FAO as "annual leguminous crops yielding from one to 12 grains or seeds of variable size, shape and color within a pod." The term "pulses" is limited to crops harvested solely for dry grain and excludes crops harvested green for food (green peas, green beans, etc.) and for oil extraction (soybeans and groundnuts), and leguminous crops that are used exclusively for sowing purposes (seeds of clover and alfalfa). Leguminous plants have the ability to fix nitrogen through symbiosis with soil bacteria of the genus *Rhizobium*, thereby enriching soils [41].

In many regions of the world, pulses are an important source of dietary protein and other nutrients, although they show, like cereals, wide variations in protein content due to genetic, environmental, and agronomic factors. Nutritionally, pulses contain approximately an average of 10% moisture, 21–25% crude protein, 1–1.5% lipids, 60–65% carbohydrates, and 2.5–4% ash.

Pulses are deficient in sulfur-containing amino acids (Table 5.9) but, as mentioned above, this deficiency can be compensated if they are combined with cereals in the diet. They may also contain some antinutritional compounds such as hydrolase inhibitors and lectins, which are part of the defensive mechanism of the seed.

5.4.1.3 Sustainable production of plant-based protein

Agriculture efficiency is likely to be affected by climate change; thus, scientific evidence shows that under conditions of elevated levels of carbon dioxide, the concentrations of minerals in some crops (e.g., wheat, rice, and soybeans) can be up to 8%

lower than normal. Protein concentrations may also be lower, while carbohydrates can be higher [42]. This situation must be considered when making long-term predictions concerning supply of protein and other nutrients from plant sources.

Although plant-based protein is preferable to animal-based proteins from an environmental perspective, high intensity production of plant protein is also subjected to criticisms concerning land and water use, soil degradation, and pollution. Feeding animals with plant protein is itself a cause of controversy (see below); actually, some authors claim that if all available edible plant proteins were directly consumed by humans, there would be a sufficient supply of dietary protein for the global population in the future. These authors demand a major shift toward a reduction in animal-based (especially beef) energy and protein intake [43]. Plant-based protein is considered as a fundamental part of the sustainable universal healthy reference diet proposed by the EAT-Lancet Commission [44] and supported by the FAO, the International Fund for Agricultural Development, the United Nations International Children's Emergency Fund, the World Food Programme, and the WHO [45].

5.4.2 Animal-based protein

The most important sources of animal-based proteins are meat, milk and its derivatives, eggs, and fish (Table 5.10).

Table 5.10: Average protein content in g/100 g and % of food energy from protein of some animal-derived raw foods.

Animal-derived foods	Protein content (g/100 g)	Energy from protein (%)
Beef (average)	16.9	48
Chicken	20.2	44
Fish (weighted average)[a]	18.9	53
Eggs (without shell)	12.6	38
Cheese (average)	23.4	34
Pork (average)	18.9	39
Milk (3.5% fat, boiled)	3.1	19
Yogurt (plain, 2.5% fat)	3.0	22

Source: [46].
[a]Pike, perch, vendace (lake), baltic herring, salmon fillet, tuna, rainbow trout fillet, saithe, powan (frozen), whitefish, pollan, lavaret, zander, pike-perch.

5.4.2.1 Meat

Meat is defined by the Codex Alimentarius[1] as "all parts of an animal that are intended for, or have been judged as safe and suitable for, human consumption" [47]. The term "meat" usually refers to skeletal muscle and associated tissues (broadly categorized into red meat, poultry, fish, and processed meat [48]) although it may include other edible parts like offal. Histologically, skeletal muscles contain muscle fibers associated with connective, adipose, vascular, and nervous tissues. Raw meat contains 20–25% protein depending on source and fat content, which can correspond to 28–36% in cooked meat due to water loss [48, 49].

From the biochemical point of view, muscle proteins can be grouped into three fractions based on location and solubility: sarcoplasmic, stromal (connective tissue), and myofibrillar proteins. Myofibrillar proteins, mainly constituted by myosin and actin and soluble in high-salt solutions, are the main component of the skeletal muscle, accounting for 50–60% of total protein content. Sarcoplasmic proteins (29% of the total) are localized in the sarcoplasm (cytoplasm) of the muscle fibers and are soluble in water, whereas stromal proteins (around 10% of total protein), such as collagen and elastin, are insoluble in aqueous solutions [50, 51]. Meat quality is basically determined by muscle fibers, intramuscular connective tissue, and intramuscular fat. In beef, the most important sensory characteristic for consumers is tenderness, which is generally considered to be determined by collagen; however, in pigs and chickens, this protein seems to have a limited impact on quality. In any case, meat quality is influenced by many factors such as species, genotypes, nutritional and environmental factors, slaughtering conditions, and postmortem processing [49]. Nutritionally, meat proteins are an excellent source of EAAs (Table 5.11), bioavailable minerals, and vitamins (Chapters 6 and 7), presenting a high net protein utilization and digestibility (Table 5.4).

In terms of food chain, animals like cows and, to a lesser extent, poultry and fishes convert protein and fibrous material from photosynthetic organisms into protein with a high nutritional value; however, there are health and environmental concerns about using animals as a major source of dietary protein:

- Health drawbacks of meat. Overconsumption of meat can lead to negative health impacts; thus, the International Agency for Research on Cancer has suggested that processed meat should be classified as *carcinogenic to humans (Group 1)* and red meat as *probably carcinogenic to humans (Group 2A)* [52]. In relation to cardiovascular disease (CVD), although data are more inconsistent, a recent publication reported that chronic dietary red meat increases systemic levels of the atherogenic metabolite trimethylamine *N*-oxide (TMAO) [53]. Meat production

---

1 The Codex Alimentarius (CA) is a collection of standards, guidelines, and codes of practice adopted by the Codex Alimentarius Commission (CAC), a central part of the Joint FAO/WHO Food Standards Programme to protect consumer health and promote fair practices in food trade.

Table 5.11: Recommended daily allowances (RDAs, for a 70 kg man) and their composition in essential amino acids (EAAs), of sample animal foods.

	Egg	Milk	Beef	Pig	Chicken	Sea bass	RDA
Grams of protein in 100 g of edible food part (or 100 ml for milk)	12.1	3.3	22	20.7	23.3	21.3	
Essential amino acid (mg)							
Lysine	1,001	272	2,002	1,737	2,246	2,021	**2,100**
Histidine	322	93	849	647	937	552	**700**
Threonine	674	164	898	919	1,160	967	**1,050**
Cysteine + methionine	740	118	871	780	974	897	**1,050**
Valine	896	233	1,063	1,243	1,384	1,044	**1,820**
Isoleucine	741	192	950	1,080	1,153	914	**1,400**
Leucine	748	355	1,892	1,624	1,955	1,655	**2,730**
Phenylalanine + tyrosine	1,247	318	1,677	1,166	1,776	1,531	**1,750**
Tryptophan	228	50	246	183	273	249	**280**
Total EAAs (mg)	**6,597**	**1,795**	**10,448**	**9,379**	**11,858**	**9,830**	**12,880**

Source: [40].

and processing also raise concerns with respect to food-borne zoonotic diseases [54].

– Environmental concerns. Global meat consumption increased by almost 60% between 1990 and 2009, based on the FAO Food Balance Sheet (FBS), and is expected to increase up to 76% by 2050. Meat production exerts a major impact on land and water availability and releases high levels of greenhouse gases, especially in the case of ruminants. Energy consumption and the use of chemicals like fertilizers and antibiotics also reflect negatively on the environmental footprint of meat production. Indirectly, the use of plant-based protein for animal feed increases environmental pressures; for example, around 85% of soy production, associated with deforestation and habitat loss in South America, is used to feed animals. There are also ethical issues: around 12 million hectares of land outside Europe are required to produce protein-rich meal for European livestock production. Increased efficiencies in production practices and improvements throughout the food chain will be needed to develop a sustainable system for meat production [55].

5.4.2.2 Milk and dairy products
Milk can be defined as "an opaque white fluid rich in fat and protein, secreted by female mammals for the nourishment of their young" [56]. In general terms, milk is a major source of dietary animal-based energy, protein, and fat, and its contribution being on average 134 kcal of energy/capita/day, 8 g of protein/capita/day and 7.3 g of fat/capita/day in the period between 2009 and 2011. Globally, cow milk accounts for

about 83% of milk production. Milk consumption greatly varies when different geographic regions are considered; thus, it provides only 6–7% of dietary protein supply in Asia and Africa compared with 19% in Europe. One major reason for this variation is the prevalence in many ethnic groups of milk hypersensitivities, especially primary lactase deficiency; actually, approximately 70% of the world's population has been estimated to develop this condition at a certain point of life. In any case, in some human groups, like Northern Europeans and a few African and Indian communities, lactase persistence predominates, in contrast to some Asian countries where lactose deficiency reaches almost 100% [57].

The role of milk and dairy products in human nutrition is controversial, and many people argue that a product "naturally designed" for babies should not be consumed by adults; moreover, most of dairy products come from cow (*Bos taurus*) milk that is supposed to provide optimal nourishment for calves and not for humans. However, cow milk can help to improve the nutritional status of children by supplying not only energy, proteins, and fat, but also vitamin B_{12}, a nutrient deficient in plant-based diets, as well as calcium, magnesium, selenium, riboflavin, and pantothenic acid [57]. Table 5.12 shows data of human and cow milk: the latter contains more protein and minerals, especially calcium and phosphorus, than human milk. Several proteins in cow milk have been shown to be allergenic for humans (Table 5.13), αS1-casein being the most important allergen in the casein fraction and α-lactalbumin and β-lactoglobulin in the whey fraction [58]. Therefore, like most foods, cow milk has advantages and disadvantages for human diet. FAO's view is that milk is a complex food, containing numerous nutrients, sometimes with conflicting health effects; however, it can be included as part of a healthy diet, provided that adequate amounts are consumed and no contraindications exist.

Table 5.12: Proximate composition of human and cow milk (per 100 g of milk).

	Human	Cow	
	Average	Average	Range
Energy (kJ)	291	262	247–274
Water (g)	87.5	87.8	87.3–88.1
Total protein (g)	1.0	3.3	3.2–3.4
Total fat (g)	4.4	3.3	3.1–3.3
Lactose (g)	6.9	4.7	4.5–5.1
Ash	0.2	0.7	0.7–0.7

Source: [57].

Milk subjected to different treatments (pasteurization, UHT, skimmed milk, semi-skimmed milk, powder milk, etc.), milk derivatives (yoghurt, kefir, cheese, butter, whey products, etc.), and milk-based products offered by food industry are available nowadays. Many of these products can have distinctive nutritional properties, includ-

Table 5.13: Protein composition of human and cow milk.

Protein	Breast (mg/mL)	Cow (mg/mL)
Casein fraction		
α-s1-Casein*	0	11.6
α-s2-Casein	0	3.0
β-Casein	2.2	9.6
κ-Casein	0.4	3.6
Y-Casein	0	1.6
Whey fraction		
α-Lactalbumin*	2.2	1.2
Immunoglobulins	0.8	0.6
Lactoferrin	1.4	0.3
β-Lactoglobulin*	0	3.0
Lysozyme	0.5	Traces
Serum albumin	0.4	0.4
Other	0.8	0.6

Source: [59].
Caseins can be separated (coagulated) from the rest of milk proteins (whey) by acidification with mineral acids, such as hydrochloric acid or sulfuric acid, or by microbial acidification using lactic acid bacteria. Alternatively, caseins can be coagulated by addition of rennet. *Proteins with high allergenicity in humans.

ing differences in the concentration and quality of proteins that have been comprehensively reviewed by FAO's experts [57].

Dairy production at farm level has many of the negative environmental impacts associated with meat production due to its requirement for land and water and production of greenhouse gases.

5.4.2.3 Eggs

In strict scientific terms, an egg (ovum) is the reproductive cell produced by animal females that remains a single cell until the single cell (nucleus) of the male sperm fertilizes. This definition can be applied to all animal species; however, in normal speech, an egg is a complex structure made and designed by oviparous animals, especially birds, to nourish and protect the embryo growing from the zygote [60].

Hen eggs represent 92% of the global primary egg production for human consumption. Eggs offer a protein of excellent quality with moderate energy content (150 kcal/100 g); moreover, they are inexpensive and provide lipids, several vitamins and minerals, and bioactive compounds (Table 5.14) [61]. They are composed of three main parts:

– Shell, a porous rigid structure that protects the embryo from mechanical damage and regulates gas exchange with the external environment. It also prevents contamination by bacteria and other pathogens and supplies nutrients like calcium to the developing embryo. For the egg industry, the eggshell provides a perfect package [62].

- Egg white mainly serves as a defense mechanism against pathogens [63], although it also provides water, protein, and other nutrients to the embryo [64]. There are two layers of egg white: the thick white layer (nearest to the yolk) and the thin white layer (nearest to the shell). Egg white contains 10.5% protein, 88.5% water, riboflavin and other B vitamins, and fat traces.
- Egg yolk basically provides nutrients for the embryo. The color of the egg yolk is due to the presence of carotenes and colorings added to a hen's feed. The nutritional content of egg yolk is: 16.5% protein, 33% fat, 50% water, fat-soluble vitamins A, D, E, and K, mineral elements, including iron, and lecithin (an emulsifier). Proteins in egg yolk are mainly associated with lipids forming lipoproteins that can be classified as high-density lipoproteins (HDLs), low-density lipoproteins (LDLs), and very-low-density lipoprotein (VLDL). The HDL fraction consists of phosvitin and the livetins (α-, β-, and χ-livetin), which contain less than 10% lipids, and the LDL compounds are α- and β-lipovitellin and contain about 20% lipids, whereas the VLDL lipovitellin contains about 40% lipid [65].

Table 5.14: Nutritional composition of hen eggs.

Component (unit)	Amount	Component (unit)	Amount
Egg shell (%)	10.5	Calcium (mg)	56.0
Egg yolk (%)	31	Magnesium (mg)	12.0
Egg white (%)	58.5	Iron (mg)	2.1
Water (g)	74.5	Phosphorus (µg)	180.0
Energy (kcal)	162	Zinc (mg)	1.44
Protein (g)	12.1	Thiamine (mg)	0.09
Carbohydrates (g)	0.68	Riboflavin (mg)	0.3
Lipids (g)	12.1	Niacin (mg)	0.1
Saturated fatty acids (g)	3.3	Folic acid (µg)	65.0
Monounsaturated fatty acids (g)	4.9	Cyanocobalamin (µg)	66.0
Polyunsaturated fatty acids (g)	1.8	Pyridoxine (mg)	0.12
Cholesterol (mg)	410	Retinol equivalents (µg)	227.0
Iodine (µg)	12.7	Potassium (mg)	147
Tocopherols (µg)	1.93	Carotenoids (µg)	10
Selenium (µg)	10	Cholecalciferol (µg)	1.8

Quantities represent an edible portion of about 100 g. Source: [61].

Some undesirable effects of egg consumption have been reported:

- High level of cholesterol and saturated fats. Although this has been used as an excuse to limit the amount of eggs that can be consumed, current evidence suggests that the link between fat intake and levels of atherogenic lipoproteins is rather complex and other factors are also involved (Chapter 4).

- The presence of antinutritional factors. Some of them are proteins, like ovomucoid (OVM, a trypsin inhibitor) and avidin (that binds biotin); however, they are denatured upon cooking. Therefore, they do not cause detrimental effects unless eggs are eaten raw.
- Some proteins in eggs are allergenic, including OVM, ovalbumin, ovotransferrin, and lysozyme (Table 5.15). OVM seems to be the dominant allergen. After cow's milk, hen egg allergy is the second most common food allergy in children [66].

Table 5.15: Major proteins in hen egg white.

Protein	% of egg white proteins
Ovalbumin	54
Ovotransferrin	12
Ovomucoid	11
Ovomucins	3.5
Lysozyme	3.4
G2 globulin	4.0
G3 globulin	4.0
Ovoinhibitor	1.5
Ovoglycoprotein	1.0
Ovoflavoprotein	0.8
Ovomacroglobulin	0.5
Cystatin	0.05
Avidin	0.05

Source: [67].

In a recent study, the intake of one egg per day, starting early in complementary feeding from 6 to 9 months and continuing for 6 months, significantly improved linear growth and reduced stunting in Andean children; moreover, the beneficial effects on the breast milk composition have also been reported. Therefore, eggs are considered an excellent affordable source of proteins and other important nutrients during infancy [63, 68]. For the rest of the population, some guidelines suggest that eating one egg per day is perfectly safe for healthy people [61].

Meat, milk, and eggs, with their products, along with food from hunting and wildlife farming and insects and insect products are collectively known as "terrestrial animal source food (TASF)," a term that includes not only protein but other nutrients supplied by these foods such as minerals, vitamins, and fatty acids. The contribution of TASF to healthy diets in the context of the Sustainable Development Goals has been recently reviewed by FAO [69].

5.4.2.4 Fish

Fish is a rich source of easily digested, high-quality proteins containing all EAAs, which accounted for about 17% of the global population's intake of animal protein (exceeding

50% in many least developed countries) in 2013. In 2014, a major shift in fish production occurred when the farmed sector's contribution to the supply of fish for human consumption surpassed that of wild-caught fish for the first time. China has played a major role in this growth, representing over 60% of world aquaculture production [70].

In addition to protein, fish provides essential fats such as the long-chain omega-3 fatty acids docosahexaenoic acid and eicosapentaenoic acid, vitamins (D, A, and B), and minerals (including calcium, iodine, zinc, iron, and selenium). Fish consumption has been reported to have positive effects on human health due to both the lipid and the protein/peptide composition. Many of the mechanisms are not fully explored, and more research is still needed to completely understand these effects [71]. However, negative impacts on human health due to exposure to chemical compounds in fish and shellfish have also been observed. An example is methylmercury that can adversely affect a baby's growing brain and nervous system if his mother consumes intoxicated seafood. Other contaminants include organochlorine pesticides, organotin compounds, phthalates, brominated flame retardants, polyflourinated compounds, polycyclic aromatic hydrocarbons, dioxins, dioxin-like PCBs and non-dioxin-like PCBs, heavy metals (mercury, cadmium, and lead), radionuclides, and arsenic. In order to minimize possible toxic effects, regulation (EC) no. 1907/2006 of the European Parliament and of the Council of 18 December 2006 establishes the maximum level of contaminants in foodstuffs, including fish, in the European Union [72].

5.4.3 New and emerging sources of protein (see Chapter 16)

5.4.4 Protein energy malnutrition

Malnutrition can be defined in a broad sense as an abnormal physiological condition caused by inadequate, unbalanced, or excessive intake of macronutrients (protein, carbohydrates, and fat) and/or micronutrients (electrolytes, minerals, and vitamins). This condition is characterized by physiological inefficiencies in the utilization of nutrients and may manifest in a variety of ways. According to FAO, there are different forms of malnutrition such as stunting or chronic malnutrition, wasting or acute malnutrition, underweight, micronutrient deficiencies or hidden hunger, anemia, overweight and obesity, and diet-related noncommunicable diseases [69]. More specifically, widespread forms of malnutrition that involve inadequate or unbalanced intake of nutrients include protein energy malnutrition (PEM), iron deficiency, vitamin A deficiency, and iodine deficiency. There are three major forms of PEM: kwashiorkor, marasmus, and marasmic kwashiorkor. Marasmus is defined as severe wasting and kwashiorkor as malnutrition with edema, whereas marasmic kwashiorkor shows symptoms of both phenotypes [73]. The traditional view that kwashiorkor is the result of protein deficiency and nutritional marasmus is the result of energy deficiency is nowadays considered an oversimplification; actually, kwashiorkor has been classified as a disease of obscure pathogenesis. Among the factors related to the occurrence of PEM are staple diets of low energy density and low in protein, EAAs, and fat content (maize, cassava, etc.), but also inappropriate weaning practices, infections (viral, bacterial, and parasitic), and external circumstances like droughts, natural disasters, and wars [73–75].

5.5 Hydrolysis and denaturing of proteins

The peptide bond links amino acid residues (Figure 5.4), thus forming a polypeptide chain; however, this is just the first level of protein structure, as polypeptides must fold in order to reach a biologically active three-dimensional structure known as "native conformation." The latter is maintained by a series of physical and chemical interactions:

- Disulfide bonds
- Electrostatic interactions
- Hydrogen bonds
- Van der Waals interactions
- Hydrophobic interactions

The breakage of interactions other than peptide bonds may produce a total or partial disruption of the native conformation of a protein, thereby producing loss of biological function. This process is known as denaturation and may be induced by high temperatures, extreme pH, or chemicals such as organic solvents, certain detergents, urea, and guanidine hydrochloride. Protein denaturation can be reversible or irreversible.

Breakage of the peptide bonds of a polypeptide to yield smaller peptides and/or free amino acids is termed protein hydrolysis or proteolysis. If all the peptide bonds in a polypeptide are broken (hydrolyzed), hydrolysis is termed as total, otherwise, partial. A standard classical procedure to achieve total hydrolysis in vitro is by treating polypeptides with 6 M HCl at 110 °C for 24 h, although this treatment degrades some amino acids. A number of proteins known as proteases, peptidases, or proteinases catalyze the hydrolysis of peptide bonds in vivo and in vitro.

Food processing often induces protein denaturation due to treatments like heating, evaporation, drying, fermentation, and irradiation. Heating is one of the most common treatments, being used both at industrial and domestic levels. Food heating may cause beneficial or detrimental effects; however, when moderately applied (up to 100 °C), benefits outweigh undesirable effects, as many proteins can be denatured while no toxic chemicals are produced (see below). Protein denaturation has several advantages in nutritional terms [76]:

- Facilitates digestion because proteases of the GIT are usually more efficient when acting on denatured proteins than on native proteins.
- Inactivates enzymes such as polyphenoloxidase, lipoxygenases, and lipases that may alter nutritional and organoleptic properties of foods.

- Inactivates toxins of microbes such as *Clostridium botulinum* and *Staphylococcus aureus*.
- Inactivates trypsin and chymotrypsin inhibitors in legumes.
- Inactivates OVMs and avidin in eggs.

Drastic thermal treatments (above 200 °C) such as roasting, grilling, baking, or direct application of fire may induce pyrolysis of amino acids, thus producing mutagens like carbolines, imidazoquinolines, acrylamide, and other products of the Maillard reaction (Chapter 3). Food processing may also involve chemical treatments like addition of alkali, oxidant agents (hydrogen peroxide or sodium hypochlorite), nitrites, or sulfites. These treatments may lower the nutritional value of food proteins by chemically altering EAAs or even producing toxic compounds like nitrosamines, in the case of nitrite treatment [76].

5.6 Functional properties of proteins

Classically, the term "functionality" referred to physicochemical properties related to sensory attributes (texture, flavor, color, and appearance) of foods. These attributes, which may be altered during preparation, processing, storage, and/or consumption, are the result of complex interactions among different food components, including proteins. Some properties of proteins involved in the functional characteristics of foods are: solubility, viscosity, cohesion, adhesion, elasticity, foaming, emulsification, capacity of gel formation, or capacity of binding water, fats, or flavors (Table 5.16).

Table 5.16: Properties of proteins with relevance in food properties.

Function	Food	Protein origin
Solubility	Beverages	Whey
Viscosity (swelling)	Soups, gravies, and salad dressings, deserts	Gelatin
Water binding	Meat sausages, cakes, and breads.	Muscle, egg
Gel formation	Meats, gels, cakes, bakeries, cheese	Muscle, egg, milk
Cohesion–adhesion	Meats, sausages, pasta, baked goods	Muscle, egg, whey
Elasticity	Meats, bakery	Muscle, cereal
Emulsification	Sausages, bologna, soup, cakes, dressings	Muscle, egg, milk
Foaming	Whipped toppings, ice cream, cakes, desserts	Egg, milk
Fat and flavor binding	Low-fat bakery products, doughnuts	Milk, egg, cereals

Source: [77].

In practical terms, three main properties seem to be essential for protein functionality: (1) hydration properties; (2) protein surface-related properties; and (3) hydrodynamic/rheological properties. These properties depend on molecular characteristics like size, charge, hydrophobicity, three-dimensional structure, flexibility, or ability to interact with other components. However, studies on the functionality of proteins have been performed with model systems, where the behavior of proteins can be very different to that in real foods, due to the interaction with other components like fats or sugars. This complicates the task of predicting how a protein or a mixture of proteins may alter the attributes of a food. In any case, both animal-based and plant-based proteins have been routinely used as additives by the food industry for decades [20].

A recent view includes some biological properties of proteins as important aspects of their functionality. These include enzymatic activity (that alter properties like color – polyphenoloxidases – or texture – hydrolytic enzymes), bioactive peptides, and allergenicity [78]. This approach integrates traditional food chemistry with current advances in biochemistry, molecular biology, and biotechnology, which might help to develop new functional foods with additional nutritional properties and/or without detrimental effects [79].

A number of procedures in food processing, some of them already mentioned like gluten involvement in dough formation (Section 5.4.1), are based on the functional properties of proteins:

– Gelatin/collagen gelation. Gelatin is a soluble mixture of proteins obtained from collagen, the main fibrous protein of bones, cartilages, and skins. Before extraction, collagen is normally heated in water at temperatures higher than 45 °C. This pretreatment breaks most noncovalent bonds, thus producing swelling and increasing solubilization. Subsequent heating cleaves hydrogen and covalent bonds, resulting in helix-to-coil transition and conversion into soluble gelatin. The most important properties of both collagen and gelatin are either associated with their gelling behavior (gel[2] formation, texturizing, thickening, and water binding capacity) or related to their surface behavior (emulsion, foam formation and stabilization, adhesion and cohesion, protective colloid function, and film-forming capacity). Enzymatic hydrolysis of collagen or gelatin has been proposed as a source of bioactive peptides, whereas adverse effects have also been investigated; thus, bovine and porcine gelatin have been reported to be more allergenic than fish gelatin [81].

– Protein hydrolysis has been used in food processing in order to enhance aspects related to the nutritional value and functional properties of foods. These include improved digestibility, modifications of sensory quality (such as texture or flavor), and health benefits, such as the improvement of antioxidant capability or reduction in allergenic compounds. Many of these processes can be achieved by protein hydrolysis using specific proteases that preserve other chemical species present in foods [82].

---

2 A gel can be described as a continuous solid network containing and supporting a continuous liquid phase [80].

Among food products elaborated using protein hydrolysis are: infant formulas for breast milk substitution [83], beer [84], and cheese (Table 5.13).

– Alkaline treatment is used to facilitate extraction of proteins from certain foods such as oilseeds and aquatic foods. This treatment improves protein solubility, enhances some functional properties (emulsification and/or foaming capacity), inactivates enzymes, and destroys antinutritional compounds like toxins, enzyme inhibitors, and allergens. However, alkaline treatment may also reduce the nutritional value of food proteins by causing racemization, cross-linking, or destruction of certain amino acids (lysine, serine, etc.). Some of the products of these chemical alterations may produce toxic compounds like lysinoalanine [85].

– Texturization, applied to foods, involves transformation of proteins from a globular state to a structure that provides a sensation of eating meat. Typical structures are fibers, shreds, chunks, bits, slices, films, and granules. Texturized protein products are expected to possess adequate chewiness, elasticity, softness, and juiciness. Vegetable proteins are often the preferred protein source for texturization (texturized vegetable protein, TVP), although protein of animal origin can also be subjected to this treatment. Many texturizing methods are available for the food industry, such as extrusion cooking, spinning (dry, wet, and spinneretless), freeze texturization, high-pressure texturization, and chemical or enzymatic texturization [20, 86]. The general principles involved in some of these methods are thermal or alkaline denaturation of proteins, realignment of the denatured proteins to obtain a fibrous network, binding of the fibers using a protein binder, and flavoring of the final product. TVP is increasingly being used as meat analog ("mock meat" or "imitation meat") [20].

– Food preservation by drying. Proteins derived from various natural sources are converted into dry powder form to enhance their stability and for long-term storage. Spray drying and freeze-drying (lyophilization or cryodessication) are the preferred methods. Spray drying is a well-established method for drying milk; however, a considerable amount of protein may be inactivated or denatured during the procedure [87]. Freeze-drying, a procedure based on eliminating the water by crystallization at a low temperature and subsequent sublimation under vacuum yields dried materials with features typical of raw material; however, it is expensive and time-consuming [88]. Other food drying methods and their effects on proteins and other components have been comprehensively reviewed elsewhere [89, 90].

5.7 Protein deterioration by microorganisms

Food proteins can be hydrolyzed by proteases that may belong to the food itself or to microorganisms present in the latter (filamentous fungi, yeasts, and bacteria), thus releasing peptides and free amino acids. Microorganisms can subsequently use amino acids as nutrients and produce metabolic end products with additional odors and tastes, thereby altering the nutritional value and the organoleptic properties of the food.

5.7.1 The Ehrlich pathway

A common biochemical pathway used by microorganisms for the catabolism of branched chain (leucine, isoleucine, and valine), sulfur-containing (cysteine and methionine), and aromatic (phenylalanine, tyrosine, and tryptophan) amino acids is the Ehrlich pathway, which starts with the transamination of the amino group and the formation of α-keto acid. These keto acids are subsequently decarboxylated to the corresponding aldehydes and, finally, depending on the redox state of the cell, they can be reduced to alcohols, or oxidized to acids [91, 92]. All these biochemical changes may give desirable flavors, like those produced during cheese ripening; however, uncontrolled microbial metabolism produces high levels of organic acids, sulfur compounds, such as mercaptans and hydrogen sulfide, indole, phenols, and ammonia. Most of these chemicals have a displeasing odor and some of them can be toxic. This process, known as putrefaction, is quite common in meats and other protein-rich foods at temperatures above 15 °C [93].

5.7.2 Production of biogenic amines

Amines produced by microbial decarboxylation of amino acids in foods belong to a family of molecules known as biogenic amines (BAs). BAs are nitrogenous compounds of low molecular weight, essential at low concentrations for natural metabolic and physiological functions in animals, plants, and microorganisms; however, high concentrations of these compounds can have toxic effects such as hypertension, cardiac palpitations, headache, nausea, diarrhea, flushing, localized inflammation, and even death. Formation of BA can occur during food processing and storage as a result of bacteria that have amino acid decarboxylase activities. The most common BAs in foods are histamine, putrescine, cadaverine, tyramine, tryptamine, 2-phenylethylamine, spermine and spermidine, and polyamines [94, 95].

5.7.3 Stickland reaction

The Stickland reaction involves the oxidative deamination, followed by decarboxylation, of one amino acid (usually alanine, valine, leucine, or isoleucine) coupled with the reductive deamination of another amino acid (proline, glycine, hydroxyproline, or the non-proteinogenic amino acid ornithine). The final products are organic acids, ammonium, and carbon dioxide [96]. As an example, the overall Stickland reaction between alanine and glycine would be:

$$CH_3\text{-}CH(NH_2)\text{-}COOH + 2NH_2\text{-}CH_2\text{-}COOH + 2H_2O \rightarrow 3CH_3\text{-}COOH + 3NH_3 + CO_2$$

The Stickland reaction is a basic bioenergetic pathway of some bacteria of the genus *Clostridium*, some of them implicated in food poisoning like *Clostridium botulinum*, associated with botulism, a disease caused by proteinaceous toxins produced by several strains of this bacterium. Botulism is characterized by flaccid descending paralysis that begins with cranial nerve palsies and might progress to extremity weakness and respiratory failure. Seven antigenically distinct toxins have been identified (A, B, C, D, E, F, and G), although only serotypes A, B, E, and (more rarely) F cause the disease in humans [97]. Botulinum toxin (mainly serotype A) has clinical applications in areas like ophthalmology (to treat strabismus among others), neurology (in dystonias, spasticity, tics, cerebral palsy, hyperhidrosis, Parkinsonian syndromes, motor neuron disease, etc.), urology, otorhinolaryngology, gastroenterology (obesity treatments, for example), aesthetic medicine, and even psychiatry (treatment of depression) [98].

5.8 Amino acid metabolism: interconversion and degradation

Amino acids and small peptides produced during the digestion of dietary proteins in the GIT are taken up by enterocytes, the absorptive cells in the lining of the intestinal mucosa. Enterocytes are differentiated epithelial cells with apical microvilli facing the intestinal lumen; microvilli increases the luminal surface area of the cell by 14- to 40-fold, thereby optimizing nutrient absorption [99] (Chapter 13).

Amino acids are subsequently exported to blood capillaries and travel to all tissues where they will be taken up (or exported, when necessary) by a numerous group of transporters. The human genome contains about 50 different amino acid transporters, many of them with specialized roles in selected cell types. Amino acid homeostasis is of paramount importance for all cells because they need to maintain optimal levels of the 20 proteinogenic amino acids for protein biosynthesis, and also to synthesize bioactive molecules or to obtain energy from their carbon skeletons [100]. In hepatocytes, the main cells in the liver that maintain nutrient homeostasis at organismal level, amino acid and protein metabolism are especially active; actually, the turnover rate of liver enzymes (i.e., the rate at which they are synthesized and degraded) is 5 to 10 higher than in other tissues [4]. Amino acids taken up by the hepatocyte have several possible fates (Figure 5.5):

- Protein synthesis: The liver does not only synthesize its own proteins but it is also the site of biosynthesis of most plasma proteins. In order to accomplish this task, amino acids must be covalently attached to the correct transfer RNA (tRNA). This step, critical for the translation of mRNA into a polypeptide, is mediated by a family of enzymes known as aminoacyl-tRNA synthetases (or aminoacyl-tRNA ligases) and requires energy in the form of adenosine triphosphate (ATP) [101]. Protein

synthesis is an extremely complex biological process that is thoroughly covered in biochemistry textbooks [4, 102].
- Export to the bloodstream in order to maintain a proper supply of amino acids for the rest of tissues and organs.
- Synthesis of nitrogen-containing compounds like nucleotides, hormones, and neurotransmitters.
- Interconversion of amino acids in a process known as transamination.
- Degradation by deamination of surplus amino acids to yield ammonium and carbon skeletons. This process is essential for nitrogen homeostasis because proteins and amino acids cannot be stored. Ammonium is converted to urea and eliminated through the kidneys, whereas carbon skeletons can be converted to pyruvate, acetyl-CoA, or intermediates of the citric acid cycle.

The excess nitrogen from other tissues is exported to the bloodstream as alanine (skeletal muscle) or glutamine (rest of tissues) and transported to the liver, where they are taken up and degraded.

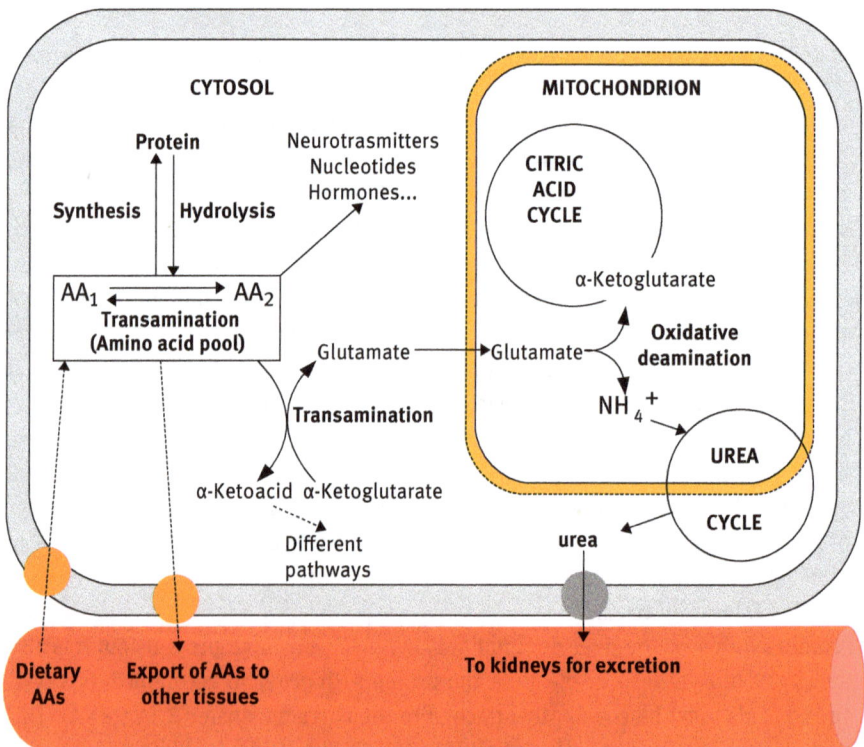

Figure 5.5: Destinations of amino acids (AAs) in hepatocytes.

5.8.1 Transamination

In transamination reactions, an amino acid acts as a donor of its α-amino group and an α-keto acid acts as an acceptor. The donor converts into the corresponding α-keto acid and the acceptor into a new amino acid:

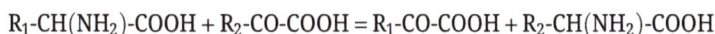

$$R_1\text{-CH}(NH_2)\text{-COOH} + R_2\text{-CO-COOH} = R_1\text{-CO-COOH} + R_2\text{-CH}(NH_2)\text{-COOH}$$

These reactions, freely reversible within the cells, are catalyzed by a family of enzymes known as aminotransferases or transaminases, most of them located in the cytosol. Cells contain several aminotransferases that use 2-oxoglutarate as the α-amino group acceptor, albeit specific for the donor α-amino acid. All aminotransferases require pyridoxal 5′-phosphate, the active form of vitamin B6 (Chapter 6), as a coenzyme for activity. The chemical mechanisms of these reactions have been studied in detail [103].

Elevated levels of aminotransferases, especially alanine transaminase (ALT) and aspartate transaminase (AST, also known as glutamic oxaloacetic transaminase, EC 2.6.1.1), in the bloodstream may indicate inflammation or damage in the liver: hepatocytes lyse releasing their contents, including enzymes, that can be detected in blood tests. High levels of ALT and AST have also been linked with an increased risk of type 2 diabetes and CVD [104]. However, unlike ALT, found predominantly in hepatocytes, AST may be derived from other tissues, such as heart, red blood cells, and muscle. High levels of serum AST alone may also indicate mononucleosis, heart problems, or pancreatitis.

5.8.2 Deamination

Deamination is the enzymatic removal of the amino groups from amino acids that normally takes place in the mitochondrial matrix of hepatocytes. Previously, the amino groups of the amino acids destined for degradation are transferred to 2-oxoglutarate in the cytosol to yield glutamate in a series of reactions mediated by aminotransferases. This also includes glutamine and alanine coming with excess nitrogen from other tissues that have to be eliminated in the liver. Deamination proceeds in a series of steps:

1. Glutamate is transported into the mitochondria in a process mediated by a glutamate/H^+ symporter [105].
2. In the mitochondrial matrix, glutamate is deaminated in a redox reaction that produces 2-oxoglutarate, ammonium, and reducing power as NAD(P)H (Figure 5.6), a reaction catalyzed by the enzyme glutamate dehydrogenase (EC 1.4.1.2 and EC 1.4.1.3).
3. 2-Oxoglutarate enters the citric acid cycle to be converted into metabolites that can be used for the synthesis of biomolecules like sugars, fatty acids, and choles-

terol, or oxidized to obtain energy. Ammonium enters the urea cycle for detoxification and elimination. It must be pointed out that, in aqueous solution, ammonium exists in a pH-dependent dynamic equilibrium with ammonia molecules:

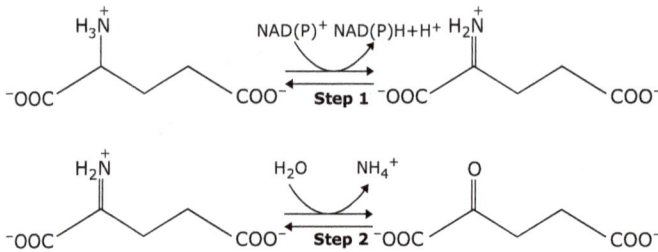

$$NH_3 + H_2O = NH_4^+ + OH^-$$

Glutamate + NAD(P)$^+$ + H$_2$O $\rightleftharpoons$ 2-oxoglutarate + NAD(P)H + NH$_4$$^+$

Figure 5.6: Oxidative deamination of glutamate mediated by glutamate dehydrogenase in two steps (adapted from J-hussain/Public domain. https://commons.wikimedia.org/wiki/File:FINAL_GLUTAMATE.GIF).

5.9 Metabolic fate of ammonium: the urea cycle

The urea cycle is a biochemical pathway elucidated by Sir Hans Krebs that begins inside the mitochondria of hepatocytes and continues in the cytosol. The overall reaction can be written as follows:

$$2NH_4^+ + HCO_3^- + 3ATP^{4-} + H_2O \rightarrow H_2N\text{-}CO\text{-}NH_2 + 2ADP^{3-} + 4Pi^{2-} + AMP^{2-} + 5H^+$$

Three molecules of ATP are spent per molecule of urea, in principle; however, in bioenergetics terms, the hydrolysis of one of these molecules to produce adenosine monophosphate (AMP) and two phosphates (Pi) is equivalent to hydrolyzing two molecules of ATP in the "standard" way (ATP + H$_2$O $\rightarrow$ ADP + Pi). Therefore, cells spend the equivalent to four molecules of ATP to produce one of urea (H$_2$N-CO-NH$_2$), an expensive, albeit necessary, pathway to convert the toxic ammonium into a very soluble, innocuous molecule. The urea cycle proceeds through five steps (Figure 5.7):
1. Reaction of ammonium with CO_2 (HCO$_3^-$) to produce carbamoyl phosphate catalyzed by the enzyme carbamoyl phosphate synthetase 1 (CPS1).
2. Combination of carbamoyl phosphate and the non-proteinogenic amino acid ornithine to form citrulline. This step is mediated by the enzyme ornithine transcarbamoylase (OTC). Citrulline is then transported from hepatocyte mitochondria into the cytoplasm by the ornithine translocase.
3. In the cytosol, citrulline reacts with aspartate to form argininosuccinate in an ATP-dependent reaction catalyzed by the enzyme argininosuccinate synthetase

(ASS). Aspartate is the source of the second amine group of urea and results from the transamination of oxaloacetate and glutamate mediated by AST.
4. Argininosuccinate is converted into arginine via argininosuccinate lyase. This reaction also produces fumarate, which is converted to malate and transported into the mitochondria where it will enter the citric acid cycle.
5. Arginine is hydrolyzed by arginase to form urea and ornithine.

Urea is exported from the hepatocytes by facilitated diffusion mediated by a group of proteins known as urea transporters and transported to the kidney for excretion, although part of the urea can also be eliminated by sweat [106].

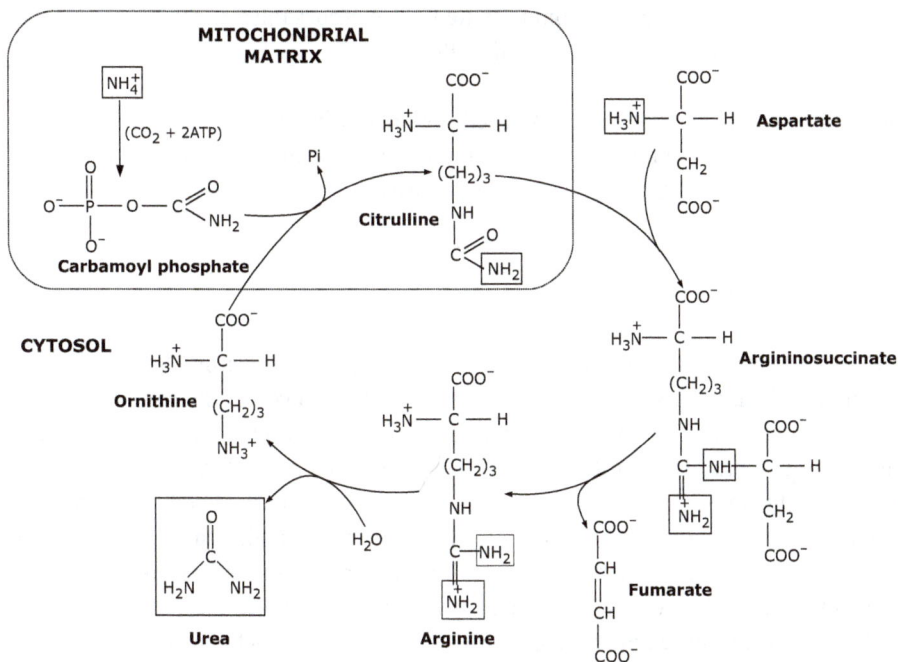

Figure 5.7: The urea cycle. See main text for more details (adapted from OpenStax College/CC BY (https://creativecommons.org/licenses/by/3.0) https://commons.wikimedia.org/wiki/File:2518_Urea_Cycle.jpg).

5.10 Proteins and adverse reactions to foods

Adverse reactions to foods can be broadly divided into several groups [107, 108]:
– Reactions with an immune basis can be subdivided into:
  – immunoglobulin E (IgE)-mediated, known as allergies or type I hypersensitivities;
  – non-IgE-mediated;

-   mixed IgE and non-IgE-mediated;
-   T-cell-mediated or type IV hypersensitivity, like CD. (Hypersensitivities are classified into four major types of allergic reactions based on pathogenesis mechanisms. There is no conclusive evidence that supports the occurrence of type II and type III hypersensitivities in food allergies. See reference [109] for further information).

- Reactions without an immune basis, also known as food intolerances, can be caused by some chemical substances present in foods (caffeine, tyramine, histamine, etc.) or enzyme deficiencies (like lactose intolerance, described in Chapter 3). Some food intolerances are of unknown origin.
- Reactions due to metabolic errors are shown below.
- Adverse reactions to food contaminated by microorganisms may result in toxic products from amino acid metabolism (Section 5.7).

Other adverse reactions are only indirectly related to foods. These include those due to malabsorption problems, such as abetalipoproteinemia and pancreatic insufficiency, or inflammatory diseases like Crohn's disease.

5.10.1 Food adverse reactions with an immune basis

– Food allergies: In strict terms, an allergen is defined as a macromolecule to which IgE binds, which means that a "food allergy" is that produced by substances, most of them proteins, that elicit an immune response mediated by IgE, a special type of antibody. This response is normally generated as part of the normal immune reaction to parasitic infections. However, for reasons not completely understood, it can also be generated by environmental agents, such as pollen, dusts, and foods [110]. There are also a number of non-IgE-mediated (or cell-mediated) food allergies that commonly affect only the GIT in a subacute or chronic way. Disorders in this category include food protein-induced enterocolitis, food protein-induced proctitis/proctocolitis, and enteropathy [111].

Foods most commonly known to cause food allergies include cow milk, hen eggs, wheat, soy, peanuts, tree nuts, fish, and seafood; however, in the Annex II of the Regulation (EU) 1169/2011 of the European Parliament, a more thorough list of 14 substances or products causing the most prevalent intolerances or allergies is included [112]. Approximately 400 proteins from more than 170 foods have been reported to cause IgE-mediated allergic reactions. They are published by the International Union of Immunological Societies in a list that is regularly updated (http://www.allergen.org). Several attempts have been made to classify allergenic proteins based on their sequence homology and domain architecture; however, more research is needed to elucidate the underlying biochemical mechanisms of food allergy. A cure for this type of adverse reactions is currently unavailable, and avoidance of the offending food is the best option for sensitive individuals. Processing may also be of major importance in food allergenicity that may increase, decrease, or remain unchanged depending on the treatment and the allergen [111, 113]. New techniques of food processing aimed at eliminating allergens by using biotechnological approaches are currently under study (Section 5.6).

The frequency, severity, and type of allergic manifestations seem to be affected by patients' genetics and a number of environmental, cultural, and behavioral factors. Nowadays, the prevalence and severity of food allergies seem to be increasing, and some ideas to explain this fact have been proposed. Thus, the hygiene hypothesis suggests that improvements in personal hygiene contribute to the increased prevalence of IgE-mediated allergies, whereas insufficient exposure to dietary and bacterial metabolites might be contributing to increases in inflammatory disorders [109].

– **Type IV hypersensitivities**: One of the main examples is coeliac (or celiac) disease that (unlike IgE-dependent wheat allergy) is an autoimmune disorder that causes immune responses in genetically predisposed individuals. In CD, the immunological response is mainly triggered by partially hydrolyzed gluten fraction from wheat and other cereals (Section 5.4.1) and mediated by the $CD4^+$ T cells of the small intestine. Some environmental factors also seem to be involved in the onset of this disease. Type IV hypersensitivity reactions also might be involved in food protein-induced enterocolitis [109, 113].

5.10.2 Amino acid disorders

Amino acid disorders or aminoacidopathies are a group of inborn diseases caused by inherited defects that result in a total or partial loss of enzymatic activities involved in the metabolism of amino acids. Amino acid disorders are biochemically characterized by abnormal levels of one or several amino acids and their downstream plasma and/or urine metabolites. Treatment of many of these pathologies often includes restriction of dietary protein. Some examples of amino acid disorders are:

– Phenylketonuria (PKU). This is characterized by deficiency of phenylalanine hydroxylase, an enzyme that catalyzes the conversion of phenylalanine to tyrosine. Phenylalanine and phenylpyruvate accumulate in blood and tissues and excreted in the urine. It causes mental retardation unless diagnosed early in the infancy. Treatment involves restriction of natural protein intake: diet must supply the minimum phenylalanine and tyrosine to meet the needs of protein synthesis. PKU can also be caused by a defect in the regeneration of tetrahydrobiopterin (BH_4), the cofactor of phenylalanine hydroxylase [4].

– Disorders of tyrosine metabolism. Three main disorders of tyrosine metabolism (tyrosinemia I, II, and III) have been described, all of them characterized by high levels of plasma tyrosine (hypertyrosinemia) and urine excretion of tyrosine-derived metabolites. Symptoms of these diseases include hepatorenal dysfunction, mental retardation, and/or convulsions [114]. Tyrosinemia I and II are successfully treated by a combination of NTBC (2-[2-nitro-4-trifluoromethylbenzoyl]-1,3-cyclohexanedione) and restricted dietary therapies starting by 1 month of age [115]. Tyrosinemia type III is a very rare disorder: only a few patients have been described and the clinical phenotype remains variable and unclear [116].

– Maple syrup urine disease (MSUD) is a disorder of branched-chain amino acid metabolism caused by a deficiency of branched-chain α-keto acid dehydrogenase complex. MSUD is characterized by elevated levels and abnormal ratios of leucine, isoleucine, valine, and allo-isoleucine in plasma. The disease is managed by dietary leucine restriction [114].

– Methylmalonic acidemia. This metabolic disorder is characterized by elevated levels of methylmalonic acid (MMA) in the blood usually caused by the loss of methylmalonyl-CoA mutase (MUT) activity. This enzyme is involved in the conversion of methylmalonyl-CoA to succinyl-CoA, an essential step in the catabolism of branched-chain amino acids and odd-numbered fatty acids, including propionic acid produced by gut microbiota. Mutations in enzymes required to produce the adenosylcobalamin cofactor (Chapter 6) of MUT can also lead to this disorder. MMA seems to produce harmful effects in mitochondria of several organs, although toxic effects by other related metabolites may also be involved. Current strategies of treatment include dietary restriction of branched-chain amino acids or adenosylcobalamin supplementation. A second form of this disease involves the additional loss of methionine synthase, leading to accumulation of homocysteine along with MMA [117].

– Disorders of sulfur amino acid metabolism. Homocystinuria is an example of this type of disorders, whose main biochemical finding is accumulation of a sulfur-containing nonproteinogenic amino acid homocysteine and its metabolites in the blood and urine. The classic homocystinuria is caused by cystathionine β-synthase deficiency, an enzyme involved in the metabolism of methionine [114].

– Nonketotic hyperglycinemia (NKH) is a severe disorder of glycine metabolism. Most of the affected individuals die within few months of life or survive with significant intellectual disabilities. The main finding in NKH is elevated levels of glycine in plasma and cerebrospinal fluid [114].

– Urea cycle disorders (UCDs). Several inherited disorders of the urea cycle have been described. All these defects, characterized by hyperammonemia and disordered amino acid metabolism, present autosomal recessive inheritance, except OTC deficiency, which is X-linked [118] (see below). Patients with primary UCDs typically present with neurological abnormalities such as abnormal posturing, vomiting, ataxia, confusion, and irritability during the neonatal period. If appropriate intervention does not take place, the patients can develop seizures, enter into coma, and die. In milder cases, clinical symptoms include loss of appetite, cyclical vomiting, lethargy, behavioral abnormalities, sleep disorders, delusions, hallucinations, and psychosis. Liver impairment is also a frequent finding in UCDs that may result in acute liver failure but also in long-term complications such as cirrhosis and hepatocellular carcinoma [119]. UCDs include:

- Deficiency of CPS1, caused by mutations in the CPS1 gene, leads to frequently fatal hyperammonemia. CPS1 catalyzes the first step of the urea cycle that converts ammonia to carbamoyl phosphate (Figure 5.7). Unless promptly treated, this disease can result in encephalopathy, coma and death, or intellectual disability in surviving patients. Current treatment includes protein restriction and supplementation with citrulline and/or arginine and ammonia scavengers [120, 121]. CPS1 deficiency is the most severe of the UCDs: individuals with complete CPS1 deficiency rapidly develop hyperammonemia in the newborn period. Children who are successfully rescued from crisis are chronically at risk for repeated bouts of hyperammonemia [122].

- Deficiency of OTC is an X-linked recessive disorder and, therefore, is much more severe in males. The classic presentation of OTC deficiency in hemizygous males is that of a catastrophic illness in the first week of life. In symptomatic heterozygous females and in males with partial OTC deficiency, symptoms may occur at any time from infancy to adulthood. Patients present hyperammonemia, hyperglutaminemia, reduced or absent citrulline in plasma, and increased urinary orotic acid [123].

- Deficiency of ASS. Metabolic disturbances associated with impairment in the function of the ASS gene are named citrullinemia type I. Biochemically, ASS deficiency is characterized by neonatal or intermittent onset of hyperammonemia, low plasma arginine, elevated plasma and urine citrulline levels, and orotic aciduria. In symptomatic cases, dietary protein restriction, substitution of arginine and EAAs, and the use of drugs allowing the alternative detoxification of surplus nitrogen is the recommended treatment [124].

- Deficiency in argininosuccinate lyase (also known as argininosuccinic aciduria), the enzyme that cleaves argininosuccinic acid to produce arginine and fumarate in the urea cycle, may present as a severe neonatal-onset form and a late-onset form. It results in elevated ammonia and citrulline concentrations in plasma, and elevated argininosuccinic acid in plasma or urine. Treatment involves rapid control of hyperammonemia during metabolic decompensations and long-term management to help prevent episodes of hyperammonemia and long-term complications. Treatment includes dietary restriction of protein and dietary supplementation with arginine is the mainstays in long-term management. Oral nitrogen-scavenging therapy can be considered for patients who do not respond to dietary measures [125].

- Deficiency in arginase. Rare disease that causes abnormalities in the development of the nervous system. It accumulates and excretes arginine. It is successfully treated with a diet rich in EAAs but excluding arginine [4].

- Deficiency in *N*-acetylglutamate synthetase results in the absence of the normal activator of CPS. This condition can be treated by administering carbamoyl glutamate, an analogue of *N*-acetylglutamate that can activate CPS [4].
- Other inborn errors affect transporters of some of the intermediates of the urea cycle [123].

UCDs illustrate the toxicity of ammonia, a small molecule that easily diffuses through the blood–brain barrier. Hyperammonemia is defined as ammonia blood levels above 110 μmol/L (198 μg/dL) in the neonatal period (including preterm) and above 50 μmol/L (90 μg/dL) from that age onward. Updated guidelines for the diagnosis and therapeutic management of hyperammonemia along with useful information about the causes of this condition (including UCDs) can be found elsewhere [126]. As far as the biochemical basis of ammonia toxicity is concerned, glutamine overproduction and the so-called glutamate excitotoxicity seem to play a major role in the brain; however, the deleterious effect of ammonia in non-brain cells is not completely understood although it apparently involves the aberrant activation of multiple signaling pathways [117].

5.11 Other nitrogen-containing nutrients: nucleotides

5.11.1 Definition and general concepts

Nucleotides are a group of biomolecules that have three main components: a heterocyclic ring of carbon and nitrogen atoms derived from pyrimidine or purine, known as nitrogenous base, a pentose sugar in ring form, and phosphate groups. The compounds obtained by linking a nitrogenous base with the pentose sugar via an *N*-glycosidic bond are known as nucleosides. Nucleotides are formally obtained from nucleosides by addition of one, two, or three phosphate groups (Figure 5.8). Nucleoside monophosphates (NMPs) are the basic units of nucleic acids; however, nucleoside triphosphates (NTPs) are the actual precursors of NMPs in vivo, that is, they are the substrates of DNA polymerases and RNA polymerases, the enzymes that catalyze replication and transcription. On the other hand, NTPs, especially ATP, are considered the "energy currency" of cells, being implicated in most of the biochemical reactions that require energy. Intracellular levels of ATP, adenosine diphosphate (ADP), and AMP are considered major determinants of the energy status, thereby affecting virtually all cellular processes. Nucleotides are also involved in signal transduction (cyclic AMP), formation of enzyme cofactors, and many PTMs of proteins.

Purines

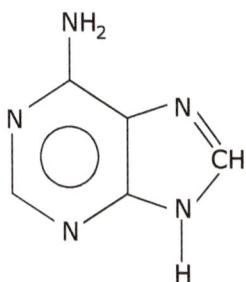

**Adenine
(DNA and RNA)**

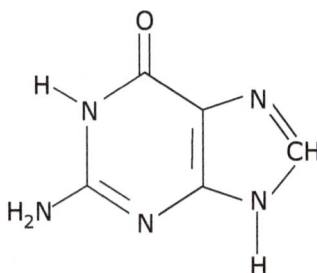

**Guanine
(DNA and RNA)**

Pyrimidines

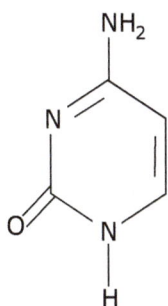

**Cytosine
(DNA and RNA)**

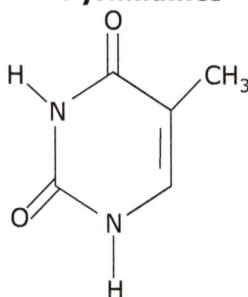

**Thymine
(DNA)**

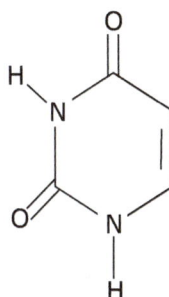

**Uracyl
(RNA)**

Pentoses

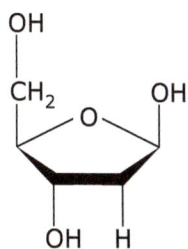

**β-D-2'-deoxyribose
(DNA)**

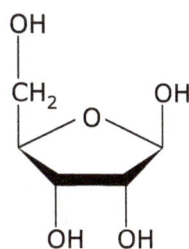

**β-D-ribose
(RNA)**

A nucleoside

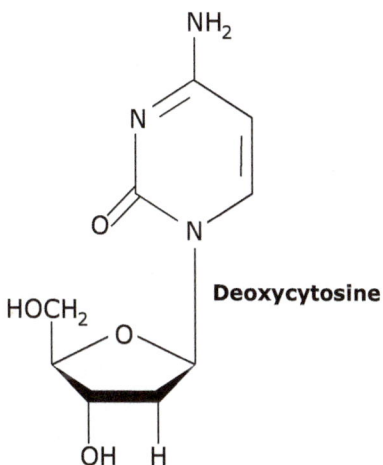

Deoxycytosine

Figure 5.8: Chemical formulae of nucleotides, precursors, and derivatives.

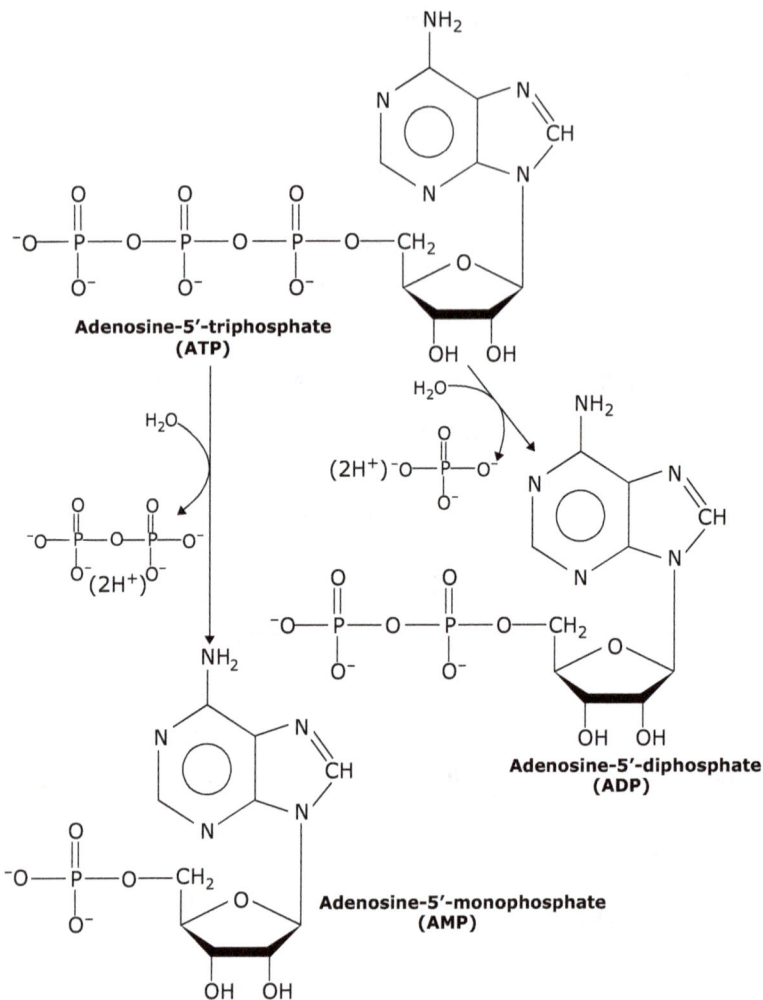

Figure 5.8 (continued)

There are two major types of nucleic acids in living organisms that differ in their chemical composition and biological functions:

- DNA is the repository of genetic information in most organisms, being the "book of instructions" of cells and responsible for passing this information to subsequent generations. In order to accomplish this task, DNA has the ability to make copies of itself in a process known as replication (Figure 5.1). Nitrogenous bases in DNA are adenine, guanine, cytosine, and thymine, and the pentose sugar is 2-deoxyribose (Figure 5.8). In vitro, DNA molecules are composed of two antiparallel chains of 3′→5′ phosphodiester-linked NMPs, held together mainly by hydrogen bonds established between complementary nitrogenous bases of both chains: A and T (two hydrogen bonds), and G and C (three hydrogen bonds) (Figure 5.9). The double chain is twisted forming a helical structure as proposed by Watson and Crick in 1953 [127]. It must be mentioned that, although this is the most popular vision of DNA structure, other in vitro structures have been identified. The situation is quite different in vivo: in the nuclei of eukaryotic cells, DNA is tightly associated with a group of proteins (histones) forming a fibrous substance known as chromatin, responsible for the high structural complexity of chromosomes [4].

– RNA has a similar chemical composition to that of DNA, except that it contains uracyl instead of thymine and ribose instead of 2-deoxyribose (Figure 5.8). Unlike DNA, RNA molecules are usually single chains of NMPs which may fold in complex three-dimensional structures. There are several types of RNA that perform different functions within the cell; the most important in quantitative terms are:
  – mRNA, which carries information contained in the DNA to the translation machinery. DNA information is transferred to RNA molecules in a process known as transcription.
  – tRNA are responsible for transporting the amino acids that will be linked to synthesize polypeptides. They play a pivotal role in the translation of the genetic code in collaboration with a set of enzymes, aminoacyl-tRNA synthetases that attach each amino acid to its corresponding tRNA .
  – Ribosomal RNA, along with several proteins, constitutes ribosomes, the translation factory. Ribosomes bind mRNA molecules and provide the optimal environment for the interaction with tRNAs so that amino acids are placed in the right sequence before the peptide bond is formed. It is the most abundant type of RNA in all cells.

Less than 2% of the DNA sequence contained in human cells is actually translated into proteins; however, it has been reported that at least 90% of the genome is transcribed. This evidence points at the existence of other types of RNA, the so-called noncoding RNAs (ncRNAs), involved in the regulation of gene expression and, consequently, in many physiological processes. Several types of ncRNAs have been identified: microRNAs, long ncRNAs, and circular RNAs, among others [128].

A more detailed description of the biological roles of DNA and RNA as well as the processes of replication, transcription, and translation can be found elsewhere [4, 102].

5.11.2 The role of nucleotides in nutrition

Human cells have three main sources of nucleotides: *de novo* synthesis from amino acids, recycling through the so-called "salvage" pathway and diet. In healthy adults, sufficient supply of nucleotides seems to be provided by synthesis and recycling; however, requirements may increase during growth, tissue damage, infections, and diseases. Many studies show that under conditions of physiological stress, nucleotides

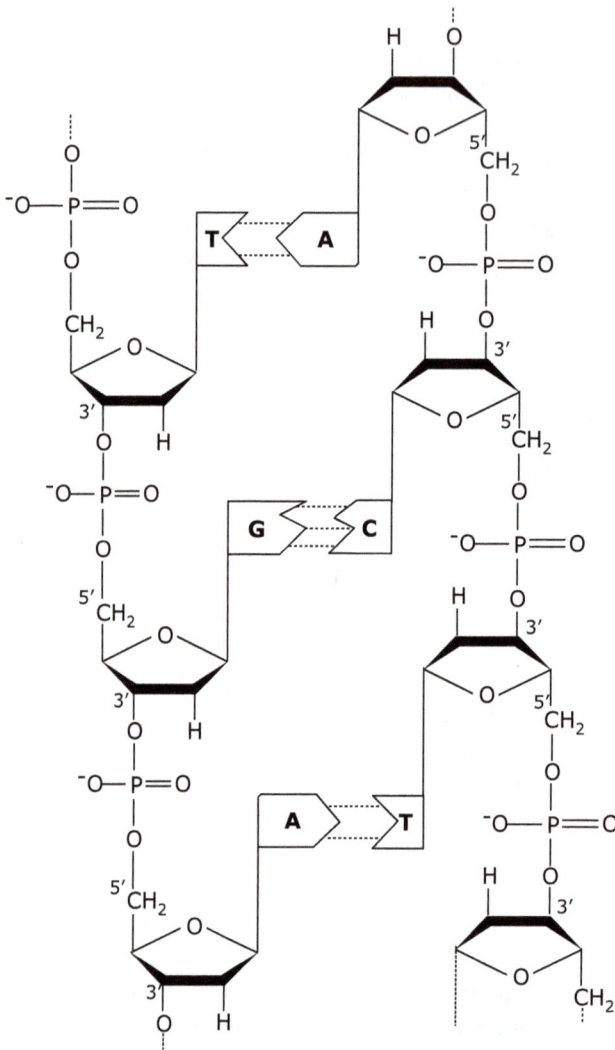

Figure 5.9: Basic structural unit of a DNA molecule. Two antiparallel chains of 3'→5' phosphodiester-linked nucleoside monophosphates are held together mainly by hydrogen bonds established between complementary nitrogenous bases of both chains: A and T (two hydrogen bonds), and G and C (three hydrogen bonds).

from the diet are required for processes that imply rapid cell growth. Nucleotides are therefore considered as conditional essential nutrients [129, 130].

The concentration of free nucleotides in human milk is higher than in bovine milk; thus, infant formulas, which are usually based in cow milk, are supplemented with nucleotides. Several studies have shown that these formulas have beneficial effects in infants, especially in those with severe intrauterine growth retardation. These evidences further suggest that requirements for nucleotides are elevated in infants and young children compared to adults [131].

Foods and food products usually contain nucleic acids as well as a mixture of nitrogenous bases, nucleosides, nucleotides, and other nucleotide-containing biomolecules (NADH, FADH$_2$, coenzyme A, etc.) that can mutually convert or wither in fresh foods such as seafood, vegetables, and fruits. Therefore, establishing the actual capacity of a given food to act as a nucleotide source is not always an easy task [132].

Dietary nucleotides have been proposed to be involved in a number of processes [129–131]:
- Maintaining the structure and growth of the intestinal tract
- Enhancing recovery of the liver and small intestine after a variety of insults
- Keeping the microbiota in the GIT healthy
- Maintaining the humoral as well as the cellular immune system in infants and young children
- Proper functioning of the central nervous system (CNS)
- Modulation of gene expression

5.11.3 Nucleotide disorders

Nucleotides are subjected to a continuous turnover of synthesis and degradation to meet cellular requirements. As mentioned above, nitrogen bases (purines and pyrimidines) may be synthesized *de novo* or recycled by a salvage pathway from normal catabolism. In the case of purines, the end product of complete catabolism is uric acid (Figure 5.10). There are a number of disorders related to nucleotides that can be classified according to the metabolic pathway involved. The severity of some of these disorders illustrates the importance of the synthesis and salvage pathways of nucleotides for human health [133]:
- Purine catabolism disorders
    - Deficiency of myoadenylate deaminase (or muscle adenosine monophosphate deaminase deficiency), the enzyme that converts AMP to inosine and ammonia. This deficiency may be asymptomatic or it may cause exercise-induced myalgias or cramping.
    - Deficiency of adenosine deaminase, which converts adenosine and deoxyadenosine to inosine and deoxyinosine, causes one form of severe combined immunodeficiency.
    - Purine nucleoside phosphorylase deficiency is characterized by immunodeficiency with severe T-cell dysfunction and often neurologic symptoms.
    - Deficiency of xanthine oxidase, the enzyme that catalyzes uric acid production from xanthine and hypoxanthine, may produce precipitation of xanthine in the urine, causing hematuria, urinary colic, and urinary tract infections.
- Purine nucleotide synthesis disorders

- Phosphoribosyl pyrophosphate synthetase superactivity is an X-linked recessive disorder that causes purine overproduction. Excess purine is degraded, resulting in hyperuricemia and gout and neurologic and developmental abnormalities.
    - Adenylosuccinase deficiency, an autosomal recessive disorder, causes profound intellectual disability, autistic behavior, and seizures. There is no effective treatment for this disorder.
- Purine salvage disorders
    - Lesch–Nyhan syndrome is a rare X-linked recessive disorder caused by deficiency of hypoxanthine-guanine phosphoribosyl transferase, which results in failure of the salvage pathway for hypoxanthine and guanine. Purines are degraded to uric acid that predisposes to gout (see below) and its complications. The disease progresses to CNS involvement with intellectual disability, spastic cerebral palsy, involuntary movements, and self-mutilating behavior. CNS dysfunction has no known treatment.
    - Adenine phosphoribosyltransferase deficiency is a rare autosomal recessive disorder that results in the inability to salvage adenine for purine synthesis. Accumulated adenine is oxidized and precipitates in the urinary tract, causing renal colic, frequent infections, and, if diagnosed late, renal failure. Treatment includes dietary purine restriction, high fluid intake, and avoidance of urine alkalinization.
- Uridine monophosphate synthase deficiency (hereditary orotic aciduria), a disorder related to pyrimidine metabolism, produces an accumulation of orotic acid that causes megaloblastic anemia, orotic crystalluria and nephropathy, cardiac malformations, strabismus, and recurrent infections. Treatment is with oral uridine supplementation.

5.11.4 Hyperuricidemia: gout

Hyperuricemia is defined as an elevated uric acid level in the blood. The normal upper limit is 6.8 mg/dL, and symptoms can occur at concentrations above 7 mg/dL. Hyperuricemia can lead to gout and nephrolithiasis (formation of kidney stones) and may be the result of increased production, decreased excretion of uric acid, or a combination of both processes. One of the major causes of increased production of uric acid is a purine-rich diet.

Gout involves accumulation of uric acid in the blood and tissues so that precipitation of urate monohydrate crystals occurs producing a painful inflammation. Precipitation is enhanced in acidic and cold environments, being more likely in peripheral joints, such as the great toe. Gout has a male predominance.

Contrary to some views, the association between hyperuricemia and dietary factors remains unclear because experimental evidence is limited. A high-protein diet

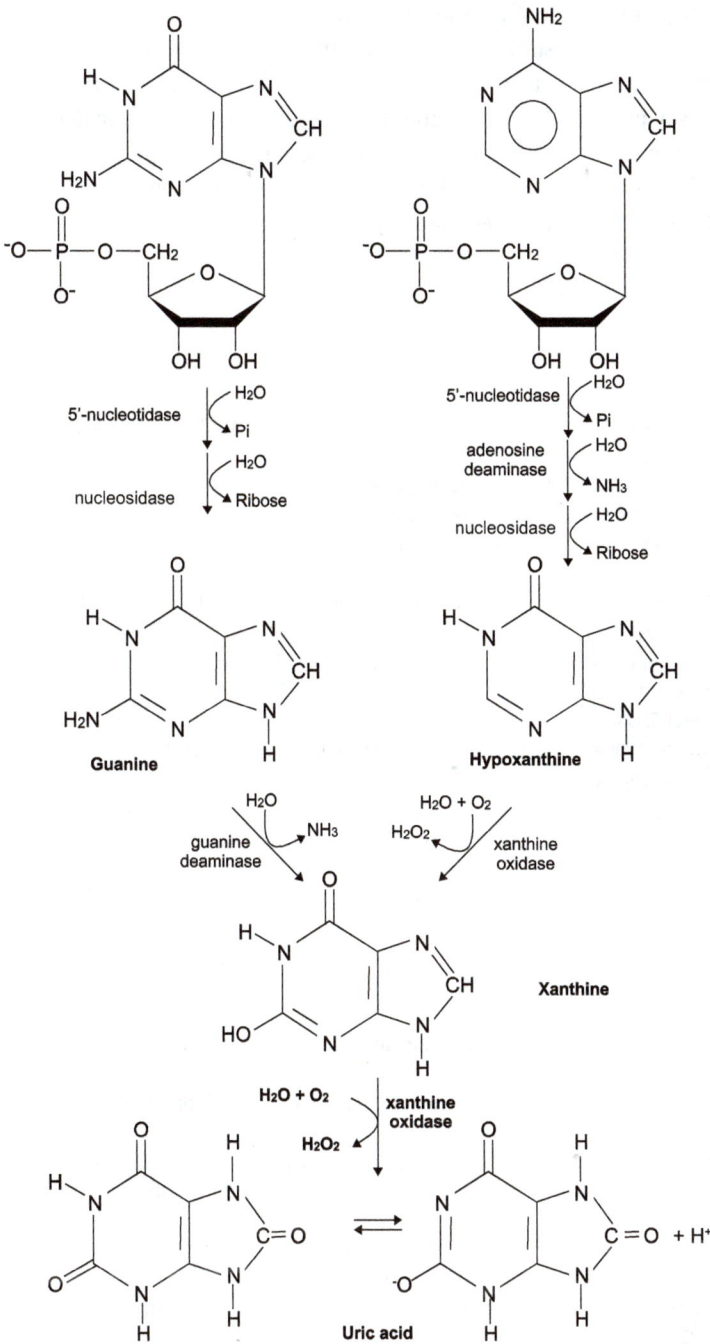

Figure 5.10: Catabolism of purines (adapted from [134]).

typically contains large quantities of purines; however, such diets may also increase excretion of uric acid in urine, resulting in lower serum acid levels. Meat and seafood have been associated with higher uric acid levels in some studies, while dairy intake and consumption of soy products have been found to decrease plasma uric acid levels. No associations between total protein intake and uric acid have been reported. Hyperuricemia has been implicated as an indicator for diseases like metabolic syndrome, diabetes mellitus, CVD, and chronic renal disease [135–137].

References

[1]   Crick FH. On protein synthesis. Symp Soc Exp Biol 1958;12:138–63.

[2]   Ohama T, Inagaki Y, Bessho Y, Osawa S. Evolving genetic code. Proc Jpn Acad Ser B Phys Biol Sci 2008;84:58–74. https://doi.org/10.2183/pjab.84.58.

[3]   Trivedi R, Nagarajaram HA. Intrinsically disordered proteins: An overview. Int J Mol Sci 2022;23:14050. https://doi.org/10.3390/ijms232214050.

[4]   Nelson DL, Cox MM, Lehninger AL. Lehninger principles of biochemistry. New York: W.H. Freeman; 2013.

[5]   Hatfield DL, Gladyshev VN. How selenium has altered our understanding of the genetic code. Mol Cell Biol 2002;22:3565–76. https://doi.org/10.1128/MCB.22.11.3565-3576.2002.

[6]   Vickery HB. The origin of the word protein. Yale J Biol Med 1950;22:387–93.

[7]   IUPAC Gold Book n.d. http://goldbook.iupac.org/ (accessed March 7, 2019).

[8]   Wilkins MR, Sanchez J-C, Gooley AA, Appel RD, Humphery-Smith I, Hochstrasser DF, et al. Progress with proteome projects: Why all proteins expressed by a genome should be identified and how to do it. Biotechnol Genet Eng Rev 1996;13:19–50.

[9]   Aslam B, Basit M, Nisar MA, Khurshid M, Rasool MH. Proteomics: Technologies and their applications. J Chromatogr Sci 2017;55:182–96. https://doi.org/10.1093/chromsci/bmw167.

[10] Baek M, DiMaio F, Anishchenko I, Dauparas J, Ovchinnikov S, Lee GR, et al. Accurate prediction of protein structures and interactions using a three-track neural network. Science 2021;373:871–6. https://doi.org/10.1126/science.abj8754.

[11] Jumper J, Evans R, Pritzel A, Green T, Figurnov M, Ronneberger O, et al. Highly accurate protein structure prediction with AlphaFold. Nature 2021;596:583–9. https://doi.org/10.1038/s41586-021-03819-2.

[12] Baek M, Baker D. Deep learning and protein structure modeling. Nat Methods 2022;19:13–4. https://doi.org/10.1038/s41592-021-01360-8.

[13] Müller S. Special issue: Ribozymes and RNA catalysis. Mol J Synth Chem Nat Prod Chem 2017; 22. https://doi.org/10.3390/molecules22050789.

[14] Ponomarenko EA, Poverennaya EV, Ilgisonis EV, Pyatnitskiy MA, Kopylov AT, Zgoda VG, et al. The size of the human proteome: The width and depth. Int J Anal Chem 2016; 2016. https://doi.org/10.1155/2016/7436849.

[15] Joint Expert Consultation on Protein and Amino Acid Requirements in Human Nutrition, Weltgesundheitsorganisation, FAO, United Nations University, editors. Protein and amino acid requirements in human nutrition: report of a joint WHO/FAO/UNU Expert Consultation ; [Geneva, 9 – 16 April 2002]. Geneva: WHO; 2007.

[16] Dietary reference values for nutrients Summary report. EFSA Support Publ 2017;14:e15121E. https://doi.org/10.2903/sp.efsa.2017.e15121.

[17] Reeds PJ. Dispensable and indispensable amino acids for humans. J Nutr 2000;130:1835S-1840S.

[18] Wu G. Amino acids: Metabolism, functions, and nutrition. Amino Acids 2009;37:1–17. https://doi.org/10.1007/s00726-009-0269-0.

[19] Food and agriculture organization of the United Nations, editor. Dietary protein quality evaluation in human nutrition: report of an FAO expert consultation, 31 March-2 April, 2011, Auckland, New Zealand. Rome: Food and Agriculture Organization of the United Nations; 2013.

[20] Damodaran S, Parkin K, Fennema OR, editors. Fennema's food chemistry. 4th edition. Boca Raton: CRC Press/Taylor & Francis; 2008.

[21] Hoffer LJ. Human protein and amino acid requirements. J Parenter Enter Nutr 2016;40:460–74. https://doi.org/10.1177/0148607115624084.

[22] Berrazaga I, Micard V, Gueugneau M, Walrand S. The role of the anabolic properties of plant-versus animal-based protein sources in supporting muscle mass maintenance: A critical review. Nutrients 2019;11:1825. https://doi.org/10.3390/nu11081825.

[23] Deglaire A, Moughan PJ. Animal models for determining amino acid digestibility in humans – a review. Br J Nutr 2012;108:S273–81. https://doi.org/10.1017/S0007114512002346.

[24] Arentson-Lantz E, Clairmont S, Paddon-Jones D, Tremblay A, Elango R. Protein: A nutrient in focus. Appl Physiol Nutr Metab Physiol Appl Nutr Metab 2015;40:755–61. https://doi.org/10.1139/apnm-2014-0530.

[25] Wu G. Dietary protein intake and human health. Food Funct 2016;7:1251–65. https://doi.org/10.1039/C5FO01530H.

[26] Soultoukis GA, Partridge L. Dietary protein, metabolism, and aging. Ann Rev Biochem 2016;85:5–34. https://doi.org/10.1146/annurev-biochem-060815-014422.

[27] Wu G. Functional amino acids in nutrition and health. Amino Acids 2013;45:407–11. https://doi.org/10.1007/s00726-013-1500-6.

[28] Paudel S, Wu G, Wang X. Amino acids in cell signaling: Regulation and function. In: Wu G, editor. Amino Acids Nutr. Health, vol. 1332. Cham: Springer International Publishing; 2021, p. 17–33. https://doi.org/10.1007/978-3-030-74180-8_2.

[29] Food and Agriculture Organization of the United Nations, editor. The future of food and agriculture: Trends and challenges. Rome: Food and Agriculture Organization of the United Nations; 2017.

[30] The State of Food Security and Nutrition in the World 2020. FAO, IFAD, UNICEF, WFP and WHO; 2020. https://doi.org/10.4060/ca9692en.

[31] FAOSTAT n.d. http://www.fao.org/faostat/en/#data/FBS (accessed May 13, 2023).

[32] Average protein content in g/100 g and % of food energy from protein in plant-derived raw foods (edible parts only) | Knowledge for policy n.d. https://knowledge4policy.ec.europa.eu/health-promotion-knowledge-gateway/dietary-protein-plant-1b_en (accessed May 20, 2023).

[33] Osborne TB. The proteins of the wheat kernel. Carnegie institution of Washington; 1907.

[34] Fermented cereals a global perspective. Chapter 1. n.d. http://www.fao.org/3/x2184e/x2184e04.htm (accessed January 2, 2020).

[35] Ortolan F, Steel CJ. Protein characteristics that affect the quality of vital wheat gluten to be used in baking: A review: Vital wheat gluten quality in baking. Compr Rev Food Sci Food Saf 2017;16:369–81. https://doi.org/10.1111/1541-4337.12259.

[36] Schalk K, Lexhaller B, Koehler P, Scherf KA. Isolation and characterization of gluten protein types from wheat, rye, barley and oats for use as reference materials. PLoS ONE 2017;12. https://doi.org/10.1371/journal.pone.0172819.

[37] Ortiz-Sánchez JP, Cabrera-Chávez F, Calderón de la Barca AM. Maize prolamins could induce a gluten-like cellular immune response in some celiac disease patients. Nutrients 2013;5:4174–83. https://doi.org/10.3390/nu5104174.

[38] Cavazos A, Gonzalez de Mejia E. Identification of bioactive peptides from cereal storage proteins and their potential role in prevention of chronic diseases: Bioactive peptides from cereal proteins. Compr Rev Food Sci Food Saf 2013;12:364–80. https://doi.org/10.1111/1541-4337.12017.

[39] Malaguti M, Dinelli G, Leoncini E, Bregola V, Bosi S, Cicero AFG, et al. Bioactive peptides in cereals and legumes: Agronomical, biochemical and clinical aspects. Int J Mol Sci 2014;15:21120–35. https://doi.org/10.3390/ijms151121120.

[40] Tessari P, Lante A, Mosca G. Essential amino acids: Master regulators of nutrition and environmental footprint? Sci Rep 2016;6:26074.

[41] N.d. http://www.fao.org/es/faodef/fdef04e.htm (accessed April 18, 2019).

[42] Myers SS, Zanobetti A, Kloog I, Huybers P, Leakey ADB, Bloom AJ, et al. Increasing CO2 threatens human nutrition. Nature 2014;510:139.

[43] Ranganathan J, Vennard D, Waite R, Dumas P, Lipinski B, Searchinger T. Shifting diets for a sustainable food future. Installment 11 of Creating a Sustainable Food Future. Washington, DC: World Resources Institute. 2016.

[44] Willett W, Rockström J, Loken B, Springmann M, Lang T, Vermeulen S, et al. Food in the anthropocene: The EAT–Lancet commission on healthy diets from sustainable food systems. The Lancet 2019;393:447–92. https://doi.org/10.1016/S0140-6736(18)31788-4.

[45] FAO, IFAD, UNICEF, WFP and WHO, editor. The state of food security and nutrition in the World 2020. Transforming food systems for affordable healthy diets. Rome: FAO; 2020.

[46] Average protein content in g/100 g and % of food energy from protein in animal-derived raw foods | Knowledge for policy n.d. https://knowledge4policy.ec.europa.eu/health-promotion-knowledge-gateway/dietary-protein-animal-1a_en (accessed May 20, 2023).

[47] Home | CODEXALIMENTARIUS FAO-WHO n.d. http://www.fao.org/fao-who-codexalimentarius/home/en/ (accessed April 20, 2019).

[48] Gifford CL, O'Connor LE, Campbell WW, Woerner DR, Belk KE. Broad and inconsistent muscle food classification is problematic for dietary guidance in the U.S. Nutrients 2017;9. https://doi.org/10.3390/nu9091027.

[49] Listrat A, Lebret B, Louveau I, Astruc T, Bonnet M, Lefaucheur L, et al. How muscle structure and composition influence meat and flesh quality. Sci World J 2016; 2016. https://doi.org/10.1155/2016/3182746.

[50] Malva A, Albenzio M, Santillo A, Russo D, Figliola L, Caroprese M, et al. Methods for extraction of muscle proteins from meat and fish using denaturing and nondenaturing solutions. J Food Qual 2018;2018:1–9. https://doi.org/10.1155/2018/8478471.

[51] Belitz H-D, Grosch W, Schieberle P. Food chemistry. 4th rev. and extended edition. Berlin: Springer; 2009.

[52] IARC Working Group on the Evaluation of Carcinogenic Risks to Humans, International Agency for Research on Cancer. Red meat and processed meat. 2018.

[53] Wang Z, Bergeron N, Levison BS, Li XS, Chiu S, Jia X, et al. Impact of chronic dietary red meat, white meat, or non-meat protein on trimethylamine N-oxide metabolism and renal excretion in healthy men and women. Eur Heart J 2019;40:583–94. https://doi.org/10.1093/eurheartj/ehy799.

[54] Food-borne zoonotic diseases. Eur Food Saf Auth n.d. https://www.efsa.europa.eu/en/topics/topic/food-borne-zoonotic-diseases (accessed April 19, 2019).

[55] Henchion M, Hayes M, Mullen AM, Fenelon M, Tiwari B. Future protein supply and demand: Strategies and factors influencing a sustainable equilibrium. Foods 2017;6. https://doi.org/10.3390/foods6070053.

[56] Milk | Definition of milk in English by Oxford Dictionaries. Oxf Dictionaries Engl n.d. https://en.oxforddictionaries.com/definition/milk (accessed April 20, 2019).

[57] Muehlhoff E, Bennett A, MacMahon D, Milk and dairy products in human nutrition. Rome: Food and Agriculture Organization of the United Nations; 2013.

[58] Hochwallner H, Schulmeister U, Swoboda I, Spitzauer S, Valenta R. Cow's milk allergy: From allergens to new forms of diagnosis, therapy and prevention methods. San Diego, CA 2014;66:22–33. https://doi.org/10.1016/j.ymeth.2013.08.005.

[59]   Rangel AH do N, Sales DC, Urbano SA, Galvão Júnior JGB, Andrade Neto JCD, Macêdo CDS. Lactose intolerance and cow's milk protein allergy. Food Sci Technol 2016;36:179–87. https://doi.org/10.1590/1678-457X.0019.

[60]   Phillip Clauer. The avian egg. Penn State Ext n.d. https://extension.psu.edu/the-avian-egg (accessed April 25, 2019).

[61]   Miranda J, Anton X, Redondo-Valbuena C, Roca-Saavedra P, Rodriguez J, Lamas A, et al. Egg and egg-derived foods: Effects on human health and use as functional foods. Nutrients 2015;7:706–29. https://doi.org/10.3390/nu7010706.

[62]   Hunton P. Research on eggshell structure and quality: An historical overview. Rev Bras Ciênc Avícola 2005;7:67–71. https://doi.org/10.1590/S1516-635X2005000200001.

[63]   Lutter CK, Iannotti LL, Stewart CP. The potential of a simple egg to improve maternal and child nutrition. Matern Child Nutr 2018;14:e12678. https://doi.org/10.1111/mcn.12678.

[64]   Mann K. The chicken egg white proteome. Proteomics 2007;7:3558–68. https://doi.org/10.1002/pmic.200700397.

[65]   Stadelman WJ. EGGS | Structure and composition. Encycl Food Sci Nutr Elsevier; 2003, p. 2005–9. https://doi.org/10.1016/B0-12-227055-X/00387-4.

[66]   Caubet J-C, Wang J. Current understanding of egg allergy. Pediatr Clin North Am 2011;58:427–43. https://doi.org/10.1016/j.pcl.2011.02.014.

[67]   Ecy L-C, Powrie WD, Nakai S. The chemistry of eggs and egg products. In: Stadelman WJ, Cotteril OJ, editors. Science and technology. The Haworth Press. Inc., NY; 1995.

[68]   Iannotti LL, Lutter CK, Stewart CP, Gallegos Riofrío CA, Malo C, Reinhart G, et al. Eggs in early complementary feeding and child growth: A randomized controlled trial. Pediatrics 2017;140. https://doi.org/10.1542/peds.2016-3459.

[69]   Contribution of terrestrial animal source food to healthy diets for improved nutrition and health outcomes. FAO; 2023. https://doi.org/10.4060/cc3912en.

[70]   FAO, editor. Contributing to food security and nutrition for all. Rome: 2016.

[71]   Khalili Tilami S, Sampels S. Nutritional value of fish: Lipids, proteins, vitamins, and minerals. Rev Fish Sci Aquac 2018;26:243–53. https://doi.org/10.1080/23308249.2017.1399104.

[72]   Seafood contaminants – Marine – Environment – European Commission n.d. http://ec.europa.eu/environment/marine/good-environmental-status/descriptor-9/index_en.htm (accessed April 25, 2019).

[73]   Soriano JM, Rubini A, Morales-Suarez-Varela M, Merino-Torres JF, Silvestre D. Aflatoxins in organs and biological samples from children affected by kwashiorkor, marasmus and marasmic-kwashiorkor: A scoping review. Toxicon 2020;185:174–83. https://doi.org/10.1016/j.toxicon.2020.07.010.

[74]   Dhaked R, Gupta CM. A critical review of malnutrition and its management. Int J Med Health Res 2016;2:43–7.

[75]   Semba RD. The rise and fall of protein malnutrition in global health. Ann Nutr Metab 2016;69:79–88. https://doi.org/10.1159/000449175.

[76]   Badui S. Química De Los Alimentos. Pearson Educación de México, SA de CV; 2006.

[77]   Kinsella JE, Damodaran S, German B. New protein foods seed storage proteins. Academic Press, New York; 1985, p. 107–79.

[78]   Foegeding EA, Davis JP. Food protein functionality: A comprehensive approach. Food Hydrocoll 2011;25:1853–64. https://doi.org/10.1016/j.foodhyd.2011.05.008.

[79]   Yada RY, editor. Proteins in food processing. Boca Raton : Cambridge, Eng: CRC Press ; Woodhead Pub; 2004.

[80]   Borlaf M, Moreno R. Colloidal sol-gel: A powerful low-temperature aqueous synthesis route of nanosized powders and suspensions. Open Ceram 2021;8:100200. https://doi.org/10.1016/j.oceram.2021.100200.

[81] Gómez-Guillén MC, Giménez B, López-Caballero ME, Montero MP. Functional and bioactive properties of collagen and gelatin from alternative sources: A review. Food Hydrocoll 2011;25:1813–27. https://doi.org/10.1016/j.foodhyd.2011.02.007.

[82] Tavano OL. Protein hydrolysis using proteases: An important tool for food biotechnology. J Mol Catal B Enzym 2013;90:1–11. https://doi.org/10.1016/j.molcatb.2013.01.011.

[83] Wu S-L, Ding D, Fang A-P, Chen P-Y, Chen S, Jing L-P, et al. Growth, gastrointestinal tolerance and stool characteristics of healthy term infants fed an infant formula containing hydrolyzed whey protein (63%) and intact casein (37%): A randomized clinical trial. Nutrients 2017;9. https://doi.org/10.3390/nu9111254.

[84] Gomaa AM. Application of enzymes in brewing. J Nutri Food Sci Forecast 2018; 1002.

[85] Pomeranz Y. Functional properties of food components. 2nd edition. San Diego: Academic Press; 1991.

[86] González-Pérez S, Arellano JB. Vegetable protein isolates. Handb Hydrocoll Elsevier; 2009, p. 383–419. https://doi.org/10.1533/9781845695873.383.

[87] Haque MA, Adhikari B. Drying and denaturation of proteins in spray drying process. Handb Ind Dry 2015:971–85.

[88] Ciurzyńska A, Lenart A. Freeze-drying-application in food processing and biotechnology – a review. Pol J Food Nutr Sci 2011;61:165–71.

[89] Afolabi IS. Moisture migration and bulk nutrients interaction in a drying food systems: A review. Food Nutr Sci 2014;05:692–714. https://doi.org/10.4236/fns.2014.58080.

[90] Mediani A, Hamezah HS, Jam FA, Mahadi NF, Chan SXY, Rohani ER, et al. A comprehensive review of drying meat products and the associated effects and changes. Front Nutr 2022;9:1057366. https://doi.org/10.3389/fnut.2022.1057366.

[91] Spinnler H-E. Flavors from amino acids. In: Food Flavors, vol. 20116950. CRC Press; 2011, p. 121–36. https://doi.org/10.1201/b11187-7.

[92] Mas A, Guillamon JM, Torija MJ, Beltran G, Cerezo AB, Troncoso AM, et al. Bioactive compounds derived from the yeast metabolism of aromatic amino acids during alcoholic fermentation. BioMed Res Int 2014;2014:1–7. https://doi.org/10.1155/2014/898045.

[93] Amit SK, Uddin MdM, Rahman R, Islam SMR, Khan MS. A review on mechanisms and commercial aspects of food preservation and processing. Agric Food Secur 2017;6:51. https://doi.org/10.1186/s40066-017-0130-8.

[94] Suzzi G, Torriani S. Editorial: Biogenic amines in foods. Front Microbiol 2015;6. https://doi.org/10.3389/fmicb.2015.00472.

[95] Linares DM, del Río B, Ladero V, Martínez N, Fernández M, Martín MC, et al. Factors influencing biogenic amines accumulation in dairy products. Front Microbiol 2012;3. https://doi.org/10.3389/fmicb.2012.00180.

[96] Nisman B. The Stickland reaction. Bacteriol Rev 1954;18:16–42.

[97] Rao AK, Sobel J, Chatham-Stephens K, Luquez C. Clinical guidelines for diagnosis and treatment of botulism, 2021. MMWR Recomm Rep 2021;70:1–30. https://doi.org/10.15585/mmwr.rr7002a1.

[98] Dressler D. Clinical applications of botulinum toxin. Curr Opin Microbiol 2012;15:325–36. https://doi.org/10.1016/j.mib.2012.05.012.

[99] Enterocytes – MeSH – NCBI n.d. https://www.ncbi.nlm.nih.gov/mesh/68020895 (accessed May 3, 2019).

[100] Bröer S, Bröer A. Amino acid homeostasis and signalling in mammalian cells and organisms. Biochem J 2017;474:1935–63. https://doi.org/10.1042/BCJ20160822.

[101] Rajendran V, Kalita P, Shukla H, Kumar A, Tripathi T. Aminoacyl-tRNA synthetases: Structure, function, and drug discovery. Int J Biol Macromol 2018;111:400–14. https://doi.org/10.1016/j.ijbiomac.2017.12.157.

[102] Berg JM, Tymoczko JL, Stryer Lubert. Biochemistry. Basingstoke: W.H. Freeman; 2012.

[103] Liang J, Han Q, Tan Y, Ding H, Li J. Current advances on structure-function relationships of pyridoxal 5′-phosphate-dependent enzymes. Front Mol Biosci 2019;6. https://doi.org/10.3389/fmolb.2019.00004.

[104] Kunutsor SK, Bakker SJL, Kootstra-Ros JE, Blokzijl H, Gansevoort RT, Dullaart RPF. Inverse linear associations between liver aminotransferases and incident cardiovascular disease risk: The PREVEND study. Atherosclerosis 2015;243:138–47. https://doi.org/10.1016/j.atherosclerosis.2015.09.006.

[105] Fiermonte G, Palmieri L, Todisco S, Agrimi G, Palmieri F, Walker JE. Identification of the mitochondrial glutamate transporter. Bacterial expression, reconstitution, functional characterization, and tissue distribution of two human isoforms. J Biol Chem 2002;277:19289–94. https://doi.org/10.1074/jbc.M201572200.

[106] Keller RW, Bailey JL, Wang Y, Klein JD, Sands JM. Urea transporters and sweat response to uremia. Physiol Rep 2016;4:e12825. https://doi.org/10.14814/phy2.12825.

[107] Turnbull JL, Adams HN, Gorard DA. Review article: The diagnosis and management of food allergy and food intolerances. Aliment Pharmacol Ther 2015;41:3–25. https://doi.org/10.1111/apt.12984.

[108] Cox AL, Sicherer SH. Classification of adverse food reactions. J Food Allergy 2020;2:3–6. https://doi.org/10.2500/jfa.2020.2.200022.

[109] Valenta R, Hochwallner H, Linhart B, Pahr S. Food allergies: The basics. Gastroenterology 2015;148:1120-1131.e4. https://doi.org/10.1053/j.gastro.2015.02.006.

[110] Mills ENC, Moreno J, Sancho A, Jenkins JA, Wichers HJ. Processing approaches to reducing allergenicity in proteins. In: Yada RY, editor. Proteins food process. Cambridge: Woodhead; 2004, p. 396–418.

[111] EFSA Panel on Dietetic Products N and A (NDA). Scientific Opinion on the evaluation of allergenic foods and food ingredients for labelling purposes. EFSA J 2014;12:3894.

[112] Council EP and Regulation (EU) No 1169/2011 of the European Parliament and of the Council of 25 October 2011 on the provision of food information to consumers, amending Regulations (EC) No 1924/2006 and (EC) No 1925/2006 of the European Parliament and of the Council, and repealing Commission Directive 87/250/EEC, Council Directive 90/496/EEC, Commission Directive 1999/10/EC, Directive 2000/13/EC of the European Parliament and of the Council, Commission Directives 2002/67/EC and 2008/5/EC and Commission Regulation (EC) No 608/2004. J Eur Union 2011;50:18–63.

[113] Sathe SK, Liu C, Zaffran VD. Food allergy. Ann Rev Food Sci Technol 2016;7:191–220. https://doi.org/10.1146/annurev-food-041715-033308.

[114] Sandlers Y. Amino acids profiling for the diagnosis of metabolic disorders. Clin Biochem Fundam Med Lab Sci Work Title, IntechOpen; 2019. https://doi.org/10.5772/intechopen.84672.

[115] Chinsky JM, Singh R, Ficicioglu C, van Karnebeek CDM, Grompe M, Mitchell G, et al. Diagnosis and treatment of tyrosinemia type I: A US and Canadian consensus group review and recommendations. Genet Med 2017;19. https://doi.org/10.1038/gim.2017.101.

[116] Najafi R, Mostofizadeh N, Hashemipour M. A case of tyrosinemia type III with status epilepticus and mental retardation. Adv Biomed Res 2018;7:7. https://doi.org/10.4103/2277-9175.223740.

[117] Lee N, Kim D. Toxic metabolites and inborn errors of amino acid metabolism: What one informs about the other. Metabolites 2022;12:527. https://doi.org/10.3390/metabo12060527.

[118] Leonard JV. Disorders of the urea cycle and related enzymes. In: Fernandes J, Saudubray J-M, van den Berghe G, Walter JH, editors. Inborn Metab Dis Diagn Treat. Berlin, Heidelberg: Springer Berlin Heidelberg; 2006, p. 263–72. https://doi.org/10.1007/978-3-540-28785-8_20.

[119] Matsumoto S, Häberle J, Kido J, Mitsubuchi H, Endo F, Nakamura K. Urea cycle disorders – update. J Hum Genet 2019;64:833–47. https://doi.org/10.1038/s10038-019-0614-4.

[120] Díez-Fernández C, Hu L, Cervera J, Häberle J, Rubio V. Understanding carbamoyl phosphate synthetase (CPS1) deficiency by using the recombinantly purified human enzyme: Effects of CPS1

mutations that concentrate in a central domain of unknown function. Mol Genet Metab 2014;112:123–32. https://doi.org/10.1016/j.ymgme.2014.04.003.

[121] Diez-Fernandez C, Häberle J. Targeting CPS1 in the treatment of Carbamoyl phosphate synthetase 1 (CPS1) deficiency, a urea cycle disorder. Expert Opin Ther Targets 2017;21:391–9. https://doi.org/10.1080/14728222.2017.1294685.

[122] Ah Mew N, Simpson KL, Gropman AL, Lanpher BC, Chapman KA, Summar ML. Urea cycle disorders overview. In: Adam MP, Ardinger HH, Pagon RA, Wallace SE, Bean LJ, Stephens K, et al., editors. GeneReviews®, Seattle (WA): University of Washington, Seattle; 1993.

[123] Nagamani SCS, Lichter-Konecki U. Inborn errors of urea synthesis. Swaimans Pediatr Neurol Elsevier; 2017, p. 298–304. https://doi.org/10.1016/B978-0-323-37101-8.00038-2.

[124] Engel K, Höhne W, Häberle J. Mutations and polymorphisms in the human argininosuccinate synthetase (ASS1) gene. Hum Mutat 2009;30:300–7. https://doi.org/10.1002/humu.20847.

[125] Nagamani SCS, Erez A, Lee B. Argininosuccinate lyase deficiency. In: Adam MP, Ardinger HH, Pagon RA, Wallace SE, Bean LJ, Stephens K, et al., editors. GeneReviews®. Seattle (WA): University of Washington, Seattle; 1993.

[126] Bélanger-Quintana A, Arrieta Blanco F, Barrio-Carreras D, Bergua Martínez A, Cañedo Villarroya E, García-Silva MT, et al. Recommendations for the diagnosis and therapeutic management of hyperammonaemia in paediatric and adult patients. Nutrients 2022;14:2755. https://doi.org/10.3390/nu14132755.

[127] Watson JD, Crick FH. Molecular structure of nucleic acids; a structure for deoxyribose nucleic acid. Nature 1953;171:737–8.

[128] Panni S, Lovering RC, Porras P, Orchard S. Non-coding RNA regulatory networks. Biochim Biophys Acta BBA – Gene Regul Mech 2020;1863:194417. https://doi.org/10.1016/j.bbagrm.2019.194417.

[129] Sanchez-Pozo A, Gil A. Nucleotides as semiessential nutritional components. Br J Nutr 2002;87:S135–7.

[130] Hess JR, Greenberg NA. The role of nucleotides in the immune and gastrointestinal systems. Nutr Clin Pract 2012;27:281–94. https://doi.org/10.1177/0884533611434933.

[131] Stein HH, Mateo CD. Nucleotides in nutrition: The importance in infant and childhood diets. Nutr. Biotechnol. Feed Food Ind. Proc. Alltechs 21st Annu. Symp. Lexingt. Ky. USA 22–25 May 2005, Alltech UK, p. 147–50.

[132] Ding T, Song G, Liu X, Xu M, Li Y. Nucleotides as optimal candidates for essential nutrients in living organisms: A review. J Funct Foods 2021;82:104498. https://doi.org/10.1016/j.jff.2021.104498.

[133] Overview of purine and pyrimidine metabolism disorders – pediatrics – MSD manual professional edition n.d. https://www.msdmanuals.com/professional/pediatrics/inherited-disorders-of-metabolism/overview-of-purine-and-pyrimidine-metabolism-disorders (accessed May 18, 2019).

[134] Benn CL, Dua P, Gurrell R, Loudon P, Pike A, Storer RI, et al. Physiology of hyperuricemia and urate-lowering treatments. Front Med 2018;5:160. https://doi.org/10.3389/fmed.2018.00160.

[135] Villegas R, Xiang Y-B, Elasy T, Xu WH, Cai H, Cai Q, et al. Purine-rich foods, protein intake, and the prevalence of hyperuricemia: The Shanghai Men's Health Study. Nutr Metab Cardiovasc Dis 2012;22:409–16. https://doi.org/10.1016/j.numecd.2010.07.012.

[136] Dalbeth N, Gosling AL, Gaffo A, Abhishek A. Gout. The Lancet 2021;397:1843–55. https://doi.org/10.1016/S0140-6736(21)00569-9.

[137] George C, Minter DA. Hyperuricemia. StatPearls, Treasure Island (FL): StatPearls Publishing; 2023.

6 Vitamins

6.1 Concept and classification

Vitamins are defined as organic substances that are required in small quantities in the diet of higher animals (vertebrates other than fish), since their organisms are unable to synthesize them, as a consequence, their deficiency can result in specific diseases or symptoms. Vitamins have no structural relationship with one another and are not energetic nutrients for humans, that is, they are not degraded to simpler molecules in catabolic reactions in order to obtain metabolically usable energy.

Vitamins can be classified according to several criteria, thus, there are **water-soluble vitamins** that include vitamin C (L-ascorbic acid, Chapter 3, Section 3.3) and complex B (that includes vitamins B_1, B_2, B_3, B_5, B_6, B_7, B_9, and B_{12}), and **fat-soluble vitamins**, known as A, D, E, and K. This classification has not only chemical but also physiological implications, as the solubility of vitamins determines their form of transport, storage, and excretion. According to the *in vivo* function, we can distinguish between vitamins that act as precursors of enzyme cofactors (complex B, vitamin K), antioxidants (vitamins A, C, and E), and those with hormonal roles, like vitamin D.

6.2 History

More than 3000 years ago, the ancients (old Egyptians, the Babylonians, the Greeks, and the Arabs) not only knew night blindness but also used animal liver for successfully cure this disease [1]. Obviously, they did not know vitamins, but this demonstrates that disorders caused by a lack of certain nutrients in the diet, the so-called deficiency diseases, have been known for a long time.

In 1753, Lind, a Royal Navy surgeon, discovered that citrus fruits (oranges, lemons, and limes) could prevent scurvy; however, major advances in vitamins began only in the nineteenth century, when the germ theory of disease, proposed by Pasteur and Koch, was dominant and dogma held that only four nutritional factors were essential: proteins, carbohydrates, fats, and minerals. During the last decades of the nineteenth century and the beginning of the twentieth century, scientists recognized scurvy, beriberi, rickets, pellagra, and xerophthalmia as specific nutrient deficiencies, rather than diseases due to infections or toxins. Experiments with animal models like mice, dogs, and guinea pigs played a fundamental role in this research. By mid-twentieth century, the role of vitamins as essential nutrients had been established and all of them had been characterized, purified, and, in some cases, synthesized. Some milestones in the discovery of vitamins, their physiological roles, and chemical compositions are [2–4]:

– In the early 1880s, Max Kassowitz, a pediatrician in Vienna, introduced cod liver oil and phosphorus for the treatment of rickets in children.

https://doi.org/10.1515/9783111111872-006

- Between 1896 and 1904, Japanese physician Masamichi Mori concluded that xerophthalmia was not an infectious disease but was caused by the lack of fat in the diet.
- In the last decade of the nineteenth century, Christiaan Eijkman observed that polyneuritis, the equivalent of human beriberi, developed in chickens fed with rice devoid of bran. In 1897, he showed that beriberi could be prevented in rice-eating populations from South-East Asia by not removing the bran from the rice. The initial Eijkman's idea was that bran had a substance that neutralized a toxin contained in rice starch; however, Gerrit Grijns concluded that rice bran contained a vital component in human diet and convinced Eijkman that beriberi was caused by a nutritional deficiency.
- In 1906, Hopkins proposed that, in addition to sugars, fats, proteins, and minerals, other compounds are required in the diet. This idea had been previously hinted by scientists like Nicolai Lunin, Gustav von Bunge, Carl A. Socin, and Jean Baptiste Dumas.
- In 1907, Holst and Frölich showed that fresh cabbage or fresh potatoes prevented scurvy in guinea pigs.
- In articles published in 1909 and 1911, Wilhem Stepp concludes that a fat-soluble substance was essential for life after a series of experiments performed with mice.
- In 1912, Hopkins says that milk contains "accessory factors" that supported life. In the same year, Casimir Funk proposed the term "vitamine" for these accessory factors, as he presumed that these compounds, essential for life (*vita* in Latin), were amines. In 1920, Jack Drummond proposed the term to be shortened to "vitamin." McCollum had previously suggested a nomenclature of "fat-soluble A" and "water-soluble B," thus recognizing the existence of at least two different types of these nutrients.
- In 1926, Jansen and Donath crystallized "water-soluble B" and identified it as thiamin(e), whose chemical description and synthesis was achieved by Robert Williams in 1936.
- In 1928, Albert Szent-Györgi isolated a substance that was later confirmed as vitamin C. Haworth described the chemical structure of vitamin C and its synthesis in 1933.
- In June 1931, the first International Conference on Vitamin Standards was held in London. In the same year, Paul Karrer described the chemical structure of vitamin A.
- In 1936, Windaus described the structures of vitamins D_2 and D_3.
- Holmes and Corbet crystallized vitamin A in 1937 and Isler synthesized it in 1947.
- In 1938, Elvehjem isolated nicotinamide and showed that it was the "anti-pellagra vitamin."

Vitamins continue to be a very active field of research nowadays and many of their biochemical and physiological aspects are yet to be established.

6.3 Structure and *in vivo* function of water-soluble vitamins

6.3.1 Vitamin B₁ (thiamin or thiamine)

6.3.1.1 Structure

Thiamin is chemically defined by IUPAC as 2-[3-[(4-amino-2-methylpyrimidin-5-yl) methyl]-4-methyl-1,3-thiazol-3-ium-5-yl]ethanol. It is structurally composed of a thiazole and a pyrimidine ring linked by a methylene group [5] (Figure 6.1A).

(A)

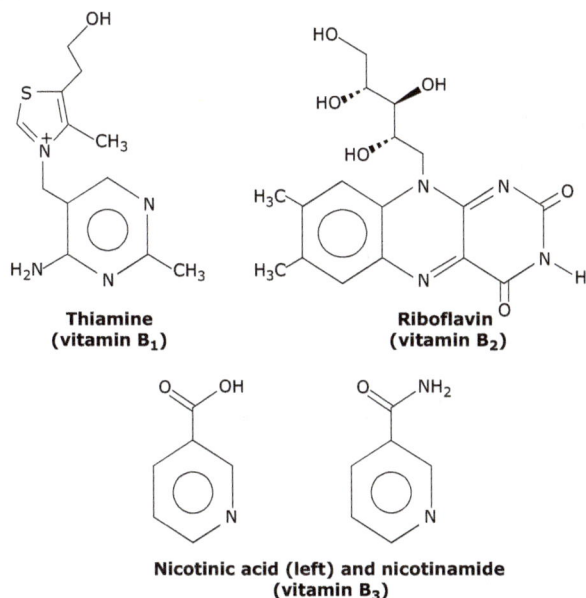

Thiamine (vitamin B₁)

Riboflavin (vitamin B₂)

Nicotinic acid (left) and nicotinamide (vitamin B₃)

Figure 6.1: Structures of water-soluble vitamins. (A) Vitamins B₁ (thiamine), B₂ (riboflavin), and B₃ (nicotinamide). (B) Vitamins B₅ (pantothenic acid) and B₆ (pyridoxine). (C) Vitamins B₇/H (biotin) and B₉ (folic acid). (D) Vitamin B₁₂.
Positions 5, 6, 7, and 8 of folic acid are reduced by NADPH *in vivo* to produce the active form 5,6,7,8-tetrahydrofolate; one-carbon units bind to positions 5, 10, or both. Formulae were drawn with structural information retrieved from the National Center for Biotechnology Information PubChem Database (accessed on June 10, 2023):
Thiamine, CID = 1130, https://pubchem.ncbi.nlm.nih.gov/compound/thiamine
Riboflavin, CID = 493570, https://pubchem.ncbi.nlm.nih.gov/compound/riboflavin
Nicotinic acid, CID = 938, https://pubchem.ncbi.nlm.nih.gov/compound/nicotinic-acid
Nicotinamide, CID = 936, https://pubchem.ncbi.nlm.nih.gov/compound/nicotinamidePantothenic acid, CID = 6613, https://pubchem.ncbi.nlm.nih.gov/compound/pantothenic-acid
Pyridoxine, CID = 1054, https://pubchem.ncbi.nlm.nih.gov/compound/pyridoxine
Biotin, CID = 171548, https://pubchem.ncbi.nlm.nih.gov/compound/biotin
Folic acid, CID = 6037, https://pubchem.ncbi.nlm.nih.gov/compound/6037
Vitamin B12: Cyanocobalamine, CID = 54605677, https://pubchem.ncbi.nlm.nih.gov/compound/54605677.

(B)

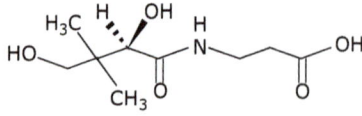

**Pantothenic acid
(vitamin B)₅**

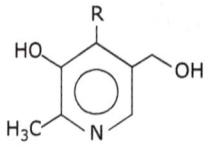

R= CH$_2$OH, pyridoxine
R= CH$_2$-NH$_2$, pyridoxamine
R= CHO, pyridoxal

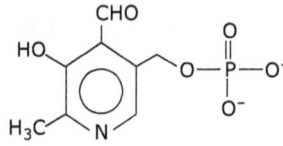

Pyridoxal-phosphate
(active form *in vivo*)

Vitamin B$_6$

(C)

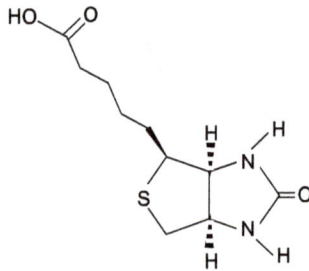

**Biotin
(Vitamin B$_7$/H)**

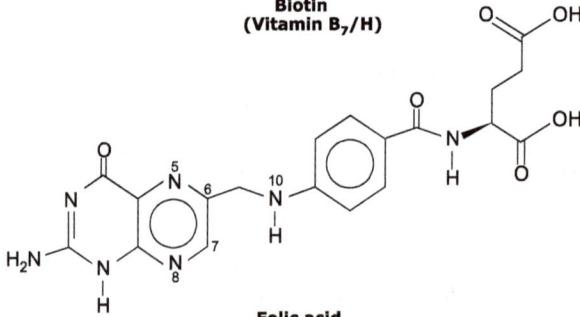

**Folic acid
(Vitamin B$_9$)**

Figure 6.1 (continued)

(D)

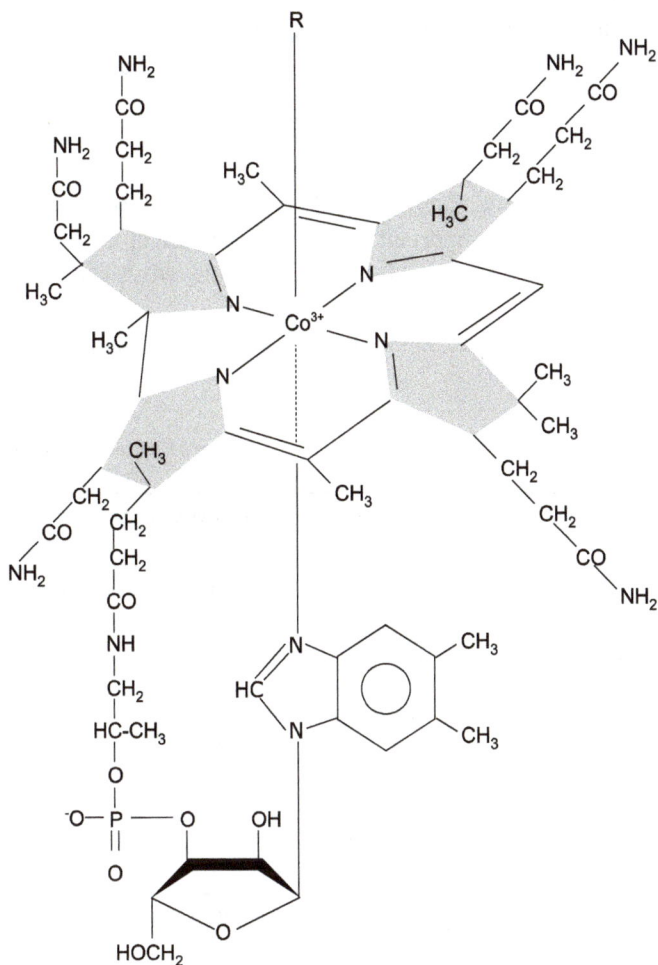

Figure 6.1 (continued)

6.3.1.2 *In vivo* function

In living organisms, thiamine is present both in free form and also as phosphorylated derivatives: thiamine monophosphate, diphosphate, triphosphate, and adenosine thiamine triphosphate. Thiamine diphosphate (TDP, also known as thiamine pyrophosphate) in cells occurs in the largest concentrations (70–90% of the total). The role of thiamine and its derivatives in the metabolism of the cells can be summarized in three major points [5–7]:

–   It is a cofactor of enzymes involved in bioenergetics (pyruvate dehydrogenase, α-ketoglutarate dehydrogenase), amino acid metabolism (α-ketoacid dehydrogenase), and transformation of various carbohydrates, including pentoses, neces-

sary for the synthesis of nucleotides (transketolase) (Table 6.1). TPP is also a cofactor of pyruvate decarboxylase (PDC, EC 4.1.1.1), a yeast enzyme that decarboxylates pyruvate to acetaldehyde, involved in alcoholic fermentation and, thus, with applications in food and beverage industry.
– It is involved in allosteric regulation of enzymes and transmission of nerve signals in synapses.
– Many studies suggest that thiamine, its phosphorylated derivatives, and thiamine-dependent enzymes play an important role in the response of microorganisms, animals, and plants to environmental factors like oxidative stress and pathogens.

TDP may be further phosphorylated to TTP whose role is still incompletely understood [5, 8].

Table 6.1: Human enzyme complexes that require thiamine pyrophosphate as a cofactor (adapted from [6]).

Enzyme	Abbreviation & EC number	Metabolic pathway and/or catalyzed reaction
Pyruvate dehydrogenase	PDH, EC 1.2.4.1	Oxidative decarboxylation of pyruvate to acetyl-CoA
2-oxoglutarate dehydrogenase	OGDH, EC 1.2.4.2	Tricarboxylic acid cycle: oxidative decarboxylation of 2-oxoglutarate
Branched chain 2-oxoacid dehydrogenase	BCOADH, EC 1.2.4.4	Catabolism of branched chain amino acids (valine, leucine and isoleucine): oxidative decarboxylation of the branched chain 2-oxo acids
Transketolase (glycolaldehyde transferase)	TK, EC 2.2.1.1	Penthose phosphate pathway
2-hydroxyphytanoyl-CoA lyase (2-hydroxyacyl-CoA lyase)	HACL, EC 4.1.2.n2	Peroxisomal alpha oxidation of 3-methyl-branched fatty acids

6.3.1.3 Sources and bioavailability

Thiamine is synthesized by microorganisms, fungi, and plants, but is an essential nutrient in animals, that require an appropriate supply of thiamine for the proper functioning of nervous, cardiovascular, and locomotive systems. Thiamine has also been reported to be very important for cancer cells development, although the relationship between vitamin B_1 and the initiation and development of cancer remains unknown [6].

Thiamine is stored in the body for a short time before it is excreted; therefore, a regular dietary intake is necessary to maintain proper blood levels. Food sources rich in thiamine include whole-grains, brown rice, pork, poultry, soybeans, nuts, dried beans, and peas (Table 6.2a), occurring mainly in phosphorylated forms in animal products, and in free form in foods of plant origin. Food processing contributes to significant loss of this vitamin (Section 6.3.1.4). Upon ingestion, thiamin phosphate esters are hydrolyzed in the intestinal lumen mainly the alkaline phosphatase associated with brush-border membranes of enterocytes and taken up through the mucosal membrane by a specific saturable transport system (see Chapter 13 and Table 13.2). Chronic alcohol consumption impairs the intestinal absorption of thiamin, possibly through the inhibition of thiamin transporters. Bioavailability of dietary thiamin can also be impaired by different types of anti-thiamin factors that degrade or modify thiamin so that it cannot be absorbed or loses its function. Sulfites, thiaminases present in some raw or fermented fish, ferns and insects, some molecules of plant origin like caffeic acid, chlorogenic acid, and tannic acid, quercetin, and rutin are considered as thiamin antagonists [5, 9]. Thiamine is commercially available as the hydrochloride and mononitrate salts that are widely used for food fortification and as nutritional supplements [10]. Gut microbiota may also be a significant source of this vitamin (Section 6.3.9).

More than 80% of total thiamin in the blood is found in erythrocytes in the form of TDP and TTP, lower amounts being present in plasma as free thiamin, TMP, and protein-bound TDP. Thiamin is taken up by cells of the blood, liver, heart and other tissues, including the placenta and brain, by active transport. Total thiamin content of the adult body has been estimated to be about 25–30 mg, located mostly in the skeletal muscles, heart, brain, liver and kidneys. Thiamin is excreted in urine (as free thiamin and other derived metabolites) and feces, although the latter seems to be due to the activity of gut microbiota; sweat is not a significant route of loss. Breast milk also contains thiamin, mostly as TMP (about 70%) and free thiamin (about 30%), with a negligible amount of TDP [5].

Measurement of erythrocyte transketolase (ETKA) is a functional test of thiamin status and can be carried out without or with added TDP. The degree to which ETKA rises in response to the addition of TDP is known as αETK (sometimes expressed as percentage of activation when TDP is added). A value of αETK < 1.15 (that is, < 15% increase in ETKA upon addition of TDP) is generally considered to reflect an adequate thiamin status; however, several factors may affect the specificity of these assays and confound their interpretation. Consequently, a combination of ETKA and αETK with other biomarkers (total thiamin or TDP concentrations in blood and erythrocytes or urinary thiamin excretion) is required to reliably assess the thiamin status of individuals [5].

6.3.1.4 Stability

Thiamine stability in aqueous solution is relatively low, being influenced by pH, temperature, ionic strength, and metal ions. The enzyme-bound form is less stable than free thiamine. Sulfites and alkaline pH cause rapid decomposition of thiamine, whereas practically no degradation occurs in acidic medium (e.g., lemon juice). Nitrites inacti-

Table 6.2a: Average content of some water-soluble vitamins in selected foods.

Food type	Food	Vit. C (mg/100 g)	Vit. B₁ (mg/100 g)	Vit. B₂ (mg/100 g)	Vit. B₃ (mg/100 g)	Vit. B₅ (mg/100 g)	Vit. B₆ (mg/100 g)	Vit. B₉ (µg/100 g)
Vegetables (raw)	Carrot	2.05	0.028	<0.01	<0.1	0.20	0.093	59.4
	Sweet pepper (green, yellow, red)	121	0.041	0.041	0.74	0.12	0.38	40.7
	Tomato	15.5	0.039	0.019	0.65	0.21	0.082	22.7
	Spinach	41.1	0.087	0.21	0.71	0.18	0.21	207
	Potato, peeled	18.9	0.068	0.048	1.33	0.34	0.25	26
Legumes	Lentil, dried	4.5	0.69	0.21	2.3	1.75	0.55	257
Fruits	Orange, pulp, raw	47.5	0.015	<0.01	0.37	0.16	<0.01	25.9
	Apple, pulp and peel, raw	6.25	0.016	0.019	0.091	0.079	0.048	6
	Grape, red, raw	10.8	0.045	0.01	0.15	0.075	0.04	5
Nuts	Almond (with peel)	<0.5	0.15	0.29	1.97	0.62	0.07	120
Seeds	Sunflower seed	1.4	1.98	0.16	4.8	0.83	1.24	254
Cereals	Bread, French bread, baguette	<0.5	0.17	<0.01	2.88	0.46	<0.01	26.8
	Wholemeal or integral bread	<0.5	0.11	<0.01	<0.1	0.44	0.087	35.6
	Wheat flour, type 150 (whole grain)	<0.5	0.34	0.11	3.48	1.02	0.26	52.4
	Wheat flour, type 55 (for bread)	<0.5	0.15	<0.01	<0.1	0.56	0.054	15.4
	Rice flour	<0.5	0.33	<0.01	<0.1	0.97	0.13	25.1
	Maize/corn flour	0	0.2	0.084	4.18	0.052	0.21	34
Meat	Meat, cooked (average)	0.69	0.21	0.17	7.45	1.01	0.37	8.38
	Poultry, cooked (average)	1.15	0.095	0.17	9.82	1.37	0.38	8.1
	Beef, sirloin steak, raw	0	0.059	0.15	5.77	0.73	0.48	9.6
	Liver, young cow, cooked	20	0.33	3.43	18	8.5	1.4	536
	Pork loin, cooked	0.6	0.99	0.31	5.57	0.76	0.52	6
	Chicken, meat, raw	2.3	0.084	0.15	8.15	1.05	0.41	14.1

Fish	Salmon, smoked	0	0.23	0.1	10	0.71	1	26
	Salmon, steamed	0	0.22	0.11	7	1.09	0.49	11.7
	Tuna, roasted/baked	–	0.052	0.062	16.4	0.13	0.47	40.6
	Salmon, raw, farmed	1.8	0.21	0.076	8.25	0.95	0.58	20.8
	Tuna, raw	1.37	0.13	0.12	11	0.66	0.54	13.6
	Salmon, raw, wild	0	0.21	0.24	8.83	1.48	0.49	13.1
	Shrimp or prawn, cooked	0.83	0.019	0.04	1.77	0.31	0.1	14.6
	Mussel, boiled/cooked in water	5.87	<0.04	0.29	1.16	0.28	0.044	48
Eggs	Egg, hard-boiled	0	0.066	0.51	0.064	1.4	0.12	44
	Egg, raw	0	0.055	0.45	0.063	1.57	0.15	34
Milk and milk products	Milk, whole, UHT	<0.5	0.041	0.17	<0.1	0.43	0.02	<2.5
	Milk, semi-skimmed, UHT	1.5	0.06	0.18	0.093	0.41	0.037	9.49
	Milk, skimmed, UHT	0.89	0.049	0.19	0.09	0.42	0.041	5.11
	Cheese (average)	0.057	0.035	0.41	1	0.51	0.11	34.4
Baby food	Baby milk, first age, powder	–	0.47	0.87	4.75	–	0.41	77.6
	Baby milk, second age, powder	–	0.54	0.94	4.75	–	1.02	83.4

Table 6.2b: Biotin content of select foods determined by HPLC/avidin binding assay.

Food type	Foods	µg biotin/100 g food
Meat	Beef liver, cooked	41.6
	Chicken nuggets, breaded, fried	1.34
	Chicken strips, breaded, fried	0.43
	Chicken liver, cooked	187.2
	Pork chop, cooked	4.5
Fish	Salmon, pink, canned in water	5.9
	Tuna, canned in water	0.682
Eggs	Egg, whole, cooked	21.4
Milk and milk products	2% milk	0.113
	American cheese	3.1
	Cheddar cheese, mild	1.4
	Skim milk	0.131
	Whole milk	0.091
Vegetables	Broccoli, fresh	0.943
	Carrots, canned	0.622
	Cauliflower, fresh	0.161
	Green beans, canned	0.007
	Mushrooms, canned	2.16
	Spinach, frozen	0.705
	Sweet potato, cooked	1.45
	Whole kernel corn, canned	0.047

Fruits and berries	
Apple, fresh	0.02
Apple juice, canned, from concentrate	0.052
Avocado, fresh	0.961
Banana, fresh	0.133
Orange, fresh	0.049
Orange juice, canned, from concentrate	0.413
Raspberries, fresh	0.178
Strawberries, fresh	1.5
Tomatoes, fresh	0.701
Nuts and seeds	
Almonds, roasted, salted	4.407
Peanuts, roasted, salted	17.5
Sunflower seeds, roasted, salted	7.8
Walnuts, fresh	2.59
Bread and grains	
Grilled toast	1.23
Hamburger bun	0.289
Oatmeal	0.191
Roll, dinner	0.048
Whole wheat bread	0.074
Beverages	
Tea, sweet	0.142
White wine	0.117
Beer	0.114

Note: The HPLC/avidin-binding assay confirmed previous assessments that meat, fish, poultry, egg, some cheeses, and some vegetables are rich dietary sources of biotin. This assay further showed that the richest sources of biotin are liver, eggs, and mushrooms, as well as some cheeses, while smaller amounts were contained in lean meat, fruit, cereals, and bread. Most previously published values were higher than those measured by the HPLC/avidin-binding assay (Comment by EFSA [11]).

Table 6.2c: Average content of Vitamin B_{12} in selected foods.

Food	Vit. B_{12} (µg/100 g)
Liver, lamb, cooked	60
Liver, calf, cooked	52,6
Kidney, all types, cooked	31,2
Meat, cooked (average)	1,34
Poultry, cooked (average)	0,65
Salmon, smoked	3,35
Salmon, steamed	3,05
Tuna, roasted/baked	2,57
Salmon, raw, farmed	3,95
Tuna, raw	3,79
Salmon, raw, wild	4,84
Shrimp or prawn, cooked	1,64
Mussel, boiled/cooked in water	17,6
Egg, hard-boiled	1,11
Milk, whole, UHT	0,24
Milk, semi-skimmed, UHT	0,38
Milk, skimmed, UHT	0,39
Cheese (average)	1,33

Source for Tables 6.2a and 6.2c: French Agency for Food, Environmental and Occupational Health & Safety. Anses. 2020. Ciqual French food composition table. Retrieved on June 10, 2023, from the Ciqual homepage: https://ciqual.anses.fr/ [12]. Source for Table 6.2b [13].

vate thiamine, probably through reaction with the amino group attached to the pyrimidine ring. Thermal degradation of thiamine is involved in the formation of meat-like aroma in cooked food [14]. More detailed information on thiamine stability and mechanisms of degradation can be found elsewhere [10, 14].

6.3.1.5 Dietary Reference Values (DRVs): hypo- and hypervitaminosis

The European Food Safety Agency published in 2017 a Summary Report with Dietary Reference Values for nutrients. This report was updated in September 2019 and includes Dietary Reference Values (DRVs) for most nutrients, including thiamin [15]. Milled rice and grains have little amounts of thiamine as processing involves removal of bran, that contains most of it. Certain food products such as tea, coffee, raw fish, and shellfish contain enzymes that destroy thiamine (thiaminases) [16].

Worldwide, thiamine deficiency is primarily due to inadequate dietary intake, specifically in diets comprised mainly of polished rice and grains. It can also be related to diets that are rich in thiaminase, a natural thiamin-degrading enzyme, which is abundantly present in some raw or fermented fish, ferns, and insects consumed primarily in Africa and Asia. In Western countries, it is most common in patients suffering from alcoholism or chronic illness. Individuals also at risk for thiamine deficiency include pregnant women, those requiring parental feeding, individuals who have undergone bariatric surgery, those with overall poor nutritional status, and patients on chronic diuretic therapy.

Thiamin deficiency usually presents with symptoms of peripheral neuritis, cardiac insufficiency and a tendency for edemas and may be accompanied by extreme fatigue, irritability, forgetfulness, poor coordination, gastrointestinal disturbances, constipation, labored breathing, loss of appetite and weight loss. The two most common complications of extreme thiamin deficiency are Wernicke-Korsakoff syndrome (WKS) and beriberi [5, 17].

- WKS affects the central nervous system (brain and spinal cord). It is most commonly caused by alcohol abuse and poor nutrition, although it can be observed in other individuals at risk for thiamin deficiency. WKS consists of two different syndromes: Wernicke encephalopathy, characterized by non-inflammatory brain lesions, that may present symptoms like ataxia, ophthalmoplegia, punctate hemorrhages in the brain, altered mental status, and balance abnormalities. If untreated, Wernicke encephalopathy can evolve to include Korsakoff psychosis that causes delirium and permanent memory loss. WKS is treated with a therapy that includes high doses of thiamine and glucose.
- Beriberi is most frequently seen in people who abuse alcohol but can also be due to other etiologies resulting in thiamin deficiency. First symptoms of beriberi are non-specific and include constipation, appetite suppression, nausea, mental depression, fatigue, peripheral neuropathy, anorexia, and weight loss.

With progression, chronic symptoms can begin to manifest as either wet beri-beri or dry beriberi.

-   Dry beriberi is predominately a neurological disorder with a sensory and motor peripheral neuropathy. Complications involve the peripheral nervous system. Symptoms include paresthesia, foot drop, muscle wasting, numbness, and absent ankle reflexes.
-   Wet beriberi is the term used for thiamin deficiency that, in addition to the presence of peripheral neuropathy, involves manifestations that include edema, an enlarged heart, congestive heart failure, cardiomegaly and tachy-cardia, warm extremities, pleural effusions, and pulmonary edema. A rapidly developing form of wet beriberi refers to the acute fulminant cardiovascular beriberi (Shoshin beriberi), or acute pernicious beriberi
-   Infantile beriberi can occur in breastfed infants of thiamin-deficient mothers at the age of 2–6 months and may be characterized by both neurologic and cardiac signs with lethal outcome due to heart failure.

As far as hypervitaminosis is concerned, the Scientific Committee for Food (SCF) of EFSA has noted that data on adverse effects of oral intake of thiamin in humans are limited and that thiamin absorption declines for an intake higher than 5 mg/day, while absorbed thiamin is actively excreted in the urine; therefore, no tolerable upper intake level (UL) has been set for this vitamin [5].

6.3.2 Vitamin B$_2$ (riboflavin)

6.3.2.1 Structure

Riboflavin is a heterocyclic isoalloxazine derivative with a ribitol side chain whose IUPAC systematic name is 7,8-dimethyl-10-[(2S,3S,4R)-2,3,4,5-tetrahydroxypentyl]benzo-[g]pteridine-2,4-dione. An alternative name is 7,8-dimethyl-(N-10-ribityl)isoalloxazine (Figure 6.1A). In older literature, riboflavin is referred to as vitamin G because of its green fluorescent properties in UV light, actually, it can be conveniently assayed using its natural yellow-green fluorescence spectra. Solutions of riboflavin exhibit four characteristic UV/visible spectral absorption bands at 220, 266, 375, and 447 nm [18].

6.3.2.2 *In vivo* function

Riboflavin is the parent precursor to the coenzymes flavin mononucleotide (riboflavin 5'-phosphate or FMN) and flavin adenine dinucleotide (FAD). *In vivo*, FMN and FAD normally occur tightly bound to certain proteins, collectively known as flavoproteins. The majority of flavoproteins are capable of mediating oxidation and reduction reactions (redox) due to the capacity of the isoalloxazine ring to undergo reversible reduction, by accepting either one or two electrons plus one or two protons, respectively.

Flavoproteins participate in a wide array of cellular processes involving electron transport, metabolism of lipids. other vitamins, drugs, and xenobiotics, as well as cell signaling and protein folding. The fully reduced forms of FAD and FMN are abbreviated as $FADH_2$ and $FMNH_2$. Approximately 10% of flavoproteins participate in other types of biochemical processes such as transferase, lyase, isomerase, and ligase reactions [7, 19]. Riboflavin has been reported to support anti-oxidant defense mechanisms (it is involved in the reduction of oxidized to reduced glutathion), to act as an anti-oxidant by its own oxidation, and to be involved in protection against ischemia/ reperfusion (I/R) oxidative injuries [20].

6.3.2.3 Sources and bioavailability
Significant dietary sources of riboflavin in developed countries include plant foods as well as animal sources, namely, meat, poultry, fish, and dairy products, particularly eggs, milk, and cheese (Table 6.2a). In developing countries, plant sources contribute nearly to all of the dietary riboflavin intake. Green vegetables, such as broccoli and turnips, are also reasonably good sources of riboflavin, whereas natural grain products tend to be relatively low in this vitamin, unless they are fortified during processing. In general terms, high-quality, protein-rich foods are excellent sources of riboflavin as well as of other B vitamins [18]. Breast milk is also a good source of riboflavin for babies, the average concentration in milk of unsupplemented women being around 360 µg/L [21].

Dietary FMN and FAD covalently bound to food protein are hydrolyzed to free riboflavin, whereas acidification in the stomach releases more loosely bound molecules, which are also hydrolyzed to free riboflavin by non-specific phosphatases of the brush border and basolateral membranes of enterocytes in the upper small intestine. Absorption of free riboflavin mainly takes place in the proximal small intestine through a carrier-mediated, saturable transport process. A carrier-mediated absorption of riboflavin is also present in the colon (Table 13.2). Prevalence of riboflavin deficiency is high in chronic alcoholics, as ethanol consumption apparently inhibits the release of riboflavin from dietary FMN and FAD as well as its absorption. In the bloodstream, riboflavin is transported mainly in erythrocytes but also in the plasma, bound to albumin and immunoglobulins. There is a positive transfer of riboflavin from the pregnant woman to the fetus. Uptake of riboflavin into the cells of organs is facilitated and may require specific carriers. Urine is the main route for elimination of this vitamin [21].

The EFSA Panel on Dietetic Products, Nutrition and Allergies considers that 24 h (preferably) or fasting urinary excretion of riboflavin is a suitable biomarker of riboflavin short-term intake and of riboflavin status [21].

6.3.2.4 Stability

Riboflavin usually occurs in food sources as phosphorylated derivatives, FMN, FAD, and their covalently bound forms, with the exception of some dairy products that may contain free riboflavin. Food processing can adversely influence the amount of bioavailable riboflavin: blanching, milling, fermenting, and extruding may result in physical removal of the vitamin; however, riboflavin is relatively stable to thermal processing, although heating procedures performed for short periods of time, such as microwaving, are preferable for its preservation. Exposure to UV light would result in appreciable losses of riboflavin; therefore, milk and milk products must be perfectly protected against light. Large amounts of riboflavin are also lost during the sun-drying of fruits and vegetables. Flavin degradation in irradiated foods varies considerably because this process depends on the chemical environment surrounding the isoalloxazine ring that can be stabilized by forming intramolecular complexes with purines, pyrimidines, or aromatic amino acid moieties. Foods cooked at elevated temperatures or under slightly acidic conditions experience increased dissociation of these complexes. The addition of sodium bicarbonate (baking soda) to accentuate the green color of vegetables can result in accelerated photodegradation of riboflavin [18]. Upon light exposure riboflavin also forms highly reactive oxygen species (Chapter 8), which may lead to significant losses of other vitamins (including folate, thiamine, ascorbate (vitamin C), and vitamins A, D, and E) and oxidative damage of proteins and lipids (e.g., degradation of unsaturated lipids and formation of toxic cholesterol oxides) [22].

6.3.2.5 Dietary Reference Values (DRVs): hypo- and hypervitaminosis

Dietary Reference Values (DRVs) and recommendations for this vitamin are shown in Table 6.3.

Clinical signs of riboflavin deficiency have been reported to include sore throat, hyperemia, and edema of the pharyngeal and oral mucous membranes, cheilosis, glossitis (magenta tongue), seborrheic dermatitis, and skin lesions like angular stomatitis; however, these symptoms are unspecific and may be due to other nutrient deficiencies as well. The metabolic role of riboflavin within food sources is intimately associated with those of the other B vitamins; thus, flavin coenzymes are involved in the metabolism of vitamins like folic acid, pyridoxine, vitamin K, niacin, and vitamin D; therefore, a primary deficiency of dietary riboflavin has wide nutritional implications [19, 21].

Certain endocrine abnormalities, such as adrenal and thyroid hormone insufficiency, and rare diseases like Brown-Vialetto-Van Laere (BVVL) and Fazio-Londe syndromes may interfere significantly with vitamin utilization. Some drugs may impair riboflavin utilization by inhibiting the conversion of this vitamin into its active coenzyme derivatives and there is evidence that alcohol causes riboflavin deficiency by inhibiting both its digestion and its intestinal absorption [18, 19].

Table 6.3: Some Dietary Reference Values for vitamins. Source [15].

Average Requirements (AR), defined as: "the level of (nutrient) intake that is adequate for half of the people in a population group, given a normal distribution of requirement." See Appendix 1 for further information about Dietary Reference Values (DRVs).

Females

Age (years)	Folate [a] (μg DFE/day)	Vit. B$_3$ [b] (mg NE/MJ)	Riboflavin (mg/day)	Thiamin (mg/MJ)	Vit. A [c] (μg/day)	Vit. B$_6$ (mg/day)	Vit. C (mg/day)
7–11 months	–	1.3	–	0.072	190	–	–
1–3	90	1.3	0.5	0.072	205	0.5	15
4–6	110	1.3	0.6	0.072	245	0.6	25
7–10	160	1.3	0.8	0.072	320	0.9	40
11–14	210	1.3	1.1	0.072	480	1.2	60
15–17	250	1.3	1.4	0.072	490	1.3	75
≥18	250	1.3	1.3	0.072	490	1.3	80
Pregnancy	–	1.3	1.5	0.072	540	1.5	–
Lactation	380	1.3	1.7	0.072	1,020	1.4	145

Males

Age (years)	Folate [a] (μg DFE/day)	Vit. B$_3$ [b] (mg NE/MJ)	Riboflavin (mg/day)	Thiamin (mg/MJ)	Vit. A [c] (μg/day)	Vit. B$_6$ (mg/day)	Vit. C (mg/day)
7–11 months	–	1.3	–	0.072	190	–	–
1–3	90	1.3	0.5	0.072	205	0.5	15
4–6	110	1.3	0.6	0.072	245	0.6	25
7–10	160	1.3	0.8	0.072	320	0.9	40

(continued)

Table 6.3 (continued)

Males

Age (years)	Folate[a] (μg DFE/day)	Vit. B₃[b] (mg NE/MJ)	Riboflavin (mg/day)	Thiamin (mg/MJ)	Vit. A[c] (μg/day)	Vit. B₆ (mg/day)	Vit. C (mg/day)
11–14	210	1.3	1.1	0.072	480	1.2	60
15–17	250	1.3	1.4	0.072	580	1.5	85
≥18	250	1.3	1.3	0.072	570	1.5	90

MJ, megajoule

[a]DFE: dietary folate equivalents. For combined intakes of food folate and folic acid, DFEs can be computed as follows: μg DFE = μg food folate + (1.7 × μg folic acid)

[b]NE: niacin equivalent (1 mg niacin = 1 niacin equivalent = 60 mg dietary tryptophan)

[c]RE: retinol equivalent, 1 μg RE equals 1 μg of retinol, 6 μg of β-carotene and 12 μg of other provitamin A carotenoids

Population Reference Intakes (PRIs), defined as: "the level of (nutrient) intake that is adequate for virtually all people in a population group," and **Adequate Intake (AIs)**, defined as "the value estimated when a Population Reference Intake cannot be established because an average requirement cannot be determined." An Adequate Intake is the average observed daily level of intake by a population group (or groups) of apparently healthy people that is assumed to be adequate.

Females

Age (years)	Vit. E (mg/day)	Age (years)	Vit. B7/H (µg/day)	Vit. B12 (µg/day)	Vit. B9 [a] (µg DFE/day)	Vit. B3 [b] (mg NE/MJ)	Vit. B5 (mg/day)	Vit. B2 (mg/day)	Vit. B1 (mg/MJ)	Vit. A [c] (µg/day)	Vit. B6 (mg/day)	Vit. C (mg/day)	Vit. D [e] (µg/day)	Vit. K [g] (µg/day)
7–11 months	5	7–11 months	6	1.5	80	1.6	3	0.4	0.1	250	0.3	20	10	10
1–2	6	1–3	20	1.5	120	1.6	4	0.6	0.1	250	0.6	20	15 [f]	12
3–9	9	4–6	25	1.5	140	1.6	4	0.7	0.1	300	0.7	30	15 [f]	20
10–17	11	7–10	25	2.5	200	1.6	4	1.0	0.1	400	1.0	45	15 [f]	30
		11–14	35	3.5	270	1.6	5	1.4	0.1	600	1.4	70	15 [f]	45
		15–17	35	4.0	330	1.6	5	1.6	0.1	650	1.6	90	15 [f]	65
≥18	11	≥18	40	4.0	330	1.6	5	1.6	0.1	650	1.6	95	15 [f]	70
Pregnancy	11		40	4.5	600	1.6	5	1.9	0.1	700	1.8	105	15 [f]	70
Lactation	11		45	5.0	500	1.6	7	2.0	0.1	1,300	1.7	155	15 [f]	70

Males

Age (years)	Vit. E (mg/day)	Age (years)	Vit. B7/H (µg/day)	Vit. B12 (µg/day)	Vit. B9 [a] (µg DFE/day)	Vit. B3 [b] (mg NE/MJ)	Vit. B5 (mg/day)	Vit. B2 (mg/day)	Vit. B1 (mg/MJ)	Vit. A [c] (µg/day)	Vit. B6 (mg/day)	Vit. C (mg/day)	Vit. D [e] (µg/day)	Vit. K [g] (µg/day)
7–11 months	5	7–11 months	6	1.5	80	1.6	3	0.4	0.1	250	0.3	20	10	10
1–2	6	1–3	20	1.5	120	1.6	4	0.6	0.1	250	0.6	20	15 [f]	12
3–9	9	4–6	25	1.5	140	1.6	4	0.7	0.1	300	0.7	30	15 [f]	20

Table 6.3 (continued)

Age (years)	Vit. E (mg/day)	Age (years)	Vit. B₇/H (µg/day)	Vit. B₁₂ (µg/day)	Vit. B₉ (a) (µg DFE/day)	Vit. B₃ (b) (mg NE/MJ)	Vit. B₅ (mg/day)	Vit. B₂ (mg/day)	Vit. B₁ (mg/MJ)	Vit. A(c) (µg/day)	Vit. B₆ (mg/day)	Vit. C (mg/day)	Vit. D(e) (µg/day)	Vit. K(g) (µg/day)
10–17	**13**	**7–10**	**25**	**2.5**	**200**	**1.6**	**4**	**1.0**	**0.1**	**400**	**1.0**	**45**	**15 (f)**	**30**
		11–14	35	3.5	270	1.6	5	1.4	0.1	600	1.4	70	15 (f)	45
		15–17	35	4.0	330	1.6	5	1.6	0.1	750	1.7	100	15 (f)	65
≥18	**13**	**≥18**	**40**	**4.0**	**330**	**1.6**	**5**	**1.6**	**0.1**	**750**	**1.7**	**110**	**15 (f)**	**70**

MJ, megajoule

PRIs are presented **in bold type** and AIs in ordinary type

(a)DFE: dietary folate equivalents. For combined intakes of food folate and folic acid, DFEs can be computed as follows: µg DFE = µg food folate + (1.7 × µg folic acid)

(b)NE: niacin equivalent (1 mg niacin = 1 niacin equivalent = 60 mg dietary tryptophan)

(c)RE: retinol equivalent, 1 µg RE equals 1 µg of retinol, 6 µg of β-carotene and 12 µg of other provitamin A carotenoids

(e)for conversion between µg and International Units (IU) of vitamin D intake: 1 µg = 40 IU and 0.025 µg = 1 IU

(f)under conditions of assumed minimal cutaneous vitamin D synthesis. In the presence of endogenous cutaneous vitamin D synthesis, the requirement for dietary vitamin D is lower or may be even zero.

(g)based on phylloquinone only. The AI of 1 µg phylloquinone/kg body weight set by SCF (1993) was multiplied by reference body weights of the European population

As far as toxicity is concerned, the levels of riboflavin consumed orally from the diet or from most multivitamin supplements rarely cause side effects or exhibits toxicity. Riboflavin that is not converted to FMN or FAD can exist as free riboflavin and be excreted mainly by the kidney. Repeatedly consumed pharmacologic doses (>100 mg) have the potential to react with light, which may have adverse cellular effects [19].

6.3.3 Vitamin B$_3$ (niacin)

6.3.3.1 Structure
Niacin is a generic term for nicotinic acid (pyridine-3-carboxylic acid) and nicotinamide (niacinamide or pyridine-3-carboxamide), both compounds being identical in their vitamin function (Figure 6.1A). Nicotinic acid is a white crystalline solid moderately soluble in water and alcohol, but insoluble in ether, whereas nicotinamide is soluble both in water and in ether, thus allowing the separation of the two substances. Niacin can be produced in the liver from the amino acid tryptophan (60 mg of tryptophan yields approximately 1 mg of niacin, although this relationship is not constant); however, this may not be sufficient for human needs, especially in the case of individuals with diets low in tryptophan. In 2004, nicotinamide riboside, a molecule present in milk, was identified as another source of nicotinamide, thus becoming the third member of the vitamin B$_3$ family [23–25].

6.3.3.2 *In vivo* function
Nicotinamide is a constituent of both nicotinamide adenine dinucleotide (NAD$^+$) and nicotinamide adenine dinucleotide phosphate (NADP$^+$). NAD$^+$ is synthesized *in vivo* by a series of biochemical pathways that differ depending on the precursor molecule (nicotinic acid, nicotinamide, nicotinamide riboside, or tryptophan). NADP$^+$ is obtained from NAD$^+$ by ATP-dependent phosphorylation. Both NAD$^+$ and NADP$^+$ can accept a proton (H$^+$) and two electrons at the C-4 position on the pyridine ring of the nicotineamide moiety, thereby resulting into formation of NADH and NADPH [7]. Redox pairs NAD$^+$/NADH and NADP$^+$/NADPH are indispensable cofactors in over 400 enzymatic reactions central to most metabolic processes, by acting as electron acceptors (NAD$^+$, NADP$^+$) or donors (NADH, NADPH) [26]:

-   NAD$^+$ is reduced to NADH in many catabolic reactions and pathways, such as glycolysis, oxidative decarboxylation of pyruvate, citric acid cycle, oxidation of ethanol, β-oxidation of fatty acids, and others. NADH is subsequently oxidized back to NAD$^+$ at the electron transport chain (ETC) of the inner mitochondrial membrane. Ultimately, the electrons donated by NADH to ETC are transferred to oxygen, which is reduced to water. The energy resulting from these electron transfers is used to generate ATP by means of a complex mechanism, known as oxidative phosphorylation that involves proton electrochemical gradients and a nanoma-chine: the ATP synthase [7].

– In the cytosol of mammalian cells, NADPH is basically produced from the reduc-tion of $NADP^+$ in the pentose phosphate pathway, although other potential sour-ces exist, such as the reactions catalyzed by specific isozymes of malic enzyme (ME) and several dehydrogenases [27]. In mitochondria, NADPH is mainly pro-duced by the oxidation of isocitrate catalyzed by the enzyme isocitrate dehydro-genase 2 (IDH2), and also by the proton-translocating transhydrogenase, that mediates the conversion of $NADP^+$ to NADPH coupled to the translocation of pro-tons across the inner mitochondrial membrane using the reducing power of NADH [28, 29]. NADPH acts as a reducing agent for the synthesis of biomolecules like fatty acid or cholesterol and derivatives (bile acids and steroids), although it is also notable for its involvement in detoxification of reactive oxygen species (ROS), ethanol, and xenobiotics. Paradoxically, NADPH is also used by phagocytes to generate ROS, used as an unspecific defense against bacteria and fungi [30].

NAD^+ is also used as a substrate by three enzyme groups: mono(ADP-ribose) transfer-ases (MARTs), poly(ADP-Ribose) polymerases (PARPs, also known as ADP-ribose trans-ferases or ATRDs), and sirtuins. These enzymes are involved in a large array of biological processes, such as control of energy metabolism and mitochondrial func-tions, calcium homeostasis, control of oxidative stress, signaling and gene regulation pathways, immunological functions, aging, and cell death. More recently, NAD^+ has been reported to play a role as a neurotransmitter [26].

6.3.3.3 Sources and bioavailability

Plants, bacteria, and yeast produce niacin and nicotinamide. From there, the vitamin molecules are maintained in the food chain, being found in a wide range of foods. The main sources of niacin include liver, meat and meat products, fish, peanuts, and whole grains (Table 6.2a). Coffee and tea are also sources of niacin. Nicotinic acid and nicotinamide may be added to foods, food supplements, and infant and follow-on for-mulae. Foods with high protein content, such as milk, cheese, and eggs, are good sour-ces of the amino acid tryptophan [25].

The mean intestinal absorption of niacin ranges from about 23% (cereals) to 70% (animal products). In cereals, niacin is mostly present as nicotinic acid esterified to polysaccharides, polypeptides, and glycopeptides, which is basically unavailable after cooking; only a small part of these bound forms being hydrolyzed by gastric acid. The bioavailability of these forms of niacin can be increased by pretreatment of the food with alkali for ester bond hydrolysis, actually, a traditional Latin American food proc-essing practice involved presoaking corn in alkaline "lime-water" ($Ca(OH)_2$) prior to cooking. This tradition explained why Latin American populations had been spared from pellagra (see below), unlike other communities that also used corn as their main staple food [25, 26]. In animal food, niacin occurs mainly in the form of the nucleotides NAD^+ and $NADP^+$, that can be cleaved by the intestinal mucosal enzyme $NAD(P)^+$-glycohydrolase, thus releasing nicotinamide which is subsequently absorbed [31]. Some nicotinic acid is absorbed by passive diffusion in the stomach and the small intestine [32]; however, the mechanism of transport across the enterocyte brush border membrane is not fully clarified yet, although several transporters, such as the human organic anion transporter-10 (OAT-10) or the sodium-coupled monocarboxylate transporter (SMCT1), appear to be involved [33].

Nicotinamide is the major form of niacin found in the bloodstream. From the blood, both nicotinic acid and nicotinamide move across cell membranes by simple diffusion, although transport into kidney tubules and erythrocytes requires a carrier. Within the cell, both nicotinamide and nicotinic acid are used to synthesize NAD but utilize slightly different pathways to achieve this. NAD^+ can be converted to $NADP^+$ by reaction with ATP. Intracellular concentrations of NAD^+ are generally higher than $NADP^+$ concentrations. The mean concentration of niacin in mature human milk is about 2.1 mg/L but this concentration seems to depend on maternal niacin intake. The major pathway of catabolism of both nicotinic acid and nicotinamide is methylation in the liver and subsequent oxidation, while excretion is mainly via the urine. Urinary excretion of niacin metabolites is considered as a marker of niacin status [25].

6.3.3.4 Stability

Heat, especially under acid or alkaline conditions, converts nicotinamide to nicotinic acid without loss of vitamin activity. Niacin is not affected by light, and no thermal losses occur under conditions relevant to food processing; however, significant amounts of niacin can be lost if cooking water is discarded [10].

6.3.3.5 Dietary Reference Values (DRVs): hypo- and hypervitaminosis

The recommended daily intake of niacin is expressed as milligrams (mg) of niacin equivalents (NE = mg of preformed nicotinic acid or nicotinamide + (1/60) of mg of tryptophan). Some reference values for niacin intake published by the Scientific Committee for Food of the European Union are shown in Table 6.3. More information from the European Food Safety Agency (EFSA) can be found elsewhere [15, 25].

Long-term inadequate intake of tryptophan and niacin can lead to the development of pellagra (from the Italian "pelle agra," rough skin). Early symptoms are usually nonspecific and include weakness, loss of appetite, fatigue, digestive disturbances, abdominal pain, and irritability. Common symptoms of pellagra include photosensitive dermatitis, skin lesions, tongue and mouth soreness, vomiting, diarrhea, depression, dementia, and, ultimately, death from multiorgan failure [25]. Traditionally, pellagra was known as the disease of the four "D's": diarrhea, dermatitis, dementia, and death [26].

Primary causes of pellagra include dependence on a corn-based diet or general malnutrition, secondary causes include chronic alcoholism and general malabsorptive states such as prolonged diarrhea. Other factors that induce pellagra-like symptoms and contribute to impaired NAD synthesis include [26]:

- Altered tryptophan metabolism in diets rich in leucine.
- Hartnup disease, an autosomal recessive mutation in the gene for a neutral amino acid transporter that leads to depletion of tryptophan.
- Carcinoid tumors can cause an increase in serotonin synthesis. This depletes tryptophan, which is also a precursor for that neurotransmitter.
- Infections with human immunodeficiency virus (HIV).
- Diseases like liver cirrhosis, diabetes mellitus, as well as prolonged fevers.
- Increasing age correlates with dropping NAD^+ levels.
- Women, in particular pregnant females, are at an increased risk for niacin deficiency and pellagra compared to men.

High doses of nicotinic acid have been used as a lipid-lowering treatment due to its important effects on lipid metabolism; however, these high doses can cause an acute flushing reaction, with vasodilatation and severe itching. Nicotinamide does not have this side effect but is not useful for treatment of hyperlipidemia. The mechanism of action of nicotinic acid is complex and not completely understood but is known to involve the hydroxycarboxylic acid receptor 2, encoded by the human gene HCAR2/GPR109A [34, 35]. Hcar2 is widely expressed throughout the body including adipocytes of white and brown tissues, epithelial cells of several organs and different types of immune cells. Recent work is establishing niacin as a promising therapeutic option in a range of neurological diseases such as multiple sclerosis, Alzheimer's disease, and glioblastoma, although several questions about the mechanisms and utility of niacin still remain and are currently under study [36]. Intakes of niacin above 500 mg/day over a period of months can cause liver damage [37].

6.3.4 Vitamin B₅ (pantothenic acid)

6.3.4.1 Structure
The biologically active form of vitamin B_5 is D-pantothenic acid, whose IUPAC name is 3-[[(2R)-2,4-dihydroxy-3,3-dimethylbutanoyl]amino]propanoate. It is synthesized by microorganisms via an amide linkage of β-alanine and D-pantoic acid (Figure 6.1B).

6.3.4.2 *In vivo* function
Pantothenic acid is a component of coenzyme A (CoA) that performs multiple roles in cellular metabolism by facilitating the transfer of acetyl or acyl groups. CoA is involved in cellular processes and pathways like β-oxidation of fatty acids, oxidative degradation of amino acids, citric acid cycle, cholesterol synthesis, and many others.

CoA also participates as carrier of acetyl and acyl groups in protein acetylations and acylations, respectively; these post-translational modifications of proteins modulate their structure, location, and activity within the cells. Pantothenic acid is also a cofactor of the acyl carrier protein (ACP) domain of fatty acid synthase (Chapter 4, Section 4.10.2) [7].

6.3.4.3 Sources and bioavailability

Pantothenic acid is ubiquitous in foods and dietary deficiency is rare. The main contributors to pantothenic acid intakes include meat products, bread, milk-based products, and vegetables (Table 6.2a). Pantothenic acid may be added to foods, food supplements, and infant and follow-on formulae (as calcium and sodium salts or as dexpanthenol). The average pantothenic acid concentration in breast milk is 2.5 mg/l [38].

Dietary CoA is hydrolyzed in the intestine in a multistep process that releases pantothenic acid which is mainly absorbed via a saturable sodium-dependent carrier-mediated process (Table 13.2). Studies in rodents suggest that pantothenic acid is transported in blood in the free form (not bound to albumin) and is rapidly taken up by tissues and red blood cells. In red blood cells, pantothenic acid could be converted to 4'-phosphopantothenic acid, but not further to CoA. The uptake into different tissues is via the Sodium Dependent Multivitamin Transporter (SMVT) (Chapter 13) [22]. Mean concentrations of pantothenic acid in mature human milk typically range between 2 and 3 mg/L and correlate with maternal intake and urinary excretion of the vitamin [38].

6.3.4.4 Stability

In solution, pantothenic acid is most stable at pH 5–7. Pantothenic acid exhibits relatively good stability during food storage, especially at reduced water activity. Losses (from 30% to 80%) can occur in cooking and thermal processing in proportion to the severity of the treatment and extent of leaching [10]. Losses of 10% are experienced in milk processing [14].

6.3.4.5 Dietary Reference Values (DRVs): hypo- and hypervitaminosis

The EFSA Panel on Dietetic Products, Nutrition and Allergies (NDA) concluded in a report published in 2014 that there is insufficient evidence to derive an average requirement (AR) and a population reference intake (PRI) for pantothenic acid. An adequate intake (AI) for adults (men and women) was set at 5 mg/day based on observed intakes. This value also applies to pregnant women. For lactating women, an AI of 7 mg/day is proposed, to compensate for pantothenic acid losses through breast milk. For infants over 6 months, an AI of 3 mg/day is proposed, whereas for children and adolescents was set at 4 and 5 mg/day, respectively, based on observed intakes in the European Union (Table 6.3) [38, 39].

Deficiency symptoms have been described in subjects on a pantothenic acid antagonist and/or pantothenic acid-deficient diet and include mood changes, as well as sleep, neurological, cardiac, and gastrointestinal disturbances. The latter are due to inflammation caused by dysfunction of the epithelial barrier and the production of pro-inflammatory molecules [9]. The Scientific Committee on Food (SCF) of the European Commission has noted that pantothenic acid has a low toxicity based on evidence available from clinical studies that indicates that doses of up to 2 g/day do not represent a health risk for the general population [38].

An updated and thorough review of the biological properties of vitamins B_1, B_2, B_3, and B_5 can be found elsewhere [22].

6.3.5 Vitamin B_6

6.3.5.1 Structure

Vitamin B_6 is a generic descriptor for several derivatives of 3-hydroxy-5-hydroxymethyl-2-methyl pyridine that differ in the group located at the fourth position of the pyridine ring: hydroxymethyl (-CH_2OH) in pyridoxine (PN), aldehyde (-CHO) in pyridoxal (PL), and aminomethyl (-CH_2NH_2) in pyridoxamine (PM). The hydroxymethyl group at the fifth position of the pyridine ring can be esterified by phosphate forming pyridoxine-5'-phosphate (PNP), pyridoxal-5'-phosphate (PLP), and pyridoxamine-5'-phosphate (PMP) (Figure 6.1B). All vitamin B_6 derivatives (vitamers) can be found in foods.

6.3.5.2 *In vivo* function

Pyridoxal-5'-phosphate is the most common active form of vitamin B_6, acting as a cofactor for more than 100 enzymes involved in several types of biochemical reactions [40, 41]:

- In amino acid metabolism, PLP is a cofactor for enzymes participating in decarboxylation, transamination, and racemization reactions of amino acids reactions. The chemical role of PLP in these reactions has been well established [7].
    - Decarboxylation of amino acids produces amines, that can act as neurotransmitters or hormones (i.e. serotonin, taurine, dopamine, noradrenaline, histamine, and γ-aminobutyric acid), and other substances, like diamines and polyamines, involved in the regulation of DNA metabolism.
    - Transamination reactions, in which the amino group from one amino acid is transferred to a α-keto acid, allow the synthesis of dispensable amino acids and the interconversion and catabolism of all amino acids, with the exception of lysine.
    - Racemization reactions lead to the formation of racemic mixtures of D- and L-amino acids, which have a role in signaling during brain development.

- In one-carbon metabolism, PLP is a cofactor for both serine hydroxymethyltransferase and glycine decarboxylase, enzymes are responsible for the transfer of one-carbon units to folate derivatives, which are used for the synthesis of purine and pyrimidine nucleotides, remethylation of homocysteine to methionine and production of S-adenosylmethionine (SAM). SAM is involved in transmethylation reactions that activate a wide range of bioactive compounds including DNA. PLP is also a cofactor of cystathionine β-synthase and cystathionine γ-lyase, enzymes involved in the conversion of homocysteine to cysteine.
- In glycogenolysis, glycogen phosphorylase relies on PLP as a coenzyme in the enzymatic cleavage of glycogen that sequentially releases glucose-1-phosphate units.
- In gluconeogenesis, PLP-dependent transaminases convert gluconeogenic amino acids to α-keto acids to create substrates for the production of glucose.
- PLP is a cofactor for the δ-aminolevulinate synthase, a key enzyme in heme biosynthesis in erythrocytes. Symptoms associated with vitamin B_6 deficiency include hypochromic microcytic anemia.
- The metabolism of vitamin B_6 interacts with that of other vitamins like riboflavin, niacin, and folate. Synthesis of carnitine (Chapter 4, Section 4.9) also requires PLP, thus establishing a link between vitamin B_6 and lipid metabolism.

In addition to its coenzyme role, all forms of vitamin B_6 have been reported to be potent antioxidants and anti-inflammatory molecules, and to play a role in immune function, protein folding, and cancer [42, 43].

6.3.5.3 Sources and bioavailability

Foods rich in vitamin B_6 include grains (whole grain corn/maize, brown rice, sorghum, quinoa, wheat germ), pulses, nuts, seeds, potatoes, some herbs and spices (e.g., garlic, curry, ginger), meat and meat products (poultry, pork, liver), fish, eggs, and dairy products (Table 6.2a). Some plants contain glycosylated vitamin B_6 in the form of pyridoxine-5'-β-D-glucoside (PNG). Vitamin B_6 is used as an additive for foods (PN, PN hydrochloride, and PN dipalmitate) and food supplements (PN, PN hydrochloride, and PLP). The vitamin B_6 content of infant and follow-on formulae, processed cereal-based foods, and foods for infants and children is regulated in the European Union [41].

Intestinal absorption of vitamin B_6 is thought to take place in the jejunum by passive diffusion of the non-phosphorylated forms, although there is some evidence for a co-transport with protons. Dietary PLP, PMP, and PNP are enzymatically dephosphorylated at the brush-border membrane by alkaline phosphatase before absorption. Bioavailability of vitamin B_6 from a mixed diet is around 75%, whereas that from supplements is considered to be almost complete (around 95%) [40, 44]. The microbiota of the large intestine has also been suggested to be a significant source of vitamin B_6 for humans [9, 45]. Absorption takes place in the jejunum via diffusion, although in vitro studies have suggested that it may also be carrier-mediated by thiamine multispecific membrane transporters [46].

After intestinal absorption, vitamin B_6 derivatives (mainly PLP and PL) are transferred via the portal circulation to the liver, where they are metabolized or released back in the bloodstream for distribution to other tissues. Before secretion into the circulatory system from hepatocytes, PLP is bound to lysine residues of proteins, mostly albumin, which is the main transport protein for PLP. When reaching the target tissue, PLP disassociates from protein and is dephosphorylated by alkaline phosphatases in order to enter the cells, where it will be re-phosphorylated. Intracellular concentrations of PLP are tightly regulated [46]. Excretion of vitamin B_6 occurs mainly as 4-pyridoxic acid through the urine, loss of excreted active forms is minimized by reabsorption in the kidney tubules [41].

6.3.5.4 Stability

Thermal processing and storage can influence the vitamin B_6 content of foods. Chemical changes include interconversion of chemical forms of vitamin B_6 (mainly by non-enzymatic transamination), thermal and photochemical degradation, as well as irreversible complexation and reactions with other molecules like proteins, peptides, amino acids, or reducing sugars. Degradation rates are strongly dependent on the form of the vitamin, temperature, pH, and the presence of other compounds; however, these studies are complicated by the different reactivities exhibited by the multiple forms of vitamin B_6 and the interconversions that can occur among them. In any case, low pH values have been shown to stabilize all forms of vitamin. Exposure to water can cause leaching and consequent losses, as with other water-soluble vitamins. Commercial sterilization of evaporated milk or unfortified infant formula causes 40–60% loss of the naturally occurring vitamin B_6, probably due to the interaction of PL, the major naturally occurring form of the vitamin in milk, with proteins. This led to over 50 cases of convulsive seizures in infants in the early 1950s that were corrected by administration of PN [10]. Vitamin B_6 loss is 45% in cooking of meat and 20–30% in cooking of vegetables [14].

6.3.5.5 Dietary Reference Values (DRVs): hypo- and hypervitaminosis

The most suitable biomarker for evaluating the vitamin B_6 status is the concentration of PLP in plasma: a value of more than 30 nmol/l has been traditionally considered to be adequate in adult humans. Consequently, recommended intakes of vitamin B_6 are those that sustain a mean plasma PLP concentration above the cut-off of 30 nmol/l (Table 6.3).

Symptoms associated with vitamin B_6 deficiency include eczema, seborrheic dermatitis, cheilosis, glossitis, angular stomatitis, hypochromic microcytic anemia, hyperirritability, convulsive seizures, and neurological abnormalities. The latter are related to both the decrease in the synthesis of γ-aminobutyric acid, a major inhibitory neurotransmitter in the brain, and to the increased concentration of tryptophan metabolites in the brain that have a proconvulsant effect. Chronic vitamin B_6 deficiency can precipitate microcytic hypochromic anemia (small erythrocytes with reduced hemoglobin concentration) due to defective hemoglobin biosynthesis [41]. Various inborn errors leading to PLP deficiency manifest as vitamin B_6-responsive epilepsy [43].

The relationship between high intakes of vitamin B_6 and the development of peripheral neuropathy is well established both in humans and in animals. The SCF panel of EFSA have recently reported that there is evidence that symptoms of peripheral neuropathy may occur at supplemental vitamin B_6 intakes below the reference point of 100 mg/day previously used to establish the Tolerable Upper Intake Level (UL), actually, there is evidence that peripheral neuropathy may occur at supplemental intakes of 50 mg/day in some individuals. The possible mechanisms of vitamin B_6 intoxication concerning peripheral neuropathy have been review by Hadtstein and Vrolijk [47]. There is also some evidence that doses of 35–40 mg/day are associated with an increased risk of hip fracture, while doses of 600 mg/day may suppress lactation. With all this evidence available EFSA has recently established a new Tolerable Upper Intake Level (UL) for vitamin B_6 intake of 12 mg/day for adults, including pregnant and lactating women. For children UL values have been set as follows: 4–6 months, 2.2 mg/day; 7–11 months, 2.5 mg/day; 1–3 years, 3.2 mg/day; 4–6 years, 4.5 mg/day; 7–10 years, 6.1; 11–14 years, 8.6 mg/day; 15–17 years, 10.7 mg/day [46].

6.3.6 Biotin (vitamin B_7/H)

6.3.6.1 Structure

Biotin is a bicyclic vitamin: one ring contains an ureido group and the other contains sulfur and has a valeric acid side chain (Figure 6.1C). The ring system of biotin can exist in eight possible stereoisomers, only one of which, the (3aS, 4S, 6aR) compound, also known as D-(+)-biotin, is biologically active. The IUPAC name of D-(+)-biotin is 5-[(3aS,4S,6aR)-2-oxo-1,3,3a,4,6,6a-hexahydrothieno[3,4-d]imidazol-4-yl]pentanoic acid. The carboxyl group of biotin may form an amide bond with the ε-amino group of a lysine residue of certain proteins, this reaction (known as biotinylation) is catalyzed by the enzyme holocarboxylase synthetase (HLCS) [10, 14, 48].

6.3.6.2 *In vivo* function

In mammals, biotin is known to be the prosthetic group of five carboxylases: acetyl-CoA carboxylase (EC 6.4.1.2) isoforms I and II, pyruvate carboxylase (EC 6.4.1.1), methylcrotonyl-CoA carboxylase (EC 6.4.1.4), and propionyl-CoA carboxylase (EC 6.4.1.3) [48].

–   Acetyl-CoA carboxylase I (ACC1) is located in the cytosol and catalyzes the rate-limiting step of fatty acid synthesis (Chapter 4, Section 4.9.2).

- Acetyl-CoA carboxylase II (ACC2) resides at the outer mitochondrial membrane and seems to play a role in the control of fatty acid oxidation.
- Pyruvate carboxylase catalyzes the incorporation of bicarbonate into pyruvate to form oxaloacetate (OAA), an intermediate of citric acid cycle. In tissues such as liver and kidney, OAA can be converted to glucose via gluconeogenesis (Chapter 3, Section 3.8.2).
- Methylcrotonyl-CoA carboxylase catalyzes an essential step in the degradation of the branched-chain amino acid leucine.
- Propionyl-CoA carboxylase catalyzes the incorporation of bicarbonate into propionyl-CoA to form methylmalonyl-CoA; methylmalonyl-CoA undergoes isomerization to succinyl-CoA and enters the tricarboxylic acid cycle. This series of reactions is essential for the degradation of odd-number fatty acids [7].

Experimental evidence gathered for more than 50 years suggests that, beyond its role as a prosthetic group of carboxylases, biotin is involved in the regulation of gene expression, cell proliferation, repair of DNA damages, and in the stability of the chromatin structure. This role might explain some of the symptoms associated with biotin deficiency [48–50]. The mechanism of action of biotin at this level has been suggested to occur through the biotinylation of histones; however, some authors have proposed that histone biotinylation marks play no direct role in gene repression but they are just a side effect of HLCS being in close physical proximity to histones. There is also evidence that the number of proteins containing covalently bound biotin is larger than previously thought [51].

6.3.6.3 Sources and bioavailability

Biotin is widely distributed in plant and animal products (Table 6.2b) and deficiency is rare in healthy humans. At usual intakes, free biotin is absorbed nearly completely; however, biotin is present as a protein-bound vitamin in most foods and must be released by the pancreatic form of the enzyme biotinidase that acts on biotinylated peptides or biocytin (biotinyl-lysine). In human milk, more than 95% of the biotin is free in the skim fraction, although this concentration varies substantially in some women. In any case, it exceeds the concentration in serum by one to two orders of magnitude, which suggests a transport system into milk [48].

There is a lack of data on the level of absorption of protein-bound biotin from foods in humans. Fecal excretion of biotin has been observed to be several times higher than intakes, due to production by the intestinal microbiota; however, the extent to which this biotin is absorbed from the large intestine is uncertain [11].

Avidin, a protein found in raw egg white, has a very high affinity for biotin and prevents its absorption in the small intestine, consequently, chronic consumption of raw eggs impairs biotin absorption and can lead to deficiency. Cooking denatures avidin and eliminates its biotin-binding properties [10]. The strong interaction between avidin and biotin is the chemical basis of the most accurate and sensitive method of biotin quantification, the high-performance liquid chromatography (HPLC)/avidin-binding assay (Table 6.2b) [13].

Free biotin is transported across the brush border of enterocytes by a variety of sodium-dependent transporters, the most studied of which being the sodium-dependent multivitamin transporter (SMVT), also involved in the uptake of pantothenic acid and lipoic acid [49, 52]. Biotin is distributed around the organism via bloodstream, either free or associated with plasma proteins. Uptake of biotin by many tissues and organs seems to be mediated by SMVT [48]. See also Chapter 13.

Within the cell, biotin is transformation into its activated form, biotinyl-AMP, in an ATP-dependent reaction catalyzed by the enzyme holocarboxylase synthetase (HCS). Biotinyl-AMP is subsequently used as substrate by HCS in order to establish an amide linkage between the carboxylic group of the biotin moiety and the ε-amino group of a lysine residue. This reaction, that requires ATP, is mediated by the enzyme. Intracellular biotin can be recycled after proteolytic degradation of endogenous biotin-dependent carboxylases and subsequent hydrolysis of biotinylated peptides or biocytin. This so-called biotin cycle ensures the adequate biotinylation status of carboxylases and maintains a delicate balance between utilizing dietary biotin and recycling the endogenous vitamin [49, 53].

6.3.6.4 Stability
Biotin is stable to heat, light, and oxygen but high or low pH values can cause inactivation, probably by hydrolysis of the amide bond of the biotin ring. Oxidizing conditions may lead to formation of biologically inactive biotin sulfoxide or sulfone. During food processing and storage, losses may occur by chemical degradation or by leaching of free biotin. Little degradation of biotin occurs during low moisture storage of fortified cereal products. Overall, biotin is quite well retained in foods. The stability of biotin in human milk is also high [10].

6.3.6.5 Dietary Reference Values (DRVs): hypo- and hypervitaminosis
The EFSA Panel on Dietetic Products, Nutrition and Allergies have reported that there are biomarkers sensitive to biotin depletion. These include urinary excretion of biotin and derived metabolites like 3-hydroxyisovaleric acid (3HIA) and 3HIA-carnitine, activity of propionyl-CoA carboxylase and others; however, the ability of these markers to discriminate between biotin insufficiency and adequacy are not well known. Consequently, the panel has concluded that there is insufficient evidence to derive an AR and a PRI for biotin, alternatively AI for adults is proposed based on observed intakes in the EU (Table 6.3) [11, 39].

Dietary biotin deficiency is rare but has been clearly documented in three situations: prolonged consumption of raw egg white, parenteral nutrition without biotin supplementation, and infant feeding with an elemental formula devoid of biotin.

Symptoms of biotin deficiency include periorificial dermatitis, conjunctivitis, alopecia, ataxia, and developmental delay. Biotin deficiency during pregnancy is teratogenic in several species. In humans, a marginal degree of biotin deficiency develops in at least one-third of women during normal pregnancy. Data obtained from a multivitamin supplementation study suggest that this deficiency might be teratogenic. Biotin deficiency has also been reported or inferred in several other circumstances like chronic alcoholism, gastrointestinal diseases (perhaps through an effect on intestinal biotin uptake), Leiner disease (a severe form of seborrheic dermatitis that occurs in infancy), and renal dialysis [48].

Some biotin-dependent inherited metabolic disorders have been described [49]:

- The neonatal form of multiple carboxylase deficiency (MCD) is caused by mutations in the gene that codes for holocarboxylase synthetase (HCS), that can reduce the affinity for biotin of the mutated enzyme up to 70-fold compared to normal enzyme. In these patients, the metabolic homeostasis is severely compromised by reduction in the activities of all biotin-dependent enzymes. Onset of the disease usually occurs from within a few hours of birth until 15 months of age with symptoms similar to those of biotin deficiency. Biochemical and clinical manifestations can be successfully reversed with pharmacological doses of biotin (10–100 mg/day).
- The juvenile form or late-onset MCD, whose incidence is 1 in 61,067 births, is caused by mutations in the gene encoding biotinidase. This disease is highly heterogeneous and patients can exhibit profound or partial biotinidase deficiency. The latter are unable to release dietary protein-bound biotin or to recycle the endogenous cellular vitamin during carboxylase turnover, thus, exhibiting a severe reduction in the enzymatic activity of all biotin-dependent carboxylases. The clinical findings and biochemical abnormalities caused by biotinidase deficiency are similar to those of biotin deficiency (dermatitis, conjunctivitis, alopecia, ataxia, and developmental delay); however, some patients with biotinidase deficiency develop neurological disorders that cannot be ameliorated or reversed with pharmacological doses of biotin. These neurological disorders include mental retardation, hearing loss, optic nerve atrophy, myelopathy, and Leigh syndrome. The particularities of biotinidase deficiency are not fully understood.

The role of biotin in the development and treatment of neurological disorders such as biotin-thiamine-responsive basal ganglia disease and progressive multiple sclerosis has also been studied, although further clinical and molecular evidences are needed in order to fully understand the effect of biotin on the central nervous system [49].

As far as toxicity of biotin is concerned even though daily doses up to 200 mg orally and up to 20 mg intravenously have been given to treat biotin-responsive inborn errors of metabolism and acquired biotin deficiency, no toxicity has been reported to date [48].

6.3.7 Vitamin B$_9$ (folic acid)

6.3.7.1 Structure
Vitamin B$_9$ is also known as vitamin B$_c$, vitamin M, *Lactobacillus casei* factor, folacin, pteroyl-L-glutamic acid, folic acid, and folate. It was first discovered by Wills and Mehta in 1931 as a factor in yeast that corrected the macrocytic anemia of pregnant women in India. The factor was later isolated from spinach leaves and was given the name folic acid [54].

Folate is a generic term used for a group of compounds with a basic structure consisting of a substituted pterin (6-methylpterin) linked through a methylene bridge to *p*-aminobenzoic acid, to which one or more glutamate residues are bound by peptide bonds. The latter are required to retain folates within the cell and subcellular organelles, whereas the number of glutamate residues determines the affinity of folate cofactors for folate-dependent enzymes.

The pterin moiety can exist in three oxidation states: oxidized (shown in Figure 6.1C, IUPAC name when completely protonated: (2S)-2-[[4-[(2-amino-4-oxo-3H-pteridin-6-yl) methylamino]benzoyl]amino]pentanedioic acid), partially reduced as 7,8-dihydrofolate, and fully reduced as 5,6,7,8-tetrahydrofolate. The latter, simply known as tetrahydrofolate or THF, carries one-carbon units at different oxidation levels, so that different forms of THF are present in cells: N^5-formyl-THF; N^{10}-formyl-THF; N^5,N^{10}-methenyl-THF; N^5-formimino-THF; N^5,N^{10}-methylene-THF; and N^5-methyl-THF. These forms are interconverted through enzyme-mediated catalysis [7, 55].

6.3.7.2 *In vivo* function
THF functions as a cofactor or co-substrate in numerous one-carbon transfer reactions:
- N^{10}-formyl-THF form provides one-carbon units for the formation of purine nucleotides (adenine and guanine) for synthesis of both DNA and RNA.
- N^5,N^{10}-methylene-THF is a cofactor in the reaction that generates thymidine monophosphate, a pyrimidine nucleotide specific for DNA.
- N^5-methyl-THF can be used as a cofactor for homocysteine remethylation to methionine, in a vitamin B$_{12}$-dependent reaction catalyzed by methionine synthase (Section 6.3.8.2). This implicates THF in the metabolism of methionine, an amino acid that can be used for protein synthesis and also for the generation of S-adenosylmethionine (SAM). SAM is a potent alkylating agent that donates its methyl group to wide range of substrates such as DNA, hormones, proteins, neurotransmitters, and membrane phospholipids, all of which are regulators of important physiological processes [7, 55, 56].
- THF is involved in the catabolism of at least three proteinogenic amino acids, histidine, serine, and glycine, as well as choline [57].

Enzyme-mediated oxidation of N^{10}-formyl-THF with concomitant reduction of $NADP^+$ to NADPH has been proposed to be a significant source of NADPH both in mitochondria and in the cytosol [57]. Folate metabolism is compartmentalized among the cytoplasm, the nucleus, and the mitochondria, and requires vitamins B_2, B_3, B_5, B_6, and B_{12}, as well as choline, a nitrogen-containing compound involved in a broad range of physiological functions [54].

6.3.7.3 Sources and bioavailability

Naturally occurring folates are found in a wide variety of foods, where they occur as reduced vitamers usually containing from five to seven glutamate residues. Few foods can be considered particularly rich sources; among those are dark green leafy vegetables, legumes, orange, and grapefruit (juice), peanuts, and almonds. Another rich source of folate is baker's yeast. Most fruits and vegetables, as well as meat (with the exception of liver and kidney), contain small amounts of folate. Potatoes and dairy products are not rich sources, but they contribute to natural folate intake because they are consumed in relatively large quantities.

In European countries, folic acid and calcium-L-methylfolate may be added to foods; as a consequence, fortified foods, such as breakfast cereals and some fat spreads are major contributors to folic acid intake. The folate content of infant and follow-on formulae, processed cereal-based foods, and foods for infants and young children is regulated within the EU [55].

The bioavailability of food folate, defined as the fraction of ingested folate that is absorbed and can be used for metabolic processes or storage, has been estimated to be around 50%, whereas that of folic acid from fortified foods or from a supplement ingested with food is about 85%. However, studies of folate bioavailability are hampered by the fact that it is influenced by a high number of factors such as the source and the amount ingested, as well as individual factors such as folate status, health, age, gender, gastrointestinal function, and the use of medication and alcohol [55, 58]. Because the absorption efficiency of synthetic and natural folates varies, dietary folate equivalents (DFE) have been defined by the Institute of Medicine: 1 µg DFE = 1 µg food folate = 0.6 µg folic acid from fortified food or as a supplement consumed with food = 0.5 µg of a folic acid supplement taken on an empty stomach. For combined intakes of food folate and folic acid, DFEs can be computed as follows: µg DFE = µg food folate + (µg folic acid × 1.7). This definition is based on evidence that folic acid has a higher bioavailability than food folate [55].

Folic acid can be absorbed directly; however, the polyglutamate chain of natural folates must be removed, mainly by glutamate carboxy-peptidase II (folate conjugase) present in the brush border of the small intestine, prior to absorption and transport as a monoglutamate into the portal vein. The monoglutamate forms of folate are transported through the action of specific transporters into the enterocyte (Table 13.2), where they are reduced to di- and tetrahydrofolate (THF). THF is subsequently converted to N^5,N^{10}-methylenetetrahydrofolate and N^5-methyltetrahydrofolate and transferred to the hepatic portal vein in order to be distributed to the rest of the organism. Synthetic folic acid gains metabolic activity after reduction to tetrahydrofolate by the enzyme dihydrofolate reductase in the enterocyte, although there is also evidence that it can enter the portal vein unchanged, with reduction and methylation taking place only once it reaches the liver [58].

Serum/plasma folate concentration is considered a sensitive marker of recent dietary intake; however, a single measurement of serum/plasma folate cannot be informative of folate status. For the assessment of folate status, multiple measurements of serum folate should be taken over a period of several weeks or a single measurement should be combined with other biomarkers of folate status. Serum folate concentrations of less than 6.8 nmol/L, confirmed on consecutive occasions, indicate folate deficiency. Red blood cell folate is an indicator of long-term dietary intake and responds slowly to changes in intake. Red blood cell folate is the most reliable biomarker of folate status, as it reflects tissue folate stores. Values of this parameter below 317 nmol/L are indicative of folate deficiency. Folates are eliminated via urine and feces, normally in the form of breakdown products. Average folate concentration of breast milk is about 80 µg/L, this value not being dependent on dietary folate intake or status of the lactating women. Folates in breast milk are bound to folate-binding proteins that stimulate the absorption of this vitamin by infants and protect it from degradation and utilization by intestinal microbiota [55].

6.3.7.4 Stability

Folates are susceptible to oxidative degradation during food processing, which is enhanced by oxygen, light, and heat. Oxidation results in a splitting of the molecule into biologically inactive forms, such as *p*-aminobenzoyl-glutamate. The susceptibility of folate to oxidation is largely influenced by the pH of the medium, folate derivatives, and the food system. In addition to susceptibility to oxidative degradation, folates are readily leached from foods by aqueous cooking media. Unlike natural folates, folic acid exhibits excellent retention during the processing and storage of fortified foods and premixes, including fortified infant formulae and medical formulae [10, 58].

6.3.7.5 Dietary Reference Values (DRVs): hypo- and hypervitaminosis

Dietary Reference Values for folate intake published by the EFSA Panel on Dietetic Products, Nutrition and Allergies are summarized in Table 6.3. Values for adults are based on biomarkers of folate status; however, it must be mentioned that measuring folate levels is particularly challenging because of the multiple chemical forms of the vitamin *in vivo* and its chemical lability. Values of AR and PRI for children were extrapolated from those for adults using allometric scaling and age-specific growth factors, whereas for lactating women AR and PRI are derived considering their need to compensate for the amount of folate secreted in breast milk. AIs are proposed for pregnant women and for infants aged 7–11 months. The latter is based on folate intake from breast milk extrapolated from infants aged 0–6 months [39, 55].

Folate deficiency impairs DNA replication and cell division, which adversely affects rapidly proliferating tissues such as bone marrow and results in megaloblastic anemia, that is, the occurrence of unusually large red blood cells (macrocytic cells) with abnormal nuclear maturation. Although megaloblastic anemia is typical of folate deficiency, it can also occur as a result of cobalamin deficiency, owing to the metabolic interactions of the two vitamins (see above). It has also been reported that folate deficiency is associated with structural damage of DNA as a consequence of misincorporation of uracil instead of thymine, which might have implications for cancer development [55].

Maternal folic acid supplementation prevents up to 70% of neural tube defects (NTDs), the most common and severe of which include spina bifida. On the contrary, clinical trials have not conclusively validated observational studies indicating a preventive role for folate in cardiovascular disease and cancer. Folates can also have an important role in epigenetics. This term is often used to describe the transmission of DNA methylation patterns and other covalent chromatin modifications that may be heritable (Chapter 14). Dietary folate and intake of other B vitamins and metabolites of one-carbon metabolism might induce alterations in DNA methylation patterns, thus altering gene expression patterns. These genomic changes can result in risk phenotypes such as obesity and metabolic syndrome that increase the risk of adult-onset disease including cardiovascular disease and certain cancers [54].

Natural food folate is considered safe, and high intakes have not been associated with any adverse effects; however, based on safety concerns for high intake of folic acid, related mainly to individuals with cobalamin deficiency, the Lowest Observed Adverse Effect Level (LOAEL) was set at 5 mg/day and the Tolerable Upper Intake Level (UL) at 1 mg/day for adults by the European Scientific Committee for Food. The UL also applies to pregnant or lactating women. ULs for children were derived from the adult value on the basis of body weight, ranging from 200 µg/day (1–3 years) to 800 µg/day (15–17 years). Some observational and clinical studies have suggested that the use of folic acid supplements is inversely associated with cancer incidence, although this effect relates to intakes at or above the currently accepted UL. Concerns have also been raised regarding the presence of unmetabolized folic acid in the circulation; however, the metabolic and biological consequences of this situation are uncertain [55].

6.3.8 Vitamin B$_{12}$

6.3.8.1 Structure

Minot, Murphy, and Whipple demonstrated that the symptoms of pernicious anemia could be overcome through the addition of liver to the diet. The nutrient responsible was eventually isolated from liver after addition of cyanide and extraction into organic solvent in 1948. Dorothy Hodgkin revealed that the compound, called cyanocobalamin, was a cyanolated, cobalt-containing. Although technically the term vitamin B_{12} refers to cyanocobalamin (IUPAC Name: cobalt(3+);[(2R,3S,4R,5S)-5-(5,6-dimethylbenzimidazol-1-yl)-4-hydroxy-2-(hydroxymethyl)oxolan-3-yl] [(2R)-1-[3-[(1R,2R,3R,5Z,7S,10Z,12S,13S,15Z,17S,18S, 19R)-2,13,18-tris(2-amino-2-oxoethyl)-7,12,17-tris(3-amino-3-oxopropyl)-3,5,8,8,13,15,18,19-octamethyl-2,7,12,17-tetrahydro-1H-corrin-24-id-3-yl]propanoylamino]propan-2-yl]phosphate;cyanide), it is used to name a number of derivatives of cobalamin that have biological activity or that can be converted to biologically active compounds *in vivo*. Cobalamin is a metal complex constituted by a cobalt ion coordinated to a corrin ring (a planar tetrapyrrole, similar to that of chlorophyll and hemoglobin) via the four pyrrole nitrogen atoms. A 5,6-dimethylbenzimidazole is linked to cobalt in the α-position below the corrin plane through a phosphoribosyl moiety. The upper or β-axial ligand varies giving rise to different chemical forms of the vitamin (Figure 6.1D) [59, 60].

- Methylcobalamin and 5′-deoxyadenosylcobalamin, that present a methyl-, and adenosyl- group, respectively, are the actual coenzymes *in vivo*.
- Hydroxocobalamin (hydroxo- group) and aquocobalamin (aquo- group) are intermediates formed during the synthesis of the coenzyme forms.
- Cyanocobalamin, that has cyano-group, is a stable synthetic form used as a food additive, and in food supplements and drugs.

Other forms of cobalamin, including sulfite, nitrite and glutathionyl derivatives have also been described in human cells, although their metabolic role is unknown [61]

6.3.8.2 *In vivo* function

In contrast to the broad range of enzymes and molecular processes involving vitamin B_{12} in prokaryotic systems, in eukaryotes (mammals, birds, fish, worms, and some protists) is restricted to just methionine synthase and methylmalonyl-CoA mutase:

- Methylcobalamin acts as methyl group donor in the conversion of homocysteine to methionine, mediated by methionine synthase. Cobalamin, a product of this reaction, is reconverted to methylcobalamin by using N^5-methyl-THF (see Section 6.3.7). Mechanistically, this process depends on the ability of the central cobalt ion to form a Co(I) species. The close metabolic relationship between folates and cobalamin explains why deficiencies in vitamins B_9 and B_{12} share several symptoms. Vitamin B_{12} does not occur in organisms like plants that have a cobalamin-independent methionine synthase [59]. This has important implications for human nutrition.
- Methylmalonyl-CoA mutase (MCM) catalyzes the reversible isomerization of L-methylmalonyl-CoA to succinyl-CoA using adenosylcobalamin as a cofactor. The importance of MCM in mammals is related to its key role in the degradation of the amino acids valine, isoleucine, methionine, and threonine, odd-chain fatty acids, and cholesterol. The enzyme is synthesized in the cell cytoplasm as a larger

precursor and is subsequently exported to the mitochondrial matrix where it is processed, thus generating the mature active enzyme. The mechanism of action of this enzyme as well as the role of adenosylcobalamin has been studied in detail [62].

6.3.8.3 Sources and bioavailability

The ability to biosynthesize vitamin B_{12} is restricted to certain prokaryotes. Some mammals, especially ruminants, are able to absorb this vitamin from the enteric bacteria that live in their intestines. For this reason ruminants accumulate more of this vitamin in their tissues than monogastric animals, such as poultry and pigs; moreover, older cattle have higher concentrations because the bioaccumulation in tissues is time-dependent. In addition, it has been reported that different cuts show variable concentrations depending on their oxidative or glycolytic metabolism tendency [63].

Humans have a rich microbiota fauna mainly in the large intestine; however, their vitamin B_{12} uptake system is located in the small intestine and they have to rely on dietary sources. Some bacteria growing in foods or beverages can provide vitamin B_{12}, thus, paradoxically, hygienic measures in food processing can lower the levels of this vitamin. Some authors suggest that this process partly explain the finding that the average plasma concentration of vitamin B_{12} in urban middle-class Indian men is lower than that measured in men living in urban slums [59].

Cobalamin is not a normal constituent of plant foods unless they contain certain microorganisms, have been exposed to microbial fermentation, or have been fortified with cobalamin. The main sources of the vitamin are animal products, including meat, fish, dairy products, eggs, and liver (Table 6.2c). In the European Union, cyano- and hydroxocobalamin may be added to both foods (including those for infant and young children) and food supplements. 5′-deoxyadenosyl- and methylcobalamin can be added to food supplements only [60]. It is worth mentioning that some foods have been found to contain both active and inactive forms of cobalamin, for example, most of the edible cyanobacteria used for human supplements, like *Spirulina*, predominately contain pseudovitamin B_{12}, which is inactive in humans but may interfere with the adsorption of active forms. This has to be considered when evaluating a given foods as a source of vitamin B_{12} [63, 64].

In humans, several proteins facilitate the absorption of vitamin B_{12} from dietary sources. Among these are haptocorrin, intrinsic factor, and transcobalamin. Haptocorrin (HC) is excreted into the saliva and upper part of the gastrointestinal tract, and it is also found in blood serum. This protein has a high affinity for cobalamin and analogs and binds them as they are released from the food source in mouth and stomach. HC (and other vitamin B_{12}-bound protein complexes) is further digested in the small intestine to release the vitamin that is subsequently bound by the intrinsic factor (IF). The latter is a protein produced by the parietal cells of the stomach that is more specific for cobalamin than HC. In ileum enterocytes, the IF-cobalamin complex is taken up by receptor-mediated endocytosis into the lysosome. Here, IF is degraded and the liberated cobalamin is released into the cytosol and exported to the bloodstream via the multidrug resistance protein MRP1. In the bloodstream around 80% of cobalamin and analogs are bound to HC, whereas the remainder binds to the more specific transcobalamin (TC). Only TC is able to facilitate the uptake of cobalamin into cells of peripheral tissues (also by receptor–mediated endocytosis), whereas the role of circulating HC is not fully understood. The trafficking within cells and its distribution between cytosol and mitochondria has been properly reviewed elsewhere [59].

On the basis of some experimental data, it has been generally assumed that around 50% of cobalamin in a typical meal is actively absorbed in healthy individuals; however, fractional absorption of cobalamin seems to be highly variable and depends on factors such as dietary source, amount ingested, the ability to release cobalamin from food, and the proper functioning of the IF system. The average cobalamin content of the body in healthy adults is estimated to be 2–3 mg, half of that amount being found in the liver. On the other hand, the main route of cobalamin excretion is via the feces, and daily losses of 0.1–0.2% of the cobalamin pool have been observed [60].

6.3.8.4 Stability
The stability of vitamin B_{12} depends on a number of conditions:
- It is fairly stable at pH 4–6, even at high temperatures. Lower pH values can cause the hydrolytic removal of the nucleotide moiety, with additional fragmentation occurs as the severity of the acidic conditions increases. Exposure to acid or alkaline conditions causes hydrolysis of amides, yielding biologically inactive carboxylic acid derivatives.
- In alkaline media or in the presence of reducing agents, such as ascorbic acid or SO_2, the vitamin is destroyed to a greater extent.
- In milk, one of the most important sources of vitamin B_{12} for humans, more than 90% mean retention has been observed during HTST and UHT processing, but storage of UHT-processed milk at ambient temperature for up to 90 days causes progressive losses that can approach 50% of the concentration.
- Loss of vitamin B_{12} is dependent on the temperature and time of conventional cooking and is further affected by other food ingredients.
- Storage in light for a long period can induce degradation of both vitamins B_2 and B_{12} in various types of liquid milk. An appreciable loss of B_{12} in multivitamin/mineral supplements because of degradation by addition of a substantial amount of vitamin C in the presence of copper.
- Chloramine-T (N-chloro-p-toluenesulfonamide), an active chlorine compound with strong oxidative activity used as in the food industry as a disinfectant for plant sterilization, produces a lactone form of vitamin B_{12}, that might block mammalian B_{12} metabolism.

Cyanocobalamin, the synthetic form of vitamin B_{12} used in food fortification and nutrient supplements, exhibits superior stability to other cobalamin derivatives and is readily available commercially. Therefore, fortified foods have become a particularly valuable source of vitamin B_{12}, especially for population groups with special needs like elderly people [64].

6.3.8.5 Dietary Reference Values (DRVs): hypo- and hypervitaminosis

Humans require vitamin B_{12} for only two enzymes and our bodies tend to retain and recycle cobalamin; therefore, the daily requirement for this vitamin is very low. The EFSA Panel on Dietetic Products, Nutrition and Allergies has concluded that there is insufficient evidence to derive an AR and PRI for cobalamin. However, it sets an AI of 4 µg/day for adults, based on data on cobalamin status, collected with different biomarkers, and in consideration of observed cobalamin intakes in several EU countries [39, 60]. Values adapted for different groups of population are summarized in Table 6.3. It has been reported that the percent bioavailability of vitamin B_{12} is inversely proportional to the dose consumed due to saturation of the active absorption process which is capable of binding only 1.5–2 µg, that is, within the range of usual intake from foods. Therefore, spreading discrete intakes of vitamin throughout the day is the best way to optimize absorption [65].

Vitamin B_{12} deficiency has been traditionally associated with a type of megaloblastic anemia, known as pernicious anemia, and evaluated by measuring total serum or plasma levels, the cutoff value for serum vitamin B_{12} being originally set at 150 pmol/l by the WHO and the US Institute of Medicine. However, pernicious anemia is a particularly severe form of vitamin B_{12} deficiency and other undesirable outcomes (including NTDs, stroke, and dementia) inversely correlate with serum levels in the range of 150 to about 350 pmol/l. The brain is particularly vulnerable, especially in children, whose brain and intellectual development can be hampered by an inadequate vitamin B_{12} status [59].

Total serum/plasma concentration is a limited biomarker for vitamin B_{12} deficiency, because it measures cobalamin bound to either haptocorrin or transcobalamin. The former binds the majority of cobalamin derivatives, but up to 60% of these molecules are inactive forms; by contrast, TC only binds biologically active forms. Nowadays, several biochemical markers like serum TC-cobalamin, serum methylmalonic acid, and plasma total homocysteine are used to evaluate possible deficiencies with better accuracy [59, 60].

Recent studies suggest that, unlike severe deficiency, suboptimal vitamin B_{12} status is very common and tens of millions of people in the world, in particular pregnant women and people in less-developed countries, may be suffering harm for this reason. Several causes have been linked with vitamin B_{12} deficiency, these include: inadequate dietary intake (veganism, highly restricted diets), gastrointestinal malabsorption (due to pernicious anemia, atrophic gastritis, total or partial gastrectomy, or other gastric surgery, ileal disease or damage, parasites, etc.), nitrous oxide toxicity, and metabolic and genetic disorders. Pernicious anemia is known to be due to a lack of intrinsic factor usually caused by an autoimmune condition that targets the parietal cells [66]; however, other aspects of vitamin B_{12} deficiency/insufficiency are far from being understood and definitely require further research [59].

Long-term oral or parenteral administration of daily cobalamin doses between 1 and 5 mg given to patients with compromised cobalamin absorption have not revealed adverse effects. There is no evidence relating cobalamin to teratogenicity or adverse effects on fertility or post-natal development. Cobalamin has not been found to be carcinogenic or genotoxic *in vitro* or *in vivo*. Thus, no adverse effects were identified that could be used as a basis for deriving a Tolerable Upper Intake Level (UL) [60].

6.3.9 Vitamins of complex B and gut microbiota

Trillions of microbes are present in the guts of mammals, most of them in the large intestine. These complex communities have been thoroughly studied during the last decades due to their involvement in multiple aspects of their hosts' health. One of these aspects is their role as a source of vitamins of complex B (B-vitamins). Metagenomic (direct genetic) analyses of the genomes present in the human gut microbiota predict that many intestinal bacteria possess synthesis pathways for several B-vitamins; however, other bacteria need some vitamins although they lack the ability to synthesize them. Therefore, there may be a competition between producers and utilizers so that it is not easy to know the actual importance of the supply of vitamins from gut microbiota to the host [9, 67]. The fact that many B-vitamin transporters are expressed in colonocytes (see Table 13.2) suggests that the microbial communities that dwell in our guts are real B-vitamin providers. Moreover, experiments done with germ-free animals show that they require more vitamins than conventional animals [67]. The metabolism of dietary and microbial B-vitamins has been reported to affect host immunity [9].

6.3.10 B vitamins and mitochondrial function

Mitochondria have traditionally been considered as cells' powerhouses as they are estimated to provide over 90% of the ATP required for cell metabolism. Nowadays these organelles are known to perform multiple cellular functions beyond energy production, thus, mitochondria can regulate gene expression within the nucleus, modulate synaptic transmission within the brain, release molecules that contribute to oncogenic transformation and trigger inflammatory responses systemically, and influence the regulation of complex physiological systems. Due to their central role in the cell, mitochondria are also involved in the pathogenesis and progression of numerous human diseases [68, 69].

B vitamins are known to be essential for mitochondrial function and, consequently, for our health. A brief summary of how B vitamins are involved in mitochondrial metabolism is listed below [70]:

– Thiamine (vitamin B_1): 90% of all cellular thiamine is in the mitochondria where they act as co-factors of several enzymatic complexes: alpha-ketoglutarate (α-KG) dehydrogenase (OGDH), pyruvate dehydrogenase (PDH), and branched-chain keto-acid dehydrogenase. Besides, thiamine acts as an allosteric regulator of the malate-aspartate shuttle and acetyl-CoA metabolism and plays a pivotal role in oxidative stress (Chapter 8).

– Riboflavin (vitamin B_2): As described above riboflavin is the precursor of FAD. Five acyl-CoA dehydrogenases present in the mitochondria require FAD and catalyze the first step in each cycle of β-oxidation: isovaleryl CoA dehydrogenase, branched-chain acyl CoA dehydrogenase, 2-methyl-branched-chain acyl CoA dehydrogenase, isobutyryl CoA dehydrogenase, and short-chain acyl CoA dehydrogenase. FAD also catalyzes redox reactions, supports antioxidant defense mechanisms, and is itself an antioxidant.

– Niacin (vitamin B_3) is the precursor of NADH and NADPH. NADH serves as a central hydride donor to ATP synthesis in the so-called oxidative phosphorylation, the main mechanism of ATP production, that is located in the inner membranes of mitochondria [7]. NADPH is involved in control of oxidative stress.

– Pantothenic acid (vitamin B_5), precursor of Coenzyme A (CoA), which is essential for the mitochondrial enzymes PDH and alpha-keto glutarate dehydrogenase of the TCA cycle as well as for the beta oxidation pathway. CoA also regulates glucose and fatty acid oxidation and serves as a cofactor for the biosynthesis of ketone bodies. Its concentration in mitochondria (2.2 mM) is several orders of magnitude higher than in peroxisomes or in the cytoplasm. Pantothenic acid also protects mitochondrial constituents from oxidative damage by modulating the levels of certain metabolites and enzymatic activities.

– Pyridoxal phosphate, the biologically active form of vitamin B_6, is specifically involved in amino acid metabolism, de novo synthesis of NAD^+ and iron-sulfur (Fe-S) cluster biosynthesis. Vitamin B6, although not being classified as classical antioxidant compound, has been shown to have antioxidant properties.

– Biotin (vitamin B_7) can enhance mitochondrial biogenesis and is required as cofactor by four carboxylases found within mitochondria: pyruvate carboxylase, propionyl-CoA carboxylase, methylcrotonyl-CoA carboxylase (MCCC), and acetyl-CoA carboxylase beta (ACC2 or ACACB). The latter is localized in the mitochondrial outer membrane and is thought to regulate fatty acid oxidation. Biotin has been reported to limit mitochondrial reactive oxygen species (ROS, Chapter 8) production.

– Folate (vitamin B_9). The network of folate metabolism is compartmentalized within mammalian cells in the nucleus, the cytosol, and in the mitochondria. Folate deficiency results in the accumulation of deletions in mitochondrial DNA in model systems, which may impair mitochondrial function and energy production.

Folate deficiency has been reported to produce a huge depletion of mitochondrial reduced glutathion (GSH) pool. GSH is an antioxidant metabolite (Chapter 8).

– Cobalamin (vitamin B_{12}) is a cofactor for methylmalonyl-CoA mutase (converts methylmalonyl-CoA into succinyl-CoA) in mitochondria. Cobalamin deficiency leads to accumulation of methyl-malonic acid (MMA), that has been reported to impair mitochondrial function. Several studies have reported the effectiveness of physiologically relevant concentration of vitamin B_{12} against mitochondrial oxidative stress.

6.4 Structure and *in vivo* function of lipid-soluble vitamins

6.4.1 Vitamin A

6.4.1.1 Structure

The term vitamin A comprises *all-trans*-retinol (simply called retinol, IUPAC Name: (2E,4E,6E,8E)-3,7-dimethyl-9-(2,6,6-trimethylcyclohexen-1-yl)nona-2,4,6,8-tetraen-1-ol) and the family of naturally occurring molecules associated with its biological activity (retinal, retinoic acid, and retinyl esters), as well as provitamin A carotenoids (β-carotene, α-carotene, and β-cryptoxanthin, Chapter 4) that are dietary precursors of retinol. *Retinoids* is another widely used term that refers to vitamin A and structurally related compounds, including its metabolites and synthetic analogs. The major synthetic forms of vitamin A are retinyl palmitate and retinyl acetate that are produced commercially and used in the production of animal feeds, nutritional supplements, and food fortification. Many synthetic analogs have pharmacologic activity, such as Am80/580, hydroxyphenyl-retinamide (fenretinide), and acitretin. Chemically, retinol is composed of a β-ionone ring, a polyunsaturated side chain and a polar end group (Figure 6.2A). This chemical structure makes it poorly soluble in water but easily transferable through membrane lipid bilayers. There are variations of retinol with "partial" vitamin A activity, thus, α-retinol, found in tropical oils such as red palm oil and in carrots, has approximately one half of the bioactivity of retinol, whereas vitamin A_2 (3,4-didehydroretinol), present in freshwater fish, is a metabolite of vitamin A in human skin, and has approximately 40% the bioactivity of retinol [71].

6.4.1.2 *In vivo* function

Vitamin A and its derivatives, mainly retinoic acid, are involved in the regulation of processes like reproduction, embryogenesis, vision, growth, cellular differentiation and proliferation, maintenance of epithelial cellular integrity, and immune function. Several biochemical and molecular studies have shown that the mechanism of action of retinoic acid is similar to that of the steroid hormones (Chapter 15); moreover, the

(A)

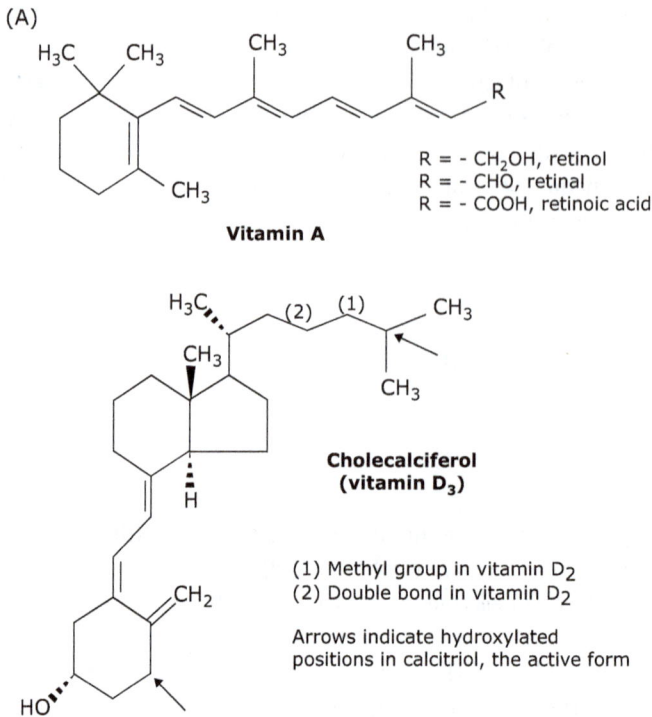

R = - CH₂OH, retinol
R = - CHO, retinal
R = - COOH, retinoic acid

Vitamin A

**Cholecalciferol
(vitamin D₃)**

(1) Methyl group in vitamin D₂
(2) Double bond in vitamin D₂

Arrows indicate hydroxylated
positions in calcitriol, the active form

Figure 6.2: Structures of fat-soluble vitamins. (A) Vitamins A and D3. (B) Vitamins E and K.
Structural information obtained from the National Center for Biotechnology Information. PubChem
Database (accessed on June 16, 2023):
Vitamin A, CID = 445354, https://pubchem.ncbi.nlm.nih.gov/compound/Retinol: Retinol
Vitamin D, CID = 5280795, https://pubchem.ncbi.nlm.nih.gov/compound/CholecalciferolCholecalciferol::
Vitamin E, CID = 14985, https://pubchem.ncbi.nlm.nih.gov/compound/Vitamin-E
Vitamin K, CID = 5280483, https://pubchem.ncbi.nlm.nih.gov/compound/Phylloquinone.

role of vitamin A in stem cells, cancer, and immunity seems to occur via epigenetic
mechanisms [72]. Some reported functions of vitamin A are:

– Ocular metabolism involves two major roles of vitamin A: on the one hand, *11-cis-*
retinal is involved in the processes of photoisomerization and signal transduction in
the retina, on the other, retinoic acid promotes cell differentiation, normal morphol-
ogy, and the barrier function of conjunctival membranes and the cornea. The role
of *11-cis-*retinal has been studied in detail for several decades, thus, it is known that
the human retina is lined with millions of photoreceptors expressing rhodopsins, a
group of proteins (opsins) bound to *11-cis-*retinal, that acts as the visual chromophore. Absorption of a photon by opsin in the photoreceptors causes isomerization
of *11-cis-*retinal to an *all-trans* configuration, thereby triggering a signal transduction
cascade that leads to transmission of a visual signal to the brain [7]. This process
leaves opsins insensitive to further light stimulation, thus, sustained vision requires

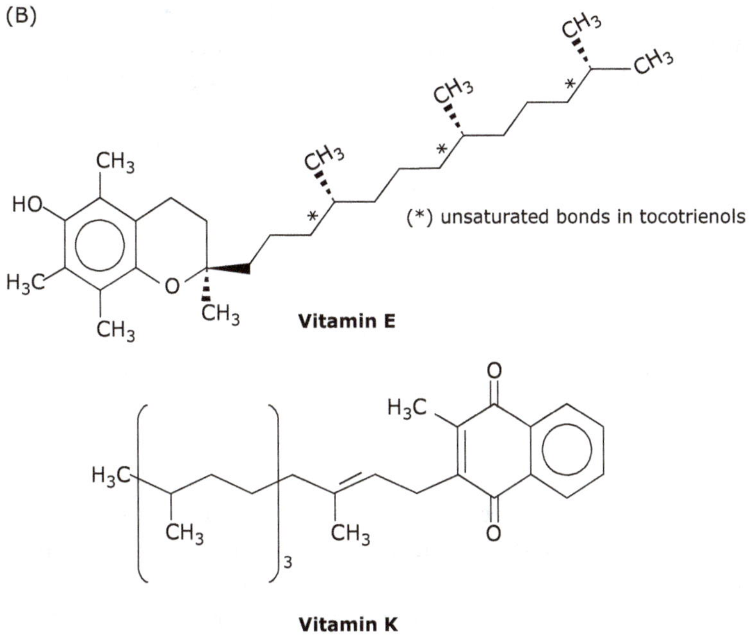

(B)

(*) unsaturated bonds in tocotrienols

Vitamin E

Vitamin K

Figure 6.2 (continued)

continuous renewal of the chromophore upon light exposure. The whole process, from the absorption of the photon to the regeneration of functional rhodopsin, is achieved by an enzymatic pathway known as the visual cycle [71, 73, 74]. In the cornea and conjunctiva, vitamin A is essential for cell differentiation and the structural integrity of these tissues. The cornea receives vitamin A in tear fluid associated with a retinol-binding protein, synthesized and secreted by the lacrimal gland [71].

– Prenatal and postnatal development. The role of vitamin A compounds in development has been studied since the 1930s. Nowadays, it is known that retinoid signaling begins soon after gastrulation and that appropriate levels of retinoic acid are critical for the correct development of heart, lungs, eyes, gonads, urogenital tract, and other organs [71]. In recent years, the multiple roles of vitamin A/retinoic acid signaling in developmental (and adult) hematopoiesis have raised much interest among researchers [75].

– Tissue repair. Hepatic and pancreatic stellate cells function in tissue repair and organ regeneration. These cells, especially hepatic stellate cells, store between 50% and 80% of all vitamin A in the human body as retinyl esters. The ability of these cells to store and metabolize retinoids seems to be highly related to their physiological and pathophysiological functions [76].

- Immunity. Vitamin A is not only essential for first-line defense mechanisms but it also decreases susceptibility to infections, activates the initial response to substances ingested orally, and affects the development and regulation of proper lymphocytes (T and B cells) [77].
- Vitamin A and carotenoids are considered antioxidant nutrients (Chapter 8) and overall dietary intake and intake of specific carotenoids (β-carotene, lycopene) have been reported to be inversely associated with coronary heart disease, stroke, and mortality [78].

6.4.1.3 Sources and bioavailability

The amount of vitamin A in foods must be expressed in terms of equivalents because of the differences in contents of preformed and provitamin, as well as in the bioavailability of carotenoids. The international unit (IU) was used for many years (1 IU = 0.30 µg *all-trans*-retinol or 0.6 µg *all-trans*-carotene). This unit is still found on supplement labels, but it is outdated. In 1967, IU was replaced by the retinol equivalent (RE), which takes into account the differences in vitamin A bioactivity between β-carotene and other provitamin A carotenoids. One RE equals 1 µg of retinol, 6 µg of β-carotene, and 12 µg of other provitamin A carotenoids (Table 6.3). In 2001, after studies had shown that the bioavailability of carotenoids in foods is lower than previously thought, a new unit, the retinol activity equivalent (RAE), was adopted by the Institute of Medicine (IOM) in the USA: 1 µg RAE equals 1 µg of pure *all-trans*-retinol, 2 µg of pure *all-trans*-β-carotene in oil (a highly absorbable form), 12 µg of food-based *all-trans*-β-carotene in foods (from which absorption is less), and 24 µg of other, food-based, *all-trans*-provitamin A carotenoids [71].

Foods rich in retinol include kidney and meat, butter, retinol-enriched margarine, dairy products, and eggs, while foods rich in provitamin A carotenoids, in particular β-carotene, include vegetables and fruits, such as sweet potatoes, carrots, pumpkins, dark green leafy vegetables, sweet red peppers, mangoes, and melons (Table 6.4A). Currently, vitamin A (as retinol, retinyl acetate, retinyl palmitate and β-carotene) may be added to foods and food supplements.

Preformed vitamin A is efficiently absorbed in the intestine, in the range of 70–90%, whereas dietary retinyl esters must be hydrolyzed within the intestinal lumen by non-specific pancreatic enzymes to yield free retinol. The latter is taken up into the intestinal cells by protein-mediated facilitated diffusion and passive diffusion mechanisms via the action of membrane-bound lipid transporters involved in fatty acid and cholesterol uptake (Chapter 13). Retinol then binds to specific cytoplasmic retinol-binding proteins (CRBPs) and undergoes esterification with long-chain fatty acids. The resulting retinyl esters are packed, along with dietary fat and cholesterol, into chylomicrons (Chapter 4), that are secreted into the lymphatic system for delivery to the blood.

Dietary provitamin A carotenoids are absorbed via passive diffusion or taken up through facilitated transport by the enterocyte. Once inside, more than 60% of these

Table 6.4: (A) Average content of fat-soluble vitamins (except vitamin K$_2$) in selected foods.

Food	Retinol (µg/100 g)	Beta-carotene (µg/100 g)	Vit. D (µg/100 g)	Vit. E (mg/100 g)	Vit. K1 (µg/100 g)
Carrot, raw	0	8,290	0	0.27	2.96
Sweet pepper, green, yellow or red, Raw	<2	834	<0.5	1.44	–
Tomato, raw	0	449	0	0.66	7.9
Spinach. raw	0	5,630	0	2.47	521
Sweet potato, raw	0	8,510	0	0.26	1.8
Orange, pulp, raw	0	<5	0	0.19	<0.8
Watermelon, pulp, raw	0	1,220	0	<0.08	<0.8
Grapefruit, pulp, raw	0	539	0	<0.08	<0.8
Almond (with peel)	0	7.33	<0.25	22.3	<0.8
Sunflower seed	0	30	0	42.3	0
Wheat flour, type 55 (for bread)	0	1	<0.25	0.4	<0.8
Rice, brown, raw	0	<50	0	0.2	1.9
Corn or maize grain, raw	–	–	–	1.7	–
Breakfast cereals, sweet (average)	–	–	1.19	6.44	
Muesli (average)	1.2	1.54	0.2	1.76	
Meat, cooked (average)	7.52	0.52	0.22	0.22	1.41
Poultry, cooked (average)	11.1	0.83	0.24	0.2	1.84
Offal, cooked (average)	3,450	34.3	0.77	0.39	1.4
Liver, young cow. Cooked	7,730	182	0.51	0.46	3.9
Kidney, all types. Cooked	111	0	0.44	0.57	0.44
Salmon, smoked	15	<8	5.45	2.85	0
Salmon, steamed	5	0	8.7	2.05	–
Tuna, roasted/baked	8	0	1.83	0.32	–
Cod, salted, dried	24.5	0	4	1.67	0.4
Crustacean or mollusk (average)	35	0	0.18	1.91	0.13
Egg, hard-boiled	61.5	11	1.12	1.03	0.3
Egg, raw	182	0	1.88	1.43	0.3
Milk, whole. UHT	31.4	21.9	<0.25	0.089	<0.8
Milk, semi-skimmed. UHT	19.2	9.45	<0.5	0.13	0.2
Cheese (average)	226	–	0.29	0.55	–
Thick cream 30% fat, refrigerated	195	73	<0.2	0.5	0
Orange juice, home-made	0	33	0	0.04	0.1
Butter, 82% fat, unsalted	675	158	1.12	2.11	7
Palm oil, refined	0	0	–	15.9	8
Almond oil	0	0	0	39.2	7
Avocado oil	0	<50	–	45.3	–
Maize/corn oil	0	0	0	13.2	1.9
Olive oil, extra virgin	0	210	<0.25	22.3	58.1
Soy oil	0	0	0	6.1	362
Sunflower oil	0	<5	<0.25	57.3	<0.8
Cod liver oil	30,000	–	250	30	0
Baby milk, first age, powder	–	–	8.11	8.89	–
Baby milk, second age, powder	–	–	8.59	9.15	–

Table 6.4: (B) Average content of vitamin K$_2$ in selected foods.

Name	Vit. K$_2$ (µg/100 g)
Liver, goose, raw	369
Edam cheese, from cow's milk	47,5
Chicken, leg, meat, raw	34,3
Egg yolk, raw	32,1
Goose, meat, raw	31
Butter, unknown fat content, unsalted (average)	15
Butter, 82% fat, unsalted	15
Salami	9
Chicken, breast, without skin, raw	8,9
Fermented milk drink, plain, whole milk	8,15
Liver, young cow, raw	7,94
Beef, minced steak, 10% fat, raw	6,7
Emmental cheese, from cow's milk	5,23
Pork tenderloin, lean, raw	3,7
Duck, meat, raw	3,6
Liver, pork, raw	3,44
Rainbow trout, raw, farmed	3,39
Pork, chop, raw	3,22
European plaice, raw	2,2
Dark chocolate bar, less than 70% cocoa	1,5
Milk, whole, UHT	0,9
Egg white, raw	0,9
Pike-perch, raw	0,78
Venison (roebuck), roasted/baked	0,7
Yogurt, fermented milk or dairy specialty, plain	0,68
Yoghurt, plain (average)	0,61
Salmon, raw, farmed	0,5
Mackerel, raw	0,4
Atlantic herring, raw	0,21
Yogurt, fermented milk or dairy specialty, plain, fat free	0,1

Source: French Agency for Food, Environmental and Occupational Health & Safety. Anses. 2020. Ciqual French food composition table. Retrieved June 19, 2023, from the Ciqual homepage https://ciqual.anses.fr/

molecules are cleaved at their central double bond into *all-trans*-retinal. This compound can be oxidized to retinoic acid, reduced to retinol or bound to CRBP in order to be incorporated into chylomicrons, along with the rest of absorbed provitamin A carotenoids (less than 40%) that are left intact. Overall, absorption of β-carotene appears to be highly variable (5–65%) and depends on factors like the type of food and diet, and the genetics and health of the individual. Data on absorption of the other provitamin A carotenoids are more limited (see Chapter 13 for further details).

Chylomicrons deliver lipid nutrients, including retinyl esters and intact carotenoids, to the liver, where they are taken up and transferred into hepatic stellate cells, that constitutes the primary sites of vitamin A storage with up to ~80–85% of total retinyl ester and retinol stores in the body (adipose tissue stores ~ 15–20%). Liver also plays a central role in the distribution and homeostasis of retinoids by means of several biochemical systems: very low-density lipoprotein (VLDL), low-density lipoprotein (LDL), and high-density lipoprotein (HDL) (Chapter 4); retinol bound to retinol-binding proteins, such as retinol-binding protein 4 (RBP4, that circulates as a complex with transthyretin); retinoic acid bound to albumin; and the water-soluble β-glucuronides of retinol and retinoic acid [79, 80].

Within cells, retinol can be oxidized to retinal and retinoic acid by several dehydrogenases. The oxidation of retinol to retinal is generally considered rate limiting for retinoic acid production, concentration of retinal being very low in most tissues. The oxidation of retinal to retinoic acid is an irreversible process that seems to be mediated by several enzymes [71]. Retinoic acid is transported from the cytosol to the nucleus where it forms complexes with nuclear retinoic acid receptors (RAR) and members of the retinoid X receptor (RXR) family. These complexes subsequently bind to specific response elements of the DNA, thereby initiating the transcription of target genes. Transport to the nucleus is mediated by the cellular retinoic acid binding protein type II (CRABP2) [81].

6.4.1.4 Stability

Food processing and storage can lead to 5–40% destruction of vitamin A and carotenoids. In the presence of oxygen, oxidative degradation leads to a series of products, some of which are volatile. The degradation of retinoids and vitamin A-active carotenoids generally parallels the oxidative degradation of unsaturated lipids, thus, factors that promote oxidation of unsaturated lipids (oxygen partial pressure, water activity, temperature, etc.) enhance degradation of vitamin A, either by direct oxidation or by indirect effects of free radicals (Chapter 4, Section 4.5.2). Dehydrated foods are particularly sensitive to oxidative degradation.

In the absence of oxygen and at higher temperatures, as experienced in cooking or food sterilization, the main reactions are fragmentation and isomerization of *all-trans* forms of retinoids and carotenoids to various *cis* isomers. Isomerization implies loss of vitamin A activity and can be induced by exposure to light, acid, chlorinated solvents (e.g., chloroform), and dilute iodine. Fragmentation of β-carotene, either by oxidative oxidation or by thermal treatment, can have a significant effect on food flavor [10, 14].

6.4.1.5 Dietary Reference Values (DRVs): hypo- and hypervitaminosis

One of the biomarkers for vitamin A status is plasma/serum retinol concentration: values below 0.7 μmol/l are considered indicative of vitamin A deficiency, whereas values below 0.35 would indicate severe deficiency. For pregnant and lactating women, plasma concentrations below 1.05 μmol/l are considered low [77]. However, the EFSA Panel on Dietetic Products, Nutrition, and Allergies considers that plasma/serum retinol concentration is under tight homeostatic control and does not reflect vitamin A status until body stores are very low. The Panel proposes the measurement of the total body retinol content or liver retinol concentration as a more specific and sensitive markers of vitamin A status. Thus, a liver concentration below 20 µg (or 0.07 µmol) of retinol/g, either in free form or as retinyl esters, would indicate vitamin A deficiency. Concentrations above this value are considered to maintain adequate plasma concentrations, prevent clinical signs of deficiency, and reflect adequate vitamin A status [79].

The EFSA has concluded that AR and PRI for vitamin A in healthy adults can be derived from the vitamin A intake, in RE, required to maintain a concentration of 20 µg retinol/g of liver. Values for infants and children were derived by using specific age-related values for reference body weight and liver/body weight ratio. Values for pregnant and lactating women were adapted by considering the estimated amount of retinol accumulated in the fetus over the course of pregnancy and the amount of retinol secreted in breast milk, respectively (Table 6.3).

Vitamin A deficiency (VAD) is a serious and widespread public health problem and has been associated with preventable blindness in children and increased infectious morbidity and mortality, including respiratory infection and diarrhea, especially in low-income countries. Major symptoms of VAD are intrauterine and post-natal growth retardation and a large array of congenital malformations collectively referred to as the fetal "vitamin A deficiency syndrome." Other health consequences include keratinization of epithelial cells and poor wound healing, nervous disorders, defective reproduction, disorders in bone formation, increased risk for cysts formation in endocrine glands, higher prevalence for formation urinary calculi and nephritis, loss of appetite, impaired ability to sense flavor, intestine malfunction, inflammatory conditions, anemia due to interferences in iron metabolism, alterations in the immune system and the microbioma, and alterations in the vision and eye development. The most specific clinical consequence of VAD is xerophthalmia, which includes night blindness, due to slow regeneration of rhodopsin, early keratinizing metaplasia (Bitot's spots), impaired production of tears, conjunctival xerosis, corneal xerosis, and corneal ulceration and scarring. Xerophthalmia can be reversible but may result in blindness if untreated. Conversely, night blindness, the first ocular symptom of deficiency, responds rapidly to an increase in vitamin A intake [77, 79, 81].

Hypervitaminosis A has been reported to cause severe headache, blurred vision, dizziness, increased spinal fluid pressure, congenital birth defects and malfunctions of the eye, skull, lungs and heart, reduced bone mineral density, drowsiness and irritability, portal hypertension, elevation of blood calcium levels, skin disorders, nausea, vomiting, disorders of the musculoskeletal system, and liver damage. In infants, bulging fontanelle and increased intracranial pressure are also classical adverse effects of vitamin A toxicity. The teratogenic effect of excessive intake of vitamin A or specific retinoids in both animals and humans is also well documented. The Scientific Committee for Food (SCF) of EFSA has set a Tolerable Upper Intake Level (UL) for preformed vitamin A of 3,000 µg RE/day for adults, including women during pregnancy and lactation. ULs for children were extrapolated from the UL for adults, based on allometric scaling: 800 µg RE/day for children aged 1–3 years; 1,100 µg RE/day for children aged 4–6 years; 1,500 µg RE/day for children aged 7–10 years; 2,000 µg RE/day for children aged 11–14 years; and 2,600 µg RE/day for children aged 15–17 years [79, 81].

Vitamin A has been reported to play important roles in cardiometabolic health, including the regulation of adipogenesis, energy partitioning and homeostasis, and lipoprotein metabolism; however, while these roles are strongly supported by animal and in vitro studies, they remain poorly understood in human physiology and disease [80, 81].

6.4.2 Vitamin D

6.4.2.1 Structure
The Joint Commission on Biochemical Nomenclature of the International Union of Pure and Applied Chemistry (IUPAC) and the International Union of Biochemistry and Molecular Biology (IUBMB) recommended in 1981 that the term vitamin D should be used as a general term to describe all steroids that qualitatively exhibit the biological activity of calciol, also known as cholecalciferol or vitamin D_3 (IUPAC Name: (1S,3Z)-3-[(2E)-2-[(1R,3aS,7aR)-7a-methyl-1-[(2R)-6-methylheptan-2-yl]-2,3,3a,5,6,7-hexahydro-1H-inden-4-ylidene]ethylidene]-4-methylidenecyclohexan-1-ol). The term vitamin D_2 refers to ergocalciferol or ercalciol, obtained from organisms like fungi. The use of abbreviations "D" or "D_3" or "1,25-(OH)$_2$-D" and "1,25-(OH)$_2$-D$_2$" for derivatives are strongly discouraged by the IUPAC and IUBMB [82], although they are often found in scientific literature.

6.4.2.2 *In vivo* function
In the skin, the 9,10-bond of 7-dehydrocholesterol is opened by a process involving UVB light (wavelengths 290–315 nm) to give an intermediate that is subsequently isomerized by heat to give vitamin D_3 (Figure 6.2A). In the bloodstream, this compound binds to the vitamin D-binding protein (DBP) or to albumin, in order to be carried to storage tissues or to the liver.

In the liver, vitamin D_3 undergoes a first step of activation, namely, 25-hydroxylation, thus being converted into the main circulating form, 25-hydroxycholecalciferol, or calcidiol. Several enzymes have been reported to have 25-hydroxylase activity, the most important being encoded by the gene CYP2R1. Calcidiol is converted, mainly in the kidney, into the biologically active metabolite 1α,25-dihydroxycholecalciferol, also called calcitriol. This reaction is mediated by the enzyme 1α-hydroxylase, encoded by the gene CYP27B1. Other tissues, including various epithelial cells, cells of the immune system, and the parathyroid gland have been shown to contain this enzymatic activity. Dietary vitamins D_2 and D_3 (see below) also undergo this activation process. From the biochemical point of view, calcitriol is a classical steroid hormone that enters cells crossing the plasma mem-

brane, either in a free form or by receptor-mediated endocytosis. Then, it strongly interacts with a vitamin D-binding receptor (VDR) inside the nucleus. The complex thus
formed regulates the transcription of many genes in the cells of the main target tissues:
intestine, kidneys, and bones [83, 84]. These genes are mainly involved in maintaining
calcium and phosphorus homeostasis in the body, together with parathyroid hormone
(PTH) and fibroblast growth factor, thus:

– In the intestine, activated genes include those coding for proteins involved in calcium and phosphorus absorption by active transport.
– In the kidneys, calcitriols stimulate the tubular reabsorption of calcium dependent and decreases the activity of the enzyme 1α-hydroxylase.
– In the bone, calcitriols interact with PTH to activate the cells responsible for bone
resorption, osteoclasts, that release hydrochloric acid and hydrolytic enzymes to dissolve the bone matrix, thus releasing calcium and phosphorus into the bloodstream.
– In parathyroid cells, calcitriols inhibits cell proliferation and suppress the expression of the gene encoding PTH.

Other functions of calcitriols include cell differentiation and antiproliferative actions in
various cell types, such as bone marrow, cells of the immune system, skin, breast, and
prostate epithelial cells, muscle and intestine. In addition to these roles, there is an ex-
panding volume of data regarding associations between vitamin D deficiency and medical pathologies such as autoimmune disorders (multiple sclerosis, type 1 diabetes and
rheumatoid arthritis), cardiovascular disease, hypertension, type 2 diabetes, neurocognitive dysfunction (Alzheimer's disease), infectious diseases (including COVID-19), myopathy, skin diseases, several types of cancer, as well as liver disease [83, 85–87]. The
widespread effects of vitamin D are not surprising since essentially all organs and cells
in the body are now known to express vitamin D-binding receptors [87].

6.4.2.3 Sources and bioavailability

As described in the previous section, vitamin D_3 can be synthesized in the skin by the
action of UV light from the sun, therefore, in strict terms, is not a vitamin. However,
endogenous synthesis may be negligible in certain countries during the winter, when
UVB light does not penetrate the atmosphere and can also be affected by ecological
factors, such as UVB-absorbing aerosols in the environment, and by individual factors
such as dark skin pigmentation, conservative clothing habits, limited sun exposure,
and liberal use of sunscreen [88]. In these circumstances, dietary sources become critical. The major food sources for naturally occurring vitamin D_3 include animal foods
such as fatty fish, offal (particularly liver), meat and meat products, and egg yolks
(Table 6.4A). Vitamin D_2 is produced in fungi and yeasts by UV-B exposure of ergosterol (also called provitamin D_2) and the content depends on the amount of UV-B light
exposure and time of exposure. Currently, vitamins D_3 and D_2 may be added to both
foods and food supplements, so that further sources of dietary vitamin D are fortified

foods (milk, margarine and/or butter, and breakfast cereals), and dietary supplements [85, 86]. Vitamin D in food or supplements may be quantified in either international units (IU) or micrograms (µg), whereby 1 IU = 0.025 µg for both vitamin D_3 and D_2 [86].

There seems to be no significant differences between vitamins D_3 and D_2 as far as digestion and absorption is concerned; therefore, the term *vitamin D* normally includes both vitamers. However, the absorption efficiency of calcidiol and calcitriol by the enterocyte is around threefold those of ergocalciferol and cholecalciferol; this is probably due to the higher water solubility of the former molecules [89]. The first step of the process of digestion is the solubilization of dietary vitamin D in the fat phase of the meal, which is emulsified into lipid droplets in the stomach and duodenum, and transferred to the mixed micelles generated by lipolysis of dietary fat. Mixed micelles are composed of phospholipids, cholesterol, lipid digestion products (such as free fatty acids, monoacylglycerols, and lysophospholipids), and bile salts. It has been suggested that the higher the percentage of fat-soluble vitamin incorporation into micelles (the so-called bioaccessibility), the higher their absorption efficiency. Bioaccessibility depends on factors like the chemical structure of the vitamin and the nature of the other lipids consumed during the same meal. Altogether, absorption efficiency of vitamin D varies between 55% and 99% in humans. Maximal uptake occurs in enterocytes of the jejunum and ileum by mechanisms that are not fully elucidated, although it has been reported that three cholesterol transporters are involved in this process when vitamin D is given at dietary doses, whereas passive diffusion occurs mainly at pharmacological doses. The transport of vitamin D within enterocytes may involve binding proteins, but none has yet been clearly identified. Vitamin D in its free form is incorporated into chylomicrons and secreted in the lymph, although other secretion pathways may exist [90] (see also Chapter 13).

Within hours of ingestion or synthesis in the skin (see above), vitamin D is distributed to the liver for conversion or delivered to the storage tissues, that include mainly the adipose tissue, muscle, and liver. Vitamin D from dietary sources is released from the chylomicrons by action of the enzyme lipoprotein lipase upon arrival in the tissues, whereas that synthesized in the skin, mainly associated to DBP, is taken up by receptor-mediated endocytosis. After lipolysis, the chylomicron remnants still contain a fraction of vitamin D bioavailable, which could remain in the bloodstream for its subsequent metabolism. It is worthwhile pointing out that there is evidence of vitamin D transfer from chylomicrons to DBP [86, 89, 91].

Dietary lipids (amounts, types of phospholipids and lipids emulsification) could influence vitamin D absorption, unlike the total amount of fat ingested, that appears to have no significant effect. Long-chain fatty acids have been reported to hinder vitamin D absorption, while a high monounsaturated fatty acids diet might improve it. High polyunsaturated fatty acids diet may reduce the effectiveness of cholecalciferol supplements in healthy older adults. In any case, more studies are needed to confirm or deny these claims. On the other hand, not enough data are available to reach a conclusion about the effect of fiber on the bioavailability of vitamin D. Several drugs (anticonvulsants, antiretrovirals, glucocorticoids, antiobesity drugs, and others) might reduce vitamin D absorption, whereas β-cyclodextrin and thiazide diuretics might have the opposite effect [89].

6.4.2.4 Stability
Vitamin D is susceptible to degradation by light and by oxidation, therefore, storage away from light, under anaerobic conditions and/or with an antioxidant content, such as vitamin E (see below), seems to stabilize vitamin D in foods [10, 92].

6.4.2.5 Dietary Reference Values (DRVs): hypo- and hypervitaminosis
Serum concentrations of 25-hydroxyvitamin D (that include the hydroxylated derivatives of both vitamin D_3 and D_2) can be used as a biomarker of vitamin D intake and status at population level. The EFSA Panel on Dietetic Products, Nutrition, and Allergies considers that a serum 25-hydroxyvitamin D concentration of 50 nmol/l is a suitable target value for all population groups; this concentration range has been associated with many of the noncalcemic health benefits of vitamin D, including reduced risk for cardiovascular disease, neurocognitive dysfunction, several cancers, and infectious diseases. In order to reach this target, the Panel sets an AI for vitamin D at 15 µg/day for all population groups with the only exception of infants aged 7–11 months (10 µg/day). In the presence of endogenous cutaneous vitamin D synthesis, the requirement for dietary vitamin D is lower or may even be zero (Table 6.3). Vitamin D intakes are sometimes expressed in International Units (IU), that are equal to 0.025 µg of ergocalciferol and/or cholecalciferol [39, 86, 87].

Vitamin D deficiency is conventionally defined as serum concentrations of 25-hydroxyvitamin D below 30 nmol/l [88]. Major manifestations of vitamin D deficiency are rickets in children and osteomalacia in adults.
– Nutritional rickets is a disorder of defective chondrocyte differentiation and mineralization of the growth plate caused by low vitamin D status, inadequate calcium intake in children, or both. Clinical diagnosis is usually obtained by physical examination and biochemical testing, and confirmed radiographically. Biochemical features of nutritional rickets include elevated levels of bone alkaline phosphatase (the most sensitive biomarker), low serum phosphorus and calcium, and elevated PTH. Symptoms and complications of rickets include impaired linear growth, chest wall deformity predisposing to pneumonia, fractures, bone pain, leg deformities, and potentially lethal outcomes like hypocalcemic seizures, cardiomyopathy, and cardiac arrest. Vitamin D–deficient mothers give birth to babies who are themselves deficient in vitamin D and are thus at risk of hypocalcemia and congenital rickets. There is also evidence suggesting that vitamin supplementation during pregnancy may reduce the risk of pre-eclampsia, preterm birth, and low birth weight, although these results have not been conclusively confirmed by sub-

sequent studies. In any case, prenatal vitamin D supplementation seems to have beneficial effects for pregnant women and newborns [88].

– Osteomalacia is a bone disease in adults that results from defective mineralization due to inadequate calcium or phosphorus availability or excessive calcium resorption from bone, most commonly due to severe vitamin D deficiency. Serum calcium concentration is often normal (2.25–2.6 mmol/L) despite the undermineralization of bones. Clinical symptoms of vitamin D deficiency in adults are less pronounced than in children and may include diffuse pain in muscles and bone and specific fractures. Muscle pain and weakness (myopathy) that accompany the skeletal symptoms in older adults may contribute to poor physical performance, increased risk of falls/falling and a higher risk of bone fractures. The role of vitamin D in adult bone health has been a major focus of research in high-income countries; however, there is inconsistent evidence of the benefits of vitamin D in the prevention of falls or fractures in older adults. Prolonged vitamin D insufficiency may lead to low bone mineral density and may dispose older subjects, particularly post-menopausal women, for osteoporosis, a situation characterized by a reduction in bone mass, reduced bone quality, and an increased risk of bone fracture, predominantly in the forearm, vertebrae, and hip [86, 88].

Over the past two decades, laboratory and epidemiological studies have also suggested that low vitamin D status may be associated with a variety of health risks (see above). However, more experimental evidence is needed to support the effects of vitamin D on health outcomes other than osteomalacia [88, 93].

Serum concentrations of 25-hydroxyvitamin D above 220 nmol/L can lead to hypercalcemia, which may eventually cause soft tissue calcification and resultant renal and cardiovascular damage. These levels cannot be obtained by endogenous synthesis or by diet, but only by ingestion of pharmacological doses (e.g., 125–1,000 µg/day) of vitamin D over a period of at least 1 month. In this respect, it must be mentioned that these extremely high doses have been effective in treating autoimmune disorders including psoriasis, multiple sclerosis, rheumatoid arthritis, and vitiligo. The toxicity associated with these doses of vitamin D can be mitigated by having the patients on an extremely low calcium diet and maintaining good hydration [87]. A Tolerable Upper Intake Level (UL) of 100 µg/day has been set by the EFSA for adolescents (11–17 years) and adults, including pregnant and lactating women. For children aged 1–10 years and infants, ULs of 50 µg/day and 25 µg/day, respectively, were selected [86].

The one-hundred-year anniversary of the discovery of vitamin D was celebrated in 2022, and, despite the advances achieved since 1922, many aspects of the biology and health benefits of this vitamin are yet to be established and are currently under active research.

6.4.3 Vitamin E

6.4.3.1 Structure

The term vitamin E was formerly used as the generic descriptor for those tocol and tocotrienol derivatives exhibiting the biological activity of 5,7,8-trimethyltocol (α-tocopherol) in qualitative terms. The systematic IUPAC nomenclature of tocol is 2-methyl-2-(4,8,12-trimethyltridecyl)-3,4-dihydrochromen-6-ol (Figure 6.2B), whereas the term tocotrienol refers to the unsaturated derivative 2-methyl-2-[(3E,7E)-4,8,12-trimethyltrideca-3,7,11-trienyl]-3,4-dihydrochromen-6-ol. Naturally occurring derivatives of α-tocopherol include β-tocopherol (5,8-dimethyltocol), γ-tocopherol (7,8-dimethyltocol), and δ-tocopherol (8-methyltocol) and their tocotrienol homologues [94].

α-Tocopherol has three stereogenic centers, at position 2 on the ring and at positions 4′ and 8′ in the side chain; thus, there are eight possible stereoisomers, that appear in equal proportions in the synthetic commercially available forms of α-tocopherol (*all-rac*-α-tocopherol): RRR-, RRS-, RSR-, RSS- and their enantiomers SSS-, SSR-, SRS-, SRR-.

RRR-α-tocopherol is the only naturally occurring isomer and is considered to be the physiologically active vitamer, due to its preferential binding to the α-tocopherol transfer protein (α-TTP) in blood. Bioactivity of each stereoisomer has been determined in rats and ranges from 21% for the SSR isomer to 90% for the RRS isomer, compared with RRR-α-tocopheryl acetate. Other tocopherols and tocotrienols possess antioxidant activity to a different degree.

Contents of vitamin E can be found in the literature in mg, μmol, α-TEs or in International Units (IU) in a rather confusing way. Currently, international institutions like the EFSA set dietary reference values (see below) considering only 2 R-α-tocopherol stereoisomers (that is, RRR-, RSR-, RRS-, and RSS-α-tocopherol) and, in accordance with the Institute of Medicine (IOM) of the USA, the following conversion rules are followed:

–   The difference in relative activity of commercial *all-rac*-α-tocopherol compared with RRR-α-tocopherol is 50%, thus, 1 mg *all-rac*-α-tocopherol is equal to 0.5 mg RRR-α-tocopherol.
–   1 IU of *all-rac*-α-tocopherol or its esters is equal to 0.45 mg 2 R-stereoisomeric forms of α-tocopherol.
–   1 IU of pure RRR-α-tocopherol or its esters is equal to 0.67 mg of 2 R-α-Tocopherol [95].

6.4.3.2 *In vivo* function

α-Tocopherol is part of the antioxidant defense system, which is a complex network including endogenous and dietary antioxidants, antioxidant enzymes and repair mechanisms (Chapter 8). It is a potent peroxyl radical scavenger and especially protects polyunsaturated fatty acids (PUFAs) within membrane phospholipids and plasma lipoproteins. It has been reported that peroxyl radicals (ROO·) react 1,000 times faster with

vitamin E (Vit E-OH) than with PUFA (RH) and form the tocopheroxyl radical (Vit E–O·),
thus:

$$ROO· \rightarrow ROOH + VitE - O·$$

The tocopheroxyl radical (Vit E–O·) is relatively stable (the unpaired electron reso-
nates across the phenolic ring system [10]) and emerges from the lipid bilayer into the
aqueous domain, thus reacting with reductants that serve as hydrogen donors (AH)
and returning to its reduced state.

$$VitE - O· + AH \rightarrow VitE - OH + A·$$

Biologically important hydrogen donors include vitamin C, coenzyme Q and thiols, especially glutathione. Regeneration of tocopherol from its radical by vitamin C appears
to be a physiologically relevant mechanism *in vivo* and suggests that the antioxidant
function of the vitamin E radical is continuously restored by other antioxidants and
by the metabolic activity of cells [96].

By protecting PUFAs within membrane phospholipids, α-tocopherol preserves intracellular and cellular membrane integrity and stability, plays an important role in
the stability of erythrocytes and the conductivity in central and peripheral nerves,
and prevents hemolytic anemia and neurological symptoms occurring in α-tocoph-
erol-deficient individuals (see below) [95, 97]. It has been reported that some cellular
and plasma proteins can prevent its chemical oxidation by binding to vitamin E [98].

Experimental evidence obtained with animals and in vitro studies suggests that
many cell responses, including inflammatory responses, cell proliferation, programmed
cell death, and lipid homeostasis can be affected by vitamin E; however, studies in humans have yielded conflicting results so far [96].

6.4.3.3 Sources and bioavailability

The main dietary sources of vitamin E include vegetable oils, fat spreads from vegetable
oils, nuts and seeds, some fatty fish, egg yolk, and whole grain cereals. The proportions
of the four tocopherols (α-, β-, γ- and δ-tocopherol) and their tocotrienol homologues
vary according to the food source. In particular, vegetable oils vary in their content of
the different tocopherol forms: wheat germ, sunflower, olive, and rapeseed oils are
good sources of α-tocopherol, wheat germ oil of β-tocopherol, soybean, corn, and rapeseed oils of γ-tocopherol and soybean oil of δ-tocopherol. Foods that have been fortified
with *all-rac*-α-tocopheryl acetate include some breakfast cereals, tomato juice, orange
juice, and milk (Table 6.4A and references [95, 97]).

The absorption of tocopherols and tocotrienols is thought to occur in parallel to
that of vitamin D and other lipid compounds (see above). Tocopherol esters must be
hydrolyzed in the duodenum by pancreatic hydrolases. The average efficiency of vitamin E absorption is below 50%, the rest being excreted in the feces, and increases
with the fat content of foods. The bioavailability of vitamin E varies greatly between

individuals and can be affected by several factors, such as different forms of the vitamin, chemical modifications (e.g., esterification), the amount consumed at once, and interactions with other nutrients or drugs consumed in parallel, as well as vitamin E status, health conditions, and genetic factors [96].

Unlike other fat-soluble vitamins, which have their own specific plasma transport proteins, vitamin E is transported by the different lipoproteins in the plasma (Chapter 4, Section 4.13), thus, newly absorbed vitamin E is incorporated into chylomicrons, secreted in the lymph and then in the bloodstream to be distributed to other organs and tissues. The liver discriminates among various vitamin E forms, so that 2R-α-tocopherols are selectively secreted into plasma associated to very-low-density lipoproteins (VLDLs). The α-tocopherol transfer protein (α-TTP), an enzyme expressed in hepatocytes, seems to play a role in this selection of α-tocopherol, although the mechanisms involved are not understood. It has been reported that the different lipoproteins exchange vitamin E in a process mediated by the phospholipid transfer protein (PLTP).

Vitamin E is not accumulated in the liver to toxic levels, all forms being metabolized by ω-hydroxylation mediated by cytochrome P-450s (CYPs). The metabolites that result from CYPs action are the so-called long-chain metabolites (LCMs), namely, 13'-tocopherol-hydroxychromanol (13'-OH) and 13'-tocopherol-carboxychromanol (13'-COOH). The latter has emerged as a derivative of vitamin E with many regulatory functions, revealing superior and sometimes even different effects compared to its metabolic precursor; moreover, LCMs occur in human serum and accumulate, for example, in primary immune cells. These findings support the idea that vitamin E may require metabolic activation before it can carry out all of its physiological functions [99]. β-oxidation shortens the side-chain of LCMs yielding several intermediate-chain metabolites and, finally, short-chain metabolites of vitamin E. All metabolites are excreted with feces or urine following conjugation with taurine, glycine, glucuronide, or sulfate. Tocotrienols follow almost the same metabolic degradation route, except for one additional step required for removing the double bonds [96, 97].

Physiological interactions between vitamin E and other micronutrients, such as vitamin C, β-carotene, selenium, and zinc have been reported; however, more studies in humans are required to better understand the complexity of these interactions and their possible contribution to metabolic diseases [96].

6.4.3.4 Stability

Vitamin E compounds exhibit reasonably good stability in the absence of oxygen and oxidizing lipids. Anaerobic treatments in food processing, such as retorting of canned foods, have little effect on vitamin E activity. In contrast, the rate of degradation increases in the presence of molecular oxygen and can be especially rapid when free radicals are also present. Oxidative degradation of vitamin E is strongly influenced by the same factors that influence oxidation of unsaturated lipids. Losses may occur in vegetable oil processing into margarine and shortening and also in dehydrated or deep fried foods. The use of intentional oxidative treatments, such as the bleaching of flour, can lead to large losses of vitamin E. Vitamin E compounds can also contribute indirectly to oxidative stability of other compounds by scavenging singlet oxygen. They also reduce formation of nitrosamines in the curing of bacon, apparently by quenching nitrogen free radicals ($NO^•$, $NO_2^•$) [10, 14].

6.4.3.5 Dietary Reference Values (DRVs): hypo- and hypervitaminosis

The IOM determined that a plasma concentration lower than 12 µmol α-tocopherol/l was associated with evidence of α-tocopherol inadequacy; plasma concentrations in physiologically normal subjects being approximately 20 µmol/l. The EFSA considers that plasma/serum α-tocopherol concentration is not a sensitive marker of dietary α-tocopherol intake and, although concentrations below about 12 µmol/l may be indicative of deficiency, there is a lack of data to set a precise cut-off value above which α-tocopherol status may be considered as adequate. The EFSA Panel on Dietetic Products, Nutrition, and Allergies proposes adequate intake values (AIs) for α-tocopherol based on observed intakes in European countries for children and adults. For infants aged 7–11 months, the Panel proposes AIs based on estimated intakes in fully breastfed infants and upward extrapolation by allometric scaling. The AI set for pregnant or lactating women is the same as for adults (Table 6.3) [95].

Vitamin E deficiency occurs only rarely in humans and usually as a result of genetic abnormalities in α-TTP or various fat malabsorption syndromes, such as biliary obstruction, cholestatic liver disease, pancreatitis, or cystic fibrosis. Primary manifestations of deficiency include spinocerebellar ataxia, skeletal myopathy, and pigmented retinopathy. The progression of neurologic symptoms appears to depend on the level of oxidative stress accompanying the α-tocopherol deficiency [97, 100].

The chemical name "tocopherol" derives from its essentiality for normal reproduction in animals, although this function has never been clearly demonstrated in humans. However, it has been reported one case of a woman with recurrent spontaneous abortions, that had a healthy baby after administration of 300 mg/day of tocopherol nicotinate [100].

The Scientific Committee on Food (SCF) of the European Commission has set a Tolerable Upper Intake Level (UL) for adults of 270 mg α-TE/day, rounded to 300 mg α-TE/day using an uncertainty factor of 2. This UL also applies to pregnant and lactating women as there was no indication from animal studies of a specific risk for these population groups. The ULs for children were derived from the adult UL and ranged from 100 mg α-TE/day (1–3 years) to 260 mg α-TE/day (15–17 years) [95].

6.4.4 Vitamin K

6.4.4.1 Structure

The term vitamin K refers to a family of fat-soluble compounds with the common chemical structure 3-substituted 2-methyl-1,4-napthoquinone that includes:

- Phylloquinone (also called phytonadione, phytomenadione or vitamin K_1, IUPAC Name: 2-methyl-3-[(E,7R,11R)-3,7,11,15-tetramethylhexadec-2-enyl]naphthalene-1,4-dione), from plant origin, that contains a phytyl group and is the primary dietary form of vitamin K (Figure 6.2B) [101]. Phylloquinone serves as an electron acceptor during photosynthesis, forming part of the electron transport chain of photosystem I. The function of vitamin K_1 in other cellular compartments is also associated with its redox properties [102].
- Menaquinones (vitamin K_2), a group of compounds with unsaturated side chains of varying length (MK-n, "n" indicating from 4 to 13 isoprenyl units) at the 3-position of the 2-methyl-1,4-napthoquinone group and found in animal products such as meat, cheese, and egg. Menaquinones, with the exception of MK-4, are produced by certain obligate and facultative anaerobic bacteria. MK-4 is formed in animals from vitamin K_1 [102].
- Menadione (unsubstituted 2-methyl-1,4-napthoquinone, also called vitamin K_3) is a water-soluble synthetic form of vitamin K that plays a role as an intermediate in the metabolic conversion of phylloquinone to MK-4.
- Menadiol sodium phosphate (also called vitamin K_4) is a synthetic water-soluble form derived from menadione by reduction.
- Dihydrophylloquinone present in foods made with partially hydrogenated fat-like hydrogenated soybean oil.

6.4.4.2 *In vivo* function

Vitamin K participates in the carboxylation of specific glutamate residues that are converted into gamma-carboxyglutamate (Gla). This post-translational modification is required for the biological function of numerous proteins, known as vitamin K-dependent proteins, and occurs at specific domains responsible for high-affinity binding of calcium ions. Glutamate carboxylation is catalyzed by the enzyme gamma-glutamyl carboxylase (GGCX), which uses a reduced form of vitamin K (KH2), carbon dioxide, and oxygen as cofactors. In each glutamate modification, KH2 is oxidized to the 2,3-epoxide form (KO), that is converted back to KH2 through a two-step reduction process mediated by the multimeric protein vitamin K oxidoreductase (VKORC1) [103]. This pathway is known as the vitamin K cycle.

Vitamin K has been traditionally linked with coagulation; however, it seems to have many other roles in human physiology. Vitamin K is necessary for the posttranslational modification of 18 or 19 proteins, among which 11 or 12 are not related to coagulation. In particular, its involvement in connective tissue calcification has stimulated intense research, although the data are still inconclusive. Vitamin K-dependent proteins include [102, 104–106]:

- Prothrombin, the circulating zymogen of the procoagulant protein thrombin.
- Clotting factors, such as factor (F) II, FVII, FIX, and FX.
- Natural anticoagulants, such as protein C, protein S, and protein Z.
- Transthyretin, a homo-tetrameric carrier protein, which transports thyroid hormones in the plasma and cerebrospinal fluid. It is also involved in the transport of retinol in the plasma.
- Osteocalcin (also known as bone-Gla protein, BGP), a protein produced by osteoblasts, important for bone formation. Partially carboxylated and uncarboxylated osteocalcin is considered by some authors to be hormone that may affect glucose and lipid metabolism and modulate fertility [102].
- Bone matrix Gla protein (MGP), which functions as a strong inhibitor of vascular calcification and connective tissue mineralization. MGP has also been described as a critical regulator of endothelial cell function, implicated in both physiological and tumor-related angiogenesis.
- Proteins whose metabolic significance is poorly understood: Gas6 (growth arrest-specific protein 6), PRGPs (proline-rich Gla proteins), and TMGs (transmembrane Gla proteins).
- Data from animal tests indicate that vitamin K is involved in the down-regulation of expression of genes involved in acute inflammatory response; however, the relevance of this role in humans is unclear [101]. Some researchers have suggested that vitamin K might directly bind to intracellular receptors, interfere with enzymes, or have antioxidant activity [102].

6.4.4.3 Sources and bioavailability

Phylloquinone (vitamin K_1) is the predominant dietary form of vitamin K in the human diet, the primary sources of phylloquinone including dark green leafy vegetables (e.g., spinach, lettuce, and other salad plants) and Brassica (flowering, head or leafy). Other sources of phylloquinone include some seed oils (soybean or canola), spreadable vegetable fats, and blended fats/oils, the latter depending on the source of fat or oil (Table 6.4A).

Menaquinones (vitamin K_2) are found in animal-based foods, in particular in liver products, meat and meat products (Table 6.4B). Poultry products are particularly rich in MK-4, as poultry feed is a rich source of menadione, subsequently converted to MK-4 in certain tissues of these animals. Menaquinones are also present in some cheese and other dairy products. In eggs (in particular in egg yolk), the most abundant menaquinone is MK-4. A fermented soybean product, natto, that is consumed mainly in the Japanese market contains nearly 1,000 µg/100 g of MK-7. Currently, phylloquinone (phytomenadione) and menaquinone (mainly as MK-7 and, to a minor extent, MK-6) may be added to foods and food supplements [101].

Substantial amounts of vitamin K in the form of long-chain MKs are known to be present in the human gut. Relatively few of the bacteria that comprise the normal intestinal flora are major producers of MKs. Obligate anaerobes of the *Bacteroides* (*B. fragilis*), *Eubacterium*, *Propionibacterium*, and *Arachnia* genera, as well as facultatively anaerobic organisms, such as *Escherichia coli*, are major producers of menaquinones MK-6 or higher, but not of the short-chain MK-4. In breastfed infants, the production of menaquinones by gut microbiota is probably low, as most bacteria of their microbiota, including *Bifidobacterium*, *Lactobacillus*, and *Clostridium* species, do not produce menaquinones [101, 105].

The EFSA Panel on Dietetic Products, Nutrition, and Allergies considers that it is not possible to estimate precisely an average absorption of phylloquinone and menaquinones from the diet. Some data suggest that absorption might be enhanced by dietary fat by up to threefold and that there are no sex or age differences in phylloquinone absorption in adults; however, there are differences between people with different types of diet. Dietary phylloquinone is less easily absorbed than menaquinones, although the bioavailability of the former increases if administered in pure form, probably because phylloquinone molecules are tightly bound to plant tissue. It is possible that vegetable oil offers more easily absorbable vitamin K_1 than whole vegetables. The absorption of dietary vitamin K_2, in particular long-chain MKs, is excellent and may even be complete due to the co-presence of fat. As far as menaquinones produced by gut bacteria are concerned, it is unlikely that the forms with large side chains are absorbed in the distal colon, but they can be absorbed in the terminal ileum, where microflora is still present, and absorption can be facilitated by bile salts. In any case, their contribution to vitamin K status is unclear [101, 102, 105]. Vitamin K absorption can be decreased by certain drugs, such as cholestyramine, rifampicin, and orlistat [102].

The mechanism of absorption is likely to be similar to that of other lipid compounds like vitamins D and E (Chapter 13), therefore, absorption is decreased in patients with biliary insufficiency or various malabsorption syndromes. Phylloquinone in plasma is predominantly carried by the triglyceride-rich lipoprotein fraction containing very-low-density lipoproteins and chylomicron remnants, although some is located in the low-density lipoprotein and high-density lipoprotein fractions. Plasma phylloquinone concentrations in a physiologically normal population range from 0.3 to 2.6 nmol/l. The major route of entry of phylloquinone into tissues appears to be through clearance of chylomicron remnants by apolipoprotein E (ApoE) receptors. The secretion of phylloquinone from the liver and the process by which the vitamin moves among organs are not yet understood. The total human body pool of phylloquinone is very small, and turnover is rapid. Long-chain MKs, rather than phylloquinone, are the major source of the vitamin in liver and turns over at a much lower rate. The catabolism of vitamin K basically involves an initial ω-hydroxylation mediated by members of the cytochrome P450 family, followed by shortening of the polyisoprenoic side chain via β-oxidation to carboxylic acids which are glucuronidated and excreted in feces (via the bile) and urine [102, 105].

6.4.4.4 Stability

Little is known about the reactions of vitamin K in foods but it seems to be more stable during processing than other vitamins. Some naturally occurring vitamin K can be found in oils that are resistant to heat and moisture during cooking, but it is diminished by acids, bases, light, and oxidizers. Freezing may also decrease the level of vitamin K [107].

The process of hydrogenation to convert plant oils to solid margarines or shortening converts some of the phylloquinone to 2′,3′-dihydrophylloquinone with a completely saturated side chain. The biologic activity of this form of the vitamin is lower than that of phylloquinone but has not been accurately determined [105].

6.4.4.5 Dietary Reference Values (DRVs): hypo- and hypervitaminosis

The EFSA Panel on Dietetic Products, Nutrition and Allergies considers that none of the biomarkers of vitamin K intake or status is suitable by itself to derive dietary reference values for vitamin K. Consequently, it maintains the value proposed by European Union Scientific Committee for Food (SCF) in 1993 for phylloquinone adequate intakes (AIs) of 1 µg phylloquinone/kg body weight per day for all age and sex population groups (Table 6.3) [101].

In adults, vitamin K deficiency is clinically characterized by a bleeding tendency in relation to a low activity of the blood coagulation factors, actually, antagonists of vitamin K, like warfarin (a competitive inhibitor of VKORC1), are used as a therapy for thrombotic disease. Symptoms of deficiency include spontaneous cutaneous purpura, epistaxis (nosebleed), gastrointestinal, genitourinary, gingival, or other bleeding. Although vitamin K deficiency is not common in adults, it can occur for several reasons: (a) poor vitamin K dietary content, (b) some pathological states (e.g., liver disease, cholestasis, cystic fibrosis, alcoholism, malabsorption states (including inflammatory bowel disease), and bariatric surgical intervention), and (c) pharmacotherapy with several drugs [102].

Oral or parenteral administration of vitamin K immediately following birth is the standard cure for the hemorrhagic disease of the newborn, also called early vitamin K deficiency bleeding (VKDB). This disease occurs during the first week of life in healthy-appearing neonates and may be due to low placental transfer of phylloquinone, low clotting factor levels, a sterile gut, and/or low vitamin K content of breast milk. Although the incidence is low, the mortality rate from intracranial bleeding is high. Exclusively breast-fed infants are more susceptible to the development of vitamin K hypovitaminosis than formula-fed children. Vitamin K deficiency has also been suggested to cause alterations in bone mineral density and aortic calcification; however, more experimental evidence is needed to sustain these suggestions [102, 105].

Several studies in humans have shown no evidence of adverse effects associated with vitamin K supplementation up to 10 mg/day for 1 month. This supports studies performed in animals, which showed no adverse effect after daily administration of 2,000 mg/kg body weight for 30 days. Considering these limited data, the SCF has concluded that there is no appropriate evidence to derive a tolerable upper intake level (UL) for vitamin K [101].

6.5 Vitamin-like or conditionally essential nutrients (CENs)

Conditionally essential nutrients can be defined as "those that can usually be synthesized in adequate amounts endogenously, but may require exogenous supplementation during some circumstances" [108], that mainly include physiological (e.g., pregnancy, lactation, aging) or pathological conditions. Choline, carnitine, taurine, and some amino acids (Chapter 5) are considered as conditionally essential nutrients.

– Choline is a quaternary amine (IUPAC Name: 2-hydroxyethyl(trimethyl)azanium, formerly 2-hydroxyethyl-N,N,N-trimethylammonium) present in food in free and esterified forms. The main forms present in foods are phosphatidylcholine (PC, aka lecithin), which is also the main form present in animal tissues, free choline, phosphocholine (PChol), glycerophosphocholine (GPC), and sphingomyelin. Choline is an integral part of some phospholipids, essential components of biological membranes, and has been reported to be involved in physiological processes such as signal transduction, DNA and histone methylation, and nerve myelination. Choline is also a precursor of the neurotransmitter acetylcholine, and of betaine, an osmoregulator. Phosphatidylcholine plays an important role in the metabolism and transport of lipids and cholesterol by lipoproteins. Dietary deficiency of choline can cause fatty liver or hepatic steatosis that may result in non-alcoholic fatty liver disease (NAFLD) and liver and muscle damage. Premenopausal women have an enhanced capacity for *de novo* biosynthesis of choline; however, many of them (around 45%) develop organ dysfunction when deprived of this nutrient [109–111]. The best source of choline is eggs (raw egg yolk: about 670 mg/100 g food, whole raw fresh egg: about 290 mg/100 g food) followed by meats and fish, whole grains, vegetables and fruit, and fats and oils (median content of fats and oils: about 5 mg/100 g food). The EFSA has set an AI of choline for all adults of 400 mg/day, 160 mg/day for infants aged 7–11 months, and, for children, values ranging from 140 mg/day (1–3 years) to 400 mg/day (15–17 years) depending on body mass. For pregnant women, the Panel set an AI of 480 mg/day, and 520 mg/day for lactating women [110]. Nutritional and biochemical properties of choline have been reviewed elsewhere [109, 110].

– L-Carnitine [IUPAC Name: (3R)-3-hydroxy-4-(trimethylazaniumyl)butanoate] is a zwitterionic quaternary amino acid involved in the translocation of long-chain fatty acid from the cytosol to the mitochondria in order to be subjected to β-oxidation (Chapter 4, Section 4.9.1). L-carnitine also seems to be involved in the regulation of cellular energy and phospholipid metabolism. It is abundant in most food products of animal origin. Carnitine is not a required nutrient for children and adults. Deficiency may occur due to genetic disorders or the use of certain drugs. Population groups that may be most vulnerable to nutritional carnitine deficiency are strict vegetarians and newborn infants. Strict vegetarians have somewhat lower plasma carnitine concentrations compared with omnivores, but no indication of clinically relevant carnitine deficiency has been reported. Infant formulas are fortified with carnitine, based of reported biochemical differences when infants were fed carnitine-free diets, compared with similar diets with carnitine. More information in [112]. Recent studies suggest that prolonged L-carnitine supplementation, especially in combination with carbohydrates, may increase total carnitine content in skeletal muscle, thereby affecting physical performance. However, this treatment elevates fasting plasma levels of trimethylamine N-oxide (TMAO), compound supposed to be pro-atherogenic (Chapter 5) [113].

– Taurine [IUPAC Name: 2-aminoethanesulfonic acid] is formed from the amino acid cysteine by removal of the carboxyl group and oxidation of the sulfur to form a sulfonic acid group. Taurine has multiple functions and plays an important role in several physiologic processes, but many of these are poorly understood, the best known function being its role in bile acid conjugation. Taurine is considered to be conditionally essential during infant development and probably for adults in some special circumstances. Taurine-supplemented beverages (containing 1,000 mg per 250 mL) are popular in many countries. Consumption of taurine-supplemented beverages dramatically increases taurine intake to eight or more times the typical intakes in a population. There is a lack of evidence to conclude that the amounts of taurine in these beverages have either therapeutic benefits or adverse effects [114]. Concentrations of circulating taurine decline with aging in mice, monkeys, and humans. Taurine supplementation has been recently reported to increase the health and life span in mice and health span in monkeys. The authors concluded that taurine deficiency may be a driver of aging in these animals. In humans, lower taurine concentrations correlated with several age-related diseases, while taurine concentrations increased after acute endurance exercise. Clinical trials in humans are needed to support these claims [115].

252 —— 6 Vitamins

References

[1]   Al Binali HAH. Night blindness and ancient remedy. Heart Views Off J Gulf Heart Assoc
       2014;15:136–9. https://doi.org/10.4103/1995-705X.151098.
[2]   Semba RD. The discovery of the vitamins. Int J Vitam Nutr Res 2012;82:310–5. https://doi.org/
       10.1024/0300-9831/a000124.
[3]   Mozaffarian D, Rosenberg I, Uauy R. History of modern nutrition science – implications for current
       research, dietary guidelines, and food policy. BMJ 2018:k2392. https://doi.org/10.1136/bmj.k2392.
[4]   The long, rocky road to understanding vitamins. World Rev. Nutr. Diet., vol. 104. Basel: S. Karger;
       2012, p. 65–105. https://doi.org/10.1159/000338592.
[5]   EFSA Panel on Dietetic Products, Nutrition and Allergies (NDA), Turck D, Bresson J, Burlingame B,
       Dean T, Fairweather-Tait S, et al. Dietary reference values for thiamin. EFSA J 2016;14. https://doi.
       org/10.2903/j.efsa.2016.4653.
[6]   Tylicki A, Łotowski Z, Siemieniuk M, Ratkiewicz A. Thiamine and selected thiamine antivitamins –
       biological activity and methods of synthesis. Biosci Rep 2018;38:BSR20171148. https://doi.org/
       10.1042/BSR20171148.
[7]   Nelson DL, Cox MM, Hoskins AA, Lehninger AL. Lehninger principles of biochemistry. 8th edition.
       New York, NY: Macmillan International, Higher Education; 2021.
[8]   Lonsdale D. Thiamin. Adv. Food Nutr. Res., vol. 83. Elsevier; 2018, p. 1–56. https://doi.org/10.1016/
       bs.afnr.2017.11.001.
[9]   Yoshii K, Hosomi K, Sawane K, Kunisawa J. Metabolism of dietary and microbial vitamin B family in
       the regulation of host immunity. Front Nutr 2019;6:48. https://doi.org/10.3389/fnut.2019.00048.
[10]   Damodaran S, Parkin K, Fennema OR, editors. Fennema's food chemistry. 4th edition Boca Raton:
       CRC Press/Taylor & Francis; 2008.
[11]   EFSA Panel on Dietetic Products, Nutrition and Allergies (NDA). Scientific opinion on dietary
       reference values for biotin. EFSA J 2014;12. https://doi.org/10.2903/j.efsa.2014.3580.
[12]   Ciqual Table de composition nutritionnelle des aliments n.d. https://ciqual.anses.fr/
       (accessed January 3, 2020).
[13]   Staggs CG, Sealey WM, McCabe BJ, Teague AM, Mock DM. Determination of the biotin content of
       select foods using accurate and sensitive HPLC/avidin binding. J Food Compost Anal 2004;17:767–76.
       https://doi.org/10.1016/j.jfca.2003.09.015.
[14]   Belitz H-D, Grosch W, Schieberle P. Food chemistry. 4th rev. and extended ed. Berlin:
       Springer; 2009.
[15]   European Food Safety Authority (EFSA). Dietary Reference Values for nutrients Summary report.
       EFSA Support Publ 2017;14. https://doi.org/10.2903/sp.efsa.2017.e15121 (amended September 2019).
[16]   Wiley KD, Gupta M. Vitamin B1 (Thiamine) deficiency. StatPearls, Treasure Island (FL): StatPearls
       Publishing; 2023.
[17]   Martel JL, Kerndt CC, Doshi H, Franklin DS. Vitamin B1 (Thiamine). StatPearls, Treasure Island (FL):
       StatPearls Publishing; 2023.
[18]   Pinto J, Rivlin R. Riboflavin (Vitamin B2). In: Zempleni J, Suttie J, Iii J, Stover P, editors. Handb. Vitam.
       5th edition. CRC Press; 2013, p. 191–266. https://doi.org/10.1201/b15413-7.
[19]   Pinto JT, Zempleni J. Riboflavin12. Adv Nutr 2016;7:973–5. https://doi.org/10.3945/an.116.012716.
[20]   Janssen JJE, Grefte S, Keijer J, De Boer VCJ. Mito-nuclear communication by mitochondrial
       metabolites and its regulation by B-vitamins. Front Physiol 2019;10:78. https://doi.org/10.3389/
       fphys.2019.00078.
[21]   Turck D, Bresson J-L, Burlingame B, Dean T, Fairweather-Tait S, Heinonen M, et al. Dietary reference
       values for riboflavin. EFSA J 2017;15. 10.2903/j.efsa.2017.4919.

[22]   Hrubša M, Siatka T, Nejmanová I, Vopršalová M, Kujovská Krčmová L, Matoušová K, et al. Biological
       properties of vitamins of the B-complex, Part 1: Vitamins B1, B2, B3, and B5. Nutrients 2022;14:484.
       https://doi.org/10.3390/nu14030484.

[23]   Zempleni J, editor. Handbook of vitamins. 4th edition. Boca Raton: Taylor & Francis; 2007.

[24]   Bieganowski P, Brenner C. Discoveries of nicotinamide riboside as a nutrient and conserved NRK
       genes establish a preiss-handler independent route to NAD+ in fungi and humans. Cell
       2004;117:495–502. https://doi.org/10.1016/S0092-8674(04)00416-7.

[25]   Scientific opinion on dietary reference values for niacin. EFSA J n.d. https://doi.org/10.2903/j.efsa.
       2014.3759.

[26]   Kirkland JB, Meyer-Ficca ML. Niacin. Adv. Food Nutr. Res., vol. 83, Elsevier; 2018, p. 83–149.
       https://doi.org/10.1016/bs.afnr.2017.11.003.

[27]   Lewis CA, Parker SJ, Fiske BP, McCloskey D, Gui DY, Green CR, et al. Tracing compartmentalized
       NADPH metabolism in the cytosol and mitochondria of mammalian cells. Mol Cell 2014;55:253–63.
       https://doi.org/10.1016/j.molcel.2014.05.008.

[28]   Han SJ, Jang H-S, Noh MR, Kim J, Kong MJ, Kim JI, et al. Mitochondrial NADP$^+$-dependent isocitrate
       dehydrogenase deficiency exacerbates mitochondrial and cell damage after kidney ischemia-
       reperfusion injury. J Am Soc Nephrol 2017;28:1200–15. https://doi.org/10.1681/ASN.2016030349.

[29]   Rydström J. Mitochondrial NADPH, transhydrogenase and disease. Biochim Biophys Acta BBA –
       Bioenerg 2006;1757:721–6. https://doi.org/10.1016/j.bbabio.2006.03.010.

[30]   Agledal L, Niere M, Ziegler M. The phosphate makes a difference: Cellular functions of NADP. Redox
       Rep 2010;15:2–10. https://doi.org/10.1179/174329210X12650506623122.

[31]   Combs GF, McClung JP. Niacin. The Vitamins, Elsevier; 2017, p. 331–50. https://doi.org/10.1016/B978-
       0-12-802965-7.00013-7.

[32]   Cornell L, Arita K. Water soluble vitamins: B1, B2, B3, and B6. In: Pitchumoni CS, Dharmarajan TS,
       editors. Geriatr. Gastroenterol. Cham: Springer International Publishing; 2021, p. 569–96.
       https://doi.org/10.1007/978-3-030-30192-7_21.

[33]   Gasperi V, Sibilano M, Savini I, Catani M. Niacin in the central nervous system: An update of
       biological aspects and clinical applications. Int J Mol Sci 2019;20:974. https://doi.org/10.3390/
       ijms20040974.

[34]   Bodor ET, Offermanns S. Nicotinic acid: An old drug with a promising future. Br J Pharmacol
       2008;153:S68–75. https://doi.org/10.1038/sj.bjp.0707528.

[35]   HCAR2 hydroxycarboxylic acid receptor 2 [Homo sapiens (human)] – Gene – NCBI n.d. https://www.
       ncbi.nlm.nih.gov/gene/338442 (accessed June 7, 2019).

[36]   Wuerch E, Urgoiti GR, Yong VW. The promise of niacin in neurology. Neurotherapeutics 2023.
       https://doi.org/10.1007/s13311-023-01376-2.

[37]   Bender DA, Oxford University Press. A dictionary of food and nutrition. Oxford; New York: Oxford
       University Press; 2009.

[38]   EFSA Panel on Dietetic Products N and A (NDA). Scientific opinion on dietary reference values for
       pantothenic acid. EFSA J 2014;12:3581. https://doi.org/10.2903/j.efsa.2014.3581.

[39]   Dietary Reference Values for nutrients Summary report. EFSA Support Publ 2017;14:e15121E.
       https://doi.org/10.2903/sp.efsa.2017.e15121.

[40]   Da Silva, Vanessa R., Mackey, Amy D., Davis, Steven R., Gregory, Jesse F. Vitamin B6. Mod. Nutr.
       Health Dis. 11th edition. Philadelphia: Wolters Kluwer Health/Lippincott Williams & Wilkins; 2014,
       p. 341–50.

[41]   EFSA Panel on Dietetic Products, Nutrition and Allergies (NDA). Dietary reference values for vitamin
       B6. EFSA J 2016; 14. https://doi.org/10.2903/j.efsa.2016.4485.

[42]   Bird RP. The emerging role of vitamin B6 in inflammation and carcinogenesis. Adv Food Nutr
       Res., vol. 83, Elsevier; 2018, p. 151–94. https://doi.org/10.1016/bs.afnr.2017.11.004.

[43] Wilson MP, Plecko B, Mills PB, Clayton PT. Disorders affecting vitamin B6 metabolism. J Inherit Metab Dis 2019. https://doi.org/10.1002/jimd.12060.

[44] Said HM. Recent advances in carrier-mediated intestinal absorption of water-soluble vitamins. Annu Rev Physiol 2004;66:419–46. https://doi.org/10.1146/annurev.physiol.66.032102.144611.

[45] Magnúsdóttir S, Ravcheev D, de Crécy-Lagard V, Thiele I. Systematic genome assessment of B-vitamin biosynthesis suggests co-operation among gut microbes. Front Genet 2015;6. https://doi.org/10.3389/fgene.2015.00148.

[46] EFSA Panel on Nutrition, Novel Foods and Food Allergens (NDA), Turck D, Bohn T, Castenmiller J, de Henauw S, Hirsch-Ernst K, et al. Scientific opinion on the tolerable upper intake level for vitamin B6. EFSA J 2023; 21. https://doi.org/10.2903/j.efsa.2023.8006.

[47] Hadtstein F, Vrolijk M. Vitamin B-6-induced neuropathy: Exploring the mechanisms of pyridoxine toxicity. Adv Nutr 2021;12:1911–29. https://doi.org/10.1093/advances/nmab033.

[48] Mock, Donald M. Biotin. Mod. Nutr. Health Dis. 11th edition. Philadelphia: Wolters Kluwer Health/Lippincott Williams & Wilkins; 2014, p. 390–8.

[49] León-Del-Río A. Biotin in metabolism, gene expression, and human disease. J Inherit Metab Dis 2019:jimd.12073. https://doi.org/10.1002/jimd.12073.

[50] Jungert A, Ellinger S, Watzl B, Richter M, The German Nutrition Society (DGE). Revised D-A-CH reference values for the intake of biotin. Eur J Nutr 2022;61:1779–87. https://doi.org/10.1007/s00394-021-02756-0.

[51] Mock DM. Biotin: From nutrition to therapeutics. J Nutr 2017;147:1487–92. https://doi.org/10.3945/jn.116.238956.

[52] Prasad PD, Wang H, Kekuda R, Fujita T, Fei YJ, Devoe LD, et al. Cloning and functional expression of a cDNA encoding a mammalian sodium-dependent vitamin transporter mediating the uptake of pantothenate, biotin, and lipoate. J Biol Chem 1998;273:7501–6. https://doi.org/10.1074/jbc.273.13.7501.

[53] León-Del-Río A, Valadez-Graham V, Gravel RA. Holocarboxylase synthetase: A moonlighting transcriptional coregulator of gene expression and a cytosolic regulator of biotin utilization. Annu Rev Nutr 2017;37:207–23. https://doi.org/10.1146/annurev-nutr-042617-104653.

[54] Stover, Patrick J. Folic acid. Mod. Nutr. Health Dis. 11th edition. Philadelphia: Wolters Kluwer Health/Lippincott Williams & Wilkins; 2014, p. 358–66.

[55] EFSA Panel on Dietetic Products, Nutrition and Allergies (NDA). Scientific opinion on dietary reference values for folate. EFSA J 2014;12. https://doi.org/10.2903/j.efsa.2014.3893.

[56] Mato JM, Martínez-Chantar ML, Lu SC. Methionine metabolism and liver disease. Annu Rev Nutr 2008;28:273–93. https://doi.org/10.1146/annurev.nutr.28.061807.155438.

[57] Brosnan ME, MacMillan L, Stevens JR, Brosnan JT. Division of labour: How does folate metabolism partition between one-carbon metabolism and amino acid oxidation? Biochem J 2015;472:135–46. https://doi.org/10.1042/BJ20150837.

[58] Naderi N, House JD. Recent developments in folate nutrition. Adv Food Nutr Res., vol. 83, Elsevier; 2018, p. 195–213. https://doi.org/10.1016/bs.afnr.2017.12.006.

[59] Smith AD, Warren MJ, Refsum H. Vitamin B12. Adv Food Nutr Res., vol. 83, Elsevier; 2018, p. 215–79. https://doi.org/10.1016/bs.afnr.2017.11.005.

[60] EFSA Panel on Dietetic Products, Nutrition, and Allergies (NDA). Scientific opinion on dietary reference values for cobalamin (vitamin B12). EFSA J 2015;13. https://doi.org/10.2903/j.efsa.2015.4150.

[61] Hannibal L, Axhemi A, Glushchenko AV, Moreira ES, Brasch NE, Jacobsen DW. Accurate assessment and identification of naturally occurring cellular cobalamins. Clin Chem Lab Med 2008;46:1739–46. https://doi.org/10.1515/CCLM.2008.356.

[62] Takahashi-Iñiguez T, García-Hernandez E, Arreguín-Espinosa R, Flores ME. Role of vitamin B12 on methylmalonyl-CoA mutase activity. J Zhejiang Univ Sci B 2012;13:423–37. https://doi.org/10.1631/jzus.B1100329.

[63] Rizzo G, Laganà AS. A review of vitamin B12. Mol Nutr, Elsevier; 2020, p. 105–29. https://doi.org/10.1016/B978-0-12-811907-5.00005-1.

[64] Watanabe F, Yabuta Y, Tanioka Y, Bito T. Biologically active vitamin B $_{12}$ compounds in foods for preventing deficiency among vegetarians and elderly subjects. J Agric Food Chem 2013;61:6769–75. https://doi.org/10.1021/jf401545z.

[65] Allen LH. Bioavailability of Vitamin B12. Int J Vitam Nutr Res 2010;80:330–5. https://doi.org/10.1024/0300-9831/a000041.

[66] Toh B-H, van Driel IR, Gleeson PA. Pernicious anemia. N Engl J Med 1997;337:1441–8. https://doi.org/10.1056/NEJM199711133372007.

[67] Uebanso T, Shimohata T, Mawatari K, Takahashi A. Functional roles of B-vitamins in the gut and gut microbiome. Mol Nutr Food Res 2020;64:2000426. https://doi.org/10.1002/mnfr.202000426.

[68] Javadov S, Kozlov AV, Camara AKS. Mitochondria in health and diseases. Cells 2020;9:1177. https://doi.org/10.3390/cells9051177.

[69] Picard M, Wallace DC, Burelle Y. The rise of mitochondria in medicine. Mitochondrion 2016;30:105–16. https://doi.org/10.1016/j.mito.2016.07.003.

[70] Mukherjee S, Banerjee O, Singh S. The role of B vitamins in protecting mitochondrial function. Mol Nutr Mitochondria, Elsevier; 2023, p. 167–93. https://doi.org/10.1016/B978-0-323-90256-4.00001-1.

[71] Ross, A. Catharine. Vitamin A. Mod. Nutr. Health Dis. 11th edition, Philadelphia: Wolters Kluwer Health/Lippincott Williams & Wilkins; 2014, p. 260–74.

[72] Bar-El Dadon S, Reifen R. Vitamin A and the epigenome. Crit Rev Food Sci Nutr 2017;57:2404–11. https://doi.org/10.1080/10408398.2015.1060940.

[73] Daruwalla A, Choi EH, Palczewski K, Kiser PD. Structural biology of 11- *cis*- retinaldehyde production in the classical visual cycle. Biochem J 2018;475:3171–88. https://doi.org/10.1042/BCJ20180193.

[74] Tsin A, Betts-Obregon B, Grigsby J. Visual cycle proteins: Structure, function, and roles in human retinal disease. J Biol Chem 2018;293:13016–21. https://doi.org/10.1074/jbc.AW118.003228.

[75] Cañete A, Cano E, Muñoz-Chápuli R, Carmona R. Role of vitamin A/retinoic acid in regulation of embryonic and adult hematopoiesis. Nutrients 2017;9. https://doi.org/10.3390/nu9020159.

[76] Carmona R, Barrena S, Muñoz-Chápuli R. Retinoids in stellate cells: Development, repair, and regeneration. J Dev Biol 2019;7:10. https://doi.org/10.3390/jdb7020010.

[77] Wiseman EM, Bar-El Dadon S, Reifen R. The vicious cycle of vitamin a deficiency: A review. Crit Rev Food Sci Nutr 2017;57:3703–14. https://doi.org/10.1080/10408398.2016.1160362.

[78] Aune D, Keum N, Giovannucci E, Fadnes LT, Boffetta P, Greenwood DC, et al. Dietary intake and blood concentrations of antioxidants and the risk of cardiovascular disease, total cancer, and all-cause mortality: A systematic review and dose-response meta-analysis of prospective studies. Am J Clin Nutr 2018;108:1069–91. https://doi.org/10.1093/ajcn/nqy097.

[79] Scientific opinion on dietary reference values for vitamin A. EFSA J n.d. https://doi.org/10.2903/j.efsa.2015.4028.

[80] Yadav AS, Isoherranen N, Rubinow KB. Vitamin A homeostasis and cardiometabolic disease in humans: Lost in translation? J Mol Endocrinol 2022;69:R95–108. https://doi.org/10.1530/JME-22-0078.

[81] Chen G, Weiskirchen S, Weiskirchen R. Vitamin A: Too good to be bad? Front Pharmacol 2023;14:1186336. https://doi.org/10.3389/fphar.2023.1186336.

[82] Vitamin D n.d. https://www.qmul.ac.uk/sbcs/iupac/misc/D.html#r2 (accessed June 26, 2019).

[83] Duffy, Valerie B. Nutrition and the chemical senses. Mod. Nutr. Health Dis. 11th edition. Philadelphia: Wolters Kluwer Health/Lippincott Williams & Wilkins; 2014, p. 574–88.

[84] Bikle D. Vitamin D: Production, metabolism, and mechanisms of action. In: Feingold KR, Anawalt B, Boyce A, Chrousos G, Dungan K, Grossman A, et al., editors. Endotext. South Dartmouth (MA): MDText.com, Inc.; 2000.

[85] Keane JT, Elangovan H, Stokes RA, Gunton JE. Vitamin D and the liver – correlation or cause? Nutrients 2018; 10. https://doi.org/10.3390/nu10040496.

[86] EFSA Panel on Dietetic Products, Nutrition and Allergies (NDA). Dietary reference values for vitamin D. EFSA J 2016;14. https://doi.org/10.2903/j.efsa.2016.4547.

[87] Holick MF. The one-hundred-year anniversary of the discovery of the sunshine vitamin D3: Historical, personal experience and evidence-based perspectives. Nutrients 2023;15:593. https://doi.org/10.3390/nu15030593.

[88] Roth DE, Abrams SA, Aloia J, Bergeron G, Bourassa MW, Brown KH, et al. Global prevalence and disease burden of vitamin D deficiency: A roadmap for action in low- and middle-income countries. Ann N Y Acad Sci 2018;1430:44–79. https://doi.org/10.1111/nyas.13968.

[89] Meza-Meza MR, Ruiz-Ballesteros AI, De La Cruz-Mosso U. Functional effects of vitamin D: From nutrient to immunomodulator. Crit Rev Food Sci Nutr 2022;62:3042–62. https://doi.org/10.1080/10408398.2020.1862753.

[90] Reboul E. Intestinal absorption of vitamin D: From the meal to the enterocyte. Food Funct 2015;6:356–62. https://doi.org/10.1039/C4FO00579A.

[91] Rowling MJ, Kemmis CM, Taffany DA, Welsh J. Megalin-mediated endocytosis of vitamin D binding protein correlates with 25-hydroxycholecalciferol actions in human mammary cells. J Nutr 2006;136:2754–9.

[92] Hemery YM, Fontan L, Moench-Pfanner R, Laillou A, Berger J, Renaud C, et al. Influence of light exposure and oxidative status on the stability of vitamins A and D₃ during the storage of fortified soybean oil. Food Chem 2015;184:90–8. https://doi.org/10.1016/j.foodchem.2015.03.096.

[93] Liu D, Meng X, Tian Q, Cao W, Fan X, Wu L, et al. Vitamin D and multiple health outcomes: An umbrella review of observational studies, randomized controlled trials, and Mendelian randomization studies. Adv Nutr 2022;13:1044–62. https://doi.org/10.1093/advances/nmab142.

[94] Tocopherols n.d. https://www.qmul.ac.uk/sbcs/iupac/misc/toc.html (accessed June 29, 2019).

[95] Scientific opinion on dietary reference values for vitamin E as α-tocopherol. EFSA J 2015;13:4149. https://doi.org/10.2903/j.efsa.2015.4149.

[96] Liao S, Omage SO, Börmel L, Kluge S, Schubert M, Wallert M, et al. Vitamin E and metabolic health: Relevance of interactions with other micronutrients. Antioxidants 2022;11:1785. https://doi.org/10.3390/antiox11091785.

[97] Traber, Maret G. Vitamin E. Mod. Nutr. Health Dis. 11th edition, Philadelphia: Wolters Kluwer Health/Lippincott Williams & Wilkins; 2014, p. 293–304.

[98] Ciarcià G, Bianchi S, Tomasello B, Acquaviva R, Malfa GA, Naletova I, et al. Vitamin E and non-communicable diseases: A review. Biomedicines 2022;10:2473. https://doi.org/10.3390/biomedicines10102473.

[99] Schubert M, Kluge S, Schmölz L, Wallert M, Galli F, Birringer M, et al. Long-chain metabolites of vitamin E: Metabolic activation as a general concept for lipid-soluble vitamins? Antioxidants 2018;7:10. https://doi.org/10.3390/antiox7010010.

[100] Harada M, Kumemura H, Harada R, Komai K, Sata M. Scleroderma and repeated spontaneous abortions treated with vitamin E-A case report. Kurume Med J 2005;52:93–5. https://doi.org/10.2739/kurumemedj.52.93.

[101] Turck D, Bresson J-L, Burlingame B, Dean T, Fairweather-Tait S, Heinonen M, et al. Dietary reference values for vitamin K. EFSA J 2017;15:e04780. https://doi.org/10.2903/j.efsa.2017.4780.

[102] Mladěnka P, Macáková K, Kujovská Krčmová L, Javorská L, Mrštná K, Carazo A, et al. Vitamin K – sources, physiological role, kinetics, deficiency, detection, therapeutic use, and toxicity. Nutr Rev 2022;80:677–98. https://doi.org/10.1093/nutrit/nuab061.

[103] Rishavy MA, Hallgren KW, Wilson LA, Usubalieva A, Runge KW, Berkner KL. The vitamin K oxidoreductase is a multimer that efficiently reduces vitamin K epoxide to hydroquinone to allow vitamin K-dependent protein carboxylation. J Biol Chem 2013;288:31556–66. https://doi.org/10.1074/jbc.M113.497297.

[104] Tie J-K, Stafford DW. Structural and functional insights into enzymes of the vitamin K cycle. J Thromb Haemost JTH 2016;14:236–47. https://doi.org/10.1111/jth.13217.

[105] Suttie, John W. Vitamin K. Mod. Nutr. Health Dis. 11th edition, Philadelphia: Wolters Kluwer Health/Lippincott Williams & Wilkins; 2014, p. 305–16.

[106] Gamma-carboxyglutamic acid-rich (GLA) domain (IPR000294) < interpro < embl-ebi n.d. https://www.ebi.ac.uk/interpro/entry/ipr000294 (accessed june 30, 2019).

[107] Marcus JB. Vitamin and mineral basics: The ABCs of healthy foods and beverages, including phytonutrients and functional foodsCulin Nutr., Elsevier; 2013, p. 279–331. https://doi.org/10.1016/B978-0-12-391882-6.00007-8.

[108] Mischley LK. Conditionally essential nutrients: The state of the science. J Food Nutr 2014;1:1–4.

[109] Zeisel, Steven H. Choline. Mod. Nutr. Health Dis. 11th edition, Philadelphia: Wolters Kluwer Health/Lippincott Williams & Wilkins; 2014.

[110] Dietary reference values for choline. EFSA J 2016. https://doi.org/10.2903/j.efsa.2016.4484.

[111] Kansakar U, Trimarco V, Mone P, Varzideh F, Lombardi A, Santulli G. Choline supplements: An update. Front Endocrinol 2023;14:1148166. https://doi.org/10.3389/fendo.2023.1148166.

[112] Rebouche, Charles J. Carnitine. Mod. Nutr. Health Dis. 11th edition, Philadelphia: Wolters Kluwer Health/Lippincott Williams & Wilkins; 2014, p. 440–6.

[113] Sawicka AK, Renzi G, Olek RA. The bright and the dark sides of L-carnitine supplementation: A systematic review. J Int Soc Sports Nutr 2020;17:49. https://doi.org/10.1186/s12970-020-00377-2.

[114] Stipanuk, Martha H. Cysteine, taurine, and homocysteine. Mod. Nutr. Health Dis. 11th edition, Philadelphia: Wolters Kluwer Health/Lippincott Williams & Wilkins; 2014, p. 447–63.

[115] Singh P, Gollapalli K, Mangiola S, Schranner D, Yusuf MA, Chamoli M, et al. Taurine deficiency as a driver of aging. Science 2023;380:eabn9257. https://doi.org/10.1126/science.abn9257.

7 Minerals

7.1 Definition and classification

Minerals can be defined as the constituents that remain as ash after the combustion of living matter. This definition includes elements other than C, H, O, and N, which are present in organic molecules and water and constitute about 99% of the total number of atoms in living systems (see previous chapters) [1, 2].

About 25 of the 90 chemical elements that occur naturally in the earth's crust are known to be present in cells (Figure 7.1), normally in the form of their most stable ions. Historically, they have been classified as either major or trace elements [2]:

–   Major elements (also known as macroelements or macrominerals) include calcium (Ca^{2+}), phosphorus (as PO_4^{3-}), magnesium (Mg^{2+}), sodium (Na^+), potassium (K^+), and chloride (Cl^-)
–   Trace elements (micro-elements or micro-minerals) include iron ($Fe^{2+/3+}$), iodine (I^-), zinc (Zn^{2+}), selenium (Se^{2-}), cobalt ($Co^{2+/3+}$), molybdenum ($Mo^{4+/5+/6+}$), manganese ($Mn^{2+/3+}$), copper ($Cu^{1+/2+}$), and fluorine (F^-)

H																	He
Li	Be											B	C	N	O	F	Ne
Na	Mg											Al	Si	P	S	Cl	Ar
K	Ca	Sc	Ti	V	Cr	Mn	Fe	Co	Ni	Cu	Zn	Ga	Ge	As	Se	Br	Kr
Rb	Sr	Y	Zr	Nb	Mo	Tc	Ru	Rh	Pd	Ag	Cd	In	Sn	Sb	Te	I	Xe
Cs	Ba	La	Hf	Ta	W	Re	Os	Ir	Pt	Au	Hg	Tl	Pb	Bi	Po	At	Rn
Fr	Ra	Ac															

Figure 7.1: Periodic table highlighting the elements that are known to be essential for life. Basic elements of macronutrients (i.e., carbohydrates, lipids, and proteins) are shown on orange background, macroelements on yellow background, and microelements on blue background. Some additional elements have been tested in model organisms but their functions and/or essentiality has not been tested in humans.

Some authors also mention "ultra-trace elements," that include some quantitatively minor elements whose essentiality has been tested in experiments with model organisms, although their functions in the human body have not been clearly elucidated in some cases. These include aluminum (Al^{3+}), barium (Ba^{2+}), bismuth (Bi^{3+}), boron (B^{3+}), bromine (Br^-), cadmium (Cd^{2+}), cesium (Cs^+), molybdenum (Mo^{n+}), lithium (Li^+), lead ($Pb^{2+/4+}$), rubidium (Rb^+), silicon (Si), strontium (Sr^{2+}), and wolfram (W^{n+}), among others [1].

Although mineral elements are present in relatively low concentrations in living organisms (hence, in foods), they play key functional roles in vivo [2, 3]:

https://doi.org/10.1515/9783111111872-007

- Participation in the structural and functional properties of a myriad of proteins (including many enzymes) and hormones, thus being involved in metabolism and its regulation
- Strength and rigidity of bones and teeth
- Transport of oxygen (O_2) and carbon dioxide (CO_2) in the bloodstream
- Cell adhesion and division
- Signal transduction
- Transmission of nerve impulses
- Protection against oxidative stress
- Cobalt is a component of Vitamin B_{12}, its biological role being associated to that of this vitamin (Section 6.3.8)

Minerals can also be toxic and there are many documented cases of severe injury and even death from exposure to minerals. On the other hand, they not only contribute to the nutritional value of foods but they are also responsible for interferences with their quality and visual appearance causing discoloration of fruit and vegetable products, catalyzing oxidation of essential nutrients like ascorbic acid (Chapter 3), and being responsible for taste defects or off-flavors. Metal ions can be derived from the food itself or acquired during food processing and storage [1].

The chemistry of mineral elements both in food and within our bodies, as well as their bioavailability, depends on factors such as pH, solubility, and interaction with other molecules.

7.2 Macroelements

7.2.1 Electrolytes: Na^+, K^+ and Cl^-

An electrolyte can be defined as "a chemical compound that dissociates into ions and hence is capable of transporting electric charge," that is, it is an electric conductor in which the flow of charge is due to the movement of ions instead of electrons [4]. Electrolytes can be solids, liquids, or solutions; however, only the latter are normally considered in biological systems; thus, salts such as sodium chloride and potassium chloride are electrolytes because they dissociate into Na^+, Cl^- and K^+ in our internal fluids. By extension, the term "electrolyte" also includes the constituent negative and positive ions. Na^+, Cl^- and K^+ are the major electrolytes in living organisms, although ions such as Ca^{2+}, Mg^{2+}, and $H_2PO_4^-$/HPO_4^{2-} are also of physiological importance [5]. Electrolyte solutions are electrically neutral, that is, the total number of positive charges balances the number of negative charges; however, cells have devised molecular mechanisms to selectively move ions across biological membranes, thus generating charge imbalances that are essential for survival.

Sodium (chemical symbol Na, atomic number 11, atomic mass 22.99 Da) and potassium (chemical symbol K, atomic number 19, atomic mass 39.10 Da) are alkali metals, solid at normal temperature and pressure. Both are highly reactive in water and air and are naturally found as cations Na^+ and K^+, respectively, forming different salts that are usually soluble in water. Typical total body content of Na^+ is 1.3–1.5 g/kg of body weight, equivalent to a total of 85–96 g for a 70 kg adult. Around 95% of Na^+ is in the extracellular fluid (ECF), being the major determinant of its volume. Total body content of K^+ in humans is about 1.6–2.2 g/kg body weight, which corresponds to 112–156 g for a 70 kg adult, around 98% occurring within the cells, making potassium the most abundant intracellular cation [6, 7].

Chlorine (chemical symbol Cl, atomic number 17, atomic mass 35.45 Da) is a halogen element that can exist in the oxidation states −1 (Cl^- anion or chloride), 0 (Cl_2, chlorine gas), +1, +3, +5, and +7, although the most important chemical form in living organisms is Cl^-. Total body content of chloride has been reported to range from 85 to 115 g in an adult, 88% of which being in the extracellular fluid [8].

7.2.1.1 Biological roles

The biological roles of sodium, potassium, and chloride are interrelated: Na^+ is the dominant cation in ECF with Cl^- as its accompanying extracellular anion. Together, they are the major contributors of extracellular osmolality of water (Chapter 2). By contrast, potassium (K^+), chloride, and low molecular weight organic metabolites are the major contributors to intracellular osmotic activity. The ECF sodium concentration approximates to 135–145 mmol/L and that of potassium is 3.5–5.5 mmol/L, whereas within cells sodium and potassium concentrations approximate 15 mmol/L and 150 mmol/L, respectively. Therefore, there is a significant concentration gradient of Na^+ and K^+ between the intracellular and the extracellular spaces, albeit maintaining an osmotic equilibrium (i.e., similar water potential – Chapter 2). The extracellular concentration of chloride is approximately 115 mmol/L, whereas intracellular concentrations can range from approximately 70 mmol/L in red blood cells to 3 mmol/L in muscle tissue. Modulating and controlling the intertwined homeostases of Na^+, K^+, and Cl^-, among other minor ions and osmolytes, is essential for water movements, thereby maintaining the volumes of both intracellular and extracellular fluids and, consequently, the cellular volume. Moreover, the main protein that drives Na^+ and K^+ movements, the Na^+/K^+-ATPase, pumps three Na^+ out of the cell while introducing only two K^+ in each catalytic cycle, thus generating a difference of electrical potential of approximately 60 millivolts (positive outside the cell). The concentration gradient of Na^+ across the plasma membrane along with the polarization of the latter, the so-called electrochemical gradient, pushes Na^+ to flow back into the cell. This movement is enabled and regulated by specific membrane transporters that couple this flow to the influx of solutes, such as amino acids and monosaccharides, and efflux of waste products. The importance of Na^+ and K^+ movements across the plasma membrane of cells is illustrated by the fact that the Na$^+$/K$^+$-ATPase is estimated to spend 20–30% of the total amount of the energy (ATP) generated by human metabolism (Figure 7.2). This protein is present in most cells but its action is particularly evident during the transmission of electrical signal in the nervous system, muscle, and heart [6, 9].

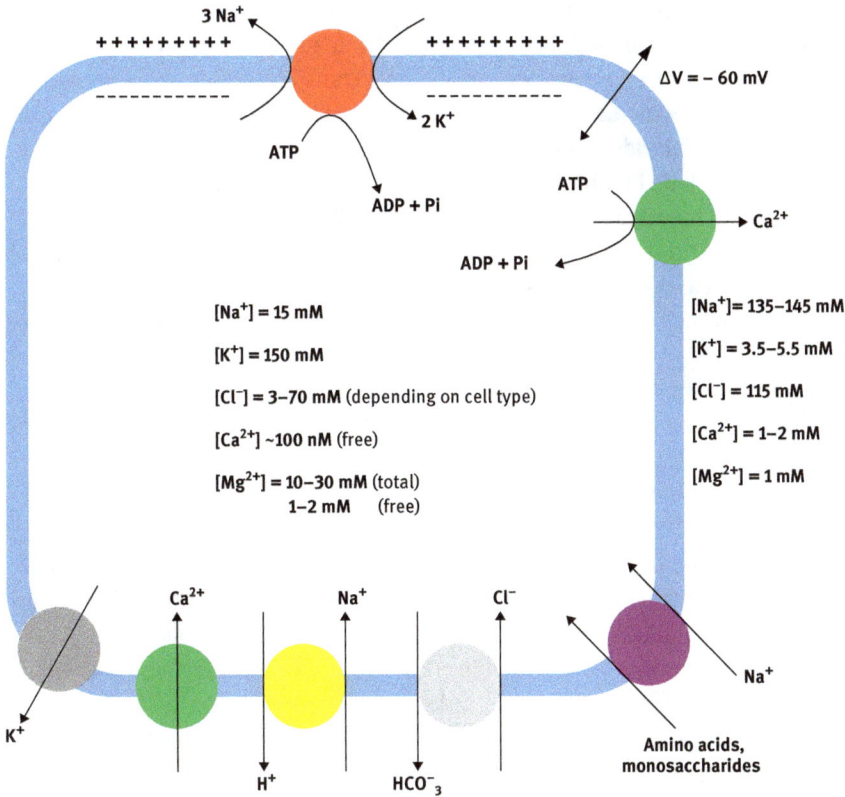

Figure 7.2: Transport of electrolytes across the plasma membrane of most human cells.

The intracellular potassium concentration is about 30 times higher than that of plasma and interstitial fluid, being the most abundant cation and the predominant osmotically active element inside cells, thereby contributing to the distribution of fluids inside and outside cells and the regulation of the acid–base balance. As in the case of Na$^+$, the Na$^+$/K$^+$-ATPase plays an important role in the strict control of intra- and extracellular concentrations of K$^+$, which are essential for neural transmission, muscle contraction, and vascular tone. Potassium is also crucial for the excretion of sodium in the kidneys and the regulation of blood pressure. Transport mechanisms of K$^+$ across biological membranes also involve the so-called potassium channels, many of them being able to open or close in response to specific stimuli (voltage, ATP, Ca^{2+}

concentration, hormones, neurotransmitters, etc.). These channels exhibit great diversity and have been implicated in functions such as secretion of saliva, gastric acid, and bile; digestion and absorption of proteins and carbohydrates; and transduction of taste signals. Potassium also has a role in cell metabolism, participating in energy transduction, hormone secretion, regulation of protein, and glycogen synthesis, as well as acting as a cofactor for a number of enzymes [7].

Chloride represents about 0.15% of total weight in an adult human body (115 g on average) and 70% of its total negative ion content. Transport of this anion across biological membranes is mediated by a group of ubiquitous pore-forming membrane proteins, known as chloride channels, that are involved in a wide variety of physiological roles. These include [8, 10]:

- Modulation of membrane polarity coordinated with the movements of Na^+ and K^+
- Osmotic and acid–base balance between the cytoplasm and both intracellular compartments and extracellular fluid
- Generation of electrical signals in muscle and peripheral and central nervous systems
- Transport of solutes across membranes
- Secretion and resorption of fluids. For example, chloride transport by the cystic fibrosis transmembrane conductance regulator (CFTR) is crucial to moisten mucus, thus providing its adequate fluidity in the lungs. Genetic defects in the gene coding for CFTR cause cystic fibrosis.
- Generation of HCl in the parietal (oxyntic) cells of the gastric mucosa (Chapter 12)
- A Cl^-/HCO_3^- exchange protein located in erythrocyte membranes facilitates uptake of oxygen and release of carbon dioxide in the lungs. Chloride ions induce a conformational change in hemoglobin (Hb) within the erythrocyte, so that Hb increases its affinity for O_2 (this effect is known as "chloride shift"). The same protein works in reverse in peripheral tissues, that is, it mediates the release of oxygen and the uptake of carbon dioxide (as HCO_3^-) by erythrocytes.

Chloride also plays a major role in innate host defense against infection by serving as a substrate for the generation of the biocide hypochlorous acid (HClO) by neutrophils. Some authors have suggested that chloride may be involved in cell cycle and apoptosis [8, 11].

7.2.1.2 Sources, bioavailability, and homeostasis

Sodium is a ubiquitous cation: the content of unprocessed raw meat and fish is typically between 30 mg and 150 mg/100 g, whereas fruits and vegetables generally contain less than 50 mg/100 g. It is also present in variable amounts in water, especially if treated with sodium fluoride, sodium bicarbonate, and/or sodium hypochlorite. Sodium concentrations in tap water sampled in several European countries was found to range between 4.3 mg and 20.0 mg/L, while bottled mineral water contained from 1 mg to

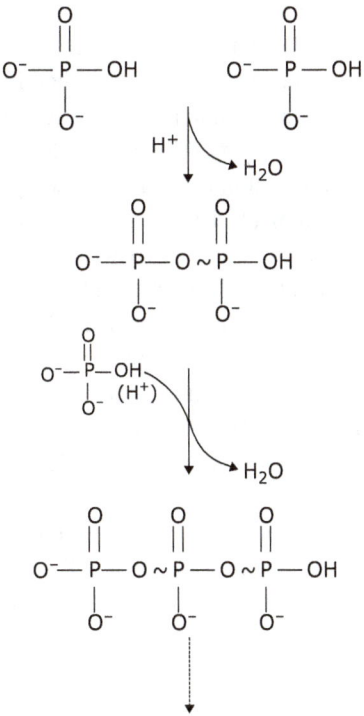

Figure 7.3: Formation of phosphate polymers. The symbols (~) indicate those bonds whose hydrolysis releases energy under cellular conditions.

1,400 mg/L; therefore, the source and quantity of drinking water significantly determine its contribution to dietary sodium intake. On the other hand, sodium (mainly as sodium chloride (NaCl), the main constituent of table salt) is profusely used in food processing and also appears in additives such as sodium bicarbonate in bakery, sodium nitrate in processed meat, monosodium glutamate (a flavor enhancer), and many others [6].

Foods of plant origin are good sources of potassium, especially starchy roots or tubers, vegetables, fruits, whole grains, and coffee. Dairy products are also rich in potassium; however, food processing may cause substantial losses. Drinking water and many food additives also contain potassium, although they do not represent major sources [7].

The major source of dietary chloride in Western diets is common salt added during food processing and preservation, as well as during discretionary use in households. Unprocessed foods contain low levels of chloride: the content of unprocessed meat and fish may be up to 4 mg/g, whereas fruit and vegetables contain generally less than 1 mg/g. Some food additives may contain chloride associated to cations other than sodium. In general terms, chloride content is substantially higher than sodium content in fruit and vegetables, whereas foods from animal sources contain approximately equimolar concentrations of sodium and chloride, or more sodium than chloride [8, 12].

NaCl is physiologically the most important salt in the diet and is already detected within the oral cavity where it generates what we call "salty taste" that guides and maintains the consumption of this salt. The mechanism of salty taste is described in Chapter 9. The information on the entry of Na^+ in the body is rapidly transmitted to the central nervous system (CNS); in this respect, it must be mentioned that NaCl-sensitive receptors have been identified in the gastrointestinal mucosa. The consumption of water and NaCl are controlled by neuroanatomical circuits that are not completely elucidated but appear to include numerous interconnected structures distributed throughout the CNS, particularly the hypothalamus and brainstem [13].

Sodium is absorbed in the distal small bowel and the colon by absorptive cells of the epithelium (Section 13.2). In the small intestine Na^+ absorption is transcellular, specific cotransporters located at the brush border of the enterocytes couple the flow of Na^+ into the cells to the uptake of low molecular solutes and micronutrients (e.g., Na^+/glucose and Na^+/amino acids cotransporters or Na^+/H^+ exchangers (NHEs)). The tendency of sodium ions to enter the enterocyte is generated and maintained by the action of the Na^+/K^+-ATPase located at the basolateral membrane. In the distal bowel, other carriers are responsible for the uptake of sodium. These include Na^+/H^+ antiporters of the SLC9 family, coupled to Cl^-/HCO_3^- exchangers to maintain electroneutrality, and electrogenic Na^+ channels (ENaC). The intestinal segment in which Na^+ absorption occurs varies, thus, it seems to occur preferentially in the small intestine after meals (coupled to nutrient absorption, as described above), whereas it takes place mainly in the ileum and colon during interdigestive periods [13, 14].

Dietary K^+ is mainly absorbed in the small intestine (around 90% of total intake) by a passive paracellular mechanism that depends on the fluxes of Na^+ and nutrients like glucose and amino acids. Therefore, factors that decrease Na^+ and water absorption may also decrease K^+ absorption. By contrast, the colon has the capacity to absorb and secrete K^+, thereby playing an important role in K^+ homeostasis [12, 15]. Daily K^+ intake can be equal to the amount of K^+ in the entire extracellular fluid (ECF); this creates a critical challenge to the body that has developed complex mechanisms of homeostatic control [16] (see below).

Chloride is absorbed from the intestinal lumen by three distinct mechanisms: (a) paracellular (passive) pathway; (b) electroneutral pathway that involves Na^+/H^+ antiporters coupled to Cl^-/HCO_3^- exchange; and (c) HCO_3^--dependent Cl^- absorption (Chapter 13). The passive pathway is predominant in the small intestinal epithelium, whereas electroneutral exchange pathway is the main route of Cl^- absorption in the ileum and colon. Several anion exchangers (AEs) have been reported to be involved in the latter [17].

Movements of Na^+, K^+, and Cl^- in the gastrointestinal tract are currently accepted to be major determinants of the driving forces that control water absorption and secretion (see Chapter 13 for further details on the absorption of water and electrolytes).

On average, small intestine and colon absorb about 9 l of water per day, that derives both from the diet (~2 L/day) and from secretion of digestive juices (~7 L/day).

About 84% of the water is absorbed by the small intestine and the remaining 16% by the large intestine, so that only 100 ml of water are excreted with feces. Two interconnected pathways for water transport occur in gut epithelia: the paracellular route, through the spaces between cell junctions, and the transcellular route, through apical and the basolateral cell membranes. The latter involves three different mechanisms: (a) passive diffusion through the phospholipid bilayer, (b) cotransport with ions and nutrients, and (c) diffusion through water channels called aquaporins (Chapters 2 and 13) [18].

Following absorption, sodium ions are distributed by portal and systemic circulations, where their concentrations are maintained between 135 and 145 mmol/L, approximately. A large proportion of body sodium is present in bones, skin, and muscles, which may act as a reserve among other roles. Potassium is transported in blood mainly as free ions (10–20% are bound to proteins) with serum concentrations ranging from 3.5 to 5.5 mmol/L, while plasma concentrations are about 0.3–0.4 mmol/L smaller, due to the release of potassium that occurs during clot formation. Nearly 70% of total potassium in the body is located in the muscles, with lower amounts present in the bones, liver, skin, and red blood cells. Chloride anions are freely transported in the blood, serum concentration being approximately 97–107 mmol/L [6–8].

The homeostasis of sodium is of paramount importance for the control of extracellular body fluids because changes in Na^+ concentrations would change osmotic pressures that could lead to swelling or shrinking of cells, thus altering their functioning. The hypothalamus exerts a complex neurohormonal control of Na^+ (and hence water) homeostasis, while the kidney is the main organ mediating its excretion and retention. Other components of this mechanism of control include renal sympathetic nerves, arterial pressure, and endocrine factors such as the renin–angiotensin system (RAS), aldosterone, cardiac natriuretic peptides, the antidiuretic hormone (ADH), and oxytocin. Other paracrine (that act locally) factors like endothelin I, bradykinin and nitric oxide might also be implicated in Na^+ homeostasis. The final outcome is that kidneys excrete sodium (along with water) in response to high dietary intakes and salvage sodium when dietary intake is low. It must be pointed out that there is evidence of complex extrarenal regulatory mechanisms that involve Na^+ stores in other organs, such as muscle and skin, that are subjected to weekly and monthly infradian rhythmicity. During pregnancy, the volume of ECF expands producing a fall in both plasma osmolality and plasma sodium concentrations. As a consequence, there are increases in the volume of the kidneys, renal blood flow, and glomerular filtration rate, which significantly alter Na^+ and water homeostasis. Progesterone has a major influence on these changes [6, 13, 19–21].

It has been recently reported that sodium and water handling by the body under circumstances of excessive intake of salt during very long periods (several months) may be different to the short-term mechanisms just described. In these conditions kidneys can increase significantly the sodium concentration in the urine in order to excrete the excess salt minimizing water loss. This mechanism requires the production of high quantities of osmolytes (mainly urea) by multiple organs and implies significant changes in energy metabolism [22].

Sodium loss through feces is relatively stable and normally limited to a few mmol/day, whereas the amount lost through sweat can vary widely, depending on factors like environmental conditions or levels of physical activity. Average sodium content in mature breast milk has been estimated to be 150 mg/L in women from Western countries, that implies a loss of 120 mg per day assuming a milk production of 0.8 L/day [6].

Dietary potassium intake increases excretion via kidneys and sequestration mainly in liver and skeletal muscle; however, renal excretion also has a circadian rhythm that is independent of food intake. At cellular level, the extra- and intracellular concentrations of K^+ are linked to those of sodium and are generated and maintained against electrochemical gradients by the Na^+/K^+-ATPase and other transporters. Potassium metabolism and homeostasis are influenced by a variety of endogenous and exogenous factors. The evidence available suggests that, right at the beginning of the meal, K^+ is sensed in the gastrointestinal tract and signals are sent to the periphery and the kidneys. This triggers K^+ transport from extracellular (ECF) to intracellular fluids. Additionally, the pancreas is stimulated to release insulin, that activates muscle Na^+-K^+-ATPase. As the meal is absorbed, the rise in plasma concentration of K^+ is a signal that targets skeletal muscle and kidney (increasing K^+ secretion and excretion, and depressing ammoniagenesis), and stimulate the biosynthesis and release of aldosterone in the adrenal gland. Aldosterone induces K^+ secretion and excretion at the renal distal nephron, colon, sweat and salivary glands. The main goal of this system is the tight control of K^+ concentration in the ECF; however, much remains to be learned about the mechanisms of K^+ homeostasis [16, 23].

Fecal potassium excretion is usually about 10–25 mmol/day, which constitutes 10–20% of total potassium elimination from the body. Potassium concentration in feces is highly variable (ranging from 20 to 200 mmol/L) and net absorption in the colon only takes place when there are large gradients of concentration between this organ and the blood. Fecal potassium excretion increases with fiber intake and may considerably increase in pathological situations, especially in cases of diarrhea or renal insufficiency. Sweat potassium concentration stays relatively constant ranging from 3 to 7 mmol/L, representing a loss of 2–3.5 mmol/day at moderate physical activity performed around thermoneutrality. In breast milk, potassium concentration shows both a decline over the first weeks of lactation and also diurnal variations. The European Food Safety Authority (EFSA) considers an approximate midpoint of potassium concentration in mature breast milk of 500 mg/L in women from Western countries, which means 400 mg/day assuming a production of 0.8 L of milk per day [7].

The overall regulation of chloride balance is linked to that of sodium through hormonal control by the renin–angiotensin–aldosterone system and cortisol. The kidney is the main route of chloride excretion, although some losses are associated to feces, sweat (around 20–40 mmol/L in healthy adults), and breast milk (11.3 mmol/L on average). Fecal losses are generally small (a few mmol/day) and relatively constant but they can become significant when chloride intakes are low or in rare cases of chloride malabsorption [8].

7.2.1.3 Electrolytes status, Dietary Reference Values (DRVs)

The EFSA considers that there is no biomarker of sodium status that can be used for setting DRVs for sodium in the general population. As far as a biomarker for sodium intake is concerned, urinary sodium excretion is considered to be more reliable than estimates of intake based on dietary assessments. In any case, the adaptation mechanisms triggered by neural and hormonal signals maintain sodium balance over a wide range of sodium intakes, consequently, balance studies cannot be used to determine sodium requirements. Based on empirical evidence about the relationship between sodium intake, blood pressure, and risk of cardiovascular disease (see below), as well as on data from balance studies, the EFSA has established a safe and adequate intake (AI) of sodium of 2.0 g/day for the general population of adults, including pregnant and lactating women. Values for children are extrapolated from values for adults based on the energy requirements of the age groups and applying a growth factor (Table 7.1) [6].

Table 7.1: Summary of dietary reference values for sodium. Table reproduced from reference [6].

Age	Safe and adequate intake (g/day)
7–11 months	0.2 (a)
1–3 years	1.1
4–6 years	1.3
7–10 years	1.7
11–17 years	2.0
≥18 years (c)	2.0

(a): Adequate intake
(c): Including pregnant and lactating women

Urinary potassium excretion is considered a reliable biomarker of dietary intake in adults on a population basis. Potassium daily intakes can be derived from the former values by multiplying by a factor of 1.30; however, no biomarker is considered reliable and accurate to establish potassium status. Adequate intakes (AI) for potassium have been set based of evidence on the relationships between potassium intake and blood pressure and risk of stroke (Table 7.2) [7].

The EFSA considers that a single 24 h excretion of chloride can be used as a valid marker for groups' average intake of chloride. No sensitive marker exists for chloride status. Safe and adequate intakes proposed for chloride are shown in Table 7.3 [8].

Table 7.2: Summary of dietary reference values for
potassium. Table reproduced from reference [7].

Age	Adequate intake (mg/day)
7–11 months	750
1–3 years	800
4–6 years	1,100
7–10 years	1,800
11–14 years	2,700
15–17 years	3,500
≥18 years	3,500
Pregnancy	3,500
Lactation	4,000

Table 7.3: Summary of dietary reference values for
chloride. Table reproduced from reference [8].

Age	Safe and adequate intake for chloride (a) (g/day)
7–11 months	0.3 (b)
1–3 years	1.7
4–6 years 1.7	2.0
7–10 years	2.6
11–17 years	3.1
≥18 years (c)	3.1

(a): Derived by multiplying the reference values for
sodium (Table 7.1) by 35.5/23
(b): Adequate intake
(c): Including pregnant and lactating women

7.2.1.4 Sodium, potassium, and chloride deficiency

Dietary sodium deficiency is rare in healthy people due to a tight regulation of water
and sodium homeostasis mediated by the antidiuretic hormone vasopressin and the
sensation of thirst. Low serum sodium concentration (below 135 mmol/L), known as
hyponatremia, usually arises when the intake of water exceeds its excretion rate (dilutional disorders) or when a rapid depletion of sodium occurs (depletional disorder).
Causes of dilutional disorders include renal failure, a syndrome of inappropriate antidiuretic hormone secretion (SIADH), diseases of the neuroendocrine system, hyperglycemia, heart failure, hepatic disease (mainly cirrhosis), or fatal complications of
diuretic therapy. Another cause of hyponatremia is the rare inherited Bartter's syndrome. Depletional disorders may arise from abnormal sweating or as a result of
vomiting and diarrhea usually caused by bacterial enterotoxins or enteroviruses [21,
24]. Symptoms of hyponatremia depend on the velocity of decrease in plasma sodium

concentration, thus, chronic hyponatremia (present for >48 h or of unknown duration), symptoms and signs are gradual in onset and often non-specific, for example lethargy, forgetfulness or unsteadiness, whereas acute hyponatremia, defined as being present for up to 48 h, may lead to raised intracranial pressure with the risk of cerebral herniation, hypoxia, and even death [21, 25].

Serum potassium concentrations lower than 3.5 mmol/L (hypokalemia) is the criterion usually accepted for assessing potassium deficiency; however, patients are not typically symptomatic until the K^+ concentration is <2.5 mmol/L or it has fallen rapidly; significant muscle weakness tends to occur at concentrations below 2.0 mmol/L. True hypokalemia can be the result of intracellular shift of potassium without decreasing total body content into cells, or to renal or gastrointestinal (diarrhea, vomiting, burns) losses, rather than low dietary intake. One of the most frequent causes of hypokalemia is loss of potassium via urine from patients treated with diuretics. Potassium loss also occurs as a result of excessive sweating, folic acid deficiency, heavy alcohol consumption, chronic kidney diseases, eating disorders such as anorexia or bulimia and starvation. Hypokalemia is associated with increased morbidity and mortality, especially from cardiac arrhythmias or sudden cardiac death. Other adverse consequences include polyuria, muscle weakness, decreased peristalsis, mental depression, and respiratory paralysis [7, 21, 24].

7.2.1.5 Electrolytes excess

Hypernatremia, defined as a serum sodium concentration higher than 145 mmol/L, is usually a consequence of dehydration rather than of excessive sodium intake. Dehydration is usually due to solute-free water loss either from the gastrointestinal tract (vomiting, osmotic diarrhea), from urine (mainly in diabetes insipidus or osmotically driven diuresis secondary to glycosuria or hypertonic solution infusion) or by excessive insensible water loss (from lung or skin). Water loss is a strong trigger for thirst and should not induce hypernatremia unless there is an associated lack of thirst (e.g. after a head injury and damage to the hypothalamus) or no ready access to water [21].

High intake of sodium, normally as sodium chloride, is generally regarded as a significant contributor to hypertension (systolic blood pressure ≥140 mm Hg and/or diastolic blood pressure ≥90 mm Hg). Hypertension is in turn associated with the development of cardiovascular and renal diseases; however, individuals respond differently to dietary salt intake, thus, some exhibit an increase in blood pressure with increasing dietary salt (salt-sensitive individuals), while others show no significant changes (salt-insensitive). Salt sensitivity of blood pressure is related to several physiological, environmental, genetic, and demographic factors. It has been shown that postmenopausal women, black populations, and older individuals are the groups with higher risk of developing salt sensitivity [26]. The underlying mechanisms that promote salt sensitivity are complex: the classic proposal was that a high-salt diet produces sodium accumulation, thirst, water consumption and, therefore, volume expansion of body fluids that elevates arterial pressure. The kidneys act by increasing the excretion of sodium and water until blood volume is reduced and arterial pressure returns to control values. According to this proposal, hypertension would be the result of kidney malfunction; however, recent studies suggest that sodium homeostasis and salt sensitivity are also related to salt accumulation in the skin interstitium and to alterations of endothelial cell function. Moreover, some new findings cannot be explained by the classical scheme, thus, excess salt intake has been reported to exert pathological effects on the vasculature that are independent of blood pressure and excretion of large quantities of salt by the kidneys with minimal water losses has been described [27–29].

Various observational and interventional studies have shown that K^+ intake is inversely related to blood pressure even with high dietary Na^+ intake; moreover, K^+ supplementation has been reported to decrease blood pressure and protect against cardiovascular events. Several mechanisms have been proposed to explain the antihypertensive effect of dietary K^+, including natriuresis (rapid urinary excretion of Na^+ induced by K^+), decreased sympathetic tone, and direct effects on vascular tone. Consequently, high K^+ diets have been proposed to control hypertension and relieve cardiovascular disease, although this approach requires a careful control of K^+ intake, especially in patients with disorders such as chronic kidney disease [30]. Other cations like Ca^{2+}, Mg^{2+}, and Zn^{2+} have also been reported to influence the occurrence and development of hypertension [31].

The gut microbiota is another factor that has been related to salt sensitivity and hypertension development, thus, a high salt intake has been associated with major changes in the gut microbiome (modifications in bacterial taxa, dysbiosis) that are in turn related to salt sensitivity in hypertension. The mechanisms involved are not fully elucidated [13].

Recent studies show that a moderate or normal sodium consumption (3 to 6 g/day) appears optimal with regard to reducing hypertension and cardiovascular risk, compared not only to high-sodium diets but also to low-sodium diets. Actually, an aggressive sodium restriction might have deleterious effects even in heart failure patients [32]. In any case, the World Health Organization (WHO) currently recommends that adults consume less than 5 g of salt (NaCl, equivalent to 2 g of sodium) per day and that this amount is adjusted downward for children aged 2–15 years based on their energy requirements relative to those of adults [33]. Among dietary factors, high intake of sodium has been reported to be the leading risk for mortality, accounting for an estimated 3.20 million deaths in 2017 [34].

Hyperkalemia is defined as a serum potassium concentration greater than approximately 5.5 mmol/L in adults, and it is often asymptomatic. Clinical manifestations of mild to moderate hyperkalemia include non-specific symptoms such as generalized weakness, paralysis, nausea, vomiting, and diarrhea. Severe hyperkalemia may lead to life-threatening cardiac arrhythmias. Hyperkalemia due to excessive dietary intake of potassium is rare because of the effective homeostasis of this cation in the body, therefore, the majority of cases occur from impaired renal function, excessive intake of oral potassium supplements or parenteral administration, and a potassium shift from cells (see reference [35] for a more exhaustive list of causes of hyperkalemia). No tolerable upper intake level (UL) has been set for potassium by the EFSA [7].

Chloride excess secondary to dietary intake is uncommon and hyperchloremia (defined as a serum chloride concentration above 107 mmol/L) is usually caused by loss of bicarbonate in the feces due to severe diarrhea, although it may occur with several other conditions associated with abnormal losses of water (skin, renal, or extra-renal), extracellular fluid volume depletion, increase in the tubular chloride re-absorption, or as a result of excessive administration of salts (e.g., NaCl, NH_4Cl, $CaCl_2$) or intake of certain medications (e.g., cortisone preparations, acetazolamide). The EFSA has not set a UL for chloride but noted that current levels of NaCl intake among European populations exceeded amounts required for normal function, which may result in hypertension (see above) [8].

7.2.1.6 Acid–base metabolism and electrolytes

In aqueous solutions, acidity is usually measured by the pH value, defined as the logarithm to the base 10 of the reciprocal of hydrogen ions (H^+ or protons) activity.

$$pH = log[1/a_{H^+}]$$

The activity is a measure of the effective concentration of protons in solution (i.e., considering possible interactions with other components of the mixture), and it may be assumed to be equal to the concentration expressed in moles/Liter, therefore:

$$pH \sim log(1/[H^+]) = -log[H^+]$$

In this scale, neutral aqueous solutions have a pH value of 7, acidic solutions have pH < 7 and alkaline (basic) solutions have pH > 7. Acid–base homeostasis and pH regulation are critical for both normal physiology and cell metabolism and function. The organs involved in the maintenance of acid-base balance are mainly the lungs and kidneys, in addition to a complex system of buffers. The latter are a variety of chemical systems that prevent radical changes in fluid pH by capturing excess hydrogen ions or releasing them if needed. The interaction and proper function of these factors are required to maintain the extracellular pH in optimal ranges (7.35–7.45). The condition where the pH value is less than 7.35 is called acidemia, whereas a value greater than 7.45 is called alkalemia [36, 37]. There are several processes that can alter the acid-base balance within our bodies [12, 37, 38]:

- Aerobic degradation of carbon skeletons of nutrients (mainly monosaccharides, fatty acids, and amino acids), which produces carbon dioxide (CO_2) and water
- Catabolism of methionine, cysteine and taurine, which produces hydrogen ions and sulfate (SO_4^{2-}).
- Catabolism of phospholipids and nucleic acid, which produces protons, hydrogen-phosphate (HPO_4^{2-}), and dihydrogenphosphate ($H_2PO_4^-$)
- Organic acids, like lactic or citric acids, can also be produced if nutrients are not completely metabolized
- Dietary cations like K^+, Mg^{2+}, and Ca^{2+} are regarded as precursors of alkali

Carbon dioxide is called volatile acid because it is eliminated by alveolar ventilation, whereas other metabolic products, known as nonvolatile or fixed acids and less abundant, must be excreted in an aqueous medium, mainly in the urine. In blood plasma the acid–base pair carbonic acid/bicarbonate is the main buffer system:

$$H^+ + HCO_3^- \rightleftharpoons H_2CO_3 \rightleftharpoons CO_2 + H_2O$$

Other extracellular buffers include phosphate, certain organic acids, and plasma proteins. Proteins are also the main buffer systems inside the cells.

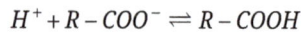

$$H^+ + HPO_4^{2-} \rightleftharpoons H_2PO_4^-$$

$$H^+ + R-COO^- \rightleftharpoons R-COOH$$

Net acid excretion by the kidneys occurs associated with urinary buffers (mainly phosphate) or with ammonium (NH_4^+), the latter being responsible for one-half to two-thirds of net acid excretion. Increase in pH is corrected by avoiding reabsorption of HCO_3^-. The movements of NH_4^+ and HCO_3^- in the kidneys and the rest of organs and tissues are closely linked to those of Na^+, K^+, and Cl^-; that is, the major electrolytes are involved in the acid–base balance of the body [36]. Disorders in acid–base balance are rather common and can be classified according to the causative underlying process (Table 7.4) [39].

Table 7.4: Definition and causes of acid–base disorders. Reproduced from [39].

Metabolic acidosis	**Process that primarily reduces bicarbonate:** Excessive H^+ formation, e.g., lactic acidosis, ketoacidosis Reduced H^+ excretion, e.g., renal failure Excessive HCO_3^- loss, e.g., diarrhea
Metabolic alkalosis	**Process that primarily raises bicarbonate:** Extracellular fluid volume loss, e.g., due to vomiting or diuretics Excessive potassium loss with subsequent hyperaldosteronism
Respiratory acidosis	**Process that primarily causes elevation in partial pressure of CO_2 (pCO_2)** Reduced effective ventilation, e.g., many chronic respiratory diseases or drugs depressing the respiratory center
Respiratory alkalosis	**Process that primarily causes reduction in pCO_2** Increased ventilation, e.g., in response to hypoxia or secondary to a metabolic acidosis

Diet plays an essential role in the maintenance of acid-base balance. Foods can be categorized by the potential to produce nonvolatile acids, the so-called potential renal acid load (PRAL), thus, grain products, meats, dairy products, and fish have high acid loads, whereas fruits, vegetables, fruit juices, and potatoes have a negative acid load. PRALs are calculated by a formula published by Remer and Manz in 1994, which takes into consideration factors such as the content of methionine, cysteine, and phosphate in the food, as well as the alkali content that is mostly present as Na^+, K^+, Ca^{2+}, and Mg^{2+} salts of organic acids [12, 37, 40, 41]. Chronic consumption of diets with high acid load reportedly favors the development of low grade metabolic acidosis, that is linked to disorders such as osteoporosis, development of kidney stones, chronic kidney disease, insulin resistance, sarcopenia, hypertension, and cardiovascular risk [37].

The so-called alkaline diets (aka low PRAL diets or LPD) propose the consumption of foods with negative acid loads and minimizing those with high acid loads. These diets have been promoted as a way of reducing morbidity and mortality from chronic diseases and cancer. Recent reports show that, although individuals on an LPD met more nutritional goals than individuals on high PRAL diets, the former may have some difficulties to meet the dietary recommendations of nutrients like vitamin D and calcium. Therefore, the nutritional adequacy of LPDs for the general population and especially for groups like pregnant women or patients with cancer clearly deserves further studies. Some authors claim that LPDs may result in a number of health benefits simply because they are based on the intake of foods of plant origin [41, 42].

7.2.2 Calcium

Calcium (chemical symbol Ca, atomic number 20, atomic mass 40.08 Da) is the fifth most abundant element in the biosphere after iron, aluminum, silicon, and oxygen, occurring in limestone and marble, coral, pearls, shells (from shellfish and eggs), antlers, bones, and teeth [43]. The total calcium content in the body of an individual increases from ~25 to 30 g at birth to ~1,000–1,500 g in a well-fed adult, 99% of which is found in the skeleton mainly as calcium hydroxyapatite $[Ca_{10}(PO_4)_6(OH)_2]$. The remaining 1% of Ca^{2+} is found in extracellular fluids, intracellular structures, and cell membranes. Bone is a living and constantly remodeling tissue that, among other functions, can act as a reservoir for calcium (and other inorganic ions), thereby participating in mineral homeostasis through the processes of resorption and formation: the old or damaged bone is resorbed by specialized cells called osteoclasts that release calcium, whereas new bone is constructed by osteoblasts that deposit calcium and decrease its concentration in extracellular fluids. These processes are tightly controlled by complex biochemical mechanisms (see below) [44–46].

7.2.2.1 Biological roles
Calcium serves a key role in bone and teeth structure, blood coagulation, platelet adhesion, endocrine and exocrine secretory functions, neuromuscular activity, and electrophysiology of the heart and smooth muscles [47]. Ca^{2+} ions can form coordination bonds with up to 12 oxygen atoms, a capacity that has been used by living organisms: within the cells, hundreds of proteins can bind Ca^{2+} with different affinities, thereby

changing their composition and structure and, thus, functional properties. The changes in the conformations of so many proteins exert an impact in nearly every aspect of cellular life. Cells usually maintain a 10,000 to 20,000-fold gradient between the cytoplasmic (<100 nM free) and extracellular (1–2 mM) concentrations of Ca^{2+}; however, concentrations inside specific intracellular organelles such as mitochondria and the endoplasmic reticulum are high. Following an appropriate signal, Ca^{2+} release from intracellular stores, or rapid influx from extracellular milieu increase cytosolic calcium concentration to approximately 100 µM. This triggers a huge number of cellular reactions and processes, such as muscle contraction, enzyme activation, cell differentiation, immune response, programmed cell death, and neuronal activity [45, 48, 49].

7.2.2.2 Sources, bioavailability, and homeostasis

Dietary sources of calcium include dairy products, selected vegetables (such as spinach, chard, endive, and broccoli), legumes, nuts, fish with soft bones (e.g., tinned sardines), and calcium-fortified foods. Tap water may also contribute significantly to the daily intake in hard water areas. The main compounds used for fortification are calcium carbonate, gluconate, glycerophosphate, lactate, sulfate, and chloride, salts of citric acid and orthophosphoric acid, as well as calcium hydroxide and oxide. Calcium salts are generally water soluble, with the exception of calcium sulfate, carbonate, and phosphates, which are soluble in acids, that is, they solubilize in the acidic environment of the stomach. Dairy products provide about 75% of dietary calcium in the United States and about 58% in the Netherlands, while in China only 6.7% of calcium comes from dairy products and the main sources of calcium in this country are vegetables (30.2%), legumes (16.7%) and cereals (14.6%) [43, 44, 46]. Diet affects the bioavailability of Ca^{2+}, so that the percentage of absorption can range from <10 to >50 depending on the source. These differences may be related to the interactions of Ca^{2+} with other components in the food matrix, which are in turn affected by several factors:

- Milk, a product designed to be the sole source of nutrition in the early stages of life in mammals, should be able to deliver sufficient calcium in a bioavailable form; however, significant variations in Ca^{2+} absorption among different dairy products have been observed in humans [46].
- Oxalic acid (HOOC-COOH) and its partially or fully ionized salts, oxalates (HOOC-COO⁻ and ⁻OOC-COO⁻), are the most potent inhibitors of Ca^{2+} absorption due to their capacity to form insoluble complexes with this cation. Oxalates are found in high concentrations in spinach, rhubarb, and, to a lesser extent, sweet potatoes and dried beans. For example, calcium absorption from spinach is only 5% compared with an average of 27% from milk [43].
- (Phytic acid *myo*-inositol hexaphosphoric acid) is the storage form of phosphorus in seeds, being present in cereals, legumes, nuts, oilseeds, and, in lower quantities, in roots, tubers, fruits, and berries. The molecule of phytic acid contains 12 dissociable hydrogens, so that several anions (phytates) with different degrees of protonation

may be formed depending on the pH. Phytates readily bind cations such as Ca^{2+}, Mg^{2+}, Zn^{2+}, and $Fe^{2+/3+}$, making them unavailable for humans, that have very low levels of endogenous phytase; therefore, phytates are considered anti-nutritional due to their negative effects on the bioavailability of these essential minerals [50]. Some processing methods can reduce phytic acid content in certain foods, thus, fermentation that occurs during bread making results in increased calcium absorption, by virtue of the phytase present in yeast. In any case, the negative effect of phytates on mineral absorption seems to be more significant in the case of Zn^{2+}, and $Fe^{2+/3+}$ (see below), whereas only concentrated sources of phytate such as wheat bran ingested as extruded cereal or dried beans seem to have substantially reduced calcium absorption. For other plants rich in calcium (primarily the *Brassica* genus, which includes broccoli, kale, and cabbage) the bioavailability is as good as that obtained from milk [3, 43, 51].

- Other food components like fiber, tannins, and lectins have been suggested to form strong complexes with ions like Ca^{2+}, Mg^{2+}, Zn^{2+}, $Fe^{2+/3+}$, and $Cu^{1+/2+}$; however, some authors point at the phytic acid associated with fiber-rich foods as the component that affects calcium absorption [43].
- Some nutrients have been suggested to enhance absorption of Ca^{2+}; these include lactose in infants as well as some nondigestible carbohydrates and proteins. Addition of soluble dietary fiber to dairy infant formulas has been reported to affect calcium, iron, and zinc availabilities in positive and negative ways, depending on the type of the dietary fiber used. The most positive effects have been obtained with inulin; however, more experimental evidence is needed to sustain these claims [43, 52].
- Heat treatments such as boiling, drying, frying, pressure cooking, steaming, and sterilization can significantly decrease some macro- and micro-minerals, including calcium, apparently by leaching into hot cooking fluids. Microwave cooking has been reported to be a more convenient method to minimize the loss of minerals like Ca^{2+} from lentils than boiling and autoclaving, whereas blanching of spinach leaves seems to improve HCl-extractable Ca^{2+} and Zn^{2+} [3].
- Heat-induced chemical reactions like Maillard-browning reaction (Chapter 3) can form compounds that bind minerals tightly, thus decreasing the absorption and bioavailability of Ca^{2+} [3, 44].
- Other nutrients like vitamin D and K (Chapter 6), and moderate alcohol consumption have been associated with changes in calcium balance that can positively influence bone calcium content. Absorption of calcium from food supplements depends on when they are consumed and the dose; smaller doses taken with meals are better absorbed [44].

The physicochemical aspects of calcium absorption as well as the differences observed in this respect among different food products have been reviewed elsewhere [46].

Calcium absorption in the body occurs by two pathways: a transcellular active metabolically driven transport and a paracellular (i.e., between adjacent epithelial cells) passive process (Chapter 13). In general terms, transcellular transport is the major route of absorption, paracellular transport being responsible for an estimated 8–23% of total calcium absorbed. When calcium intake is high, paracellular transport becomes more important. Studies in rodents and humans suggest that, although most calcium is absorbed in the ileum, it can also be taken up in the colon via the pathways just mentioned [53]. Calcium absorption in the gastrointestinal tract is basically governed by three key aspects: the concentration of free ("ionized") calcium that depends on salt equilibria and pH, the rate of absorption of ionized calcium, and the transit time of the material through the intestine. A number a physiological factors can affect these aspects and, hence, Ca^{2+} absorption [44, 46]:

- It has been shown to be low in patients with vitamin D deficiency, although there is uncertainty about the serum concentration of this vitamin required for optimal absorption.
- It is inversely related to the concentration of calcium present in the gut lumen and dietary load.
- It varies throughout the lifespan, being higher during periods of rapid growth and lower in old age:
  - Calcium absorption is high in infancy (absorption efficiency of about 60%) and decreases during childhood, from around 45% in children aged 1–3 years to 30% in children aged about 6 years.
  - During puberty, absorption increases in line with skeletal growth: 35% at 7–10 years, 40% at 11–14 years, and 45% in boys aged 15–17 years. In post-pubertal girls aged 15–17 years, absorption is 35% (data obtained with children consuming from 800 to 1,800 mg per day of dietary calcium). The absorption values reported in the literature differ depending on factors like the type of population, habitual calcium intake, and stage of puberty.
  - In adults, dietary calcium absorption is approximately 25%, being lower in post-menopausal women and in men over 60 years of age, probably as a result of a developing resistance to the action of vitamin D. Menopause is associated with a significant fall in calcium absorption, possibly as a result of lower estrogen levels affecting receptors in the small intestine. Some studies show a continuous reduction in absorption from the age of 60 years both in men and women.
  - Absorption increases approximately twofold during pregnancy. By 2–3 months postpartum, calcium absorption returns to values close to those observed in early gestation or prior to conception.

After absorption, calcium is distributed in the organism through the bloodstream where it can be found as free ions, bound to proteins, mainly albumin (about 45%), and complexed to anions like citrate, phosphate, sulfate, and carbonate (about 10%). Calcium concentration in blood plasma is tightly regulated at around 2.5 mM by calcium-sensing receptors (CaSRs) through the modulation of parathyroid hormone, calcitonin, and calcitriol (1α,25-dihydroxycholecalciferol or 1,25-dihydroxy vitamin D$_3$, Chapter 6) secretion [4, 8]:

- A decrease in serum concentrations of Ca^{2+} activates CaSRs in parathyroid glands, which further promotes the secretion of PTH. PTH increases blood calcium levels by the direct activation of calcium reabsorption in kidney and calcium release in bone. PTH also promotes the production and secretion of calcitriol in kidney cells. Calcitriol regulates the intestinal calcium absorption, kidney calcium reabsorption, and bone calcium release. CaSRs expressed in bone, kidney, and intestine cells are also involved in the regulation of calcium homeostasis [45].
- An increase in serum concentrations of Ca^{2+} inhibits PTH secretion and stimulates calcitonin secretion by the thyroid gland. The major target site for calcitonin is bone, where it inhibits bone resorption by osteoclasts [54].
- Calcium metabolism is further regulated by other hormones such as estrogen and testosterone, adrenal steroids, glucocorticoids, growth hormone, and thyroid hormones. Moreover, factors such as ethnicity, individual health state, vitamin D status, genetic polymorphisms of the vitamin D receptor (VDR) (Section 6.4.2.2) and certain conditions such as achlorhydria (reduced production of hydrochloric acid in the stomach) have also been reported to affect calcium absorption and metabolism [44, 49].

Calcium deposition into bone occurs during periods of growth, with a marked increase during puberty. In the fetus, serum calcium, phosphorus, and ionized calcium are higher than maternal values. This is achieved by means of physiological changes in the mother, which increase the efficiency of absorption and decrease bone mineral. Calcium is actively transported across the placenta to the fetus, thus producing a fall in maternal serum calcium concentrations, although ionized serum calcium are maintained within the normal range.

Bone mass increases substantially during the first two decades of life, reaching a plateau, referred to as peak bone mass (PBM), when bone mineral density becomes stable. Some studies estimate that overall PBM is reached at 18.8 years in women and 20.5 years in men, with maximal growth occurring at 11.8 years in girls and 13.5 years in boys. However, PBM of some parts of the skeleton, like lumbar spine, seems to occur much later: at 33–40 years in women and 19–33 years in men. Other studies indicate that calcium continues to be accrued in bones in young adults, with men having PBM at a later age than women. Bone constantly undergoes remodeling, and almost the entire adult skeleton is remodeled over a 10-year cycle [44].

Calcium is continuously excreted from the body via urine, feces, and skin and sweat (dermal losses), whereas in lactating women, breast milk is another route of calcium loss.

- Urinary excretion is a function of the balance between calcium load filtered by the kidneys and the efficiency of absorption by the renal tubules. Urinary calcium comprises absorbed calcium that is lost from the body after the requirements for bone and endogenous fecal and dermal excretion have been met. During periods of rapid growth, the principal determinants of urinary calcium excretion are body weight and age, a value of 2 mg/kg body being accepted for obligatory urinary calcium losses in children. No consensus exists for other groups of population.

– Fecal calcium is derived from a mixture of unabsorbed calcium and endogenous losses (shed mucosal cells and intestinal secretions). Endogenous losses vary with body size (and possibly calcium intake), but are unrelated to age or sex. Average values for endogenous fecal losses of calcium have been estimated at 1.5 mg/kg body weight per day and 2.1 mg/kg body weight per day for children and adults, respectively.

– Dermal losses are difficult to measure accurately but seem to be very variable, depending on factors like body surface, climate, and level of physical activity. In the absence of exercise and with minimal sweating, a value of 40 mg/day is accepted as dermal losses in adults. No data are available for children.

– Calcium in breast milk (post colostrum) is relatively constant for the first three months of lactation, with a concentration of 200–300 mg/L (5.0–7.5 mmol/L), and from then on it progressively declines. A series of compensatory physiological changes maintain the calcium supply to the infant, including increased maternal efficiency of absorption in the later stages of lactation, enhanced renal reabsorption, and reduced bone mineral density, although no long-term effects on bone have been attributed to lactation.

A more detailed account of calcium homeostasis can be found in references [44, 49].

7.2.2.3 Calcium status, Dietary Reference Values (DRVs)

The role of the skeleton as a huge reserve that maintains calcium at appropriate levels for normal body functions complicates the assessment of calcium status [43, 44], consequently, dietary reference values are established by measuring calcium balance, that is, by determining if losses are met by intake. Calcium requirement is thus defined as the amount of dietary calcium required to replace losses in the urine, feces, and sweat, plus that needed for bone accretion during periods of growth [43].

The European Food Safety Authority (EFSA) Panel on Dietetic Products, Nutrition and Allergies derived values of average requirement (AR) and population reference intake (PRI) for calcium in adults based on balance data from North America, adding an allowance for dermal losses of calcium. The PRI for young adults (18–24 years), who still accumulate calcium in bones, is the intermediate value between adolescents aged 15–17 years and adults ≥25 years. For infants (7–11 months), an adequate intake (AI) was derived by extrapolating the average amount of calcium absorbed by exclusively breast-fed infants using isometric scaling and taking into account the percentage of calcium absorption. For children, ARs were estimated based on factorial calculation of losses and considering the need for calcium accretion in bone; the per-

Table 7.5: Summary of dietary reference values for calcium. Table reproduced from reference [44].

Age	Adequate intake (mg/day)	Average requirement (mg/day)	Population reference intake (mg/day)
7–11 months	280	–	–
1–3 years		390	450
4–10 years		680	800
11–17 years		960	1,150
Adults 18–24 years (a)		860	1,000
Adults ≥25 years (a)		750	950

(a): Including pregnancy and lactation

centage of calcium absorption at different ages was also considered in this case. PRIs for children and young adults were estimated based on a coefficient of variation of 10%. The AR for adult women aged 18–24 years and ≥25 years, respectively, also applies to pregnant and lactating women [44]. Data are summarized in Table 7.5.

7.2.2.4 Calcium deficiency

Insufficient dietary supply, low absorption, and/or high losses of calcium induce resorption from the skeleton in order to maintain blood concentrations within the range required for normal cellular and tissue functions. This can reduce bone mass eventually leading to osteopenia (a lower than normal bone mineral density (BMD)) and osteoporosis, characterized by a very low BMD. In any case, the origin of osteoporosis is multifactorial. Other skeletal disorders like rickets and osteomalacia are associated with suboptimal bone mineralization and are usually caused by vitamin D deficiency (Chapter 6), although a low intake of calcium often coexists with vitamin D deficiency and both can independently cause nutritional rickets. An inadequate supply of calcium during bone development leads to stunted growth and bowing of long bones. Older adults will have a reduced bone mass, which leads to impaired bone strength. Bone loss is strongly related to genetic factors and specific environmental influences, such as lifestyle, that become more important with increasing age [44].

Disorders that involve regulation of Ca^{2+} levels in extracellular fluids are not nutritional but usually due to problems in the parathyroid gland function. Magnesium deficiency, usually due to alcoholism or malabsorption, can also affect calcium metabolism by impairing PTH release [44].

7.2.2.5 Calcium excess

Hypercalcemia is defined as serum calcium concentrations higher than 2.75 mmol/L (11 mg/dL) and affects approximately 1% of the worldwide population. Mild hypercalcemia, defined as total calcium of less than <3 mmol/L (12 mg/dL) is usually asymptomatic but may be associated with symptoms such as fatigue and constipation in approximately 20% of people. Severe hypercalcemia, defined as total calcium of 3.5 mmol/L or greater (>14 mg/dL) or that develops rapidly over days to weeks, can cause nausea, vomiting, dehydration, confusion, somnolence, and coma.

The most common causes of hypercalcemia include malignant tumors and hyperparathyroidism of different etiology. Additional causes include granulomatous diseases such as sarcoidosis, endocrinopathies such as thyroid disease, immobilization, genetic disorders, medications such as thiazide diuretics and supplements such as calcium, vitamin D, or vitamin A. Hypercalcemia has also been associated with SARS-CoV-2, ketogenic diets, and extreme exercise, but these account for less than 1% of causes. Chronic hypercalcemia may also lead to calcification of soft tissues (e.g., nephrocalcinosis and vascular calcification), particularly when phosphorus concentration in the blood is also high. The decrease in renal function with age increases the sensitivity of older people to excess calcium intake. The underlying cause of hypercalcemia should be identified and treated. Serum intact parathyroid hormone (PTH) is the most important initial test to evaluate hypercalcemia. A Tolerable Upper Intake Level (UL) of 2,500 mg of calcium per day has been proposed for adults, including pregnant and lactating women, by different health authorities worldwide. No data allow the setting of a UL for infants, children, or adolescents [44, 55].

7.2.3 Phosphorus

Phosphorus (chemical symbol P, atomic number 15, atomic mass 30.97 Da) is the eleventh most abundant element in the earth's crust and the sixth most abundant element in the human body, which contains around 400 g in women and 500 g in men (approximately 1.35 g per 100 g of fat-free tissue). At birth, total phosphorus reaches an average of 20 g (0.5 g/100 g fat-free tissue). As in the case of calcium, approximately 85% of the total phosphorus of our bodies is found in bone, mainly in the form of hydroxyapatite (see above), although it also appears as amorphous calcium phosphate; approximately 14% is located in soft tissues, muscle, and viscera, usually as phosphate esters and, to a lesser extent, as phosphoproteins and free phosphate ions. Finally, a small fraction (~1%) is found in the extracellular space bound to proteins, complexed to sodium, calcium, and magnesium, or as inorganic phosphate [56, 57].

7.2.3.1 Biological roles

Unlike other primordial bioelements like carbon or oxygen, phosphorus does not change its oxidation state (+5) in metabolic reactions, being found as (ortho)phosphate (PO_4^{3-}) or linear polymers (polyphosphates) formed by the condensation of phosphate units with concomitant release of water molecules. This process results in the linkage of phosphorus atoms by the so-called phosphoanhydride bonds (Figure 7.3). Hydrolysis of phosphoanhydride bonds is thermodynamically favorable, albeit kinetically slow. This property is of paramount importance in living cells, which can use phosphate polymers as a store of energy that can be released in a controlled manner (usually by means of appropriate enzymes) where and when necessary [58, 59]. Phosphate and polyphosphates can also be reversibly incorporated to biomolecules, thereby changing properties of biological significance. The most relevant functions of phosphorus in living organisms are [56]:

- Along with calcium and other ions, it has a key role in bone and teeth structure
- It forms the backbone of nucleic acids (Chapter 5)
- It is an essential component of the phospholipids of membrane bilayers (Chapter 4).
- Phosphorylation (and polyphosphorylation) of proteins, enzymes, and sugars is frequently used by organisms to change properties such as mass, charge, structure, and the ability to interact with cations like calcium or magnesium. This process determines the activity and function of molecules involved in an enormous variety of biological roles.
- It is an integral component of the universal energy currency, adenosine triphosphate (ATP) (Figure 5.9b). Other phosphorylated molecules, like creatine phosphate in muscle, serve as a rapid source of phosphate for ATP production.
- A phosphorylated molecule, 2,3-diphosphoglycerate (or 2,3-bisphosphoglycerate), modulates the affinity of hemoglobin toward oxygen.
- Phosphate is an important intracellular buffer and is therefore essential for pH regulation in the human body.
- Many intracellular signaling processes depend on phosphorus-containing compounds such as cyclic adenosine monophosphate (cAMP), cyclic guanine monophosphate (cGMP), and inositol polyphosphates (e.g., inositol triphosphate or IP3).

7.2.3.2 Sources, bioavailability, and homeostasis

Dietary phosphorus comes from three main sources:

- Naturally occurring phosphorus as component of cell membranes, tissue structures, phosphoproteins, and other biomolecules (see above), often referred to as "organic phosphorus." This is found in most foods, with approximately 20–30% of the dietary phosphorus for most adults coming from milk and milk products and 20–30% from meat, poultry and fish, grain products, and legumes. There is a relationship between high protein foods and natural phosphorus content of a mean total intake of 15–17 mg of phosphorus for every gram of protein. In any case, the rate of phosphorus absorption and the bioavailability of "organic phosphorus" vary with the food source and the available information is limited [56, 60].
- Inorganic phosphorus added as an ingredient in some medications, dietary supplements and food enrichment/fortification, and water. Phosphorus-containing salts added to foods and food supplements include glycerophosphates of calcium, magnesium, manganese and potassium, ferric sodium diphosphate, ferrous am-

monium phosphate, ferric diphosphate, calcium, magnesium, potassium, and sodium salts of orthophosphoric acid, riboflavin 5′-phosphate (sodium salt), and pyridoxine 5′-phosphate. Other salts only used in food supplements include ferrous
phosphate, sodium monofluorophosphate, thiamine monophosphate chloride, thiamine pyrophosphate chloride, and pyridoxal 5′-phosphate [61].
– Additives used in the processing of many different foods, ranging from baked
   goods and restructured meats to cola beverages, for specific approved functions.
   These include emulsification, acidification, leavening, moisture binding, color stability, and iron binding, among others [60, 61].

The amount of phosphorus contributed by the use of additives in processed and prepared foods is difficult to quantify, as the food industry is not required to publish
these amounts on food labels. This makes the assessment of dietary phosphorus a
complicated task, although it has been estimated that these additives may increase
phosphorus intake by as much as 1,000 mg/day [56].

Phosphorus absorption from the diet varies over a wide range. It typically ranges
from 65% to 90% of intake in infants and from 55% to 80% in adults, with a tendency
to decrease with aging. The organic forms of phosphate are hydrolyzed by phosphatases within the gut lumen, thus releasing inorganic phosphate that is absorbed by
the epithelium of the small intestine (duodenum and jejunum) via both a paracellular
passive diffusional process and a transcellular active sodium-dependent process
(Chapter 13). The former depends on the amount of phosphorus in the gut, whereas
the latter is increased by calcitriol. In turn, a decrease in serum phosphate concentration leads to an increase in the synthesis of calcitriol. The phosphate incorporated
into the enterocytes is transferred to the bloodstream by poorly understood mechanisms [57, 61].

The ability to absorb and use phosphorus is affected by the total amount of phosphorus in the diet and also by the type of phosphorus (organic versus inorganic) and
the food origin (animal- versus plant-derived). Dietary phosphate also interacts with
other minerals present in food, such as calcium, sodium, and magnesium; thus, phosphate absorption decreases with dietary calcium and magnesium and requires an adequate luminal concentration of sodium. Changes in intestinal phosphate absorption
affect renal phosphate handling, either indirectly, through alterations in serum phosphate concentration, or through the production of intestinally derived peptides or
proteins [57].

Regulators of phosphate homeostasis include parathyroid hormone, calcitriol,
and a number of polypeptidic hormones collectively known as the "phosphatonins,"
fibroblast growth factor-23 (FGF-23) being the most studied. These substances coordinate mineral metabolism in the gut, bone, kidney, and parathyroid gland, with the
goal of maintaining calcium and phosphate homeostasis:
– In addition to its effects on calcium (see above), PTH is one of the best characterized hormonal regulators of plasma phosphate concentration. The dominant ef-

fect of this hormone on phosphorus is to lower circulating levels by inhibiting phosphate reabsorption in the kidneys [56, 57].

- Calcitriol and PTH have opposing effects, thus, PTH promotes calcitriol production, while calcitriol inhibits the synthesis of PTH. The final consequence is that calcitriol increases both phosphate intestinal absorption and tubular reabsorption, the latter by suppressing PTH. This mechanism balances phosphate levels while maintaining calcium homeostasis [57].
- The most prominent hormone involved in phosphorus regulation, however, is FGF23, that is predominantly produced in the bones and participates in endocrine networks to counteract the multiple functions of calcitriol, including regulation of mineral homeostasis. FGF23 suppresses calcitriol production, thereby creating a bone/kidney axis to protect against vitamin D toxicity and to coordinate phosphate handling by the kidneys with bone mineralization and turnover [49, 62]. FGF23 also lowers serum PTH levels [63].

The regulation of phosphate homeostasis maintains a normal concentration of phosphate in human serum/plasma of 0.8–1.5 mmol/L (2.5 to 4.6 mg/dL) in adults (4 to 7 mg/dL in children). Around 85–90% of serum phosphate is in the form of anions HPO_4^{2-} and $H_2PO_4^-$, which exist in a pH-dependent equilibrium, whereas 10–15% is bound to proteins. By contrast, in the blood as a whole, the majority of phosphate is associated with the phospholipids of blood cell membranes and plasma lipoproteins [61]. Serum concentrations of phosphorus and calcium are interrelated, so that the increase of phosphorus concentration following ingestion depresses Ca^{2+}. This in turn stimulates the secretion of PTH that re-equilibrates serum levels of both minerals (see above).

Phosphorus is eliminated from the body mainly via urine and feces [61].

- Under normal conditions, about 15% of the phosphorus filtered by the kidneys is ultimately excreted (1–1.5 g/24 h). The rest is reabsorbed as inorganic phosphate coupled to sodium transport mediated by specific cotransporters. As described above, the reabsorption of phosphorus in the kidneys is subjected to hormonal regulation.
- The total fecal excretion of phosphorus has been estimated between 300 to 600 mg/day. This represents non-absorbed dietary phosphorus as well as losses of endogenous phosphorus from digestive secretions that have not been reabsorbed.

Around 140 mg/L (4.5 mmol/L) of phosphorus is secreted with mature human milk in lactating women with a correct intake of phosphorus (see below).

7.2.3.3 Phosphorus status, Dietary Reference Values (DRVs)

Like in the case of calcium, there is currently no reliable biomarker of phosphorus intake and status. As far as phosphate balance is concerned, few studies are available in comparison with other minerals, such as calcium, partly because phosphorus isotopes cannot be safely used for kinetic studies. On the other hand, the EFSA does not consider serum phosphate to be a reliable biomarker for phosphorus nutritional adequacy and has proposed adequate intakes (AI) for phosphorus from DRVs established for calcium (Table 7.6). In this proposal, the EFSA accepts a calcium to phosphorus molar ratio in the whole body of 1.4:1 for all population groups. No additional dietary phosphorus is required for pregnant and lactating women due to the existence of physiological adaptive processes that ensure sufficient phosphorus for fetal growth and breast milk production (Table 7.6) [61].

Table 7.6: Summary of adequate intakes for phosphorus. Table reproduced from reference [61].

Age	Adequate intake (mg/day)
7–11 months	160
1–3 years	250
4–10 years	440
11–17 years	640
Adults ≥18 years[a]	550

[a]Including pregnant and lactating women

7.2.3.4 Phosphorus deficiency

Hypophosphatemia is defined as phosphorus concentrations in serum below 0.80 mmol/L (2.5 mg/dL) in adults. It can be further classified as mild (<2.0–2.5 mg/dL), moderate (<1.0–2.0 mg/dL), or severe (<1.0 mg/dL) [64]. In general terms, hypophosphatemia can occur by inadequate phosphate intake, increased phosphate excretion, or shift from extracellular phosphate into the intracellular space.

– Inadequate intake of phosphorus mainly occurs in cases of chronic and severe malnutrition or phosphorus malabsorption, because most diets contain sufficient phosphate. Phosphorus malabsorption is often secondary to inflammatory states, surgery, or diarrhea. Chronic alcoholism is a frequent cause of hypophosphatemia by inadequate phosphate intake. Antacids containing Mg^{2+} and Al^{3+} can also interfere with phosphate absorption [65].

– Increased excretion of phosphate can occur due to acquired (primary hyperparathyroidism, vitamin D deficiency, renal tubular disease, osmotic diuresis, renal transplant, and administration of bicarbonate, corticosteroids, or diuretics) and hereditary conditions (Fanconi's syndrome, X-chromosome–linked hypophosphatemia, autosomal-dominant hypophosphatemic rickets, type 1 distal renal tubular acidosis) [65].

– Phosphate shifts can be associated with insulin administration in treatment of diabetic ketoacidosis or hyperglycemic hyperosmolar nonketotic coma, administration of glucose or total parenteral nutrition in malnourished patients, acute respiratory alkalosis, hungry bone syndrome, or oncogenic hypophosphatemic osteomalacia [65].

Patients with chronic obstructive pulmonary disease (COPD) show lower average serum phosphate levels than controls; however, further investigations are needed to identify the pathogenesis of hypophosphatemia associated with COPD [66].

Clinical symptoms of hypophosphatemia depend on the severity and duration of low plasma phosphate levels. Patients with moderate hypophosphatemia (1.5–2.5 mg/dL) are generally asymptomatic. The effects of severe hypophosphatemia include anorexia, anemia, muscle weakness, bone pain, rickets, and osteomalacia, increased susceptibility to infection, paresthesia, ataxia, confusion, and even death [61].

7.2.3.5 Phosphorus excess

Some adverse effects of excessive phosphorus intake have been reported in animal studies; however, such effects have not been observed in humans, except in patients with end-stage renal disease. No significant changes either in serum PTH concentration or in markers of bone remodeling have been demonstrated in long-term studies with dosages of up to 3,000 mg/day. On the other hand, although some gastrointestinal symptoms (diarrhea, nausea, and vomiting), were observed in some healthy subjects taking phosphate supplements with dosages higher than 750 mg/day, this has not been considered a reliable basis for establishing a UL for phosphorus by the EFSA [61].

7.2.4 Magnesium

Magnesium (chemical symbol Mg, atomic number 12, atomic mass 24.30 Da) is the eighth most abundant element in the earth's crust and the eleventh most abundant element in the human body. It is also the second most abundant cation in sea water (average concentration 55 mM), only after sodium. Like calcium, its most stable form is as cation Mg^{2+}, both having similar charge and chemical properties; however, in aqueous solution Mg^{2+} binds water molecules tighter than calcium (Ca^{2+}), potassium (K^+), or sodium (Na^+). This property allows Mg^{2+} ions to be surrounded by two hydration shells, making them harder to dehydrate than other cations. This basic property explains the often antagonistic behavior of Mg^{2+} and Ca^{2+} *in vivo* and the fact that Mg^{2+} binding to proteins and other biological molecules tends to be weaker than that of Ca^{2+} [67, 68].

7.2.4.1 Biological roles

The human body contains approximately 24 g of Mg^{2+}, 99% of which being stored in bone (53%), muscle (27%), and other soft tissues (19%). Only 1% of total Mg^{2+} is in the extracellular fluid. In bones, Mg^{2+} ions are located at the surface of the hydroxyapatite crystals and increase the solubility of the latter; one-third of these cations are exchangeable, serving as a reservoir for maintaining physiological levels [67, 68]. Mg^{2+} is the second most abundant intracellular cation (after potassium) with total concentrations between 10 and 30 mM; however, since most of the intracellular Mg^{2+} is bound to different biomolecules, the concentration of freely available Mg^{2+} falls to 0.5–1.2 mM, which is in the range of the concentration in extracellular fluid [67, 69]. Mg^{2+} is involved in practically every major metabolic and biochemical process within the cell due to its capacity to interact with macromolecules and metabolites:

–   Nucleic acid structure. Mg^{2+} is an essential component of the RNA and DNA tertiary structures. The most studied Mg^{2+}-RNA interaction is that of transfer RNA (tRNA), where Mg^{2+} stabilizes the structure; however, this cation is also crucial for the stabilization of tertiary structures present in other types of RNA: messenger, ribosomal, RNA, catalytic self-splicing, and viral genomic. It also stabilizes the natural DNA conformation, known as B-DNA [67].

–   Enzymatic activity. Traditionally, Mg^{2+} has been suggested to be cofactor of more than 300 enzymatic reactions, acting either as a cosubstrate or on the enzyme itself as a structural or catalytic component; however, many Mg^{2+}-dependent enzymes have been described in the past three decades. Nowadays, more than 600 enzymes for which Mg^{2+} serves as cofactor have been identified, and in many more (around 200) it can act as an activator. For most enzymatic reactions that involve ATP^{4-} (Chapter 5) the true substrate is $MgATP^{2-}$, a key component in the bioenergetics of cell; consequently, magnesium is essential for the viability of major metabolic processes, such as the syntheses of carbohydrates, lipids, nucleic acids, and proteins, as well as for glucose metabolism. Moreover, regulatory proteins like kinases, guanylyl cyclases, and adenylyl cyclases also use $MgATP^{2-}$ as substrate [67].

–   Structure and function of biomembranes and transporters. Mg^{2+} has been reported to stabilize biomembranes, thereby affecting their fluidity and permeability. This process influences membrane-bound proteins like ion channels, transporters, and signal transducers [70]. Major ion transporters across biomembranes, the so-called membrane-bound ATPases, including the ATP-synthase (aka F-type ATPase or F_1F_0-ATPase), the splendid molecular machine that synthesizes ATP in the mitochondrial inner membrane, require Mg^{2+} for function [9].

–   Cell signaling. Mg^{2+} acts as a physiological Ca^{2+} antagonist within cells, so that the Mg^{2+}/Ca^{2+} ratio is of major importance for the activity of proteins modulated by calcium [67].

–   Muscle contraction. The antagonistic role of Mg^{2+} with respect to Ca^{2+} also plays an important role in the regulation of muscle contraction [71].

–   Bone structure. Magnesium affects the synthesis and secretion of hormones that regulate skeletal homeostasis and bone cell function, and influences hydroxyapatite crystal formation and growth [72].

–   Neuronal magnesium is involved is the regulation of excitatory synaptic transmission and neuronal plasticity in learning and memory [73, 74].

7.2.4.2 Sources, bioavailability, and homeostasis

Foods high in magnesium include almonds, bananas, legumes (especially soybeans), broccoli, brown rice, cashews, green vegetables (spinach), berries, nuts, oatmeal, seeds (flaxseed, pumpkin, sesame, sunflowers), sweet corn, tofu, whole grains and grain products, fish and seafood, and some coffee and cocoa beverage preparations. Tap or bottled water can also make a significant contribution to intake [75, 76]. Nowadays, crops and vegetables are not as rich in magnesium and other minerals as they used to be, due to the depletion of minerals in soils; moreover, it has been estimated that 80–90% of Mg^{2+} can be lost during food processing and the diet in Western countries contains increasing amounts of refined grains and processed food. For these reasons, many authors suggest that a significant part of the population in these countries is consuming less than the required amount of magnesium (see below) [67, 75].

Many magnesium additives are currently used in foods and food supplements, these include inorganic (carbonate, chloride, phosphates, hydroxide, oxide, and sulfate) and organic (lactate, citrate, L-ascorbate, bisglycinate, L-lysinate, gluconate, glycerophosphate, and more) salts. The magnesium content in infant and follow-on formulae, as well as the maximum content of processed cereal-based foods and foods for infants and young children is regulated by the EFSA [76].

Intestinal Mg^{2+} absorption occurs predominantly in the small intestine (mainly in the distal jejunum and ileum) by simple diffusion via a paracellular pathway, regulated by the tight junctions, whereas smaller amounts are absorbed in the colon, mainly via a transcellular pathway (Chapter 13). At usual intakes, the percentage of Mg^{2+} absorption is considered to be 40–50%; however, the fractional absorption of Mg^{2+} seems to decrease with increasing intakes [76]. Dietary Mg^{2+} uptake depend on a number of exogenous and endogenous factors. Among the exogenous factors are solubility, dose, and the presence of enhancing and inhibiting factors in the food; endogenous factors include age, health, hormones, and magnesium status. Food components like oxalic acid, phytic acid, high doses of phosphates, Ca^{2+} or Zn^{2+}, aluminum, excess ethanol, soft drinks, and coffee, wheat bran, cellulose, lignin, hemicellulose, pectin, and potato starch have been reported to inhibit Mg^{2+} absorption; whereas proteins, peptides from casein or whey, lipids, vitamins D and B6, and some low or indigestible carbohydrates (inulin, oligosaccharides, and resistant starch, among others) are regarded as enhancing factors. In the case of food supplements and pharmaceutical preparations, the type of magnesium salt used seems to exert only a slight effect on absorption. In any case, more studies are needed in order to deepen the understanding of Mg^{2+} absorption and its influencing factors [73, 77].

Mg^{2+} ions are distributed throughout the organism by the blood. Approximately 0.3% of body magnesium is in the serum: 54% as free cations, 33% as a protein bound form (most of it to albumin), and around 13% as anion complexes. Magnesium is concentrated in blood cells with respect to the serum [76].

The homeostasis of magnesium is regulated both at intracellular and extracellular levels:

- Within the cells, there is a compartmentalization of Mg^{2+}, which can be found in nucleus, mitochondria, endoplasmic/sarcoplasmic reticulum, and cytoplasm. Mg^{2+} concentration in the cytoplasm is maintained relatively constant, ranging from 10 to 30 mM or from 5 to 20 mM for total concentration, according to different authors, and from 0.5 to 1.2 mM for free Mg^{2+}. As previously mentioned, this is due to the binding capacity of this cation. Maintenance of intracellular concentrations of magnesium requires the presence of multiple transport systems across the different biomembranes. All these systems are controlled at different levels (enzymatic, hormonal, etc.) in a complex way not completely understood. (More information in reference [69].)
- Extracellular homeostasis of magnesium requires the concerted involvement of different organs (small and large intestines, kidneys, and bones) in order to maintain the balance between intake and losses.
- At intestinal level, no hormone or factor has been clearly established as a regulator of absorption, which seems to depend basically on dietary intake, thus, absorption can rise from 30% to 50% of dietary Mg^{2+} to <80% if intake is low. Experiments in mice have shown that low and high dietary Mg^{2+} affects Ca^{2+} balance via the kidney (by increasing reabsorption and elimination, respectively), although the mechanisms responsible for these phenomena are unknown [69, 78].
- Kidneys are critical for the regulation of Mg^{2+} homeostasis: around 10% of total magnesium in the body (~2,400 mg) is filtered daily through the renal glomeruli, only 3–5% being excreted in the urine. In this organ, the activity of the main magnesium channel (TRPM6) is regulated by the epidermal growth factor, whereas the expression of the corresponding gene appears to be regulated by plasma Mg^{2+} levels and estrogens [69, 78]. Many other factors have been reported to increase or decrease renal absorption of magnesium, including hypermagnesemia and hypercalcemia [79].
- Bones are the largest Mg^{2+} stores in the human body, thus providing a large exchangeable pool (one-third of skeletal magnesium) to buffer changes in serum concentration. Bone surface levels of Mg^{2+} are similar to those of serum, which suggests a continuous exchange of this cation between bone and blood [68, 78].

Magnesium can be lost not only via kidneys (~100 mg/day, see above), but also through feces and sweat. In the case of fecal losses, a large, albeit variable, proportion comes from unabsorbed dietary Mg^{2+}, although there are also endogenous losses of magnesium contained in bile, pancreatic, and intestinal juices, as well as in intestinal cells, that is not reabsorbed. Overall, an average 260 mg of magnesium can be lost per day through feces. Different values have been reported for magnesium sweat losses, the highest values being after intense exercise and/or in a hot environment, with values in the range of 1–5 mg/day. Magnesium losses through menstruation in women are considered negligible, whereas a secretion of 25 mg/day has been estimated in breast milk during the first 6 months of lactation [68, 76].

7.2.4.3 Magnesium status, Dietary Reference Values (DRVs)

Magnesium concentrations in urine, feces, serum, and erythrocytes have been used for the assessment of magnesium status, with serum magnesium concentration being the most frequently used marker; however, all these biomarkers have significant limitations, therefore, adequate intakes (AIs) based on observed intakes, instead of average requirements (ARs) or population reference intakes (PRIs), have been proposed for magnesium by the EFSA. AI values are shown in Table 7.7.

Table 7.7: Adequate intakes for magnesium. Source [76].

Age	Adequate intake (mg/day)	
	Males	Females
7–11 months	80	80
1–<3 years	170	170
3–<10 years	230	230
10–<18 years	300	250
≥18 years[a]	350	300

[a]Including pregnant and lactating women.

7.2.4.4 Magnesium deficiency

Although total serum and plasma values are considered unreliable indicators of Mg^{2+} depletion, magnesium deficiency is commonly associated with hypomagnesemia, generally defined as serum Mg^{2+} levels below 0.7 mM. Patients with this condition present nonspecific symptoms such as depression, tiredness, muscle spasms, and muscle weakness, whereas severe depletion (<0.4 mM) may lead to cardiac arrhythmias, tetany, seizures, and disturbances in K^+ and Ca^{2+} handling. Several studies have associated low serum Mg^{2+} values with osteoporosis and supplementation with magnesium seems to increase bone mineral density, although the effects are relatively small (1–3%) [67]. Mg^{2+} deficiency has been associated with numerous clinical disorders, such as migraine, chronic pain, neurological disorders (epilepsy, Alzheimer's, Parkinson's, and stroke), anxiety, depression, diabetes mellitus (see below), cardiovascular diseases, and cancer [73].

Hypomagnesemia is not a rare condition and can be caused by a number of factors:

– Nutritional deficiency. It has been estimated that up to 60% of the population in the United States does not meet daily Mg^{2+} requirements, a situation that may eventually lead to hypomagnesemia. Alcoholism and pathological conditions such as anorexia nervosa can also be a cause of magnesium deficiency [67].

- Gastrointestinal disorders, such as Crohn's disease and ulcerative colitis, celiac disease, acute and chronic diarrhea, regional enteritis, intestinal and biliary fistulas, or steatorrhea [69, 73]. In the case of people suffering from celiac disease, besides the absorption inefficiency, gluten free-diets have been found to be poor in fiber and micronutrients such as Mg^{2+} [73].
- Renal disorders [69].
- Endocrine and metabolic disorders like diabetes mellitus [69].
- Genetic mutations (see reference [67] for a complete list of genes involved)
- The use of certain drugs, such as diuretics and proton pump inhibitors, among others [67].
- Chronic magnesium deficiency is common in the elderly, because they absorb less Mg^{2+} in the gut and excrete more Mg^{2+} in the kidneys. This situation is probably exacerbated by estrogen deficit, which occurs in aging women and men [73].
- Mg^{2+} deficiency is often observed in people affected by type 2 diabetes, moreover, Mg^{2+} is reportedly an insulin sensitizer; however, it still remains unclear if Mg^{2+} deficiency is a cause or a consequence of this pathology [73].

Hypomagnesemia is generally treated by oral Mg^{2+} supplementation (360 mg/day), although this may cause diarrhea; intravenous supplementation may be a more effective alternative. In any case, the underlying cause of magnesium deficiency must be established in order to prevent its occurrence and/or minimize its severity [67, 69].

7.2.4.5 Magnesium excess

Hypermagnesemia a is generally defined as serum Mg^{2+} levels above 1.1 mM. Symptoms include nausea, vomiting, lethargy, headaches, and/or flushing. Mg^{2+} levels in serum above 3.0 mM may even cause severe cardiac defects, whereas extreme hypermagnesemia (above 5 mM) can result in coma, asystole, and death by cardiac arrest. In any case, hypermagnesemia is rare and no genetic causes for it have been identified so far. It may be caused by administration of certain drugs and renal failure [67, 69].

A Tolerable Upper Intake Level (UL) for magnesium of 250 mg/day has been established by European authorities for adults, including pregnant and lactating women, and children from 4 years of age. This UL does not include magnesium normally present in foods and beverages, but refers exclusively to magnesium salts and compounds in nutritional supplements, water, or food additives. No UL value has been established for children aged 1–3 years, due to the lack of data [76].

7.3 Microelements

7.3.1 Iron

Iron (chemical symbol Fe, atomic number 26, atomic mass 55.85 Da) is the most common element on earth by mass, forming much of the outer and inner core of our planet, and the fourth most common element in the crust. Its most relevant roles in living organisms are as ferrous (Fe^{2+}) and ferric (Fe^{3+}) cations. These states can readily interconvert by donating (Fe^{2+}) and accepting electrons (Fe^{3+}), which allows iron to participate in oxidation–reduction reactions that are essential for a number of fundamental biologic processes [35, 36]; moreover, iron's redox potential can be modulated by binding to different ligands. By exploiting these properties biological systems can adjust the chemical reactivity of this element to suit physiological needs. Iron usually form complexes with nitrogen-containing ligands, like in the porphyrin ring of heme group, and with sulfur, forming iron–sulfur clusters [80–82].

7.3.1.1 Biological roles

Total iron body content is estimated to be ~3–5 g, most of it being found as heme iron, the prosthetic group of hemoglobin in erythrocytes (~2 g), and myoglobin in muscles (~300 mg). The rest is distributed among macrophages of the spleen, liver, and bone marrow (~600 mg), associated with ferritin in the liver parenchyma (~1,000 mg) and other cellular iron-containing proteins and enzymes (~8 mg) of iron [82].

Iron-containing proteins can be classified in hemoproteins (also known as haemoproteins, haemproteins, or hemeproteins), proteins containing iron–sulfur clusters, and non-heme, non-iron–sulfur, iron-containing proteins (see reference [83] for a more comprehensive review):

- Three types of hemoproteins have been identified so far: oxygen carriers (hemoglobins and myoglobins), activators of molecular oxygen (cytochrome oxidases, peroxidases, and cytochrome P-450s), and electron transport proteins (cytochromes *a*, *b*, and *c*)
- Iron–sulfur proteins contain iron atoms bound to inorganic sulfur or with the sulfur atoms of cysteine residues in the protein, or both. Iron–sulfur proteins can be classified as either simple or complex, with the latter containing additional cofactors, such as flavins, molybdenum, or heme group. Fe–S clusters can have redox potentials ranging from −500 mV to +300 mV, thereby acting as donors and acceptors of electrons in many biological reactions like those mediated by mitochondrial respiratory complexes I–III.
- Non-heme, non-sulfur iron-containing proteins are a heterogeneous collection of proteins that can be further classified into three categories:
  - Mononuclear non-heme iron enzymes, like the Fe^{2+}-2-oxoglutarate-dependent (2-OGDDs) family of dioxygenases, that catalyze the hydroxylation of prolyl

and lysyl-residues in collagen biosynthesis (Chapter 3, Section 3.3.2), among other reactions.

- Dinuclear non-heme iron enzymes, like the ferroxidase activity of human ferritin, that oxidizes Fe^{2+} to Fe^{3+}, so that the latter can be subsequently stored.
- Proteins of iron storage, transport, and metabolism, such as transferrin and hemosiderin.

7.3.1.2 Sources, bioavailability, and homeostasis

Dietary iron can be absorbed as either heme or non-heme iron. The former mainly comes from hemoglobin or myoglobin and is contained in foods such as red meats, fish, and poultry. Only small amounts of heme iron are present in some plants and fungi. Non-heme iron (mainly in the form of ferritin) is present in some animal and plant foods, particularly liver and legume seeds. Although non-heme iron is the predominant form in the diet, heme iron is more bioavailable: approximately 15% to 35% of heme iron is estimated to be assimilated, compared with 2% to 20% non-heme iron absorption. The proportion of heme iron with respect to total iron has been reported to vary considerably depending on the origin of meat: 69% for beef; 39% for pork and veal, 26% for chicken and fish, and 21% for liver [80, 84].

Iron deficiency is one of the most prevalent micronutrient deficiencies in the world (see below), therefore, a number of iron-containing compounds are added to foods as fortificants in many countries. However, compounds with the best bioavailability tend to interact strongly with food constituents, thus producing undesirable organoleptic changes. When selecting an iron fortificant, the main goal is obtain the best absorbability with the fewest changes to the taste, color, or texture of the food. Iron compounds used as food fortificants can be divided into three categories [85]:

- Water soluble compounds. They are usually the preferred choice because of their high bioavailability, although they are more likely to have adverse effects on the color and flavor of foods (rancidity and subsequent off-flavors). In case of multiple fortification, some vitamins can be oxidized due to the occurrence of free iron. Among these compounds are ferrous sulfate, gluconate, lactate, and bisglycinate, as well as ferric ammonium citrate and sodium iron (Fe^{3+}) EDTA. Ferrous sulfate and gluconate are also used as iron supplements.
- Compounds poorly soluble in water but soluble in dilute acid. These compounds cause fewer sensory problems in foods than water soluble compounds, while being reasonably well absorbed from food, as they are soluble in the gastric acids produced in the stomach of healthy adults and adolescents. More studies are needed to assess absorption rates in infants. Ferrous fumarate and ferric saccharate are the most commonly used compounds in this group. Ferrous fumarate is also a popular iron supplement.
- Water insoluble and poorly soluble compounds in dilute acid. These compounds have been widely used by the food industry because they have less effect on the

sensory properties of foods and because they are cheaper than the other types of compounds. This group includes ferric orthophosphate and pyrophosphate, as well as elemental iron. The bioavailability of the latter depends on its solubility, which in turn is very dependent on the size, shape, and surface area of the iron particles added to the food.

Several of the abovementioned compounds are currently available in encapsulated form. This technology improves the bioavailability of iron, reduces organoleptic alterations, and decreases gastrointestinal problems. An active research is currently under way to determine the best combination of iron compounds and coatings to optimize their efficacy as iron fortificants. Some of these preparations are being used in dry infant formulas and cereals, mainly in industrialized countries. A major drawback of use of encapsulated forms of iron compounds is their high cost [85, 86].

Iron follows an almost cyclic pathway within the body, with few new inputs and few losses. In blood, most iron is found within erythrocytes associated to hemoglobin, only 5 mg of iron being linked to the transport protein transferrin (see below) that receives iron from two sources: dietary absorption and recycled iron that derives from senescent erythrocytes subjected to phagocytosis in the spleen and other tissues. The amount of dietary iron absorbed must balance losses to maintain homeostasis, but, in any case, this amount is much lower than recycled iron (1–2 mg/day vs. 20 mg/day) [87].

During digestion, iron is released from foods in three main forms: non-heme iron, heme iron, and ferritin, each of them having specific mechanisms of absorption, which occur mainly in the enterocytes of the duodenum and proximal small intestine as Fe^{2+} (Chapter 13). In humans, iron absorption must be finely regulated due to lack of an active mechanism of excretion; consequently, the rate of absorption is modulated by the iron status of the body. This prevents toxic accumulation of iron while compensating the small amounts lost in feces, urine, and sweat. On the other hand, there are a number of enhancing and inhibiting factors of iron absorption [80, 84]:

- Enhancers include ascorbic and citric acids, fructose, dietary protein (muscle tissue from animals), and certain amino acids like cysteine, histidine, lysine, and methionine
- Inhibitors include oxalic acid and oxalates, tannins, phytic acid and phytates, polyphenols, carbonate, phosphate, fiber, and other metal ions like calcium.

There is controversy about the actual role of enhancers and inhibitors of iron absorption, thus, in both cases the effects are more pronounced in single-meal absorption studies than in longer term interventions. In any case, body iron status seems to be the key determinant of absorption efficiency [80].

Enterocytes can rapidly export the absorbed Fe^{2+} cations across the basolateral membrane into the bloodstream, a process mediated by ferroportin (Chapter 13). The protein hephestin subsequently converts Fe^{2+} to the Fe^{3+} form, which is bound to plasma transferrin and distributed throughout the organism by the blood. Each transferrin molecule binds up to two Fe^{3+} cations and, under normal circumstances, only around 30% of the binding sites are occupied at a given time. This provides a considerable buffering capacity against an excess of iron. If iron is not immediately required by the body, it becomes sequestered in the cell, bound to ferritin, and can be lost from the body when the enterocyte is sloughed after a few days [88].

Transferrin delivers iron into the cells by binding to a receptor on the plasma membrane, followed by the internalization of the transferrin -receptor complex by clathrin-mediated endocytosis. Other forms of iron (non–transferrin -bound iron and ferritin) can also be taken up by most cells. Free iron in solution is quite toxic; therefore, it is mainly sequestered within the cells by proteins like ferritin and some enzymes. Ferritin, a large protein of 24 subunits that can hold up to 4,500 Fe^{3+} cations per molecule, is the major intracellular iron-storage protein. The levels of ferritin increase when cellular iron concentrations rise, a situation that may lead to the aggregation of iron-laden molecules. These aggregates fuse with lysosomes being degraded to hemosiderin. Iron can be mobilized from both ferritin and hemosiderin if required. A small proportion of ferritin, that correlates with the concentration of intracellular iron, can be secreted out of the cell. This association makes serum ferritin concentrations a good indicator of body iron stores [88]. Intracellular iron transport is mediated by iron chaperones, in this respect, poly(rC)-binding proteins (PCBPs), which were originally identified as RNA/DNA-binding molecules, have been recently identified as key intracellular iron chaperones. PCBPs are not only cytosolic iron carriers, but also regulators of iron transport and recycling. Two isoforms have been described, PCBP1, involved in the iron storage pathway that involves ferritin, and PCBP2, involved in processes such as iron transfer from the iron importer, iron export, and heme degradation [89].

Iron must be adequately supplied to organs avoiding both iron deficiency and iron overload; therefore, a tight coordinated regulation of the available iron is necessary at both cellular and systemic levels. Regulation of intracellular iron metabolism is complex and involves transcriptional, posttranscriptional, translational, and posttranslational mechanisms, whereas regulation at organismal level is mediated by hepcidin, a 25-amino acid peptide mainly synthesized by the liver in response to increased body iron. Hepcidin action results in reduced dietary iron absorption (Chapter 13) and higher iron retention in macrophages of the reticuloendothelial system. Conversely, if iron levels are low, hepcidin secretion is reduced, thus increasing intestinal absorption and iron mobilization from the stores. Hepcidin synthesis is also induced during inflammatory responses [84, 90].

7.3.1.3 Iron status, Dietary Reference Values (DRVs)

There are a number of parameters than can be analyzed to assess iron status [80, 84]:
- Tests of anemia due to iron deficiency:
    - Hemoglobin and hematocrit

- Reticulocyte hemoglobin content (CHr)
- Mean cell volume (MCV), mean cell hemoglobin (MCH) and the red cell distri-
  bution width (RDW)
- Hypochromic red blood cells (HYPO)
- Erythrocyte zinc protoporphyrin (ZPP)
- Tests for iron depletion in the body:
  - Serum iron, total iron-binding capacity (TIBC) and transferrin saturation (TSAT)
  - Serum ferritin
  - Soluble serum transferrin receptor (sTfR)
  - Ratio of sTfR (R) to ferritin (F). The ratio has been shown to be more reliable
    than either parameter alone for the identification of iron deficiency The
    EFSA Panel on Dietetic Products, Nutrition and Allergies has derived ade-
    quate requirements (ARs) and Population Reference Intake (PRIs) for iron
    [80, 91]. These values are shown in Table 7.8.

7.3.1.4 Iron deficiency

Around 30% of world's population, that is, around 2,000 million people, are estimated
to suffer from anemia; many of them due to iron deficiency (ID). Although this condition is frequent (and exacerbated by infectious diseases) in low-income areas, it is also
prevalent in industrialized countries [92]. Despite this high prevalence, diagnosing and
treating ID is a challenge in clinical practice and there is a remarkable heterogeneity in
the guidelines published by professional associations worldwide about its management.
One of the reasons for this situation is that ID can be masked by conditions that stimulate hepcidin production, such as many types of cancer, chronic kidney disease, and inflammatory disorders associated with atherosclerosis, obesity, diabetes, and metabolic
syndrome. Iron deficiency can be classified according to different criteria [93]:
- Based on etiology and pathophysiology:
  - Absolute ID: iron stores are reduced or depleted even in the absence of an
    obvious anemia. Patients with absolute ID have low tissue iron stores, low
    bone marrow iron stores, and low plasma iron (transferrin saturation) [94].
  - Relative ID: total body iron is not decreased but iron is lacking in specific tis-
    sues/organs.
- Based on laboratory and clinical results: ID (without anemia) and iron deficiency
  anemia (IDA).
- Based on inheritance: a genetic form of iron deficiency anemia, recognized as a
  rare disease, has been identified and there seems to exist a genetic predisposition
  toward ID.

During absolute ID or periods of increased iron demand, hepcidin suppression upre-
gulates iron absorption and recycling to optimize iron supply. Inflammation (acute
and chronic infection, autoimmune conditions, cancer, recent surgery, and heart fail-

ure) is associated with a type of anemia whose predominant mechanism is the so-called functional ID, in which inflammation-mediated increases in hepcidin prevent cellular iron export (especially from macrophages) to the plasma. This results in reduced transferrin saturation, iron deficient erythropoiesis, and anemia, even with sufficient body iron stores [94].

The gold standard test for absolute ID is the finding of absent stainable bone marrow iron; however, this analysis is invasive and rarely done routinely, although it remains useful in complex cases. Iron deficiency is usually diagnosed by blood biomarkers, such as serum ferritin concentration of transferrin saturation (see reference [94] for a more comprehensive list).

Although ID does not always progress to anemia, the latter is a major sign of this deficiency. Anemia due to iron deficiency (IDA) is associated with impaired erythropoiesis due to the shortage of iron. Other tissues or organs, such as skeletal muscles, heart, or brain, can also be deficient in iron; however, these situations are not recognizable by using traditional tests. Other symptoms associated with iron deficiency include fatigue, muscle weakness, reduced physical performance, cognitive impairment, and restless leg syndrome. ID is a predictor of mortality in chronic heart failure and chronic kidney disease, and has also been associated with aging [93, 95]. It must be mentioned that, even when asymptomatic, ID can promote suboptimal outcomes, including impaired physical exercise performance and child neurocognitive development, as well as adverse pregnancy consequences [94].

Major causes of ID and IDA are increased iron requirements (growth, menses, and pregnancies), reduced iron intake, chronic blood loss, and defective absorption. IDA is widespread both in developing countries, with prevalences reaching 40%, 30%, and 38% in children, young females, and pregnant women, respectively, and also in high-income countries, especially in children and pregnant females. In many cases, diagnosing ID/IDA leads to the discovery of underlying gastrointestinal disorders that result in blood loss from the gut or reduced iron absorption [93].

Dietary recommendations to prevent iron deficiency include consumption of foods with high iron availability and with the ability to enhance absorption, as well as avoidance of foods that reduce absorption (see above). The strategies based on iron fortification and iron supplements are controversial, thus, fortification has been successful only in developed countries and supplementation has been shown to be beneficial for pregnant women, preterm or low birth weight infants, young children, women in their reproductive years, as well as for individuals with kidney failure, inflammatory bowel diseases, or hookworm infection. By contrast, the use of supplements may be contraindicated in anemia associated with chronic diseases, whereas iron fortification have been reported to produce an excess of iron levels in people with susceptibility for genetic or acquired iron loading (see below). The benefits of iron supplementation in pregnant women are also controversial. The use of other micronutrients commonly provided with iron supplements, like folate, vitamin A, and riboflavin, must also be carefully evaluated to optimize the beneficial effects of supplementation [84].

7.3.1.5 Iron excess

Iron excess, usually known as iron overload, is nearly impossible to attain from dietary sources with normal intestinal function. Some adverse gastrointestinal effects have been reported after short-term ingestion of high doses of non-heme iron preparations (50–60 mg/day), particularly if taken without food. The Institute of Medicine of the United States has set a UL at 45 mg/day for men and women aged 14 years and older, including pregnant and lactating women. For infants and children, the UL was set at 40 mg/day [96].

Iron overload is usually due to genetic diseases caused by mutations in genes that codify for proteins involved in hepcidin regulation (HFE, transferrin receptor 2, hemojuvelin, hepcidin, matriptase 2) or in the interaction of hepcidin with its effector (ferroportin), in cellular iron handling [divalent metal transporter 1 (DMT1)], iron transport (transferrin), or release of iron to transferrin (ceruloplasmin) [97]. Repeated blood transfusions and excessive supplementation of parenteral iron are also causes of acquired overload [87]. A particular case of iron overload (previously called Bantu siderosis) occurs in Southern and Central Africa due to the combination of an uncharacterized genetic defect with increased exposure to iron from food and beer prepared in iron utensils [80].

Iron toxicity is associated with non-transferrin bound iron (NTBI), a form of iron captured and accumulated primarily by hepatocytes, which can appear in the blood when transferrin saturation rises above 45%. When transferrin saturation is greater than 75–80%, a new form called labile plasma iron or reactive plasma iron (RPI) appears. Both NTBI and RPI generate radical oxygen species that can damage cell membranes, intracellular organelles, and DNA [87].

7.3.2 Zinc

Zinc (chemical symbol Zn, atomic number 30, atomic mass 65.39 Da) is the 24th most abundant element in the earth's crust, usually occurring as a stable divalent cation (Zn^{2+}). Zinc compounds, unlike those of other transition metals such as copper or iron, are colorless due to its completely filled set of d-orbitals. This feature renders Zn^{2+} diamagnetic and prevents it from undergoing oxidation or reduction reactions, thus facilitating its safe transport within the body; nevertheless, Zn^{2+} can bind to organic molecules that are redox sensitive, especially thiolate and amine electron donors. The coordination number of Zn^{2+} in its complexes varies from two to eight, but most frequently are found to be four, five and six. Zn^{2+} complexes have the ability to adopt transient coordination geometries, which makes them effective catalysts. On the other hand, zinc has the capacity for fast ligand exchange, a property that is believed to be important for some biological functions [24, 98, 99].

7.3.2.1 Biological roles

The total content of zinc in an adult human body ranges from 1.5 g (females) to 2.5 g (males), most of it (85%) being located in muscles and bones, although some tissues (e.g., the prostate) have a high overall concentration. There are also regional differences in zinc abundance within organs that are believed to be functionally relevant. Approximately 95% of zinc in the body is intracellular, most of it found in the cytosol, although smaller amounts can reside in vesicles. Nucleic acids, protein thiols, and nitrogen ligands have a high binding affinity toward Zn^{2+}, therefore, concentrations of free cations within the cell are believed to be very small, in the picomolar range, whereas the total cellular zinc concentration is in the range of hundreds of micromolar [98, 100].

By using bioinformatics approaches, around 3,000 proteins (2,800 ± 400) have been estimated to bind Zn^{2+} in humans, a number that could be even larger if additional regulatory functions of zinc are considered (see below). The coordination chemistry of Zn^{2+} in proteins and peptides involves N, O, and S atoms of residues such as histidine, glutamate, aspartate, and/or cysteine, and shows a remarkable flexibility as the number of ligands range from three to six. Moreover, ligands may not stem from a single protein but from up to four proteins [101, 102]; therefore, the interaction with Zn^{2+} allows proteins and peptides to modulate their structures and, hence, their functional properties. Zinc functions *in vivo* can be classified in catalytic, structural, and regulatory:

–   Catalytic function. In general terms, an enzyme is considered a zinc metalloenzyme if the removal of the cation(s) causes loss of activity without irreversibly altering the polypeptide, so that reconstitution with zinc restores the activity. More than 300 zinc metalloenzymes of all six enzyme classes have been identified so far in different organisms. Zn^{2+} binding is a posttranslational protein modification, although the mechanism of incorporation to the polypeptide is not completely understood. Examples of zinc metalloenzymes are RNA polymerases I, II, and III, alkaline phosphatases, and carbonic anhydrases [98].

–   Structural functions. Zinc ions are essential to establish and maintain an extremely important structure that appears in the so-called zinc-finger proteins (ZNFs). In these structures, Zn^{2+} coordinates cysteine and histidine in different combinations. ZNFs are one of the most abundant groups of proteins (around 2,500 ZNFs have been identified in humans) and have a wide range of molecular functions, due to their interactions with DNA, RNA, poly-ADP-ribose, and other proteins. ZNFs are involved in the regulation of cellular processes, such as transcriptional regulation, ubiquitin-mediated protein degradation, signal transduction, actin targeting, DNA repair, cell migration, and many others [103].

- Regulatory functions. Three major regulatory roles at different levels have already been ascribed to Zn^{2+}: (1) gene expression via MTF1 (metal-response element–binding transcription factor 1), (2) cell signaling pathways, and (3) synaptic transmission in the central nervous system [98]. Recent findings suggest that zinc is a major regulatory cation with a more important role in cellular biology than previously thought; therefore, new discoveries about its mechanisms of action are likely to emerge in the near future [100].

Zinc has also been reported to modulate cell-mediated immunity and to act as an antioxidant and anti-inflammatory agent [104].

7.3.2.2 Sources, bioavailability, and homeostasis

Meat, legumes, eggs, fish (including molluscs, especially oysters, and crustaceans), and grains and grain-based products are rich dietary sources of zinc. Salts such as zinc acetate, bisglycinate, chloride, citrate, gluconate, lactate, carbonate, and sulfate, as well as zinc oxide may be added to both foods and food supplements. Zinc salts derived from amino acids (L-lysine, L-methionine, and L-aspartic acid) and organic acids (L-ascorbic, L-malic, L-pidolic, and picolinic) can be added to food supplements only. Breakfast cereals are usually fortified with zinc, being one of the major sources of this micronutrient in countries like the United States. Fruits and vegetables are low in zinc content [98, 99].

Zn^{2+} bioavailability depends on its solubility and stability of the respective complexes in the intestinal lumen, which is affected by the diet. Phytate forms stable complexes with Zn^{2+} at intestinal pH, which diminishes its availability for enterocytes, whereas dietary proteins have a positive effect on absorption due to the release of amino acids and peptides upon degradation. In general, plant-based diets contain high phytate levels; therefore, they provide less intestinally available zinc than meat-based diets. Zn^{2+}-phytate interactions are stronger in the presence of Ca^{2+}, which suggests that the latter might aggravate the inhibition of zinc absorption; however, this has not been confirmed in several human dietary studies. Negative effects of both heme-iron and inorganic iron on zinc absorption have been reported by several studies, while copper has shown no impact on zinc absorption. In contrast, supra-physiological zinc doses critically impair intestinal copper absorption. Fibers, such as cellulose, and ascorbic acid have no effect on intestinal Zn^{2+} bioavailability; citrate, on the other hand, positively influences zinc availability. Concentrations of Zn^{2+}-citrate complexes are higher in human milk than in cow's milk, which might explain the higher Zn^{2+} absorption from the former. Chemical and physical food processing also affect zinc bioavailability, thus, the formation of heat-derived Zn^{2+}-binding ligands, such as Maillard browning products, decreases Zn^{2+} absorption, whereas fermentation or germination have positive effects [105].

A joint FAO/WHO experts committee classified diets according to the potential bioavailability of their zinc content in high (50% absorption), moderate (30% absorption), and low (15% absorption) bioavailability diets [99, 106]. In any case, inhibition of zinc absorption by phytates still have some controversial points, thus, several studies have shown that zinc bioavailability can be relatively high in diets rich in phytates, especially those based on legumes [85].

The majority of dietary zinc is absorbed in the upper small intestine, although small quantities may be absorbed throughout the entire gastrointestinal tract. Stable isotope studies suggest that zinc is released from food components (proteins, phytate) in the acidic environment of the stomach, being subsequently bound to a variety of other organic ligands in the alkaline medium of the distal duodenum. Fractional absorption of dietary zinc in humans is typically in the range of 16–50%, being inversely related to oral zinc intake, while net absorption is regulated by body zinc homeostasis and, thus, depends on the individual zinc status; therefore, the gastrointestinal tract plays a primary role in maintaining zinc homeostasis at organismal level. Adjustments in absorption probably involve up-regulation and down-regulation of zinc transporters (the molecular mechanisms of intestinal absorption are described in Chapter 13), whereas the mechanisms that regulate pancreatic/intestinal secretion are not well understood. Studies in experimental animals suggest that absorption increases during late pregnancy and lactation and may decline with aging. Zn^{2+} absorption is also affected by the form in which it is administered, being higher from orally administered aqueous solutions than from meals. Moreover, internal factors such as serum albumin or hepcidin seem to influence Zn^{2+} absorption/export by enterocytes, which suggest that the liver participates in the regulation of Zn^{2+} homeostasis via these proteins [98, 99, 105].

In the bloodstream, 70% to 80% of the zinc is associated to blood cells, while plasma zinc represents 20% to 30% (0.1% of total body zinc) and is mainly bound to albumin. Plasma zinc can be rapidly exchanged with that located in the liver and soft tissues other than muscle. Bones have been reported to release Zn^{2+} in times of depletion, albeit at a slower rate than liver, and many cells have zinc stored in vesicles that may serve as a transient source in times of need [98, 99].

The major route of zinc excretion is secretion into the gastrointestinal (GI) tract, that combines pancreatic and intestinal secretions, as well as sloughing of mucosal cells. GI zinc excretion is directly related to dietary zinc intake: 7 to 15 mg of ingested zinc per day correspond to 3.0 to 4.6 mg/day of zinc excreted through the GI tract. Other losses of zinc include urine (<1 mg/day), integument (1 mg/day), semen (1 mg/ejaculate), menstruation (0.1 to 0.5 mg total), and parturition (100 mg/fetus, 100 mg/placenta). Lactation losses range from 2.2 mg/day at 4 weeks to 0.9 mg/day at 35 weeks [98].

7.3.2.3 Zinc status, Dietary Reference Values (DRVs)
The efficient regulation of zinc homeostasis both at organismal and cellular levels complicates the assessment of zinc status. Plasma zinc concentration, which normally ranges from 12 to 18 µmol/L (0.8 to 1.2 µg/mL), is typically used to evaluate zinc status, because it reflects dietary zinc intakes over the long term and responds to zinc supple-

mentation. The main disadvantage of this method is that plasma zinc levels are sensitive to situations not directly related to zinc status, thus, they decline with food intake, infections, and increased steroidal hormone levels that can occur during pregnancy and oral contraceptive therapy. Plasma zinc concentration also increases with muscle catabolism during illnesses or weight loss [98].

The EFSA has derived Population Reference Intakes (PRIs) for zinc in adults taking into consideration the amount of absorbed zinc necessary to match endogenous losses, body weight, and the inhibitory effect of phytate on absorption. For infants and children, adequate requirements (ARs) were calculated considering losses and an estimation needs during growth. For pregnant and lactating women, increases in physiological requirement were estimated based on the demand for new tissue, primarily by the conceptus, and for the production of breast milk, respectively [91, 99]. Values are summarized in Table 7.8.

7.3.2.4 Zinc deficiency

Nearly 17.3% of the global population has been estimated to be at risk of developing zinc deficiency, with a prevalence of inadequate zinc intake that ranges from 7.5% in high-income regions to 30% in South Asia. The major factor contributing to zinc deficiency in the developing world is high phytate-containing cereal intake [104, 107]. Zinc deficiency is characterized as a type II nutrient deficiency because it stops growth, the nutrient is avidly conserved, and, if necessary, weight is lost to make the nutrient internally available in order to maintain adequate concentration in the tissues. In adults, depletion of zinc is a slow process due to the mechanisms of homeostatic control explained above, whereas repletion appears to be rapid [98].

Classical symptoms of zinc deficiency include growth retardation, male hypogonadism in adolescents, rough skin, poor appetite, mental lethargy, delayed wound healing, cell-mediated immune dysfunctions, and abnormal neurosensory changes. In males, adequate zinc content is necessary for the formation and maturation of sperm and hormonal regulation, being also tied to the prostate gland, where zinc concentration is much higher than in other tissues. This accessory sex organ requires an adequate amount of zinc for its proper function. In females, the demand for zinc has an acute increase during pregnancy, as a consequence, Zn^{2+} deficiency may result in spontaneous abortion, extended pregnancy or prematurity, malformations, and retarded growth of fetus. Some neuropsychiatric disorders such as Alzheimer's disease, Parkinson's disease, brain aging, and depression have been linked to Zn^{2+} deficiency, most likely because zinc acts as the component of many key enzymes or serves as the activator or regulator of them. Besides, the zinc status is critical to the gastrointestinal (GI) system and to the normal development of bone. Even a mild deficiency of zinc in humans has been shown to affect clinical, biochemical, and immunological functions adversely (decreased serum testosterone level, oligospermia, hyperammonemia, decreased dark adaptation and sense of taste, among others). Acute severe zinc deficiency results from malabsorption of

Table 7.8: Dietary Reference Values (DRVs) for micronutrients. Data obtained from reference [91].

(A) Males
Average requirements (ARs)

Age group (years)	Iron (mg/day)	LPI (mg/day)	Zinc (mg/day)
7–11 mo (a)	8	(c)	2.4
1–3	5	(c)	3.6
4–6	5	(c)	4.6
7–10	8	(c)	6.2
11–14	8	(c)	8.9
15–17	8	(c)	11.8
≥18	6	300	7.5
		600	9.3
		900	11.0
		1,200	12.7

LPI, level of phytate intake; mo, months
(a): the second half of the first year of life (from the beginning of the 7th month to the 1st birthday)
(b): an AI was set for infants (see Table 7.5)
(c): The fractional absorption of zinc considered in setting ARs for children was based on data from mixed diets expected to contain variable quantities of phytate; therefore, no adjustment for phytate intake has been made.

Table 7.8 (continued)

Population reference intakes (PRIs) and adequate intakes (AIs)

Age (years)	F (mg/day)	I (µg/day)	Mn (mg/day)	Mo (µg/day)	Se (µg/day)	Fe (mg/day)	Zn (mg/day) LPI (mg/day)	Zn (mg/day)	Age (years)	Cu (mg/day)
7–11 mo (a)	0.4	70	0.02–0.5 (b)	10	15	**11**	(c)	**2.9**	7–11 mo (a)	0.4
1–3	0.6	90	0.5	15	15	**7**	(c)	**4.3**	1–2	0.7
4–6	1.0	90	1.0	20	20	**7**	(c)	**5.5**	3–9	1.0
7–10	1.5	90	1.5	30	35	**11**	(c)	**7.4**	10–17	1.3
11–14	2.2	120	2.0	45	55	**11**	(c)	**10.7**		
15–17	3.2	130	3.0	65	70	**11**	(c)	**14.2**		
≥18	3.4	150	3.0	65	70	**11**	300	**9.4**	≥18	1.6
							600	11.7		
							900	14.0		
							1,200	16.3		

PRIs are presented in bold type and AIs in ordinary type

LPI, level of phytate intake; mo, months

(b): In view of the wide range of manganese intakes that appear to be adequate, a range is set for the AI of this age group.

(c): The fractional absorption of zinc considered in setting PRIs for children was based on data from mixed diets expected to contain variable quantities of phytate; therefore, no adjustment for phytate intake has been made.

(B) Females

Average requirements (ARs)

Age (years)	Fe (mg/day)	Age (years)	LPI (mg/day)	Zn (mg/day)
7–11 mo (a)	8	7–11 mo (a)	(c)	2.4
1–3	5	1–3	(c)	3.6
4–6	5	4–6	(c)	4.6
7–11	8	7–10	(c)	6.2
12–14	7	11–14	(c)	8.9
15–17	7	15–17	(c)	9.9
≥18		≥18	300	6.2
Premenopausal	7		600	7.6
Postmenopausal	6		900	8.9
			1,200	10.2
Pregnancy	7			+1.3 (d)
Lactation	7			+2.4 (d)

Table 7.8 (continued)

Population reference intakes (PRIs) and adequate intakes (AIs)

Age (years)	F (mg/day)	I (µg/day)	Mn (µg/day)	Mo (µg/day)	Se (µg/day)	Zn (mg/day) LPI (mg/day)	Age (years)	Zn	Fe (mg/day)	Age (years)	Cu (mg/day)
7–11 mo (a)	0.4	70	0.02–0.5 (b)	10	15	(c)	7–11 mo (a)	**2.9**	**11**	7–11 mo (a)	0.4
1–3	0.6	90	0.5	15	15	(c)	1–3	**4.3**	**7**	1–2	0.7
4–6	0.9	90	1.0	20	20	(c)	4–6	**5.5**	**7**	3–9	1.0
7–10	1.4	90	1.5	30	30	(c)	7–11	**7.4**	**11**	10–17	1.1
11–14	2.3	120	2.0	45	45	(c)	12–14	**10.7**	**13**		
15–17	2.8	130	3.0	65	65	(c)	15–17	**11.9**	**13**		
≥18	2.9	150	3.0	65	65	300	≥18	**7.5**	**16 (d)**	≥18	1.3
						600	Prem.	**9.3**	**11**		
						900	Postm.	**11.0**			
						1,200		**12.7**			
Pregnancy	2.9	200	3.0	65	70			**+1.6 (e)**	**16 (d)**		1.5
Lactation	2.9	200	3.0	65	85			**+2.9 (d)**	**16 (d)**		1.5

d, day; LPI, level of phytate intake; mo, months. **PRIs are presented in bold type** and AIs in ordinary type.

(b): In view of the wide range of manganese intakes that appear to be adequate, a range is set for the AI of this age group.

(c): The fractional absorption of zinc considered in setting PRIs for children was based on data from mixed diets expected to contain variable quantities of phytate; therefore, no adjustment for phytate intake has been made.

(d): The PRI covers the requirement of approximately 95% of premenopausal women.

(e): in addition to the PRIs for non-pregnant, non-lactating women.

zinc caused by a mutation in ZIP4, the main intestinal zinc transporter (see above). This genetic disease, known as acrodermatitis enteropathica, is a lethal, autosomal, recessive trait whose major symptoms include bullous pustular dermatitis of the extremities and oral, anal, and genital areas around the orifices, paronychia, and alopecia, along with ophthalmic and neuropsychiatric signs, weight loss, growth retardation, male hypogonadism, and increased susceptibility to infections [104, 108].

Conditioned deficiency of zinc has been observed in patients with alcoholism, malabsorption syndrome, liver disease, chronic renal disease, sickle cell disease, and other chronic illnesses. On the other hand, zinc has been very successfully used as a therapeutic modality for the management of acrodermatitis enteropathica, acute diarrhea in children, Wilson's disease, common cold, blindness in patients with age-related dry type of macular degeneration, and is very effective in decreasing the incidence of infection in the elderly [98, 99, 104]. Zinc deficiency has also been associated with type II diabetes, although this association and the possibility to use zinc as a therapy require further research [109].

7.3.2.5 Zinc excess

Prolonged intakes of zinc in the range of approximately 100–300 mg/day, attainable with zinc-containing supplements and oral zinc medicines, result in copper deficiency, alterations in the activity of superoxide dismutase (an antioxidant enzyme), alterations in lipid metabolism, impairment of the pancreatic function, and decreased serum ferritin and hematocrit concentrations, especially in women, the latter due to a competitive interaction between zinc and iron. Some studies also suggest that zinc is a neurotoxin; actually, Zn^{2+} overload has been reported to be involved in the pathogenesis of Alzheimer's disease. The US Food and Nutrition Board has set a tolerable upper intake level (UL) at 40 mg of zinc/day for adults older than 19 years [110, 111].

7.3.3 Copper

Copper (chemical symbol Cu, atomic number 29, atomic mass 63.55 Da) plays a significant role in biology due to its capacity to exist in two oxidation states: Cu^+ or Cu^{2+}. Copper is rarely found as a free ion in biological systems, but is normally chelated to free amino acids and polypeptides, because free copper ions are highly toxic for cells due to their ability to catalyze reactions such as generation of reactive oxygen species. Cu^+ has an affinity for thiol and thioether groups (like those found in cysteine or methionine), whereas Cu^{2+} exhibits a preference for linking to oxygen or imidazole nitrogen groups (found in aspartic and glutamic acid, or histidine, respectively), consequently, both ions can participate in a wide spectrum of interactions with proteins [112].

7.3.3.1 Biological roles

Copper has been reported to be essential for connective and bone tissue formation and integrity, iron metabolism, the physiology of the central nervous system (including brain development), production of melanin, cardiac function, cholesterol metabolism, and immune function [113].

A total of 54 copper-binding proteins were identified in humans in 2016, which is a tiny part of the whole human proteome. About half of these proteins are enzymes that act as oxidases, mono- and dioxygenases, superoxide (O_2^-) decomposing enzymes, and nitrogen oxide (NOx) reductases. Many copper-binding proteins are associated with copper trafficking, being involved in storage, transport across different cell membranes, and transfer of copper ions (chaperones). The subcellular localization of many copper-binding proteins has also been established and 23 of them are reportedly extracellular [114, 115]. Examples of copper-dependent proteins include cytochrome c oxidase, critical to aerobic respiration and oxidative phosphorylation; dopamine monooxygenase and tyrosinase, involved in the synthesis of catecholamines and melanin, respectively; peptidyl glycine alpha hydroxylating monooxygenase (PAM), which modifies neurohypophyseal peptide hormones; lysyl oxidase, necessary for maturation of extracellular collagen and elastin; and some superoxide dismutases, which help to neutralize reactive oxygen species [116].

7.3.3.2 Sources, bioavailability, and homeostasis

Foods differ widely in their natural copper content, especially those of plant origin, influenced by factors such as season (concentration is higher in greener portions), soil quality, geography, water source and the use of fertilizers. Rich dietary sources of copper are liver, meats, shellfish (oysters), cocoa products, nuts (particularly cashew), lentils, and seeds [117, 118]. Drinking water can be another source of copper, although the mineral content depends on the natural concentration, the pH and the occurrence of copper in the plumbing system. Foods are estimated to account for 90% or more of copper intake in adults when the copper content in drinking water is low (<0.1 mg/L), otherwise (copper content > 1–2 mg/L), water may account for up to 50% of total intake. Foods can be fortified with copper lysine complex, cupric carbonate, cupric citrate, cupric gluconate and cupric sulfate, whereas food supplements can contain, in addition to those compounds, copper L-aspartate, copper bisglycinate and copper (II) oxide [118]. The primary site of dietary copper uptake is the upper small intestine where dietary copper, usually in the form of Cu^{2+} ions, is thought to be reduced to Cu^+ at the brush border prior to the transport across the membrane into the cell. Several chaperones and transporters seem to be involved in the intracellular trafficking of copper. The proposed mechanism is described in more detail in Chapter 13.

A limited number of studies on the copper absorption in humans are available and in many of them endogenous losses are not considered, therefore, results must be carefully interpreted. In adults, apparent copper absorption (in %) tends to be higher with omnivorous diets than with lacto-ovo-vegetarian diets, albeit this difference is compensated for by the higher copper content of the latter. On average, absorption of copper from a mixed diet is estimated to be around 50% [118]. Several factors, such as certain amino acids and proteins, iron, zinc, molybdenum, vitamin C, and carbohydrates, have been reported to exert adverse effects on the bioavailability of dietary copper. These effects are likely to be more pronounced in neonates [113]. High doses of iron supplements may increase the risk of copper depletion in humans, however, the interactions between the metabolisms of copper and iron are complex and are currently being investigated [119]. It is also well established that high levels of dietary zinc can affect copper absorption; actually, chronic high zinc intake can result in severe neurological diseases attributable to copper deficiency [118]. Ascorbic acid supplementation has been reported to induce copper deficiency in experimental animals, an effect that might also occur in humans [119].

The copper transferred from the enterocytes to the portal circulation is mainly transported to the liver and kidneys bound to the amino acid histidine, and the proteins albumin and transcuprein. After being taken up by liver cells, copper ions are delivered to different subcellular compartments in order to be incorporated to copper-dependent enzymes. The rest can be stored in metallothionein, incorporated into ceruloplasmin to be exported to the bloodstream, or excreted in the bile. Most of the copper in blood (from 80% to 95%) is bound to ceruloplasmin, a ferroxidase implicated in iron efflux from hepatocytes and other cells, that has six non-exchangeable copper ions, with a seventh ion that may be loosely bound [118]. Ceruloplasmin has been reported to have several functions other than the ferroxidase activity: neutralization of reactive oxygen species, oxidative inactivation of nitric oxide (NO), and it has recently been identified as a circulating source of copper for many tissues, particularly for heart, placenta, and the fetus [116]. The remainder of copper in plasma, mainly bound to histidine or albumin, has been traditionally considered the pool for transfer to tissues. In most cells, like in enterocytes, copper is taken up as Cu^+ ions through Ctr1, being subsequently bound to a series of chaperones, that are responsible for their incorporation to other proteins or to the delivery to other subcellular compartments [113, 118].

Copper homeostasis in mammalian cells is achieved through the regulation of genes coding for proteins involved in copper influx and detoxification. An example is Ctr1 protein that is removed from the cell surface in response to elevated copper levels and recycled back to the plasma membrane when the extracellular copper concentration is reduced. This regulation may explain the observation that copper depletion increases the absorption efficiency of copper, while chronic dietary consumption of copper shows the opposite effect. It has also been reported that high concentrations of cellular copper ions can enhance transcription of genes that code for metallothioneins, proteins that scavenge the excess of copper ions. When cellular concentrations of copper and other metal ions are low, the metal–free metallothioneins are highly vulnerable to proteolytic processes. Many studies further demonstrate that mammals have developed an exquisite ability to sense and retain copper in the organs; however, the underlying mechanisms are still not clearly understood [120].

The total amount of copper in an adult is estimated to be about 90–110 mg, with bones and skeletal muscles containing about 47% and 27% of this amount, respectively. Liver and brain contain about 8–11% of the total body copper and the organs with the highest concentrations are liver, brain, kidney, and heart [121].

Copper is primarily excreted through the bile, where it forms complexes with bile salts that prevent its reabsorption in the gut, thus passing through to the feces. Urinary excretion is very small, being usually ignored in most balance studies, whereas copper losses in sweat and skin may be significant, but the results obtained in different studies are variable and subjected to many confounding factors. Copper content in mature breast milk during the first 6 months of lactation has been reported to be variable, ranging about 100–1,000 µg/L in Western countries. An average value of 350 µg/L of copper in breast milk, which implies a loss of 280 µg/day, has been estimated by the EFSA [118].

The average intake of copper is 1.3 mg/day, half of which is absorbed on average, and the estimated amount of copper lost in feces is 1–1.5 mg/day. These fluxes are regulated to maintain homeostatic control of body copper levels by modulating absorption in the intestine and excretion in the liver (via bile), but the mechanisms involved have not been elucidated to date [113, 119, 121].

Two genetic diseases are associated with impairments in copper homeostasis:

- Menkes disease, an X-linked recessive disorder that affects gene *ATP7A*, that encodes a P-type Cu-ATPase involved in copper transport at the Golgi apparatus in most cells. This disease results in defective copper elimination from cells and low absorption of dietary copper. The symptoms appear by two months of age and include hypothermia, neuronal degeneration, mental retardation, abnormalities in hair, bone fragility, and aortic aneurysms. These clinical manifestations are attributable to a malfunction of copper-requiring enzymes. The disease is usually fatal by the age of 3 years, although there are milder forms with longer life expectancy [113, 122, 123].

- Wilson disease (WD), an autosomal recessive disease that affects gene *ATP7B*, that encodes another isoform of the P-type Cu-ATPase. Unlike *ATP7A*, that is expressed in almost all cell types and tissues, *ATP7B* is predominantly expressed in the adult liver. In WD both the production of ceruloplasmin and normal biliary excretion of copper are impaired. This results in toxic accumulations of copper in liver, brain, and other tissues and organs that may lead to neurological damage and cirrhosis if the condition is left untreated. Acute hepatitis, hemolytic crisis, and hepatic failure may also occur. In any case, over 600 pathogenic mutations have been detected in the *ATP7B* gene and clinical presentations vary widely. Treatments are based on removal of copper stores with chelation therapy and inhibition of intestinal copper absorption with zinc. The prognosis is good, especially if treatment is initiated at early stages of the disease [113, 124]. New

molecular details of WD progression and metabolic signatures of WD phenotypes
have begun to emerge from studies with WD patients and animal models. These
studies have revealed the contributions of non-parenchymal liver cells and extrahepatic tissues to the liver phenotype, and pointed to dysregulation of nuclear receptors (NR), epigenetic modifications, and mitochondria dysfunction as important
hallmarks of WD pathogenesis [125].

Several neurological disorders such as Alzheimer's, Parkinson's, and Huntington's diseases are characterized by disturbed copper homeostasis, although the role of copper
in these pathologies is not fully understood. The most common form of amyotrophic
lateral sclerosis (ALS) is due to mutations in the Cu,Zn superoxide dismutase (SOD1)
enzyme. There is plenty of evidence suggesting that dysregulation of copper homeostasis in the central nervous system is a crucial underlying event in ALS pathophysiology [24, 126].

7.3.3.3 Copper status, Dietary Reference Values (DRVs)

The most commonly used method to assess copper status in humans has been the
quantification of copper levels and the activity of several copper-binding enzymes in
blood, especially ceruloplasmin. This is based on the fact that significant reductions in
plasma copper and ceruloplasmin activity have been observed in severely copper de-
ficient-humans. However, the utility of this method is limited by the finding that both
parameters can also be increased by several physiologic alterations, such as the acute
phase response to infection and inflammation, pregnancy and other hormonal pertur-
bations, and some carcinogenic phenotypes [113].

The EFSA Panel on Dietetic Products, Nutrition and Allergies has proposed Adequate Intakes (AIs) for all groups of population (Table 7.8) [91, 118]. In North America,
it has been estimated that 40% of the population is in a state of functional copper deficit, probably caused by insufficient dietary copper intake. An optimal dietary intake
of 2.6 mg/day for adults has been proposed in order to correct this situation [117].

7.3.3.4 Copper deficiency

Usual pathophysiologic features of copper deficiency include anemia, leukopenia (a
decrease in the number of leukocytes), neutropenia (low concentration of neutrophils), and, in infants and children, osteoporosis, scoliosis and scorbutic-like changes.
Moderate copper deficiency as a result of long-term low copper intakes may result in
additional manifestations including arthritis, arterial disease, depigmentation, myocardial disease, and neurologic abnormalities. Additional effects of marginal copper
deficiency might include cardiac arrhythmias, increased serum cholesterol levels, and
glucose intolerance [113, 118].

7.3.3.5 Copper excess

Copper toxicity is uncommon in humans due to the existence of a precise homeostatic control of copper in response to the capacity of the free metal to generate reactive oxygen species. The Scientific Committee for Food (SCF) of the European Union has set a No Observed Adverse Effect Level (NOAEL) of 10 mg of copper per day based on some scientific studies and established a Tolerable Upper Intake Level (UL) of 5 mg/day for adults (although not for pregnant and lactating women, due to the absence of adequate data). For children, the UL of adults was extrapolated based on body weight [118, 127].

Excessive copper accumulation have been observed in very special cases of cirrhosis in India and the Austrian Tyrol, however, although copper may have some role to play in the development of these cases, a genetic predisposition is likely to be involved [118].

A novel mechanism of copper toxicity, known as cuproptosis, has been recently identified. Cuproptosis occurs by direct binding of copper to lipoylated components of the tricarboxylic acid (TCA) cycle, which leads to protein aggregation, proteotoxic stress, and ultimately cell death. This finding may explain the need for ancient copper homeostatic mechanisms [128].

7.3.4 Iodine

Iodine (chemical symbol I, atomic number 53, atomic mass 126.9 Da) can exist in the oxidation states −1 (I^- anion or iodide) to +7 (IO_4^- anion or periodate). It is distributed throughout oceans, land, and air, although the majority occurs in seawater. Inorganic iodine in seawater can be accumulated by marine organisms, thereby entering the food chain; however, some of these organisms, such as macroalgae (e.g., kelp), microalgae and marine bacteria can methylate iodine, thus producing an organic volatile compound: methyl iodide (CH_3I). Other volatile iodine-containing organic compounds are diiodo-methane (CH_2I_2), chloroiodomethane (CH_2ClI), and ethyl iodide (CH_3CH_2I). Iodine evaporation has also been observed in land environments (rice fields and peatlands). In the atmosphere, evaporated organic iodine is transformed by photolysis to inorganic forms like molecular iodine (I_2), hypoiodous acid (HIO, oxidation state +1), and iodic acid (HIO_3, oxidation state +5), that can precipitate to grounds and oceans with rain. Iodides and iodates (IO_3^- anion) in rocks and soils may also be liberated by weathering and erosion and leached by rainwater into surface water and the sea. On land, small amounts of iodine are taken up by plants that are subsequently ingested by herbivores, thereby entering the terrestrial food chain. Altogether, these complex interactions comprise the so-called biogeochemical cycle of iodine [129].

7.3.4.1 Biological roles

Iodine in the form of iodide anion (I^-) is an essential micronutrient for life, in general, and mammals in particular, that require iodine as a mandatory structural and functional element of thyroid hormones triiodothyronine (T_3) and thyroxine (T_4). I^- is oxidized by specialized peroxidase enzymes to generate these hormones, that have an essential role in energy-yielding metabolism and on the expression of genes that impact many physiological functions, including embryogenesis and growth, and the development of neurological and cognitive functions [130, 131]. Iodine also exerts autoregulatory effects on the thyroid gland to maintain an adequate thyroid hormone production [132] (see below).

Due to its reducing properties, I^- is also considered an important scavenger of reactive oxygen species *in vivo* and recent evidence indicates that it may be an antibacterial, antiviral and antifungal agent. Radioactive iodine therapy with ^{131}I has been used for the treatment of thyroid cancer for decades, whereas new studies indicate that I^- may be antineoplastic, anti-proliferative, and cytotoxic in human cancer. On the other hand, radioiodide isotopes resulting from nuclear accidents can cause thyroid dysfunction and thyroid cancer [130].

In healthy humans the thyroid gland produces predominantly the prohormone L-thyroxine (T_4) together with a small amount of the bioactive hormone 3,3',5-triiodo-L-thyronine (T_3). Most T_3 is produced by enzymatic outer ring deiodination of T_4 in peripheral tissues. Inner ring deiodination of thyroid hormones may also occur but this is considered an inactivating pathway. Three enzymes that catalyze these deiodinations have been identified, called type 1 (D1), type 2 (D2) and type 3 (D3) iodothyronine deiodinases. All of them are selenium-dependent; therefore, production and metabolism of thyroid hormones depend on two trace elements, namely, iodine and selenium [133]. Other metabolites of L-thyroxine have been identified such as 3,5-diiodo-L-thyronine, 3,3',5'-triiodo-L-thyronine (reverse-T_3), thyronamines, thyroacetic acids, as well as various sulfated or glucuronidated metabolites. All T_4 derivatives are bound in the blood to a series of distributor proteins that deliver them to the different tissues and organs. In their target cells, thyroid hormones have two distinct signaling pathways: the non-genomic transcription-independent pathway that occurs outside the nucleus, and the genomic transcription-dependent pathway that occurs inside the nucleus. The latter pathway promotes transcription of thyroid hormone-responsive genes, whereas nongenomic effects have been considered to be mainly related to homeostasis (actions on plasma membrane ion transporters or maintenance of the cytoskeleton). Recent evidence supports the existence of crosstalk between non-genomic and genomic effects of the hormone [134, 135].

7.3.4.2 Sources, bioavailability, and homeostasis

Iodine occurs in food and water mainly as iodide in highly variable concentrations. The richest sources are marine products (fish, shellfish, molluscs, seaweed), eggs and milk, as well as their derivatives and iodized salt. Iodine content of milk and eggs is influenced by feeding and hygienic practices, thus, the content of dairy products is often relatively high because of the supplements given to dairy cows. Iodine can also enter the food chain, in amounts difficult to control, *via* sanitizing solutions and iodophores. Sodium iodide, sodium iodate, potassium iodide, and potassium iodate may be added to foods and food supplements. Iodine binds to double bonds in fatty acids, allowing the preparation of iodized oils that are used for supplementation in certain countries. In Europe, iodine fortification of salt has been implemented in four countries, being mandatory in thirteen, voluntary in sixteen and not regulated in the remaining countries. The amount of iodine added mainly ranges between 15 and 30 mg/kg of salt. In countries like Japan and Korea, iodine intake is usually much higher than the internationally recommended levels, due to the consumption of kelp products. Multivitamins may contain around 150 µg of iodine per tablet [131, 132].

Dietary iodine occurs as iodate and other forms that have to be digested and/or reduced to I^- in order to be absorbed, along with iodine recycled in the saliva and stomach, also as I^-, that results from deiodination in peripheral tissues. The mechanism of iodide absorption in the small intestine is described in Chapter 13.

The absorbed iodide enters the circulating pool of inorganic iodide and is concentrated in the thyroid gland or excreted by the kidney. Thyroid uptake of iodine depends both on intake (low after excessive iodine intake and high in iodine deficiency), and also on the functional status of the thyroid (low in non-functional thyroid, high in stimulated thyroid). In pregnant women, there is a transport of iodine to the fetus through the placenta, whereas in breast-feeding women, a considerable fraction of the circulating iodide is taken up by the mammary glands and excreted in breast milk. The iodine taken up by the thyroid is used for the synthesis and subsequent secretion of thyroid hormones. Thyroid activity is mainly regulated by the thyroid-stimulating hormone (thyrotropin or TSH) and iodide by autoregulation. TSH is secreted by the pituitary, being in turn regulated by T_4 and T_3 and hypothalamic signals [132].

Excess iodine not taken up by the thyroid is excreted in the urine (more than 90% of dietary iodine), with partial reabsorption occurring in the renal tubules. In healthy subjects renal iodide clearance is 30–36 ml/min with an average concentration of 100 µg/L, and is significantly decreased either in impaired renal function or in myxedema. An increase in fluid volume intake can lead to additional iodine losses. Iodine losses may also occur via feces (between 10 and 30 µg/day) and sweat (35–40 µg/L). Concentration of iodine in breast milk is 20 to 50 times higher than in plasma, that is, up to 150–180 µg/L in iodine sufficiency. More than 80% of iodine in breast milk is in the form of inorganic iodide (I^-), whereas T_4 is <2 µg/L and T_3 is <0.05 µg/L [131].

7.3.4.3 Iodine status, Dietary Reference Values (DRVs)
Urinary iodine (UI) excretion is an excellent indicator of recent iodine intake, because it represents more than 90% of dietary iodine intake. In adult populations, an average UI concentration of 100 µg/L corresponds to an iodine intake of 150 µg/day. Bio-

markers of iodine status include urinary iodine concentration, serum concentration of thyroid hormones (such as TSH, T_3, and T4), and thyroid volume [131].

The EFSA has set Adequate Intakes (AI) of iodine based on data on the relationship between iodine intakes and UI excretion in population groups without signs of thyroid dysfunction evidenced by a minimal prevalence of goitre (Table 7.8). Similar values are proposed by other institutions [91, 131, 132].

7.3.4.4 Iodine deficiency

Iodine deficiency disorders are the result of insufficient intakes iodine, which lead to insufficient thyroid function. On a population level, iodine intake can be assessed by measuring urinary iodine (UI) concentration, and the following criteria based on UI concentration in school-aged children have been suggested [136]:

– Median UI < 20 µg/L, insufficient iodine intake and severe iodine deficiency
– Median UI 20–49 µg/L, insufficient intake and moderate iodine deficiency
– Median UI 50–99 µg/L, insufficient intake and mild iodine deficiency
– Median UI 100–199 µg/L, adequate iodine intake

Depending on the time and severity of thyroid hormone deficiency, and the stage of development different scenarios may occur [131, 132]:

– Maternal iodine deficiency during pregnancy results in fetal iodine deficiency. It is accompanied by higher rates of stillbirths, abortions, and congenital abnormalities. It constitutes a threat to early brain development with consequent physical and mental retardation and lower cognitive and motor performance in later life.
– Severe iodine deficiency in pregnancy and fetal life and during the first years results in cretinism. The most common type, neurological cretinism, is characterized by cognitive impairment, deaf mutism, and spastic diplegia (a form of cerebral palsy). The myxedematous type, less common, shows symptoms such as apathy, hypothyroidism, puffy features, growth retardation, delayed bone maturation, retarded sexual maturation, and dwarfism. Cretinism can be endemic in areas with severe iodine deficiency.
– Hypothyroidism (myxedema) may also be observed with severe iodine deficiency, and results from hormone deficiency. It is associated with reduced metabolic rate, cold intolerance, weight gain, puffy face, edema, hoarse voice, and mental sluggishness.
– Chronic iodine deficiency may lead to compensatory thyroid hypertrophy/hyperplasia with enlarged thyroid gland (goitre). Goitre may subsequently cause hyperthyroidism and also increases the risk of thyroid cancer. A large goitre may cause obstruction of the trachea and the esophagus.

Iodine deficiency is estimated to affect around 2.2 billion people (35–45% of the world's population), being the most common cause of goiter, although not all goitres are the result of an iodine deficiency. The incidence of goiter is based on the degree of iodine deficiency, thus, with mild iodine deficiency, the incidence of goiter is 5% to 20%, with moderate deficiency, the incidence is 20% to 30%, and with severe iodine deficiency, the incidence is greater than 30%. Although global iodine nutrition status has improved in the last decades, areas of deficiency are common, and not necessarily in developing countries: in 2003, 16% of the European population had goitre and, in 2011, 44% was estimated to have insufficient iodine, being a public health problem in 14 European countries [131, 137]. Changes in agriculture and industry as well as in consumer practices in some countries may be contributing to a decline of iodine content in the food supply, these include [137, 138]:

- Reduced use of iodate dough conditioners, that may affect iodine content in store-bought breads and baked goods,
- Reduced use of iodine supplemented feed for livestock, that may be contributing to lower iodine content of dairy milk, meat, and eggs
- Reduced use of iodophors as sanitizing agents in milk processing
- Veganism and some forms of vegetarianism, that avoid dairy intake and promote consumption of plant-based alternative milk sources. Soy-based dairy alternatives interfere with iodine absorption, and a number of milk alternative products have been shown to contain much less iodine than cow's milk. Data from several industrialized countries also show lower iodine concentration in organic milk than in conventional milk.
- Certain foods are sources of natural goitrogens which interfere with thyroid metabolism, thus, cruciferous vegetables (i.e., broccoli, cabbage, kale, cauliflower) contain glucosinolates that produce thiocyanate and isothiocyanate, known to compete with thyroid iodine uptake. Thiocyanate can also be a product of the metabolism of cyanogenic glucosides, a group of goitrogens found in cassava, sweet potatoes, maize, lima beans, bamboo shoots, linseed, and sorghum. Cassava is a staple in many developing countries and has been linked to the etiology of endemic goiter in Africa and Malaysia. Goitrogenic flavanoids in soy and millet may interfere with enzymatic activity involved in iodine metabolism.
- The use of iodized salt is an effective public health measure; however, up to 20% of iodine in salt may be lost during processing, and another 20% lost during food preparation. Furthermore, only 53% of table salt sold in countries like the USA is iodized.
- Lower dietary salt use at the table and in cooking due to public health messages linking high sodium intake to hypertension are a contributing factor to reduced iodized salt intake, in addition to use of non-iodized salt in processed and restaurant foods.

7.3.4.5 Iodine excess
Chronic excessive iodine supply can also lead to goiter and may accelerate the development of subclinical thyroid disorders to hypothyroidism or hyperthyroidism, increase the incidence of autoimmune thyroiditis, and increase the risk of thyroid

cancer. The Scientific Committee for Food (SCF) of the European Union has set a value of 600 µg/day as a Tolerable Upper Intake Level (UL) for iodine in adults including pregnant and lactating women. This is half of the value of 1,100 µg/day adopted by the Institute of Medicine (IOM) of the United States [131].

7.3.5 Selenium

Selenium (chemical symbol Se, atomic number 34, atomic mass 78.96 Da) has similar chemical properties than sulfur, its group superior homologue in the periodic table. Selenium usually forms inorganic and organic compounds where it occurs with oxidation states −2, +4, and +6. Selenites and selenates (SeO_3^{2-} and SeO_4^{2-}, respectively) are the most common inorganic compounds, being found in water solution. Selenium also forms stable bonds with carbon in organic compounds, thus producing selenides (R–Se–R), such as dimethylselenide, and selenium amino acids (L-selenomethionine and L-selenocysteine), that have structures similar to L-methionine and L-cysteine, except that selenium replaces sulfur (Chapter 5).

7.3.5.1 Biological roles

Selenium is important for the proper functioning of the central nervous system, the male reproductive biology, the endocrine system, muscle function, the cardiovascular system, and the immune system. Many effects of selenium are attributable to the insertion of this element into a family of proteins called selenoproteins; however, several small molecular weight selenium-containing metabolites, like hydrogen selenide and methylseleninic acid, among others, have been identified. These compounds have been suggested to affect DNA repair and epigenetics and to exert chemopreventive effects [139].

Twenty five genes coding for selenoproteins exhibiting a wide variety of tissue distribution and functions have been identified in humans. These proteins are characterized by the occurrence of selenocystein (Sec). Sec constitutes a specific amino acid residue that requires a specialized machinery for its insertion into the polypeptide. By contrast, selenomethionine seems to unspecifically replace methionine residues in proteins without exerting significant effects on their properties [140].

Selenoproteins are involved in many biological functions such as redox reactions, inactivation of free radicals (glutathione peroxidases), regeneration of reduced thioredoxin, thyroid function (iodothyronine deiodinases, that convert inactive thyroxine to active thyroid hormone, triiodothyronine, are selenium-dependent -Section 7.3.4-), protein quality control, redox regulation, phospholipid biosynthesis, calcium flux in immune cells, ER associated degradation, regulation of body weight and energy metabolism, among others (a complete list of proteins with subcellular localization and functions in vivo can be found in [139]).

7.3.5.2 Sources, bioavailability, and homeostasis

The richest food source of selenium are Brazil nuts (*Bertholletia excelsa*). Mean concentrations between 5 and 72 µg/g dry weight (up to 400 µg/nut) have been reported in samples collected in Brazil and the Amazon Basin, indicating large variability depending on their region of production. Other rich foods of selenium include offals and sea foods (0.4 to 1.5 µg/g fresh weight), followed by muscle meats (0.1 to 0.4), cereals and grains (<0.1 to >0.8), dairy products (<0.1 to 0.3), and fruits and vegetables (<0.1). In European populations, the main food groups contributing to selenium intake were found to be milk and dairy products, meat and meat products, grains and grain-based products and fish and fish products [141]. Selenium content of cereals and grains varies depending on how much soil selenium is available for uptake. Foods from animal sources vary somewhat in selenium content, but the degree of variation is less than in plants because of the homeostatic control of selenium metabolism in animals [142]. Wheat, other grains, and soya beans contain predominantly selenomethionine with smaller amounts of selenocysteine and selenate. Other organic selenium-containing compounds in foods from plant sources, such as garlic and onion, are Se-methyl-selenocysteine and γ-glutamyl-Se-methyl-selenocysteine. Data on the forms of selenium in animal foods are limited, and the content according to the diet of the animals. When inorganic selenium is given to animals, selenocysteine is the main selenocompound formed. When animals consume selenium from plant sources, they incorporate selenomethionine in their proteins in place of methionine. Selenotrisulfide, glutathione selenopersulfide and selenides have also been reported to occur in tissues. Selenate and selenite have been detected in fish and selenoneine (2-selenyl-Nα,Nα,Nα-trimethyl-L-histidine) has been identified as the major selenium compound in swordfish, tuna, mackerel and sardine. Water is not considered to significantly contribute to selenium intake. Selenium-enriched (selenized) food items may be produced by enrichment of fertilizers (selenized garlic, onion, broccoli, wheat), feed (selenized milk or eggs) or growth medium (selenized yeast). Food and food supplements can be enriched in selenium by addition of sodium selenate, sodium hydrogen selenite, sodium selenite, L-selenomethionine and selenium-enriched yeast [140]. In general terms, selenium absorption is usually in the range of 50% to 100% (the EFSA accepts an absorption efficiency of 70% from usual diets) and is not affected by selenium nutritional status [140, 142]. The molecular mechanisms of selenium absorption at the enterocytes are described in Chapter 13.

A number of factors are generally accepted to affect overall selenium bioavailability [143]:
– Selenium concentration and types of selenium-containing species in a foodstuff. Organic forms, especially selenomethionine, are more capable of increasing blood selenium levels than inorganic selenates and selenites; however, organic and inorganic forms of selenium were equally effective in raising another biomarker: blood glutathione peroxidase activity.

- Other components of foodstuff, either major (carbohydrate, fat, protein, and fiber) or minor (oligoelements and toxic metals) components, and the synergies and antagonisms that may be established among them.
- Food processing treatments like cooking, boiling, ripening, ageing or fermentation

In any case, bioavailability studies of selenium and selenocompounds in foods are limited by the lack of a generally accepted in vivo and in vitro protocols, as results significantly depend on experimental conditions, such as the selection of healthy status of examined people (in in vivo humans approaches) and the selection of animal model (in in vivo animals approaches), among others [143].

Following absorption, the different selenium compounds are mostly transported to the liver, the principal site of selenium metabolism. The metabolites are then widely distributed to other organs and tissues such as pancreas, nervous system, skin and hair, bone, both skeletal and cardiac muscle, lungs and kidneys. These organs are variably affected by deficient and excessive intakes of selenium. According to animal and in vitro studies, in the liver, all selenium compounds, except selenomethionine, Se-methyl-selenocysteine and γ-glutamyl-Se-methyl-selenocysteine, are metabolized to hydrogen selenide (H_2Se), that can be incorporated into selenoproteins, used in the formation of selenosugars or methylated for excretion [141].

In blood plasma, the selenoprotein P and the extracellular glutathione peroxidase account for approximately 30–60% and 10–30% of total selenium, respectively. The rest of plasma selenium consists of selenomethionine in albumin and other proteins and a minor fraction in small molecular compounds such as selenosugars. The relative distribution of selenium among these different compounds is affected by the amount and chemical nature of selenium in the diet. Selenium also occurs in platelet and red blood cells in glutathione peroxidase GPx1 [140].

The liver regulates whole-body selenium by producing excretory metabolites and by distributing it to other tissues via secretion of Sepp1 into the plasma. Extrahepatic tissues acquire selenium primarily by endocytosis of Sepp1. Within cells, synthesis of individual selenoproteins is regulated by the availability of selenium, so that when selenium is limiting it is supplied for synthesis of certain selenoproteins at the expense of others, thus creating a hierarchy of selenoproteins. The essential molecular components of selenoprotein synthesis have been identified although its regulation is incompletely understood. Selenium is also recycled inside the cell to support selenoprotein synthesis when supply is limited. Under conditions of excess, selenium is metabolized to small-molecule excretory forms to maintain homeostasis [144].

Selenium elimination primarily occurs by urinary excretion, the main human urinary metabolites being selenosugars 1 and 3, trimethylselenonium ion (TMSe) and, depending on the selenium compound consumed, selenate and selenium-methylselenoneine, the latter originating most likely from methylation of selenoneine in fish. Part of excreted selenium is in the form of yet unidentified metabolites [141].

Selenium in the feces consists of unabsorbed selenium and some endogenous excretion from the turnover of intestinal mucosal cells, which contain selenium in the form of selenoproteins; however, most results have been obtained in mice and require further elucidation in humans. In breast milk, selenium concentration reflects maternal intake and increases mainly in response to organic selenium intake from supplements. Selenium appears as a component of specific selenoproteins, mainly as glutathione peroxidases (15–30% of total milk selenium) and seleno-amino acids in milk proteins, while inorganic species are undetectable. The EFSA considers an average selenium concentration of 15 µg/L in mature breast milk in the European Union; this means that the amount of selenium secreted in breast milk during the first six months of lactation is 12 µg/day, assuming a milk volume of 0.8 l/day [140].

7.3.5.3 Selenium status, Dietary Reference Values (DRVs)

Markers of selenium intake or status include concentrations of selenium in whole blood, plasma and blood cells (red blood cells, platelets), hair, nails or urine, as well as concentration of selenoproteins like Sepp1 in plasma or activity of selenoenzymes, mainly glutathione peroxidase, in plasma, platelets, or whole blood. Each one of these biomarkers has its own limitations, however, the EFSA considers that plasma Sepp1 concentration is the most informative biomarker of selenium function, therefore, it has been used by this organism to establish Adequate Intakes (AI) for selenium (Table 7.8) [91, 140, 141].

7.3.5.4 Selenium deficiency

Deficiency of selenium usually results from insufficient supply in the diet and has been confirmed in humans inhabiting regions with low contents of this element in soils, like China and Central and Eastern Siberia. At molecular level, selenium deficiency leads to a decreased expression of selenoproteins, thereby affecting multiple biological processes. Symptoms of severe deficiency are disorders related to heart muscle and joints, however, even moderate deficiencies may also have a negative impact on human health, for example, increasing the risk of infertility in men, prostate cancer, nephropathy, or neurological diseases. Selenium deficiency is also involved in Keshan disease and Kashin-Beck disease. Keshan disease is an endemic cardiomyopathy occurring mainly in children and young women in regions of China with particularly low selenium intake (around 15 µg/day). Infection with an enterovirus seems to be a cofactor in this disease, along with selenium deficiency. Kashin-Beck disease is a chronic degenerative osteochondropathy that manifests itself by rheumatoid arthritis, shortened fingers and toes, or growth disorders of the organism. It is endemic in some areas in selenium-deficient areas of China, but also in Mongolia, Siberia and North Korea. The combination of selenium and iodine deficiency constitutes a factor favoring the development of Kashin-Beck disease and other possible risk factors are the presence of mycotoxins in food, and humic and fulvic acids in drinking water [140, 145].

The immune system is also affected by dietary selenium levels and selenoprotein expression, thus, selenium deficiency can lead to impaired innate and adaptive immune responses, thereby increasing susceptibility to infections and cancers. However, the benefits of selenium supplementation to boost immunity against pathogens, vaccinations, or cancers have not provided entirely clear results [139].

7.3.5.5 Selenium excess

Chronic excess of body selenium can give rise to selenosis that occurs in population groups exposed to levels of dietary selenium above 1,000 µg/day. Symptoms include headache, loss of hair, deformation and loss of nails, skin rash, malodorous breath and skin, excessive tooth decay, and discoloration, as well as numbness, paralysis, and hemiplegia. The molecular mechanisms of selenium toxicity remain unclear. Levels of dietary exposure at which selenium becomes toxic are difficult to establish, because toxicity is affected by the selenium compounds in the food supply and probably the combination of other components of the diet and individual genotypes.

The EFSA Panel on Nutrition, Novel Foods and Food Allergens has recently established the following Tolerable Upper Intake Levels (ULs) for selenium for the European Union (females and males): 4–6 months: 45 µg/day; 7–11 months: 55 µg/day; 1–3 years: 70 µg/day, 4–6 years: 95 µg/day; 7–10 years: 130 µg/day; 11–14 years: 180 µg/day; 15–17 years: 230 µg/day; adults: 255 µg/day; pregnant women 255 µg/day; lactating women 255 µg/day [141].

7.3.6 Manganese

Manganese (chemical symbol Mn, atomic number 25, atomic mass 54.94 Da) is the 12th most abundant element on earth, where is not found as a free element, but usually exists as oxides, carbonates and silicates. Manganese can be found in oxidation states ranging from −3 to +7, the most common in living organisms being Mn^{2+} and Mn^{3+}. The former, the only form absorbed by humans, is the most stable form, while Mn^{3+} is a powerful oxidant, which usually disproportionates to give Mn^{2+} and Mn^{4+}, or forms complexes with proteins [146]. The human body contains approximately 10 to 20 mg of manganese, with 25% to 40% present in bone and 5 to 8 mg turned over on a daily basis [147].

7.3.6.1 Biological roles

Manganese is required for proper immune function, regulation of blood sugar and cellular energy, reproduction, digestion, bone growth, blood coagulation and hemostasis, and defense against reactive oxygen species (ROS). These effects are due to the incorporation of the metal into metalloenzymes, thus, manganese is an essential cofactor for superoxide dismutase (SOD), arginase, and pyruvate carboxylase. SOD protects the cell against antioxidant processes, including those associated with radiation, chemicals, and ultraviolet light, whereas arginase has a significant importance in nitrogen metabolism through the urea cycle (Chapter 5) and pyruvate carboxylase is involved in gluconeogenesis (Chapter 3). In many enzymes, including various decarboxylases, glutamine synthetase, hydrolases, kinases, and transferases, manganese acts as an activator [147–149]. Activation probably involves binding of manganese to the polypeptide, thereby inducing a conformational change, or to a substrate, like ATP. In these cases, except for glycosyltransferases, manganese is not essential because the enzymes can be activated by other metals [149].

7.3.6.2 Sources, bioavailability, and homeostasis

Natural erosion releases tons of manganese into the environment which is subsequently available for absorption by microorganisms, plants, and animals. Manganese is also widely used in various types of industry and in some pesticides and fungicides used in agriculture. As manganese is absorbed by ingestion, inhalation, dermal permeation, and even administered in intravenous injection, sufficient levels of manganese are easily attained and in fact adverse effects caused by this element have been described (see below). Dietary sources with high manganese levels include legumes, rice, nuts, and whole grains, although it is also found in seafood, seeds, chocolate, tea, leafy green vegetables, spices, and some fruits such as pineapple and acai [146, 148]. Water can also contain manganese and the European Union legislation has set a parametric value of 50 µg/L for manganese in drinking water. Several manganese salts are permitted for use in foods and food supplements, these include manganese carbonate, chloride, citrate, gluconate, glycerophosphate, and sulfate (all as Mn^{2+}). Manganese ascorbate, L-aspartate, bisglycinate, and pidolate (also as Mn^{2+}) are permitted only in food supplements. Manganese content in infant and follow-on formulae is regulated by Directive 2006/141/EC 8 in the EU [147].

Ingested manganese is absorbed in the intestine as Mn^{2+} by a diffusion mechanism and a transport mechanism, the latter probably via the divalent metal transporter 1 (DMT1). DMT1 translocates other divalent metals, including Fe^{2+} and Cu^{2+}, and is reportedly inhibited by Ca^{2+} [150], therefore, the presence of other metals can compete with Mn^{2+} absorption [146]. Iron status has also been suggested to affect manganese absorption [147].

The rate of manganese absorption in the gastrointestinal tract varies depending on age, gender, and type of food, thus, it has been reported that retention was 15.4% in premature infants, 8.0% in term newborns and 1.0% to 3.0% in adults. Absorption is better from human milk than from bovine milk or soy-based formula, which may be related to the lower concentration and increased binding to lactoferrin of manganese in human milk, the increased calcium content of bovine milk, and the relatively large amounts of phytic acid for soy-based formula. The effect on manganese absorption of other dietary components, such as ascorbic acid or polyphenols, is a matter of controversy [148, 149]. The amount of manganese absorbed is also influenced by its concentration in the diet, with low intakes resulting in increased percentages of absorption, however, the mechanisms that regulate this homeostatic control are unknown [147].

Manganese is rapidly distributed from plasma to the liver (30% of the total), kidney (5%), pancreas (5%), colon (1%), urinary system (0.2%), bone (0.5%), brain (0.1%), erythrocytes (0.02%), the remaining being delivered to soft tissues. The liver, pancreas, bone, kidney, and brain maintain the highest manganese concentrations in the body. Manganese concentration in the blood of healthy adults has been reported to range from 4 to 15 µg/L and is mainly associated with erythrocytes (66%), as well as leucocytes and platelets (30%). Plasma contains about 4%, predominantly as Mn^{2+}, that forms complexes with molecules such as albumin (84%), water, as hexahydrated ion (6%), bicarbonate (6%), citrate (2%), and transferrin (Tf) (1%). Tf can also bind Mn^{3+}, thus forming a complex that can be taken up by neuronal cells [147, 148]. Several mechanisms have been proposed for the uptake and intracellular distribution of manganese in different types of cells, although the available information is contradictory [146, 149].

Manganese is excreted into the small intestine via bile, the main route of elimination being via feces, while around 1% of dietary intake is excreted in the urine (1–8 µg/L). In breast milk, there is no correlation between maternal dietary intake and manganese concentrations, which vary from 0.8 to 30 µg/L. The concentration is substantially higher in colostrum than in mature milk, whose manganese concentration appears to be relatively constant in the first half-year of breastfeeding, but may decrease afterward [147].

7.3.6.3 Manganese status, Dietary Reference Values (DRVs)
The EFSA has concluded that the assessment of manganese intake or status using biological markers is difficult due to the rapid excretion of manganese into bile, the homeostatic control, and lack of sensitivity of biomarkers over the normal range of intakes; therefore, the available data are insufficient to derive ARs and PRIs for manganese. A proposal with Adequate Intakes (AIs) is shown in Table 7.8 for all population groups [91, 147].

7.3.6.4 Manganese deficiency
Although experiments done with animals show a variety of symptoms associated with manganese deficiency, evidence is poor in humans, where a specific deficiency syndrome for manganese has not been described [147, 149].

7.3.6.5 Manganese excess
Accumulation of excessive manganese in human body can result in severe toxicity; however, reports of these adverse effects are associated primarily with inhalation in occupational settings or consumption of contaminated water. The primary target

organ of manganese toxicity is the brain, where its excessive accumulation causes a variety of psychiatric and motor disturbances known as manganism. First symptoms include reduced response speed, irritability, mood changes, and compulsive behaviors; later on, symptoms get more prominent resembling to those of idiopathic Parkinson's disease. Impairments in liver and cardiovascular function are also found in most patients [89]. Manganese excess has also been linked to pathologies such as Alzheimer's and Huntington's diseases, amyotrophic lateral sclerosis, as well as mitochondrial dysfunction [24].

7.3.7 Molybdenum

Molybdenum (chemical symbol Mo, atomic number 42, atomic mass 95.94 Da) is a metallic element of the second transition series with similar chemical properties to chromium. The oxidation state of molybdenum can range from -II to VI, a property used by some enzymes to catalyze diverse redox reactions. Molybdenum is bioavailable as molybdate (MoO_4^{2-}) and incorporated into metal cofactors after entering the cells by complex biosynthetic machineries (see below). These cofactors are subsequently linked to different polypeptides, thus yielding active enzymes. Within these enzymes, molybdenum shuttles between three oxidation states (+4, +5, and +6), thereby catalyzing two-electron redox reactions controlled by the cofactor itself and the enzyme environment. Molybdenum is the heaviest catalytic metal in humans and is also present in all plant and animal tissues, being considered an essential micronutrient for most life forms [151, 152].

7.3.7.1 Biological roles

In most living organisms, molybdenum is bound to a pterin, thus forming a special cofactor known as Molybdenum-cofactor (Moco), which is the active compound at the catalytic site of molybdo-enzymes. The only exception to this rule known so far is bacterial nitrogenase, where molybdenum participates, along with iron, in the so-called Fe-Mo-cofactor. In humans, four different enzymes are known to require Moco for activity [151, 153].

- Sulfite oxidase (SO), that catalyzes the oxidation of sulfite to sulfate, the terminal step in the degradation of the amino acid cysteine. This enzyme is localized in the mitochondrial intermembrane space.
- Xanthine oxidoreductase (XOR), that catalyzes the last two steps in purine catabolism converting hypoxanthine to xanthine and xanthine to uric acid (Chapter 5, Section 5.12.3).
- Aldehyde oxidase (AO), an enzyme located in the cytosol of cells that oxidizes aldehydes to carboxylic acids, as well as a range of N-containing heterocyclic aromatic rings that are often present in drug compounds [154].

– The mitochondrial amidoxime-reducing component (mARC), initially described as drug-metabolizing enzyme, is now believed to be involved in human diseases, such as non-alcoholic fatty liver disease (NAFLD) and non-alcoholic steatohepatitis (NASH) [155].

SO and XOR catalyze catabolic reactions in the metabolism of cysteine and purine, being involved in key aspects of the global carbon, sulfur, and nitrogen cycles, whereas AO is involved in the metabolism of some medicines, drugs, and other xenobiotics in humans [151]. Molybdoenzymes have also been associated with alternative reactions ("moonlighting functions"), such as the reduction of nitrite to NO, a process which is well known to regulate blood pressure, hypoxic vasodilation, and other physiological and cellular processes [153]

7.3.7.2 Sources, bioavailability, and homeostasis

Foods high in molybdenum are pulses, cereal grains and grain products, offal (liver, kidney), and nuts, although this element is present in nearly all foods in trace amounts as soluble molybdates. The molybdenum content in plants varies greatly depending on the properties of the soil where they are grown, whereas concentrations in drinking water are usually below 10 µg/L (concentrations up to 200 µg/L have been reported in areas near mining sites). Potassium molybdate may be added to food supplements, while ammonium and sodium molybdate may be added to both foods and food supplements [156].

Molybdenum is absorbed throughout the gastrointestinal tract, in higher amounts in the proximal portions of the small intestine [157]. Little is known abut the actual mechanism of absorption, although the protein encoded by the gene MOT2, a member of the family of sulfate transporters, has been related to molybdenum homeostasis in several organisms, including humans [158, 159]. In addition to this mechanism, passive absorption of molybdate may also occur. Overall, the absorption efficiency of this element has been reported to be very high (around 90%), although it can be decreased by the presence of other food components. From the enterocytes, molybdate enters the circulation, where it is transported mostly as a free anion, although binding to proteins, such as albumin and α_2-macroglobulin, as well as to erythrocytes, has been suggested. Plasma concentrations range between 3 and 11 nmol/L in people with usual molybdenum intakes [156].

Liver and kidney are the organs with the highest concentrations of molybdenum (1.3–2.9 mg/kg and 1.6 mg/kg dry matter, respectively) and the total body amount has been calculated to be around 2.3 mg. Most molybdenum is excreted in the urine as molybdate, although some can also be eliminated with the bile through the feces. Its homeostasis depends on the regulation of its excretion more than on its absorption [156, 157]. In mature human milk, mean molybdenum concentration has been reported to range from 0.72 to 4 µg/L, being highest during the first few days of breast-feeding, and decreasing during the course of lactation [156].

7.3.7.3 Molybdenum status, Dietary Reference Values (DRVs)
The EFSA has concluded that that there is no useful biomarker of molybdenum status and has proposed Adequate Intakes (AIs) for adults based on mean molybdenum intakes in the European Union (Table 7.8) [91, 156].

7.3.7.4 Molybdenum deficiency
Molybdenum deficiency induced by low dietary levels has not been described in humans and only one case of a patient with Crohn's disease and short bowel syndrome, that was fed by parenteral nutrition lacking molybdenum, has been reported. This person developed nausea, rapid breathing and heart rate, vision problems, and ultimately coma. The patient's clinical condition improved upon treatment with ammonium molybdate [160].

Molybdenum cofactor deficiency (MoCD) in newborn patients present with a broad spectrum of clinical severity, with the vast majority being affected from the neonatal age. The most common presentation of severe classical MoCD is of early myoclonic encephalopathy, with poor feeding, irritability, and a distressed facial expression, that quickly progresses to myoclonic seizures, decreased consciousness, and apnea. The molecular cause of the disease is the loss of sulfite oxidase (SOX) activity, that leads to a toxic accumulation of sulfite and a secondary increase of metabolites such as S-sulfocysteine and thiosulfate, along with a decrease in cysteine and its oxidized form, cystine. Moco is synthesized by a three-step biosynthetic pathway and, depending on which synthetic step is impaired, MoCD is classified as type A, B, or C. MoCD type A can be circumvented by administering cyclic pyranopterin monophosphate; however, no specific treatments for MoCD types B/C (and SOX deficiency) are currently available [161].

7.3.7.5 Molybdenum excess
High intakes of molybdenum can induce secondary copper deficiency in certain animals; however, few data are available for humans. Individuals in areas of Armenia with estimated daily intake levels of 10 to 15 mg/d showed elevated serum uric acid and high tissue xanthine oxidase level, displaying symptoms such as aching joints, hyperuricosuria, and elevation of blood molybdenum. Workers exposed to high air concentrations in a molybdenum production plant presented similar symptoms. Molybdenum levels have also been suggested to be inversely associated with testosterone levels [162].

The Scientific Committee on Food of the EU has set a Tolerable Upper Intake Level (UL) for molybdenum of 0.01 mg/kg body weight per day, equivalent to 0.6 mg/

day in the case of adults, including pregnant and lactating women. For children from 1 year of age onward, the UL was set between 0.1 and 0.5 mg/day [156].

7.4 Elements whose essentiality has not been established in humans

7.4.1 Chromium

Chromium (chemical symbol Cr, atomic number 24, molecular weight 59) occurs in each oxidative state from −2 to +6, being Cr^{3+} and Cr^{6+} the most important in human health, however, the high energy needed to oxidize Cr^{3+} to Cr^{6+} precludes the possibility of this oxidation occurring in biological systems [163]. Compounds containing Cr^{3+} are mutagenic and carcinogenic when inhaled and potentially when ingested orally in large quantities, whereas the role of Cr^{3+} as an essential nutrient or as a pharmacological agent is a matter of controversy [164, 165].

7.4.1.1 Biological roles

Cr^{3+} was proposed to be an essential element for mammals approximately sixty years ago and postulated to be necessary for the efficacy of insulin, thus being involved in the regulation of the metabolism of carbohydrates, lipids and proteins. Cr^{3+} was supposed to be a necessary cofactor for the biologic activity of the so-called glucose tolerance factor, however, after decades of research, GTF has not been characterized and there is no agreement on its composition or structure [166]. A circulating complex of Cr^{3+} and an oligopeptide named low-molecular weight chromium-binding substance or chromodulin has also been proposed as the means by which Cr^{3+} mediates responses to insulin; the EFSA considers that "chromodulin's existence and function is unclear as is the functional essentiality of Cr(III)" [163].

In the United States and Canada chromium is still considered an essential element, given that its status has not been re-evaluated since 2001; however, leading researchers in the biochemistry of chromium have asked for a review of the Dietary Reference Intake values for chromium in view of the last scientific reports. In any case, marketing claims of chromium as an agent to reduce body mass and develop muscle are no longer allowed in the United States because these claims, like those of its essential status, are not supported by experiments [164, 165].

Recent studies performed in mice identified eight Cr^{3+}-binding proteins in mitochondria. This group of proteins is predominately associated with ATP synthesis, and includes the ATP synthase, whose activity is inhibited by Cr^{3+} via displacement of Mg^{2+} (Section 7.2.4.1). The inhibition of the ATP synthase activity by Cr^{3+} improves glucose metabolism, and rescues mitochondria from hyperglycemia-induced fragmentation.

The mode of action of Cr^{3+} also holds true in male mice suffering from type II diabetes [167]. Whether or not these results are confirmed in human cells remains to be seen.

7.4.1.2 Sources, bioavailability, and homeostasis

Chromium is ubiquitous in the diet, being abundant in meat and meat products, oils and fats, breads and cereals, fish, pulses, and spices. Levels in dietary plants are determined by soil and water conditions because, although plants do not require chromium, they can accumulate high concentrations of this element when grown in soils polluted with chromium coming from industries or fertilizers. Stainless steel pots and pans used in cooking and meat processing can also contribute to the chromium content in foods. Some Cr^{3+} salts (chloride, sulfate, picolinate, and lactate) may be added to both foods and food supplements, whereas Cr^{3+} nitrate and chromium-enriched yeast may be added to food supplements only. No minimum and maximum levels for chromium on infant and follow-on formulae have been set in the European Union [163, 166].

Dietary chromium is apparently absorbed via passive diffusion [164], with an efficiency that has been estimated to range between 0.4% and 2.5%, depending on factors such as the chemical properties of the ingested source and the presence of other dietary components, in the case of Cr^{3+} from food. Absorption efficiency of supplemental chromium was reported to be between 0.1 and 5.2% depending on the type of compound ingested. Vitamin C reportedly enhances chromium absorption in women when supplied as chromium chloride [163].

Following absorption, chromium is transported to the liver by the bloodstream bound to plasma proteins such as transferrin, that is presumably taken up by endocytosis [164]. Chromium has been found to accumulate mainly in the liver and spleen, but also in soft tissues, bone, and other organs (skin, heart, brain, kidneys, pancreas, and testes). Urine is the main excretory route for absorbed chromium, with small amounts being excreted in sweat and bile, while fecal excretion mainly consists of unabsorbed chromium. Data of chromium concentrations in mature human milk obtained from small groups of women in Europe show a high variability, ranging from 0.14 to 10.8 μg/L [163].

7.4.1.3 Chromium status, Dietary Reference Values (DRVs)

In a report published in 2014, the EFSA Panel on Dietetic Products, Nutrition and Allergies found no convincing evidence for a role of chromium in human metabolism and physiology and no proof that chromium is an essential trace element, therefore, no Dietary Reference Values for this element have been established [163]. In the United States, the Food and Nutrition Board set adequate intakes for chromium in 2001 [166].

7.4.1.4 Chromium deficiency

No symptoms of chromium deficiency have been reported in healthy humans, and no animal model of chromium deficiency has been established [163].

7.4.1.5 Chromium excess

European authorities have been unable to set a Tolerable Upper Intake Level (UL) for chromium, although no evidence of adverse effects associated with supplemental intake of chromium up to a dose of 1 mg/day have been found in a limited number of studies. The EFSA Panel on Contaminants in the Food Chain has derived a Tolerable Daily Intake of 300 µg of Cr^{3+}/kg body weight per day from data obtained in rats. On the contrary, Cr^{6+} is well established as a teratogen, genotoxin, and carcinogen [163, 166].

7.4.2 Fluorine

Fluorine (atomic number 9, atomic mass 19) is the most electronegative element of the Periodic Table; therefore, it mainly occurs as fluoride (F^-) in nature. Fluorides are ubiquitous in air, water and the lithosphere. Availability of fluoride from soils depends on the solubility of the fluoride compound, the acidity of the soil, and the presence of water. Fluorine in air may also exist in gaseous (as the diatomic molecule F_2) or particulate forms, arising from fluoride-containing soils, industry, coal fires, and volcanoes [168].

7.4.2.1 Biological roles

Fluoride is not considered an essential nutrient; however, it appears in vivo associated with calcified tissue (bone and teeth).

– Dental health. The beneficial effects of small amounts of fluoride for dental health have been known since the beginning of the twentieth century. For decades, it was considered that fluoride has to be incorporated into dental enamel during teeth development to form hydroxyfluorapatite, that is more resistant to acids (either ingested or generated by oral bacteria) than hydroxyapatite. Today, fluoride is accepted to exert its effects mainly on the cyclic de- and remineralization processes that take place at the tooth/oral fluid interface, thereby allowing maximum caries protection without ingestion. Fluoride also interferes with the metabolism of oral microbial cells by inhibiting crucial enzymes [169, 170].

– Bone health. Partial substitution of fluoride for hydroxyl groups of hydroxyapatite alters the mineral structure of the bone, thus resulting in increased density and hardness, but not necessarily in increased mechanical strength. Fluoride also appears to increase osteoblast activity, although its use as a therapy for osteoporosis is controversial because, despite producing denser bones, it does not reduce fracture risk and may increase non-vertebral fractures. Moreover, chronic intake of high doses of fluoride can result in conditions known as dental and skeletal fluorosis (see below) [168, 171].

7.4.2.2 Sources, bioavailability, and homeostasis

Major fluoride food sources are water, water-based beverages, and foods reconstituted with fluoridated water. The concentration of fluoride of both tap and bottled drinking water can vary enormously (from 0.1 to 6.0 mg/L within the EU) depending on the geological conditions of the original source. Fluoride content in food is generally low (0.1–0.5 mg/kg), unless prepared with fluoridated water, with the exception of tea, which can contain considerable amounts of fluoride, depending on the type of tea, brewing procedure and fluoride concentration of brewing water. Fluoride content of selected beverages and foods can be found at the USDA National Fluoride Database [172].

The main route of fluoride absorption is via the gastrointestinal tract. Ionic fluoride is converted to hydrogen fluoride (HF) when enters the acidic environment of the stomach lumen. The permeability coefficient of HF is about a million times higher than that of F^-, therefore, it can diffuse into the cells, not requiring a specific transport system. The majority of fluoride not absorbed from the stomach will be absorbed from the small intestine in a pH-independent manner, whereas unabsorbed F^- (an average of 10–20% of fluoride intake) will be excreted in feces. The degree of fluoride absorption, that can reach nearly 100% when supplied as NaF, is affected by the presence of cations such as Ca^{2+}, Mg^{2+}, Al^{3+}, Fe^{3+}, which form poorly soluble compounds. Other factors that affect absorption include the amount of ingested food, the presence of bile salts, and the concentrations of pepsin and pancreatin [170, 173].

Fluoride is rapidly distributed in plasma and deposited in bone and other calcified tissues that contain approximately 99% of total fluoride, the remainder being distributed between blood and soft tissues. The total fluoride content of the human body amounts to 2–5 g and depends on age and exposure to fluoride, whereas the skeleton of a newborn contains only about 5–50 mg. Fluoride not deposited in calcified tissue is mainly excreted via the kidney. Breast milk is considered a minor route of fluoride loss (less than 1% of fluoride intake) [168].

7.4.2.3 Fluoride status, Dietary Reference Values (DRVs)

Fluoride concentration in plasma, saliva, urine, milk, sweat, and bone surface have been proposed as markers of recent exposure to fluoride, whereas concentrations in inner bone, teeth (dentin), nails, and hair are markers of historic fluoride intake. Bulk enamel concentrations reflect the level of systemic exposure to fluoride during tooth formation [168, 170]. In any case, the EFSA has concluded that none of the mentioned biomarkers can be used for defining Dietary Reference Values (DRVs). Moreover, no Average Requirement (AR) for the performance of essential physiological functions can be defined because fluoride is not considered an essential nutrient. However, due to the beneficial effects of dietary fluoride on prevention and severity of caries, Adequate Intakes have been set at 0.05 mg/kg body weight per day for both children and adults [168].

7.4.2.4 Fluoride deficiency

No signs of fluoride deficiency have been identified in humans. As far as its role in dental health is concerned, it must be underlined that adequate intakes of fluoride (unlike intakes of calcium, phosphorus, or vitamins A, C, and D) are not necessary for healthy tooth development, although it may decrease the susceptibility of enamel to acid attacks after eruption [168].

7.4.2.5 Fluoride excess

Acute toxicity is characterized by gastrointestinal disturbances such as pain, nausea, vomiting, and diarrhea. In severe cases, it can lead to renal and cardiac dysfunction, coma, and ultimately death. Most cases of acute toxicity are observed in children due to ingestion of fluoride-containing toothpaste or mouthwashes [171]. Chronic toxicity is usually caused by high fluoride concentrations in drinking water or the use of fluoride supplements. This leads to dental fluorosis and, in more severe cases, to skeletal fluorosis :

– Dental fluorosis is an undesirable side effect of excessive fluoride intake during critical periods of enamel formation (amelogenesis) of both primary and secondary teeth. It is characterized by increased porosity of the teeth, with a loss of enamel translucency, and increased opacity. Mild forms of dental fluorosis are of aesthetic concern only, while in severe cases teeth are stained brown, pitted, and fragile, and may be deformed or broken. Based on its effects on dental fluorosis, the EFSA set Tolerable Upper Intake Levels (ULs) for fluoride at 0.1 mg of fluoride/kg body weight per day (1.5 mg/day for children aged 1–3 and 2.5 mg/day for children aged 4–8 years, approximately).

– Skeletal fluorosis occurs after many years of excessive fluoride intake (10–20 mg/day), mostly as a consequence of living in regions with high fluoride concentrations in drinking water. The early symptoms of skeletal fluorosis include stiffness and pain in the joints. In severe cases, the bone structure may change and ligaments may calcify, resulting in impairment of muscles, paralysis, and pain. Based on data from observational and intervention studies with regard to fractures, the UL for older children and adults was set at 0.12 mg/kg body weight per day (5 mg/day for children and adolescents aged 9–14 and 7 mg/day for children aged 15 years and older) [168].

7.4.3 Boron

Boron (B, atomic number 5, atomic mass 10.81 Da) is considered as a probably essential element by the WHO, due to its beneficial effects on the metabolism. Boric acid (H_3BO_3) forms ester complexes with several biologically important sugars, especially ribose; this results in the modification of ribose-containing metabolites, such as S-adenosylmethionine, diadenosine phosphate, NAD^+, and others. These compounds are involved in bone formation and maintenance, cardiovascular health, cancer risk, and neurological function, with which boron has been beneficially associated. Boron supplementation improved mental alertness, attention, short term memory, and motor speed and dexterity in boron-deprived humans, while low boron intakes have been associated with an increased risk of prostate cancer in men and lung cancer in women [174].

Boron is abundant in vegetables, fruits, and nuts; therefore, significant amounts of this element (1–7 mg) can be consumed every day. Boron is absorbed from gastrointestinal tract almost completely and excreted in the urine, in both cases as boric acid. Total body amounts of boron in healthy people range from 15 to 80 µg/kg. It does not tend to accumulate in tissues, although higher levels are found in bone, nails, and hair. The WHO has suggested that an acceptable safe range of population mean intake of boron for adults could be 1–13 mg/d, while 0.9 mg/day seems to be beneficial for women's bone health, and 6 mg/day would be necessary to obtain positive effects in prostate cancer. Boron deprivation has been associated with impairment of growth, abnormal bone development, decrease in blood steroid hormone levels, increase in urinary calcium excretion, and changes in macromineral status in humans and animals, as well as with Kashin-Beck disease, a bone disease with high incidence in China. Boric acid is effectively used in vaginal yeast infections, especially those caused by *Candida albicans*, and has potential as a protective agent against oxidative damage caused by the fungal toxin Aflatoxin B1. A safe upper intake of 0.4 mg/kg body weight (~28 mg/d for a 70 kg person) has been indicated by the WHO, whereas the EFSA established an Acceptable Daily Intake for boron at 0.16 mg/kg body weight (~11.2 mg/d for a 70 kg person) [166, 174, 175].

7.4.4 Silicon

Silicon (Si, atomic number 14, atomic mass 28,09 Da) is the second most abundant element in the earth's crust, where it usually occurs tetrahedrally coordinated to oxygen forming silicates, that make up more than 90% of the lithosphere. Silicon is ubiquitous in the diet, the major contributors being beverages (primarily beer), coffee, and water, followed by grain, grain products, and vegetables. Orthosilicic acid [$Si(OH)_4$] is the main silicon species in man, being gastrointestinally absorbed and transported in the blood, mainly unbound. Total amount of this element in the human is around 1–2 g and concentration in serum ranges between 24 and 31 µg/dL. Excretion occurs mainly via kidneys. The essentiality of silicon for humans is still under discussion, as specific physiological and/or metabolic function for this element has not been identified. In any case, there is evidence indicating that silicon can bind to the hydroxyl groups of glycosaminoglycans, mucopolysaccharides, and collagen, thereby influencing the formation and/or utilization of these molecules in bones and connective tissue. Silicon seems to exert beneficial effects on the immune or inflammatory response and on mental health, and it has also been shown to affect the absorption, retention, or action of mineral elements, such as aluminum, copper, and magnesium. An intake of silicon of around 25 mg/day has been proposed to assure nutritional benefits of this element. Human deficiencies of silicon have not been reported, whereas adverse effects are primarily limited to silicosis, a lung disease resulting from the inhalation of silica (SiO_2) particles. No Tolerable Upper Intake Levels (ULs) have been established for silicon in humans [166, 176–178].

References

[1]   Belitz H-D, Grosch W, Schieberle P. Food chemistry. 4th rev. and extended ed. Berlin: Springer; 2009.
[2]   Damodaran S, Parkin K, Fennema OR, editors. Fennema's food chemistry. 4th edition. Boca Raton: CRC Press/Taylor & Francis; 2008.
[3]   Gharibzahedi SMT, Jafari SM. The importance of minerals in human nutrition: Bioavailability, food fortification, processing effects and nanoencapsulation. Trends Food Sci Technol 2017;62:119–32. https://doi.org/10.1016/j.tifs.2017.02.017.
[4]   Definition of Electrolyte – Chemistry Dictionary n.d. https://www.chemicool.com/definition/electrolyte.html (accessed September 14, 2019).
[5]   Lobo DN, Lewington AJ, Allison SP. Basic concepts of fluid and electrolyte therapy. 2013. ISBN 978-3-89556-058-3. Available at: https://www.researchgate.net/publication/249625074_Basic_Concepts_of_Fluid_and_Electrolyte_Balance
[6]   Turck D, Castenmiller J, de Henauw S, Hirsch-Ernst K-I, Kearney J, Maciuk A, et al. Dietary reference values for sodium. EFSA J 2019;17:e05778. https://doi.org/10.2903/j.efsa.2019.5778.
[7]   Turck D, Bresson J-L, Burlingame B, Dean T, Fairweather-Tait S, Heinonen M, et al. Dietary reference values for potassium. EFSA J 2016;14:e04592. https://doi.org/10.2903/j.efsa.2016.4592.
[8]   Turck D, Castenmiller J, de Henauw S, Hirsch-Ernst K-I, Kearney J, Maciuk A, et al. Dietary reference values for chloride. EFSA J 2019;17:e05779. https://doi.org/10.2903/j.efsa.2019.5779.
[9]   Nelson DL, Cox MM, Hoskins AA, Lehninger AL. Lehninger principles of biochemistry. 8th edition. New York, NY: Macmillan International, Higher Education; 2021.
[10]   Berend K, van Hulsteijn LH, Gans ROB. Chloride: The queen of electrolytes? Eur J Intern Med 2012;23:203–11. https://doi.org/10.1016/j.ejim.2011.11.013.
[11]   Wang G, Nauseef WM. Salt, chloride, bleach, and innate host defense. J Leukoc Biol 2015;98:163–72. https://doi.org/10.1189/jlb.4RU0315-109R.
[12]   Bailey JL., Sands JM, Franch HA Water, electrolytes, and acid-base metabolism. Mod. Nutr. Health Dis. 11th edition. Philadelphia: Wolters Kluwer Health/Lippincott Williams & Wilkins; 2014, p. 102–32.
[13]   Bernal A, Zafra MA, Simón MJ, Mahía J. Sodium homeostasis, a balance necessary for life. Nutrients 2023;15:395. https://doi.org/10.3390/nu15020395.

[14]   Kiela PR, Ghishan FK. Physiology of intestinal absorption and secretion. Best Pract Res Clin
       Gastroenterol 2016;30:145–59. https://doi.org/10.1016/j.bpg.2016.02.007.
[15]   Rajendran VM, Sandle GI. Colonic potassium absorption and secretion in health and disease. In:
       Pollock DM, editor. Compr. Physiol., Hoboken, NJ, USA: John Wiley & Sons, Inc.; 2018, p. 1513–36.
       https://doi.org/10.1002/cphy.c170030.
[16]   McDonough AA, Fenton RA. Potassium homeostasis: Sensors, mediators, and targets. Pflüg Arch –
       Eur J Physiol 2022;474:853–67. https://doi.org/10.1007/s00424-022-02718-3.
[17]   Kiela PR, Ghishan FK. Physiology of intestinal absorption and secretion. Best Pract Res Clin
       Gastroenterol 2016;30:145–59. https://doi.org/10.1016/j.bpg.2016.02.007.
[18]   Laforenza U. Water channel proteins in the gastrointestinal tract. Mol Aspects Med 2012;33:642–50.
       https://doi.org/10.1016/j.mam.2012.03.001.
[19]   Titze J. Sodium balance is not just a renal affair: Curr Opin Nephrol Hypertens 2014;23:101–5.
       https://doi.org/10.1097/01.mnh.0000441151.55320.c3.
[20]   Lowell BB. New neuroscience of homeostasis and drives for food, water, and salt. N Engl J Med
       2019;380:459–71. https://doi.org/10.1056/NEJMra1812053.
[21]   Trepiccione F, Capasso G, Unwin R. Electrolytes and acid–base: Common fluid and electrolyte
       disorders. Medicine 2023;51:102–9. https://doi.org/10.1016/j.mpmed.2022.11.001.
[22]   Minegishi S, Luft FC, Titze J, Kitada K. Sodium handling and interaction in numerous organs. Am J
       Hypertens 2020;33:687–94. https://doi.org/10.1093/ajh/hpaa049.
[23]   Gumz ML, Rabinowitz L, Wingo CS. An integrated view of potassium homeostasis. N Engl J Med
       2015;373:60–72. https://doi.org/10.1056/NEJMra1313341.
[24]   Jomova K, Makova M, Alomar SY, Alwasel SH, Nepovimova E, Kuca K, et al. Essential metals in health
       and disease. Chem Biol Interact 2022;367:110173. https://doi.org/10.1016/j.cbi.2022.110173.
[25]   Dineen R, Thompson CJ, Sherlock M. Hyponatraemia – presentations and management. Clin Med
       2017;17:263–9. https://doi.org/10.7861/clinmedicine.17-3-263.
[26]   Pilic L, Pedlar CR, Mavrommatis Y. Salt-sensitive hypertension: Mechanisms and effects of dietary
       and other lifestyle factors. Nutr Rev 2016;74:645–58. https://doi.org/10.1093/nutrit/nuw028.
[27]   Choi HY, Park HC, Ha SK. Salt sensitivity and hypertension: A paradigm shift from kidney
       malfunction to vascular endothelial dysfunction. Electrolytes Blood Press 2015;13:7. https://doi.org/
       10.5049/EBP.2015.13.1.7.
[28]   Guyton AC. Blood pressure control–special role of the kidneys and body fluids. Science
       1991;252:1813–6. https://doi.org/10.1126/science.2063193.
[29]   Titze J, Luft FC. Speculations on salt and the genesis of arterial hypertension. Kidney Int
       2017;91:1324–35. https://doi.org/10.1016/j.kint.2017.02.034.
[30]   Gritter M, Rotmans JI, Hoorn EJ. Role of dietary k⁺ in natriuresis, blood pressure reduction,
       cardiovascular protection, and renoprotection. Hypertension 2019;73:15–23. https://doi.org/10.1161/
       HYPERTENSIONAHA.118.11209.
[31]   Xiao H, Yan Y, Gu Y, Zhang Y. Strategy for sodium-salt substitution: On the relationship between
       hypertension and dietary intake of cations. Food Res Int 2022;156:110822. https://doi.org/10.1016/j.
       foodres.2021.110822.
[32]   Hogas M, Statescu C, Padurariu M, Ciobica A, Bilha SC, Haisan A, et al. Salt, not always a
       cardiovascular enemy? A mini-review and modern perspective. Medicina (Mex) 2022;58:1175.
       https://doi.org/10.3390/medicina58091175.
[33]   Salt reduction n.d. https://www.who.int/news-room/fact-sheets/detail/salt-reduction
       (accessed June 27, 2023).
[34]   Stanaway JD, Afshin A, Gakidou E, Lim SS, Abate D, Abate KH, et al. Global, regional, and national
       comparative risk assessment of 84 behavioural, environmental and occupational, and metabolic
       risks or clusters of risks for 195 countries and territories, 1990–2017: A systematic analysis for the

Global Burden of Disease Study 2017. Lancet 2018;392:1923–94. https://doi.org/10.1016/S0140-6736 (18)32225-6.

[35] Larivée NL, Michaud JB, More KM, Wilson J-A, Tennankore KK. Hyperkalemia: Prevalence, predictors and emerging treatments. Cardiol Ther 2023;12:35–63. https://doi.org/10.1007/s40119-022-00289-z.

[36] Hamm LL, Nakhoul N, Hering-Smith KS. Acid-base homeostasis. Clin J Am Soc Nephrol CJASN 2015;10:2232–42. https://doi.org/10.2215/CJN.07400715.

[37] Osuna-Padilla IA, Leal-Escobar G, Garza-García CA, Rodríguez-Castellanos FE. Carga ácida de la dieta; mecanismos y evidencia de sus repercusiones en la salud. Nefrología 2019;39:343–54. https://doi.org/10.1016/j.nefro.2018.10.005.

[38] Seifter JL, Chang H-Y. Disorders of acid-base balance: New perspectives. Kidney Dis 2017;2:170–86. https://doi.org/10.1159/000453028.

[39] Hamilton PK, Morgan NA, Connolly GM, Maxwell AP. Understanding acid-base disorders. Ulster Med J 2017;86:161–6.

[40] Remer T, Manz F. Estimation of the renal net acid excretion by adults consuming diets containing variable amounts of protein. Am J Clin Nutr 1994;59:1356–61.

[41] Schwalfenberg GK. The alkaline diet: Is there evidence that an alkaline pH diet benefits health? J Environ Public Health 2012; 2012. https://doi.org/10.1155/2012/727630.

[42] Storz MA, Ronco AL. How well do low-PRAL diets fare in comparison to the 2020–2025 dietary guidelines for Americans? Healthcare 2023;11:180. https://doi.org/10.3390/healthcare11020180.

[43] Duffy VB. Nutrition and the chemical senses. Mod. Nutr. Health Dis. 11th edition. Philadelphia: Wolters Kluwer Health/Lippincott Williams & Wilkins; 2014, p. 574–88.

[44] Scientific opinion on dietary reference values for calcium. EFSA J 2015;13:4101. https://doi.org/10.2903/j.efsa.2015.4101.

[45] Pu F, Chen N, Xue S. Calcium intake, calcium homeostasis and health. Food Sci Hum Wellness 2016;5:8–16. https://doi.org/10.1016/j.fshw.2016.01.001.

[46] Shkembi B, Huppertz T. Calcium absorption from food products: Food matrix effects. Nutrients 2022;14:180. https://doi.org/10.3390/nu14010180.

[47] Kraft MD. Phosphorus and calcium: A review for the adult nutrition support clinician. Nutr Clin Pract 2015;30:21–33. https://doi.org/10.1177/0884533614565251.

[48] Clapham DE. Calcium signaling. Cell 2007;131:1047–58. https://doi.org/10.1016/j.cell.2007.11.028.

[49] Spiardi R, Geara AS. Normal regulation of serum calcium. In: Walker MD, editor. Hypercalcemia, Cham: Springer International Publishing; 2022, p. 1–17. https://doi.org/10.1007/978-3-030-93182-7_1.

[50] Coulibaly A, Kouakou B, Chen J. phytic acid in cereal grains: Structure, healthy or harmful ways to reduce phytic acid in cereal grains and their effects on nutritional quality. Am J Plant Nutr Fertil Technol 2011;1:1–22. https://doi.org/10.3923/ajpnft.2011.1.22.

[51] Skoglund E, Carlsson N-G, Sandberg A-S. PHYTATE. Heal. Methods, Elsevier; 2009, p. 129–39. https://doi.org/10.1016/B978-1-891127-70-0.50014-5.

[52] Bosscher D, Van Caillie-Bertrand M, Van Cauwenbergh R, Deelstra H. Availabilities of calcium, iron, and zinc from dairy infant formulas is affected by soluble dietary fibers and modified starch fractions. Nutrition 2003;19:641–5. https://doi.org/10.1016/S0899-9007(03)00063-7.

[53] Beggs MR, Bhullar H, Dimke H, Alexander RT. The contribution of regulated colonic calcium absorption to the maintenance of calcium homeostasis. J Steroid Biochem Mol Biol 2022;220:106098. https://doi.org/10.1016/j.jsbmb.2022.106098.

[54] Schenck PA, Chew DJ, Nagode LA, Rosol TJ. Disorders of calcium. Fluid electrolyte acid-base disord. Small Anim Pract, Elsevier; 2012, p. 120–94. https://doi.org/10.1016/B978-1-4377-0654-3.00013-5.

[55] Walker MD, Shane E. Hypercalcemia: A review. JAMA 2022;328:1624. https://doi.org/10.1001/jama.2022.18331.

[56] O'Brien KO, Kerstetter JE, Insogna KL. Phosphorus. Mod. Nutr. Health Dis. 11th edition. Philadelphia: Wolters Kluwer Health/Lippincott Williams & Wilkins; 2014, p. 150–8.

[57]   Penido MGMG, Alon US. Phosphate homeostasis and its role in bone health. Pediatr Nephrol 2012;27:2039–48. https://doi.org/10.1007/s00467-012-2175-z.

[58]   Müller W, Schröder HC, Wang X. The phosphoanhydride bond: One cornerstone of life. Biochemist 2019;41:22–7. https://doi.org/10.1042/BIO04104022.

[59]   Pérez-Castiñeira JR, Docampo R, Ezawa T, Serrano A. Editorial: Pyrophosphates and polyphosphates in plants and microorganisms. Front Plant Sci 2021;12:653416. https://doi.org/10.3389/fpls.2021.653416.

[60]   Calvo MS, Uribarri J. Contributions to total phosphorus intake: All sources considered: Contributions to total phosphorus intake. Semin Dial 2013;26:54–61. https://doi.org/10.1111/sdi.12042.

[61]   Scientific opinion on dietary reference values for phosphorus. EFSA J 2015;13:4185. https://doi.org/10.2903/j.efsa.2015.4185.

[62]   Quarles LD. Chapter 47 – FGF-23 counter-regulatory hormone for Vitamin D actions on mineral metabolism, hemodynamics, and innate immunity. In: Feldman D, editor. Vitam. 4th edition. Academic Press; 2018, p. 871–84. https://doi.org/10.1016/B978-0-12-809965-0.00047-1.

[63]   Shimada T, Yamazaki Y, Takahashi M, Hasegawa H, Urakawa I, Oshima T, et al. Vitamin D receptor-independent FGF23 actions in regulating phosphate and vitamin D metabolism. Am J Physiol-Ren Physiol 2005;289:F1088–95. https://doi.org/10.1152/ajprenal.00474.2004.

[64]   Rudolph EH, Gonin JM. Disorders of phosphorus metabolism. Nephrol Secrets, Elsevier; 2012, p. 551–9. https://doi.org/10.1016/B978-1-4160-3362-2.00088-9.

[65]   Pappoe LS, Singh AK. Hypophosphatemia. Decis. Mak. Med., Elsevier; 2010, p. 392–3. https://doi.org/10.1016/B978-0-323-04107-2.50138-1.

[66]   Stroda A, Brandenburg V, Daher A, Cornelissen C, Goettsch C, Keszei A, et al. Serum phosphate and phosphate-regulatory hormones in COPD patients. Respir Res 2018;19:183. https://doi.org/10.1186/s12931-018-0889-6.

[67]   de Baaij JHF, Hoenderop JGJ, Bindels RJM. Magnesium in man: Implications for health and disease. Physiol Rev 2015;95:1–46. https://doi.org/10.1152/physrev.00012.2014.

[68]   Jahnen-Dechent W, Ketteler M. Magnesium basics. Clin Kidney J 2012;5:i3–14. https://doi.org/10.1093/ndtplus/sfr163.

[69]   Rude RK. Magnesium. Mod. Nutr. Health Dis. 11th edition. Philadelphia: Wolters Kluwer Health/Lippincott Williams & Wilkins; 2014, p. 159–75.

[70]   Wolf FI, Cittadini A. Chemistry and biochemistry of magnesium. Mol Aspects Med 2003;24:3.

[71]   Potter JD, Robertson SP, Johnson JD. Magnesium and the regulation of muscle contraction. Fed Proc 1981;40:2653–6.

[72]   Rude RK, Singer FR, Gruber HE. Skeletal and hormonal effects of magnesium deficiency. J Am Coll Nutr 2009;28:131–41.

[73]   Fiorentini D, Cappadone C, Farruggia G, Prata C. Magnesium: Biochemistry, nutrition, detection, and social impact of diseases linked to its deficiency. Nutrients 2021;13:1136. https://doi.org/10.3390/nu13041136.

[74]   Kirkland A, Sarlo G, Holton K. The role of magnesium in neurological disorders. Nutrients 2018;10:730. https://doi.org/10.3390/nu10060730.

[75]   Razzaque MS. Magnesium: Are we consuming enough? Nutrients 2018; 10. https://doi.org/10.3390/nu10121863.

[76]   Scientific opinion on dietary reference values for magnesium. EFSA J 2015;13:4186. https://doi.org/10.2903/j.efsa.2015.4186.

[77]   Schuchardt JP, Hahn A. Intestinal absorption and factors influencing bioavailability of magnesium-an update. Curr Nutr Food Sci 2017;13:260–78. https://doi.org/10.2174/1573401313666170427162740.

[78]   de Baaij JHF, Hoenderop JGJ, Bindels RJM. Regulation of magnesium balance: Lessons learned from human genetic disease. Clin Kidney J 2012;5:i15–24. https://doi.org/10.1093/ndtplus/sfr164.

[79]   Blaine J, Chonchol M, Levi M. Renal control of calcium, phosphate, and magnesium homeostasis. Clin J Am Soc Nephrol 2015;10:1257–72. https://doi.org/10.2215/CJN.09750913.

[80]   Scientific opinion on dietary reference values for iron. EFSA J 2015;13:4254. https://doi.org/10.2903/j. efsa.2015.4254.

[81]   Dev S, Babitt JL. Overview of iron metabolism in health and disease. Hemodial Int Int Symp Home Hemodial 2017;21:S6–20. https://doi.org/10.1111/hdi.12542.

[82]   Pantopoulos K, Porwal SK, Tartakoff A, Devireddy L. Mechanisms of mammalian iron homeostasis. Biochemistry 2012;51:5705–24. https://doi.org/10.1021/bi300752r.

[83]   Crichton RR. Iron metabolism: From molecular mechanisms to clinical consequences. 4th edition. Chichester, West Sussex: Wiley; 2016.

[84]   Wessling-Resnick M. Iron. Mod. Nutr. Health Dis. 11th edition. Philadelphia: Wolters Kluwer Health/ Lippincott Williams & Wilkins; 2014, p. 176–87.

[85]   Allen L. Guidelines on food fortification with micronutrients. Geneva: World Health Organization [u.a.; 2006.

[86]   Durán E, Villalobo C, Churio O, Pizarro F, Valenzuela C. Encapsulación de hierro: Otra estrategia para la prevención o tratamiento de la anemia por deficiencia de hierro. Rev Chil Nutr 2017;44:234–43. https://doi.org/10.4067/S0717-75182017000300234.

[87]   Brissot P, Troadec M-B, Loréal O, Brissot E. Pathophysiology and classification of iron overload diseases; update 2018. Transfus Clin Biol 2019;26:80–8. https://doi.org/10.1016/j.tracli.2018.08.006.

[88]   Anderson GJ, Frazer DM. Current understanding of iron homeostasis. Am J Clin Nutr 2017;106:1559S–1566S. https://doi.org/10.3945/ajcn.117.155804.

[89]   Yanatori I, Richardson DR, Toyokuni S, Kishi F. The new role of poly (rC)-binding proteins as iron transport chaperones: Proteins that could couple with inter-organelle interactions to safely traffic iron. Biochim Biophys Acta BBA – Gen Subj 2020;1864:129685. https://doi.org/10.1016/j.bbagen. 2020.129685.

[90]   Rossi E. Hepcidin – the iron regulatory hormone. Clin Biochem Rev 2005;26:47–9.

[91]   Dietary Reference Values for nutrients Summary report. EFSA Support Publ 2017;14:e15121E. https://doi.org/10.2903/sp.efsa.2017.e15121.

[92]   WHO | Micronutrient deficiencies. WHO n.d. http://www.who.int/nutrition/topics/ida/en/ (accessed August 19, 2019).

[93]   Camaschella C. New insights into iron deficiency and iron deficiency anemia. Blood Rev 2017;31:225–33. https://doi.org/10.1016/j.blre.2017.02.004.

[94]   Pasricha S-R, Tye-Din J, Muckenthaler MU, Swinkels DW. Iron deficiency. Lancet 2021;397:233–48. https://doi.org/10.1016/S0140-6736(20)32594-0.

[95]   Peyrin-Biroulet L, Williet N, Cacoub P. Guidelines on the diagnosis and treatment of iron deficiency across indications: A systematic review. Am J Clin Nutr 2015;102:1585–94. https://doi.org/10.3945/ ajcn.114.103366.

[96]   Institute of Medicine (U.S.), editor. DRI: dietary reference intakes for vitamin A, vitamin K, arsenic, boron, chromium, copper, iodine, iron, manganese, molybdenum, nickel, silicon, vanadium, and zinc: a report of the Panel on Micronutrients . . . and the Standing Committee on the Scientific Evaluation of Dietary Reference Intakes, Food and Nutrition Board, Institute of Medicine. Washington, D.C: National Academy Press; 2001.

[97]   Piperno A, Pelucchi S, Mariani R. Inherited iron overload disorders. Transl Gastroenterol Hepatol 2020;5:25–25. https://doi.org/10.21037/tgh.2019.11.15.

[98]   King JC, Cousins RJ Zinc. Mod. Nutr. Health Dis. 11th edition. Philadelphia: Wolters Kluwer Health/ Lippincott Williams & Wilkins; 2014, p. 189–205.

[99]   Scientific opinion on dietary reference values for zinc. EFSA J 2014;12:3844. https://doi.org/10.2903/j. efsa.2014.3844.

[100] Maret W. Zinc in cellular regulation: The nature and significance of "zinc signals". Int J Mol Sci 2017;18. https://doi.org/10.3390/ijms18112285.

[101] Andreini C, Banci L, Bertini I, Rosato A. Counting the zinc-proteins encoded in the human genome. J Proteome Res 2006;5:196–201. https://doi.org/10.1021/pr050361j.

[102] Krężel A, Maret W. The biological inorganic chemistry of zinc ions. Arch Biochem Biophys 2016;611:3–19. https://doi.org/10.1016/j.abb.2016.04.010.

[103] Cassandri M, Smirnov A, Novelli F, Pitolli C, Agostini M, Malewicz M, et al. Zinc-finger proteins in health and disease. Cell Death Discov 2017;3:17071.

[104] Prasad AS. Discovery of human zinc deficiency: Its impact on human health and disease. Adv Nutr 2013;4:176–90. https://doi.org/10.3945/an.112.003210.

[105] Maares M, Haase H. A guide to human zinc absorption: General overview and recent advances of in vitro intestinal models. Nutrients 2020;12:762. https://doi.org/10.3390/nu12030762.

[106] World Health Organization, Food and Agriculture Organization of the United Nations, editors. Vitamin and mineral requirements in human nutrition. 2nd edition. Geneva : Rome: World Health Organization ; FAO; 2004.

[107] Hagmeyer S, Haderspeck JC, Grabrucker AM. Behavioral impairments in animal models for zinc deficiency. Front Behav Neurosci 2015;8. https://doi.org/10.3389/fnbeh.2014.00443.

[108] Duan M, Li T, Liu B, Yin S, Zang J, Lv C, et al. Zinc nutrition and dietary zinc supplements. Crit Rev Food Sci Nutr 2023;63:1277–92. https://doi.org/10.1080/10408398.2021.1963664.

[109] Fernández-Cao JC, Warthon-Medina M, H. Moran V, Arija V, Doepking C, Serra-Majem L, et al. Zinc intake and status and risk of type 2 diabetes mellitus: A systematic review and meta-analysis. Nutrients 2019;11. https://doi.org/10.3390/nu11051027.

[110] Nriagu J. Zinc toxicity in humans. Encycl Environ Health Nriagu 2007:1–7.

[111] Narayanan SE, Rehuman NA, Harilal S, Vincent A, Rajamma RG, Behl T, et al. Molecular mechanism of zinc neurotoxicity in Alzheimer's disease. Environ Sci Pollut Res 2020;27:43542–52. https://doi.org/10.1007/s11356-020-10477-w.

[112] Festa RA, Thiele DJ. Copper: An essential metal in biology. Curr Biol CB 2011;21:R877–83. https://doi.org/10.1016/j.cub.2011.09.040.

[113] Collins JF. Copper. Mod. Nutr. Health Dis. 11th edition. Philadelphia: Wolters Kluwer Health/Lippincott Williams & Wilkins; 2014, p. 206–16.

[114] Messerschmidt A. Copper metalloenzymes. Compr. Nat. Prod. II, Elsevier; 2010, p. 489–545. https://doi.org/10.1016/B978-008045382-8.00180-5.

[115] Blockhuys S, Celauro E, Hildesjö C, Feizi A, Stål O, Fierro-González JC, et al. Defining the human copper proteome and analysis of its expression variation in cancers. Met Integr Biometal Sci 2017;9:112–23. https://doi.org/10.1039/c6mt00202a.

[116] Ramos D, Mar D, Ishida M, Vargas R, Gaite M, Montgomery A, et al. Mechanism of copper uptake from blood plasma ceruloplasmin by mammalian cells. PLoS One 2016;11:e0149516. https://doi.org/10.1371/journal.pone.0149516.

[117] Pierson H, Yang H, Lutsenko S. Copper transport and disease: What can we learn from organoids? Annu Rev Nutr 2019;39:75–94. https://doi.org/10.1146/annurev-nutr-082018-124242.

[118] Scientific opinion on dietary reference values for copper. EFSA J 2015;13:4253. https://doi.org/10.2903/j.efsa.2015.4253.

[119] Doguer C, Ha J-H, Collins JF. Intersection of iron and copper metabolism in the mammalian intestine and liver. Compr Physiol 2018;8:1433–61. https://doi.org/10.1002/cphy.c170045.

[120] Chen J, Jiang Y, Shi H, Peng Y, Fan X, Li C. The molecular mechanisms of copper metabolism and its roles in human diseases. Pflüg Arch – Eur J Physiol 2020;472:1415–29. https://doi.org/10.1007/s00424-020-02412-2.

[121] Manto M. Abnormal copper homeostasis: Mechanisms and roles in neurodegeneration. Toxics 2014;2:327–45. https://doi.org/10.3390/toxics2020327.

[122] Latorre M, Troncoso R, Uauy R. Biological aspects of copper. Clin Transl Perspect WILSON Dis, Elsevier; 2019, p. 25–31. https://doi.org/10.1016/B978-0-12-810532-0.00004-5.

[123] Tümer Z, Møller LB. Menkes disease. Eur J Hum Genet 2010;18:511–8. https://doi.org/10.1038/ejhg.2009.187.

[124] Vierling JM, Sussman NL. Chapter 16 – Wilson disease in adults: Clinical presentations, diagnosis, and medical management. In: Kerkar N, Roberts EA, editors. Clin. Transl. Perspect. WILSON Dis., Academic Press; 2019, p. 165–77. https://doi.org/10.1016/B978-0-12-810532-0.00016-1.

[125] Dev S, Kruse RL, Hamilton JP, Lutsenko S. Wilson disease: Update on pathophysiology and treatment. Front Cell Dev Biol 2022;10.

[126] Gil-Bea FJ, Aldanondo G, Lasa-Fernández H, de Munain AL, Vallejo-Illarramendi A. Insights into the mechanisms of copper dyshomeostasis in amyotrophic lateral sclerosis. Expert Rev Mol Med 2017;19:e7. https://doi.org/10.1017/erm.2017.9.

[127] Pratt WB, Omdahl JL, Sorenson JR. Lack of effects of copper gluconate supplementation. Am J Clin Nutr 1985;42:681–2. https://doi.org/10.1093/ajcn/42.4.681.

[128] Tsvetkov P, Coy S, Petrova B, Dreishpoon M, Verma A, Abdusamad M, et al. Copper induces cell death by targeting lipoylated TCA cycle proteins. Science 2022;375:1254–61. https://doi.org/10.1126/science.abf0529.

[129] Kaiho T. Iodine made simple. Boca Raton: CRC Press, Taylor & Francis Group; 2017.

[130] De la Vieja A, Santisteban P. Role of iodide metabolism in physiology and cancer. Endocr Relat Cancer 2018;25:R225–45. https://doi.org/10.1530/ERC-17-0515.

[131] Scientific opinion on dietary reference values for iodine. EFSA J 2014;12:3660. https://doi.org/10.2903/j.efsa.2014.3660.

[132] Laurberg P. Iodine. Mod. Nutr. Health Dis. 11th edition. Philadelphia: Wolters Kluwer Health/Lippincott Williams & Wilkins; 2014, p. 217–24.

[133] Peeters RP, Visser TJ. Metabolism of thyroid hormone. In: Feingold KR, Anawalt B, Boyce A, Chrousos G, Dungan K, Grossman A, et al., editors. Endotext, South Dartmouth (MA): MDText.com, Inc.; 2000.

[134] Incerpi S, Davis PJ, Pedersen JZ, Lanni A. Nongenomic actions of thyroid hormones. In: Belfiore A, LeRoith D, editors. Princ. Endocrinol. Horm. Action, Cham: Springer International Publishing; 2016, p. 1–26. https://doi.org/10.1007/978-3-319-27318-1_32-1.

[135] Köhrle J. Thyroid hormones and derivatives: Endogenous thyroid hormones and their targets. In: Plateroti M, Samarut J, editors. Thyroid Horm. Nucl. Recept. Methods Protoc., New York, NY: Springer New York; 2018, p. 85–104. https://doi.org/10.1007/978-1-4939-7902-8_9.

[136] De Benoist B. World Health Organization, Nutrition for Health and Development. Iodine status worldwide: WHO global database on iodine deficiency. Geneva: Dept. of Nutrition for Health and Development, World Health Organization; 2004.

[137] Hatch-mcchesney A, Lieberman HR. Iodine and iodine deficiency: A comprehensive review of a re-emerging issue. Nutrients 2022;14:3474. https://doi.org/10.3390/nu14173474.

[138] Woodside JV, Mullan KR. Iodine status in UK–An accidental public health triumph gone sour. Clin Endocrinol (Oxf) 2021;94:692–9. https://doi.org/10.1111/cen.14368.

[139] Avery J, Hoffmann P. Selenium, selenoproteins, and immunity. Nutrients 2018;10:1203. https://doi.org/10.3390/nu10091203.

[140] Scientific opinion on dietary reference values for selenium. EFSA J 2014;12:3846. https://doi.org/10.2903/j.efsa.2014.3846.

[141] EFSA Panel on Nutrition, Novel Foods and Food Allergens (NDA), Turck D, Bohn T, Castenmiller J, de Henauw S, Hirsch-Ernst K, et al. Scientific opinion on the tolerable upper intake level for selenium. EFSA J 2023; 21. https://doi.org/10.2903/j.efsa.2023.7704.

[142] Sunde RA. Selenium. Mod. Nutr. Health Dis. 11th edition. Philadelphia: Wolters Kluwer Health/Lippincott Williams & Wilkins; 2014, p. 225–37.

[143] Moreda-Piñeiro J, Moreda-Piñeiro A, Bermejo-Barrera P. In vivo and in vitro testing for selenium and selenium compounds bioavailability assessment in foodstuff. Crit Rev Food Sci Nutr 2017;57:805–33. https://doi.org/10.1080/10408398.2014.934437.

[144] Burk RF, Hill KE. Regulation of selenium metabolism and transport. Annu Rev Nutr 2015;35:109–34. https://doi.org/10.1146/annurev-nutr-071714-034250.

[145] Kieliszek M, Błażejak S. Current knowledge on the importance of selenium in food for living organisms: A review. Molecules 2016; 21. https://doi.org/10.3390/molecules21050609.

[146] Chen P, Bornhorst J, Aschner M. Manganese metabolism in humans. Front Biosci Landmark Ed 2018;23:1655–79.

[147] Scientific opinion on dietary reference values for manganese. EFSA J 2013;11:3419. https://doi.org/10.2903/j.efsa.2013.3419.

[148] Horning KJ, Caito SW, Tipps KG, Bowman AB, Aschner M. Manganese is essential for neuronal health. Ann Rev Nutr 2015;35:71–108. https://doi.org/10.1146/annurev-nutr-071714-034419.

[149] Buchman AL. Manganese. Mod. Nutr. Health Dis. 11th edition. Philadelphia: Wolters Kluwer Health/ Lippincott Williams & Wilkins; 2014, p. 238–44.

[150] Shawki A, Mackenzie B. Interaction of calcium with the human divalent metal-ion transporter-1. Biochem Biophys Res Commun 2010;393:471–5. https://doi.org/10.1016/j.bbrc.2010.02.025.

[151] Dror Y, Stern F. Molybdenum. In: Malavolta M, Mocchegiani E, editors. Trace Elem. Miner. Health Longev., Cham: Springer International Publishing; 2018, p. 179–207. https://doi.org/10.1007/978-3-030-03742-0_7.

[152] Schwarz G, Mendel RR, Ribbe MW. Molybdenum cofactors, enzymes and pathways. Nature 2009;460:839–47. https://doi.org/10.1038/nature08302.

[153] Schwarz G. Molybdenum cofactor and human disease. Curr Opin Chem Biol 2016;31:179–87. https://doi.org/10.1016/j.cbpa.2016.03.016.

[154] Montefiori M, Jørgensen FS, Olsen L. Aldehyde oxidase: Reaction mechanism and prediction of site of metabolism. ACS Omega 2017;2:4237–44. https://doi.org/10.1021/acsomega.7b00658.

[155] Clement B, Struwe MA. The history of mARC. Molecules 2023;28:4713. https://doi.org/10.3390/molecules28124713.

[156] Scientific opinion on dietary reference values for molybdenum. EFSA J 2013;11:3333. https://doi.org/10.2903/j.efsa.2013.3333.

[157] Blanco A, Blanco G. Chapter 29 – Essential minerals. In: Blanco A, Blanco G, editors. Med. Biochem., Academic Press; 2017, p. 715–43. https://doi.org/10.1016/B978-0-12-803550-4.00029-X.

[158] Tejada-Jimenez M, Chamizo-Ampudia A, Calatrava V, Galvan A, Fernandez E, Llamas A. From the eukaryotic molybdenum cofactor biosynthesis to the moonlighting enzyme mARC. Molecules 2018; 23. https://doi.org/10.3390/molecules23123287.

[159] Tejada-Jiménez M, Galván A, Fernández E. Algae and humans share a molybdate transporter. Proceedings of National Academy of Science USA 2011;108:6420. https://doi.org/10.1073/pnas.1100700108.

[160] Abumrad NN, Schneider AJ, Steel D, Rogers L. Amino acid intolerance during prolonged total parenteral nutrition reversed by molybdate therapy. Am J Clin Nutr 1981;34:2551–9.

[161] Johannes L, Fu C-Y, Schwarz G. Molybdenum cofactor deficiency in humans. Molecules 2022;27:6896. https://doi.org/10.3390/molecules27206896.

[162] Novotny JA. Molybdenum nutriture in humans. J Evid-Based Complement Altern Med 2011;16:164–8. https://doi.org/10.1177/2156587211406732.

[163] Scientific opinion on dietary reference values for chromium. EFSA J 2014;12:3845. https://doi.org/10.2903/j.efsa.2014.3845.

[164] Vincent JB, Lukaski HC. Chromium. Adv Nutr 2018;9:505–6. https://doi.org/10.1093/advances/nmx021.

[165] Vincent JB. New evidence against chromium as an essential trace element. J Nutr 2017;147:2212–9. https://doi.org/10.3945/jn.117.255901.

[166] Eckhert CD. Trace elements. Mod. Nutr. Health Dis. 11th edition. Philadelphia: Wolters Kluwer Health/Lippincott Williams & Wilkins; 2014, p. 245–59.

[167] Wang H, Hu L, Li H, Lai Y-T, Wei X, Xu X, et al. Mitochondrial ATP synthase as a direct molecular target of chromium(III) to ameliorate hyperglycaemia stress. Nat Commun 2023;14:1738. https://doi.org/10.1038/s41467-023-37351-w.

[168] Scientific opinion on dietary reference values for fluoride. EFSA J 2013;11. https://doi.org/10.2903/j.efsa.2013.3332.

[169] Aoba T, Fejerskov O. Dental fluorosis : Chemistry and biology. Crit Rev Oral Biol Med 2002;13:155–70. https://doi.org/10.1177/154411130201300206

[170] Štepec D, Ponikvar-Svet M. Fluoride in human health and nutrition. Acta Chim Slov 2019;66:255–75. https://doi.org/10.17344/acsi.2019.4932.

[171] Aoun A, Darwiche F, Al Hayek S, Doumit J. The fluoride debate: The pros and cons of fluoridation. Prev Nutr Food Sci 2018;23:171–80. https://doi.org/10.3746/pnf.2018.23.3.171.

[172] Available N. USDA National Fluoride Database of Selected Beverages and Foods – Release 2 (2005) 2015. https://doi.org/10.15482/usda.adc/1178143.

[173] Buzalaf MAR, Whitford GM. Fluoride metabolism. In: Buzalaf MAR, editor. Monogr. Oral Sci., vol. 22, Basel: KARGER; 2011, p. 20–36. https://doi.org/10.1159/000325107.

[174] Nielsen FH, Eckhert CD. Boron. Adv Nutr 2020;11:461–2. https://doi.org/10.1093/advances/nmz110.

[175] Uluisik I, Karakaya HC, Koc A. The importance of boron in biological systems. J Trace Elem Med Biol 2018;45:156–62. https://doi.org/10.1016/j.jtemb.2017.10.008.

[176] Götz W, Tobiasch E, Witzleben S, Schulze M. Effects of silicon compounds on biomineralization, osteogenesis, and hard tissue formation. Pharmaceutics 2019;11:117. https://doi.org/10.3390/pharmaceutics11030117.

[177] Nielsen FH. Update on the possible nutritional importance of silicon. J Trace Elem Med Biol 2014;28:379–82. https://doi.org/10.1016/j.jtemb.2014.06.024.

[178] Sadowska A, Świderski F. Sources, bioavailability, and safety of silicon derived from foods and other sources added for nutritional purposes in food supplements and functional foods. Appl Sci 2020;10:6255. https://doi.org/10.3390/app10186255.

8 Oxidative stress and antioxidants in nutrition

Victoria Valls-Bellés

8.1 Oxygen and its toxicity

Molecular oxygen is one of nature's most abundant gaseous elements that occur in the form of diatomic gas (O_2). Despite containing an even number of electrons, molecular oxygen behaves like a paramagnetic molecule, that is, two of its electrons are located in different orbitals and spinning in the same direction with parallel spins. This implies certain connotations of interest in relation to its mechanism of molecular action and more specifically in its oxidative function.

Oxygen has a tendency to complete its electronic structure in its last layer with eight electrons (octet law). If an oxygen molecule acquires an extra electron in the course of a reaction, it is transformed into another negatively charged paramagnetic and monoradical species known as superoxide anion ($O_2^{\bullet -}$). This is one of the most important free radicals or reactive oxygen species (ROS), responsible for the so-called oxygen toxicity. Its metabolic importance is of particular interest since the discovery of the enzyme superoxide dismutase (SOD), by McCord and Fridovich in 1969 [1], which provided the first in vivo evidence of the superoxide anion and the subsequent identification of antioxidant defenses. This enzyme is responsible for the dismutation of the superoxide anion to hydrogen peroxide.

However, regardless of evolutionary theories and their selection mechanisms, it is evident that oxygen was selected by nature to act as a terminal acceptor of cellular oxidations, thus playing a key role in its reaction with cytochrome oxidase, representing the last link in the electronic transport chain. In this way, cellular metabolism is the result of a series of oxidation–reduction reactions whose ultimate purpose is to obtain biological or chemical energy to achieve the performance of different types of biological work, such as macromolecule synthesis, active transport, muscle movement, and secretory mechanisms.

One of the first known bibliographic reviews on oxygen toxicity is due to Lavoisier, who in 1785 already suggested that excess oxygen administration could be as dangerous as its defect. Later, in the eighteenth century, other oxygen-derived products, such as H_2O_2 and ozone, were discovered, and their effects were shown to be clearly harmful from the beginning of their synthesis [2].

The free radical hypothesis began to be considered as responsible for oxygen toxicity with Michaelis, who published the participation of free radicals as intermediaries of organic oxidations in 1939. Both Michaelis in 1946 and Gilbert and Gerschamn in 1954 provided the first theoretical and scientific basis to explain oxygen toxicity

Victoria Valls-Bellés, Unitat Predepartamental de Medicina (Area: Fisiologia), Universitat Jaume I, Castelló de la Plana, Spain

https://doi.org/10.1515/9783111111872-008

mediated by highly reactive species [3, 4]. Since the 1950s, the participation of reactive species in most of the biochemical processes of biological systems has been accepted. From then on, these species and the proposal of others that would be discovered later acquired a leading role in the mechanisms of cytotoxic action of oxygen. The implications of this toxicity can be considered in biochemical, toxic-metabolic, and pathophysiological processes, whose molecular bases obey or are a consequence of a general mechanism known as oxidative stress. Despite the existence of metabolizing and defensive enzymes against oxygen and its free radicals, oxidative stress remains a threatening and continuing danger to living cells [5].

8.2 Formation of reactive oxygen species (ROS) and other free radicals in vivo

8.2.1 The concept of free radical

A free radical is defined as any chemical species (atom, molecule, or ion) that contains at least one missing electron on an energy level and that is, in turn, capable of existing independently (hence, the free term). Free radicals are highly reactive, needing to "steal" or "donate" an electron to another atom or molecule, which is transformed into another free radical, thus generating a chain reaction and producing the so-called oxidative stress.

Free radicals are also known as reactive oxygen species, and reactive nitrogen species (RNS). The combination of these free radicals gives rise to other nonradical reactive species, also called reactive oxygen metabolites (ROMs) or active oxygen (AO). Among ROS, the ones shown in Table 8.1 are worth mentioning.

Table 8.1: Reactive oxygen species (ROS) and reactive nitrogen species (RNS).

	Radicals	Nonradicals
Reactive oxygen species (ROS)	Superoxide ($O_2^{\bullet-}$)	Hydrogen peroxide (H_2O_2)
	Hydroxyl ($\bullet OH$)	Hypochlorous acid (HOCl)
	Peroxyl (ROO$\bullet$)	Ozone (O_3)
	Alkoxyl (RO$\bullet$)	Singlet oxygen ($^1O_2$)
	Hydroperoxyl ($HO_2\bullet$)	
Reactive nitrogen species (RNS)		Nitrous acid (HNO_2)
		Dinitrogen tetroxide (N_2O_4)
	Nitric oxide (NO)	Dinitrogen trioxide (N_2O_3)
	Nitric dioxide (NO_2)	Peroxynitrite ($ONOO^-$)
		Peroxynitrous acid (ONOOH)
		Nitronium cation (NO_2^+)
		Alkyl peroxynitrites (ROONO)

As Halliwell indicated in 1996 [6], both ROS and RNS are global terms that, in English, include both radicals and some nonradicals that are oxidizing agents of oxygen and nitrogen and/or are easily converted into radicals, that is, they are reactive species whether or not free radicals. Along with the oxygen and nitrogen radicals, there are other derivatives, some centered on atoms of hydrogen, carbon, sulfur, chlorine, and so on, which undoubtedly contribute to the propagation and maintenance of new reactions that lead to the formation of radicals. There are many oxygen species that act as biological oxidants. From the chemical point of view, the capacity of each reactive oxygen radical or species is determined by four basic characteristics such as (a) reactivity, (b) specificity, (c) selectivity, and (d) diffusivity [7].

Superoxide ($O_2^{\bullet-}$) is the most potent reducing agent, and the simple addition of a proton results in the formation of $HO_2^{\bullet}$, thus becoming a very active, selective, and specific oxidizing agent. $O_2^{\bullet-}$ is not particularly reactive with lipids, carbohydrates, or nucleic acids, and exhibits limited reactivity with certain proteins. This evidence confirms that $O_2^{\bullet-}$ reacts with proteins that contain metals in its prosthetic group. The hydroxyl radical ($^{\bullet}OH$), however, reacts with any molecule that is nearby, without any specificity so that the danger lies in the functional importance of the cell compartment in which it originates or the molecule to which it attacks. In this way, if it attacks DNA, it can produce or generate serious alterations. On the contrary, if the production of the radical takes place in an environment such as plasma and the damaged molecule is an enzyme that is present in large quantities, the actual biological damage will be practically imperceptible.

The three components with the greatest diffusion capacity are $O_2^{\bullet-} < H_2O_2^{\bullet} < {}^{\bullet}OH$, being able to react with molecules that are far from the place of origin and even with the ability to pass through cell membranes (Figure 8.1).

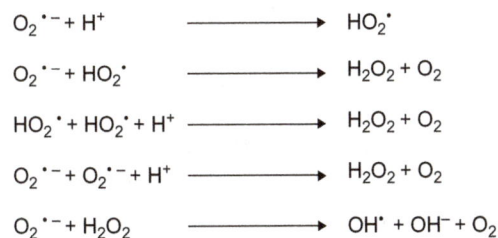

$$O_2^{\bullet-} + H^+ \longrightarrow HO_2^{\bullet}$$

$$O_2^{\bullet-} + HO_2^{\bullet} \longrightarrow H_2O_2 + O_2$$

$$HO_2^{\bullet} + HO_2^{\bullet} + H^+ \longrightarrow H_2O_2 + O_2$$

$$O_2^{\bullet-} + O_2^{\bullet-} + H^+ \longrightarrow H_2O_2 + O_2$$

$$O_2^{\bullet-} + H_2O_2 \longrightarrow OH^{\bullet} + OH^- + O_2$$

Figure 8.1: Reactivity of oxygen species.

The production of free radicals has to be continuously monitored and maintained at very low concentrations. This is done by the different mechanisms that exist in the organisms forced to live in aerobic environments. For this reason, oxygen species occur in environments sufficiently enclosed to prevent their diffusion or, alternatively, where they can be controlled by the action of defensive enzymes or free radical scavengers synthesized by aerobic cells that are responsible for their rapid metabolization to more stable or harmless species [8–10].

8.2.2 Sources of free radicals

Free radicals can be both endogenous, generated by the cell itself, or exogenous. They occur naturally as intermediaries or as a product of the numerous oxidative reactions of the cells, as well as through various physical–chemical or biotransformation processes. ROS can also be produced through exposure to environmental oxidants, toxicants, and heavy metals that can disturb the balance between cell oxidation–reduction reactions, thus altering the normality of biological functions [8–10].

8.2.2.1 Endogenous sources

The main endogenous sources of free radicals are:

– Electronic transport chain. In general, it is assumed that mitochondria provide the greatest source of ROS in most cells [11–13]. The main function of the mitochondria is to generate energy, the electronic transport chain being the last link in the combustion of nutrients for obtaining energy in the form of ATP. About 95% of the oxygen we breathe is reduced to H_2O by the action of cytochrome oxidase a3, the last link in the electronic transport chain, through a mechanism in which four redox centers participate, providing, in addition, the main source of energy (ATP) to the organism. However, the monovalent reduction of the oxygen molecule results in the formation of three highly reactive species responsible for oxygen toxicity. These species are the superoxide radical ($O_2^{\bullet-}$), hydrogen peroxide (H_2O_2), and the hydroxyl radical ($^{\bullet}OH$). Their toxicity is a consequence of their extreme reactivity, which in turn follows from their physical–chemical condition, characteristic of most paramagnetic species. The exception in this case is the hydrogen peroxide (H_2O_2), which, not being a radical, becomes part of important redox reactions to reduce or oxidize to species of greater reactivity (Figure 8.2) [14]. In certain circumstances, these ROS can also be produced at the level of complex I and the quinone–semiquinone–ubiquinol complex (Q_{10}) acting as electron acceptors (Figure 8.2) [15]. This situation usually occurs in healthy mitochondria; however, this production can be much higher in certain physiological processes such as aging. Any physiological situation that involves an increase in mitochondrial respiration will lead to an increase in the formation of free radicals, as occurs during physical exercise. Mitochondria can generate more than 85% of ROS in skeletal muscle tissue [12], but between 1% and 2% of the O_2 consumed by the organism is naturally converted into $O_2^{\bullet-}$ that will lead to the formation of other ROS after dismutation. It has been estimated that a human being can produce about 2 kg of superoxide in the body every year and that individuals with chronic inflammatory processes can generate much more [17–21].

– Microsomal electronic transport (hydroxylation reactions). The endoplasmic reticulum, that is, the microsomal fraction of the cell, contains the non-phosphorylating electronic transport system (different from the mitochondrial electronic transport that is phosphorylating). These systems participate in various hydroxylation and desaturation reactions that produce ROS. A hydroxylation system is constituted by the microsomal enzymes of cytochrome P-450, which are responsible for metabolizing xenobiotics. The hepatic microsomal system (Figure 8.3) consists of a flavoprotein called NADPH-cytochrome P450 reductase, and a microsomal cytochrome, P450. An electronic equivalent is transferred from NADPH to flavoprotein that contains a FAD prosthetic group, which is completely reduced. Subsequently, the electrons are transferred from the reduced flavoprotein to the oxidized form of cytochrome P450 (Fe^{3+}) to give the reduced form P450 (Fe^{2+}), which reacts with O_2 forming the superoxide ion [22].

– Phagocytic cells. A common fact to all types of inflammation is the infiltration into the affected tissue of cells capable of moving freely. These are mainly leukocytes, neutrophils, monocytes, or macrophages. These cells are activated and carry out phagocytosis through an oxygen consumption mechanism, using the NADPH oxidase system directly generating O_2. This consumption can be up to 20–30 times higher than that existing prior to activation. On the other hand, phagocytic cells also generate nitric oxide (NO), by the action of nitric oxide synthase (NOS) on intracellular arginine, as a defense mechanism. The combination of $O_2^{\bullet-}$ with NO results in the formation of $ONOO^-$ that is capable of inducing lipid peroxidation in lipoproteins, thus destroying cell membranes [23] (Figure 8.4).

– Autoxidation of reduced carbon compounds such as amino acids, proteins, lipids, carbohydrates, and nucleic acids also results in the formation of $O_2^{\bullet-}$.

– The catalytic activation of various enzymes of the intermediate metabolism, such as hypoxanthine and xanthine oxidase, aldehyde oxidase, monoamine oxidase, cyclooxygenase, lipoxygenase, NOS, is also a source of free radicals. An example is the deamination of dopamine by the monoamine oxidase that generates H_2O_2 in some neurons and has been implicated with the etiology of Parkinson's disease [24]. Another example is the enzyme NOS (types I, II, and III) that produces one of the most important radicals in biological regulation in cells: nitric oxide (NO). Another of the most relevant radicals is the superoxide produced by the NAD(P)H oxidase [25]. Xanthine oxidase is another enzyme that participates in the production of ROS, generating superoxide by the oxidation of hypoxanthine to xanthine and then to uric acid [26].

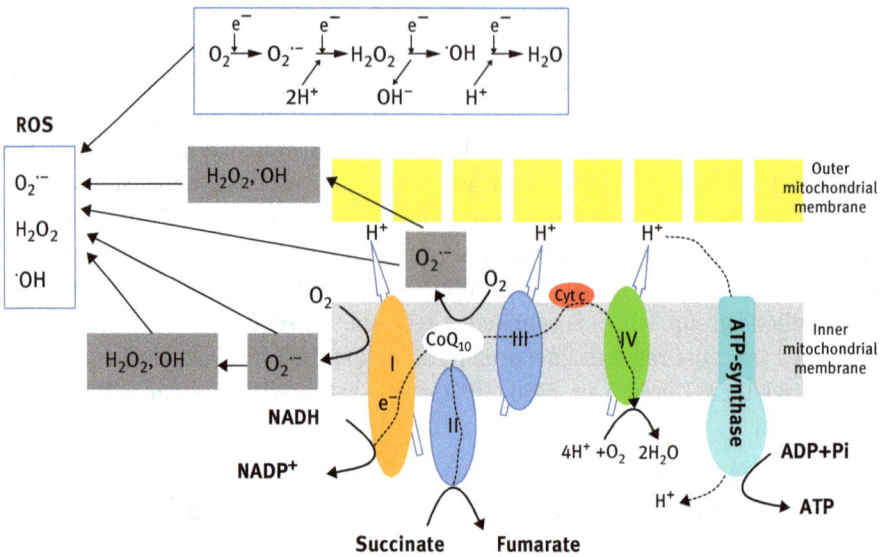

Figure 8.2: Generation of reactive oxygen species (ROS) in the monovalent reduction of oxygen and in the electronic transport chain [16].

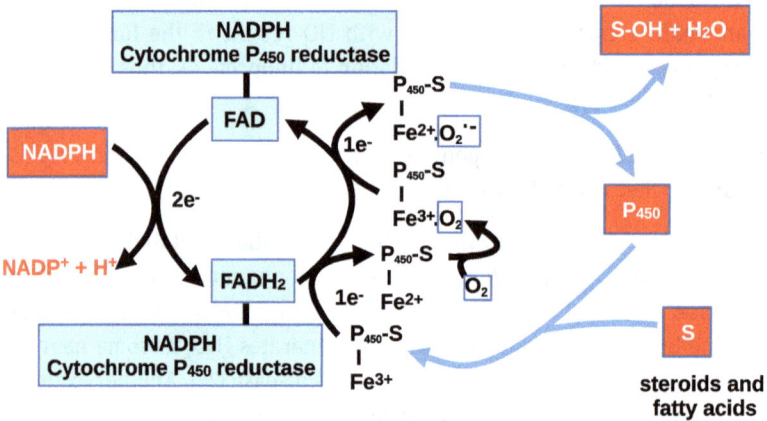

Figure 8.3: Hepatic microsomal hydroxylating system (adapted from [27]).

8.2.2.2 Exogenous sources

As previously mentioned, ROS can also be generated by exogenous sources:

– Environmental (electromagnetic radiation, sunlight, ozone, and tobacco). Free radicals can be produced in response to electromagnetic radiation, such as gamma rays, which can cleave water and produce hydroxyl radicals [28]. Nitrogen oxides in cigarette smoke cause the oxidation of macromolecules and the re-

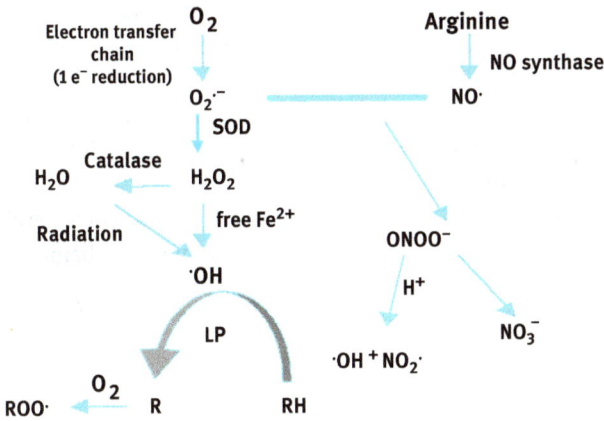

Figure 8.4: Antioxidant system in phagocyte cells [16].

duction of antioxidant levels, which contributes to the appearance of pathologies in the smoker, such as cardiovascular processes and a variety of related cancers, especially lung cancer [28–32].

– Pharmacological (xenobiotics, drugs, etc.). This is the case of anthracyclines, which interact with complex I of the electronic transport chain and induce the formation of free radicals [16, 33–35].

– Nutritional. ROS are produced in the presence of food contaminants, additives, PUFA (polyunsaturated fatty acid), and so on. Iron and copper salts promote the formation of free radicals generating H_2O_2. When an individual absorbs a significant amount of dietary iron due to a genetic defect, particularly heme iron, it becomes a risk factor for cardiovascular disease (CVD) and certain types of cancer [36].

8.3 Biological damage by ROS

Although ROS have traditionally been observed from a negative point of view for cell function and viability, they can play an important role in the origin of life and biological evolution with beneficial effects on organisms. In recent years, the roles of ROS in signaling and gene expression modulation have been recognized and re-evaluated. In fact, it is not easy to categorize ROS or free radicals as beneficial or harmful molecules. According to Jackson and collaborators 2002 [37], everything depends on the cellular process that is analyzed. These authors take two examples; thus, in cell death by necrosis due to ischemia–reperfusion mechanisms, ROS are not beneficial, while in cell death by apoptosis, they can be seen as harmful or beneficial.

The role of ROS in inflammation processes can also be ambivalent. It has been clearly seen as beneficial in the proinflammatory role since it provides an improvement in the immune response following the infection, but in disorders such as rheumatoid arthritis, the inappropriate inflammatory response generated by ROS must be suppressed.

Researchers have concluded that ROS play an important role, which can be harmful or beneficial, in changes of modulation of gene expression and cellular function. These changes can be used as biomarkers of oxidative stress.

ROS cause oxidation to biomolecules such as polyunsaturated lipids, cholesterol molecules, carbohydrates, proteins, and nucleic acids that are susceptible to being attacked in vivo by free radicals (Figures 8.5 and 8.6).

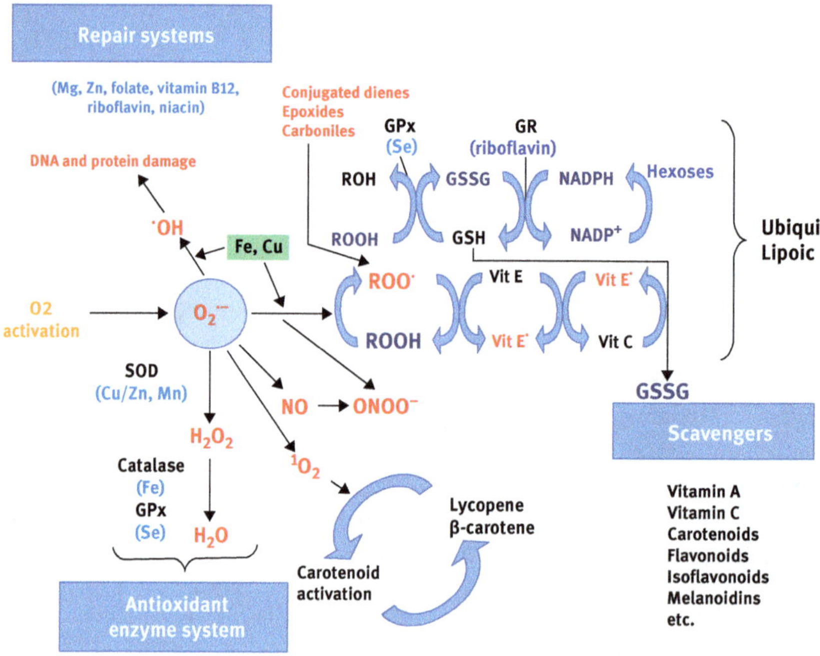

Figure 8.5: Scheme of ROS generation and antioxidant defense [16].

8.3.1 Damage to lipids

PUFAs are highly susceptible to being altered by free radicals, producing lipid peroxidation. Since they have C = C double bonds of the *cis*-type, the divinyl-methane structure being repeated within them, each double bond is separated from the successive by an allylic CH_2, which makes them particularly susceptible to be attacked by free radicals.

The chain of reactions that free radicals produce in fatty acids consists of three essential stages: initiation, propagation, and termination (Figure 8.6).

Initiation

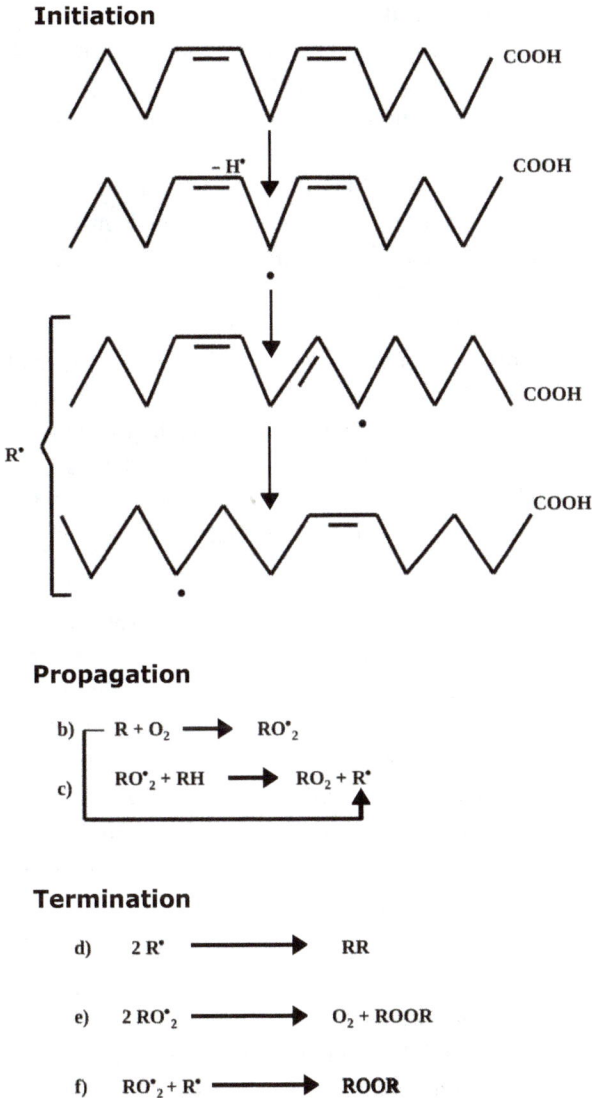

Propagation

b) ⎯ $R + O_2 \longrightarrow RO{}^{\bullet}_2$

c) $RO{}^{\bullet}_2 + RH \longrightarrow RO_2 + R{}^{\bullet}$

Termination

d) $2\,R{}^{\bullet} \longrightarrow RR$

e) $2\,RO{}^{\bullet}_2 \longrightarrow O_2 + ROOR$

f) $RO{}^{\bullet}_2 + R{}^{\bullet} \longrightarrow ROOR$

Figure 8.6: Scheme of lipid peroxidation (Valls-Bellés V.).

– Initiation reactions. Any species capable of sequestering a hydrogen atom from a carbon chain in a fatty acid (LH) will give rise to a radical located in the corresponding carbon atom (C${}^{\bullet}$ or L${}^{\bullet}$) of this structure (Figure 8.6). In the fatty acid molecule, a restructuring takes place to form a conjugated diene, a change in the arrangement of the double bonds, together with the carbonyl radical shift (C) to be placed on the next adjacent atom. In lipid peroxidation, there are other initiating mechanisms, such as those triggered by the products of this peroxidation, since they are activated species, although they can also be considered as propagating agents in the sense of generating and attacking neighboring lipid structures. The role of electronic transfer from metal ions is important, since redox reactions between transition metal ions and peroxide compounds play an important role in the formation of free radicals in vivo. This is mainly the case of copper and iron, although the reaction of initiation of lipid peroxidation has also been observed with other transition metals. The most important mechanism of action from the point of view of in vivo production is the Fenton reaction [38]. Undoubtedly, the species responsible for this initiation in the presence of metal ions is the hydroxyl radical. Both these species and others derived from lipid peroxidation are capable of abstracting a hydrogen atom by themselves [39]. There is another mechanism for initiation of peroxidation reactions, such as the breakdown of chemical bonds by photolytic action [40]. Finally, it is necessary to mention the role of toxic reactions triggered by various xenobiotics such as carbon tetrachloride or antitumor drugs such as adriamycin. The formation of activated molecular species is involved in the pharmacological mechanism of action of adriamycin; however, the cellular selectivity of this molecule or its effect over normal cells remains to be resolved [33].

– Propagation reactions. Unlike what happens in the processes of initiation and termination, in propagation reactions, the number of participating free radicals remains constant. After the extraction of a hydrogen and the molecular readjustment, the radical created has a high reactivity toward O_2 molecules, giving rise to the peroxide radical ($RO^\bullet_2$). On the other hand, the peroxide radical can induce the extraction of another hydrogen from a neighboring fatty acid molecule, thus initiating another peroxidation cycle. The reaction will end when there is a consumption of substrates or by the conversion of paramagnetic species to radicals or molecules of greater stability and lower reactivity.

– Termination reactions. Termination reactions involve a process in which more than one type of free radicals participates. There are essentially three termination reactions, which lead to the disappearance of the most reactive species, to gradually replace them with molecules of greater chemical stability. Among these three reactions, two types are distinguished: (d) homotermination and (e) and (f) cross-termination (Figure 8.6). The radicals $RO^\bullet$ and $O^\bullet_2$ will combine with each other to form the compound ROOR, which can undergo further oxidation. The choice of one or another reaction path will depend, among other things, on the viscosity, as well as on the concentration of oxygen in the medium. Another factor that determines which reaction will occur is the structure of the substrate. The connotations of termination reactions can be important from the cellular pathophysiological point of view. In no way should it be assumed that the evolution of termination mechanisms ends the distorting danger of free radicals, but quite the opposite. Indeed, in reaction (d), newly formed links with various biomolecules can be highly destructive for those biological systems in which they are systematically, permanently, and definitively established. As a consequence of lipid peroxidation, there is an alteration in the conductivity, fluidity, permeability, and transport of the membranes. If we consider the critical role of PUFAs as major components of cell membranes, these alterations can be very important [41, 42].

8.3.2 Damage to proteins

Proteins and amino acids are also attacked by free radicals. This attack causes changes in cell function, chemical fragmentation, and an increase in susceptibility to proteolytic attack. An example is the reversible oxidation of the –SH groups, which is closely linked to oxidative stress in many aspects. There are other types of oxidations of reactive groups of the reversible type; for example, the oxidation of methionine to methionine sulfoxide and its enzymatic reduction back to methionine. However, irreversible oxidation through damage of some amino acid groups is also possible, and we find an example in the rupture of the imidazole ring of histidine or tryptophan [43, 44].

The possibility of a protein being attacked by a free radical or an oxygen species depends on its composition of amino acids and the accessibility of the oxidizing species to them. Certain amino acids strongly react with free radicals, such as methionine and cysteine, both present in some enzymes (lysozyme, pepsin, etc.). This is associated with the loss of biological activity of these enzymes. Proline is another target of oxidative stress, especially as a mechanism for breaking peptide bonds. One of the susceptible proteins to this type of destruction is the collagen molecule.

In addition to the oxidation of amino acids, the oxidative stress of proteins is also closely related to the reversible oxidation–reduction of thiol groups; thus, the alteration of the thiol/disulfide status has been shown to have biological consequences such as changes in the kinetic constants and maximum speed of various enzymes.

An example of the importance of protein oxidation is represented by low-density lipoproteins (LDLs), where histidines and lysines are modified by oxidation, thus causing an alteration in the recognition by receptors. In addition, protein oxidation processes frequently introduce new functional groups such as hydroxyl groups and carbonyl groups, which contribute to altering mobility and protein function. An improvement in the characterization of these effects has allowed to identify several secondary processes that include fragmentation, cross-linking, and splitting, which can accelerate or prevent proteolysis mediated by proteasomes according to the severity of oxidative damage [45].

8.3.3 DNA damage

The attack of free radicals on DNA generates a series of injuries, which include the breaking of chains and the modification of bases that produce mutagenesis and carcinogenesis. Oxidative alterations interrupt transcription and replication, increasing the number of mutations. The hydroxyl radical ($^{\bullet}$OH) attacks DNA, which leads to a large number of changes in purine and pyrimidine bases. Some of these modified bases are considered potentially harmful for genome integrity [16, 46–54].

8.3.4 Cholesterol damage

The oxidation of cholesterol is of particular biological interest, since cholesterol hydroperoxides and a family of oxidized oxysterols on the β-ring of the sterol are produced, as well as derivatives of oxidized cholesterol that are involved in atherosclerosis and CVD (Figure 8.7). Currently, there is a lot of evidence indicating that free radicals, lipid peroxidation, and oxidative modification of LDL are involved in the atherosclerosis initiation process. Oxidized LDL is more easily captured by macrophages, which results in the formation of foam cells and induces the proliferation of smooth muscle cells [16] (Chapter 4. Section 4.15). It has also been postulated that the presence of oxysterols in the blood may be the result of an effective antioxidant mechanism in vivo. This system is based on the possible interaction at the level of blood and various tissues of different oxidizing elements with cholesterol, and oxidized derivatives thereof would be excreted via biliary–fecal route [55–60].

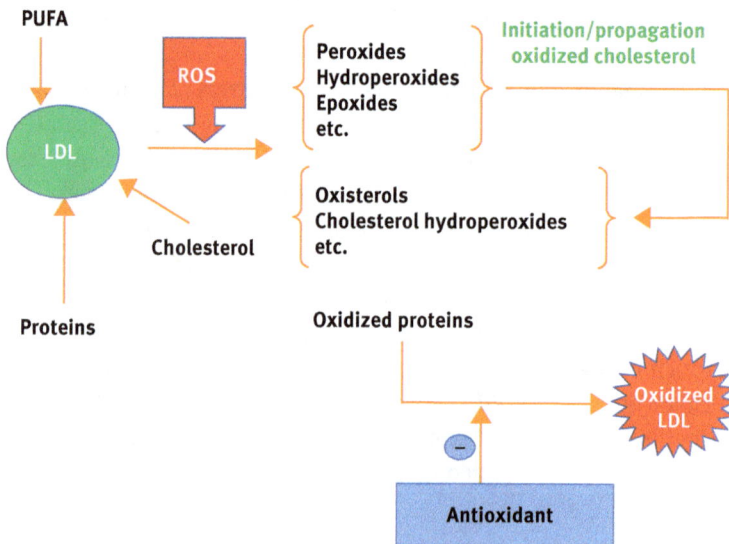

Figure 8.7: LDL oxidation scheme (Valls-Belles V).

8.4 Natural defenses

At low concentrations, free radicals are necessary for good cell functioning, being able to act as second messengers stimulating cell proliferation, and/or acting as mediators for cell activation. However, an excess of them can accumulate to toxic levels resulting in various actions on the metabolism of the immediate principles, which may be the origin of cellular damage. In order to counteract the toxic action of free radicals, organisms have developed numerous antioxidant defense mechanisms that allow their elimination or transformation into stable molecules [61].

According to Sies et al. [10], the biological systems are in a state of approximate equilibrium between prooxidant forces and their antioxidant capacity. The imbalance in favor of the prooxidant action is what is known as oxidative stress. In fact, oxidative damage only occurs when oxidizing mechanisms exceed the capacity of defense systems. Therefore, the survival of aerobic cells requires mechanisms that counteract the negative effects of free radicals. According to Halliwell and Gutteridge [11], an antioxidant is any substance that, in the presence or at low concentrations with respect to the oxidizable substrate, significantly delays or inhibits the oxidation of the latter. A good antioxidant is characterized by its high effectiveness, its operational variability, and versatility to be able to combine with an important variety of ROS. There are antioxidant systems at both physiological and biochemical levels:

– At physiological level, the function of the microvascular system is to maintain tissue levels of O_2, always within relatively low partial pressures.
– At biochemical level, the antioxidant defense can be enzymatic or nonenzymatic; in addition, there are also molecule repair systems.

8.4.1 Antioxidant enzyme system

Aerobic organisms have developed antioxidant enzymes such as SOD, catalase (CAT), glutathione peroxidase (GPx), and DT-diaphorase. SOD is responsible for the $O_2^{\bullet-}$ to H_2O_2 dismutation reaction, which will be detoxified by a subsequent reaction catalyzed by CAT or GPx, producing H_2O and O_2. CAT is found mainly in peroxisomes, and its main function is to eliminate the H_2O_2 generated in the beta-oxidation of fatty acids, while GPx will degrade cytoplasmic H_2O_2 [62].

There are other important enzymes that also participate in the defense system, including in the regeneration reactions of glutathione (GSH), such as GSH reductase, or NADPH-quinone oxidoreductase (DT-diaphorase) [63]:

– SOD (EC.1.15.1.) catalyzes the dismutation reaction of the superoxide ion to hydrogen peroxide, which can be subsequently reduced by CAT or GPx (Figure 8.8). Another function of SOD is to protect dehydratases (dihydroxide dehydratase acid, aconitase, 6-phosphogluconate dehydratase, and fumarate) against inactivation by the superoxide free radical. Four classes of SOD that have been identified contain copper, zinc, iron, manganese, or nickel, as cofactors. In humans, there are three forms of SOD: cytosolic Cu/Zn-SOD, Mn-SOD mitochondrial, and extracellular SOD [18]. SOD catalyzes the dismutation of $O_2^{\cdot-}$ by successive oxidations and reductions of the transition metal ion at its active site by a ping-pong mechanism with a high reaction rate [64, 65]. There is a fourth Ni-SOD enzyme that has been purified from the cytosolic fraction of the mycobacterium *Streptomyces* sp. and *Streptomyces coelicolor*. It is composed of four identical subunits, and its amino acid composition is different to those of the three previous SODs [66].

– CAT (EC 1.11.1.6) acts similarly to SOD [18]. It reacts at any concentration with H_2O_2 to form molecular oxygen and water, and also shows peroxidase activity with hydrogen donors (methanol, ethanol, formic acid, phenol, etc.), thus protecting the cells from internal peroxides (Figure 8.9). CAT is a tetrameric enzyme with four identical 60 kDa subunits that contain a ferriprotoporphyrin per subunit, its molecular mass being over 240 kDa. It is so efficient that can be saturated by H_2O_2 at any concentration [66, 67]. It is abundant in mammalian cells and is located in peroxisomes, where it destroys the H_2O_2 generated by oxidases within these organelles [6]. Although CAT is not essential under normal conditions, it plays an important role in the acquisition and tolerance of oxidative stress and cellular response in some cell types [66]. The increased sensitivity of cells enriched with CAT to adriamycin, bleomycin, and paraquat is attributed to the ability of CAT to prevent drug-induced consumption of O_2, either by eliminating H_2O_2, or by direct interaction with the drug [66].

– GPx (EC 1.11.1.8) is probably the major H_2O_2 scavenger in mammalian cells. In higher organisms, GPx appears largely to supplant the need for CAT. It contains active selenium centers, involved not only in the elimination of H_2O_2 but also in the metabolism of lipid peroxides (ROOH and H_2O_2). This enzyme uses GSH as a co-substrate to catalyze the reduction of lipid peroxides, which in turn acts as a nonenzymatic antioxidant (Figure 8.10) [18]. At least five isoenzymes of GPx have been found in mammals, and their expression levels vary depending on the type of tissue. Cytosolic and mitochondrial GPx reduces lipid peroxides and H_2O_2 at the expense of glutathione (CGPx or GPx1); similarly, glutathione phospholipid hydroperoxidase (GPx4 or PHGPx) is found in most tissues. The latter is located both in the cytosol and in the membrane, and can directly reduce lipid hydroperoxides, fatty acids, and cholesterol hydroperoxides that occur in the peroxidation of oxidized membranes and lipoproteins. It is mostly expressed in renal epithelial cells. Cytosolic GPx2 and extracellular GPx3 are poorly detected in most tissues except in the gastrointestinal tract and kidney. Another isoenzyme, GPx5, is specifically expressed in mouse [68]. GPx is considered the main antioxidant enzyme that detoxifies H_2O_2 in animal cells, especially in human erythrocytes, since CAT has a lower affinity for H_2O_2 than GPx. It is also known that cells with decreased GPx are more sensitive to the toxicity of paraquat and adriamycin [69]. GPx and other selenoproteins containing selenocysteine or selenomethionines also function in the maintenance of the defense against peroxynitrite-mediated oxidations [69]. This enzyme can reduce lipid peroxides, as well as hydrogen peroxide, being a very important enzyme for the maintenance of the structure and function of biological membranes [70].

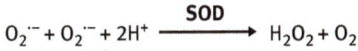

$$O_2{}^{\cdot-} + O_2{}^{\cdot-} + 2H^+ \xrightarrow{\text{SOD}} H_2O_2 + O_2$$

Figure 8.8: Reaction of SOD with the superoxide radical.

$$2\ H_2O_2 \xrightarrow{\text{CAT}} 2\ H_2O\ +\ O_2$$

$$ROOH\ +\ AH_2 \xrightarrow{\text{CAT}} H_2O\ +\ ROH\ +\ A$$

Figure 8.9: Reaction of catalase with hydrogen peroxide.

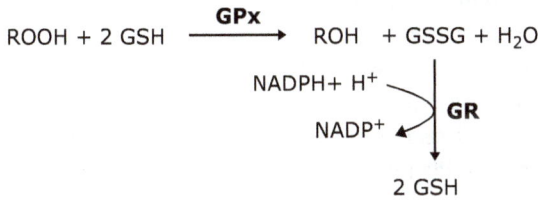

$$ROOH + 2\ GSH \xrightarrow{\text{GPx}} ROH\ +\ GSSG + H_2O$$

NADPH+ H$^+$ → NADP$^+$ → GR → 2 GSH

Figure 8.10: Reduction of lipid peroxide (ROOH) by glutathione peroxidase (GPx). GR, glutathione reductase.

8.4.2 Nonenzymatic antioxidant systems

A second group of antioxidants are those of nonenzymatic nature, among which various types of molecules of both hydrophilic and lipophilic nature are grouped, whose defensive action will depend in some cases on a direct interaction on the reactive species to yield stable or less reactive complexes. Among them are endogenous antioxidants such as GSH, uric acid, and certain plasma proteins such as ceruloplasmin and ferritin.

Reduced GSH, a tripeptide composed of cysteine, glutamic acid, and glycine (γ-Glu -CysH-Gly), is the largest intracellular antioxidant component. Its distribution is universal as it is present in both plants and animals and plays an important role in cellular protection against the toxic effects of free radicals. It occurs mainly in its reduced form in the cells, and a large part of its functions are due to the presence of the reduced thiol group conferred by cysteine. It has an important role as an antioxidant in the cellular defense against free radicals, being able to interact and stabilize hydroxyl, superoxide, and peroxide radicals, in addition to participating in the reduction of other antioxidants such as α-tocopherol and donate hydrogens to repair damaged DNA. On the other hand, it can act as a co-substrate of antioxidant enzymes such as GPx, as we mentioned earlier. The most important characteristics of GSH are:

- GSH is an exogenous and endogenous antioxidant. The GSH of the diet can be absorbed in the small intestine and can be synthesized again.
- Although the GSH radical formed by the oxidation of GSH is a prooxidant radical, it can react with another GSH radical (GS•) producing GS-SG that is reduced to GSH by the NADPH-dependent glutathione reductase.
- GSH can react with electrophilic components of xenobiotics in a reaction catalyzed by glutathione-S-transferase.
- The GSH can be conjugated with NO, thus forming nitrosylated GSH, which will release GSH and NO through a thiol protein system.
- GSH interacts with thiol proteins (glutaredoxin and thioredoxin) that can play an important role in the regulation of homeostasis of the cell's oxidation–reduction system.

GSH is synthesized from glutamate, cysteine, and glycine in two stages. In the first stage, glutamic acid and cysteine bind to form γ-glutamylcysteine. This reaction is catalyzed by γ-glutamylcysteine synthetase. This step is limited by the availability of cysteine. In the second stage, γ-glutamylcysteine reacts with glycine to form the tripeptide, a reaction catalyzed by glutathione synthetase (Figure 8.11).

Total GSH is regulated by a feedback reaction by γ-glutamylcysteine synthase. The dietary availability of sulfur amino acids may have an influence on the concentrations of cellular GSH [71].

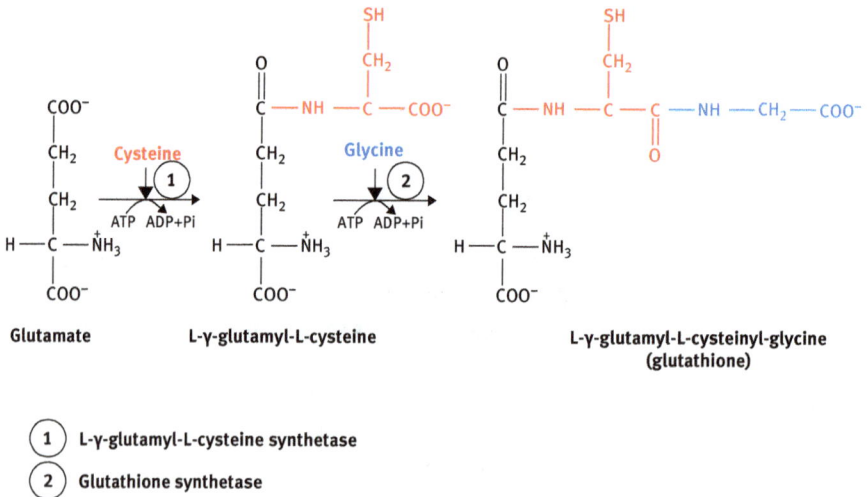

1 L-γ-glutamyl-L-cysteine synthetase
2 Glutathione synthetase

Figure 8.11: Synthesis of glutathione (GSH) (Valls-Bellés, V.).

8.5 Antioxidant nutrients

A second group of nonenzymatic antioxidants are those of exogenous origin, mainly from the diet such as vitamins (C, E), carotenoids, flavonoids, melanoidins, and selenium [72, 73].

8.5.1 Vitamin E

Vitamin E is considered as the main antioxidant sequestering agent of lipophilic radicals in vivo in lipid phase and in the external part of lipoproteins. Eight homologues, alpha, beta, gamma, delta-tocopherols, and tocotrienols, are known and their antioxidant action is given by the group –OH of the aromatic ring of the vitamin that can undergo oxidation reactions. One of its most important functions is the inhibition of lipid peroxidation, acting as a "scavenger" of the peroxyl radical ($ROO^•$). Another important reaction is the reduction of the alpha-tocopherol radical by other antioxidants such as vitamin C, CoQ, and GSH. This reaction is very important as it regenerates or saves vitamin E and also reduces the prooxidant character of the vitamin E radical [74, 75].

Vitamin E protects cells from peroxidation of membranes and their subsequent degeneration, and prevents oxidative damage of LDL, cellular proteins, and DNA [76]. A diet deficient in vitamin E is related to a reduction in the activity of hepatic CAT, GSH peroxidase, and glutathione reductase. It induces lipid peroxidation in the liver and causes cardiovascular and neurological disorders [77, 78].

8.5.2 Vitamin C

Vitamin C or ascorbic acid is a water-soluble vitamin that is found in a very high concentration in numerous tissues and plasma. It is one of the most potent antioxidants in the aqueous phase, which acts at an extracellular and cytosolic level. It reacts with $O_2^{•-}$, H_2O_2, $ROO^•$, $^•OH$, and $^1O_2$ by oxidizing to dehydroascorbate [79]. It acts synergistically with other scavengers such as vitamin E or urate to regenerate them, reducing them by returning them to their active state. Absorption is a function of intake, greater intake less absorption and vice versa. It can also act as a prooxidant in the presence of transition metals (Cu and Fe), generating the hydroxyl radical. This prooxidant effect of ascorbic acid does not normally take place in vivo since in nonpathological situations there is no copper or free iron in extracellular fluids (Figure 8.12) [80–82] (see also Chapter 3).

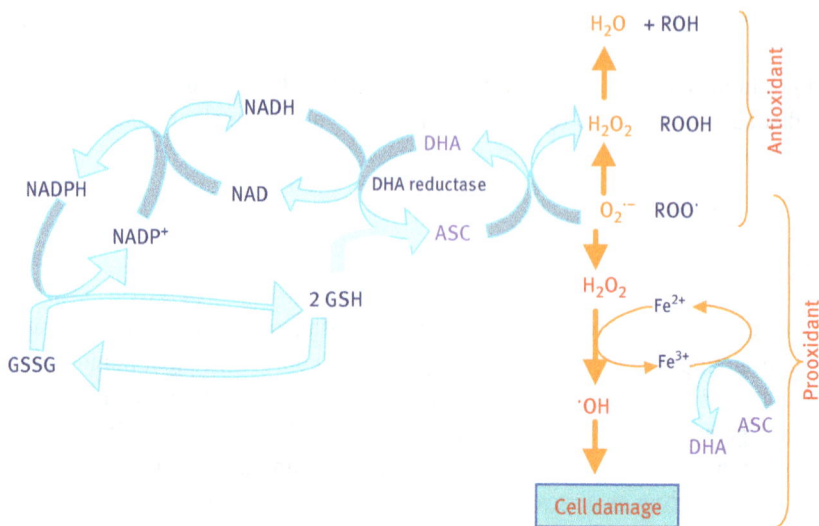

Figure 8.12: Scheme of the antioxidant and prooxidant activities of vitamin C [6].

8.5.3 Carotenoids

Carotenoids are divided into two groups, carotenes and xanthophylls. Carotenes such as α- or β-carotene or lycopene contain only carbon and hydrogen atoms, while xanthophylls such as cryptoxanthin, cantaxanthin, and lutein have at least one oxygen atom in their structure (Chapter 4). Due to the presence of multiple conjugated bonds, the carotenoids and in particular the 4-oxo derivatives such as astaxanthin and canthaxanthin act as free radical scavengers. They are efficient antioxidants against singlet oxygen and peroxide radical, thus contributing to the body's lipophilic antioxidant defense system. In the presence of peroxyl radicals, the oxidative chain ends, as long as the partial O_2 pressures are kept low; otherwise, the oxidative process continues. Therefore, physiological conditions determine the antioxidant or prooxidant character of these compounds (Figure 8.13) [83, 84].

The antioxidant capacity of lycopene has long been known as demonstrated by the protection exerted against DNA damage produced by ROS [85]. This action, in any case, is also explained by the synergistic effect exerted by the different antioxidants present in foods rich in lycopene such as tomatoes; however, it has been shown that high doses of purified lycopene (30 mg/day) have the same protective effect on DNA [85]. Likewise, this antioxidant role is reinforced by the observation that oxidative stress markers, which usually increase after a meal rich in fat, are attenuated in healthy subjects after ingestion of lycopene [86].

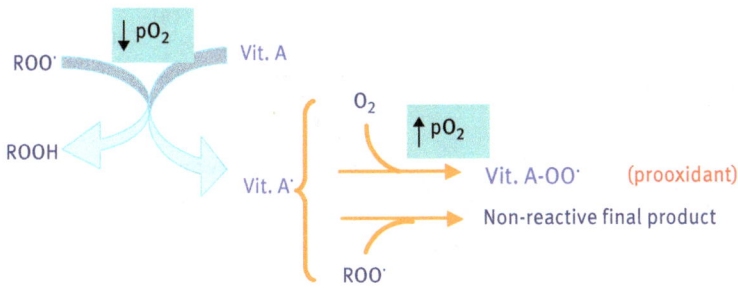

Figure 8.13: Scheme of the antioxidant and prooxidant activities of vitamin A [6].

8.5.4 Phenolic compounds

Phenolic compounds, including flavonoids, are a group of natural substances found in the plant kingdom, in fruits, vegetables, seeds, stems, and flowers, and are, therefore, important constituents of the human diet. The most abundant flavonoids in the diet are flavanols (catechins and proanthocyanidins), anthocyanins, and oxidation products derived from them [87–88]. Flavanols are the basic units of the so-called "condensed tannins" (see also Section 9.2.7).

A mixed western diet provides approximately 1 g of flavonoids per day. Structurally, flavonoids are benzo-γ-pyrone derivatives, their basic structure being diphenylpyranes: two benzene rings linked through a ring-shaped pyrone or heterocyclic pyran [89]. The chemical structure of phenolic compounds is what gives them their ability to act as radical scavengers. The type of compound, the degree of methoxylation, and the number of hydroxyl groups are some of the parameters that determine its antioxidant activity. Thus, according to Rice-Evans et al. [90], the compounds with the highest activity are those with two hydroxyl groups in ortho-position in ring B, which gives high stability to the radical that is formed after the free radical capture reaction: those containing a double 2,3-bond in conjugation with the 4-oxo (C=O) in the C ring, and those compounds that have OH– groups in 3 and 5 and the oxo group (C=O) in 4 in rings A and C. They combine with sugars to form glycosides, the most common way in which they are found in nature. However, it is observed that the aglycone shows a higher antioxidant activity than their corresponding glycosides (Figure 8.14) [91, 92]. They can act as hydrogen donors or chelate metal ions such as iron and copper, preventing the oxidation of LDLs, which are involved in the pathogenesis of coronary heart disease, inhibit aggregation platelet, and protect DNA from oxidative damage. In animal models, the effect of epicatechin has been studied in rat hepatocytes and it has been observed that they inhibit lipid peroxidation and increase cell viability. Likewise, catechin prevents the toxic effects of the antineoplastic adriamycin, and the flavonoid fraction of beer, both blonde and black, decreases the oxidation of lipids and proteins, while increasing cell viability, in liver cells after induction of oxidative stress [16].

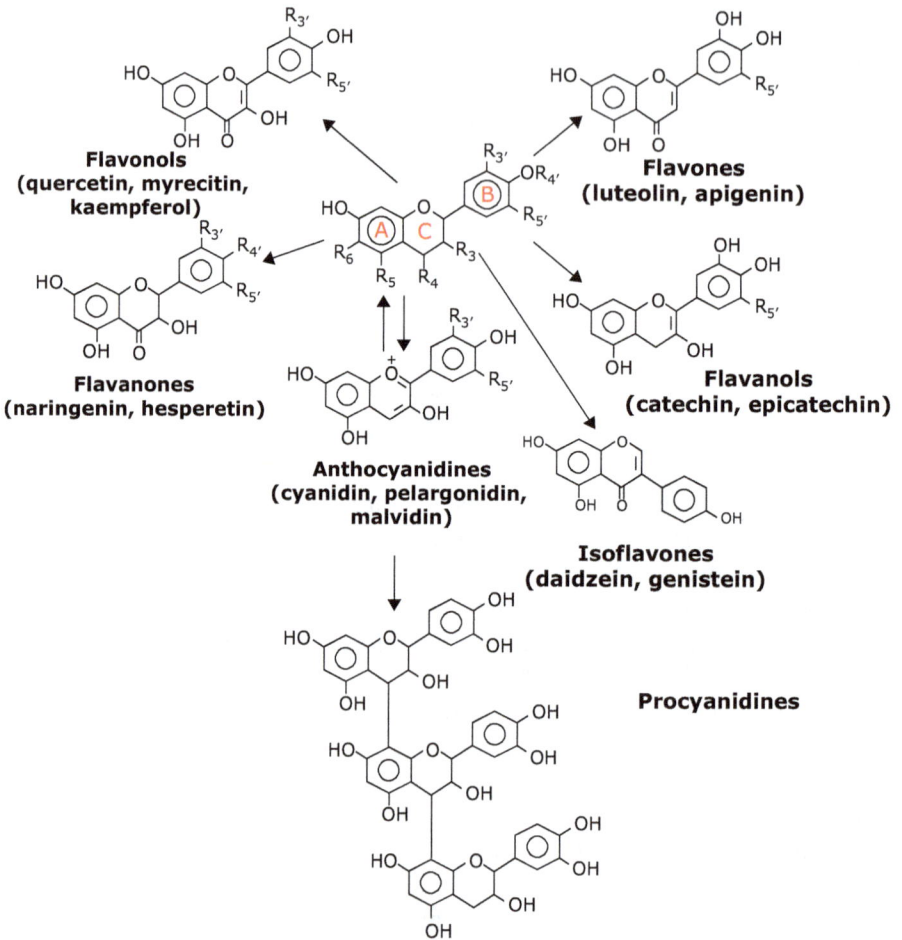

Figure 8.14: Chemical structure of flavonoids (Valls-Bellés, V.).

On the other hand, these compounds present in the diet can favor the defense of endogenous antioxidants through the antioxidant response elements (AREs) found in the promoters of some genes, which are inducible by oxidative stress. Some carcinogenic chemopreventives are thought to act through ARE, by increasing antioxidants and detoxification [93, 94].

It has also been proven that flavonoids in combination have greater antioxidant power than in isolation. Recently, it has been observed that flavonoids stimulate P-glycoprotein, which participates in the mechanism of cellular defense against the action of xenobiotics [51, 57, 95–97].

8.6 Repair systems

8.6.1 Direct

Reduction of the groups (S–S) of sulfur amino acids in proteins by specific enzymes such as disulfide reductase and sulfoxide reductase.

8.6.2 Indirect

In these repair systems, the molecular damage is first recognized and eliminated or degraded, and then the eliminated part is synthesized. This occurs both in the oxidized and lipid peroxide proteins of the carbon chains and in the oxidations of DNA and RNA:

- The oxidized proteins are recognized by proteases and completely degraded to amino acids, and proteins are replaced by de novo synthesis. The oxidized protein may contain two or three oxidized amino acids and probably the rest will be reused in the synthesis. If the protein is well oxidized, proteolytic degradation may be inadequate and cross-linking may occur.
- Lipid peroxidation begins after the extraction of a hydrogen atom in the hydrocarbon chain of PUFAs, whose vinyl-methane structure represents the target and place of initiation of the peroxidation process. In an aerobic environment, interaction of the carbonyl radical ($R^•$) with O_2 occurs, leading to the formation of $ROO^•$. Subsequently, a new hydrogen (sequential reaction) can be extracted and may give rise to the ROOH that will form the alkoxy radical ($RO^•$) by decomposition. This first process is followed by a series of propagation and termination reactions to finally give more stable products, such as malondialdehyde and other carbon products that are removed from the cells (Figure 8.5).
- DNA oxidation. The genetic material is also vulnerable to oxidative damage. Oxidative alterations interrupt transcription, translation, and replication by increasing the number of mutations; on the other hand, the increase in oxidative damage is a natural process that under normal physiological conditions produces a modification in the bases, the ratio being of 1/130,000 bases in nuclear DNA and 1/8,000 in mitochondrial DNA, since the latter is closer to the places where ROS are generated. The DNA repair system is made up of endonucleases and glyosylases [98].

8.7 Oxidative stress and health

There are many pathophysiological processes that are currently associated with the production of free radicals, such as mutagenesis, cell transformation, cancer, diabetes, atherosclerosis, myocardial infarction, ischemia/reperfusion processes, neonate disease (neonatal retinopathy), inflammatory diseases (rheumatoid arthritis and lupus), disorders of the central nervous system (Parkinson's disease and Alzheimer's disease) [99], and aging. In numerous pathologies, reduced levels of antioxidant enzymes or total antioxidants have been observed [100, 101].

8.7.1 Oxidative stress and cardiovascular pathology

CVDs are the leading cause of morbidity and mortality in developed countries. In the year 2020, according to the data from the World Health Organization (WHO), it will be the leading cause of mortality in the world. Cardiovascular pathology is a multifactorial process where obesity, diabetes, hypertension, genetics, dyslipidemia, free radicals, lifestyle, and so on are involved, with hyperlipidemia being one of the greatest risks in atherosclerotic processes. Atherosclerosis is a very complex process in which LDL and cell proliferation are involved in the endothelium [102–103].

There are different hypotheses to explain the processes associated with the development of atherosclerosis; today one of the most accepted is oxidative modification. LDL is trapped in the subendothelial space where it is susceptible to oxidative modification by resident vascular cells, such as smooth muscle cells, endothelial cells, and macrophages. Oxidized LDL stimulates monocytic chemotaxis, prevents monocytic outflow, and supports the formation of foam cells. Once formed, oxidized LDL also causes endothelial dysfunction and damage, and foam cells become necrotized due to the accumulation of oxidized LDL (Figure 4.15) [104].

Currently, there is much evidence to indicate that free radicals, lipid peroxidation, and oxidative modification of LDL are involved in the process of initiation of atherosclerosis [105, 106], giving greater validity to the last hypothesis described [107–109].

8.7.1.1 Oxidation of LDL

In atherosclerotic lesions, the existence of different oxidative modifications that take place on lipoprotein has been observed, in addition to the production of ROS and RNS by vascular cells. Therefore, atherosclerosis has been represented as a state of high oxidative stress characterized by lipid and protein oxidation in the vascular wall. Thus, recent studies have established the presence of oxidized lipids [110], such as peroxides, hydroperoxides, and epoxides, in those lesions. Similarly, when reactive species act on cholesterol molecules, they produce cholesterol hydroperoxides and oxysterols. In addition, there is evidence of protein oxidation in such lesions [111].

Atherosclerosis and its vascular pathological consequences have been shown to be accentuated with oxidative stress. In cases in which these oxidation phenomena occur on plasma lipoproteins (LDLs), they significantly increase their atherogenic power [112]. These oxidized LDLs are responsible for a series of effects as follows:

- They are captured more easily by macrophages, which produce their enrichment in cholesterol esters and the formation of foam cells.
- They inhibit macrophage motility in the arterial wall.
- They alter gene expression, inducing cytokine production.
- They can adversely alter coagulation processes, such as the alteration of platelet aggregation.

8.7.2 Oxidative stress and inflammation processes

There has been a growing interest for a long time to establish the exact role that oxidative processes play in the pathogenesis of inflammation, rheumatoid arthritis, asthma, psoriasis, and contact dermatitis among other diseases, all of them with a possible common link with oxidative stress [113].

The set of processes associated with the inflammatory response are very complex [114] and often involve the action of ROS. Numerous mediators have been described that could begin to amplify the inflammatory response, such as histamine, serotonin, cytokines, and tumor necrosis factor. Inflammation plays an important role in the development of numerous pathologies among which we can mention type II diabetes as well as its associated pathologies such as obesity [115].

In fact, the relationship between inflammation, diabetes, and diet has been verified in different studies carried out both on animals and with people. Thus, blood markers of inflammation, such as C-reactive protein and interleukin-6, are often considered as predictive parameters of diabetes. A specific case that illustrates the complexity of these processes is prostate inflammation, which may contribute to the appearance of cancer in this gland. This risk decreases after ingestion of anti-inflammatory drugs and antioxidant substances [116]. According to some authors, antioxidants may reduce inflammation in the respiratory tract, as well as the increased reactivity that exists in asthmatics [117]. Thus, some studies have shown how antioxidants can reduce the expression of certain types of cytokines (interleukin-18) in these asthmatic processes, inhibiting the activity of NF-kappa B, thereby suggesting that ROS could regulate interleukin expression [115, 117, 118].

There is also a growing interest in the role of antioxidants in the control of other diseases with an inflammatory base, such as allergy. It is believed that the antioxidant status of the individual is associated with an increased immune response, although there is no evidence that a lower response to allergens is associated with higher levels of antioxidants [119]. The possible action of antioxidants on immune function has also aroused the interest of researchers on the possible effect of these compounds on pathologies linked to immunological disorders such as multiple sclerosis [120]. Quercetin and other antioxidant polyphenols have a direct relationship with the decrease in inflammation, an effect that would be mediated by the inhibition of proinflammatory cytokines such as tumor necrosis factor [121].

8.8 Mechanisms of ROS elimination

Diet composition plays an important role in oxidative stress as it can contribute to both oxidative damage and antioxidant defense [122]. This partially explains the relationship between diet and some chronic diseases such as atherosclerosis and cancer. More than 2,000 epidemiological studies show that most of the protective effects against a variety of mainly CVDs and cancer correlate with a high intake of fruits and vegetables. Traditionally, nutrition has been recognized as an important factor in the modulation of different diseases and longevity. Diet, particularly through fruit, vegetables, nuts, and beverages made from vegetables, such as beer [16] and wine, provides antioxidants, such as vitamins and other phytochemicals, which are an important exogenous source capable of increasing the cellular response to oxidative stress [123, 124]. In epidemiological studies by Gey et al. (WHO/Proyecto Mónica) [125], where they determined plasma antioxidants (alpha-tocopherol, ascorbate, vitamin A, carotenoids, and selenium) in 16 European populations, the incidence of mortality from ischemic heart disease shows an inverse relationship with plasma levels of alpha-tocopherol ($P = 0.002$). Studies on the intake of fruits and vegetables in Europe show the high difference in consumption between Northern and Southern Europe. These studies have shown a lower incidence of CVDs and cancer in Southern European countries, where a Mediterranean diet is consumed, with respect to Nordic countries.

The Mediterranean diet (Chapter 15) is mainly constituted by a high consumption of fruits, vegetables, legumes, unrefined cereals [126, 127], low levels of meat and dairy products, and a moderate consumption of wine, beer, and fish. Olive oil is a major contributor to this diet as a source of fat. The Mediterranean diet is considered a diet with high antioxidant activity [128–131], and nowadays it is known that fruits, vegetables, olive oil, and so on contain other compounds, even with greater antioxidant activity than vitamins and flavonoids [132, 133].

The Rotterdam study investigated the relationship between dietary intake of flavonoids and antioxidant vitamins and the risk of ischemic cerebrovascular accident (CVA) for an average of 6 years, observing that a high dietary intake of antioxidants is associated with a lower risk of CVA. It has been epidemiologically proven that a high dietary intake of fruits and vegetables produces a 50% reduction in the risk of digestive and respiratory tract cancers [12]. These studies support the hypothesis that natural antioxidants from food can protect cells from oxidative stress [134, 135].

In addition, we have to keep in mind that not all effects are due to antioxidant vitamins and flavonoids, but other compounds found in foods can indirectly contribute to the reduction of these pathologies. For example, high plasma levels of homocysteine are a risk factor for CVDs, while folates reduce homocysteine levels in plasma; therefore, folate from the diet indirectly contributes to reducing the risk of CVDs [134].

To reduce or eliminate free radicals, you must follow a diet rich in fruits and vegetables, and include a minimum of four different colors every day, in order to obtain a wide variety of compounds with antioxidant activity [135].

Dietary antioxidants, which include catechins, flavonoids, anthocyanins, stilbenes, and carotenoids, demonstrate benefits in the prevention and/or support of therapy in chronic diseases. Antioxidants reshape DNA methylation patterns through multiple mechanisms, including regulation of epigenetic enzymes and chromatin remodeling complexes. These effects may further contribute to the antioxidant properties of the compounds [135–138].

At present, and due to their stability, the most effective antioxidants are polyphenols with different chemical structures, which give them different antioxidant activities [139, 140]. They present antioxidant and anti-inflammatory activity against a wide variety of pathologies that are very typical in developed countries [141–143]. Recent studies have shown that lycopene reduces serum lipid levels, endothelial dysfunction, inflammation, and blood pressure, and increases antioxidant potential [144]. Likewise, it has been observed that these effects are significantly greater if a tomato paste (rich in lycopene) is mixed with extra virgin olive oil. These natural antioxidants, which can also improve the nutritional value of food, can lead to new forms if used in food [145]. CVD is one of the leading causes of morbidity and mortality, and atherosclerosis is the common root of most CVDs. Oxidative stress is one of the most important factors driving atherosclerosis and its complications [146, 147].

On the other hand, several components, natural bioactives, such as polyphenols, have anticancer properties [148]. In a recent review on the anticancer, antiprogressive, and apoptotic effects of lycopene, it has been reported that this carotenoid exerts a powerful positive effect in prostate cancer [149].

In the process of aging, the dietary contribution of certain foods rich in antioxidants is reduced, due to different circumstances. In addition, the mitochondrial function is closely related to the processes of cell aging. If we manage to adapt a diet with combinations of certain foods rich in antioxidants, we will achieve healthier aging, with lower risks of associated pathologies [150, 151].

References

[1] McCord JM, Fridovich I. Superoxide dismutase. An enzymic function for erythrocuprein (hemocuprein). J Biol Chem 1969;244(22):6049–55.

[2] Bannister JV. Foreword. Handbook of methods of oxygen radical research. Boca Ratón, FL, USA: RA Greenwald ed. CRC Press; 1986.

[3] Michaelis L. Fundamentals of oxidations and reduction. In: Green DE, editor. En currents in biochemical research. New York: Interscience; 1946, p. 207–27.

[4] Gilbert DL. Oxygen and living processes. An interdisciplinary approach. New York: Springer Verlag; 1981.

[5] Jomova K, Jenisova Z, Feszterova M, Baros S, Liska J, Hudecova D, Rhodes CJ, Valko M. Arsenic: Toxicity, oxidative stress and human disease. J Appl Toxicol 2011;31(2):95–107.

[6] Halliwell B. Antioxidants in human health and disease. Ann Rev Nutr 1996;16:33–50.

[7] Kehrer JP, Klotz LO. Free radicals and related reactive species as mediators of tissue injury and disease: Implications for Health. Crit Rev Toxicol 2015;45(9):765–98.

[8] Auten RL, Davis JM. Oxygen toxicity and reactive oxygen species: The devil is in the details. Pediatr Res 2009;66(2):121–27.

[9] Cadenas E, Packer L, Traber MG. Antioxidants, oxidants, and redox impacts on cell function – A tribute to Helmut Sies. Arch Biochem Biophys 2016;595:94–98.

[10] Sies H, Berndt C, Jones DP. Oxidative stress. Ann Rev Biochem 2017;86:715–48.

[11] Halliwell B, Gutteridge JM. Free radicals in biology and medicine. 2nd edition. New York: Oxford University Press; 1988.

[12] Lindsay DG, Astley SB. European research on the functional effects of dietary antioxidants-EUROFEDA. Mol Aspects Med 2002;23:1–38.

[13] Fang YZ, Yang S, Wu G. Free radicals, antioxidants and nutrition. Nutrition 2002;18(10):872–78.

[14] Ames BN, Viguie CA, Frei B, Shinenaga MK, Paches L, Brooks GA. Antioxidant status and indexes of oxidative stress during consecutive days of exercise. J Appl Physiol 1993;75(2):566–75.

[15] Onukwufor JO, Berry BJ, Wojtovich AP. Physiologic implications of reactive oxygen species production by mitochondrial complex I reverse electron transport. Antioxidants (Basel) 2019;8(8).

[16] Valls- Bellés V, Codoñer-Chanch P, Gonzaléz San-José ML, Muñiz Rodríguez P. Biodisponibilidad de los flavonoides de la cerveza. Efecto antioxidante "in Vivo". Centro de Información Cerveza y Salud. DL- M-36370-2005. Monografía 14:105.

[17] Halliwell B, Gutteridge JMC. Free radicals in biology and medicine. 3rd edition. Oxford University Press; 1999.

[18] McCord JM. The evolution of free radicals and oxidative stress. Am J Med 2000;108:652–59.

[19] Fang YZ, Yang S, Wu G. Free radicals, antioxidants and nutrition. Nutrition 2002;18(10):872–78.

[20] Costa RA, Romagna CD, Pereira JL, Souza-Pinto NC. The role of mitochondrial DNA damage in the cytotoxicity of reactive oxygen species. J Bioenerg Biomembr 2011;43(1):25–28.

[21] Xiao M, Zhong H, Xia L, Tao Y, Yin H. Pathophysiology of mitochondrial lipid oxidation: Role of 4-hydroxynonenal (4-HNE) and other bioactive lipids in mitochondria. Free Radic Biol Med 2017;111:316–27.

[22] Hrycay EG, Bandiera SM. Involvement of cytochrome P450 in reactive oxygen species formation and cancer. Adv Pharmacol 2015;74:35–84.

[23] Nathan C, Xien QW. Regulation of biosynthesis of nitric oxide. J Biol Chem 1994;269(19):13725–28.

[24] Fahn S, Cohen G. The oxidant stress hypothesis in Parkinson´s disease: Evidence supporting it. Ann Neurol 1992;32(6):804–12.

[25] Dröge W. Free radicals in the physiological control of cell function. Physiol Rev 2002;82:47–95.

[26] Schmidt HM, Kelley EE, Straub AC. The impact of xanthine oxidase (XO) on hemolytic diseases. Redox Biol 2019;21:101072.

[27] Mataix-Albert B. 2005. Ph. D. thesis. University of Granada (Spain). Available at: https://digibug.ugr.es/bitstream/handle/10481/762/15743226.pdf?sequence=1

[28] Betteridge J. What is de oxidative stress?. Metabolism 2000;49(2):3–8.

[29] Muñiz P, Saez P, Iradi A, Viña J, Oliva MR, Saez GT. Differences between cysteine and homocysteine in the induction of deoxyribose degradation and DNA damage. Free Radic Biol Med 2001;30 (4):354–62.

[30] Smith RW, Wang J, Schültke E, Seymour CB, Bräuer-Krisch E, Laissue JA. Proteomic changes in the rat brain induced by homogenous irradiation and by the bystander effect resulting from high energy synchrotron X-ray microbeams. Int J Rad Biol 2012:118–27.

[31] Asavei T, Bobeica M, Nastasa V, Manda G, Naftanaila F, Bratu O, Mischianu D, Cernaianu MO, Ghenuche P, Savu D, Stutman D, Tanaka KA, Radu M, Doria D, Vasos PR. Laser-driven radiation: Biomarkers for molecular imaging of high dose-rate effects. Med Phys 2019;46(10):e726–e734.

[32] Kelly FJ, Fussell JC. Role of oxidative stress in cardiovascular disease outcomes following exposure to ambient air pollution. Free Radic Biol Med 2017;110:345–67.

[33] Valls V, Castellucio C, Fato R, Genova ML, Bovina C, Sáez G, Marchetti M, Parenti-Castelli G, Lenaz G. Protective effect of exogenous coenzyme Q against damage rat liver. Biochem Mol Biol Inter 1996;33(4):633–42.

[34] Valls V, Peiró C, Muñiz P, Sáez GT. Age-related changes in antioxidant status and oxidative damage to lipid and DNA in mitochondria of rat liver. Process Biochem 2005;40:903–08.

[35] Valls-Belles V, Torres C, Muñiz P, Beltran S, Martinez-Alvarez JR, Codoñer-Franch P. Effect of grape seed polyphenols before adriamycin toxicity in rat hepatocytes. European J Nutr 2006;10:1–7.

[36] Kim SM, Hwang KA, Choi KC. Potential roles of reactive oxygen species derived from chemical substances involved in cancer development in the female reproductive system. BMB Rep 2018;51(11):557–62.

[37] Jackson MJ, Papa S, Bolaños J, Bruckdorfer R, Carlsen H, Elliott RM, Flier J, Griffiths HR, Heales S, Holst B, Lorusso M, Lund E, Øivind Moskaug J, Moser U, Di Paola M, Polidori MC, Signorile A, Stahl W, Viña-Ribes J, Astley SB. Antioxidants, reactive oxygen and nitrogen species, gene induction and mitochondrial function. Mol Aspects Med. 2002;23:209–85. doi: 10.1016/s0098-2997(02)00018-3

[38] Wink DD, Wink CB, Nims RW, Ford PC. Oxidizing intermediates generated in the Fenton reagent. Kinetic arguments against the intermediary of the hydroxyl radical. Environ Health Perspect 1994;102S(3):11–15.

[39] Halliwell B, Gutteridge JM. Biologically relevant metal iron-dependent hydroxyl radical generation, An update. FEBS Lett 1992;307(1):108–12.

[40] Elgendy FM, Abou-Seif MA. Photolysis and membrane lipid peroxidation of human erythrocytes by m-chloroperbenzoic acid. Photochem Photobiol 1998;277(1):1–11.

[41] Shadyro O, Lisovskaya A. ROS-induced lipid transformations without oxygen participation. Chem Phys Lipids 2019;221:176–83.

[42] Peña-Bautista C, Baquero M, Vento M, Cháfer-Pericás C. Free radicals in Alzheimer's disease: Lipid peroxidation biomarkers. Clin Chim Acta 2019;491:85–90.

[43] Okumura H, Ishii H, Pichiorri F, Croce CM, Mori M, Huebner K. Fragile gene product, Fhit, in oxidative and replicative stress responses. Cancer Sci 2009;100(7):1145–50.

[44] Hauck AK, Huang Y, Hertzel AV, Bernlohr DA. Adipose oxidative stress and protein carbonylation. J Biol Chem 2019;294(4):1083–88.

[45] Griffiths HR, Moller L, Bartosz G, Bast A, Bertoni-Freddari C, Collins A, et al. Biomarkers. Mol Asp Med 2002;23:101–208.

[46] Floyd RA. The role of 8-hydroxyguanine in carcinogenesis. Carcinogenesis 1990;11(9):1447–50.

[47] Dizdaroglu M. Chemical determination of oxidative DNA damage by gas chromatography-mass spectometry. Methods Enzymol 1994;234:3–6.

[48] Collins AR. Measuring oxidative damage to DNA and its repair with the comet assay. Biochim Biophys Acta 2014;1840(2):794–800.

[49] Valls-Belles V, Torres MC, Boix L, Muñiz P, Codoñer-Franch P. α-Tocopherol, MDA–HNE and 8-OHdG levels in liver and heart mitochondria of adriamycin-treated rats fed with alcohol-free beer. *Toxicology* 2008;249:97–101.

[50] Fiotakis K, Valavanidis A. Comparative study of the formation of oxidative damage marker 8-hydroxy-2′-deoxyguanosin (8-OhdG) adduct from the nucleoside 2′-deoxyguanosine by transition of particulate matter in relation to metal content and redox activity. Free Radic Res 2005:39–1071–1081.

[51] Beard WA, Batra VK, Wilson SH. DNA polymerase structure-based insight on the mutagenic properties of 8-oxoguanine. Mutat Res 2010;703(1):18–23.

[52] Monzo-Beltran L, Vazquez-Tarragón A, Cerdà C, Garcia-Perez P, Iradi A, Sánchez C, Climent B, Tormos C, Vázquez-Prado A, Girbés J, Estáñ N, Blesa S, Cortés R, Chaves FJ, Sáez GT. (One-year

follow-up of clinical, metabolic and oxidative stress profile of morbid obese patients after laparoscopic sleeve gastrectomy. 8-oxo-dG as a clinical marker. Redox Biol 2017:389–402.

[53]   Osawa T. Development and application of oxidative stress biomarkers. Biosci Biotechnol Biochem 2018;82(4):564–72.

[54]   Beetch M, Harandi-Zadeh S, Shen K, Lubecka K, Kitts DD, O'Hagan HM, Stefanska B. Dietary antioxidants remodel DNA methylation patterns in chronic disease. Br J Pharmacol 2019. 10.1111/bph.14888.

[55]   Martínez Álvares JR, Villarino Marín A, Valls Bellés V, Codoñer Franch P, Lopez Jaén AB, Yao Lee S, Ambrós Marigómez MC. El lúpulo contenido en la cerveza, su efecto antioxidante en un grupo controlado de población. In: Martinez Alvares JR, Villarino MA, Valls Bellés V, editors. Monografía. Editada por Centro de Información Cerveza y Salud; 2007, p. 113.

[56]   Yoshida H, Kisugi R. Mechanisms of LDL oxidation. Clin Chim Acta 2010;411(23–24):1875–82.

[57]   Gradinaru D, Borsa C, Ionescu C, Prada GI. Oxidised LDL and NO synthesis–Biomarkers of endothelial dysfunction and ageing. Mech Ageing Dev 2015;151:101–13.

[58]   Kiokias S, Proestos C, Oreopoulou V. Effect of natural food antioxidants against LDL and DNA oxidative changes. Antioxidants (Basel) 2018;7(10):E133. 10.3390/antiox7100133.

[59]   Choi SH, Sviridov D, Miller YI. Oxidised cholesteryl esters and inflammation. Biochim Biophys Acta Mol Cell Biol Lipids 2017;1862(4):393–97.

[60]   Miyoshi N, Iuliano L, Tomono S, Ohshima H. Implications of cholesterol autoxidation products in the pathogenesis of inflammatory diseases. Biochem Biophys Res Commun 2014;446(3):702–08.

[61]   Murdolo G1, Bartolini D2, Tortoioli C3, Piroddi M2, Iuliano L4, Galli F2. Lipokines and oxysterols: novel adipose-derived lipid hormones linking adipose dysfunction and insulin resistance. Lipokines and oxysterols: Novel adipose-derived lipid hormones linking adipose dysfunction and insulin resistance. Free Radic Biol Med 2013;65:811–20.

[62]   Valko M, Rhodes CJ, Moncol J, Izakovic M, Mazur M. Free radicals, metals and antioxidants in oxidative stress-induced cancer. Chem Biol Interact 160(1):1–40.

[63]   Fang YZ, Yang S, Wu G. Free radicals, antioxidants and nutrition. Nutrition 2002;18(10):872–79.

[64]   Muñiz P, Saez P, Iradi A, Viña J, Oliva MR, Saez GT. Differences between cysteine and homocysteine in the induction of deoxyribose degradation and DNA damage. Free Radic Biol Med 2001;30 (4):354–62.

[65]   Benov L, Sztejnberg L, Fridovich I. Critical evaluation of the use of hydroethidine as a measure of superoxide anion radical. Free Radic Biol Med 1998;25(7):826–31.

[66]   Mates JM, Pérez-Gómez C, Núñez de castro I. Antioxidant enzymes and human diseases. Clin Biochem 1999;32(8):595–603.

[67]   Miao L, Clair DK. Regulation of superoxide dismutase genes: Implications in disease. Free Radic Biol Med 2009;47(4):344–56.

[68]   Lleídas F, Rangel P, Hansberg W. Oxidation of catalase by singlet oxygen. J Biol Chem 1998:10630–37.

[69]   Ding L, Liu Z, Zhu Z, Luo G, Zhao D, Ni J. Biochemical characterisation of selenium-containing catalytic antibody as a cytosolic glutathione peroxidase mimic. Biochem J 1998;332:251–55.

[70]   Taylor S, Davenport LD, Speranza MJ, Mullenbach GT, Linch RE. Glutathione peroxidase protects cultured mammalian cells from the toxicity of adriamycin and paraquat. Arch Biochem Biophys 1993;305:600–05.

[71]   Morillas Ruiz Juana M. Los antioxidantes en la prevención del estrés oxidativo en la actividad física. Editorial Planeta SA; 2010.

[72]   Gebicki JM, Nauser T, Domazou A, Steinmann D, Bounds PL, Koppenol WH. Reduction of protein radicals by GSH and ascorbate: Potential biological significance. Amino Acids 2010;39(5):1131–37.

[73]   Surai PF, Kochish II, Fisinin VI, Kidd MT. Antioxidant defence systems and oxidative stress in poultry biology: An update. Antioxidants (Basel) 2019;8(7):E235. 10.3390/antiox8070235.

[74] Ashor AW, Siervo M, Lara J, Oggioni C, Afshar S, Mathers JC. Effect of vitamin C and vitamin E supplementation on endothelial function: A systematic review and meta-analysis of randomised controlled trials. Br J Nutr 2015;113(8):1182–94.

[75] Abudu N, Miller JJ, Attaelmannan M, Levinson SS. Vitamins in human arteriosclerosis with emphasis on vitamin C and vitamin E. Clin Chim Acta 2004;339(1–2):11–25.

[76] Schubert M, Kluge S, Schmölz L, Wallert M, Galli F, Birringer M, Lorkowski S. Long-chain metabolites of vitamin E: Metabolic activation as a general concept for lipid-soluble vitamins?. Antioxidants (Basel) 2018;7(1):E10. 10.3390/antiox7010010.

[77] Codoñer-Franch P, López-Jaén AB, Muñiz P, Sentandreu E, Valls-Bellés V. Mandarin juice improves the antioxidant status of hypercholesterolemic children. J Pediatr Gastroenterol Nutr 2008;47 (3):349–55.

[78] Birringer M, Lorkowski S. Vitamin E: Regulatory role of metabolites. IUBMB Life 2019;71(4):479–86.

[79] Cobley JN, McHardy H, Morton JP, Nikolaidis MG, Close GL. Influence of vitamin c and vitamin e on redox signaling: Implications for exercise adaptations. Free Radic Biol Med 2015;84:65–76.

[80] Dhremer E, Valls V, Muñiz P, Cabo J, Sáez GT. 8-Hydroxydeoxyguanosine and antioxidant status in rat liver fed with olive and corn oil diets. Effect of ascorbic acid supplementation. J Food Lipids 2001;8:281–94.

[81] Lebel M, Massip L, Garand C, Thorin E. Ascorbate improves metabolic abnormalities in Wrn mutant mice but not the free radical scavenger catechin. Ann N Y Acad Sci 2010;1197:40–44.

[82] Traber MG, Stevens JF. Vitamins C and E: Beneficial effects from a mechanistic perspective. Free Radic Biol Med 2011;51(5):1000–13.

[83] Kamiloglu S, Toydemir G, Boyacioglu D, Beekwilder J, Hall RD, Capanoglu E. A review on the effect of drying on antioxidant potential of fruits and vegetables. Crit Rev Food Sci Nutr 2016;56(1): S110–28.

[84] Martínez-Tomás R, Larqué E, González-Silvera D, Sánchez-Campillo M, Burgos MI, Wellner A, Parra S, Bialek L, Alminger M, Pérez-Llamas F. Effect of the consumption of a fruit and vegetable soup with high in vitro carotenoid bioaccessibility on serum carotenoid concentrations and markers of oxidative stress in young men. Eur J Nutr 2012;51(2):231–38.

[85] Young AJ, Lowe GL. Carotenoids – antioxidant properties. Antioxidants (Basel) 2018;7(2):28.

[86] Müller L, Caris-Veyrat C, Lowe G, Böhm V. Lycopene and its antioxidant role in the prevention of cardiovascular diseases-a critical review. Crit Rev Food Sci Nutr 2016;56(11):1868–78.

[87] Denniss SG, Haffner TD, Kroetsch JT, Davidson SR, Rush JW, Hughson RL. Effect antioxidants and biomarkers of endothelial health in young, healthy individuals. Vasc Health Risk Manag 2008;4 (1):213–22.

[88] Joseph SV, Edirisinghe I, Burton-Freeman BM. Fruit polyphenols: A review of anti-inflammatory effects in humans. Crit Rev Food Sci Nutr 2016;56(3):419–44.

[89] Bravo L. Polyphenols: Chemistry, dietary sources, metabolism, and nutritional significance. Nutr Rev 1998;56(11):317–33.

[90] Rice-Evans CA, Miller NJ, Paganga G. Structure-antioxidant activity relationships of flavonoids and phenolic acids. Free Radic Biol Med 1996;20(7):933–56.

[91] Weseler AR, Bast A. Masquelier's grape seed extract: From basic flavonoid research to a well-characterised food supplement with health benefits. Nutr J 2017;16(1):5.

[92] Zhang YJ, Gan RY, Li S, Zhou Y, Li AN, Xu DP, Li HB. Antioxidant phytochemicals for the prevention and treatment of chronic diseases. Molecules 2015;20(12):21138–56.

[93] Mattoo AK, Shukla V, Fatima T, Handa AK, Yachha SK. Genetic engineering to enhance crop-based phytonutrients (nutraceuticals) to alleviate diet-related diseases. Adv Exp Med Biol 2011;698:122–43.

[94] Murray M, Dordevic AL, Ryan L, Bonham MP. An emerging trend in functional foods for the prevention of cardiovascular disease and diabetes: Marine algal polyphenols. Crit Rev Food Sci Nutr 2018;58(8):1342–58.

[95] Valls-Belles V, Torres MC, Muñiz P, Codoñer-Franch P. Changes in mitochondrial rat liver and heart enzymes (complex I and complex IV) and coenzymes Q9 and Q10 levels induced by alcohol-free beer consumption. Eur J Nutr 2010;49(3):181–87.

[96] Medina-Remón A, Tresserra-Rimbau A, Pons A, Tur JA, Martorell M, Ros E, Buil-Cosiales P, Sacanella E, Covas MI, Corella D, Salas-Salvadó J, Gómez-Gracia E, Ruiz-Gutiérrez V, Ortega-Calvo M, García-Valdueza M, Arós F, Saez GT, Serra-Majem L, Pinto X, Vinyoles E, Estruch R, Lamuela-Raventos RM. PREDIMED study investigators. Effects of total dietary polyphenols on plasma nitric oxide and blood pressure in a high cardiovascular risk cohort. The PREDIMED randomized trial. Nutr Metab Cardiovasc Dis 2015;25(1):60–67.

[97] Martín-Peláez S, Covas MI, Fitó M, Kušar A, Pravst I. Health effects of olive oil polyphenols: Recent advances and possibilities for the use of health claims. Mol Nutr Food Res 2013;57(5):760–71.

[98] Sedelnikova OA, Redon CE, Dickey JS, Nakamura AJ, Georgakilas AG, Bonner WM. Role of oxidatively induced DNA lesions in human pathogenesis. Mutat Res 2010;704(1–3):152–58.

[99] Srinivas Bharath MM. Post-translational oxidative modifications of mitochondrial complex I (NADH: Ubiquinone Oxidoreductase): implications for pathogenesis and therapeutics in human diseases. J Alzheimer's Dis 2017;60(s1):S69–S86.

[100] Conti E, Musumeci MB, De Giusti M, Dito E, Mastromarino V, Autore C, Volpe M. IGF-1 and atherothrombosis: Relevance to pathophysiology and therapy. Clin Sci (Lond) 2011;120(9):377–402.

[101] Naik E, Dixit VM. Mitochondrial reactive oxygen species drive proinflammatory cytokine production. J Exp Med 2011;208(3):417–20.

[102] Sugamura K, Keaney JF Jr. Reactive oxygen species in cardiovascular disease. Free Radic Biol Med 2011;51(5):978–92.

[103] Cadenas S. ROS and redox signaling in myocardial ischemia-reperfusion injury and cardioprotection. Free Radic Biol Med 2018;117:76–88.

[104] Tsimikas S, Miller YI. Oxidative modification of lipoproteins: Mechanisms, role in inflammation and potential clinical applications in cardiovascular disease. Curr Pharm Des 2011;17(1):27–37.

[105] Codoñer-Franch P, Bataller Alberola A, Domingo Camarasa JV, Escribano Moya MC, Valls Bellés V. Influence of dietary lipids on the erythrocyte antioxidant status of hypercholesterolemic children. Eur J Pediatrics 2009;168:321–27.

[106] Codoñer-Franch P, Murria-Estal R, Tortajada-Girbés M, Castillo-Villaescusa C, Valls-Bellés V, Alonso-Iglesias E. New factors of cardiometabolic risk in severely obese children: Influence of pubertal status. Nutr Hosp 2010;25(5):845–51.

[107] De Rosa S, Cirillo P, Paglia A, Sasso L, Di Palma V, Chiariello M. Reactive oxygen species and antioxidants in the pathophysiology of cardiovascular disease: Does the actual knowledge justify a clinical approach?. Curr Vasc Pharmacol 2010;8(2):259–75.

[108] Saji N, Francis N, Schwarz LJ, Blanchard CL, Santhakumar AB. Rice bran derived bioactive compounds modulate risk factors of cardiovascular disease and Type 2 diabetes mellitus: An updated review. Nutrients 2019;11(11):E2736. 10.3390/nu11112736.

[109] Reverri EJ, Morrissey BM, Cross CE, Steinberg FM. Inflammation, oxidative stress, and cardiovascular disease risk factors in adults with cystic fibrosis. Free Radic Biol Med 2014;76:261–77.

[110] Yoshida H, Kisugi R. Mechanisms of LDL oxidation. Clinica Chimica Acta 2010;411:1875–82.

[111] Tsimikas S, Miller YI. Oxidative modification of lipoproteins: Mechanisms, role in inflammation and potential clinical applications in cardiovascular disease. Curr Pharm Des 2011;17(1):27–37.

[112] Itabe H, Obama T, Kato R. The dynamics of oxidised LDL during atherogenesis. J Lipids 2011;2011:418313.

[113] Liang N, Kitts DD. Role of chlorogenic acids in controlling oxidative and inflammatory stress conditions. Nutrients 2015;8(1):E16. 10.3390/un8010016.

[114] Geronikaki AA, Gavalas AM. Antioxidants and inflammatory disease: Synthetic and natural antioxidants with anti-inflammatory activity. Comb Chem High Throughput Screen 2006;9 (6):425–42.

[115] Gupta SC, Kim JH, Kannappan R, Reuter S, Dougherty PM, Aggarwal BB. Role of nuclear factor κB-mediated inflammatory pathways in cancer-related symptoms and their regulation by nutritional agents. Exp Biol Med (Maywood) 2011;236(6):658–71.

[116] Zenkel M, Lewczuk P, Jünemann A, Kruse FE, Naumann GO, Schlötzer-Schrehardt U. Proinflammatory cytokines are involved in the initiation of the abnormal matrix process in pseudoexfoliation syndrome/glaucoma. Am J Pathol 2010;176(6):2868–78.

[117] Lee KS, Kim SR, Park SJ, Min KH, Lee KY, Jin SM, Yoo WH, Lee YC. Antioxidant down-regulates IL-18 expression in asthma. Mol Pharmacol 2006;70(4):1184–93.

[118] Navarrete-Reyes AP, Montaña-Alvarez M. Inflammaging. Aging inflammatory origin. Rev Invest Clin 2009;61(4):327–36.

[119] Dunstan JA, Breckler L, Hale J, Lehmann H, Franklin P, Lyonso G, Ching SY, Mori TA, Barden A, Prescott SL. Associations between antioxidant status, markers of oxidative stress and immune responses in allergic adults. Clin Exp Allergy 2006;36(8):993–1000.

[120] Libby P. Role of inflammation in atherosclerosis associated with rheumatoid arthritis. Am J Med 2008;121(10):S21–31.

[121] Papaconstantinou J. The role of signaling pathways of inflammation and oxidative stress in development of senescence and aging phenotypes in cardiovascular disease. Cells 2019;8(11):E1383. 10.3390/cells8111383.

[122] Harasym J, Oledzki R. Effect of fruit and vegetable antioxidants on total antioxidant capacity of blood plasma. Nutrition 2014;30(5):511–17.

[123] Riccioni G, Bazzano LA. Antioxidant plasma concentration and supplementation in carotid intima media thickness. Expert Rev Cardiovasc Ther 2008;6(5):723–28.

[124] Bulló M, Lamuela-Raventós R, Salas-Salvadó J. Mediterranean diet and oxidation: Nuts and olive oil as important sources of fat and antioxidants. Curr Topics Med Chem 2011;11(14):1797–810.

[125] Gey KF, Puska P, Jordan P, Moser UK. Inverse correlation between plasma vitamin E and mortality from ischemic heart disease in cross-cultural epidemiology. Am J Clin Nutr 1991;53(1):326S–334S.

[126] Masisi K, Beta T, Moghadasian MH. Antioxidant properties of diverse cereal grains: A review on in vitro and in vivo studies. Food Chem 2016;196:90–97.

[127] Wang Y, Chun OK, Song WO. Plasma and dietary antioxidant status as cardiovascular disease risk factors: A review of human studies. Nutrients 2013;5(8):2969–3004.

[128] Hoffman R, Gerber M. Food processing and the Mediterranean diet. Nutrients 2015;7(9):7925–64.

[129] Ditano-Vázquez P, Torres-Peña JD, Galeano-Valle F, Pérez-Caballero AI, Demelo-Rodríguez P, Lopez-Miranda J, Katsiki N, Delgado-Lista J, Alvarez-Sala-Walther LA. The fluid aspect of the Mediterranean diet in the prevention and management of cardiovascular disease and diabetes: The role of polyphenol content in moderate consumption of wine and olive oil. Nutrients 2019;11(11):E2833. 10.3390/nu11112833.

[130] Soory M. Relevance of nutritional antioxidants in metabolic syndrome, ageing and cancer: Potential for therapeutic targeting. Infect Disord Drug Targets 2009;9(4):400–14.

[131] Martínez-González MA, Salas-Salvadó J, Estruch R, Corella D, Fitó M, Ros E. PREDIMED investigators. Benefits of the Mediterranean diet: Insights From the PREDIMED Study. Prog Cardiovasc Dis 2015;58(1):50–60.

[132] Singh MD, Thomas P, Owens J, Hague W, Fenech M. Potential role of folate in pre-eclampsia. Nutr Rev 2015;73(10):694–722.

[133] Luo H, Chiang HH, Louw M, Susanto A, Chen D. Nutrient sensing and the oxidative stress response. Trends Endocrinol Metab 2017;28(6):449–60.

[134] Afrin S, Gasparrini M, Forbes-Hernandez TY, Reboredo-Rodriguez P, Mezzetti B, Varela-López A, Giampieri F, Battino M. Promising health benefits of the strawberry: A focus on clinical studies. J Agric Food Chem 2016;64(22):4435–48.

[135] Beetch M, Harandi-Zadeh S, Shen K, Lubecka K, Kitts DD, O'Hagan HM, Stefanska B. Dietary antioxidants remodel DNA methylation patterns in chronic disease. Br J Pharmacol 2019. 10.1111/bph.14888.

[136] Vetrani C, Costabile G, Di Marino L, Rivellese AA. Nutrition and oxidative stress: A systematic review of human studies. Int J Food Sci Nutr 2013;64(3):312–26.

[137] Özen AE, Bibiloni M, Pons A, Tur JA. Consumption of functional foods in Europe; a systematic review. Nutr Hosp 2014;29(3):470–78.

[138] Forman HJ, Davies KJ, Ursini F. How do nutritional antioxidants really work: Nucleophilic tone and para-hormesis versus free radical scavenging in vivo. Free Radic Biol Med 2014;66:24–35.

[139] Tenore GC, Caruso D, Buonomo G, D'Avino M, Ciampaglia R, Maisto M, et al. Lactofermented Annurca apple puree as a functional food indicated for the control of plasma lipid and oxidative amine levels: Results from a randomised clinical trial. Nutrients 2019;11(1):122.

[140] Esmaeilinezhad Z, Barati-Boldaji R, Brett N, de Zepetnek J, Bellissimo N, Babajafari S, et al. The effect of synbiotics pomegranate juice on cardiovascular risk factors in PCOS patients: A randomized, triple-blinded, controlled trial. J Endocrinol Invest 2020;43(4):539–48.

[141] Ohishi T, Fukutomi R, Shoji Y, Goto S, Isemura M. The beneficial effects of principal polyphenols from green tea, coffee, wine, and curry on obesity. Molecules 2021;26(2):453.

[142] Shen N, Wang T, Gan Q, Liu S, Wang L, Jin B. Plant flavonoids: Classification, distribution, biosynthesis, and antioxidant activity. Food Chem 2022;383:132531.

[143] Gorzynik-Debicka M, Przychodzen P, Cappello F, Kuban-Jankowska A, Marino Gammazza A, Knap N, Wozniak M, Gorska-Ponikowska M. Potential health benefits of olive oil and plant polyphenols. Int J Mol Sci 2018;19(3):686.

[144] Khan UM, Sevindik M, Zarrabi A, Nami M, Ozdemir B, Kaplan DN, Selamoglu Z, Hasan M, Kumar M, Alshehri MM, Sharifi-Rad J. Lycopene: Food sources, biological activities, and human health benefits. Oxid Med Cell Longev 2021;2021:2713511.

[145] Martínez Álvarez JR, Lopez Jaen AB, Cavia-Saiz C, Muñiz P, Valls-Belles V. Beneficial effects of olive oil enriched with lycopene on the plasma antioxidant and anti-inflammatory profile of hypercholesterolemic patients. Antioxidants 2023;12(7):1458.

[146] Violi F, Nocella C, Loffredo L, Carnevale R, Pignatelli P. Interventional study with vitamin E in cardiovascular disease and meta-analysis. Free Radic Biol Med 2022;178:26–41.

[147] Castro C. Editorial: Natural plant antioxidants and cardiovascular disease. Front Physiol 2022;13:848497.

[148] Maiuolo J, Gliozzi M, Carresi C, Musolino V, Oppedisano F, Scarano F, Nucera S, Scicchitano M, Bosco F, Macri R, Ruga S, Cardamone A, Coppoletta A, Mollace A, Cognetti F, Mollace V. Nutraceuticals and Cancer: Potential for Natural Polyphenols. Nutrients 2021;13(11):3834.

[149] Mirahmadi M, Azimi-Hashemi S, Saburi E, Kamali H, Pishbin M, Hadizadeh F. Potential inhibitory effect of lycopene on prostate cancer. Biomed Pharmacother 2020;129:110459.

[150] Lippi L, Uberti F, Folli A, Turco A, Curci C, d'Abrosca F, De sire A, Invernizzi M. Impact of nutraceuticals and dietary supplements on mitochondria modifications in healthy aging: A systematic review of randomized controlled trials. Aging Clin Exp Res 2022;34(11):2659–74.

[151] Demirci-Çekiç S, Özkan G, Avan AN, Uzunboy S, Çapanoğlu E, Apak R. Biomarkers of oxidative stress and antioxidant defense. J Pharm Biomed Anal 2022;209:114477.

9 The biochemistry of flavor perception

9.1 Definition of flavor

Human senses can be classified according to the type of external signal that elicits their responses; thus, sight (elicited by light), hearing (by vibrations), and touch (by mechanical stimuli) can be regarded as "physical senses," whereas taste and smell are specialized in sensing chemicals. For this reason, smell and taste are involved in what is called "chemosensation," which would include other perceptions such as chemesthesis (see below) [1].

The acceptance of a food depends on properties such as color, appearance, taste, aroma, texture, and even the sound generated during chewing [2]. These properties are often termed as organoleptic, which means "of, relating to, or involving the use of sense organs or senses, esp. of smell and taste" [3]. Although all our senses are admittedly involved in food acceptance, it is clear that the sensations elicited in the mouth significantly drive our dietary choices. These sensations are basically determined by the so-called *flavor*, which can be defined as the combination of gustatory (taste) and olfactory (aroma) stimuli produced in the mouth by foods, with the addition of somatosensory sensations, such as texture, temperature, astringency, or irritation [4–6]. The amazing complexity of taste has led some authors to propose "that there is no taste and no smell in nature [. . .] but a vast and blurred variety of modes of chemical communication that could be collectively called "chemosensation," which always starts from the interactions between ligands and receptors . . ." [1].

Flavor perception is affected by sight, hearing, touch, internal (state of organism, hunger, satiety, emotion, expectation, etc.), and external factors (ambient lighting, background music, and/or noise), as well as social and cultural circumstances [2, 5]. All these sensory inputs are integrated in the brain by means of complex mechanisms in order to produce an adequate response: pleasure, disgust, food rejection. For these reasons, flavor perception is regarded by some authors as "the most multisensory of our everyday experiences" [5].

9.2 Taste

Taste is mediated by a small number of receptors that sense either simple essential nutrients or potentially toxic substances, that is, the task of this system is to solve nutritional problems that are of capital importance for survival [6, 7]. The taste system does not only drive food preferences leading to ingestion or rejection but some of its components, especially taste receptors (see below), can also be found in organs such as the brain, heart, blood, respiratory tract, pancreas, urogenital tract, testis, thyroid gland, thymus, and gastrointestinal tract [8]. This suggests that nutrient sensing

https://doi.org/10.1515/9783111111872-009

is involved in processes like nutrient absorption; regulation of metabolic parameters in the brain, pancreas, gut, and thyroid gland; and even immune responses and fertility [8, 9]. Classically, sweet, salty, bitter, and sour had been considered the basic tastes and, more recently, "umami," a savory taste elicited by certain L-amino acids, was added to this list, while there is an increasing agreement on the existence of a "fat" taste [10].

The organs responsible for taste perception are the taste buds, which are located mainly in the tongue epithelium and palate, although they are also found on the pharynx, larynx, and epiglottis [4, 9]. In the tongue, taste buds are enclosed within small structures with different shapes and distribution known as papillae, which can be of three types: (1) fungiform papillae, shaped like mushrooms with up to 1 mm in diameter and concentrated at the tips and edges of the tongue; (2) circumvallate papillae, circular structures located on the dorsal surface of the tongue at the junction of the oral and pharyngeal cavities; (3) foliate papillae, which appear as a series of clefts along the lateral margins of the tongue [6, 11] (Figure 9.1). A fourth type of papilla, known as filiform papillae, are devoid of taste buds and believed to be responsible for texture perception [12]. Taste buds are compact clusters of 50–100 elongated cells that resemble a garlic bulb. Four types of cells have been identified in these organs (Figure 9.1):

– Type I cells that comprise half the total number of cells in taste buds seem to function as supporting and protective cells ("glial-like" cells), although some authors suggest that they might also be involved in the perception of salty flavor [9, 13]. They express proteins involved in neurotransmitter clearance [8].
– Type II cells form three distinct subpopulations responsible for sensing sweet, umami, and bitter tastes, respectively, which represent about one-third of the cells in a taste bud. Type II cells do not possess a synaptic machinery, a fact that determines the way they transmit sensory signals [8, 9, 14]. In these cells, taste compounds are detected by proteins that belong to the superfamily of G-protein-coupled receptors (GPCRs), the largest and most diverse group of membrane receptors in eukaryotes. Humans have nearly a thousand GPCRs involved in a wide variety of processes. GPCRs can be classified into five classes and further divided into subfamilies based on sequence similarities [15].
– Type III cells only represent 2–20% of the cells in a taste bud. They are acid-sensing (sour taste) cells that have neuronal properties with visible synaptic structures [14]. Type III cells have been proposed to contain the taste receptors mediating high concentrations of salt [15, 16].
– Type IV cells are stem cells located at the bottom of the taste buds that replace the other cell types when they die [8].

Individual taste cells are specialized in sensing one of the five basic tastes; however, every taste bud can have several types of cells. Therefore, the existence of preferential

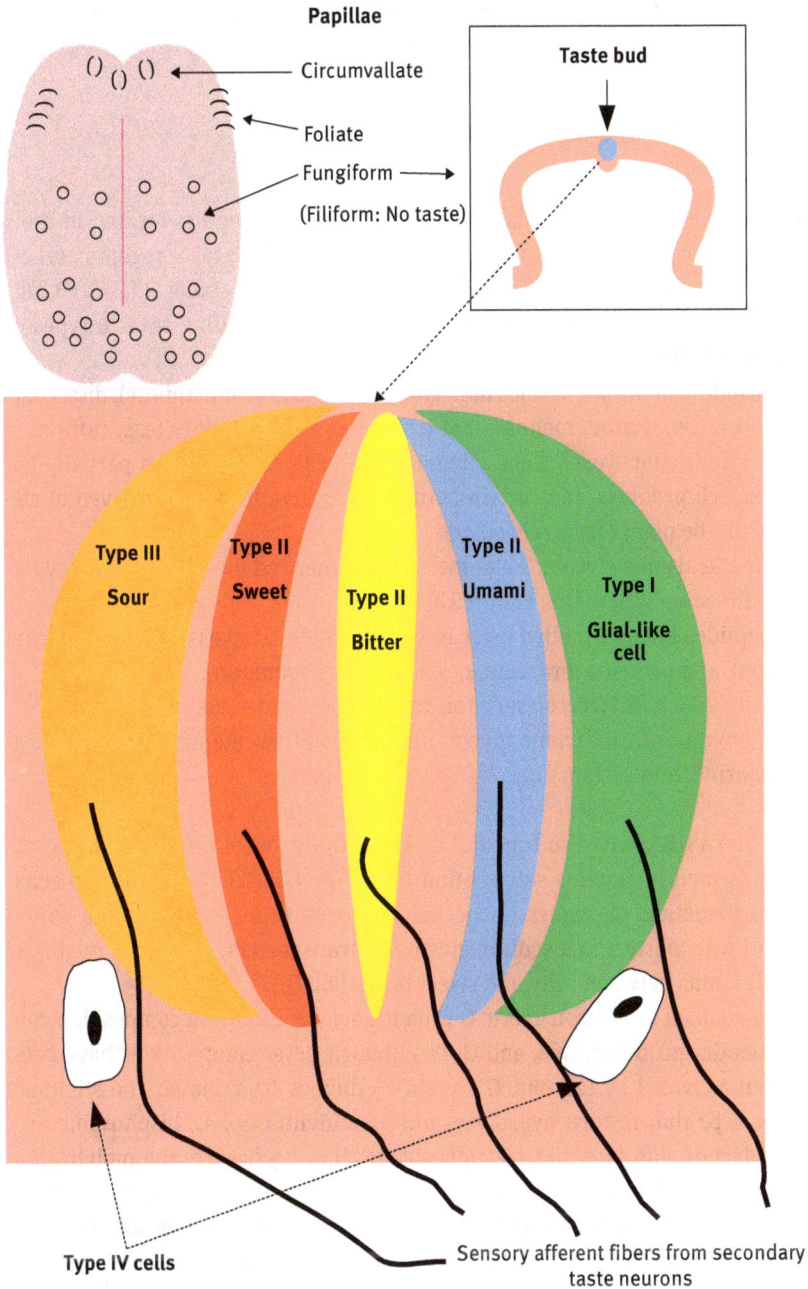

Figure 9.1: Anatomy of taste buds.
(adapted from Version 8.25 of the Textbook OpenStax Anatomy and Physiology. OpenStax – https://cnx.
org/contents/FPtK1zmh@8.25:fEI3C8Ot@10/Preface, CC BY 4.0, via https://commons.wikimedia.org/w/
index.php?curid=30147989)

regions in the mouth for the detection of the different tastes, the so-called tongue map, is an idea that is nowadays regarded as false [6].

9.2.1 Sweet taste

Sweetness is detected in type II cells that express specific receptors located in their apical tips. These are dimers comprising two different GPCRs: taste receptor type I member 2 (TAS1R2, formerly T1R2) and taste receptor type I member 3 (TAS1R3, formerly T1R3).[1] TAS1R2–TAS1R3 heterodimers are capable of binding several types of molecules (Figure 9.2) [7, 17]:

- Sugars, including monosaccharides (e.g., glucose, fructose, and mannose), disaccharides (e.g., sucrose, lactose, maltose, and trehalose), trisaccharides (e.g., raffinose), tetrasaccharides (stachyose), some oligosaccharides (e.g., polycose, a partially hydrolyzed starch product), and certain natural glycosides, like those derived of steviol, found in the plant *Stevia rebaudiana.*
- Some amino acids like glycine, L-alanine, L-threonine, and the D-isomers of tryptophan and histidine (Chapter 5. Figure 5.3).
- Certain peptides like the methyl ester of the dipeptide L-aspartyl-L-phenylalanine (aspartame), and proteins (brazzein, monellin, and thaumatin).
- Some alcohols such as glycerol, sorbitol, and xylitol (Figure 3.6).
- Synthetic compounds (ethylene glycol) and artificial sweeteners such as sucralose, saccharine, and cyclamate.

Homodimers of TAS1R3 may also function as low-affinity sweet receptors, while experiments performed in rodents suggest that there are TAS1R3-independent mechanisms for the detection of sugars and other sweet substances. The latter would involve glucose transporters and sodium/glucose co-transporters, a fact that might explain the interactions between salty and sweet tastes [9, 17].

GPCRs are coupled to heterotrimeric G proteins, whose canonical composition consists of subunits Gα gustducin, Gβ3, and Gγ13, although other compositions have been reported. When activated by tastants, G protein βγ dimers are released and stimulate phospholipase C β2 that in turn hydrolyzes phosphatidylinositol-4,5-bisphosphate releasing diacylglycerol and inositol-1,4,5-triphosphate (IP$_3$). IP$_3$ triggers the mobilization of Ca^{2+} from intracellular stores, which leads to the opening of the cation-permeable channel TRPM5, which initiates the depolarization of the plasma membrane. This activates voltage-gated Na$^+$ channels (VGNa$^+$), which rapidly increase the depolarization, thus triggering the action potential. The combined action of elevated Ca^{2+} and membrane depolarization results in ATP efflux via the CALHM1/CALHM3 heterohexameric

---

1 Both name types can be found in the literature. The new notation (i.e., TASxRy) will be used here.

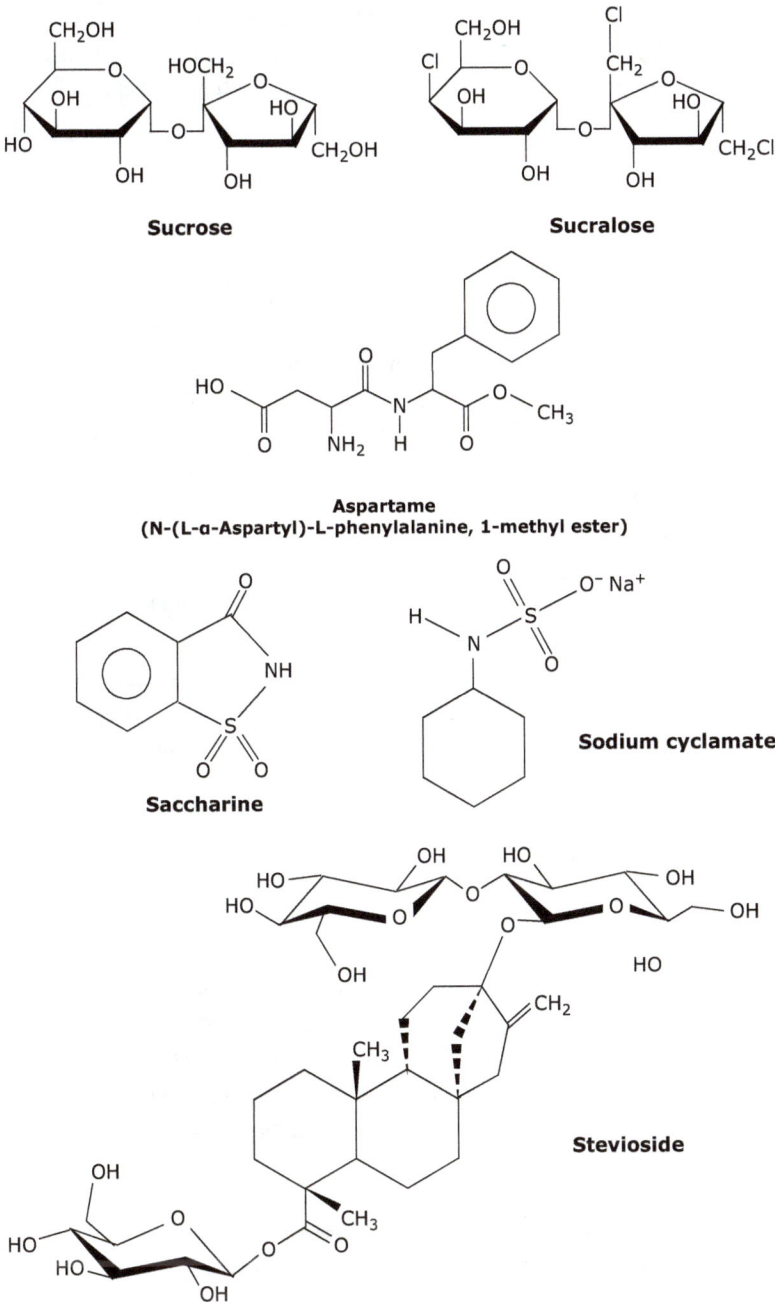

Figure 9.2: Chemical formulae of some sweeteners.
Except for stevioside (see below), chemical formulae were elaborated and adapted using the structural information available at National Center for Biotechnology Information. PubChem Database (accessed on July 8, 2023):

ion channel [8, 18]. The ATP is provided by unusual, large mitochondria, closely opposed to clusters of CALHM1 channels within the plasma membrane of type II taste cells [19]. ATP secreted into the extracellular spaces stimulates afferent fibers that send the signal to the brain [9] (Figure 9.3).

9.2.2 Umami taste

Multiple receptors associated with umami taste are present in specialized type II cells. The most studied of these receptors are heterodimers of GPCR type I member 1 (TAS1R1) and TAS1R3, which are activated by the amino acids L-glutamate and L-aspartate (Chapter 5), although the prototypical stimulus for umami taste is monosodium glutamate (MSG). The intracellular signaling pathway triggered in this case is identical to that described for sweet taste. The second type of umami taste receptors is glutamate receptors GRM1 to GRM4 (previously called mGLUR1 to 4), which are activated by glutamate and analogs such as L-(+)-2-amino-4-phosphonobutyrate. The GRMs that mediate umami tasting are short less sensitive versions of the glutamate neuroreceptors present in the central nervous system, especially in the brain [9, 15].

In humans, guanosine monophosphate and inosine monophosphate were demonstrated to strongly enhance umami taste in the presence of glutamate already in the 1960s. Other purine 5'-ribonucleotides have also shown this effect to varying degrees. Interestingly, the synergism between MSG and nucleotides is stimulated by NaCl, a well-known taste enhancer. The molecular mechanisms underlying the umami taste are reasonably well established [15].

9.2.3 Bitter taste

An enormous variety of substances are capable of stimulating bitter taste, such as caffeine, quinine, or denatonium benzoate [22]; however, the goal of this system is not distinguishing among potential toxic substances but just recognizing them in order to transmit this information to the brain so that the foods containing these substances can be rejected [6]. In order to accomplish this task, humans have a set of around 25

Figure 9.2 (continued)
Sucrose: CID = 5988, https://pubchem.ncbi.nlm.nih.gov/compound/5988
Aspartame: CID = 134601, https://pubchem.ncbi.nlm.nih.gov/compound/Aspartame
Sucralose: CID = 71485, https://pubchem.ncbi.nlm.nih.gov/compound/Sucralose
Saccharin. CID = 5143, https://pubchem.ncbi.nlm.nih.gov/compound/Saccharin
Sodium cyclamate. CID = 23665706, https://pubchem.ncbi.nlm.nih.gov/compound/Sodium-cyclamate
Stevioside: Formula elaborated using information from Yikrazuul – Own work; ISBN 3-540-40291-8,
Dominio público, https://commons.wikimedia.org/w/index.php?curid=9149312, and http://www.chm.bris.
ac.uk/motm/stevioside/steviosideh.htm

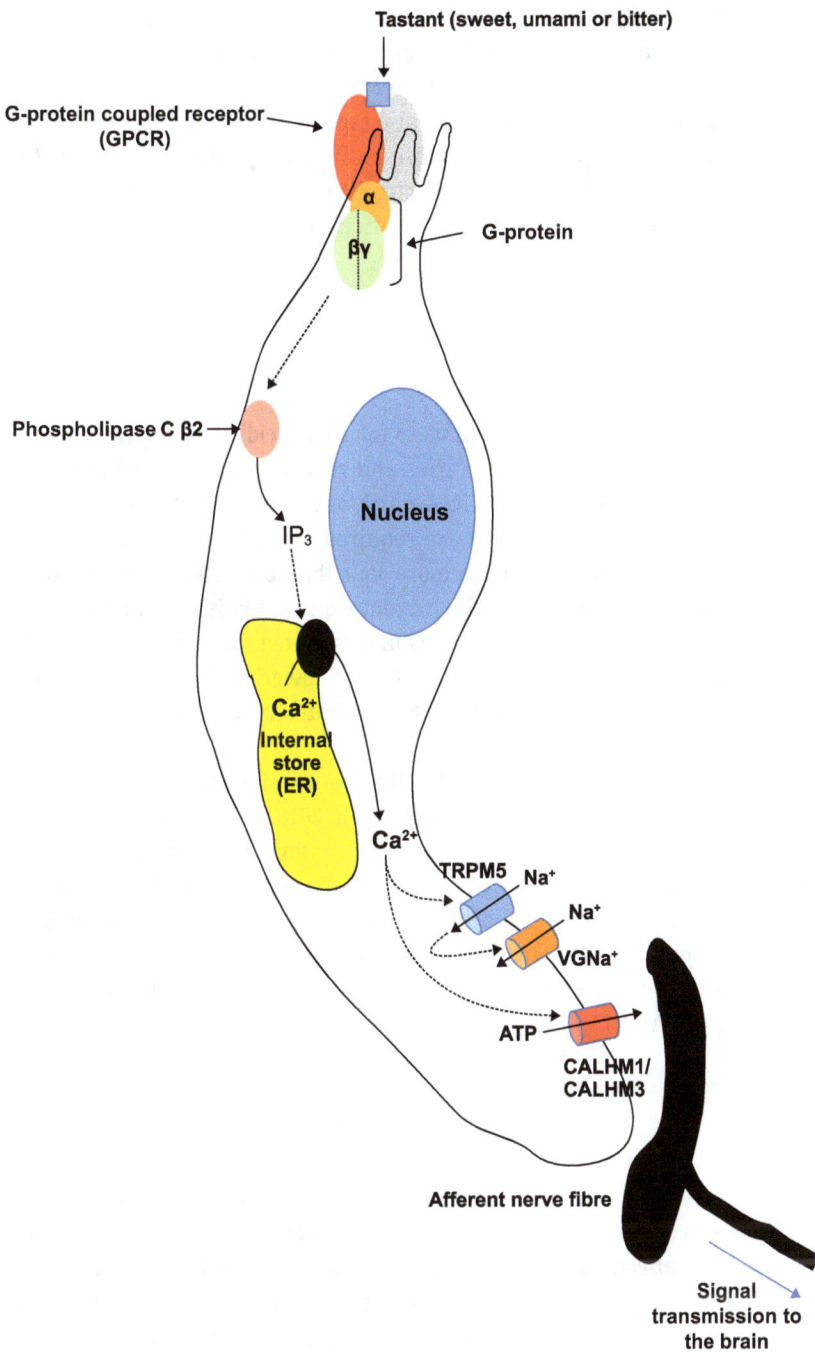

Figure 9.3: Signal transduction of sweet, bitter, and umami tastes (elaborated and adapted from references [15, 20, 21]).

different receptors with variable affinities toward a wide range of substances. These receptors belong to a family of GPCRs distinct to those involved in the sweet and umami tastes. They are known as type 2 taste receptors, or TAS2Rs, and are generally considered to act as monomers, although they also seem to form homodimers and heterodimers [9]. Receptors of bitter taste are also located in specific type II taste bud cells and share the intracellular signaling mechanism with sweet and umami tastes (Figure 9.3). TAS2Rs also trigger a second transduction cascade, which plays a modulatory role [23].

9.2.4 Sour taste

The first studies on the physical–chemical basis for sour (acid) taste appeared in the late nineteenth and early twentieth centuries and revealed that hydrochloric acid (HCl) as well as organic acids such as citric, tartaric, and malic acids tasted sour. This provided the first evidence that protons are required to elicit sour taste; however, weak (organic) acids were shown to taste more sour than predicted based solely on the concentration of the hydrogen ion. The current explanation is that the sour taste is mainly elicited by weak acids that diffuse as undissociated forms into type III cells and dissociate inside, thus releasing protons (H^+) and lowering the pH. In any case, the reason why weak organic acids taste more sour than strong acids remains unanswered [24].

Protons produced outside the cells by the dissociation of acids are also taken up via Otopetrin 1, a proton channel that was identified in 2018 by Tu et al. [25]. Cytosolic acidification blocks the Kir2.1 potassium channel and, thus, the resting K^+ current, thereby amplifying the signal elicited by the intracellular protons [26]. The accumulation of positive charges inside the cell produces the depolarization of the basolateral membrane, which opens voltage-gated Ca^{2+} channels. Ca^{2+} influx triggers neurotransmitters release (5-hydroxytryptamine (serotonin), acetylcholine, noradrenaline, and γ-aminobutyric acid (GABA)) by exocytosis [9, 17] (Figure 9.4).

9.2.5 Salty taste

Salt (NaCl) induces two different behaviors both in insects and mammals: at low concentrations, it acts as an attractant, whereas at high concentrations, it acts as a repellant. This probably reflects, on the one hand, the necessity to detect an essential nutrient required for various physiological processes and, on the other hand, a protection against hypernatremia and dehydration. Studies in rodents suggest that the amiloride-sensitive Na^+-channel EnaC is involved in the perception of salty taste, along with a second amiloride-insensitive system that may mediate aversive taste at high salt concentrations.

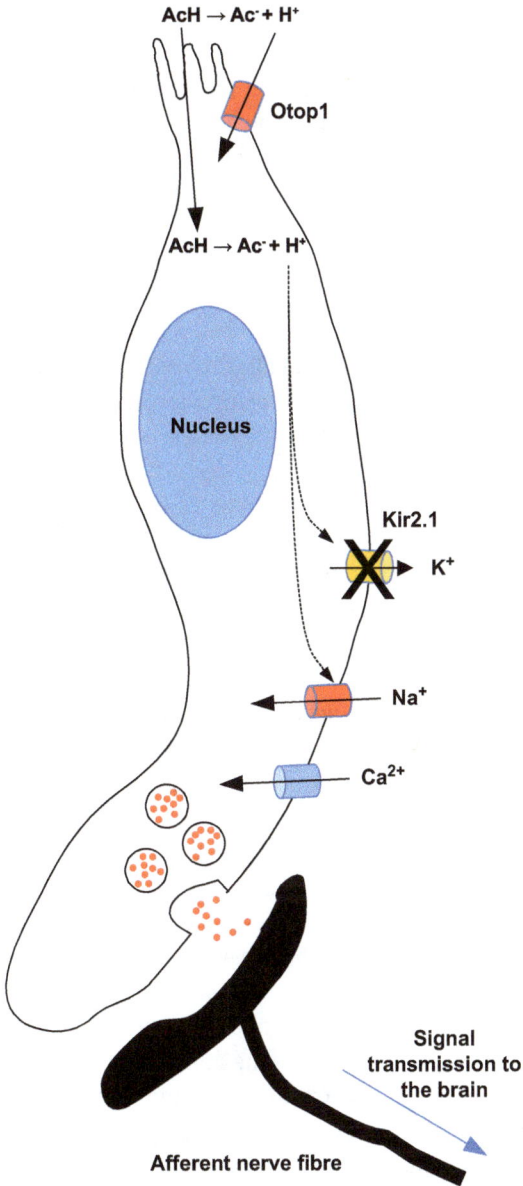

Figure 9.4: Signal transduction of sour taste. Red dots represent neurotransmitter molecules (adapted from references [20, 21, 24]).

Several studies support the involvement of EnaC in the perception of salty taste in humans; however, there are many conflicting results regarding amiloride's inhibitory effect on perceived saltiness. Moreover, amiloride has a more limited effect in humans than in rodents. There are several hypotheses to explain this evidence but none has been demonstrated to date. Therefore, significant questions remain unanswered about the perception of salty taste in humans. These include the true composition of ENaC, the molecular identity of high salt taste sensors, neural circuit mechanisms involved, and others. The identification of cells and receptors responsible for human perception of NaCl should pave the way to design artificial salt substitutes that could help to tackle hypertension [16, 27, 28].

9.2.6 Fat taste

Traditionally, lipids were accepted to elicit only texture and olfactory sensations; however, recent evidences suggest that type II cells express specific receptors for fatty acids: GPR120, another member of the GPCR family expressed both in taste buds and the surrounding gustatory epithelium, and CD36, a protein that belongs to the class B scavenger receptor family, which might also be expressed in type III cells. These two receptors might play a nonoverlapping role in the detection of fat taste. The transduction mechanism of the signal elicited by fatty acids in taste bud cells seems to be similar to that of bitter taste [10, 29]. GPR120 is also highly expressed in adipocytes and macrophages where it has been reported to mediate insulin-sensitizing and antidiabetic effects by repressing macrophage-induced inflammation [30]. The structure of this receptor bound to different fatty acids has been determined by cryoelectron microscopy [31], representing an important breakthrough that may help to deepen into the different functions of GPR120 in vivo.

9.2.7 Other tastes

Foods can trigger additional sensory perceptions that have been less studied, and these include:

– Chemesthesis. This term refers to the direct activation of somatosensory nerves (like the trigeminal nerve), by chemical stimuli. Within the mouth, chemesthesis is associated with sensations such as pungency, freshness, tingling, sharpness, cooling, and even pain. Common chemesthetic agents include capsaicin (chili peppers), ethanol, menthol (peppermint), zingerone (ginger), allyl isothiocyanate (mustard oil, horseradish, and wasabi), piperine (black peppers), various acids, and carbonated water (carbon dioxide) (Figure 9.5). These compounds are known to interact with specific cation channels (such as TRPV1 and TRPA2) expressed in nociceptors, that is, the receptors responsible for the perception of irritation, pain, and temperature [9, 32]. Intraoral chemesthetic sensations are assumed to influence taste perceptions; however, it has been suggested that some of them, like that elicited by capsaicin, might also have a genuine taste component [9, 33].

- Astringency. This sensation is described as "a feeling of puckering, rough and drying sensation plus a slight bitter taste on the tongue and membranes of the oral cavity" [34]. Tannic acid (whose structure formally contains ten 3,4,5-trihydroxyphenyl (galloyl) units surrounding a glucose center [35]) is one of the common chemicals that produce astringency, along with other tannins. The latter are generally defined as soluble, astringent complex phenolic substances of plant origin mainly derived from gallic acid (hydrolyzable tannins) or from flavan-3,4-diol (a flavanol, Figure 8.14) (condensed tannins) [36]. Whether astringency is a taste or an oral somatosensation remains in dispute: on the one hand, astringent compounds have been shown to interact with taste receptors in animal studies and to activate taste nerves; on the other hand, astringent perception has been reported to be dependent on lingual nerve (trigeminal) function. Recent studies suggest that the human brain might recognize astringency as a real taste [34].
- Kokumi taste. Addition of some nearly tasteless substances to salty, sweet, and/or umami-eliciting foods enhances mouthfulness and complexity and induces a long-lasting savory sensation on the tongue. This effect is known as "kokumi taste" and seems to be associated with some Ca^{2+}-sensing receptors expressed in certain taste bud cells. Glutathione (Figure 8.11) is the prototypical kokumi taste substance: it is tasteless by itself, but synergistically reinforces umami tastants such as MSG and inosine monophosphate. Other substances, including γ-glutamyl peptides, are also known to elicit the kokumi taste [37–40].
- Calcium taste. Ca^{2+} has been shown to interact with TAS1R3, also involved in sweet and umami tastes (see above), in humans; however, it is not known whether this interaction produces a perception of taste that drives an appetite for this nutrient [9, 41].
- Water taste. Recent evidence suggests that acid-sensing taste bud receptor cells (type III cells) also mediate taste responses to water [42].
- Metallic taste. Salts like $CuSO_4$, $ZnSO_4$, and $FeSO_4$, which produce a metallic taste sensation, have been reported to activate receptor TRPV1. Moreover, synthetic sweeteners, such as saccharin, aspartame, acesulfame-K, and cyclamate, also seem to interact with this receptor, thus explaining the metallic aftertaste exerted by these compounds [43]. Divalent salts have also been proposed to produce a bitter sensation by activating several TAS2Rs, while receptor TAS2R7 appears to respond to trivalent salts such as aluminum sulfate [44].

9.2.8 The neuroendocrinology of taste

As mentioned previously, type II cells, responsible for the detection of sweet, bitter, and umami tastes, secrete ATP when stimulated by the appropriate tastants. ATP is detected by the so-called purinergic receptors located in nearby afferent fibers, that transmit the appropriate signal to the central nervous system. Purinergic receptors

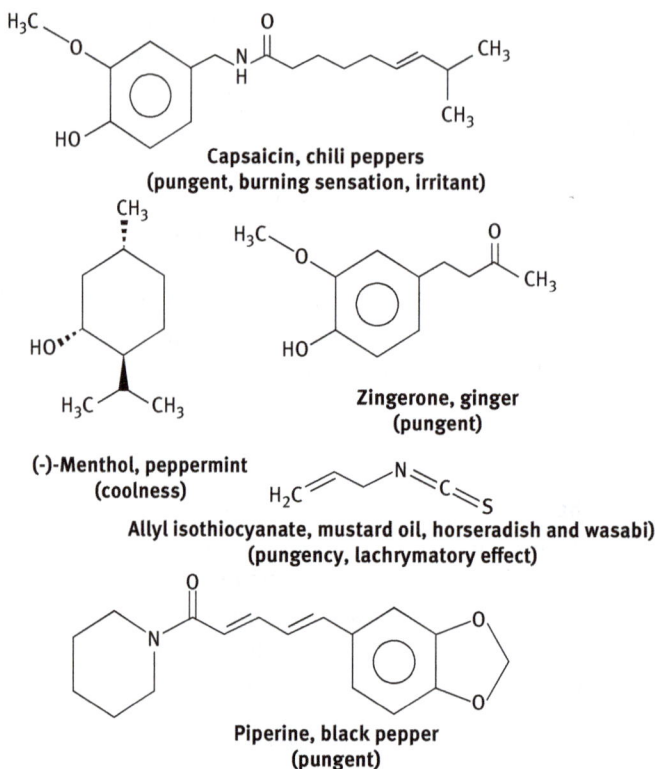

Figure 9.5: Chemical formulae of some chemesthetic agents.
Chemical formulae were elaborated and adapted using the structural information available at National Center for Biotechnology Information. PubChem Database (accessed on July 8, 2023):
Capsaicin, CID = 1548943, https://pubchem.ncbi.nlm.nih.gov/compound/Capsaicin
Menthol, CID = 1254, https://pubchem.ncbi.nlm.nih.gov/compound/Menthol
Zingerone, CID = 31211, https://pubchem.ncbi.nlm.nih.gov/compound/Zingerone
Allyl isothiocyanate, CID = 5971, https://pubchem.ncbi.nlm.nih.gov/compound/Allyl-isothiocyanate
Piperine, CID = 638024, https://pubchem.ncbi.nlm.nih.gov/compound/Piperine

are also present in other taste bud cells, so that ATP can amplify its own release (an autocrine effect) and stimulate type III cells (a paracrine effect). In addition to ATP, type II cells secrete acetylcholine and locally produced hormones. Acetylcholine seems to have an autocrine effect that increases ATP secretion, while hormones mediate the communication with neighboring cells, thereby altering taste perception and modulating the intensity of the taste signal (see below). A feedback mechanism is established in type III cells in response to the ATP released by type II cells: the former secrete 5-hydroxytryptamine (5-HT) and GABA, inhibitors of ATP secretion [9, 22].

Results obtained mainly in rodents show that all types of taste bud cells express and/or detect a wide variety of peptidic hormones and neuropeptides (Table 9.1). This allows TBCs to communicate among one another, thus modulating taste perception according to the complex mixtures of tastants that can appear during food processing in the mouth. The presence of receptors for hormones involved in the regulation of energy metabolism in these cells further indicates the importance of the taste perception system beyond food acceptance/rejection. Conversely, taste receptors occur along the gastrointestinal tract, the thyroid gland and the pancreas, thus demonstrating the strong interconnection between nutrient perception and absorption, and the regulation of metabolism. Taste receptors have also been identified in the hypothalamus, the region of the brain that controls energy homeostasis and regulates hunger/satiety [8, 22].

Table 9.1: Peptidic hormones and neuropeptides expressed or detected by taste bud cells (TBCs).

Hormone/ neuropeptide	Main production site	General function	TBC type E[1]	TBC type R[2]	Function in taste system
Cholecystokinin (CCK)	Enteroendocrine I cells of duodenum and jejunum	Reduction of appetite (anorexigenic)	II	II	The balance between CCK and NPY adjusts bitter sensing to the energy content of the food (sensed by the presence of sweet and umami tastants)
Neuropeptide Y (NPY)	Brain	Regulation of energy metabolism	II	II	
Peptide tyrosine tyrosine (PYY)	Enteroendocrine L cells	Anorexigenic	II	II[3]	Involved in fat taste perception
Insulin	Pancreas (β-cells)	Regulation of blood glucose levels and energy metabolism	–	II/ III	Unknown
Glucagon-like peptide 1 (GLP-1)[4]	Enteroendocrine L cells	Reduction of blood glucose levels by stimulating insulin secretion (incretin)	II/ III	III	Combined with NPY contributes to differentiate between sweet, umami, bitter and sour tastes
Glucagon-like peptide 2 (GLP-2)[4]	Enteroendocrine L cells	Regulation of gut function	II	–	Unknown
Glucagon[4]	Pancreas (α-cells)	Antagonist to insulin	II	II	Involved in sweet taste signaling
Vasoactive intestinal peptide (VIP)	Produced mainly by neurons, endocrine, and immune cells. Widespread distribution	Wide variety of biological functions [20]	II	II	Decreases perception of sweet and bitter tastants

Table 9.1 (continued)

Hormone/ neuropeptide	Main production site	General f unction	TBC type E[1]	TBC type R[2]	Function in taste system
Ghrelin	Stomach	Feeding behavior and energy homeostasis	I/ II/ III/ IV	I/ II/ III/ IV	Fat and salt perception
Oxytocin	Hypothalamus	Regulation of reproductive behavior and mother–infant interaction Psychosocial function Eating behavior [21]	–	I	Regulation of salt appetite and natriuresis
Galanin	Central nervous system and gut	Regulation of food intake, gut motility, and hormone secretion	II/ III	II	Regulation of fat consumption
Leptin	Adipose tissue	Regulation of adipose tissue mass	–	II	1. Stimulation of ATP release with increasing temperatures 2. Regulation of sweet taste perception

Receptors for CCK, GLP-1, and NPY are also present in afferent nerve fibers. Table elaborated with data obtained mainly in rodents and published in references [8, 22].
[1]Expression of hormone/neuropeptide.
[2]Expression of the receptor for the hormone/neuropeptide.
[3]Interacts with a NPY receptor.
[4]These hormones are three different cleavage products generated from a single pro-glucagon peptide.

Many authors have suggested that pleasant tastes and aromas, that is, the hedonic aspect of eating, may cause overeating by suppressing the homeostatic mechanisms that control body weight. According to this view, the availability of energy-dense hyperpalatable foods would drive the obesity epidemic. The presence of taste receptors in organs crucial for the control of metabolism, eating behavior, and energy homeostasis would be a support for this idea; however, a recent proposal suggests that the energy content of food (detected by sweet, umami, and fat taste receptors) can also drive intake in an unconscious manner [45].

A strict control of metabolism and energy homeostasis is essential for survival; therefore, it is not surprising that it depends on extremely complex mechanisms, including the taste system. However, much more research is required in order to reach a deep understanding of these mechanisms. This research will undoubtedly shed light on the factors that drive eating behavior and will help implement strategies to tackle important social problems such as obesity and other metabolic disorders.

9.3 Olfaction

Olfaction is a much more complex system than taste: volatiles (aroma compounds or odors, i.e., compounds that are in gaseous form) can arise from a source external to the body and stimulate the olfactory epithelium upon inhalation through the nostrils; this is known as "orthonasal olfaction." Alternatively, volatiles may arise from inside the mouth when foods are chewed and swallowed, the so-called retronasal olfaction. Retronasal (but not orthonasal) olfactory input has been reported to be processed by the brain region responsible for taste processing, where they are presumably integrated [6, 46].

Perception of volatiles is due to the interaction of these compounds with the olfactory receptor neurons (ORNs) that lie in the olfactory epithelium, which occupies a 3.7 cm^2 zone in the upper part of the nasal cavity. Humans have around 12 million ORNs in each epithelium (right and left), each ORN having at one end 20–30 cilia bathing in mucus and containing one type of olfactory receptors (ORs). ORs bind to odor molecules and give an electrical response that is transferred through the axon situated at the other end of the ORNs to nerve fibers situated at the back of the nasal cavity. The axons of all the ORNs containing the same type of odor receptor are grouped in the same glomerulus. More than 5,500 glomeruli are present in humans, forming the olfactory bulb, where the signal is treated and transmitted to different zones in the brain, such as the thalamus, amygdala, and orbitofrontal cortex [47].

Mammalian ORs are predicted to contain seven transmembrane domains and have either been shown to be coupled to G proteins or are expected to do so based on their sequence similarity to GPCRs. The predominant receptors of the main olfactory epithelium are encoded by the approximately 400 OR genes that occur in the human genome, although pseudogenes, variations in copy number, and single nucleotide polymorphisms have been identified in many individuals [48, 49]. Mammalian ORs are classified into class I ORs, which bind primarily hydrophilic odorants, and class II ORs, which bind hydrophobic odorants. The odorant must cross a hydrophilic mucus, where the dendrites of olfactory neurons are immersed; therefore, hydrophobic odorants need to be transported. This transport is believed to be the role of the so-called odorant-binding proteins [50]. How odorants are recognized by ORs remains unclear; however, the structure of an active human odorant receptor (OR51E2) bound to the fatty acid propionate has been recently determined by cryoelectron microscopy. This result provides an insight into how ORs enable our olfactory sense [51].

Traditionally, it has been accepted that humans can discriminate 10,000 odors; however, more recent calculations estimate that at least 1 trillion (10^{12}) olfactory stimuli may be detected [52]. How the combinatorial activation of the ORs encodes odor perception is currently unknown [49]. Similarly to taste receptors, ORs have been identified outside the nose and are currently considered as general chemoreceptors involved in physiological and pathophysiological processes throughout the human body [53].

Most flavor perception is due to retronasal olfaction; however, there are complex interactions between olfactory, taste, and somatosensory stimuli. Moreover, these interactions appear to be largely learned; therefore, understanding how flavor perception functions at physiological and biochemical levels is an extremely difficult task that will require an important research effort during the next years [6, 54].

9.4 Flavor formation in foods

An important aspect of flavor is the chemistry of the molecules associated with the perception of taste, olfaction, and somatosensory sensations and how they are generated both within the natural sources of food and during processing. More than 25,000 molecules associated with flavor have been reported and classified in the comprehensive database FlavorDB (http://cosylab.iiitd.edu.in/flavordb); 2,000 of these molecules have been reported to be found in 936 natural entities/ingredients, whereas nearly 14,000 have been identified as synthetic. For around 9,500 molecules, no specific source has been ascertained so far. Information for 33 taste receptors related to sweet, bitter, sour, and umami tastes, as well as for 1,068 odor receptors is also included [55]. FlavorDB gathers information from other databases such as FooDB (http://foodb.ca/), Flavornet (http://www.flavornet.org/), SuperSweet (http://bioinformatics.charite.de/sweet/), and BitterDB (http://bitterdb.agri.huji.ac.il/dbbitter.php), as well as from sources such as *Fenaroli's Handbook of Flavor Ingredients* [56] and other scientific literature. The information concerning the chemical aspects of flavor is of paramount importance not only for academic reasons but also for the food industry that tries to develop new products that attract customers.

In natural sources, the molecules responsible for flavor can be generated in a wide variety of chemical and biochemical reactions. Taste is usually associated with nonvolatile molecules, such as sucrose, glucose, or capsaicin, whereas compounds that become volatile into the mouth can stimulate both taste and smell. Tastants usually derive from the hydrolysis of proteins, carbohydrates, lipids, ribonucleotides, and pigments, whereas volatiles are generated by the oxidation of lipids, Maillard reaction, caramelization, degradation, and the interactions among the products of these reactions. The fact that each constituent of fresh and processed foods is a potential substrate for chemical, enzymatic, or microbiological transformation widens the number of flavor compounds that can be generated (acids, alcohols, aldehydes, ketones, esters, ethers, terpenes, etc.) [2]. Traditionally, four types of basic mechanisms were considered to be responsible for the generation of flavor molecules [57]:

- Biosynthetic. This includes molecules formed directly by the biosynthetic processes that occur in natural sources of food, such as terpenoids and ester compounds from plants like mint, citrus, pepper, and banana.
- Direct enzymatic reaction. Molecules formed by enzymes acting on specific precursors, like the formation of onion flavor by the action of allinase on sulfoxides.
- Indirect enzymatic reaction. This type of mechanisms produces flavor molecules by oxidation of flavor precursors with compounds previously generated in enzyme-catalyzed reactions. For example, during the formation of black tea aroma, flavonols are oxidized by oxygen in a reaction catalyzed by catechol oxidase. Oxidized flavonoids can subsequently react with other biomolecules giving rise to a variety of flavor compounds.
- Pyrolytic. During thermal processing and heating of foods (i.e., frying, cooking, grilling, etc.), hundreds of compounds can be generated by lipid degradation (Section 4.6.2), Maillard reaction (Section 3.6 and reference [58]), or Strecker degradation. The latter typically involves the oxidative deamination and decarboxylation of an α-amino acid in the presence of an α-dicarbonyl compound yielding an α-aminoketone and a Strecker aldehyde, which has one carbon less than the initial α-amino acid. The Strecker degradation is sometimes considered as a subset of the Maillard reaction [59]. The compounds produced by these processes contribute to the overall flavor not only by themselves but also by interacting with each other [60].

Some authors suggest that, essentially, there are only two major types of mechanism of flavor formation: biosynthetic, which includes all chemical transformations mediated by enzymatic systems and microorganisms, and pyrolytic, which comprises the changes undergone by foods during treatments such as boiling, frying, or sterilization [2]. A more comprehensive coverage of reactions that generate flavor compounds can be found elsewhere [2, 61, 62].

9.4.1 The role of saliva in flavor formation and perception

Saliva is a complex mixture synthesized and secreted by salivary glands (Chapter 12), whose flow and composition depend on endogenous (circadian rhythms, age, sex, and several diseases) or exogenous factors (diet and pharmacological agents). Basically, saliva is composed of water, salts, and proteins, as well as microorganisms, cellular debris, and food residues. Saliva plays different functions, including lubrication, predigestion, protection against microorganisms, detoxification, teeth mineralization, and transport of taste compounds to chemoreceptors located in the taste buds [63, 64].

Mucins, the major salivary proteins, are responsible for the gel nature of the mucous layer that covers epithelial surfaces throughout the body; moreover, they are important for food oral processing and digestion [65]. Mucins comprise a large family of glycoproteins, encoded by about 20 genes, which present several common structural features: protein backbones with multiple tandem repeats of serine and threonine, where oligosaccharides are covalently O-linked, as well as cysteine-rich domains at the N- and C-terminal ends that can lead to dimerization and further multimerization via S–S bonds [66]. Protection of the oral cavity involves the secretion of proteins such as immunoglobulins or enzymes that regulate the production of reactive oxygen species and reactive nitrogen species (Section 8.2) [64]. Other important components of saliva are hydrolytic enzymes (amylases and lipases) or salivary lipids (fatty acyls, glycerolipids, glycerophospholipids, sphingolipids, and sterol lipids) [67].

Many studies have been carried out on the possible role of saliva in the formation and perception of gustatory, olfactory, and chemesthetic sensations. It seems plausible that many different chemical and enzymatic reactions that occur during the interaction between saliva and food may lead to the formation of flavor compounds that are not present in the initial composition of the food. However, obtaining general conclusions from these studies is very difficult due to a number of considerations that can be summarized as follows [63, 68]:

- The flow and composition of saliva may vary under different conditions of stimulation or as a consequence of variations on the physiological status (illnesses or the aging process). In this sense, it must be mentioned that several illnesses (e.g., obesity and cancer) might modify the salivary parameters, which in turn will affect food behavior and nutritional status.
- Saliva can modify flavor, but the flavor stimuli can also modify saliva secretions.
- Saliva controls the oral health environment; thus, unbalanced saliva secretions might compromise oral microbiota or tooth status, which in turn will affect food intake.
- There exist great interindividual differences in saliva composition and oral microbiota among humans.

References

[1]   Mollo E, Boero F, Peñuelas J, Fontana A, Garson MJ, Roussis V, et al. Taste and smell: A unifying chemosensory theory. Q Rev Biol 2022;97:69–94. https://doi.org/10.1086/720097.
[2]   Badui S. Química De Los Alimentos. Pearson Educación de México, SA de CV; 2006.
[3]   Organoleptic, adj. : Oxford English Dictionary n.d. https://www-oed-com.us.debiblio.com/view/Entry/236328?redirectedFrom=organoleptic#eid (accessed September 25, 2019).
[4]   Duffy, VB. Nutrition and the chemical senses. Mod. Nutr. Health Dis. 11th edition. Philadelphia: Wolters Kluwer Health/Lippincott Williams & Wilkins; 2014, p. 574–88.
[5]   Spence C. Multisensory flavor perception. Cell 2015;161:24–35. https://doi.org/10.1016/j.cell.2015.03.007.

[6] Bartoshuk LM. Taste. In: Wixted JT, editor. Stevens Handb. Exp. Psychol. Cogn. Neurosci. Hoboken, NJ, USA: John Wiley & Sons, Inc.; 2018, p. 1–33. https://doi.org/10.1002/9781119170174.epcn203.

[7] Yarmolinsky DA, Zuker CS, Ryba NJP. Common sense about taste: From mammals to insects. Cell 2009;139:234–44. https://doi.org/10.1016/j.cell.2009.10.001.

[8] Behrens M, Meyerhof W. A role for taste receptors in (neuro)endocrinology? J Neuroendocrinol 2019;31:e12691. https://doi.org/10.1111/jne.12691.

[9] Roper SD, Chaudhari N. Taste buds: Cells, signals and synapses. Nat Rev Neurosci 2017;18:485–97. https://doi.org/10.1038/nrn.2017.68.

[10] Khan AS, Murtaza B, Hichami A, Khan NA. A cross-talk between fat and bitter taste modalities. Biochimie 2019;159:3–8. https://doi.org/10.1016/j.biochi.2018.06.013.

[11] Simpson KL. Chapter 23 – Olfaction and taste. In: Haines DE, Mihailoff GA, editors. Fundam. Neurosci. Basic Clin. Appl. 5th edition, Elsevier; 2018, p. 334–45.e1. https://doi.org/10.1016/B978-0-323-39632-5.00023-2.

[12] Lauga E, Pipe CJ, Le Révérend B. Sensing in the mouth: A model for filiform papillae as strain amplifiers. Front Phys 2016;4. https://doi.org/10.3389/fphy.2016.00035.

[13] Gravina SA, Yep GL, Khan M. Human biology of taste. Ann Saudi Med 2013;33:217–22. https://doi.org/10.5144/0256-4947.2013.217.

[14] Oka Y. Opening a "wide" window onto taste signal transmission. Neuron 2018;98:456–58. https://doi.org/10.1016/j.neuron.2018.04.020.

[15] Diepeveen J, Moerdijk-Poortvliet Tcw, Van Der Leij FR. Molecular insights into human taste perception and umami tastants: A review. J Food Sci 2022;87:1449–65. https://doi.org/10.1111/1750-3841.16101.

[16] Taruno A, Gordon MD. Molecular and cellular mechanisms of salt taste. Annu Rev Physiol 2023;85:25–45. https://doi.org/10.1146/annurev-physiol-031522-075853.

[17] Roper SD. Signal transduction and information processing in mammalian taste buds. Pflugers Arch 2007;454:759–76. https://doi.org/10.1007/s00424-007-0247-x.

[18] Ma Z, Taruno A, Ohmoto M, Jyotaki M, Lim JC, Miyazaki H, et al. CALHM3 is essential for rapid ion channel-mediated purinergic neurotransmission of GPCR-mediated tastes. Neuron 2018;98:547–61. e10. https://doi.org/10.1016/j.neuron.2018.03.043.

[19] Romanov RA, Lasher RS, High B, Savidge LE, Lawson A, Rogachevskaja OA, et al. Chemical synapses without synaptic vesicles: Purinergic neurotransmission through a CALHM1 channel-mitochondrial signaling complex. Sci Signal 2018;11:eaao1815. https://doi.org/10.1126/scisignal.aao1815.

[20] Kinnamon SC. Neurosensory transmission without a synapse: New perspectives on taste signaling. BMC Biol 2013;11:42. https://doi.org/10.1186/1741-7007-11-42.

[21] Chaudhari N, Roper SD. The cell biology of taste. J Cell Biol 2010;190:285–96. https://doi.org/10.1083/jcb.201003144.

[22] Calvo SS-C, Egan JM. The endocrinology of taste receptors. Nat Rev Endocrinol 2015;11:213–27. https://doi.org/10.1038/nrendo.2015.7.

[23] Wooding SP, Ramirez VA, Behrens M. Bitter taste receptors: Genes, evolution and health. Evol Med Public Health 2021;9:431–47. https://doi.org/10.1093/emph/eoab031.

[24] Turner HN, Liman ER. The cellular and molecular basis of sour taste. Annu Rev Physiol 2022;84:41–58. https://doi.org/10.1146/annurev-physiol-060121-041637.

[25] Tu Y-h, Cooper AJ, Teng B, Chang RB, Artiga DJ, Turner HN, et al. An evolutionarily conserved gene family encodes proton-selective ion channels. Science 2018;359:1047–50. https://doi.org/10.1126/science.aao3264.

[26] Ye W, Chang RB, Bushman JD, Tu Y-H, Mulhall EM, Wilson CE, et al. The K^+ channel K_{IR} 2.1 functions in tandem with proton influx to mediate sour taste transduction. Proc Natl Acad Sci 2016;113:E229–38. https://doi.org/10.1073/pnas.1514282112.

[27] Kaushik S, Kumar R, Kain P. Salt an essential nutrient: Advances in understanding salt taste detection using *Drosophila* as a model system. J Exp Neurosci 2018;12:117906951880689. https://doi.org/10.1177/1179069518806894.

[28] Roebber JK, Roper SD, Chaudhari N. The role of the anion in salt (NaCl) detection by mouse taste buds. J Neurosci 2019;39:6224–32. https://doi.org/10.1523/JNEUROSCI.2367-18.2019.

[29] Besnard P, Passilly-Degrace P, Khan NA. Taste of fat: A sixth taste modality? Physiol Rev 2016;96:151–76. https://doi.org/10.1152/physrev.00002.2015.

[30] Oh DY, Talukdar S, Bae EJ, Imamura T, Morinaga H, Fan W, et al. GPR120 is an omega-3 fatty acid receptor mediating potent anti-inflammatory and insulin-sensitizing effects. Cell 2010;142:687–98. https://doi.org/10.1016/j.cell.2010.07.041.

[31] Mao C, Xiao P, Tao X-N, Qin J, He Q-t, Zhang C, et al. Unsaturated bond recognition leads to biased signal in a fatty acid receptor. Science 2023;380:eadd6220. https://doi.org/10.1126/science.add6220.

[32] Slack JP. Molecular pharmacology of chemesthesis. Chemosens. Transduct. Elsevier; 2016, p. 375–91. https://doi.org/10.1016/B978-0-12-801694-7.00021-4.

[33] Braud A, Boucher Y. Intra-oral trigeminal-mediated sensations influencing taste perception: A systematic review. J Oral Rehabil 2019:joor.12889. https://doi.org/10.1111/joor.12889.

[34] Zhu Y, Thaploo D, Han P, Hummel T. Processing of sweet, astringent and pungent oral stimuli in the human brain. Neuroscience 2023;520:144–55. https://doi.org/10.1016/j.neuroscience.2023.03.011.

[35] Tannic acid. Am Chem Soc n.d. https://www.acs.org/molecule-of-the-week/archive/t/tannic-acid.html (accessed July 8, 2023).

[36] Swanson BG. Tannins and polyphenols. Encycl. Food Sci. Nutr., Elsevier; 2003, p. 5729–33. https://doi.org/10.1016/B0-12-227055-X/01178-0.

[37] Dunkel A, Köster J, Hofmann T. Molecular and sensory characterization of γ-glutamyl peptides as key contributors to the kokumi taste of edible beans (*Phaseolus vulgaris* L.). J Agric Food Chem 2007;55:6712–19. https://doi.org/10.1021/jf071276u.

[38] Maruyama Y, Yasuda R, Kuroda M, Eto Y. Kokumi Substances, Enhancers of basic tastes, induce responses in calcium-sensing receptor expressing taste cells. PLoS One 2012;7:e34489. https://doi.org/10.1371/journal.pone.0034489.

[39] Ueda Y, Sakaguchi M, Hirayama K, Miyajima R, Kimizuka A. Characteristic flavor constituents in water extract of garlic. Agric Biol Chem 1990;54:163–69. https://doi.org/10.1271/bbb1961.54.163.

[40] Ohsu T, Amino Y, Nagasaki H, Yamanaka T, Takeshita S, Hatanaka T, et al. Involvement of the calcium-sensing receptor in human taste perception. J Biol Chem 2010;285:1016–22. https://doi.org/10.1074/jbc.M109.029165.

[41] Tordoff MG, Alarcón LK, Valmeki S, Jiang P. T1R3: A human calcium taste receptor. Sci Rep 2012;2. https://doi.org/10.1038/srep00496.

[42] Zocchi D, Wennemuth G, Oka Y. The cellular mechanism for water detection in the mammalian taste system. Nat Neurosci 2017;20:927–33. https://doi.org/10.1038/nn.4575.

[43] Riera CE, Vogel H, Simon SA, Coutre J le. Artificial sweeteners and salts producing a metallic taste sensation activate TRPV1 receptors. Am J Physiol-Regul Integr Comp Physiol 2007;293:R626–34. https://doi.org/10.1152/ajpregu.00286.2007.

[44] Wang Y, Zajac AL, Lei W, Christensen CM, Margolskee RF, Bouysset C, et al. Metal ions activate the human taste receptor TAS2R7. Chem Sens 2019;44:339–47. https://doi.org/10.1093/chemse/bjz024.

[45] De araujo IE, Schatzker M, Small DM. Rethinking food reward. Annu Rev Psychol 2019. https://doi.org/10.1146/annurev-psych-122216-011643.

[46] Blankenship ML, Grigorova M, Katz DB, Maier JX. Retronasal odor perception requires taste cortex, but orthonasal does not. Curr Biol 2019;29:62–69.e3. https://doi.org/10.1016/j.cub.2018.11.011.

[47] Genva M, Kenne Kemene T, Deleu M, Lins L, Fauconnier M-L. Is it possible to predict the odor of a molecule on the basis of its structure? Int J Mol Sci 2019; 20. https://doi.org/10.3390/ijms20123018.

[48] Su C-y, Menuz K, Carlson JR. Olfactory perception: Receptors, cells, and circuits. Cell 2009;139:45–59. https://doi.org/10.1016/j.cell.2009.09.015.

[49] Trimmer C, Keller A, Murphy NR, Snyder LL, Willer JR, Nagai MH, et al. Genetic variation across the human olfactory receptor repertoire alters odor perception. Proc Natl Acad Sci 2019;116:9475–80. https://doi.org/10.1073/pnas.1804106115.

[50] Malliou F, Pavlidis P. Current theories in odorant binding. Curr Otorhinolaryngol Rep 2022;10:405–10. https://doi.org/10.1007/s40136-022-00437-y.

[51] Billesbølle CB, De March CA, Van Der Velden WJC, Ma N, Tewari J, Del Torrent CL, et al. Structural basis of odorant recognition by a human odorant receptor. Nature 2023;615:742–49. https://doi.org/10.1038/s41586-023-05798-y.

[52] Bushdid C, Magnasco MO, Vosshall LB, Keller A. Humans can discriminate more than 1 trillion olfactory stimuli. Science 2014;343:1370–72. https://doi.org/10.1126/science.1249168.

[53] Maßberg D, Hatt H. Human olfactory receptors: Novel cellular functions outside of the nose. Physiol Rev 2018;98:1739–63. https://doi.org/10.1152/physrev.00013.2017.

[54] Mainland JD. Olfaction. In: Wixted JT, editor. Stevens Handb. Exp. Psychol. Cogn. Neurosci., Hoboken, NJ, USA: John Wiley & Sons, Inc.; 2018, p. 1–46. https://doi.org/10.1002/9781119170174.epcn204.

[55] Garg N, Sethupathy A, Tuwani R, Nk R, Dokania S, Iyer A, et al. FlavorDB: A database of flavor molecules. Nucleic Acids Res. 2018;46:D1210–6. https://doi.org/10.1093/nar/gkx957.

[56] Burdock GA. Fenaroli's handbook of flavor ingredients. CRC press; 2016.

[57] Sanderson GW, Grahamm HN. Formation of black tea aroma. J Agric Food Chem 1973;21:576–85.

[58] Gao Y, Miao J, Lai K. Study on Maillard reaction mechanism by quantum chemistry calculation. J Mol Model 2023;29:81. https://doi.org/10.1007/s00894-023-05484-w.

[59] Resconi VC, Escudero A, Campo MM. The development of aromas in ruminant meat. Molecules 2013;18:6748–81. https://doi.org/10.3390/molecules18066748.

[60] Shahidi F, Hossain A. Role of lipids in food flavor generation. Molecules 2022;27:5014. https://doi.org/10.3390/molecules27155014.

[61] Belitz H-D, Grosch W, Schieberle P. Food chemistry. 4th rev. and extended edition. Berlin: Springer; 2009.

[62] Damodaran S, Parkin K, Fennema OR, editors. Fennema's food chemistry. 4th edition. Boca Raton: CRC Press/Taylor & Francis; 2008.

[63] Muñoz-González C, Feron G, Canon F. Main effects of human saliva on flavour perception and the potential contribution to food consumption. Proc Nutr Soc 2018;77:423–31. https://doi.org/10.1017/S0029665118000113.

[64] Schwartz M, Neiers F, Feron G, Canon F. The relationship between salivary redox, diet, and food flavor perception. Front Nutr 2021;7. DOI=10.3389/fnut.2020.612735

[65] Çelebioğlu HY, Lee S, Chronakis IS. Interactions of salivary mucins and saliva with food proteins: A review. Crit Rev Food Sci Nutr 2020;60:64–83. https://doi.org/10.1080/10408398.2018.1512950.

[66] Bansil R, Turner BS. The biology of mucus: Composition, synthesis and organization. Adv Drug Deliv Rev 2018;124:3–15. https://doi.org/10.1016/j.addr.2017.09.023.

[67] Matczuk J, Żendzian-Piotrowska M, Maciejczyk M, Kurek K. Salivary lipids: A review. Adv Clin Exp Med 2017;26:1023–31. https://doi.org/10.17219/acem/63030.

[68] Canon F, Neiers F, Guichard E. Saliva and flavor perception: Perspectives. J Agric Food Chem 2018;66:7873–79.

10 Food additives

María Montaña Durán Barrantes

10.1 Introduction

Food additives, although present/listed in foods as ingredients, are not characteristic to them. That is, the food will be the same with or without them. Food additives are of utmost importance in the current food industry as, since the turn of the millennium, more than two-thirds of all consumed food belong to the category of "prepared," which could not exist without the use of additives [1]. Consumers face a dilemma: on the one hand, foods free from food additives are desired, while also wanting long-lasting foods, which are pleasant and appealing, with good texture and taste, as can be seen in Table 10.1.

Table 10.1: Examples of possible groups for processed foods based on type of use [2].

Food group	Possible subgroups	Comments
Fast foods	Kebabs, tacos, hamburgers, fried chicken, pizza	
Infant foods	Infant formulas, prepared infant foods	
Special dietary foods	Reduced energy foods, diabetic foods, low-sodium foods	Including parenteral and enteral feeds, therapeutic meal replacements
Manufactured foods	Processed meals (dried, salted, smoked and canned fish, processed meat, egg-based and milk-based products, etc.), snack foods, packet mixes, soups, sauces, gravies, sugar, and syrups (confectionery, desserts, jams, jellies, vegetable and fruit preserves . . .)	
Prepared foods	Institutional meals (restaurant meals), domestic meals, recipe-based meals	
Beverages	Teas, coffees, cordials, soft drinks, fruit-flavored drinks	Including carbonated drinks but excluding milk and fruit and vegetable juices

Usage of stabilizers or modifiers of the food characteristics has its roots in ancient history; sesame oil was used in Mesopotamia to preserve fish and meat, cane sugar for fruit preserves in India, potassium nitrate for Roman brine, as well as spices and food coloring used during ancient Egyptian times, to put a few examples. There is,

María Montaña Durán Barrantes, Área Tecnología de Alimentos, Departamento de Ingeniería Química, Universidad de Sevilla, Spain

https://doi.org/10.1515/9783111111872-010

however, also a negative side to the use of additives: when they are used to trick the consumer. Only after nineteenth century, scientists could begin to detect alterations of coloring in wine, vegetables (copper sulfate), cheese (red lead), chocolate (brick dust), etc., thanks to the development of chemical analysis. Food regulations emerge as a response to these fraudulent activities at the beginning of the twentieth century.

10.2 Definition of food additive

Everything that becomes part of a food is considered an ingredient, and as such, it may be indispensable and not replaceable, or optional and replaceable. The group of replaceable ingredients includes food additive, which is understood as any substance intentionally added to the food during its elaboration in order to perform a specific technological function by remaining in food. Food is therefore the same with or without the additive, and the addition of the latter serves a purpose that is favorable to the consumers, or at least not harmful to them, by improving the stability of a food with the least possible modifications. It may be the case that the same technological effect (coloring, sweetening, increasing viscosity, etc.) can be achieved by more than one substance. Additives can help correct defects and are also used to stabilize or modify physical, chemical, biochemical, or sensory characteristics. For their technological use, it is not only necessary to know whether the additive is authorized for the food in question and at which dosage but also to take into consideration the operating conditions: solubility, presence of other ingredients, packaging material, processing temperature, etc. Finally, as any other ingredient, additives must be declared on the label in accordance with the relevant regulations.

According to the Regulation (EC) No 1333/2008 [3] under the European rules applicable to food additives, the following definition is applied: "food additive shall mean any substance not normally consumed as a food in itself and not normally used as a characteristic ingredient of food, whether or not it has nutritive value, the intentional addition of which to food for a technological purpose in the manufacture, processing, preparation, treatment, packaging, transport or storage of such food results, or may be reasonably expected to result, in it or its by-products becoming directly or indirectly a component of such foods." This definition specifies that, for example, in the case of using ascorbic acid as an antioxidant additive (E 300), it will not be considered as a nutrient (vitamin C), and vice versa, if it is used to increase the nutritional value of a food. On the other hand, the definition of an additive makes it clear that its incorporation is intentional, so its unintentional addition would imply its condition as a contaminant in the food.

The category of technological ingredients includes additives, aromas, enzymes, and processing aids. In the European Union, the following package of regulations affecting all of them include:

– Regulation (EC) No 1331/2008 establishing a common authorization procedure
– Regulation (EC) No 1332/2008 on food enzymes
– Regulation (EC) No 1333/2008 on food additives

- Regulation (EC) No 1334/2008 on flavorings

The aim of this legislative package is to unify all the legislation and to avoid transposition into the national laws of each EU country. Regulations affecting only the list of authorized additives would include:
- Regulation (EC) No 1331/2008 of the European Parliament and of the Council of December 16, 2008, establishing a common authorization procedure for food additives, food enzymes, and food flavorings.
- Commission Regulation (EU) No 234/2011 of March 10, 2011 implementing Regulation (EC) No 1331/2008 of the European Parliament and of the Council establishing a common authorization procedure for food additives, food enzymes, and food flavorings.
- Regulation (EC) No 1333/2008 of the European Parliament and of the Council of December 16, 2008 on food additives.
- Commission Regulation (EU) No 1129/2011 of November 11, 2011 amending Annex II to Regulation (EC) No 1333/2008 of the European Parliament and of the Council by establishing a Union list of food additives, text with European Economic Area relevance (consolidated text). Part A: This Union list includes:
  - The name of the food additive and its E number
  - The foods to which the food additive may be added
  - The conditions under which the food additive may be used
  - Restrictions on the sale of the food additive directly to the final consumer
- Commission Regulation (EU) No 257/2010 of March 25, 2010 setting up a program for the re-evaluation of approved food additives in accordance with Regulation (EC) No 1333/2008 of the European Parliament and of the Council on food additives.

Regulation (EC) No 1333/2008 lays down the conditions of use of food additives in foods, including food additives, food enzymes, and food flavorings, and the rules on labeling of food additives sold as such. The whole information is organized on the basis of 4 annexes as follows:
- Annex I: Functional classes of food additives in foods and of food additives in food additives and food enzymes.
- Annex II (Part A, B, C, D, and E): Union list of food additives approved for use in foods and conditions of use.
- Annex III (Part 1, 2, 3, 4, 5, and 6): Union list of food additives including carriers approved for use in food additives, food enzymes, food flavorings, nutrients, and their conditions of use.
- Annex IV: Traditional foods for which certain Member States may continue to prohibit the use of certain categories of food additives.
- Annex V: List of the food colors for which the labeling of foods shall include additional information.

Only food additives that are listed in the EU legislation can be added to food and this can be done only under specific conditions, which avoids different national interpretations. The additives are listed based on the categories of food of which they may be added.

Additives causing minimum toxicological concerns may be added in almost all processed foodstuffs. Examples include calcium carbonate (E 170), lactic acid (E 270), citric acid (E 330), pectin (E 440), fatty acids (E 570), and nitrogen (E 941). For other additives the use is more restricted, for example:

– Natamycin (E 235) can only be used as preservative for the surface treatment of cheese and dried sausages.
– Erythorbic acid (E 315) can only be used as antioxidant in certain meat and fish products.
– Sodium ferrocyanide (E 535) can only be used as anti-caking agent in salt and its substitutes.

The Commission's food additive database is available on the Internet: https://food.ec. europa.eu/safety/food-improvement-agents/additives/database_en. This database enables the consumer or business operator to find out which additives are authorized in a particular food.

10.2.1 Acceptable Daily Intake (ADI)

Scientists who evaluate the safety of chemicals in food try to establish what they call "health-based guidance values" or "safe levels." These define the maximum amount of chemicals that we can safely consume each day or week, during our entire lifetime. The most common safe levels are ADI, used for additives and other substances purposely added to food. Safe levels are based on the scientific review of all the toxicological data available at the time on a specific chemical, including long-term tests on animals. From this review, a reference point is identified for the most sensitive adverse effect in animals that is considered relevant for humans too. This reference value is usually a "no observed adverse effect level," the NOAEL (usually expressed in terms of mg compound per kg body weight per day), which is the greatest concentration or amount of substance that causes no detectable adverse effect in the animals used in the test [4]. The safe level is then derived by applying an uncertain factor of 100:10 to account for the difference between humans and animals, and another 10 to account for the differences between humans, such as those between children and adults. A food additive is considered safe for its intended use if its human exposure is less than, or is approximately, the same as the ADI. Exceeding the health-based guidance values, such as ADI, on an occasional basis is not necessarily a cause for concern because they are set so as to take into account exposure to a substance over a lifetime.

Only in the case that the safe levels are exceeded persistently over time, decision makers will take action in order to protect consumers.

The procedure for the authorization of the use of food additives is laid down in Regulation (EC) No 1333/2008, following an application to the European Commission by an interested party. The substances are evaluated based on a dossier, usually provided by an applicant (normally the producer or a potential user of the food additive). This dossier must contain the chemical identifications of the additive, its manufacturing process, methods of analyses and reaction and fate in food, the case of need, the proposed uses, and toxicological data (information on metabolism, subchronic and chronic toxicity, carcinogenicity, genotoxicity, reproduction and developmental toxicity, and, if required, other studies). For new additives, the Commission will request EFSA to assess the safety of the substance. After EFSA has given its opinion (within 9 months following the request), the Commission, together with food additive experts from all Member States, will consider the possible authorization. The safety assessment, the technological need, the possibility for misuse, and the advantages and benefits for the consumer are all taken into account.

If considered appropriate, the Commission will prepare a proposal for possible authorization of the additive and present it for vote at the Standing Committee on the Food Chain and Animal Health (SCoFCAH). If SCoFCAH supports the proposal, it will be presented to the Council and the European Parliament. They can still reject it in case they consider that the authorization does not comply with the conditions of use set out in the EU legislation.

On the international front, JECFA, Joint FAO/WHO Expert Committee on Food Additives, is an international scientific expert committee administered jointly by the Food and Agriculture Organization of the United Nations (FAO) and the World Health Organization (WHO). It has been meeting since 1956 to evaluate the safety of food additives, contaminants, naturally occurring toxicants, and residues of veterinary drugs in food [5, 6].

According to Article 9 of Regulation (EC) No 1333/2008, "food additives may be assigned to one of the functional classes in Annex I on the basis of the principal technological function of the food additive," without this precluding it from being used for several functions; and Article 10 specifies that with each additive must be entered "the name of the food additive and its E number; the foods to which the food additive may be used; conditions under which the food additive may be used; and if appropriate, whether there are any restrictions on the sale of the food additive directly to the final consumer," which is listed in Annex II. This Regulation will be open to amend nonessential elements relating to additional functional classes, if necessary, as a result of scientific progress or technological development.

The presence of food additives should therefore be considered safe even for consumers that eat large quantities of foodstuffs to which the additives have been used at the maximum permitted level.

As a result of the re-evaluation program, for example, the use of three food colors was revised because EFSA decreased their ADI considering that human exposure to these colors is likely to be too high. Therefore, the maximum levels of these colors that can be used in food were lowered in early 2012. This reference concerns E 104 Quinoline yellow, E 110 Sunset Yellow, and E 124 Ponceau 4R.

10.3 Classification of food additives

The additives can be of natural origin, coming from raw food materials or from a natural product, such as E 306, an extract rich in tocopherols obtained from vegetable oils, as well as molecules produced by synthesis but of natural origin, such as carotenoid-type colorings, and natural compounds slightly modified in their composition or structure, such as modified starches or celluloses. On the other hand, there are artificial or synthetic additives that respond to chemical structures that do not exist in nature and are obtained by synthesis (more efficient, with reproducible and very uniform results), such as erythrosine (E 127).

From a functional point of view, additives are used both to avoid changes in foods, which are subject to many environmental conditions that can modify their original composition (changes in temperature, oxidation, exposure to microbes, etc.), and to achieve a greater variety of foods, which are easy to prepare, safe, nutritious, and cheap. This involves the use of additives with the aim of improving their physical properties, taste, preservation, etc., which means that they remain safe, nutritious, and tasty until consumption.

The use of food additives should serve one or more of the following purposes:

1. Ensure food safety.
2. Enhance the keeping quality or stability of a food.
3. Increase availability of food when they are out of season.
4. Preserve the nutritional quality of the food.
5. Foster consumer acceptance.
6. Provide help in manufacture, processing, preparation, treatment, packing, transport, or storage of food.
7. Keep homogeneity to the foodstuff.

In an attempt to group the set of additives (definition according to Annex I, Regulation No 1333/2008) according to the modifying or stabilizing activity of certain properties in the food, the following groups could be identified:

I. **Modifiers of organoleptic characteristics of food:**
    Color-modifying additives:
    "Colors" are substances that add or restore color in a food, and include natural constituents of foods and natural sources, which are normally not consumed as foods as such and not normally used as characteristic ingredients of food. Preparations obtained from foods and other edible natural source materials by physical and/or chemical extraction resulting in a selective extraction of the pigments relative to the nutritive or aromatic constituents are colors within the meaning of this Regulation.

    Sapid substances:
    "Acids" are substances that increase the acidity of a foodstuff and/or impart a sour taste to it.
    "Acidity regulators" are substances that alter or control the acidity or alkalinity of a foodstuff.
    "Sweeteners" are substances used to impart a sweet taste to foods or in tabletop sweeteners.
    "Flavor enhancers" are substances that enhance the existing taste and/or odor of a foodstuff.

II. **Stabilizers of physical food appearance:**
    "Anti-caking agents" are substances that reduce the tendency of individual particles of a foodstuff to adhere to one another.
    "Anti-foaming agents" are substances that prevent or reduce foaming.
    "Emulsifiers" are substances that make it possible to form or maintain a homogeneous mixture of two or more immiscible phases such as oil and water in a foodstuff.
    "Thickeners" are substances that increase the viscosity of a foodstuff.
    "Stabilizers" are substances that make it possible to maintain the physico-chemical state of a foodstuff; stabilizers include substances that enable the maintenance of a homogenous dispersion of two or more immiscible substances in a foodstuff, substances which stabilize, retain, or intensify an existing color of a foodstuff, and substances that increase the binding capacity of the food, including the formation of cross-links between proteins enabling the binding of food pieces into reconstituted food.
    "Humectants" are substances that prevent foods from drying out by counteracting the effect of an atmosphere having a low degree of humidity, or promote the dissolution of a powder in an aqueous medium.

III. **Substances to prevent from chemical or biological alterations in food:**

"Antioxidants" are substances that prolong the shelf-life of foods by protecting them against deterioration caused by oxidation, such as fat rancidity and color changes.

"Preservatives" are substances that prolong the shelf-life of foods by protecting them against deterioration caused by microorganisms and/or which protect against growth of pathogenic microorganisms.

"Sequestrants" are substances that form chemical complexes with metallic ions.

IV. **Plastic properties modifiers, capable of contributing to the most desirable texture:**

"Modified starches" are substances obtained by one or more chemical treatments of edible starches, which may have undergone a physical or enzymatic treatment, and may be acid or alkali thinned or bleached.

"Firming agents" are substances that make or keep tissues of fruit or vegetables firm or crisp, or interact with gelling agents to produce or strengthen a gel.

"Raising agents" are substances or combinations of substances which release gas and thereby increase the volume of a dough or a batter.

"Gelling agents" are substances which impart a foodstuff texture by forming of a gel.

"Emulsifying salts" are substances which convert the proteins contained in cheese into a dispersed form and thereby bring about homogeneous distribution of fat and other components.

"Foaming agents" are substances which make it possible to form a homogeneous dispersion of a gaseous phase in a liquid or solid foodstuff.

V. **Substances that carry out functions not included in the previous paragraphs:**

"Bulking agents" are substances which contribute to the volume of a foodstuff without contributing significantly to its available energy value.

"Glazing agents" (including lubricants) are substances which, when applied to the external surface of a foodstuff, impart a shiny appearance or provide a protective coating.

"Flour treatment agents" are substances, other than emulsifiers, which are added to flour or dough to improve its baking quality.

"Propellants" are gases other than air, which expel a foodstuff from a container.

"Packaging gases" are gases other than air, introduced into a container before, during, or after the placing of a foodstuff in that container.

"Carriers" are substances used to dissolve, dilute, disperse, or otherwise physically modify a food additive or a flavoring, food enzyme, nutrient, and/or other substance added for nutritional or physiological purposes to a food without altering its function (and without exerting any technological effect themselves) in order to facilitate its handling, application, or use.

10.3.1 International numbering system of food additives: E number

The category of the additive is determined by the competent authority in its geographical area of influence, establishing its inclusion for specific uses. In Europe, the competent authority is the European Food Safety Authority (EFSA), and it is the European Commission that decides whether or not to include the additive under consideration. Based on the advice of EFSA the Commission may propose a revision of the current conditions of use of the additives and if needed remove an additive from the list. In the United States it is the Food and Drug Administration (FDA) that evaluates the suitability of substances for use as additives. This corresponds in the European Union to a code known as the "E number": the letter E followed by the numbering of the International Numbering System for Food Additives (INS) proposed by the Codex Alimentarius, consisting of three or four digits; in some cases, this number is followed by an alphabetical suffix [7].

Initially, the allocation of the INS was arranged in such a way as to group together food additives with similar technological functions, as will be discussed in the following sections of this chapter. However, because of the length of the list, most of the three-digit numbers have already been allocated, and therefore the position of the additive in the list can no longer be taken as indicative of the function of the additive, although this is often the case.

The additives in foodstuff label must be mentioned in the ingredient list and designated by the name of their functional class, followed by their specific chemical name or EC number (European Community number), according to the rules set out in Regulation (EC) No 1169/2011 [8]. For instance: "color: curcumin" or "color: E 100."

As can be seen, this E number can be used in order to simplify the labeling of substances with sometimes complicated chemical names.

10.4 Food colorings

The color of food is a very important consideration when consuming it. Everyone knows the expression "to have eyes bigger than your belly." The consumer establishes an unconscious relationship between a food and its typical color, which is influenced by technological processes with factors such as:
- New shades can appear: caramelization, carbonylamine browning, etc.
- Characteristic shades may disappear: discoloration or change in color.

Colored substances naturally present in foods are usually called pigments, with the designation colorings being reserved for those, natural or artificial, which are used as additives to help correct undesirable changes. All colored substances are characterized by a high degree of electronic delocalization, with any changes to the electronic structure resulting in the modification of its color.

The use of food colors is considered acceptable for the following purposes [9]:
– To restore the original appearance of food of which the color has been affected by processing, storage, packaging, and distribution.
– To make food more visually appealing.
– To give color to food otherwise colorless.

The use of food colors must always comply with the general condition that they do not mislead the consumer. For example, the use of colors should not give the impression that it contains ingredients that have never been added.

In addition, although natural colors have been used in food since ancient times (saffron, caramel, cochineal, turmeric . . .), today there are a significant number of additives of artificial origin, whose presence in food has led to warnings in the EU through the RASFF (Rapid Alert System for Food and Feed) [10], due to their adverse effects. The trend in the food industry today is to replace these synthetic or artificial colors with natural ones, although the latter are more sensitive to alterations due to heat, light, pH, storage, and oxidizing or reducing substances in the food. Examples include curcumin (E 100), which is very stable to light but unstable to heat, and anthocyanins (E 163), whose pH must be below 3.5 to maintain their color.

According to Regulation (EC) No 1333/2008, colors from synthetic sources, and selectively extracted colors from natural sources, as well as inorganic pigments, are regulated together. The labeling of foods shall include additional information about some of them (tartrazine (E 102), carmoisine or azorubine (E 122), etc.) in order that "may have an adverse effect on the activity and attention in children." Natural colors are generally considered to be harmless and have fewer specific limitations than artificial colors, such as riboflavins (E 101), chlorophylls (E 140), vegetable carbon (E 153), lycopene (E 160d), etc.

10.5 Food stabilizers for chemical and biological alterations

Some additives act by directly preventing changes (stabilizers), while others (synergists or synergistic synergists) enhance them. A mixture of two or more substances is synergistic when the total dose of the mixture is more effective than an equal dose of each substance separately.

For a reaction to occur that chemically alters the food, the necessary reagents must be present and in contact in a suitable medium. Therefore, these alterations can be prevented by adding substances that remove the reagents, prevent interaction between them, or modify the conditions of the medium. Principal chemical alterations in food are:
– Enzymatic browning. There are enzymes (polyphenol oxidase, tyrosinase) that can introduce, in ortho, a second hydroxyl group in a phenol and oxidize the diphenols to diquinones, which are unstable, polymerized, and undergo oxidation,

giving rise to brown compounds (melanoids). These enzymes have copper (II) as a prosthetic group and their activity is maximal at a pH between 5 and 7. Therefore, acids and sequestrants, such as citric and ascorbic acid, are effective (the latter is also a reducing agent).

– Non-enzymatic browning. This alteration can occur either by degradation of ascorbic acid or by Maillard browning (carbonylamine). This occurs when there are reducing sugars together with amino acids, peptides, or proteins. It is prevented by blocking the free carbonyl group by adding bisulfite (with sulfur dioxide) and is made more difficult by acidification. Besides the darkening, characteristic aromas appear. It occurs to a greater extent in concentrated or dehydrated foods and is facilitated by temperature and a low acidic pH. (More information on browning reactions in Chapter 3, Section 3.6.)

– Oxidative reactions [11]. These alterations require the presence of oxygen, from the environment or absorbed in the food, and are therefore prevented by: inert atmospheres, vacuums, barriers, or reducing substances (ascorbic acid, erythorbic acid, sulfur dioxide . . .) applied in mass or on the surface.

  – The process of fat rancidity occurs when an unsaturated lipid absorbs radiant energy in the presence of oxygen, resulting in the formation and propagation of free radicals, peroxides and further radicals until they combine with each other in the completion phase of this auto-oxidation (Chapter 8). Transition metal ions are prooxidant substances which favor the propagation of radicals. Antioxidants are additives that prevent rancidity. They must be soluble in the food as well as stable under operating conditions and act in the initiation or propagation phase. Natural: tocopherols, gallic acid and gallates, extracts of spices and condiments; the possible technological use of flavonoids, carotenoids, etc., is being investigated. Synthetic: butylhydroxyanisole, butylhydroxytoluene, tertbutilhydroquinone. Synergists or synergists of antioxidants act by regenerating them (ascorbic acid and ascorbates . . .) or by sequestering prooxidants (citric acid and citrates, phosphates, EDTA, etc.).

Preservatives (benzoic, sorbic, propionic, sulfurous acids, etc.) are used to prevent microbial deterioration. Not all of them are equally effective against different microbes (bacteria, moulds, or yeasts) and can act by killing them (microbicides) or by preventing their growth (microbiostatics). The active form is usually undissociated acid (present in an acidic medium), so their effectiveness depends on the pH, which determines the degree of dissociation. They are used alone, mixed with another preservative or with a synergist (usually an acid, although synergistic effects of some antioxidants have also been described). Stability under operating conditions and solubility must also be considered: soluble salts (of K^+ or Na^+) are used in acidic water-based foods, and calcium salts or acids in solid foods.

There is a group of widely used preservatives, whose levels are not limited by Regulation (EU) 1129/2011, which are of natural origin: acetic acid (E 260) and its salts, lactic acid (E 270) and its salts, malic acid (E 296) and its salts, tartaric acid (E 334) and its salts, and carbon dioxide (E 290). Some are also used as acidulants, and synergists of antioxidants, and may also be obtained naturally by extraction from foods (acetic from wine, lactic from milk lactose, etc.), or by synthesis from reactions between other pure chemical compounds.

By contrast, there are also preservative additives with limitations on the quantity to be added in the food categories where their use is authorized, such as:

- Sorbic acid (E 200) and its salts, which are easily assimilated by the body, although they may react with other additives in the food, such as nitrites or sulfites, causing compounds with a mutagenic action.
- Benzoic acid (E 210) and its salts, which are found naturally in red fruits, cloves, and cinnamon, for example, but are obtained synthetically because they are cheaper.
- Sulphur dioxide (E 220) and sulfites, widely used in winemaking, but also very effective as an antioxidant in enzymatic browning, blocking the free carbonyl groups of the sugars, which are therefore unable to interact with the amino acids in the Maillard reaction (non-enzymatic browning). This additive causes adverse reactions in people with high sensitivity to sulfites, such as asthma, skin processes, and diarrhea, but without teratogenic or carcinogenic effects.
- Nitrites and nitrates (E 249 to 252), highly specific against the growth of *Clostridium botulinum*, an anerobic microorganism that produces a potent neurotoxin, for which they must be used with a pH of 5.0 to 5.5, in synergy with sodium chloride (common salt) and refrigeration temperatures. When used in meat products, it helps to stabilize the red color of the meat by reaction of the generated nitric oxide with myoglobin, thus producing nitrosomioglobin; but it can also result in nitrosamines, carcinogenic substances. Therefore, they are used together with additives that inhibit this reaction, such as ascorbic acid and various phenolic compounds.

10.6 Structural food stabilizers

A disperse system is multiphase and contains a continuous phase, within which small particles from one or more disperse phases are distributed. The greater the stability, the smaller is the difference in density between the phases or the size of the particles and the higher the viscosity of the liquid. Dispersion is also stabilized by electrostatic action (protective colloids) or by the adsorption of surface-active substances at the interface.

In physical stabilization, the viscosity of the continuous phase is increased by cooling or by the action of thickening substances (which are usually polysaccharides or proteins, naturally present or added, capable of forming hydrocolloids). A particular case is that of gelifiers, which cause the immobilization of the continuous phase

by leaving it trapped in a structure with the consistency of a gel. In fruit derivatives, only vegetable polysaccharides or polysaccharides of microbial origin (pectin, xanthan gum, etc.) are used, since gelifiers of animal origin (gelatin, "surimi") are reserved for other products. Vegetable polysaccharides may come from terrestrial or marine plants (algae) and may be found in tissues or as exudation products. The following additives are authorized under European legislation (for any given application, please refer to the lists of authorized additives):
- Algae extracts: alginic acid and alginates (E 400 to E 404), agar-agar (E 406), and carrageenan (E 407)
- Seed extracts: locust bean gum (E 410) and guar gum (E 412)
- Polysaccharides biosynthesized by bacteria: xanthan gum (E 415) and gellan gum (E 418)
- Fruit extracts: pectins (E 440)

Gelation conditions vary from one to another. Highly methylesterified pectins (fast or slow) are gelled with sugar, acid, and heat, whereas low methylester pectins are gelled at a lower temperature, in a wide range of pHs and with divalent cations. Alginates, carrageenans, and gellan gum also require certain types of ions to gel. Agar-agar and locust bean gum mixtures with xanthan gum only need to be dispersed in hot water, gelling when cooled. The choice of gelling agent should be made according to the desired characteristics of the final products: for example, the use of highly esterified pectins would not be advisable for products with low sugar content. It should also be taken into account that these substances are degraded by prolonged heating in an acidic medium, so they should be added to the rest of the previously dispersed ingredients at the end of the food processing.

In physico-chemical stabilization, the affinity of the particles dispersed by the continuous phase can be favored by the adsorption in the interface of amphiphilic substances (compounds with affinity for both phases, such as some lipids and proteins) or substances that give them an electrostatic charge (which causes the particles to repel each other, helping to keep them dispersed). Emulsifying agents are additives that allow the formation of a homogeneous mixture of two or more immiscible phases in a food, preventing the dispersed phase from becoming associated and migrating to the surface or precipitating, due to a difference in density, as occurs in cream (oil-in-water emulsion) or butter (water-in-oil emulsion). Emulsifiers are compounds that are adsorbed on the surface of the drops of the dispersed phase, which slows down the process of aggregation of the drops, thus protecting the emulsion, and may be as follows:
- Surfactants (amphiphilic molecules), such as the calcium, potassium, sodium, and magnesium salts of fatty acids (E 470), lecithin (E 322), or sorbitan esters (E 491 to E 495), used for example in margarines, chocolates, or meat products.
- Hydrophilic colloids, such as monoglycerides and diglycerides (E 471), which are used to give stability to the incorporation of air into baking or pastry doughs.

Agglomeration allows small particles to be joined together to achieve solids of a given shape and size, but the particles must be adhesive. This property can be conferred by a naturally present or added substance: there are "stabilizers" that adhere to the particles and cover them, making them stick together (they are not considered additives when they are provided by a characteristic ingredient of the formulation, but they are when are added for the purpose of agglomeration). They are usually polymers (polysaccharides, proteins) capable of forming hydrocolloids, although sometimes fatty substances, more or less viscous (such as cocoa butter), are used. There are various agglomeration techniques, such as the use of coating agents/glazing agents: this is usually done by dropping a coating mass of the appropriate viscosity onto the foodstuff while it is being transported on an endless mesh belt (the mass is applied from below by means rollers placed underneath). The product can also be immersed in a coating bath while being transported between two mesh belts.

10.7 Sweeteners

Sweeteners are substances that give sweetness to foods, such as sugars and sweetening additives. Some also affect texture, stability, taste, color, or nutritional value.

Under the name of sugars, only carbohydrates obtained from vegetables or milk serum, in which they are free, or those obtained by hydrolysis of vegetable starch [12], free of sweetening additives, can be marketed. The most commonly used sweeteners are sucrose and starch hydrolysates.

Sweetening additives are substances of natural or synthetic origin and of diverse composition that have a sweetening power equal to or greater than that of sugars, but that provide little or no energy value. Among the natural sweeteners authorized in the European Union by Regulation (EU) 1333/2008, updated by Regulation (EU) 1129/2011, include, among others:

–   Sweeteners of a glycidic nature, such as first- and second-generation polyols (sorbitol (E 420), mannitol (E 421), isomaltitol (E 953), maltitol (E 965), lactitol (E 966), and xylitol (E 967))
–   Non-glycidic sweeteners, flavonoid derivatives (neohesperidine dihydrochalcone (E 959), 1,500 times more sweetening power than sucrose), and nitrogenous derivatives (aspartame (E 951), obtained by combination of two amino acids; and thaumatin (E 957), with more than 2,000 times more sweetening power than sucrose, and plant-based.

Polyalcohols are compounds obtained by the hydrogenation of reducing sugars and, in addition to being used as sweeteners, are also used to reduce the activity of water and therefore increase the osmotic pressure in the food, solubilize flavors, give texture, etc., and can be added without restrictions on the maximum quantity (quantum satis), in accordance with good manufacturing practices. However, they have a laxative action, so their use is not allowed in baby foods.

Among the synthetic or artificial additives, the most widely used today are acesulfame K (E 950), cyclamates (E 952), saccharin (E 954), aspartame (E 954), sucralose (E 955), and neotame (E 961), all of which have a high sweetening power in comparison to sucrose, which allows their use in very low concentrations (some chemical formulae are shown in Chapter 9, Figure 9.2). Moreover, none of them have any other technological function besides sweetening, they are suitable for diabetics and they are not carcinogenic. Saccharin, the first synthetic sweetener discovered that was more powerful than sucrose (300 times more powerful), in 1879, with great stability in food, had to face a study carried out on rats at the end of the 1970s that linked it to the appearance of cancer. The FDA (Food and Drug Administration) considered banning it, but in 1991 it was concluded that this cancer threat in animals was not relevant to humans due to critical differences between species, according to the IARC (International Agency for Research on Cancer, which is part of the World Health Organization of the United Nations). Today, its use is totally safe worldwide, and it is eliminated without hardly being metabolized, like acesulfame K (E 950), although it crosses the placental barrier and is secreted through maternal milk, so it is contraindicated in pregnant and lactating women.

Mono and disaccharides (glucose, fructose, lactose), although they are sweeteners, are not considered as additives according to the mentioned Regulation.

There are countries that allow the use of other natural sweeteners (glycerin, tagatose, stevioside . . .). For example, tagatose, very similar to fructose and obtained by enzymatic treatment from lactose, is recognized in the United States as GRAS (Generally Recognized as Safe), while in the EU it is not considered as an additive, although it is used as an ingredient in foods to replace sugar, allowing the declaration on the food label of the following information "causes a lower increase in blood glucose than foods containing sugar," or "helps maintain mineralization of teeth," which refers to demineralization that occurs due to the fermentation of glucose by bacterial action in the mouth.

10.8 Flavorings and flavor enhancers

As defined by Regulation (EC) 1334/2008 [13], flavorings are products that are added to foods in order to impart or modify odor and/or taste, which includes certain food ingredients with flavoring properties for use in and on foods. On the other hand, flavor enhancers are substances that enhance the existing taste and/or odor of a foodstuff, as established by Annex III in Regulation (EC) 1333/2008, such as neotame (E 961), neohesperidine dihydrochalcone (E 959), or thaumatin (E 957).

Legislation classifies flavorings as: "flavoring substances" (defined chemical compounds: "natural," "nature identical," or "artificial," for example, amyl acetate from banana or by chemical synthesis), "flavoring preparations" (mixtures obtained from biological raw materials by physical, enzymatic, or microbiological processes, such as peppermint oil), "processing flavorings" (originating from heating, among others, amine compounds together with reducing sugars or Maillard reaction), and "smoke flavorings." Natural plant flavorings are derived from the essential oils in their juices and are mixtures of volatile (terpenes, sesquiterpenes, alcohols, aldehydes, ketones, acids, and esters) and fixed substances (camphors and waxes). They can be recovered mechanically (citrus fruits), by extraction or by distillation (steam trapping). If they are obtained by maceration and evaporation of the solvent, they can give rise to solids (concretes) or liquids (oleoresins). Separating the hydrocarbons by "deterpenation" results in products concentrated in odoriferous substances.

The principal regulations applicable to food flavorings are shown below:

– Regulation (EC) No 1331/2008 of the European Parliament and of the Council of December 16, 2008, establishing a common authorization procedure for food additives, food enzymes, and food flavorings.

– Regulation (EC) No 1334/2008 of the European Parliament and of the Council of December 16, 2008, on flavorings and certain food ingredients with flavoring properties for use in and on foods and amending Council Regulation (EEC) No 1601/91, Regulations (EC) No 2232/96 and (EC) No 110/2008 and Directive 2000/13/EC.

– Commission Regulation (EU) No 234/2011 of March 10, 2011, implementing Regulation (EC) No 1331/2008 of the European Parliament and of the Council establishing a common authorization procedure for food additives, food enzymes, and food flavorings.

– Regulation (EU) No 1169/2011 of the European Parliament and of the Council of October 25, 2011, on the provision of food information to consumers, amending Regulations (EC) No 1924/2006 and (EC) No 1925/2006 of the European Parliament and of the Council, and repealing Commission Directive 87/250/E EC, Council Directive 90/496/E EC, Commission Directive 1999/10/EC, Directive 2000/13/EC of the European Parliament and of the Council, Commission Directives 2002/67/EC and 2008/5/EC, and Commission Regulation (EC) No 608/2004.

– Commission Implementing Regulation (EU) No 872/2012 of October 1, 2012, adopting the list of flavoring substances provided for by Regulation (EC) No 2232/96 of the European Parliament and of the Council, introducing it in Annex I to Regulation (EC) No 1334/2008 of the European Parliament and of the Council and repealing Commission Regulation (EC) No 1565/2000 and Commission Decision 1999/217/EC.

10.9 Others

Hereafter, the different classes of food additives considered in Parts 1 to 6 of Annex III of Regulation (EC) 1333/2008, are briefly defined below:

–   Part 1: Carriers approved for use in food additives, food enzymes, food flavorings, nutrients whose conditions of use are "to dissolve, dilute, disperse or otherwise physically modify whichever of them, added for physiological purposes to a food without altering its function or exerting any technological effect themselves, in order to facilitate its handing, application or use," for instance: mannitol (E 421), isomalt (E 953), or xylitol (E 967), also used as sweeteners.

–   Part 2: Food additives, other than carriers, used in Food Additives preparations, such as sorbic acid-sorbates (E 200–203) or sulfur dioxide-sulfites (E 220–228) for color preparations under specific conditions; butylated hydroxyanisole (BHA) (Figure 10.1) as emulsifiers containing fatty acids; or phosphoric acid (E 338) for preparation of the color E 163 (anthocyanins).

–   Part 3: Food additives including carriers in food enzymes, which include different maximum levels depending on enzyme preparation, used in final food or in beverages: sorbic acid (E 200), benzoic acid (E 210) (Figure 10.1), sodium methyl p-hydroxybenzoate (E 219), etc.

–   Part 4: Food additives including carriers in food flavorings, specifying the flavoring categories to which the additive may be added, for example: propyl gallate (E 310), added in essential oils in 1,000 mg/kg, or in flavorings other than essential oils in 100 mg/kg, both established at a maximum level.

–   Part 5: Food additives in nutrients (those substances added for physiological or nutritional purposes) "except nutrients intended to be used in foodstuff for infants and young children." Most of them may also be used as a carrier (Part 1), such as, for example: lecithin (E 322), citric acid (E 330), ascorbic acid (E 300), etc.

As has been noted throughout this chapter, there is a significant number of food additives with different technological functions in foods, such as sorbic acid: preservative, carrier, additive in food enzyme preparations . . . which, in addition, can be of natural origin (extracted from fruits of the genus *Sorbus*) or synthetic. Therefore, Regulation (EC) 1169/2011 specifies the obligation to indicate on the label the technological application with which the additive is functioning for that particular type of food, which helps the consumer to know the maximum concentration, if any, in which it is being used. The name of its functional class is followed by its specific designation or by the code E, for example, "acidifier (citric acid)" or "acidifier (E 330)."

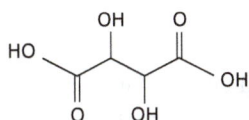

Tartaric acid (E 334)
Antioxidant

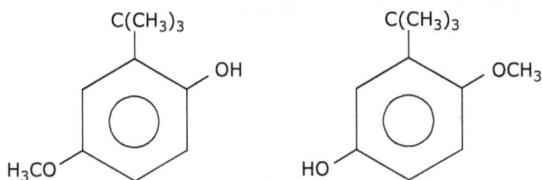

Butylhydroxyanisole (BHA) (E 320)
(mix of isomers)
Antioxidant

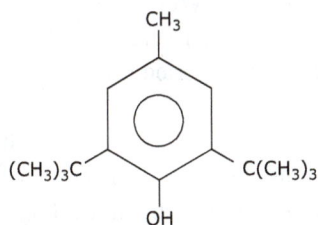

Butylhydrotoluene (BHT) (E 321)
Antioxidant

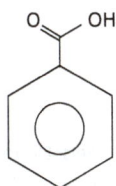

Benzoic acid (E 210)
Preservative

Figure 10.1: Chemical formulae of some compounds used as food additives. Formulae were elaborated and adapted using the structural information available at National Center for Biotechnology Information. PubChem Database. Tartaric acid, CID = 875, https://pubchem.ncbi.nlm.nih.gov/compound/Tartaric-acid. Butylhydroxyanisole, CID = 24667, https://pubchem.ncbi.nlm.nih.gov/compound/Butylhydroxyanisole. Butylated hydroxytoluene, CID = 31404, https://pubchem.ncbi.nlm.nih.gov/compound/Butylated-hydroxyto luene. Benzoic acid, CID = 243, https://pubchem.ncbi.nlm.nih.gov/compound/Benzoic-acid. (Accessed on May 25, 2020.)The structure of other compounds can be found elsewhere in this book: lycopene (E 160d), riboflavin (E 101), and anthocyanidines, the aglycone of anthocyanins (E 163), all of them natural colors, are shown in Figures 4.11 (Chapter 4), 6.1 (Chapter 6), and 8.14 (Chapter 8), respectively. In Chapter 3, several formulae of sugars and derivatives, including L-ascorbic acid (vitamin C, E 300), pectins (E 440), and gums, are shown in different figures. A list of sweeteners is shown in Figure 9.2 (Chapter 9).

10.9.1 Food enzymes

According to Regulation (EC) No 1332/2008, "food enzyme" means a product obtained from plants, animals or microorganisms or products thereof, including a product obtained by a fermentation process using microorganisms:

i. containing one or more enzymes capable of catalyzing a specific biochemical reaction; and

ii. added to food for a technological purpose at any stage of the manufacturing, processing, preparation, treatment, packaging, transport, or storage of foods.

"Food enzyme preparation" means a formulation consisting of one or more food enzymes in which substances such as food additives and/or other food ingredients are incorporated to facilitate their storage, sale, standardization, dilution, or dissolution. For instance, invertase (E 1103) is an enzyme that catalyzes the hydrolysis of sucrose into glucose and fructose, also known as inverted sugar syrup. Fondant candies include invertase to liquefy the sugar during the storage period prior to sale of the final product.

Food enzymes permitted within the European Union appear in a list that describes the enzymes and specifies any conditions governing their use, including their function in the final food, and remain subjected to the general labeling obligations. This list remains under continuous observation and re-evaluated whenever necessary as a function of changing conditions governing their use.

References

[1] Regmi A., Gehlhar M. editors. New directions in global food markets. Electronic report from the Economic Research Service; 2005 [cited 20 October 2019]. Available from: http://www.ers.usda.gov.

[2] Greenfield H, Southgate D.A.T. Chapter 3. Selection of foods. In Food composition data. Production, management and use. Technical editors: B. A. Burlingame and U. R. Charrondiere. Rome: FAO Publishing Management Service; 2003 [cited 27 April 2023]. Available from: http://www.fao.org/3/ y4705e/y4705e.pdf.

[3] Regulation (EC) No 1333/2008 of the European Parliament and of the Council of 16 December 2008 on food additives. Official Journal L 354, 31.12.2008, p. 16–33.

[4] European Food Safety Authority (EFSA). Glossary [cited 26 April 2023]. Available from: https://www. efsa.europa.eu/en/glossary/noael

[5] World Health Organization (WHO). Principles for the safety assessment of food additives and contaminants in food. Environmental Health Criteria 70. 1987 [cited 27 April 2023]. Available from: http://www.inchem.org/documents/ehc/ehc/ehc70.htm.

[6] Food and Agricultural Organization (FAO). Chemical and technical assessments of food additives (CTAs) [27 April 2023]. Available from: https://www.fao.org/food/food-safety-quality/scientific-advice/ jecfa/technical-assessments/en/.

[7] Codex Alimentarius, International Food Standards. Class Names and the International Numbering System for Food Additives. CXG 36-1989, Amended in 2021. [cited 26 April 2023]. Available from:

https://www.fao.org/fao-who-codexalimentarius/sh-proxy/en/?lnk=1&url=https%253A%252F%252Fworkspace.fao.org%252Fsites%252Fcodex%252FStandards%252FCXG%2B36-1989%252FCXG_036e.pdf.

[8]   Regulation (EU) No 1169/2011 of the European Parliament and of the Council of 25 October 2011 on the provision of food information to consumers. Latest consolidated version: 01/ 01/2018.Official Journal L 304, 22.10.2011, p. 18–63.

[9]   Guidance Notes on the Classification of Food Extracts with Colouring Properties. Standing Committee on the Food Chain Anim. Health. p. 1–18. 2013 [cited 27 April 2023]. Available from: https://www.foedevarestyrelsen.dk/SiteCollectionDocuments/25_PDF_word_filer%20til%20download/06kontor/Guidance%20notes%20on%20the%20classification%20of%20food%20extracts%20with%20colouring%20properties.pdf.

[10] Rapid Alert System for Food and Feed (RASFF) [cited 26 April 2023]. Available from: https://food.ec.europa.eu/safety/rasff_en

[11] Giese J. Antioxidants: Tools for preventing lipid oxidation. Food Technol 1996; 50 (11):73–81.

[12] Codex Alimentarius, International Food Standards. Standard for sugars. CXG 212–1999, Amended in 2019. [cited 26 April 2023]. Available from: https://www.fao.org/fao-who-codexalimentarius/sh-proxy/es/?lnk=1&url=https%253A%252F%252Fworkspace.fao.org%252Fsites%252Fcodex%252FStandards%252FCXS%2B212-1999%252FCXS_212e.pdf.

[13] Regulation (EC) No 1334/2008 of the European Parliament and of the Council of 16 December 2008 on flavouring and certain food ingredients with flavouring properties for use in and on foods. Latest consolidated version: 21/03/2023.Official Journal L 354, 31.12.2008, p. 218–234.

11 Food safety

María Montaña Durán Barrantes

11.1 Introduction

Food or foodstuff means "any substance or product, whether processed, partially processed or unprocessed, intended to be, or reasonably expected to be ingested by humans" [1]. Food is obtained from primary production – production, rearing, or growing of primary products including harvesting, milking, and farmed animal production prior to slaughter, so as hunting and fishing and the harvesting of wild products – and may or may not have undergone any transformation (action that substantially alters the final product). Food law includes, equally, requirements for feed when it is used for food-producing animals, that is, for human consumption. The main objective of food law is to strengthen the confidence of the consumer, and other entities involved, in the decision-making process underpinning food law, its scientific basis, and the structures and independence of the institutions protecting health and other interests. In this way, all aspects of the food production chain are considered, such as the production, manufacture, transport, and distribution of feed given to food-producing animals, since adulteration or fraudulent or other bad practices will lead to damages in food safety.

Measures to control food safety should be based on risk assessment, which will be carried out in an independent, objective, and transparent manner based on available scientific data. In this end, the establishment of the European Food Safety Authority (EFSA) [1] reinforces scientific and technical support. Through its Scientific Committee and Permanent Scientific Panels, EFSA contributes to the establishment and development of international food safety standards.

According to Regulation (EC) 178/2002 [1], "hazard means a biological, chemical or physical agent in, or condition of, food or feed with the potential to cause an adverse health effect." The four stages, hazard identification, hazard characterization, exposure assessment, and risk characterization, are part of the scientific-based risk assessment.

Health hazards related to food can be classified according to their origin as follows [2]:

– Biological hazards caused by microorganisms: bacteria, viruses, yeasts, molds, algae, parasitic protozoa, microscopic parasitic helminths, and their toxins and metabolites.
– Chemical hazards caused by toxic chemicals, heavy metals, environmental contamination, and different situations, both natural and man-made, resulting from an incorrect operation of handling, preparing, preserving, and presenting foods.

María Montaña Durán Barrantes, Área Tecnología de Alimentos, Departamento de Ingeniería Química, Universidad de Sevilla, Spain

https://doi.org/10.1515/9783111111872-011

- Physical hazards due to the presence of foreign material in food at the time of consumption, such as soil, glass, metal scraps, and plastic.

Whatever the hazard, it can occur at any stage of the food chain due to the misuse of chemical products, poor hygiene practices, etc., resulting in the contamination of food and the occurrence of foodborne diseases. Therefore, it is important to know both the causes and the effects of such contamination in order to be able to predict and act accordingly during the processing and distribution of food. The effects in humans can range from mild gastrointestinal disorders to severe neurological disorders (sometimes fatal) and the symptoms manifest themselves more rapidly when poisons or toxins are ingested. These effects can be prevented by avoiding toxic foods, the presence of contaminants in food (segregation and hygiene), control of pH, water activity (a_w), and temperature (greater risk between 35 and 45 °C), and so on [3].

When classifying foodborne diseases, it must be considered whether the vectors of transmission are toxic or contaminated food (externally or internally) and whether the causative agent is biological or not. Depending on the agent, the types of diseases are: intoxication (the agent is a non-bioorganic toxic, a toxin, or poison); infection (the agents are living pathogenic microbes that invade tissues or release toxic substances into the consumer); infestation (the agents are not microbes but microscopic organisms); viral infections (the agent is a virus); and foodborne diseases caused by prions.

Biological foodborne diseases can be classified as microbial and nonmicrobial depending on the origin of the contaminant [4].

11.2 Biological foodborne diseases of microbial origin

The most important biological pollutants of food are microorganisms, that is, living beings whose individuals can only be seen under a microscope. Microbes are uni- or multicellular microorganisms without tissues, which reproduce asexually (with the exception of microscopic fungi, which may also reproduce sexually), and their associations (colonies) can be seen with the naked eye. Those of food interest are chemoorganotrophs (their sources of energy and nutrient sources are organic materials) and usually belong to one of the following types [3]:

- Bacteria: Unicellular and prokaryotic microorganisms (lack membrane-bound nucleus and subcellular organelles)
- Yeasts: Eukaryotic microscopic fungi, typically unicellular
- Molds: Eukaryotic microscopic fungi, typically multicellular (filamentous)

Other microscopic organisms such as insect eggs and larvae, as well as microscopic phases of higher organisms, can also contaminate food. These organisms have tissues and are therefore not considered microbes.

Factors that have a major influence on microbial development are:

- Temperature. All microbes grow in a temperature range between a minimum and a maximum, the optimum temperature being close to it, depending on which they are classified as psychrophiles (below 10 °C), mesophiles (20–45 °C), and thermophiles (41 and 122 °C). In general, it is accepted that there is no development below −7 °C.
- Water content. Microbes can only use free water: bacteria generally need more than yeasts and the latter more than molds. When the activity of water (a_w) is less than 0.70, the most common microbes will not develop.
- pH. Microbes grow within a range of a minimum and a maximum pH value, presenting an optimal value at which growth is maximal. In general, bacteria grow better at pH values close to neutrality and most fermentative yeasts grow well between pH 4.0 and 4.5 (almost none grow at alkaline pH). Molds usually have a broader pH range of growth than most yeasts and bacteria.
- Molecular oxygen (O_2). Microbes can display a wide range of responses to O_2: *obligate aerobes* require O_2 for growth; *obligate anaerobes* do not need or use O_2, actually, this is a toxic substance, which either kills or inhibits their growth; *facultative anaerobes* (or *facultative aerobes*) can switch between aerobic and anaerobic types of metabolism; *aerotolerant anaerobes* are bacteria with an exclusively anaerobic (fermentative) type of metabolism but they are insensitive to the presence of O_2 [5].
- Some substances can have a favorable effect on microbial development (e.g., growth factors), while others do so unfavorably (microbicides and microbiostatics).

These factors, along with the nutritional characteristics and energy sources of the habitat, determine where microorganisms can grow.

Foodborne diseases of microbial origin can be classified according to the causative agent:
- Microalgae: The disease is acquired by ingesting aquatic animals that are contaminated with toxic microalgae (their toxins do not normally affect the animals that fix them):
  - Mussels. During the "Red Tides," they incorporate dinoflagellate toxins (*Gonyaulax tamarensis/G. catenella*). The neurotoxin PSP (saxitoxin) can be lethal; DSP toxin causes gastrointestinal disorders.
  - Tropical fish. *Lyngbya majuscula* ciguatoxin causes ciguatera (gastrointestinal disease).
- Molds secrete toxins when they grow on starch residues. These toxins include ergotoxins (*Claviceps purpurea*, "ergot of rye"), aflatoxins (*Aspergillus flavus* and others), ocratoxin, and patulin (*Aspergillus* and *Penicillium*). Diseases are acquired directly or through animals.
- Protozoa: The most important foodborne disease associated with protozoa is amebic dysentery (caused by *Entamoeba histolytica*, a thermosensitive organism). It reaches the food by the fecal–oral route through water, manipulators, and contaminated utensils. Other protozoa of interest are *Giardia lamblia*, *Cryptosporidium parvum*, *Toxoplasma gondii*, *Naegleria*, and *Acanthamoeba*.

- Bacteria: They are the main responsible for foodborne diseases, the most frequent being caused by mesophiles with optimal temperatures of 35–40 °C (only *Clostridium botulinum, Listeria monocytogenes,* and *Yersinia enterocolitica* grow below +5 °C). There are two main causes of foodborne diseases associated with bacteria: intoxication (poisoning) and infection [6]. Bacteria that cause intoxications include (Table 11.1):
  - Botulism. This is caused by *Clostridium botulinum* neurotoxin. The microorganism is resistant (spores), but the toxin (protein) is thermolabile. Anaerobe. Development: pH > 4.5; 3–47.5 °C (optimal temperature: 35 °C); a_w > 0.94. It causes vomiting, dizziness, pain, dryness, paralysis, and diplopia in 12–72 h. Lethality: according to the types.
  - *Staphylococcus aureus.* This is a facultative anaerobe. Growth conditions: pH > 4.0; 10–46 °C (optimal temperature: 40–45 °C); a_w > 0.83 (up to 20% salt and up to 60% sucrose). Agent: thermostable enterotoxin. It causes vomiting, diarrhea, and hypothermia within 1/2 to 8 h, with recovery at 1–2 days. Rarely lethal.
  - *Bacillus cereus.* This is aerobic and spore-forming. Growth conditions: optimal pH: 6.0–8.5; 5–50 °C. It causes vomiting after 1–5 h.

Table 11.1: Bacterial intoxications, showing limit values for bacterial growth.

Bacteria	a_w	pH	Temperature	Type of toxin
Bacillus cereus	>0.95	>4.9	5–50 °C	Enterotoxin
Clostridium botulinum	>0.94	>4.6	3.3–47.5 °C	Thermolabile neurotoxin
Staphylococcus aureus	>0.83	>4.0	6.5–46 °C	Heat-resistant enterotoxin

Infections of bacterial origin include (Table 11.2) [6]:
- *Bacillus cereus*: 8–20 h after ingesting more than 100,000 germs/gram, diarrhea. Toxicogenic.
- Salmonellosis: 6–48 h after ingesting *Salmonella*, acute gastroenteritis and fever, which occurs before 1 week (about 2 days). Intestinal. Facultative anaerobe.
- Campylobacteriosis: acute gastroenteritis (with heavy diarrhea) caused by ingestion of *Campylobacter jejuni.* Heat-sensitive microaerophile. It does not grow in refrigeration (optimal temperature: 37 °C).
- Listeriosis: by ingestion of *Listeria monocytogenes*, pseudo-flu symptoms appear, followed by a second phase with very varied symptoms. Anaerobe. Very widespread (milk). They tolerate more salt at lower temperature (30% (w/v) at 4 °C).
- *Clostridium perfringens*: after ingesting these spore-forming bacteria (8–24 h), a mild diarrheal condition occurs. Anaerobe. Tolerates less than 5% salt.
- *Vibrio*: After 1–5 days, mild fever, vomiting, and diarrhea appear. Some species produce cholera (*V. cholerae*). Optional. Seafood and waters. Heat sensitive.

- *Escherichia coli* (some strains): traveler's diarrhea, hemorrhagic diarrhea with renal failure (strain O157: H7). Aerobic or optional. Heat sensitive.
- Yersiniosis: Gastroenteritis due to *Yersinia enterocolitica* (in children, with fever and acute abdominal pain). Intestinal, it can contaminate utensils. Aerobic or facultative. Less than 5% salt.
- Shigellosis (bacillary dysentery): Emetic-diarrheal syndrome caused (7–36 h) by ingestion of *Shigella* (intestinal, contaminates utensils) and seafood. Aerobic or facultative. Heat sensitive.
- Brucellosis (Maltese fever): through milk with *Brucella*. Aerobic. Optimum pH: 6.6–7.4. Heat sensitive.

Table 11.2: Bacterial infections, showing limit values of bacterial growth and source of contamination [6].

Bacteria	a_w	pH	Temperature	Likely source of contamination
Bacillus cereus	>0.95	>4.9	5–50 °C	Soil, dust
Brucella	>0.88	>4.5	6–42 °C	Milk from infected animals
Campylobacter jejuni	>0.98	>4.9	30–47 °C	Undercooked food from infected animals
Clostridium perfringens	>0.93	>5.5	15–50 °C	Soil, raw food
Escherichia coli	>0.95	>3.6	7.1–42 °C	Human feces, direct or via water
Listeria monocytogenes	>0.90	>4.1	−0.1–45 °C	Soil or infected animals, directly or via manure
Salmonella	>0.94	>4.0	5–46 °C	Food from infected animals, human feces
Shigella	>0.94	>5.0	10–48 °C	Human feces, direct or via water
Vibrio	>0.93	>4.8	5–44 °C	Marine coastal environment
Yersinia enterocolitica	>0.95	>4.6	−0.9–45 °C	Infected animals (especially swine), contaminated water

11.3 Biological foodborne diseases of nonmicrobial origin

Symptoms are varied, mainly emetic (vomiting) and diarrhea. Sometimes a preconsumption treatment (disinfection, freezing, heating, etc.) can be effective, but it is best not to eat suspicious foods. The most common foodborne diseases of nonmicrobial origin include [7]:

- Poisons from poisonous animals, fungi, and higher plants. Typical examples are poisonous mushrooms:
  - Nervous (muscarinic) syndrome: Mushrooms containing muscarine (typically, *Amanita muscaria*). Symptoms: At 2–3 h, headache, intense sweating, and loss of consciousness. Lethality: low.
  - Degenerative syndrome (phalloid): Mushrooms containing phalloid venom (typically, *Amanita phalloides*). Symptoms: at 10–12 h, vomiting, diarrhea, oliguria, and cardiac disorders; tremors, cramps, hepatorenal lesions, hypothermia, albuminuria, heart disease, and coma. Lethality: high.

- Gastrointestinal (lividinic) syndrome: For harmless-looking mushrooms (typically, *Entoloma lividum*). Symptoms: at 1/2 h, vomiting and diarrhea (severe abdominal pain). Lethality: very rare.
- Infestations by larvae or eggs of parasitic worms that grow inside the body [8].
    - Trichinosis: By viable larvae of *Trichinella spiralis*, cystic in uncontrolled pig or wild boar. Vomiting and diarrhea when larvae colonize the intestinal mucosa; later, muscular pain and inflammation, when the encystment occurs in muscles (consequences depending on the muscle). It can be avoided by controlling the pigs, as well as their meat and by treating it properly (especially by heating). A similar disease is anisakiasis, due to *Anisakis*, caused by *Anisakis* larvae in fish.
    - Flat worms: Development in the intestine of *Taenia saginata/T. solium* (lonely). Through infested meats. Avoidance: control. Treatment: vermifuge.
    - Cylindrical worms: Development in the intestine of *Ascaris lumbricoides* (America), *Trichuris trichiura* (Europe), or *Enterobius vermicularis* (children). By ingestion of their eggs (resistant) via the fecal–oral route. Avoidance: hygiene. Treatment: vermifuge.
- Viral infections: Viruses usually enter the body with bacteria via the fecal–oral route (direct or indirect), in unheated food [9]. Some examples:
    - Hepatitis A: After incubation (15–50 days), fever, malaise, anorexia, vomiting, and diarrhea; sometimes, not always, jaundice. Origin: water, bivalve mollusks, infected handlers (via food).
    - Norwalk virus: At 12–48 h, gastrointestinal syndrome. For seafood and raw foods.

11.4 Abiotic foodborne diseases

These include contaminants produced during processing, contaminant from food contact materials, veterinary residues, and environmental contaminants [10, 11]. The symptoms are:
- Poisoning caused by a high and single dose of a toxic substance, usually with rapid onset of symptoms. It is usually accidental and its frequency is low. It is prevented with good manufacturing practices.
- Degenerative diseases. Continued ingestion of small amounts of certain substances may have long-term effects. Origin: ignorance of the presence or effects of the agent. They could be avoided with a better understanding of the effects of constituents and materials, and the products of their reactions.
- Mutagenesis. Carcinogens may be present in a very small proportion in some foods (polycyclic aromatic hydrocarbons for cooking at high temperatures, some nitrosamines in meat, beer, etc.).

11.5 Food hygiene: food safety from production to consumption

11.5.1 Legislation and reference bodies

Food hygiene implies "the measures and conditions necessary to control hazards and to ensure fitness for human consumption of a foodstuff taking into account its intended use" (Regulation (EC) 852/2004). The set of rules covering all stages of the production, processing, distribution, and placing on the market (holding the food for the purpose of sale) are listed below:

- Regulation (EC) 852/2004 of the European Parliament and the Council of April 29, 2004, on the hygiene of foodstuffs.
- Regulation (EC) 853/2004 of the European Parliament and the Council of April 29, 2004, laying down specific hygiene rules for food of animal origin.
- Regulation (EC) 854/2004 of the European Parliament and the Council of April 29, 2004, laying down specific rules for the organization of official controls on products of animal origin intended for human consumption. No longer in force from December 13, 2019.

This initial "Hygiene Package" takes into account, in particular, the primary responsibility for food safety borne by the food business operator; it also implies a general implementation of procedures based on the Hazard Analysis and Critical Control Point (HACCP) principles, which are developed in guides to good practice for hygiene or for the application of HACCP principles, as an instrument for all business operators of the food chain [12].

In a current revision of official controls on products of animal origin intended for human consumption, Regulation (EC) 854/2004 has been replaced by the requirements in Regulation (EU) 2017/625, Commission Delegated Regulation (EU) 2019/624, Commission Delegated Regulation (EU) 2019/625, Commission Implementing Regulation (EU) 2019/626, Commission Implementing Regulation (EU) 2019/627, and Commission Implementing Regulation (EU) 2019/628.

Implementing and delegated acts (former PRAC) of the hygiene package, and related key rules, are:

- Regulation (EC) 178/2002 laying down the general principles and requirements of food law, establishing the EFSA and laying down procedures in matters of food safety.
- Regulation (EC) 2073/2005 on microbiological criteria for foodstuffs.
- Regulation (EC) 2074/2005 laying down implementing measures for certain products under Regulation (EC) 853/2004 of the European Parliament and of the Council and for the organization of official controls.
- Regulation (EU) 2016/429 on transmissible animal diseases and amending and repealing certain acts in the area of animal health (animal health law).

- Regulation (EC) 2017/625 on official controls and other official activities performed to ensure the application of food and feed law, rules on animal health and welfare, plant health and plant protection products, and repealing Regulation (EC) 854/2004, among others.
- In addition, in some Member States, national guides to good practice have been developed by the food sectors and assessed by the competent authorities. These guides may be used on voluntary basis and may serve as a good and practical tool for the implementation of hygiene requirements in food businesses. Besides, in small shops, the HACCP-based procedures may be implemented in a flexible way [13, 14].

HACCP principles must be established for each specific installation and product, following a detailed analysis of the production process, in order to assess the hazards associated with each point in the production chain (ingredients, treatments, personnel, etc.). This information must be used to identify the "critical points," that is, the treatments that cancel out or reduce to acceptable limits the negative effects of the previous steps. Tolerance limits must be set and strictly controlled for each of these, and the information, corrective actions, and system responses must be documented. To minimize the number of critical points, consider whether the specific hazard that each one avoids can be corrected by a subsequent one, leaving the minimum number necessary to ensure safe production. HACCP systems can be implemented within the framework of international quality standards (ISO/EN/UNE 9000 series), which are not mandatory and are supplemented by other standards, such as environmental or risk prevention standards, and require external accreditation.

Current challenges for Food Hygiene and Safety include:
- Complex food chain
- Massification of industrial food production (concentration in a few producers)
- Frequent use of collective catering services, including new distribution agents (Glovo, Just Eat, etc.)
- Increases in the consumption of precooked, prepared, vacuum-packed dishes, and so on
- Concentration of purchases due to changes in family and social organization
- Population mobility: work trips, leisure, migratory movements, and so on
- Need for innovation in the kitchen and in the industry. Fusion of cultures
- Environmental issues and how they affect human, animal, and plant health

In any case, according to Eurobarometer data on food safety published in 2022 [15], almost half of Europeans consider food safety important and assume that the food they buy is safe, despite the global Covid-19 pandemic [16]. However, the impact of a war in Europe is the main concern because of its impact on the rising cost of food.

11.5.2 Hygienic design for food industry

General principles of hygienic design based on Regulation 852/2004/CE (food hygiene) and Directive 2006/42/EC (machines) determine that:
–   the design of the equipment and premises will allow for proper maintenance, cleaning, and disinfection;
–   construction, composition, and state of conservation should minimize the risk of contamination of food products;
–   prevent the accumulation of dirt, contact with toxic materials, and the deposit of particles in foodstuffs and the formation of condensation or undesirable mold on surfaces, and so on.

However, the legislation does not establish specific requirements to guarantee the hygienic design.

The European Hygienic Engineering and Design Group (EHEDG) [17] is a nonprofit organization without legislative functions. It offers practical guidance on the aspects of hygienic design essential for safe manufacturing and with food guarantees. The certification of products, services, and processes following the guidelines of the EHEDG is one of the most valuable tools available to companies to achieve a differentiation from the competition. Principles of hygienic design, from EHEDG, are:
–   Easy cleaning of equipment and facilities
–   Control of microorganisms: Prevent access of microorganisms to the facilities, use of impervious equipment to microorganisms, prevent growth of microorganisms, avoid areas where they can hide (cracks, gaps, etc.)
–   Pest control: Prevent pest access to facilities, avoid areas where they can hide and multiply
–   Prevent contamination with foreign particles: Use of abrasion-resistant materials, eliminate fasteners that can loosen and fall, implement preventive maintenance program, use of covers and/or covers of hygienic design, clean and inspect all new equipment before using them
–   Prevent chemical contamination: Design equipment to prevent contamination by cleaning chemicals, lubricants, thermal fluids, and so on

It is recommended to integrate the hygienic design of the facilities and equipment, within the quality management and food safety systems, such as HACCP, Globalgap, BRC (British Retail Consortium) protocol, IFS (International Food Standards) protocol, or UNE-EN ISO 22000.

11.5.3 Food alert network

Food safety must be applied in the following three phases of action:
- Risk assessment through scientific advice and data analysis
- Risk management through the application of safety standards, developing control mechanisms that ensure compliance
- Risk communication, nationally and internationally, when necessary

About risk communication of food alerts, the Regulation (CE) 178/2002 establishes the EFSA and laying down procedures in matters of food safety, about the precautionary or precautionary principle, extends the responsibility to all food chain operators, and defines rapid alert procedures as control mechanisms that are put in place as soon as a risk appears.

The International Food Safety Authorities Network (INFOSAN) is a global network of 186 national food safety authorities, managed jointly by the Food and Agricultural Organization (FAO) and the World Health Organization (WHO), to ensure rapid sharing of information during food safety emergencies to stop the spread of contaminated food.

At European level, the Rapid Alert System for Food and Feed (RASFF) [18] was implemented to provide food and feed control authorities with an effective tool to exchange information about measures taken responding to serious risks detected in relation to food or feed (see Table 11.3 as an example) [19]. This policy gives the European Union the opportunity to exchange information and coordinate administrative

Table 11.3: Notifications by hazard category and risk decision (RASFF annual report, 2018) [19].

Hazard category	Undecided	Serious	Not serious
Adulteration/fraud	10	6	101
Allergens	3	147	8
Biological contaminants (other)		42	1
Chemical contaminants (other)	1		
Composition	62	108	14
Environmental pollutants	2	48	2
Feed additives	1		3
Food additives and flavorings	9	30	86
Foreign bodies	14	78	42
Genetically modified food or feed	14		1
Industrial contaminants		2	
Labeling absent/incomplete/incorrect	7	10	12
Metals	5	144	15
Microbial contaminants (other)	6	59	62
Migration	35	48	34
Mycotoxins	5	542	2

Hazard category	Undecided	Serious	Not serious
Natural toxins (other)	1	24	1
Not determined/other	2	5	
Novel food	47	16	15
Organoleptic aspects	4	3	34
Packaging defective/incorrect	4	6	15
Parasitic infestation	2		21
Pathogenic microorganisms	40	488	111
Pesticide residues	66	180	5
Poor or insufficient controls	10	6	81
Process contaminants		1	2
Radiation		1	4
Residues of veterinary medical products	6	24	16
Transmissible spongiform encephalopathies			8

and other measures taken as appropriate for the purpose of early identification of the risks in a rapid coordinated manner within the European institutions [20].

Since March 2021, alongside the RASFF system, the AAC (Administrative Assistance and Cooperation Network) and the FFN (EU Agri-Food Fraud Network) [21] work together to make up the ACN (Alert and Cooperation Network) to facilitate the exchange of administrative information and cooperation between Member States on official controls in the agri-food chain [22]. Each network is responsible for certain aspects of the agri-food chain: noncompliance with a potential health risk (RASFF), noncompliance with a health risk (AAC), and suspected fraud (FFN) (Figure 11.1) [23]. Figure 11.2 shows the main activities of the ACN in 2022 [24].

11.5.4 Emerging risks

EFSA defines an emerging risk as: "A risk resulting from a newly identified hazard to which a significant exposure may occur, or from an unexpected new or increased significant exposure and/or susceptibility to a known hazard" [25]. As a result, emerging risk identification is a dynamic and process activity that involves a broad range of expertise and close cooperation between Member States, stakeholders, and EU and international bodies.

Through recent networking and international cooperation, 13 potential emerging issues were discussed and published in 2023 [26]. These issues were classified according to the hazard or driver identified (hazards: microbiological, chemical, microbiological and chemical, other; drivers: illegal activity, new consumer trends, new process or technology, climate change, other) (Figure 11.3). It is noted that the emerg-

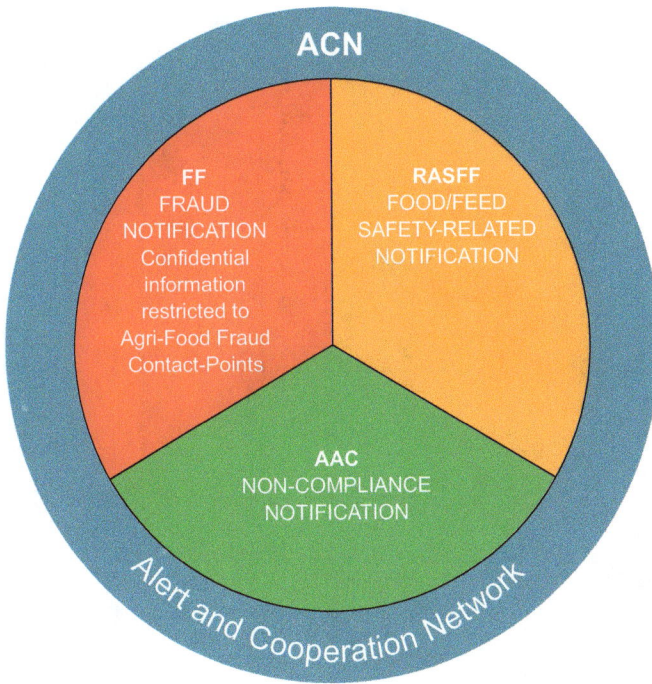

Figure 11.1: Components of the Alert and Cooperation Network [22].

ing risk landscape is constantly changing and evolving, requiring EFSA to improve the tools (operational platform) and methodologies to anticipate potential risks in the future, such as emerging chemicals, new food and feed sources and production technologies, food fraud, the circular economy, etc.

For example, among the current emerging risks food allergen control involves, on the one hand, ensuring that labeling guarantees that food allergens are accurately identified, and on the other hand, the existence of good hygiene practices to prevent transfer of an allergen from one food to another that does not contain it. Food allergies are an increasing food safety issue globally, resulting in many food recalls, as well as a number of deaths, every year. So, it is crucial to have internationally developed guidance on best practices to ensure an understanding of how to manage food allergens [27].

In conclusion, the identification of emerging risks also helps to improve EFSA's ability to estimate the value of the outputs of the process in terms of risk averted, for example, by tapping into new data sources, developing new analytical tools and methods, and widening networks of scientific knowledge. Recommendations were made in three areas [25]:

1. Further development of a food system-based approach including the integration of social sciences to improve understanding of the interactions and dynamics between actors and drivers and the development of horizon scanning protocols.

Alert and Cooperation Network in 2022

(1) RASFF - food/feed safety-related notifications
(2) AA - non compliance notifications without public health risk involved
(3) Agri-FF - non compliance notifications with suspicion of fraud
date of extraction: 19/01/2023

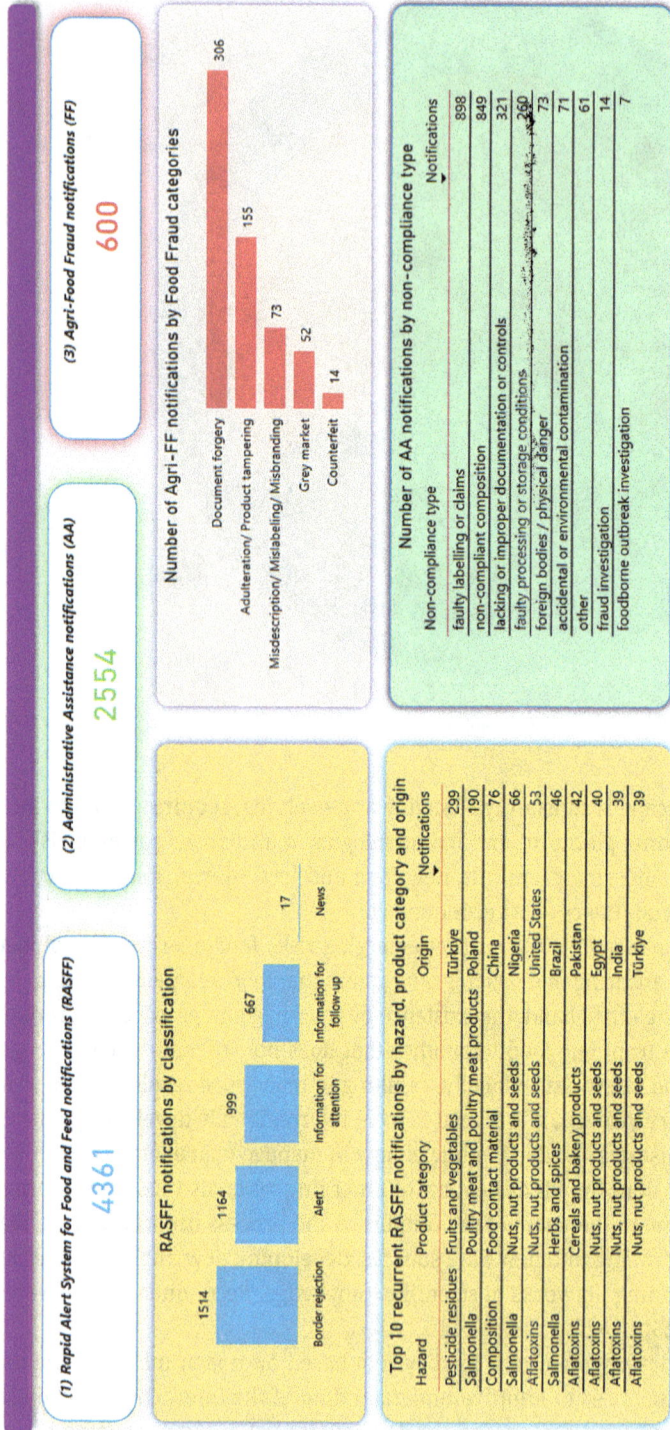

(1) Rapid Alert System for Food and Feed notifications (RASFF)

4361

RASFF notifications by classification

Border rejection	1514
Alert	1164
Information for attention	999
Information for follow-up	667
News	17

Top 10 recurrent RASFF notifications by hazard, product category and origin

Hazard	Product category	Origin	Notifications
Pesticide residues	Fruits and vegetables	Türkiye	299
Salmonella	Poultry meat and poultry meat products	Poland	190
Composition	Food contact material	China	76
Salmonella	Nuts, nut products and seeds	Nigeria	66
Aflatoxins	Nuts, nut products and seeds	United States	53
Salmonella	Herbs and spices	Brazil	46
Aflatoxins	Cereals and bakery products	Pakistan	42
Aflatoxins	Nuts, nut products and seeds	Egypt	40
Aflatoxins	Nuts, nut products and seeds	India	39
Aflatoxins	Nuts, nut products and seeds	Türkiye	39

(2) Administrative Assistance notifications (AA)

2554

(3) Agri-Food Fraud notifications (FF)

600

Number of Agri-FF notifications by Food Fraud categories

Document forgery	306
Adulteration/ Product tampering	155
Misdescription/ Mislabeling/ Misbranding	73
Grey market	52
Counterfeit	14

Number of AA notifications by non-compliance type

Non-compliance type	Notifications
faulty labelling or claims	898
non-compliant composition	849
lacking or improper documentation or controls	321
faulty processing or storage conditions	290
foreign bodies / physical danger	73
accidental or environmental contamination	71
other	61
fraud investigation	14
foodborne outbreak investigation	7

Figure 11.2: ACN (Alert and Cooperation Network) from ACN Annual Report 2022 [24].

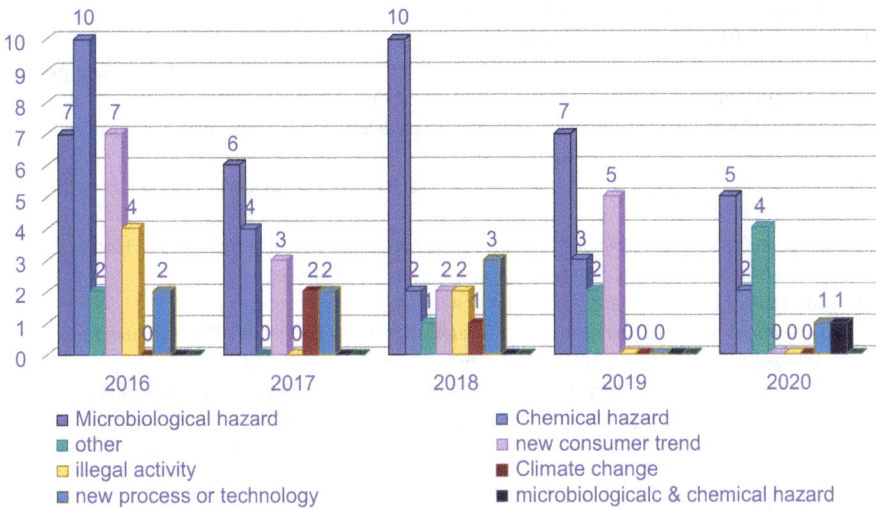

Figure 11.3: Potential emerging issues and drivers 2016–2020 (adapted from [26]).

2. Improving data processing pipelines to prepare for big data analysis, implementing a data validation system, and developing data sharing agreements to explore mutual benefits.
3. Revise EFSA's emerging risk identification process to increase transparency and improve communication.

References

[1]   Regulation (EC) No 178/2002 of the European Parliament and of the Council of 28 January 2002 laying down the general principles and requirements of food law, establishing the European Food Safety Authority and laying down procedures in matters of food safety. Official Journal L 031, 01/02/2002, 0001–0024

[2]   International Commission on Microbiological Specifications for Foods (ICMSF). El sistema de análisis de riesgos y puntos críticos. Su aplicación a las industrias de alimentos. Zaragoza: Acribia; 1991. Translated from Microorganisms in Foods 4: Application of the Hazard Analysis Critical Control Point (HACCP) System to Ensure Microbiological Safety and Quality. Oxford: Blackwell Scientific Publications; 1988. ISBN: 0632021810.

[3]   Frazier WC, Westhoff DC. Microbiología de los alimentos. 4a ed. 1a reimp. Zaragoza: Acribia; 2000.

[4]   Scientific Status Summary by the Institute of Food Technologists' Expert Panel on Food Safety and Nutrition. Foodborne illness: Role of home food handling practices. Food Technol 1995;49 (4):119–131.

[5]   Todar K. Nutrition and growth of bacteria. Todar's Online Textbook of Bacteriology. 2008–2012. [cited 14 December 2019]. Available from: http://textbookofbacteriology.net/nutgro_4.html

[6]   Scientific Status Summary by the Institute of Food Technologists' Expert Panel on Food Safety and Nutrition. Bacteria associated with foodborne diseases. Food Technol 1988;42(4):181–200.

[7]   Hodgson E. Toxins and venoms. Prog Mol Biol Transl Sci 2012;112:373–415.

[8]   Jackson GJ. Parasitic protozoa and worms relevant to the U.S. Food Technol 1990;44(5):106–112.

[9]   Scientific Status Summary by the Institute of Food Technologists' Expert Panel on Food Safety and Nutrition. Virus transmission via foods. Food Technol 1988;42(10):241–248.

[10]   Lawley R, Curtis L, Davis I. The food safety hazard guidebook. 2nd edition. Royal Society of Chemistry; 2012.

[11]   Todd EC Foodborne diseases: Overview of biological hazards and foodborne diseases. Encycl Food Safety 2014:221.

[12]   European Commission. Biological Safety [cited 29 April 2023]. Available from: https://food.ec.eu ropa.eu/safety/biological-safety/food-hygiene/guidance-platform_en

[13]   Food and Agricultural Organization of the United Nations. Food hygiene at 50: A Codex Alimentarius journey from small beginnings to stories of success. [cited 10 December 2019]. Available from: http://www.fao.org/3/CA2323EN/ca2323en.pdf

[14]   European Commission. Information on the implementation of food safety management systems covering Good Hygiene Practices and procedures based on the HACCP principles, including the facilitation/flexibility of the implementation in certain food businesses. Official Journal of the EU C355/1, 16/09/2022.

[15]   EFSA, Report Special Eurobarometer Wave EB97.2, Food safety un the EU, March-April 2022 [cited 01 May 2023]. Available from: https://www.efsa.europa.eu/sites/default/files/2022-09/EB97.2-food-safety-in-the-EU_report.pdf

[16]   Van Kerhove M. WHO conference on COVI-19 and other global health issues, 29 March 2023 [cited 01 May 2023]. Available from: https://www.who.int/multi-media/details/who-press-conference-on-covid-19-and-other-global-health-issues-29-march–2023

[17]   European Hygienic Engineering Design Group (EHEDF). Guidelines Hygienic Design Principles for Food Factories [cited 01 May 2023]. Available from: https://www.ehedg.org/

[18]   Rapid Alert System for Food and Feed (RASFF). [cited 02 July 2023]. Available from: https://food.ec. europa.eu/safety/rasff_en

[19]   Afonso A, García Matas R, Maggiore A, Merten C, Robinson YA. EFSA's activities on emerging risks in 2018. First published: 20 August 2019. Wiley Online Library https://doi.org/10.2903/sp.efsa.2019.EN-1704 [cited 10 December 2019]. Available from: https://www.foodsafetynews.com/2019/11/experts-discuss-food-hygiene-allergens-and-stec-at-codex-meeting/

[20]   RASFF Window. European Commission. [cited 02 July 2023]. Available from: https://webgate.ec.eu ropa.eu/rasff-window/screen/search

[21]   Winkler B, Maquet A, Reeves-Way E, Siegener E, Cassidy T, Valinhas De Oliveira T, Verluyten J, Jelic M, Muznik A. Fighting fraudulent and deceptive practices in the agri-food chain. Technical Report: Implementation of Article 9(2) of Regulation (EU) 2017/625. Publications Office of the European Union, Luxembourg, 2023, ISBN 978-92-68-00336-7, 10.2760/31366, JRC131525. [cited 07 May 2023]. Available from: https://publications.jrc.ec.europa.eu/repository/handle/JRC131525.

[22]   European Commission. Food Safety. The Alert and Cooperation Network. [cited 01 July 2023] Available from: https://food.ec.europa.eu/safety/acn_en

[23]   European Commission, Alert and Cooperation Network, 2021 Annual Report, Publications Office of the European Union, Luxembourg, 2022, ISBN 978-92-76-52918-7, 10.2875/328358 [cited 07 May 2023]. Available from: https://acn_annual-report_2021-final.pdf(europa.eu)

[24]   Alert and Cooperation Network (ACN) overview. European Commission. [cited 01 July 2023]. Available from: https://food.ec.europa.eu/safety/acn/reports-and-publications_en

[25]   European Food Safety authority. Emerging risks. [cited 10 June 2023]. Available from: https://www.efsa.europa.eu/en/topics/topic/emerging-risks

[26] EFSA (European Food Safety Authority), Bottex B, Gkrintzali G García MR, Georgiev M, Maggiore A, Merten C, Rortais A, Afonso A, Robinson T. Technical report on EFSA's activities on emerging risks in 2020. EFSA supporting publication 2023:EN-8024. 51 2023, 2397–8325. Available from: https://efsa.onlinelibrary.wiley.com/doi/pdf/10.2903/sp.efsa.2023.EN–8024

[27] Food Safety News (FSN). Allergens on the agenda. https://www.foodsafetynews.com/2019/11/experts-discuss-food-hygiene-allergens-and-stec-at-codex-meeting/

12 The biochemistry of digestion

Victoria Valls-Bellés

12.1 Introduction: functional organization of the digestive system and associated organs

The basic function of the digestive system is the transformation of food into nutrients, so that the body's cells have the necessary molecules for metabolic maintenance and regeneration (Figure 12.1).

The gastrointestinal tract (GIT) is a hollow cylinder divided into large functional segments, which vary along its path (10–11 m). The GIT, which begins in the mouth and ends in the anus, also has several associated glands (salivary glands, liver, and pancreas). The main parts of the GIT include:

- Mouth, or oral cavity, is the route of entry or ingestion of food, which is subjected to a mechanical process, mastication, that reduces the food to smaller pieces by the action of chewing (teeth). Food particles are then chemically processed by the enzymes contained in the saliva. In this process, starch is already transformed into maltose.
- Pharynx and esophagus are two consecutive segments that communicate the mouth with the stomach. Pharynx is the place where swallowing occurs; from here on, all digestive processes are involuntary. In the esophagus, the bolus of food passes through the upper sphincter and is propelled by a mechanical process known as peristalsis; then, it crosses the lower sphincter or cardia to reach the stomach.
- Stomach is where food is stored and subjected to the process of digestion. The partially digested food bolus is transformed into the chyme (semi-liquid) after the chemical actions of gastric juices and hydrochloric acid, which in turn destroys most of the microorganisms that may be present in the food bolus. At the same time, a series of peristaltic waves act on the food in both directions in order to reduce the particle size. Thereafter, the chyme is being gradually eliminated, in a process that depend on the waves, which allow the closure and opening of the pyloric sphincter. Some nutrients such as alcohol, water, and a small part of mineral salts are absorbed in the stomach.
- Small intestine, formed by three segments (duodenum, jejunum, and ileum), is where the final stages of digestion occur by means of the chemical action of secretions, such as biliary and pancreatic, and also by mechanical processes, since a large peristalsis is produced in this segment of the GIT. After this treatment, digested products are prepared to be absorbed, mainly in the small intestine, which

Victoria Valls-Bellés, Unitat Predepartamental de Medicina (Area: Fisiologia), Universitat Jaume I, Castelló de la Plana, Spain

https://doi.org/10.1515/9783111111872-012

Mechanical fragmentation (chewing)

Mouth

Chemical digestion

Saliva → Starch hydrolysis

Bolus formation

Epiglottis

Esophagus → Peristalsis

Cardia

Mixing, crushing, and propulsion movements

Stomach

Chemical digestion → Gastric juice → [HCl, Pepsin, Amylase, Lipase]

Bolus

Pylorus

Chyme

Small intestine

Duodenum ← Bile (fat emulsification)
Duodenum ← Pancreatic juice (hydrolytic enzymes)

Jejunum → Absorption of nutrients

Ileum → Absorption of vitamin B_{12} and bile salts

Ileocaecal valve

Absorption of water, salts, and vitamins

Colon

Microbiota →
- Barrier for germs and pathogens
- Activation of immune system
- Synthesis of vitamin K and some vitamins of group B
- Energy from fibre

Rectum → Defecation reflex

Anal sphincters

Faeces

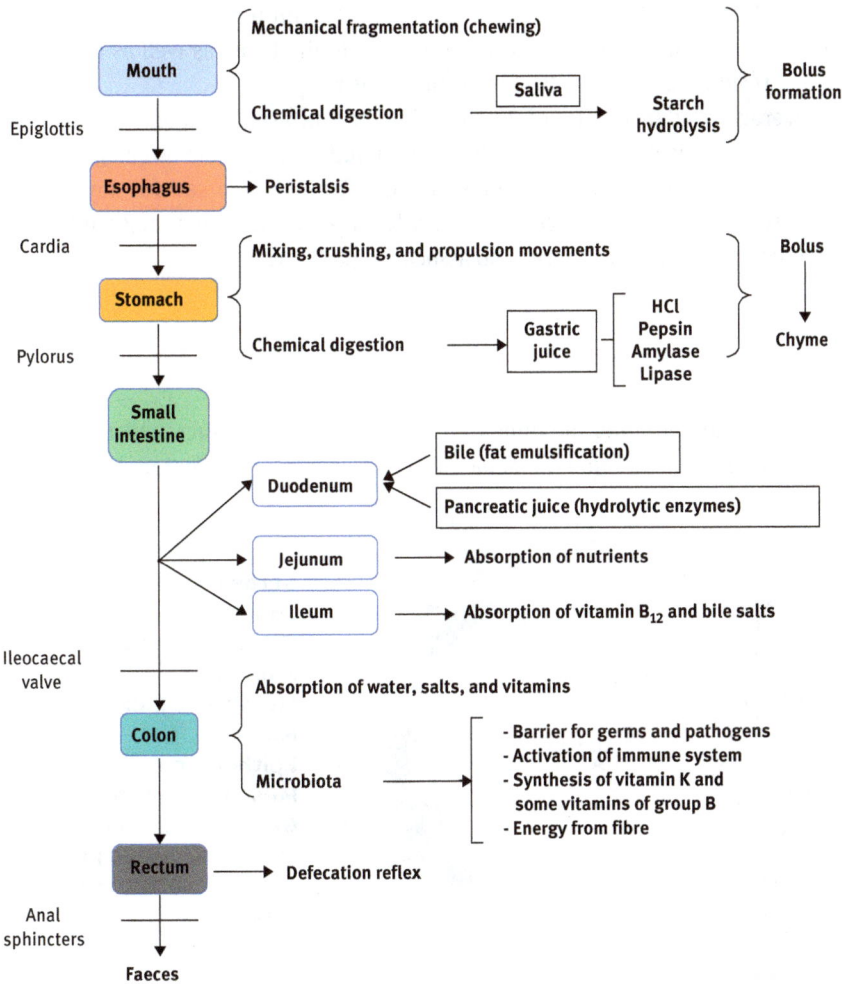

Figure 12.1: Simplified general scheme of the digestion process.

has a large internal surface due to the presence of villi and microvilli. The rest of the chyme passes into the large intestine through the ileocaecal valve.

- Large intestine is the final zone of adjustment of water and ion absorption processes. In addition, it is a storage organ of non-absorbed products that will be eliminated in the form of feces via the anus, through the anal sphincters (upper and lower).

The main function of sphincters is to prevent reflux, that is, they make sure that the remains of the bolus do not flow back up. In addition, along the tract there are glandular structures that are invaginations of the tube wall, which empty their secretions into the intestinal lumen. This is the case of Brunner glands, which secrete large amounts of HCO_3^- into the duodenum. In general, the main function of the GIT is the administration of water, electrolytes, and nutrients to the body, as well as the excretion of food residues and products of liver metabolism.

The average residence times of the food along the different segments of the digestive tract, in theoretical terms, are 1 min in the mouth, 2–3 s in the esophagus, 2–4 h in the stomach, 1–4 h in the small intestine, and from 10 h to several days in the colon. In any case, these residence times are a function of the type of macronutrients that form the food, as long as there is no associated pathology.

12.1.1 Gastrointestinal wall

This is the wall surrounding the lumen of the gastrointestinal tract, which acts as a mechanical, biological, and functional barrier between the lumen content of the intes-

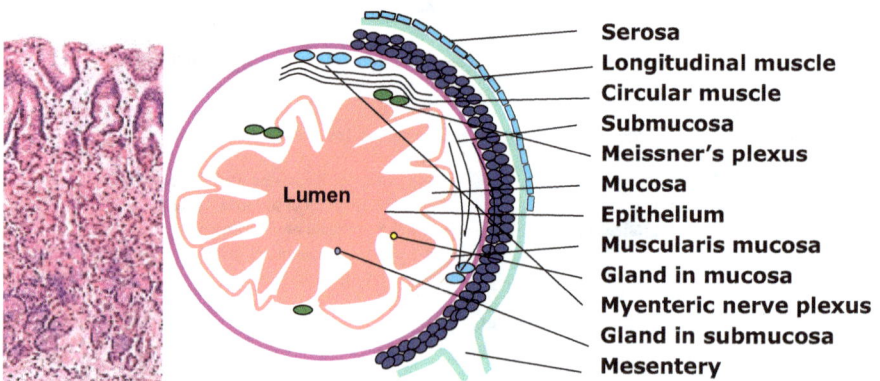

Figure 12.2: Gastrointestinal wall. Longitudinal and transverse section (Valls-Belles, V.).

tine and our internal environment. It consists of several types of tissues superimposed on concentric layers (Figure 12.2):
–   The mucosa is the inner layer of the GIT, and it surrounds the lumen and is in direct contact with digested food (chyme). The mucosa is formed by microvilli, villi, and invaginations or sacs formed in the epithelium, known as crypts. It consists of three parts:
    –   The epithelium, the innermost layer, is where most of the processes of digestion, absorption, and excretion occur.
    –   Lamina propria is a layer of connective tissue within the mucosa.
    –   Muscularis mucosae is a thin layer of smooth muscle.

Each part of the GIT and associated organs have a highly specialized mucosa.

–   The submucosa is a layer of dense disordered connective tissue located underneath the mucosa that contains blood vessels, lymphatic vessels, nerve fibers (plexuses), and prolongations of mucosal glands.

–   The muscular layer or muscularis propria is in turn composed of two layers and presents a variable thickness depending on the part of the GIT. The muscle of the inner layer is arranged in circular rings around the tract, while the muscle of the outer layer is arranged longitudinally. The stomach has an extra inner oblique layer. The coordinated contractions of these layers, known as peristalsis, drive food through the gastrointestinal tract. Peristalsis is controlled by the myenteric plexus or Auerbach's plexus, which is located between the longitudinal and circular muscular layers. Peristaltic activity is initiated by the myenteric interstitial cells of Cajal (ICCs), which are responsible for creating the slow wave bioelectric potential leading to smooth muscle contraction. ICCs are mediators of enteric neurotransmission.

–   The serosa layer is responsible for covering the intraperitoneal regions of the GIT (those parts suspended by the peritoneum). This structure consists of connective tissue covered by a simple squamous epithelium, called the mesothelium, which reduces frictional forces during digestive movements. The intraperitoneal regions include most of the stomach, first part of the duodenum, all of the small intestine, caecum and appendix, transverse colon, sigmoid colon, and rectum. These parts of the tract have mesentery. Retroperitoneal regions include the oral cavity, esophagus, pylorus, distal duodenum, ascending colon, descending colon, and anal canal.

Along the GIT there are the organs associated with the alimentary canal, such as the liver with the gallbladder and the pancreas. The liver secretes bile into the duodenum, whereas the pancreas, a gland of mixed origin, secretes pancreatic juice in the duodenum and hormones that control sugar levels into the blood.

12.1.2 Gastrointestinal blood flow: splanchnic circulation

The digestive system and associated organs have their own blood circulation, splanchnic circulation, which passes through the liver, thus allowing the reticuloendothelial cells that line sinusoids to eliminate bacteria and other particles that can penetrate the general circulation. This blood flow allows the transport of absorbed nutrients to the cells and hence to the entire organism. Water-soluble nutrients (proteins, carbohydrates) are transported to the liver through the portal vein; by contrast, lipid molecules pass to the general circulation via the lymphatic vessels. Hepatic cells (hepatocytes) temporarily take up and store between 1/2 and 2/3 of the absorbed nutrients.

The splanchnic circulation receives 25% of the cardiac output. Blood flow is proportional to gastric activity, thus, in periods between meals activity is minimal, whereas during postprandial periods the activity is maximal.

12.1.3 Neural regulation of gastrointestinal function

The gastrointestinal tract also has its own nervous system, the enteric nervous system ("the brain of the GIT"), which is localized throughout the entire tract, from the esophagus to the anus. This intrinsic innervation is composed of two intramural plexuses, the submucosal (Meissner's) plexus and the myenteric (Auerbach's) plexus, which control the movements and secretions of the GIT (Figure 12.3). The enteric nervous system regulates local function through short reflexes.

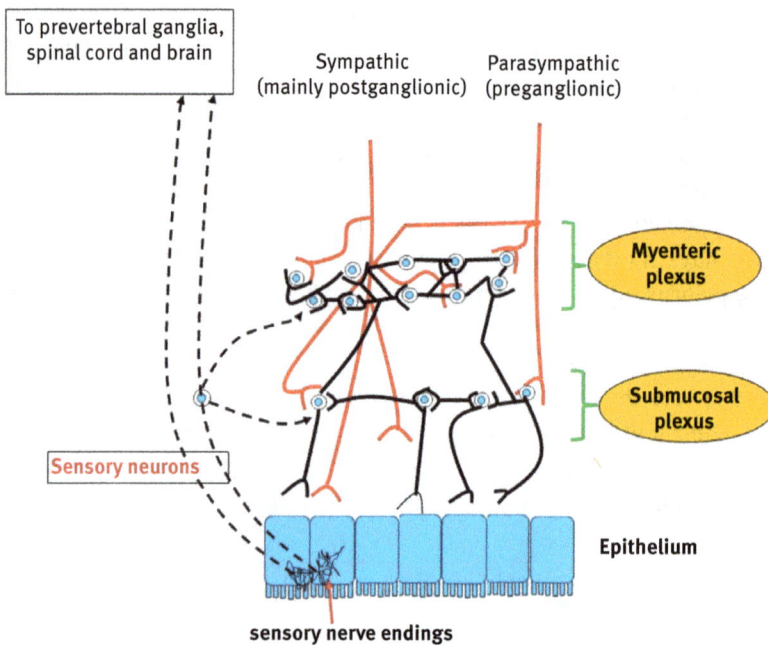

Figure 12.3: Submucosal plexus and myenteric plexus (Valls-Bellés, V.).

The submucosal plexus or Meissner's plexus increases tonic contraction of the intestinal wall, the intensity of rhythmic contractions, and the contraction rate of excitation waves along the intestine. In addition, it regulates the activity of the mucosa, playing an important role in the control of the secretions of the digestive tract and local blood flow. Some neurons of the Meissner's plexus are inhibitory (relaxation in the sphinc-

Table 12.1: Some neurotransmitters secreted by enteric neurons.

Acetylcholine	Substance P
Noradrenaline (Norepinephrine)	Vasoactive intestinal peptide
Cholecystokinin	Somatostatin
Serotonin	Leuenkephalin
Dopamine	Metenkephalin
Tryptophan	Bombesin

All of them are either inhibitory or excitatory, thus, acetylcholine stimulates gastrointestinal activity whereas norepinephrine inhibits it.

ters). Auerbach's plexus regulates much of the motility of the digestive tract, that is, muscle contractions. Enteric neurons secrete different neurotransmitters (Table 12.1). The extrinsic innervation is formed by the autonomic system (sympathetic and parasympathetic) that coordinates the function through long reflexes. Sensory nerve endings send afferent fibers to both plexuses and the central nervous system, while receiving efferent information from the autonomic nervous system. On the other hand, sensory nerve endings can trigger local reflexes inside the intestine.

Some sympathetic fibers innervate the smooth muscle (blood vessels) of the digestive tract and cause vasoconstriction while others innervate secretory cells. On the other hand, parasympathetic innervation is provided by the vagus nerve and goes from the esophagus to the transverse colon. The rest of the colon, rectum, and anus receive innervation through the pelvic nerves.

The activation of the sympathetic nerves inhibits the secretory motor activities of the digestive system and causes the contraction of the smooth musculature in the sphincters. The activation of parasympathetic nerves stimulates gastrointestinal motor and secretory activity.

12.1.4 Chemical regulation of gastrointestinal function

The gastrointestinal tract is regulated by a series of hormones that are synthesized and secreted in the digestive system and act in an autocrine, paracrine, and/or endocrine manner, regulating both its motor activity and its secretory activity (Chapter 14). Autocrine means that cells respond to molecules secreted by themselves; paracrine manner occurs when a sensor cell releases a chemical messenger or regulatory peptide that acts on nearby target cells, by diffusion through the interstitial space; endocrine describes the process in which a sensing cell responds to the stimulus by secreting a peptide or regulatory hormone that travels through the bloodstream to target cells away from the secretion point.

12.1.5 Membrane potential in the GIT

As a result of the selective permeability of biomembranes and the action of multiple membrane transporters, there is an uneven distribution of charges across the cell membranes in such a way that there are more negative charges inside the cell than outside. This results in an electric potential difference known as membrane potential (Chapter 7, Figure 7.2). There are two basic patterns of electrical activity across the membranes of smooth muscle cells, slow waves, and spike potentials (Figure 12.4):

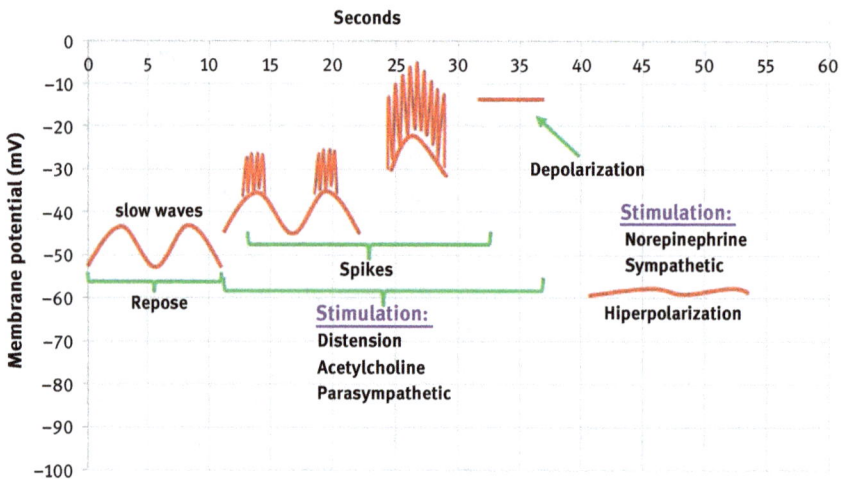

Figure 12.4: Membrane potential across the membranes of smooth muscle cells of the gastrointestinal tract.

- Slow waves are fluctuations in the resting membrane potential (see below), whose intensity vary between 5 and 15 millivolts (mV), and show a frequency of 3 waves per minute in the stomach, 12 per minute in the duodenum and 8–9 per minute in the terminal ileon. Slow waves seem to be caused by an interaction between smooth muscle cells and ICCs. They do not elicit contractions, except in the stomach.
- Spike waves are generated automatically when the membrane potential in the resting smooth muscle exceeds −40 mV. They are 10–40 longer than neuron action potentials and are produced by calcium-sodium channels. Spike waves are real action potentials.

The resting membrane potential is around −56 mV, when it becomes more positive (i.e., less negative) membrane depolarization occurs and muscle fibers become more excited. When the potential becomes more negative, hyperpolarization occurs and muscle fibers become less excitable. Parasympathetic nerves increase excitability by secreting acetylcholine, whereas sympathetic nerves decrease the excitability through the secretion of norepinephrine in its terminations.

During spike waves great quantities of calcium enter smooth muscle cells and most contractions are generated (Figure 12.5). Slow waves do not promote the entry of calcium ions (only sodium ions) into smooth muscle fibers and cannot produce muscle contraction.

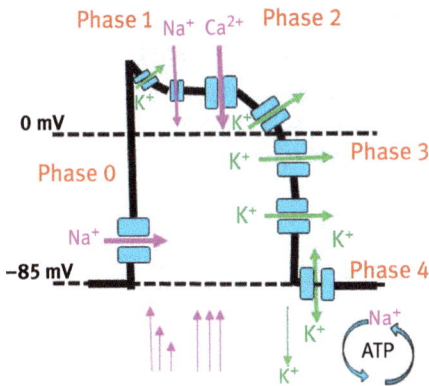

Figure 12.5: Action potentials at the membranes of smooth muscle cells (adapted by Valls-Belles).

Part of the muscle of the GIT produces tonic contractions, a state of permanent continuous semi-contraction of the muscle that is not associated with slow waves. They can last for several minutes or even several hours and may be caused by continuous repetitive spike potentials, hormones, or other factors that induce membrane depolarization (continuous entry of calcium ions into cells).

12.2 The mouth: phases of swallowing

12.2.1 Cephalic phase

Before the food is ingested there are already changes in the digestive system to prepare for the different processes, such as absorption, digestion, etc. Thus, the sight of food and the anticipation of feeding stimulates the brain cortex, whereas food aromas stimulate the hypothalamus and the spinal cord. These stimuli produce an increase in salivary secretion, a contraction in the gallbladder, and relaxation of the sphincter of Oddi, which implies a pancreatic secretion. Around 30% of gastric secretions are produced in the cephalic phase.

12.2.2 Oral phase

In this phase the food is already in contact with the gastrointestinal tract. The responses that occur in the oral cavity are the same as those initiated in the cephalic phase, plus the stimuli of taste, which involve a stimulation of the hypothalamus and spinal cord (Chapter 9). In this phase the food is crushed through the process of chewing, lubricated, and mixed to form the bolus. Absorption in the oral phase is minimal, although alcohol and some drugs can be absorbed. The lubrication process is performed by the secretions of the salivary glands that are formed by acinar and ductal cells. The former are secretory cells grouped in a spherical structure that release their contents toward the center of the acini (Figure 12.6). Ductal cells form the duct that drains the salivary secretion from the glands into the mouth.

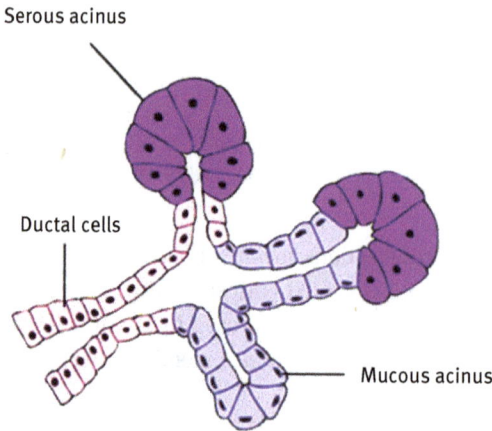

Figure 12.6: Structural organization of salivary glands (image obtained from: Sonia Gupta and Nitin Ahuja (January 10, 2019). Salivary Glands, Histology, Thomas Heinbockel and Vonnie D.C. Shields, IntechOpen, DOI: 10.5772/intechopen.81213. Available from: https://www.intechopen.com/books/histology/salivary-glands).

There are three main types of acinar cell: serous, mucinous, and seromucous (mixed)
– Serous cells produce a watery secretion
– Mucinous cells produce a much thicker secretion
– Seromucous cells produce a mixed secretion

The proportion of serous, mucous, and mixed cells varies in different salivary glands, the latter being found in the oral cavity and classified in minor glands, located in the oral mucosa, and named based on their location (palatine, lingual, buccal, and labial), and major glands (parotid, submandibular, and sublingual). Each of them participates in the production of saliva in different percentages, thus, submandibular gland produces ~70%, parotid gland ~25%, and sublingual gland ~5% under baseline conditions.

Saliva is secreted in the mouth in large quantities (1–1.5 l/day; 1 ml/min/g of gland). The composition of saliva (shown in Table 12.2) is determined by the biphasic model of salivary secretion:

- In phase 1 the cells of the acini and intercalar ducts produce a secretion with Na^+, K^+, and Cl^- concentrations close to those of plasma concentrations (135–145 mM, 3.5–5.5 mM, and 115 mM, respectively).
- In phase 2 the concentration of solutes is modified while the primary secretion runs through the ducts: Na^+ and Cl^- are extracted and K^+ and HCO_3^- are added to the saliva, thus becoming hypotonic with respect to the plasma.

Salivary secretion is regulated by the autonomic nervous system through the salivary reflexes, which can be conditioned during the cephalic phase (aromas, food sight, noise of plates and cutlery, etc.), or unconditioned reflexes during the oral phase. Both types of reflexes stimulate salivary secretion.

12.2.3 Pharyngeal phase

Once the food bolus has formed, the tongue pushes it back and enters the pharynx, it is the so-called swallowing process, the process by which the bolus continues to the esophagus, a tube of approximately 25 cm that connect the pharynx with the stomach. The voluntary phase occurs once the food is prepared for swallowing, the tongue compresses the bolus against the palate and pushes it voluntarily toward the pharynx. After this phase the rest of the processes are involuntary:

- The palate is pulled upward and palatopharyngeal folds move inward. This prevents food from going to the nasopharynx.
- Vocal cords are pushed and the pharynx moves forward and upward against the epiglottis. This prevents food from entering the trachea and helps open the upper esophageal sphincter (UES).
- The superior pharyngeal constrictor muscles contract, thereby propelling the bolus into the pharynx.
- A peristaltic wave initiated by the contraction of the muscles pushes the bolus into the esophagus and relaxes the UES.

12.2.4 Esophageal phase

The main function is to quickly drive food from the esophagus to the stomach. The bolus passes through the UES to the esophagus in less than 1 s. The peristaltic wave that begins in the pharynx continues and spreads to the esophagus (primary peristalsis), then, a secondary peristalsis occurs as a result of esophageal distension, the lower esophageal sphincter (LES) relaxes and the bolus enters the stomach.

Table 12.2: Chemical composition of some secretions of the gastrointestinal tract.

	Saliva	Gastric juice	Pancreatic juice	Pancreatic cells secretions
pH	6.0–7.0	1.0–3.5	8.1–8.5	7.5–8.0
Inorganic components	Water Na^+ HCO_3^- Cl^- K^+	Water H^+ (*) Cl^- Na^+ HCO_3^- K^+ (*) Characteristic of gastric juice	Water Na^+ HCO_3^- Cl^- K^+ Ca^{2+} Zn^{2+} $H_2PO_4^-/HPO_4^{2-}$ SO_4^{2-}	Water Na^+ HCO_3^- Cl^- K^+ Ca^{2+}
Organic components	– Hydrolytic proteins: Salivary amylase Lingual lipase Kallikreins (serine proteases) – Defense mechanisms Lysozyme Immunoglobulins ABO antigens – Mucins – Plasma proteins (incorporated by filtration) – Salivary lipids	– Pepsinogen -> Pepsin – Trefoil factors (protective function) – Histamine (enterochromaffin cells, regulates acid secretion) – Gastrin (G cells, regulates acid secretion) – Somatostatin (D cells, Regulates acid secretion) – Rennin or chymosin (only found in babies) – Intrinsic factor (absorption of vitamin B_{12})	– Mucins – Amylase – Lipase & colipase (*) – Phospholipase A (*) – Cholesterol esterase – Endopeptidases (trypsin and chymotrypsin) – Exopeptidases – Aminopeptidases – Carboxipeptidases – Ribonucleases – Deoxyribonucleases (*) Require the presence of bile salts for activity.	– Dipeptidases – Aminopeptidases – Dextrinases – Disacaridases (sucrase-isomaltase, lactase, maltase) – Enterokinase (converts trypsinogen in trypsin)

12.3 Stomach. Functional structure. Secretion. Postprandial activity and gastric motility

The stomach has three layers of smooth muscle: the longitudinal, which continues in the duodenum, and the circular and the oblique, both ending in the pyloric sphincter. On the other hand, based on secretory, electrical, and motor characteristics, the stomach can be divided into two functional regions (Figure 12.7):

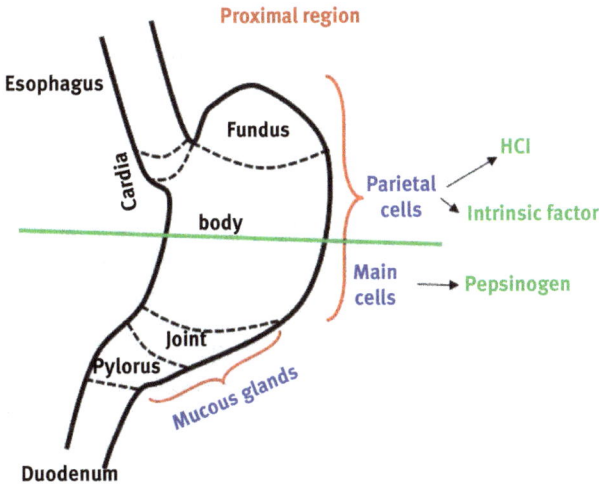

Figure 12.7: Schematic representation of the structure of the stomach.

- The proximal stomach, that anatomically coincides with the cardia, fundus and one third of the stomach body.
- The distal stomach, that anatomically coincides with the antrum, two-thirds of the body and the pylorus. The circular muscular layer is thicker in the antrum.

Glands differ depending on the region of the stomach, so that the gastric mucosa can be divided into three regions based on the structure of the glands:
- Glandular region of cardia, composed mainly of mucus-secreting cells
- Oxyntic or parietal glandular region composed of parietal cells that secrete HCl and intrinsic factor (the protein that binds vitamin B_{12}), as well as enterochromaffin cells and D cells, which secrete histamine and somatostatin, respectively.
- Body region: Chief cells (secrete pepsinogen)
- Pyloric region: G cells (secrete gastrin)

12.3.1 Gastric secretion

Gastric juice is a mixture of secretions from specialized surface cells and cells of the gastric glands. Gastric juice, whose chemical composition is shown in Table 12.2, mixes with the bolus, thus producing the chyme. Around 2–3 l of gastric juice are secreted every day. The most important characteristic of gastric juice is its low pH (as low as 1) due to its high content in HCl. The stomach has approximately one billion parietal cells that secrete 0.16 M HCl. The parietal cell is highly specialized in this operation, which depends on active transport and, consequently, requires and enormous amount of ATP to be accomplished. The process of HCl production in parietal cells is shown in Figure 12.8. Some of the main functions of HCl are:

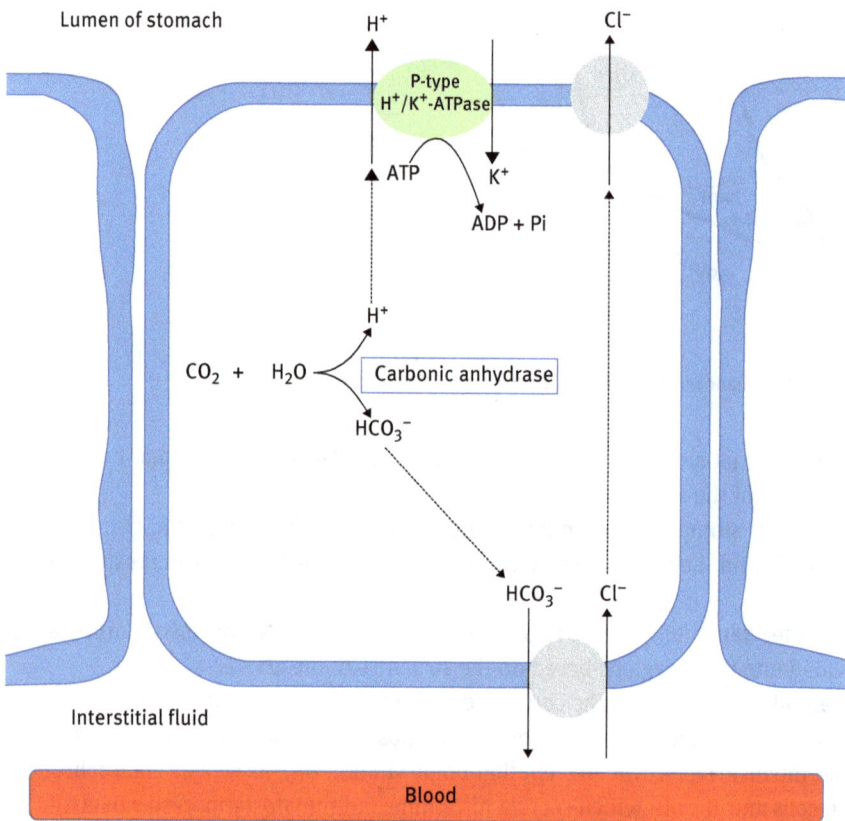

Lumen of stomach $\qquad$ H^+ $\qquad$ Cl^-

P-type H^+/K^+-ATPase

ATP $\qquad$ K^+

ADP + Pi

H^+

CO_2 + H_2O $\qquad$ Carbonic anhydrase

HCO_3^-

HCO_3^- Cl^-

Interstitial fluid

Blood

Figure 12.8: Production of hydrochloric acid (HCl) in parietal (oxyntic) cells.

– Transformation of pepsinogen into active pepsin. Gastric pepsin is actually a heterogeneous set of proteins that are responsible for the proteolytic activity of gastric juice. These are secreted in the form of inactive zymogenic precursors called pepsinogen I (PGI) and pepsinogen II (PGII), two molecular variants that differ in net load and/or molecular weight (isozymes) (Figure 12.9). At pH 1.5, pepsin exhibits about 90% of maximum activity, decreasing to about 35% of maximum activity at pH 4.5.

– It facilitates the digestion of connective tissue and muscle fibers from ingested meat.
– It solubilizes Ca^{2+} and $Fe^{2+/3+}$ salts, thus facilitating the absorption of these cations.
– It acts as a mechanism of defense by destroying bacteria.

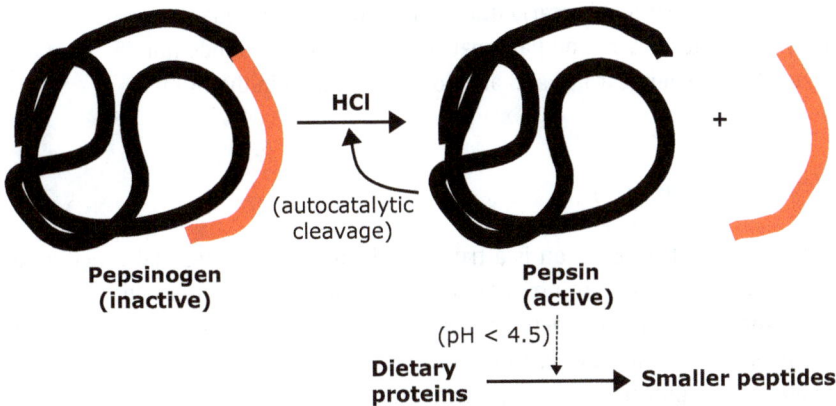

Figure 12.9: Activation of pepsin from pepsinogen in the presence of hydrochloric acid (HCl).

Surface cells secrete bicarbonate (HCO_3^-), which is trapped by the mucus to form an alkaline viscous layer to protect the walls from acidity. The mucus is made up of glycoproteins, which are extremely hydrophilic and can form gels. There are two types of gastric mucus, visible and soluble. Visible mucus is made up of mucins (glycoproteins) that form a gelatinous coating with a high concentration of bicarbonate that protects the gastric epithelium from acid and pepsin. Mucin molecules are cross-linked by disulfide bridges which, along with the oligosaccharide chains, give the mucus a highly viscous consistency that easily expands upon hydration. Soluble mucus contains mucins without disulfide bonds, therefore it has a less viscous consistency that allows the lubrication of the bolus and facilitates its mixing. Mucosal stability is increased by the presence of small peptides known as trefoil factors.

12.3.2 Gastric function: objectives

– Mechanical crushing of food
– Liquefaction of solids
– Digestion of macronutrients
– Maximal exposure of the products to enzymatic action

- Control of gastric emptying by regulating intestinal transit to a value of around 200 kcal/h (depending on whether it is solid or liquid, nutrients, etc.).
- Optimizes pressure/volume ratio thus preventing gastroesophageal reflux and accelerating emptying
- Distention of the stomach wall generates important signals for the control of the posterior segments and gives a feeling of satiety
- Preferential digestion of proteins, which requires a very acidic pH and a gastric mucosa protected by an alkaline mucus
- Cleaning of residues and bacteria during the interdigestive phase
- Acidity stimulates biliary and pancreatic secretions in the duodenum
- Unlike other nutrients, water and alcohol are absorbed in the stomach

12.3.3 Regulation of gastric secretion

The regulation of gastric secretion is a true paradigm of gastrointestinal functioning as a whole and depends on an intricate balance of chemotransmitters with simultaneous excitatory and inhibitory actions. The regulation of gastric secretion is divided into three phases: cephalic, gastric, and intestinal.

- Cephalic phase: Enteric neurons are activated via vagal route. These neurons release acetylcholine, which acts directly on parietal and enterochromaffin cells, and gastrin-releasing peptide (GRP) in the vicinity of G cells, which in turn release gastrin which activates parietal and chief cells.
- Gastric phase occurs when food has reached the stomach and causes the greatest generation of acid secretion of the three phases. The presence of food in the gastric lumen stimulates chemical and mechanical receptors, allowing amino acids and short chain peptides to stimulate the release of gastrin from G cells. Gastric distension triggers the release of acetylcholine and GRP.
- Intestinal phase occurs when the chyme reaches the duodenum.
    - Stimulation of the neuropeptide calcitonin gene–related peptide (CGRP), which acts on D cells to induce the release of somatostatin (decreases the release of gastrin and histamine).
    - In the duodenum, gastric acid is neutralized by sodium bicarbonate. This inhibits gastric enzymes that have optimal activity at low pH values.
    - Local and hormonal reflexes cause the inhibition of the secretion mechanisms of HCl.

12.3.4 Gastric motility

Mixing, crushing, and propulsion movements occur. Peristalsis begins minutes after the food has reached the pyloric part of the stomach, which is the one with the greatest muscle thickness and the greatest crushing power (Figure 12.10).

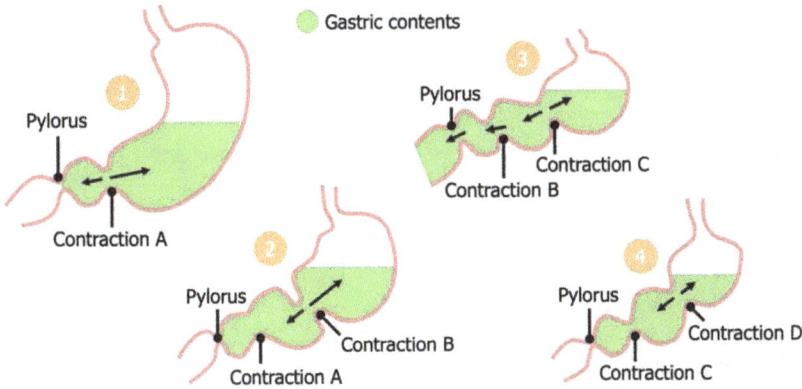

Figure 12.10: Gastric motility. Reproduced with permission from: Lecturio (https://www.lecturio.com/es/concepts/motilidad-gastrointestinal/)

- During gastric filling, a weak peristaltic wave "A" begins at the antrum toward the pylorus. The gastric contents are compressed and pushed back toward the stomach body.
- A stronger "B" wave originates in the body and squeezes the gastric contents in both directions.
- The pylorus opens when wave "B" approaches and the duodenal bulb is filled. The "C" wave begins just above the incisure of the body.
- The pylorus closes again. Wave C fails to evacuate the content and a "D" wave starts higher in the body. The bulb can contract or remain full.
- Peristaltic waves are now originated higher in the body. Gastric content is intermittently evacuated.
- 3–4 h later the stomach is almost empty. Small peristaltic waves empty the bulb with some reflux inside the stomach. There is antegrade and reverse peristalsis in the duodenum. Solid particles pass into the duodenum only if it is 0.3 mm or less in size, therefore the content is practically liquid. Gastric emptying depends on the quantity and quality of chyme that the duodenum can process.

12.3.5 Regulation of gastric emptying

Gastric emptying is regulated by strong inhibition and weak facilitation.
- Strong inhibition by enterogastric reflex (that produces the contraction of pylorus and the inhibition of the propelling contractions of the antrum) and hormones, such as cholecystokinin, secretin, gastric inhibitor peptide (GIP), and somatostatin. The inhibition of gastric emptying triggered by the presence of the chyme in the duodenum is a function of its abundance, the presence of fat and protein degradation products, acidity of the chyme, hyperosmolarity, which implies contraction of the antrum and pylorus.
- Weak facilitation is triggered by gastric distension due to increased volume of the chyme in the antrum. Motilin, a peptide secreted in the interdigestive period, also accelerates gastric emptying.

12.4 Pancreas

Pancreas is divided into several parts: head, uncinate process, neck, body, and tail. The main pancreatic duct begins at the tail and ends in the lower portion of the head. It joins the bile duct forming the hepatopancreatic ampulla or ampulla of Vater, which terminates in the lumen of the duodenum. The accessory pancreatic duct begins in the cavity of the main duct itself, crosses the head of the pancreas, and joins the duodenum through the Santorini's minor caruncle. It has no valves and can be considered as a simple way of derivation.

The pancreas is formed by exocrine and endocrine tissues. Exocrine tissue secretes digestive enzymes into a network of ducts that join the main pancreatic duct. Endocrine tissue, which is formed by the islets of Langerhans, secrete hormones into the bloodstream.

12.4.1 Physiology of exocrine pancreas

- The function of pancreatic acinar cells. The exocrine part is made up of epithelial (acinar and centroacinar) cells arranged in spherical or ovoid structures (pancreatic acini) formed by the cells. The primary function of pancreatic acinar cells is the production of large amounts of enzymatic proteins that are stored as zymogen granules after being synthesized. Cells pour their contents into the luminal space of the acini by exocytosis. Cellular contents are subsequently transferred to the duodenum through the pancreatic ducts. Acinar cell are stimulated by acetylcholine, cholecystokinin (CKK), bombesin, and substance P via membrane receptors that use inositol triphosphate (IP3) as a second messenger. Secretin and vasoactive intestinal peptide (VIP) stimulate acinar cells through adenyl cyclase. The aqueous component secreted by ductal cells is slightly hypertonic and presents a high concentration of HCO_3^-. As it progresses through the ducts, water is adjusted through the epithelium until the pancreatic juice becomes isotonic and part of the HCO_3^- is exchanged for Cl^- (Figure 12.11). Under resting conditions, the aqueous component is produced mainly in the intercalated ducts and other intralobular ducts. When the secretion is stimulated by secretin, the additional flow comes mainly from the extralobular ducts.

– Control of ionic secretion. More bicarbonate than chloride is secreted in the pancreatic duct, however, the situation is the opposite in the acini. Higher rates of secretion of pancreatic juice imply more production of bicarbonate and lower secretion of chloride, whereas concentrations of Na^+ and K^+ are maintained stable.

– Composition of pancreatic secretion. The pancreas secretes between 1.5 and 3 L per day of an alkaline liquid (pH 8.1–8.5) that contains about 20 enzymes and proenzymes. HCO_3^- is the most important ion in pancreatic juice. Between 120 and 300 mmol/day are secreted daily. Its function is to neutralize the acidic chyme that comes from the stomach, thus providing the adequate pH for pancreatic enzyme action (Table 12.2).

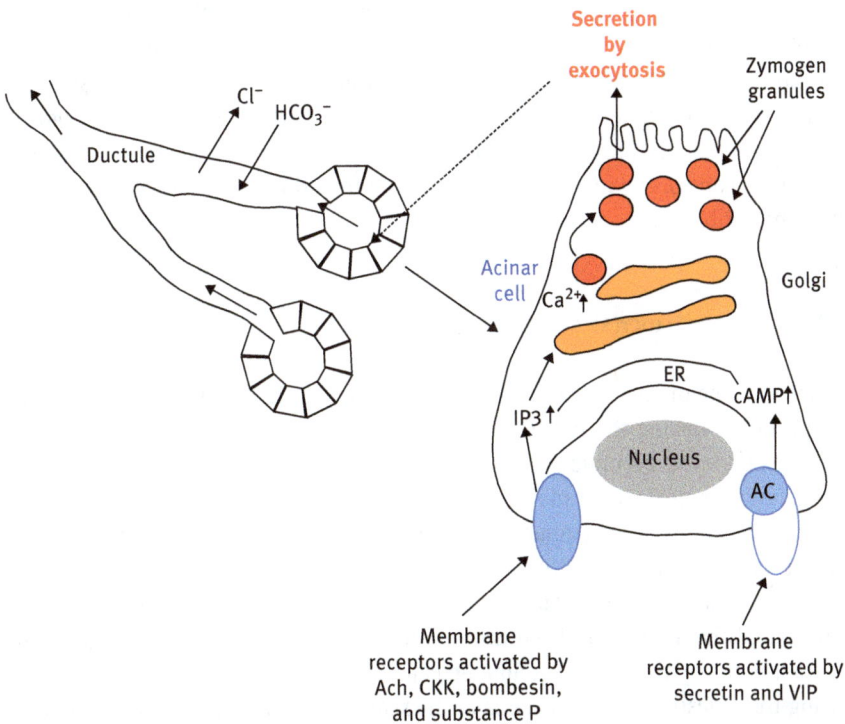

Figure 12.11: Structure of exocrine pancreas and function of pancreatic acinar cells. Abbreviations: IP3, inositol 1,4,5-trisphosphate; AC, adenyl (adenylyl or adenylate) cyclase; cAMP, cyclic adenosine monophosphate. Adapted from: Jurij Dolenšek, Viljem Pohorec, Marjan Slak Rupnik and Andraž Stožer (April 26, 2017). Pancreas Physiology, Challenges in Pancreatic Pathology, Andrada Seicean, IntechOpen, DOI: 10.5772/65895. Available from: https://www.intechopen.com/books/challenges-in-pancreatic-pathology/pancreas-physiology.

12.4.2 Physiology of endocrine pancreas

The endocrine pancreas is formed by the islets of Langerhans, which in turn are composed by different types of cells that secrete hormones directly into the blood:
- Alpha cells (α-cells) synthesize and secrete glucagon, a peptide hormone that increases the level of blood glucose (hyperglycemic hormone) by acting on different organs and tissues (Chapter 14).
- Beta cells (β-cells) produce and release insulin, a hypoglycemic hormone that regulates the level of glucose in the blood. Like glucagon, it exerts effects on multiple target organs and tissues (Chapter 14). Insulin is synthesized as "pre-proinsulin" and converted to "proinsulin" in the endoplasmic reticulum. Pro-insulin generates equimolar amounts of insulin and C-peptide through the action of proteolytic enzymes. Insulin and the unfolded C-peptide are packed in secretory granules that accumulate in the cytosol of the β cell and are released simultaneously in response to glucose.
- Delta cells (δ-cells) synthesize somatostatin, a hormone that inhibits the contraction of both the smooth muscle of the digestive system and the gallbladder when digestion is over. It also inhibits the secretion of insulin and glucagon. Somatostatin secretion is regulated by high levels of glucose, amino acids, and glucagon.
- PP cells (gamma cells or F-cells) produce and release the pancreatic polypeptide (PP), which controls and regulates the exocrine secretion of the pancreas and contracts the gallbladder.

12.4.3 Regulation of exocrine secretion

Two periods can be distinguished in the exocrine pancreatic secretion: interdigestive and postprandial. During the interdigestive period pancreatic secretion is quite limited and considered to be under the control of nervous and hormonal mechanisms. Nerve regulation is performed mainly by the parasympathetic control, with enteropancreatic connections. The sympathetic nervous system influences by inhibiting interdigestive secretion and motility. The hormones with a greater role in this period are motilin and pancreatic polypeptide, which stimulate and inhibit secretion, respectively. Interdigestive regulation is considered important to clean the upper gastrointestinal tract of food particles, desquamated cells, and intestinal flora.

During the digestive period there is an increase of exocrine pancreatic secretion induced by hormonal and nervous stimuli triggered by food. Three phases are classically distinguished during this period: cephalic, gastric, and intestinal. Pancreatic secretion must neutralize the acidic pH of the chyme, and thus the activity of gastric pepsin, which can damage the duodenal mucosa. On the other hand, a neutral pH is important to activate pancreatic enzymes and to increase the solubility of bile acids.

- Cephalic phase. The stimulation comes from the integration of stimuli such as chewing, smelling, or tasting. It is responsible for almost 50% of pancreatic secretion. Stimuli travel via the vagus nerve to acini and pancreatic ducts.
- Gastric phase: This phase is initiated by the gastric distension caused by the bolus and by the presence of amino acids and peptides in the stomach lumen. This stimulation activates efferent vagovagal reflexes and increases gastrin secretion. In this phase pancreatic enzymes are preferentially secreted compared to water and bicarbonate. Its contribution to pancreatic secretion is less than 10% of total secretion.
- Intestinal phase: Pancreatic response is regulated primarily by the hormones secretin and CCK. The action of CCK is fundamentally paracrine, exerting its effect at the level of adjacent vagal fibers. The low pH in the chyme stimulates the production of secretin that, in addition to its endocrine action, also seems to act via paracrine route.

12.4.4 Summary of pancreas functions

The pancreas has digestive and hormonal functions:
- Enzymes secreted by the exocrine tissue of the pancreas participate in the digestion of carbohydrates, fats, proteins, and nucleic acids in the duodenum. These enzymes are transported through the pancreatic duct to the bile duct as inactive zymogens, which are activated when they enter the duodenum. The exocrine tissue also secretes bicarbonate to neutralize the acid coming from the stomach to the duodenum.
- The main hormones secreted by the endocrine tissue of the pancreas are insulin and glucagon (which regulate glucose level in the blood and participate in energy homeostasis, as will be discussed in Chapter 14), as well as somatostatin.

12.5 The liver: structure. Hepatic secretion: storage and regulation

The Couinaud classification divides the liver into eight independently functional segments, each of them having its own vascular flow, venous drainage, and biliary drainage. In the center of each segment there is a branch of the portal vein, the hepatic artery, and the bile duct. On the periphery of each segment there is vascular outflow through the hepatic veins. Around 25% of the total blood passes through the liver.

At microscopic level, several structures are defined (Figure 12.12):
- Classic hepatic lobule. A morphological unit that has a hexagonal shape and is organized around the centrolobular vein. The interior consists of hepatocyte cords arranged radially around the centrolobular vein. Blood flows from the pe-

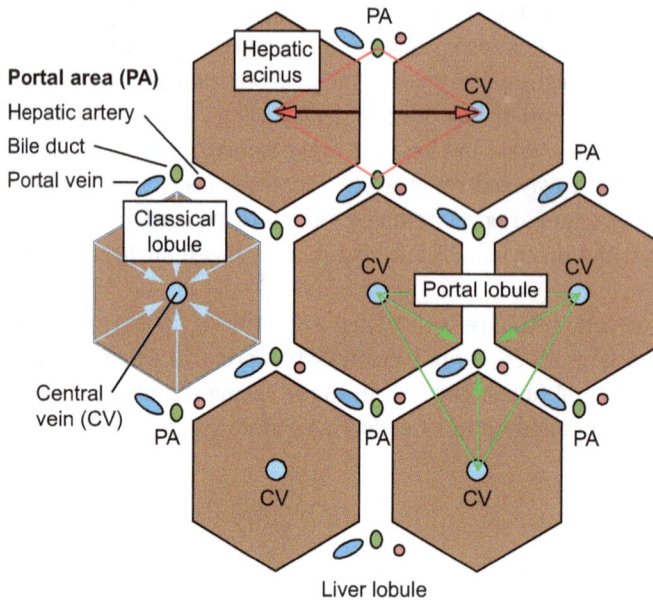

Figure 12.12: Schematic representation of the hepatic lobule. Reproduced with permission from: Gartner, Leslie P. Textbook of Histology, 4th edition. ISBN: 978-0-323-35563-6. Copyright © 2017 Elsevier, Inc.

riphery to the center of the lobule, into the centrolobular vein and bile flows to the periphery, into the bile ducts of the portal areas.

– Portal lobule. This unit is centered around the bile duct of the portal space and defined as the triangular area consisting of three hepatic lobules that are drained by the same bile duct of the portal space.

– Liver (hepatic) acinus. It has an approximate diamond shape and is formed by portions of two hepatic lobules irrigated by terminal branches of the portal vein and the hepatic artery.

– Hepatic sinusoids. They are a type of capillaries covered by a fenestrated (perforated) endothelium that separates hepatocytes from blood cells. They allow the passage of blood plasma in the space between hepatocytes and the endothelium. The blood filtered through the sinusoids comes directly from the stomach and intestines. Sinusoids also contain phagocytic cells (hepatic macrophages or Kupffer cells), which eliminate bacteria and wastes from the blood. Around the sinusoid, there are Ito cells or lipocytes that store vitamin A and lipids.

After a meal the hepatocytes:
– Absorb glucose, amino acids, iron cations, vitamins, and other nutrients in order to be metabolized or stored
– Eliminate and degrade hormones, toxins, bile pigments, and drugs

– Secrete albumin, lipoproteins, coagulation factors, angiotensinogens, and other products in the blood

During the periods between meals hepatocytes release glucose in the circulation.

The hepatic artery, which carries oxygen-rich blood into the liver, arises from the aorta, the largest artery in the body. Arteries from the aorta also give rise to the capillaries that capture nutrients from the intestine and converge in the portal vein, which enters the liver. Within the liver, the branch of the hepatic portal vein and the hepatic arteries join in the spaces between the hepatic lobes, all of them draining into the hepatic sinusoids. Therefore, there is an unusual mixture of venous and arterial blood in the sinusoids.

Substances are exchanged in the liver and blood subsequently passes to the hepatic vein, which flows directly into the vena cava, which delivers it to the heart. During interdigestive periods bile is stored in the gallbladder.

12.5.1 Bile production by the hepatocytes

Hepatocytes secrete conjugated bile acids (bile salts), conjugated bilirubin, ions, and water into the bile canaliculi that terminate in the bile ducts, where bicarbonate is secreted through the action of hormones and vagal stimulation. The bile that goes through the bile ducts is an alkaline aqueous secretion that is stored in the gallbladder. Bile is a green, bitter-tasting liquid substance produced by the liver. It is involved in digestion processes functioning as an emulsifier of fatty acids.

12.5.2 Synthesis of bile acids

Bile acids are synthesized in the liver from cholesterol by two different pathways. The primary bile acids are cholic acid (Chapter 4) and chenodeoxycholic acid.

Approximately 90–95% of bile salts (sodium and potassium salts of bile acids) are reabsorbed in the ileum, while 5–10% pass to the colon where they are modified through bacterial action, thus producing secondary bile acids. These are partially reabsorbed (some are lost in the faeces) and return to the liver to be conjugated with glycine or taurine. This process produces glycocholate, taurocholate, glycokenodeoxylate, and taurochenodeoxycholate. New bile acids are only synthesized by the liver to replenish fecal losses.

Bile salts discharged into the duodenum are absorbed in the ileum by co-transport with sodium, delivered to the liver via portal vein, and returned to the intestine with the bile. This recirculation occurs several times a day (enterohepatic recirculation).

12.5.3 Regulation of liver secretion and vesicular emptying

- Bile secretion. The substances that increase bile secretion are known as choleretics. The vagus nerve releases acetylcholine that activates bile secretion. At the same time, bile salts stimulate secretin release, which increases the secretion of bile and bicarbonate in the duodenum (Figure 12.13).
- Biliary emptying. Cholagogues stimulate vesicular emptying. The vagus nerve elicits a weak stimulation of vesicular contraction, while hormonal control by CCK produces a potent contraction of the gallbladder that pours bile into the duodenum in response to the presence of food (Figure 12.13).

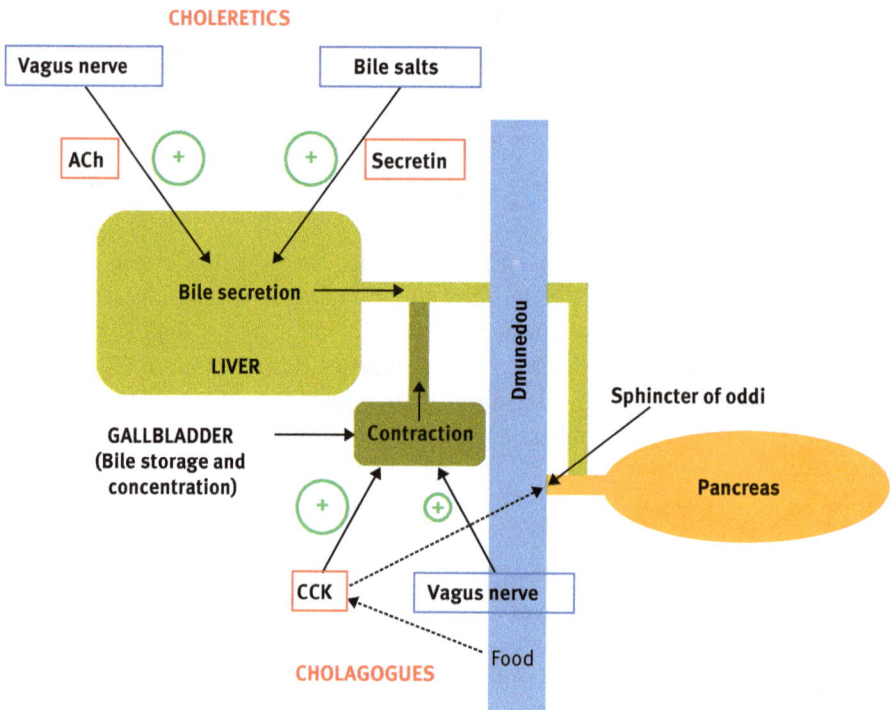

Figure 12.13: Schematic representation of the regulation of bile secretion and gallbladder emptying.

Bile is stored and concentrated in the gallbladder for several hours. Concentration is achieved by active reabsorption of Na^+, followed by passive reabsorption of Cl^- and water (Ca^{2+} is not reabsorbed). The concentrations of solutes and bile salts are therefore higher in stored bile (and with a slightly lower pH) than in the bile flowing through the hepatic duct. Bile can be concentrated 5 to 20 times.

12.5.4 Functions of bile salts

Bile salts are amphipathic: they have a hydrophilic polar groups and a hydrophobic apolar nucleus (the cyclopenta[a]-phenanthrene carbon skeleton, see Chapter 4). The main functions of bile salts are:
- Upon reaching a certain critical level of concentration, bile salts associate to form micelles with their hydrophilic part facing outward and the hydrophobic part inward. These micelles interact with the digested lipids present in the intestinal lumen, thus forming mixed micelles that deliver the lipids to the brush border of the enterocytes for absorption (Chapter 13).
- Bile salts alkalize the duodenum, along with pancreatic secretion and intestinal juice.
- They are an excretion route for some waste products: bile pigments, steroids and cholesterol, heavy metals and drugs.

12.5.5 Liver functions

(Discussed in Chapter 14, Section 14.6)

12.6 Small intestine

The basic structure and the specialized cells of the small intestine are described in Chapter 13.

12.6.1 Digestion in the small intestine

12.6.1.1 Chemical digestion
Bile and pancreatic juice are poured into the duodenum, which act on the chyme in the lumen of the intestine and form the chyle. The function of pancreatic juice is to provide enzymes that degrade carbohydrates, fats, and proteins. Intestinal juices contain enzymes that continue with the degradation of carbohydrates and proteins (Table 12.2), while bile emulsifies fats. This phase of intestinal chemical digestion, which takes place in the lumen of the organ, is known as luminal digestion.

12.6.1.2 Mechanical digestion: motility
Motility in the small intestine varies depending on the gastric phase:
- During the interdigestive phase, the pattern of motility is described by the so-called migrating motor complex (MMC) or migratory myoelectric complex, which consists of three phases:

–   Phase I lasts about 70 min. During this phase, only the basic electric rhythm and hydro-saline secretion can be observed.
–   Phase II lasts between 10 and 20 min and is characterized by a slight increase in base activity, with intermittent irregular low-amplitude contractions, along with secretion acidic and enzymatic.
–   Phase III is the motor activity phase, which presents contraction waves at a rate of 11 to 13 per minute with a duration of 1–5 min. The duration of these phases varies between individuals and even in the same individual depending on the time of day.

MMC takes about 90 min to travel the entire intestinal tube and ends upon reaching the ileocecal valve, then a new motor complex is generated in the stomach. This process is associated with two functions: secretion, during phase I and II, and propulsion, during phase III. The latter eliminates intestinal waste that might cause the growth of harmful bacteria. It has also been shown that some absorption occurs during phase I and II. The main hormonal regulator of MMC is motilin.

–   During the digestive phase the presence of circulating gastrin, cholecystokinin (CCK) and neurotensin, adapt gut motility to a digestive pattern, characterized by a segmental and random motor activity dependent on neurohormonal influences. MMC is inhibited during a variable period of time that depends on the caloric intake and on the type of nutrient, so that fats delay MMC more than carbohydrates and the latter more than proteins. The digestive pattern shows two types of movement:
    –   Segmentation movements are characterized by close contractions of the circular muscular layer, dividing different segments of the intestine into small portions that give a characteristic image of "string of sausages." Segmentations are rhythmic, with a frequency of 7 to 12 times per minute, and are produced in such a way that each time the segmentation originates at a different point. These movements allow the chyme to mix with secretions and promotes absorption of nutrients (Figure 12.14).
    –   Propulsion movements are produced by peristalsis: a contraction is followed by a relaxation of the circular muscle, which propagates in the distal direction. Contractions occur randomly at different points of the small intestine and their propagation affects segments of different lengths. This causes a very slow propulsion of the intestinal contents (1–2 cm/s).

In certain circumstances where the mucosa is threatened by mechanical or chemical damage, an intense peristaltic contraction is generated at the point of damage that quickly crosses the entire intestinal tube, in both directions (oral and caudal) from the point of origin. This movement is known as peristaltic rush and its function is to quickly evacuate the harmful contents of the intestine.

Another movement in the intestine involves irregular contractions of muscular layer of the mucosa at a frequency of 3 contractions per minute. This movement is

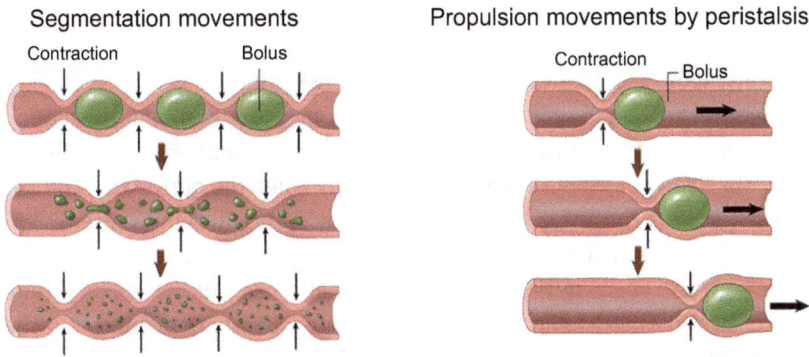

Figure 12.14: Segmentation and propulsion movements in the small intestine. Reproduced with permission from: Patton, K. T. and Thibodeau, G. A. The Human Body in Health and Disease, Seventh Edition. ISBN: 978-0-323-40211-8/978-0-323-40210-1. Copyright © 2018 by Elsevier, Inc.

under the tonic activation of the sympathetic nervous system and other chemical stimuli and leads, on the one hand, to the formation of the characteristic folds of the mucosa and, on the other, to a periodic shortening and lengthening of the intestinal villi that favors blood and lymphatic drainage.

12.6.1.3 Regulation of small intestine motility

A number of reflexes control the number, frequency, and strength of peristaltic waves. Among the most relevant are:

- Gastrointestinal reflex. The presence of food in the stomach leads to an increase in intestinal motility.
- Intestino-intestinal reflex. The presence of chyme in the intestine increases intestinal motility.
- Gastroileal reflex. The presence of food in the stomach increases motility especially at the level of the ileum in order to facilitate its emptying.
- Parasympathetic nervous system. An increase in parasympathetic stimulation increases intestinal motility.

12.6.2 Composition of intestinal secretions

The cells that line the internal surface of the small intestine release a mixture of substances known as intestinal juice, which includes water, bicarbonate, mucin, mineral salts, and a variety of enzymes, (Table 12.2). The intestinal juice has an alkaline pH to counteract the acidity of chyme and continues the digestion of macronutrients along the small intestine.

12.6.3 Small bowel emptying

Intestinal emptying occurs through the ileocaecal valve, which is contracted under basal and unstimulated conditions, generating a high pressure area and preventing the passage of the ileal content to the caecum. This optimizes nutrient uptake by the ileal mucosa. The ileocaecal valve appears to relax only when the peristaltic waves of the ileum reach it, thus causing a small spill of chyme in the caecum, whose distension causes a reflex contraction of the sphincter. This response is fundamentally nervous, so that both the vagus nerve and the sympathetic nervous system stimulate sphincter contraction. Acetylcholine is the vagal mediator and norepinephrine is the sympathetic mediator. Gastrin may also participate in this response because it stimulates the motility of the ileum and the relaxation of the sphincter. The gastroileal reflex causes an increase of the motility of the ileum and, consequently, of ileocaecal evacuation. Normal bowel emptying is about 1,500 ml/day. Overall, the functions of the ileocaecal valve are:

-   Avoid caecum overload.
-   Increase the residence time of the chyme in the ileum, favoring absorption.
-   Increase the residence time of the chyme in the large intestine, favoring the reabsorption of water and electrolytes, as well as bacterial action (see below).

12.6.4 Functions of the small intestine

-   Motor function. It allows cleaning movements during the interdigestive periods and food mixing and kneading in the digestive phase, thus facilitating nutrients absorption.
-   Secretory function. The highest proportion of intestinal secretion comes from exocrine pancreatic secretion and biliary secretion, which are poured into the duodenum through the sphincter of Oddi. The mucosa of the small intestine also secretes some digestive enzymes, the most important being enterokinase, which activates the pepsinogen secreted by the pancreas.
-   Digestive function. The intestinal mucosa contains its own enzymes at the luminal surface in addition to those from the pancreatic exocrine secretion, which are responsible for the final digestion of the chyme. Some intracellular digestion also occurs.
-   Absorption function. The intestinal mucosa is designed to perform the absorption of most of the nutrients (Chapter 13).
-   Endocrine function. A high number of hormones are produced and secreted in the small intestine, all of them aimed at regulating intestinal digestion and absorption. To accomplish this task, there is a set of chemical and mechanical receptors whose stimuli trigger specific hormonal responses. A number of reflexes are also involved.

- Protective function. This function is determined by the basicity of intestinal secretions and by a powerful immune barrier.

12.7 Large intestine. Functional structure. Absorption and secretion. Motility of the colon: peristalsis and mass movements. Defecation

The large intestine is 1.5 m long and extends from the ileocecal valve to the anus. It consists of the caecum, the appendix, the colon, the rectum, and the anal canal. The colon is in turn divided into ascending colon, transverse colon, descending colon, and sigmoid (or ileo-pelvic) colon.

The longitudinal muscle of the large intestine does not form a continuous layer, but is divided into longitudinal bands, or teniae. Since these bands are shorter than the circular muscular layer, they cause the formation of folds (haustras) in the wall of the large intestine.

The rectum is the last portion of the digestive system, being located between the sigmoid colon and the anus. It has an approximate length of 20 cm. The function of the rectum is to store the stool before being expelled through the anal opening. The rectum extends to the anus, an opening that has an internal sphincter of smooth muscle cells and an external sphincter of striated muscle.

The colon contains numerous straight tubular glands with goblet, regenerative, and enteroendocrine cells. The lining cells, the colonocytes, are specialized in the absorption of water, some vitamins, and electrolytes.

12.7.1 Motility of the colon

- Mixing or segmentation movements. They are tonic waves of contraction that move back and forth (also called antiperistaltic waves). They occur with a frequency of 3 or 4 per minute and allow prolonged contact of the intestinal contents with the mucosa.
- Propulsion movements or mass movements occur from the transverse colon to the sigmoid. They are a type of modified peristalsis: a constriction ring appears, then, 20 cm or more of the colon loses the haustrations, contracting as a unit, and pushing the stool forward. This type of movement usually occurs a few times a day (1–3), especially within the hour after breakfast and lasts for 10 to 30 min. When the stool reaches the rectum, the desire to defecate appears.

Mass movements are regulated by the gastro-colic reflex (when food enters the stomach) and duodeno-colic reflex (when food penetrates into the duodenum), thus favoring the occurrence of mass movements after meals.

12.7.2 Chemical digestion in the large intestine

Goblet cells, very abundant in the lining epithelium of the colon, secrete mucus. The purpose of this secretion is to lubricate and confer an appropriate consistency to feces, while its alkaline pH is neutralizes and protects the intestinal mucosa. The final stage of digestion in the colon is performed by bacteria (Figure 12.15).

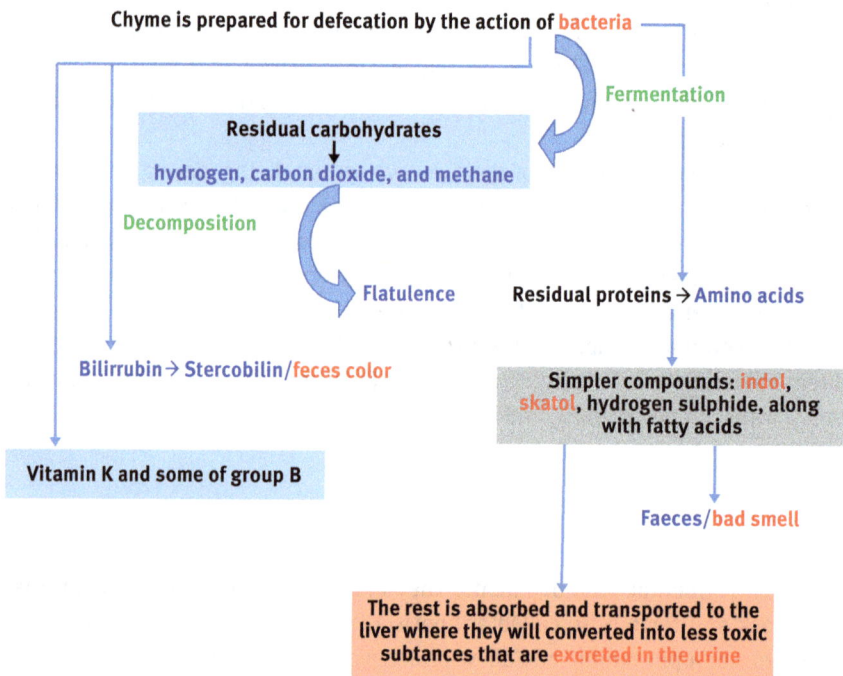

Figure 12.15: Final stages of digestion by gut microbiota.

12.7.3 The rectum

The rectum is normally empty because the intestinal content reaches the sigmoid colon and rectum only when mass movement occurs. This situation leads to the distention of the walls, thus triggering the defecation reflex. The rectum can also retain fecal content for a short period of time.

At rest, the anal canal is closed by an internal anal sphincter (involuntary tonic contraction) that acts as a first barrier, and the external anal sphincter and the puborectalis muscle (tonic contraction under voluntary control) that act as a second barrier.

When a mass movement forces stool into the rectum, a person feels like defecating.

- Afferent signals that propagate through the myenteric plexus increase peristalsis toward the rectum, driving stool toward the anus and relaxing the sphincter.
- If nerve endings of the rectum are stimulated, signals are transmitted to the spinal cord and through the parasympathetic fibers of the pelvic nerves, then, signals return to the descending colon, sigma, rectum, and anus, increasing peristalsis and relaxing the sphincter.
- Acetylcholine and substance P produce stimulation.
- Norepinephrine and peptide Y increase anal pressure.
- The sympathetic nervous system is responsible for the contraction.

12.7.4 Defecation reflex

The following sequence of events is established under the influence of the parasympathetic nervous system:

1. Peristaltic contraction of the end of the colon and rectum.
2. Contraction of the musculature of the pelvic floor.
3. Relaxation of the anal sphincters.

The reflex is usually accompanied by a strong inspiration, closure of the glottis to prevent air outflow, and contraction of the abdominal and thoracic muscles, which results in an increase in intra-abdominal and intra-thoracic pressure.

12.7.5 Feces composition

- Water: 70–80%
- Solid waste (20–30%), composed of undigested food residues, such as dietary fiber (cellulose, lignin), cellular and bacterial debris, bile compounds (stercobilin, responsible for their color), enzymes, and gases.
- Fecal fats: about 5% of ingested fats (6 g/day)
- Fecal nitrogen: 1.4–1.7% of ingested protein (1–2 g/day)

12.7.6 Functions of the large intestine

- Reabsorption of water, vitamins, and some electrolytes
- Disposal of waste substances
- Bacteria of the large intestine also make some important substances, such as vitamin K

12.8 The microbiota of the gastrointestinal tract and its functions

The gastrointestinal microbiota (formerly known as flora or microflora) is formed by a large group of more than 100 trillion (10^{14}) bacteria of more than 400 species that live in the digestive system (archaea, viruses, and fungi are also present). These microorganisms are found from the mouth to the final part of the large intestine. From the esophagus, practically aseptic, the microflora presents a gradient in quantity and variety, being scarce in the stomach and gradually increasing from the small intestine and into the colon, where it performs its main functions:

- Probiotic effect (maintenance of intestinal balance), acting as a barrier for the entry of germs and pathogens that arrive with food. This is accomplished by:
  - Competition with pathogens for nutrients and adhesion sites in the mucosa
  - Generation of a hostile environment for pathogens
  - Activation of the immune system
- Energy production from non-digestible fiber
- Synthesis of vitamin K and some vitamins of B group (biotin, cobalamin, folates, nicotinic acid, pantothenic acid, pyridoxine, riboflavin, and thiamine)
- Participation in the absorption of calcium, magnesium, sodium, and (partially) iron

Articles and textbooks used for the elaboration of Chapter 12

[1]   Zhang CX, Wang HY, Chen TX. Interactions between intestinal microflora/probiotics and the immune system. Biomed Res Int 2019; 2019:6764919. 10.1155/2019/6764919.
[2]   Molina-Tijeras JA, Gálvez J, Rodríguez-Cabezas ME. The immunomodulatory properties of extracellular vesicles derived from probiotics: A novel approach for the management of gastrointestinal diseases. Nutrients. 2019;11(5):E1038. 10.3390/nu11051038.
[3]   Guarner F. «Papel de la flora intestinal en la salud y en la enfermedad». Nutrición Hospitalaria, 2007;22 (2):212–1611. ISSN 0212-1611.
[4]   Barrett KE, Barman SM, Brooks HL, Yuan JX. Ganong's review of medical physiology. Section IV: Gastrointestinal physiology: Introduction, 26e. McGraw-Hill; 2019.
[5]   Koeppen BM, Stanton BA. Berne y Levy. Fisiología. Sección 6: fisiología digestiva. 7ª edición. Elsevier; 2018.
[6]   Hall JE. Guyton y Hall. Tratado de fisiología médica. Unidad XII: Fisiología gastro-intestinal. 14ª edición. Elservier ES; 2021. ISBN: 978-0-323-59712-8.

[7]   Stuart Ira Fox. Fisiología Humana. Capítulo 18: Sistema digestivo. 15 edición. Mac Graw Hill; 2021. ISBN: 9786071515377.

[8]   Dvorkin AD, Cardinali DP, Iermoli RB, Taylor. Bases fisiológicas de la práctica médica. Sección 5: sistema gastrointestinal. 14ª edición. Con CD-Rom: Panamericana; 2010.

[9]   Asencio Peralta C. Fisiología de la Nutrición. Tema 5: Ingreso y utilización de los alimentos en el sistema digestivo 2ª ed. Manual Moderno, Colombia; 2018.

[10]   Silverthorn, DU. Human Physiology, Chapter 21. The Digestive System. 8th Edition. Pearson; 2019. ISBN: 9781292259543.

[11]   Hershel R, Michael L. Medical Physiology: A Systems Approach. S E C T I O N VIII: GI PHYSIOLOGY – 54. Intestinal Motility 543 Kim E. Barrett. 1ª Ed. Mac Graw Hill; 2011.

13 Absorption of nutrients

13.1 Introduction

Our diet primarily consists of a macronutrient mixture, carbohydrates, lipids, and proteins, which comprise up to 51.8%, 32.8%, and 15.4%, respectively, of the total food-derived energy in the Western diet. Within the mouth, chewing of foods results in smaller pieces, while amylases and salivary lipase start breaking down the carbohydrates and triacylglycerides (TAG), respectively. The bulk of the digestive process starts in the stomach, where enzymes are secreted by chief cells as inactive forms (zymogens), upon stimulation by gastrointestinal hormones. The latter are produced after ingestion in amounts that depend on meal composition and calories. The acidic environment of the stomach triggers the transformation of zymogens into their active forms (Chapter 12). The gastric phase is followed by the small intestinal phase, where most of the food digestion and nutrient absorption occurs [1].

The small intestine is a large tube that measures 2.5 cm in diameter and 6–8 m in length in humans. It begins in the pyloric orifice and ends in the ileocaecal valve, being divided in three regions: duodenum, of around 20 cm long, jejunum, about two-fifth of the total length, and ileum. In the duodenum, the partially digested food coming from the stomach (chyme) gets in contact with secretions from the pancreas and the liver, which contain enzymes and bile salts, respectively, among other compounds. Digestion continues as the food moves along the jejunum and ileum due to the action of hydrolytic enzymes secreted by intestinal epithelial cells (IECs). The intestinal lining has circular folds, known as "plicae circulares," that in turn present millions of tiny projections termed "villi." Villi are about 1 mm in height and give the intestinal mucosa a velvety appearance. In addition, many of the epithelial cells that cover each villus present apical membranes with microscopic protrusions (microvilli) that resemble a fine brush, hence the term "brush border" is used for this side of the cell. Altogether, this organization produces a fractal-like structure of the intestinal lining, whose goal is to maximize the available absorptive surface and thus food digestion and nutrient absorption [2]. Underneath the villi, capillaries and lymph vessels mediate transport of absorbed nutrients into the body. The base of each villus is surrounded by multiple invaginations, termed crypts of Lieberkühn. IECs form a continuous single-layered sheet that separate the external environment from the internal one, being essential not only to absorb nutrients but also to avoid the entry of potentially harmful substances and microorganisms (Figure 13.1). Seven major types of differentiated IECs can be distinguished in the epithelium of the small intestine [3–5]:

– Enterocytes are the most abundant cells, representing approximately 90% of the total number of IECs, and being the main responsible for the absorption of nutrients. They also release some digestive enzymes into the intestinal lumen and express transmembrane mucins, heavily O-glycosylated proteins that cover the

https://doi.org/10.1515/9783111111872-013

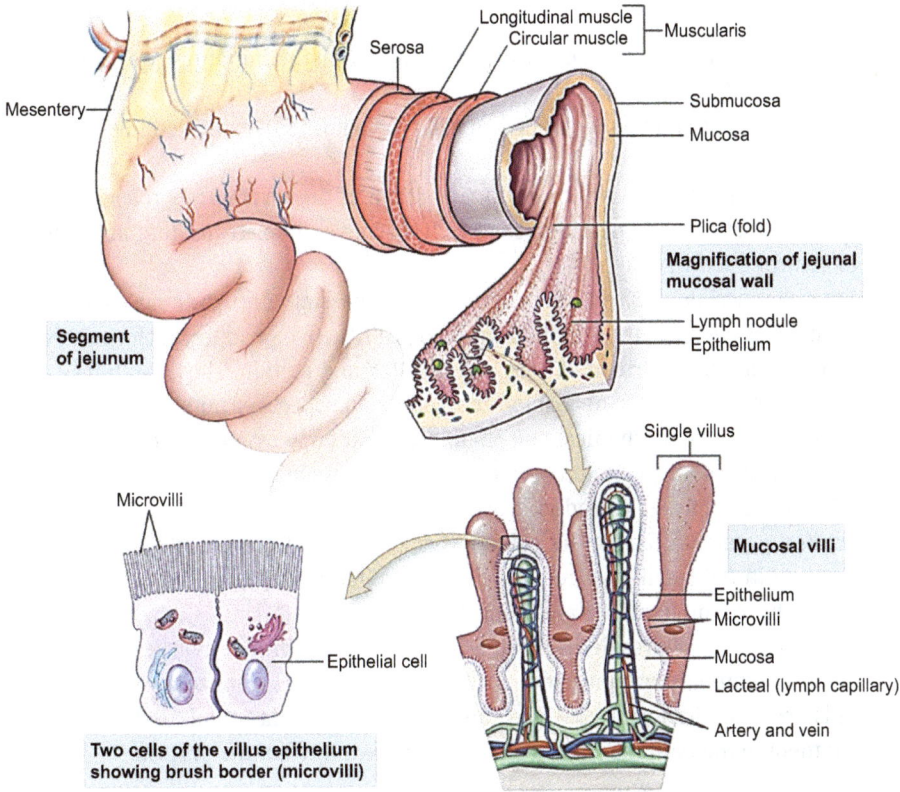

Figure 13.1: Anatomy of intestinal tissue. Reproduced with permission from: Patton, K. T. and Thibodeau, G. A. The Human Body in Health and Disease, Seventh Edition. ISBN: 978-0-323-40211-8/978-0-323-40210-1. Copyright © 2018 by Elsevier, Inc.

apical surface of these cells forming the so-called glycocalyx [6, 7]. The microvilli of enterocytes, along with the glycocalyx and the mucus secretions of other IECs, contribute to the formation of the so-called unstirred water layer, a microclimate generated by the trapping of water molecules, that presents a low pH generated by a H^+/Na^+ antiport exchange system. The unstirred water layer plays an important role in the absorption of nutrients, especially lipids [8]. Enterocytes are characterized by an intense energy metabolism that allows macromolecule synthesis and movement of nutrients in the small intestine. These cells, which are polarized, can receive their fuels from the luminal content and from the bloodstream [9].

– Goblet cells are less numerous than enterocytes and occur both on villi and in crypts. They release mucus substances, such as the gel-forming mucin MUC2,

among other proteins [6], that cover and protect the luminal surface of the intestine.
- Tuft cells promote clearance of parasitic helminths by initiating type 2 immune responses [10].
- BEST4+ cells are a recently identified cell type whose function is still being investigated [10].
- Paneth's cells are found in the bottom positions of the crypts. They release several types of enzymes and peptides, such as lysozyme and defensins, that are thought to control the microbiota.
- Enteroendocrine cells sense nutrients that are passing through the gut and in response secrete more than 20 distinct biologically active peptides, such as secretin, cholecystokinin, gastric inhibitory peptide, PYY and motilin. These peptides act in an endocrine or paracrine fashion to regulate all aspects of nutrient homeostasis including satiety, mechanical and chemical digestion, nutrient absorption, storage and utilization, or regulation of liver activity. Like goblet cells, they occur both on villi and in crypts. PYY produced by enteroendocrine cells have been reported to act in a paracrine fashion on colonocytes, to augment postprandial nutrient absorption in the small intestine and to regulate ion transport and ion-coupled nutrient absorption in mouse and human small intestine [11].
- Microfold (M) cells reside in the specialized epithelium that overlies the Peyer's patches, lymphoid accumulations that play a key role in mucosal immunity. M cells process microorganisms and molecules from the intestinal lumen and present them to the immune system cells, which are in the lamina propria.

The intestinal epithelium is renewed every 5–7 days. This high turnover rate is accomplished by intestinal stem cells (ISCs) residing at the bottom of the crypts, that generate transit-amplifying (TA) cells. The rapidly dividing TA cells differentiate into the various IEC types [10, 12, 13].

Epithelial cells are held together by the apical junctional complexes, that consist of adherent junctions and tight junctions (TJs), as well as by desmosomes. Adherent junctions and desmosomes are involved in intercellular communication and binding between adjacent IECs without altering paracellular permeability (Figure 13.2). TJs are composed of several transmembrane proteins, whose extracellular domains form a selective barrier by establishing interactions with nearby cells [14].

13.2 Transport pathways across the small intestine epithelium

IECs are polarized, that is, one of their sides differs in structure and function from the other. In the case of the small intestine epithelium, the apical membrane (aka brush border membrane) faces the intestinal lumen, whereas the basolateral membrane borders neighboring cells and the underlying basement membrane (Figure 13.2). This

LUMEN

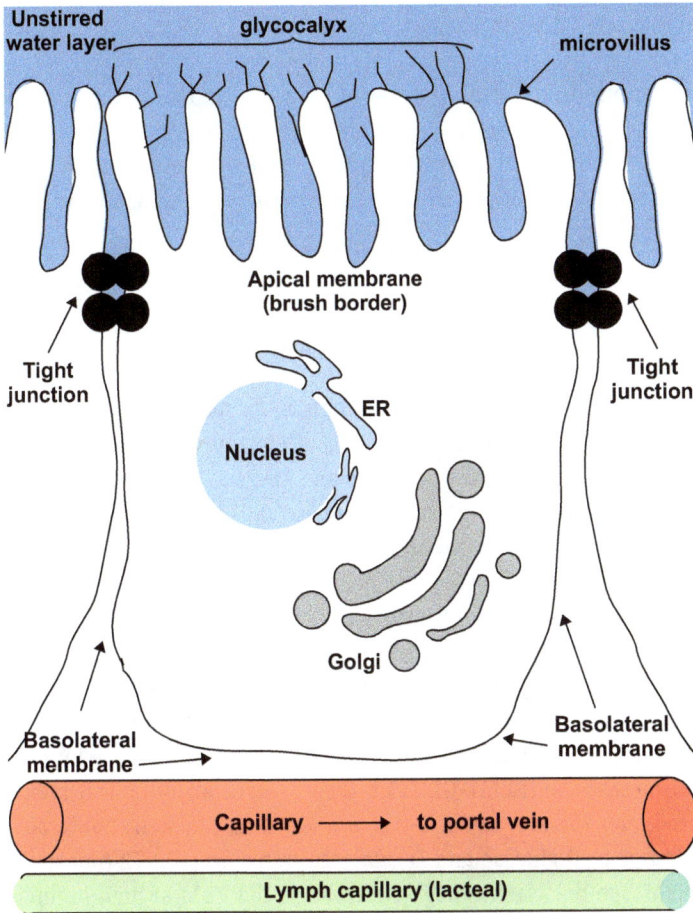

Figure 13.2: Overview of subcellular structures involved in nutrient uptake in enterocytes.

polarity is essential for barrier formation and for the uptake and vectorial transport of nutrients in enterocytes [12].

There are two general transport mechanisms across the epithelial cell layer:

– Paracellular transport that occurs through the intercellular spaces of IECs. This is regulated by intracellular processes that finely tune the permeability of TJ complexes.

– Transcellular transport that can occur via three main modes [13, 15]:

   – Passive diffusion of molecules from the apical to the basolateral side.

   – Carrier-mediated uptake and/or diffusion through the epithelial cell layer.

   – Vesicle-mediated transcellular transport or "transcytosis," used to shuttle macromolecules such as lipoproteins, antibodies, and albumin from one surface of a polarized cell to the other. This process often involves receptor-mediated endocytosis, regulated transit of the carrier vesicle through the cytoplasm, and release of the cargo via exocytosis.

The transport mechanism used for the absorption of a given compound depends on physicochemical properties like molecular weight and size, stability, charge distribution, or the ability to pass through or interact with the plasma membrane. Other factors that influence transport include intestinal motility, interactions with other molecules, and solubility in the mucus layer [13].

13.3 Absorption of carbohydrates

Starch and the disaccharides lactose and sucrose comprise the majority of digestible carbohydrates. During digestion, the combined action of enzymes like salivary and pancreatic amylases, maltase-glucoamylase, lactase, and sucrase-isomaltase, hydrolyze dietary carbohydrates to glucose, galactose, and fructose (Chapter 3). These monosaccharides are subsequently absorbed as depicted in Figure 13.3:

– Transport of glucose and galactose across the brush border of enterocytes can be mediated by the SGLT1 transporter, that drives the movement of both molecules from the lumen of the gut into the enterocyte against their chemical gradients, along with two Na^+ cations (a type of co-transport known as symport). The electrochemical gradient of Na^+ across the brush border membrane (BBM) is the main driving force, this gradient being maintained by the Na^+/K^+-ATPase located at the basolateral membrane (BLM). GLUT5 is also present at the apical membrane where it facilitates the movement D-fructose down its concentration gradient. Another facilitator of D-glucose and D-galactose transport, GLUT2 (normally localized in the BLM, see below), has been observed in the BBM at high luminal D-glucose concentrations or in some pathological conditions [16–18]. Another glucose transporter located in the BBM, GLUT1, has been reported to act by triggering the activation of GLUT2 on this membrane [19].

– Glucose, galactose, and fructose are passively transported across the BLM through the action of GLUT2 [16]. In the fasting state, the GLUT2 on the BLM may transport glucose from the blood into the enterocytes to meet the energy demands of the latter [19].

Although the major mechanisms of sugar uptake seem to be established, there are still some aspects and controversies that deserve further research, like the existence of an alternative paracellular pathway for sugar absorption or the role of other glucose transporters that have been identified in human enterocytes [18, 20].

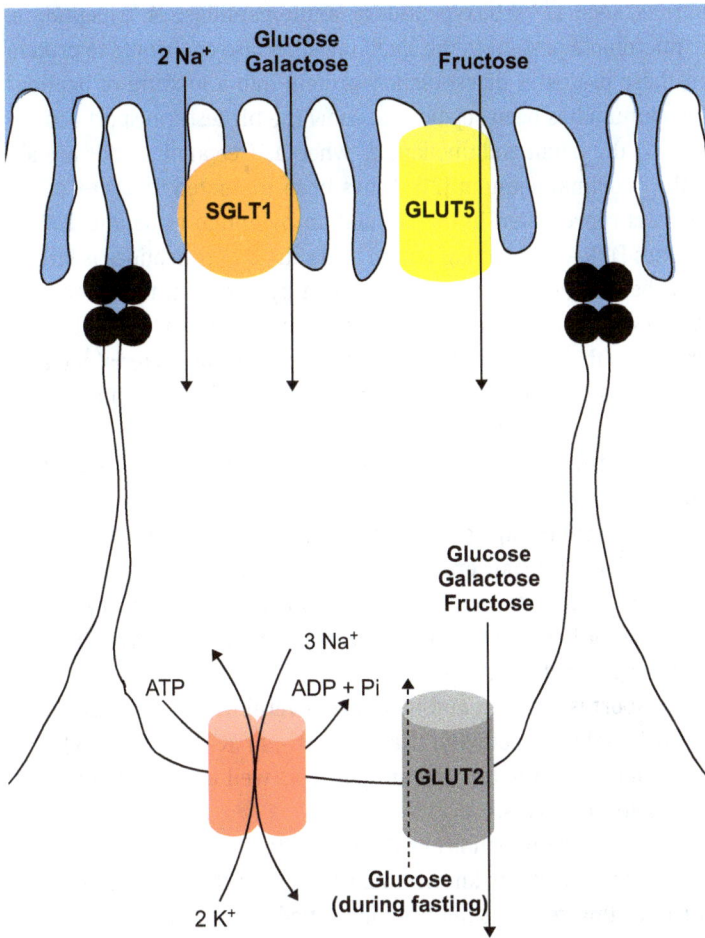

Figure 13.3: Mechanisms of absorption of monosaccharides in enterocytes. See main text for references and further details.

13.4 Absorption of amino acids and oligopeptides

The pancreatic juice secreted by the pancreas in the duodenum contains a mixture of bicarbonate (HCO_3^-) and proteases, which play an important role in protein digestion. Bicarbonate alkalizes the acidic chyme that come from the stomach, thus creating a neutral environment (pH = 6.5–7.5) ideal for the activation of the proteases. These are secreted as zymogens and activated upon enzymatic cleavage. Pancreatic proteases can be divided into endopeptidases (hydrolyzing peptides bonds in the interior of the amino acid chain) and exopeptidases (releasing the amino acids from either end of the amino acid chain). In addition, other proteolytic enzymes are produced and secreted by

IECs at the brush border, such as carboxypeptidase, aminopeptidase N, tripeptidase, and dipeptidase. The microbiota present in the small intestine also contribute to protein digestion. Altogether, these proteases digest dietary protein into a mixture of peptides and amino acids. The jejunum has been reported to show the highest transport activity of dipeptides, followed by the ileum and duodenum, whereas free amino acids are absorbed primarily in the proximal jejunum [21]. It has been estimated that 90% of absorbed dietary proteins is represented in the circulation by amino acids and 10% as dipeptides and tripeptides [17]. Absorption of amino acids, di-, and tripeptides is largely complete at the end of the small intestine, being mediated by a set of transporters that provide amino acids not only for systemic needs but also for enterocyte metabolism. The large intestine mediates the uptake of amino acids derived from bacterial metabolism and endogenous sources [22]. The general absorption mechanism for amino acids, as well as di- and tripeptides is outlined in Figure 13.4.

- Transport across the brush border of enterocytes involve several types of specific transporters (Table 13.1A):
  - Di- and tripeptides cross the apical membrane via a co-transport with protons (H^+) mediated by four transporters. This process is indirectly driven by Na^+, which is necessary to generate a proton gradient through the action of a Na^+/H^+ exchanger. Peptides of four or more amino acids in length are poorly absorbed in a non-carrier-dependent mechanism [17, 21].
  - Amino acid transport is complex and there are several transporters involved, with variations in solute specificity, (Na^+-, Cl^-, H^+-, or K^+- dependency), and mechanisms: uniports, symports, and antiports, as well as electroneutral or electrogenic transport process.
- Transport across the basolateral membrane (Table 13.1B):
  - Only one peptide transporter is known to be present at the basolateral membrane, whose encoding gene has not been identified so far [21].
  - At least six amino acid transporters are involved in the exit of amino acids through the of the enterocyte. They may work in either direction depending on the luminal amino acid concentration and cellular demand, as well as on the electrogenic driving forces [17].

Intestinal uptake of peptides may also occur by paracellular transport through the intercellular junctions and by transcytosis [23].

13.5 Absorption of lipids

The World Health Organization recommends that lipids (Chapter 4) should supply up to 30% of the total energy uptake (E) in humans, with a minimum of 15% E; however, these nutrients account for an average of 42% E in the diets of developed countries.

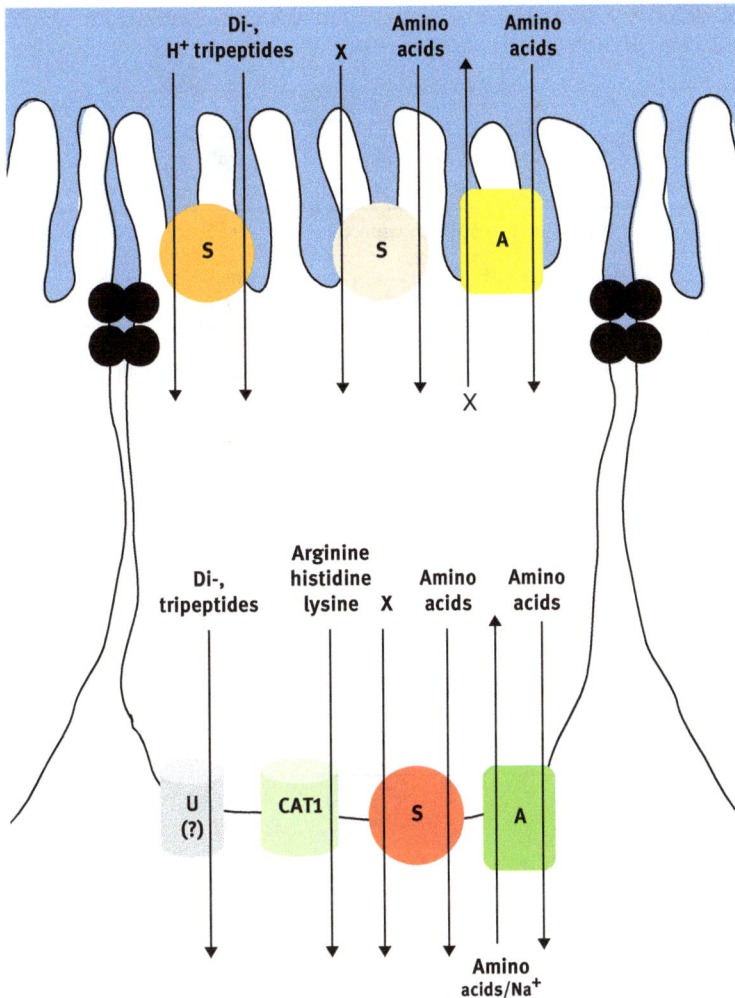

Figure 13.4: Overview of the mechanisms of absorption of amino acids and small peptides in enterocytes. A: antiporter, S: symporter, U: uniporter, X: several ions or amino acids. See Table 13.1 and main text for further details and references.

The vast majority of ingested lipids (around 95%) are triacylglycerols (TAGs), the rest being phospholipids (~4.5%) and steryl esters (~0.5%) [24, 25].

Dietary TAGs are initially digested in the stomach by the lingual and gastric lipases to form diacylglycerols (DAGs) and free fatty acids (FFAs); however, most of their digestion occurs in the duodenum through the action of pancreatic lipase, an enzyme that requires a co-lipase to efficiently exert its catalytic function. The hydrolysis of TAGs by the different lipases release FFAs and $sn2$-monoacylglycerols (MAGs), while phospholipids are hydrolyzed by pancreatic phospholipase A_2 yielding FFAs

Table 13.1: Characteristics of diverse di- and tripeptide and amino acid transporters expressed in human intestine. Source [21, 22] and https://www.ncbi.nlm.nih.gov/gene.

(A) Apical membrane

Coding gene	Transporter protein	Mechanism	Ion dependency	Substrate
Di- and tripeptide transporters				
SLC15A1	PEPT1	Symporter	H^+	Di- and tripeptide
SLC15A3	PHT2	Symporter	H^+	Histidine, di- and tripeptide
SLC15A4	PHT1	Symporter	H^+	Histidine, di- and tripeptide
CDH17	HPT-1	Symporter	H^+	Di- and tripeptide
Amino acid transporters				
SLC1A1	EAAT3/EAAC1	Antiporter	$AAA + 3Na^+ + H^+ \rightleftharpoons K^+$	Aspartic acid and glutamic acid
SLC1A5 (Expressed mainly in colon)	ASCT2/AAAT	Antiporter	$Na^+ + AA \rightleftharpoons Na^+ + AA$	Neutral amino acids primary substrates: alanine, asparagine, cysteine, glutamine, serine, and threonine
SLC7A9[1]	$b^{0,+}AT$	Antiporter	$CAA/cystine \rightleftharpoons NAA$	Cationic amino acids and cystine
SLC6A6	TauT	Symporter	Cl^- and $2Na^+$	β-alanine and taurine
SLC6A14 (Expressed in distal jejunum and colon)	$ATB^{0,+}$	Symporter	$2Cl^-$ and Na^+	Cationic and neutral amino acids
SLC6A19[2]	B^0AT1/HND	Symporter	Na^+	Neutral amino acids
SLC6A20[2]	SIT1	Symporter	Cl^- and $2Na^+$	Proline
SLC36A1	PAT1/LYAAT1	Symporter	H^+	β-alanine, glycine, and proline
SLC38A5	SN2/SNAT5	Antiporter	$AA + Na^+ \rightleftharpoons H^+$	Glutamine, asparagine, histidine, serine, alanine, and glycine

(B) Basolateral membrane

Coding gene	Transporter protein	Mechanism	Ion dependency	Substrate
Di- and tripeptide transporters				
Unknown	Basolateral peptide transporter	Uniporter	–	Di- and tripeptide

Table 13.1 (continued)

(B) Basolateral membrane

Coding gene	Transporter protein	Mechanism	Ion dependency	Substrate
Amino acid transporters				
SLC6A9	GlyT1	Symporter	Cl^- and $2Na^+$	Glycine
SLC7A1	CAT1	Uniporter	–	Arginine, histidine, and lysine
SLC7A6	y$^+$LAT2	Antiporter	$CAA \leftrightarrows NAA + Na^+$	Cationic amino acids
SLC7A7[3]	y$^+$LAT1	Antiporter	$CAA \leftrightarrows NAA + Na^+$	Cationic amino acids
SLC7A8[3]	LAT2	Antiporter	$NAA \leftrightarrows NAA$	Neutral amino acids
SLC7A10[3]	asc-1	Antiporter	$NAA \leftrightarrows NAA$	Small neutral amino acids
SLC38A2	SNAT2	Symporter	Na^+	Neutral amino acids and imino
SLC43A2	LAT4	Uniporter	–	Leucine, phenylalanine, valine and methionine
SLC16A10	TAT1	Uniporter	–	Aromatic amino acids

AAA: anionic amino acids; CAA: cationic amino acids; NAA: neutral amino acids.
(1) Forms a complex with the protein encoded by the gene SLC3A1.
(2) Requires the accessory protein angiotensin-converting enzyme 2 (ACE2) for expression and function.
(3) Form a heterotrimer with the heavy subunit transporter protein encoded by the gene SLC3A2.

and lysophospholipids (glycerophospholipids or sphingolipids – Figure 4.1 – that lack one acyl group). Sterol and FFAs are produced from steryl esters by several enzymes: cholesterol esterase, carboxylic ester hydrolase (also called carboxylic ester lipase), and sterol ester hydrolase [25].

The mechanism of digestion and adsorption of lipids (Figure 13.5) is influenced by their hydrophobicity, which forces them to organize in microscopic spherical particles in order to minimize contact with the aqueous secretions of the gastrointestinal tract (saliva, gastric juice, pancreatic juice, and others). This could hamper the action of hydrolyzing enzymes, therefore, lipid droplets are emulsified by the bile, a liver secretion that contains a family of cholesterol-derived natural detergents (bile salts), phospholipids, and free cholesterol molecules, among other substances. Bile salts are synthesized from cholesterol in the liver and account for about two-thirds of the solute mass of human bile, while phospholipids, mainly lecithin (phosphatidylcholine), comprise 15–25% of biliar lipids. It has been estimated that more than 90% of the phospholipids and cholesterol that enter the intestine comes from the bile. Emulsification not only enhances the efficiency of lipid digestion but also allows the formation of mixed micelles, which are composed of both dietary and biliary lipidic molecules, such as phospholipids, sterols, FFAs, MAGs, lysophospholipids, and fat soluble vitamins (see Chapters 4

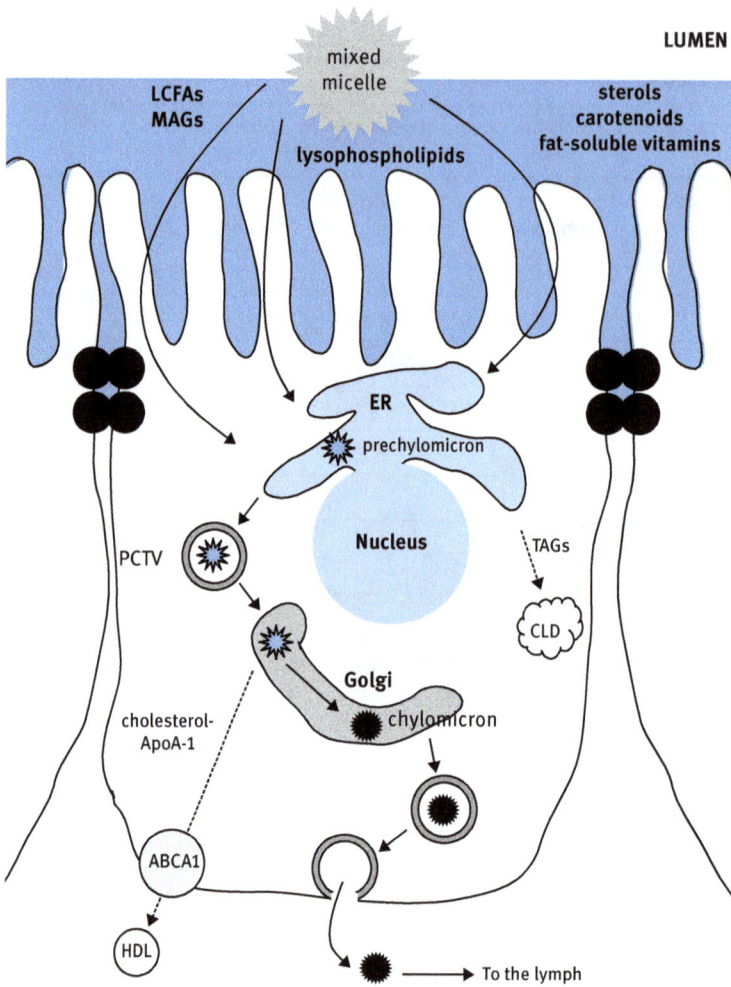

Figure 13.5: Mechanism of absorption of lipids in enterocytes. See main text for references and further details.

and 6 for chemical structures and metabolic roles). Mixed micelles move across the unstirred water layer toward the brush border (apical) membranes of enterocytes, where the acidic environment induces their dissociation and the protonation of fatty acids. This process facilitates the subsequent uptake of the different components of the micelles [8, 25].

13.5.1 Uptake of free long chain fatty acids, monoacylglycerols, and lysophospholipids by enterocytes

The side chains of fatty acids usually contain an even number of carbon atoms (Section 4.2) and can be classified into short chain (SCFAs, less than 8 carbon atoms), medium chain (MCFAs, 8–12 carbon atoms), long chain (LCFAs, 14–18 carbon atoms), and very long chain (VLCFAs, 20–26 carbon atoms) fatty acids, although slight variations in this classification can be found in the literature [10, 26, 27]. LCFAs are taken up from the intestinal lumen into enterocytes by two distinct mechanisms: (a) passive diffusion through the apical membrane when luminal concentrations are higher than those inside the cell, and (b) a saturable and probably protein-dependent mechanism, that occurs when the intracellular fatty acid concentration is higher than that in the lumen. Passive diffusion by the so-called flip-flop mechanism has been proposed to be the main route of LCFAs uptake by enterocytes and other cell types; however, some studies suggest that this process is too slow by itself and also requires protein facilitation in vivo. Altogether, several proteins have been reported to be involved in LCFAs and VLCFAs uptake in enterocytes [8, 26, 28, 29]:

- Fatty Acid Binding-Protein Plasma Membrane (FABPpm) and cluster of differentiation 36 (CD36) may act as acceptors for LCFAs prior to their diffusion across the apical membrane. CD36 is a multifunctional glycoprotein that acts as a receptor for a broad range of ligands other than LCFAs. These include certain proteins (e.g., collagen), oxidized low-density lipoproteins, anionic phospholipids, and bacterial diacylated lipopeptides [30]. Although a model whereby CD36 acts in concert with FABPpm in mediating FA uptake has been proposed, a more recent study has suggested that these two proteins act independently [26]. CD36 also seems to be implicated in the uptake of other lipidic molecules (see below).
- Fatty Acid Transport Protein 4 (FATP4) is an acyl-CoA synthetase with specificity towards VLCFAs and localized mainly at the endoplasmic reticulum (ER). Although FATP4 does not seem to be essential for FA absorption, it has been recently proposed to regulate lipid metabolism in metabolically active tissues [31]; therefore, its role in fatty acid uptake would be indirect.
- Caveolin-1 (Cav-1) is the main structural component of caveolae, that are specialized forms of lipid rafts with the membrane microdomains enriched in cholesterol and sphingolipids. Cav-1 facilitates the assembly of caveolae through interaction with cholesterol and FAs. The interaction between CD36 and Cav-1 seems to be important in the CD39-dependent internalization of LCFAs [26].

SCFAs and MCFAs can be passively diffused through enterocytes and enter the underlying portal vein as free fatty acids, where they bind to albumin; however, these types of fatty acids are not commonly found in food, therefore, this mechanism is not thought to contribute significantly to dietary fatty acids absorption. Fermentation of dietary fiber by commensal bacteria is the primary source of SCFAs, that are absorbed in the colon [10]. Monoacylglycerols seem to be taken up through similar mechanisms to those of LCFA transport, while lysophospholipids have been reported to be transported by passive diffusion as well as by caveolae-mediated endocytosis [17, 28, 32, 33].

13.5.2 Uptake of sterols and fat-soluble vitamins D, E, and K

Cholesterol in the intestinal lumen is present either as non-esterified cholesterol or as cholesteryl esters, both forms being originated from the diet or from the liver through the bile. Only non-esterified cholesterol can be absorbed by the intestine, while cholesteryl esters must be hydrolyzed in the lumen of the intestine to non-esterified cholesterol and free fatty acids. About 50% of the cholesterol in the intestine is absorbed and 50% is excreted in feces. Transport of cholesterol and other sterols into enterocytes is mediated by the apically localized Niemann-Pick C1-Like 1 (NPC1L1) permease, that is inhibited by the drug ezetimibe [34]. NPC1L1 has been proposed to function as an endocytic transporter of cholesterol; however, studies in rats suggest that endocytosis is not required for NPC1L1-mediated cholesterol uptake. Scavenger receptor class B type I (SR-BI) is also highly expressed on the apical membrane of enterocytes, although its importance in cholesterol uptake is controversial. SR-BI has an essential role in the uptake of fat-soluble vitamins and carotenoids and may also act as a lipid sensor [28]. After entering the enterocyte, unesterified cholesterol may be pumped back to the intestinal lumen by a heterodimeric sterol efflux transporter (ABCG5/ABCG8), but the majority of cholesterol is esterified in the endoplasmic reticulum and incorporated into chylomicrons (see below and Chapter 4) [35–37]. CD36 has been reported to play a role in cholesterol uptake at the proximal intestine in experiments performed in mice [38, 39].

Vitamins E and D apical transport is facilitated by membrane proteins including SR-BI, CD36 and NPC1L1. A fraction these vitamins can be effluxed back to the intestinal lumen via apical membrane transporters (SR-BI and possibly other transporters for vitamin E, and ATP binding cassette transporter B1 -ABCB1- and ABCG5/ABCG8 for vitamin D). At supplemental/pharmacological doses, these vitamins are assumed to enter the enterocyte via passive diffusion [40–43]. A recent study suggests that vitamin K absorption and excretion are regulated by NPC1L1 and ABCG5/ABCG8 [44].

13.5.3 Uptake of vitamin A and carotenoids

In the human diet, most preformed vitamin A (VA) occurs as retinyl palmitate (about 65% of total VA intake), while β-carotene is the most abundant pro-vitamin A carotenoid (30% of VA uptake). The remaining 5% are other compounds like α-carotene or β-cryptoxanthin. Dietary retinyl palmitate is incorporated into mixed micelles and hydrolyzed to all-*trans* retinol. This step seems to be necessary for absorption and is catalyzed by pancreatic lipase and pancreatic lipase-related protein-2, although phospholipase B, an enzyme associated with the brush border membrane, might also be involved. β-carotene is not significantly metabolized and stays as such in mixed micelles [45].

All-*trans* retinol is introduced into the enterocyte, at least in part, by a saturable protein-mediated passive absorption mechanism, however, the protein that mediates this process has not been identified so far. By contrast, several proteins have been reported to be involved in carotenoid intestinal absorption, these include: SR-BI (lutein, lycopene, β-carotene, α-carotene, β-cryptoxanthine, phytoene, and phytofluene), CD36 (lycopene, β-carotene, α-carotene, β-cryptoxanthine, and lutein), and NPC1L1 (α-carotene, β-carotene, β-cryptoxanthin, and lutein). Depending on the concentration in the lumen, a fraction of carotenoids might be absorbed via a passive diffusion process, while observations that some of these molecules could be effluxed back to the lumen have also been reported. In any case, many aspects of carotenoid intestinal uptake remain to be established [45, 46].

13.5.4 Intracellular processing of dietary lipids within the enterocyte

LCFAs taken up by the enterocyte are transported to the endoplasmic reticulum (ER) by intracellular fatty acyl CoA transporters FABP-1 (or liver (L)-FABP) and FABP-2 (or intestinal (I)-FABP). In the ER, they are re-esterified into TAGs, either by the monoacylglycerol or by the glycerol phosphate pathways. Free cholesterol is converted to cholesteryl ester (CE) and transported to the ER by the acyl CoA:cholesterol acyltransferase enzyme 2 (ACAT2).

In the ER, the newly synthesized protein ApoB-48 is lipidated with TAGs, phospholipids, and CE, by microsomal triglyceride transfer protein (MTP). Prechylomicrons are generated, also in the ER, by further addition of TAGs, cholesterol, and the protein apoA-IV. Prechylomicrons are delivered to the Golgi apparatus in the so-called prechylomicron transport vesicles (PCTVs) where they acquire ApoA-1. Mature chylomicrons are exported from the Golgi in vesicles and released by exocytosis at the basolateral surface of the enterocyte into the lamina propria and eventually into the circulation via the lymphatic system. Once in the circulation, chylomicrons acquire other apolipoproteins such as apoC2, apoC3, and apoE. Cholesterol can also be secreted, along with ApoA-1 and phospholipids, as a nascent high-density lipoprotein particle (HDL), by the transporter ABCA1 [34, 47]. More information on chylomicrons, HDL, and other lipoproteins can be found in Chapter 4.

Lysophospholipids, primarily lysophosphatidylcholine (LPC), that are internalized in association with caveolin-1 containing endocytic vesicles, play important roles both in intracellular lipid trafficking and in chylomicron assembly and transport. Re-esterification of LPC to phosphatidylcholine seems to be necessary to accomplish these tasks [48].

Dietary lipids may also be stored within the so-called cytoplasmic lipid droplets (CLDs), which accumulate in enterocytes in response to consumption of high levels of dietary fat, or in diseases where chylomicron synthesis or secretion are inhibited. CLDs are nowadays considered dynamic organelles that have multiple cellular functions beyond lipid metabolism [49].

Intracellular vitamin E might be associated to lipid droplets, lysosomes, ER membrane and/ or specific binding proteins. The major fraction of vitamin E is secreted in the lymph into chylomicrons, while a minor part may also be secreted at the basolateral side via ABCA1 [40]. Provitamin A carotenoids are partly metabolized within the enterocyte into retinyl-esters and incorporated to chylomicrons, along with the rest of carotenoids. Some proteins may be involved in intracellular transport of carotenoids, but none has been clearly identified. A fraction of vitamin E and polar carotenoids, like xanthophylls, may be secreted via the nascent HDL pathway mediated by ABCA1, as explained above for cholesterol [46].

A recent review on the digestion, intestinal mucosa metabolism, absorption and effects on health of fat soluble vitamins and phytochemicals can be found elsewhere [50].

13.6 Absorption of water soluble vitamins

Although many aspects of the absorption of water soluble vitamins are unknown, in some cases the molecular identity of the systems involved, their cell biology, and their regulation have been delineated. Table 13.2 summarizes the current knowledge of transporters and mechanisms involved in the intestinal absorption of these essential nutrients.

13.7 Absorption of minerals

13.7.1 Water and electrolytes (Figure 13.6)

The small intestine must absorb high quantities of water. A normal person usually drinks 1 to 2 L of dietary fluid every day, but the small intestine receives another 6 to 7 L of fluids as secretions from salivary glands, stomach, pancreas, liver, and the small intestine itself. Approximately 80% of this fluid is absorbed before reaching the large intestine.

The current view is that net movement of water across cell membranes occurs by osmosis, that is, water diffuses into the intracellular space in response to the osmotic gradient established by electrolytes, mainly sodium, chloride, and potassium, and osmolytes. Figure 13.6 shows an overview of the systems involved in water and electrolytes transport in enterocytes. Water molecules may enter the cell by simple or

Table 13.2: Transporters and molecular mechanisms involved in the intestinal absorption of water-soluble vitamins. (A) Apical membrane, (B) Basolateral membrane, and factors and conditions that negatively impact uptake. Table elaborated with data obtained from references [51–56] and https://www.ncbi.nlm.nih.gov/gene.

(A)			**Apical membrane**
Vitamin	**Sources (absorptive cell)**	**(Protein/gene)**	**Transport mechanism**
Thiamine (B$_1$)	Diet (enterocyte)	THTR1/SCL19A2	Free thiamine/H$^+$ antiport
		THTR2/SCL19A3	Free thiamine/H$^+$ antiport
		(MTPPT/SLC25A19 [1][2])	Thiamine-pyroph./? antiport
		OCT1/SLC22A1	Organic cation /H$^+$ cotransport
		OCT3/SLC22A3	Organic cation /H$^+$ cotransport
		OCTN1/SLC22A4	?
		OCTN2/SLC22A5	?
	Gut microbiota (colonocyte)	THTR1/SCL19A2	H$^+$ antiport
		THTR2/SCL19A3	H$^+$ antiport
		TPPT/SLC44A4 [2]	pH and Na+-independent, and energy-dependent transport
Riboflavin (B$_2$)	Diet (enterocyte) ——— Gut microbiota (colonocyte)	RFVT3/SLC52A3	Na$^+$-independent, H$^+$-dependent
Niacin (B$_3$)	Diet (enterocyte) ——— Gut microbiota (colonocyte)	SMCT1/SLC5A8	Na$^+$-coupled transport
		SMCT2/SLC5A12	Na$^+$-coupled transport
		MCT1/SLC16A1	H$^+$-coupled transport
Pantothenic acid (B$_5$)	Diet (enterocyte) ——— Gut microbiota (colonocyte)	SMVT/SLC5A6	Na$^+$-dependent carrier-mediated
Pyridoxine and derivatives (B$_6$)	Diet (enterocyte) ——— Gut microbiota (colonocyte)	THTR1/SCL19A2	H$^+$ antiport
		THTR2/SCL19A3	H$^+$ antiport

Table 13.2 (continued)

(A)		Apical membrane	
Vitamin	**Sources (absorptive cell)**	**(Protein/gene)**	**Transport mechanism**
Biotin (B$_7$)	Diet (enterocyte) ——— Gut microbiota (colonocyte)	SMVT/SLC5A6	Na$^+$-dependent carrier-mediated
Folate (B$_9$)	Diet (enterocyte) ——— Gut microbiota (colonocyte)	PCFT/SLC46A1 RFC/SLC19A1	Folate$^-$/H$^+$ symport Reduced folate/anion antiport
Cobalamin (B$_{12}$)	Diet (terminal ileon)	Haptocorrin/TCN1 (saliva) IF (intrinsic factor)/CBLIF Cubam complex (Cubilin/CUBN + amnionless/AMN)	Binds to vit B$_{12}$ in mouth Binds to vit B$_{12}$ in small intestine Cubilin-mediated endocytosis of cobalamin–IF complexes
Vitamin C	Diet (enterocyte)	SVCT1/ SLC23A1 (For ascorbic acid) GLUT2/SLC2A2 GLUT8/SLC2A8 (For dehydroascorbic acid)	Na$^+$-cotransport Facilitated diffusion Facilitated diffusion

(B)			Basolateral membrane	
Vitamin	**Sources (absorptive cell)**	**(Protein/ gene)**	**Transport mechanism**	**Factors and conditions that negatively impact uptake**
Thiamine (B1)	Diet (enterocyte)	THTR1/ SCL19A2 [2] RFC1/ SLC19A1	pH-dependent, electroneutral Thiamine monophosphate	– Chronic alcohol consumption – *E. coli* – Sepsis
	Gut microbiota (colonocyte)	THTR1/ SCL19A2 [2]	pH-dependent, electroneutral	

Table 13.2 (continued)

(B)			**Basolateral membrane**	
Vitamin	**Sources (absorptive cell)**	**(Protein/ gene)**	**Transport mechanism**	**Factors and conditions that negatively impact uptake**
Riboflavin (B2)	Diet (enterocyte)	RFVT1/ SLC52A1 RFVT2/ SLC52A2 [3]	Carrier-mediated	– Chronic alcohol consumption – Amiloride (Na^+/H^+ exchanger) – Chlorpromazine
	Gut microbiota (colonocyte)	RFVT1/ SLC52A1		
Niacin (B3)	Diet (enterocyte)	?	?	
	Gut microbiota (colonocyte)			
Pantothenic acid (B5)	Diet (enterocyte)	?	?	
	Gut microbiota (colonocyte)			
Pyridoxine and derivatives (B6)	Diet (enterocyte)	THTR1/ SCL19A2 (?)		
	Gut microbiota (colonocyte)			
Biotin (B7)	Diet (enterocyte)	?	Carrier-mediated, sodium-independent process	Chronic alcohol consumption Lipopolysaccharide at the outer membrane of Gram-negative bacteria Anticonvulsant drugs
Folate (B9)	Diet (enterocyte)	MDR3/ ABCB4	Multidrug-resistance-associated proteins	Chronic alcoholism
	Gut microbiota (colonocyte)			

Table 13.2 (continued)

(B)			Basolateral membrane	
Vitamin	Sources (absorptive cell)	(Protein/ gene)	Transport mechanism	Factors and conditions that negatively impact uptake
Cobalamin (B12)	Diet (terminal ileon)	MRP1/ABCC1	ATP-driven efflux of cobalamin after degradation of IF in the ER (cobalamin binds to transcobalamin in the bloodstream)	– Lack of functional IF – Atrophic body gastritis – H. pylori infections – Inhibitors of H^+/K^* ATPase – Impaired gastric function – Inflammatory bowel diseases (including Crohn's disease)
Vitamin C	Diet (enterocyte)	SVCT2/ SLC23A2	Na^+-cotransport	

(1) Located at the mitochondrial membrane.
(2) Thiamine-pyrophosphate transporter.
(3) Also localized inside intracellular vesicles.

facilitated diffusion, the latter being mediated by aquaporins, as well as by some ion channels or solute transporters (Chapter 2). Paracellular transport through the tight junctions, which is also driven by osmotic gradient, also contributes to water absorption. The latter mechanism is essential for the uptake of other nutrients by the so-called solvent drag effect, that is, the bulk movement of small soluble solutes along with the absorbed water [20, 57]. Electrical potential differences also affect the movement of ions by this mechanism.

After crossing the intestinal epithelium, water molecules and solutes can diffuse into the capillary blood within the villus.

Several mechanisms can contribute to apical Na^+ transport in both intestines [17, 57]:

–   Nutrient-coupled Na^+ absorption mediated by several families of Na^+-dependent transporters such as sugar or amino acid transporters (Sections 13.3 and 13.4), as well as other transporters like the Na^+/bile acid cotransporter ASBT (encoded by the gene SLC10A2) that mediates bile acid reabsorption in the terminal ileum, or some Na^+/vitamin cotransporters (Table 13.2).
–   Electroneutral absorption via the Na^+/H^+ exchange mechanism mediated primarily by members of the SLC9 family of Na^+/H^+ exchangers (NHE). Isoforms NHE2 and NHE3 have been identified on the enterocyte apical membrane, the latter being the most significant contributor to the intestinal Na^+ and water absorption. The same NHEs are expressed on the apical surface of colonocytes. NHE1 is found

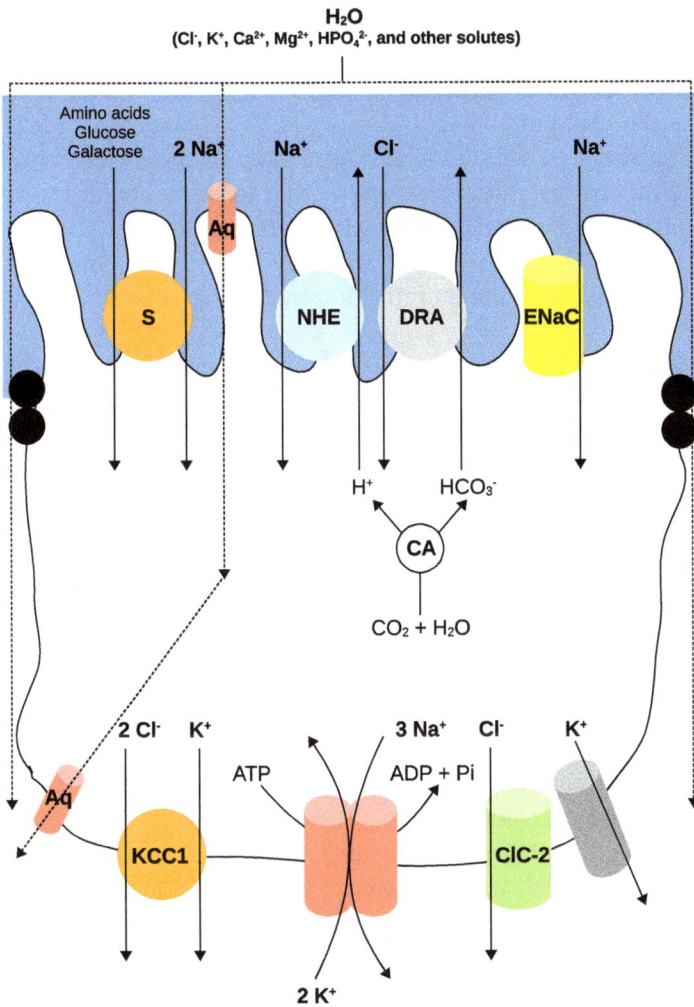

Figure 13.6: Some transporters involved in the absorption of water and electrolytes in enterocytes. Aq: aquaporin; S: Na^+/sugar or amino acid symporter; DRA: Cl^-/HCO_3^- exchanger; CA: carbonic anhydrase. Further details and references in the main text.

in the basolateral membrane of intestinal epithelial cells. This mechanism of Na^+ uptake is influenced by H^+ movements, that are in turn driven by proteins like H^+/peptide cotransporters (Table 13.1) or the H^+/monocarboxylate cotransporter MCT1 (encoded by SLC16A1).

– Electrogenic Na^+ absorption by the epithelial Na^+ channels (ENaC), composed of three subunits, α, β, and γ, encoded by the genes SCNN1A, SCNN1B, and SCNN1G, respectively.

–   Na$^+$/HCO$_3^-$ cotransporters (NBCs), found throughout the intestine, but mainly in the duodenum and colon.

Absorbed sodium is rapidly exported from the cell via the Na$^+$/K$^+$-ATPase, so that a high concentration of this cation can be achieved in the narrow space between enterocytes. A potent osmotic gradient is thus formed across apical cell membranes and their connecting junctional complexes, that osmotically drives water movement across the epithelium.

Chloride can be absorbed from the intestinal lumen via three distinct mechanisms (Figure 13.6) [17]:

–   Paracellular pathway by solvent drag, predominant in the jejunum.
–   Electroneutral pathways, the main route of absorption in the ileum and proximal colon with less prominence in the distal colon, involves coupled Na$^+$/H$^+$ and Cl$^-$/HCO$_3^-$ exchange. In this pathway, the efflux of H$^+$ alkalizes the cytoplasm, thus activating the Cl$^-$/HCO$_3^-$ exchanger. Carbonic anhydrase (CA) is the main producer of H$^+$ and HCO$_3^-$.
–   HCO$_3^-$-dependent Cl$^-$ absorption not coupled to a parallel Na$^+$/H$^+$ exchanger.

The process of Cl$^-$/HCO$_3^-$ exchange across the cell plasma membrane is mediated by multiple chloride and bicarbonate transporting anion exchangers, like DRA (down-regulated in adenoma) and PAT1, the latter being capable of several exchange modes: Cl$^-$/oxalate, SO$_4^{2-}$/oxalate, SO$_4^{2-}$/Cl$^-$, Cl$^-$/formate, and Cl$^-$/OH$^-$ [17]. Chloride export across the basolateral membrane of enterocytes and colonocytes has been proposed to be mediated by the potassium chloride cotransporter-1 (KCC1/SLC12A4) [58] and by ClC-2 chloride channels (CLCN2), with Kir 7.1 potassium channels maintaining the driving force for chloride exit [59]. Multiple isoforms of the anion exchanger (AE) gene family (also known as SLC4) are expressed in the small and large intestine, some of them with a basolateral localization. It has been proposed that these transporters may contribute to the basolateral chloride uptake by working in concert with either NBCs or NHE1 [57].

The vast majority of intestinal K$^+$ absorption occurs in the small intestine by passive paracellular diffusion. This occurs because K$^+$ concentration in the intestinal lumen is higher than in the extracellular fluids, allowing a large concentration gradient across the tight junctions. On the other hand, in apical and basolateral membranes of enterocytes, there are several K$^+$-channels and K$^+$-dependent transporters that are involved in the regulation of membrane potential and transport of other solutes. BK channels (KCNMA1 or KCa1.1) have been implicated in apical K$^+$ secretion, while calcium-dependent KCNQ channels seem to be important for colonic K$^+$ secretion, at least in rats. In the basolateral membrane KCNN4 and KCNQ1 (with its regulatory subunit, KCNE3), and perhaps others, are candidates for colonic K$^+$ channels. It is worth mentioning the importance of the Na$^+$/K$^+$-ATPase (and the H$^+$/K$^+$-ATPase in colonocytes) for nutrient absorption, that links the movements of K$^+$ in and out of intestinal cells with those of Na^+ and other electrolytes, which in turn are influenced by nutrients such as glucose, peptides, and amino acids [57, 60, 61].

The small and large intestines are also engaged in fluid secretion in order to lubricate the mucosa and prevent epithelial damage. Fluid secretion is driven predominantly by electrogenic chloride secretion, which is performed mainly by cells residing in the crypts. The secretory mechanism relies mainly on the apical chloride channel known as the cystic fibrosis transmembrane conductance regulator (CFTR), although other accessory chloride channels may also participate to a lesser degree. Chloride anions destined for secretion must be transported from the bloodstream to the intestinal lumen across the basolateral membrane in association with Na^+ and K^+ via the cotransporter NKCC1. K^+ is also recycled across the basolateral membrane by several potassium channels in order to sustain the driving force for apical chloride exit. Water and sodium ions follow chloride movements paracellularly and/or transcellularly. Alterations by endogenous and exogenous factors in the transport of water and electrolytes across the intestinal epithelium may underlie diarrheal symptoms [57, 59].

The processes described above must act in a concerted and finely tuned manner to control the fluidity of the intestinal contents, conserve the large amounts of fluid that enter the intestine each day, and maintain electrolytes homeostasis. Moreover, intestinal epithelial cells must adapt to both physiological and pathophysiological cues through a complex interplay of cell-to-cell interactions, intercellular mediators, and intracellular signal transduction cascades. An updated review of these regulatory mechanisms can be found elsewhere [57].

13.7.2 Calcium (Figure 13.7)

Calcium absorption may occur via two mechanisms [62, 63] (Figure 13.7):
- Non-saturable, paracellular diffusive pathway, which predominates at normal to high dietary calcium intake. The paracellular pathway is the major route of absorption or secretion in the jejunum and ileum, although is observed in all segments of the intestine (including the colon). A family of tight junction proteins called claudins, contribute epithelial ion selectivity via the paracellular pathway. There is some evidence that this pathway is also vitamin D sensitive in the human ileum.
- Saturable transcellular process, which occurs at lower luminal calcium levels against a concentration gradient, driven by non-voltage-gated Ca^{2+} channels expressed in the apical membranes (TRPV6 and, to a lesser extent, TRPV5) of the enterocytes. The saturable pathway is present in the proximal small intestine (duodenum and jejunum), cecum, and colon but is absent in the ileum. Another channel located in the apical membrane, $Ca_v1.3$, is thought to provide another mechanism of active transport when luminal Ca^{2+} is in abundance.

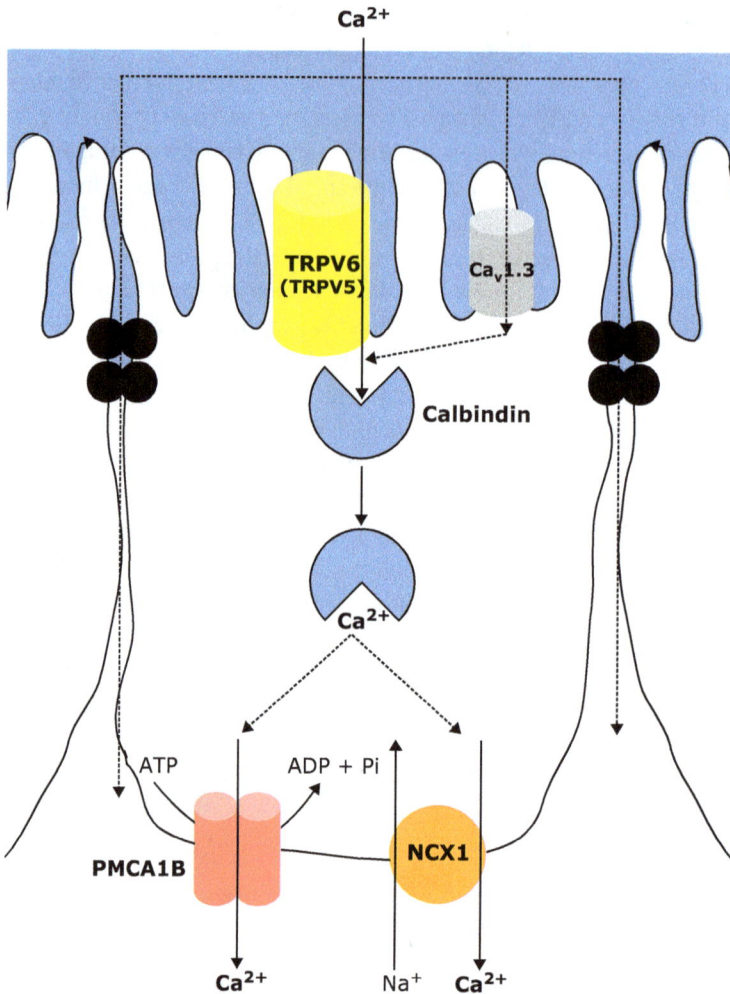

Figure 13.7: Mechanism of absorption of calcium in enterocytes. See main text for references and further details.

The relative importance of these pathways depends upon Ca^{2+} intake, thus, when it is low the saturable pathway predominates, while the bulk of absorption occurs through the paracellular pathway when Ca^{2+} intake is high.

Within the enterocyte, Ca^{2+} binds to calbindin-D9k, thus maintaining a free intracellular concentration below 10^{-7} mol/l, and preventing epithelial cell apoptosis. Calbindin carries Ca^{2+} to the basolateral membrane, where it will exit the enterocyte by means of two different mechanisms: pumping by the Ca^{2+}-ATPase PMCA1b (~80% contribution), and exchange by Na^+, a process mediated by the Na^+/Ca^{2+} exchanger NCX1 (~20% contribution) [17, 64].

Calcitriol is the major controlling hormone of intestinal Ca^{2+} transport that acts by increasing the expression of most of the proteins involved in both pathways (see Chapter 6 for further details). Other hormones have also been reported to be involved in this process, such as parathyroid hormone (PTH), estrogens, prolactin, growth hormone, and glucocorticoids, as well as prebiotics and dietary factors. Different physiological conditions, such as growth, pregnancy, lactation, and aging, adjust intestinal calcium absorption according to demands. All these factors participate in the regulation of calcium adsorption in both the small and the large intestine [62–64].

13.7.3 Phosphorus (Figure 13.8)

The majority of phosphate is absorbed in the jejunum, although the relative efficiency of absorption is greater in the duodenum. It occurs via two routes (Figure 13.8):

– Passive diffusion process through the paracellular pathway, which operates during high intake of phosphate. Inhibition of the epithelial sodium hydrogen exchanger (NHE3, see above) decreases intestinal phosphate absorption by reducing paracellular phosphate permeability. It has been recently postulated that this is the predominant pathway of intestinal phosphate absorption in humans [65].

– Transcellular, sodium-dependent, phosphate absorption is secondarily active and utilizes the sodium concentration gradient established by the Na^+/K^+-ATPase. It is driven by the apical Na^+/phosphate transporter NaPi-IIb, which belongs to a family of SLC34 solute carriers. Type III Na^+-phosphate transporters PIT1 and PIT2 may also play a minor role [17, 65].

The exit of phosphate across the basolateral membrane (BLM) is likely to occur by facilitated diffusion via unknown phosphate transporters. The human protein XPR1 has been proposed as the phosphate export transporter [66]. PIT1 and PIT2 can also mediate transport of phosphate into the enterocyte through the BLM during dietary insufficiency [17]. Dietary phosphate, calcitriol, and PTH are thought to be the most important physiological regulators of intestinal phosphate absorption. A recent study that demonstrates that phosphate is sensed by the parathyroid gland through the calcium sensing receptor (CaSR) provides a mechanism to explain the PTH dependence [67]. Other hormones, such as the epidermal growth factor (EGF), glucocorticoids, estrogens, metabolic acidosis, and fibroblast growth factor (FGF)-23, have been reported to decrease Na^+-dependent intestinal phosphate absorption [17, 68]. A variety of potential mechanisms for the maintenance of phosphate homeostasis have been postulated. These mechanisms are based on the communication between the different organ systems and cell types, actually, the hormones responsible for phosphate homeostasis come from the parathyroid gland, bone, and kidney. The liver might also be the organ that senses phosphate availability and sends signals to the other organs involved in phosphate homeostasis. In any case, our understanding of phosphate homeostasis

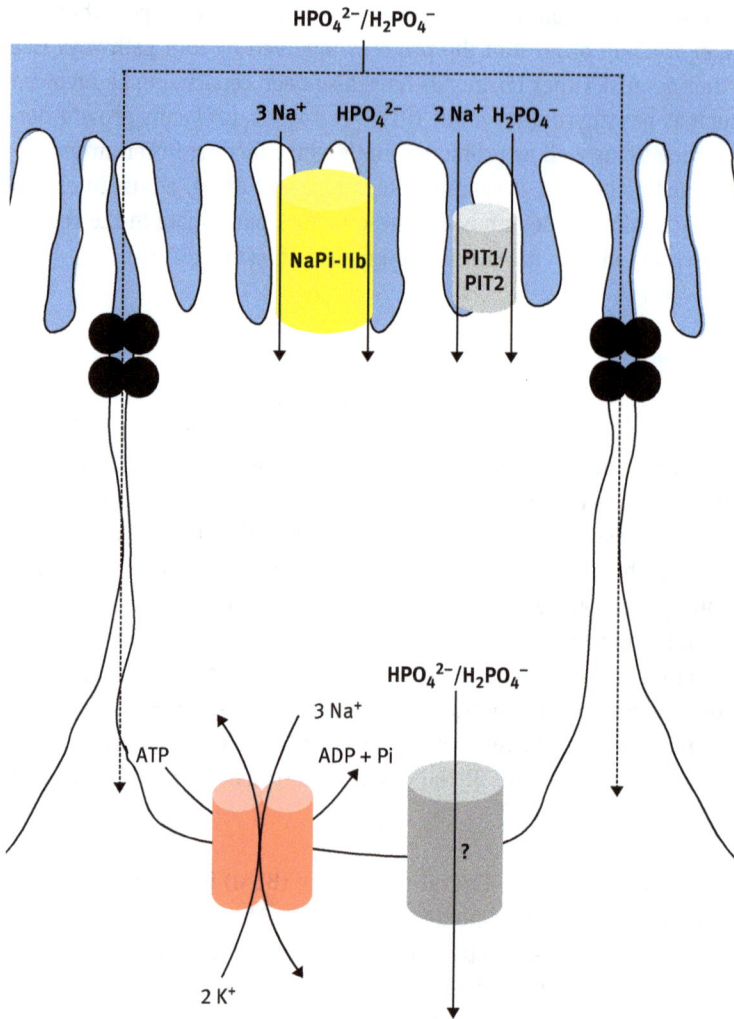

Figure 13.8: Mechanism of absorption of phosphate in enterocytes. See main text for references and further details.

and its regulation, and the transporters responsible for ensuring cellular and organism-level phosphate balance remains incomplete [66].

13.7.4 Magnesium (Figure 13.9)

Recent studies have revealed that the duodenum, jejunum, and ileum absorb Mg^{2+} through both transcellular and paracellular routes. Paracellular movement of Mg^{2+} is driven by both concentration gradient and by the solvent drag effect, which also occurs across the colonic epithelium. Transcellular Mg^{2+} absorption is driven by TRPM6 and TRPM7 homodimer channels, whose activities are negatively regulated by physiological Mg·ATP and Mg^{2+} levels. The expression of a heterodimer TRPM6/7 channel in the plasma membrane has also been reported in the duodenal and jejunal epithelium; however, heterodimer channels do not respond to physiological intracellular Mg^{2+} and Mg·ATP [17, 69, 70] (Figure 13.9).

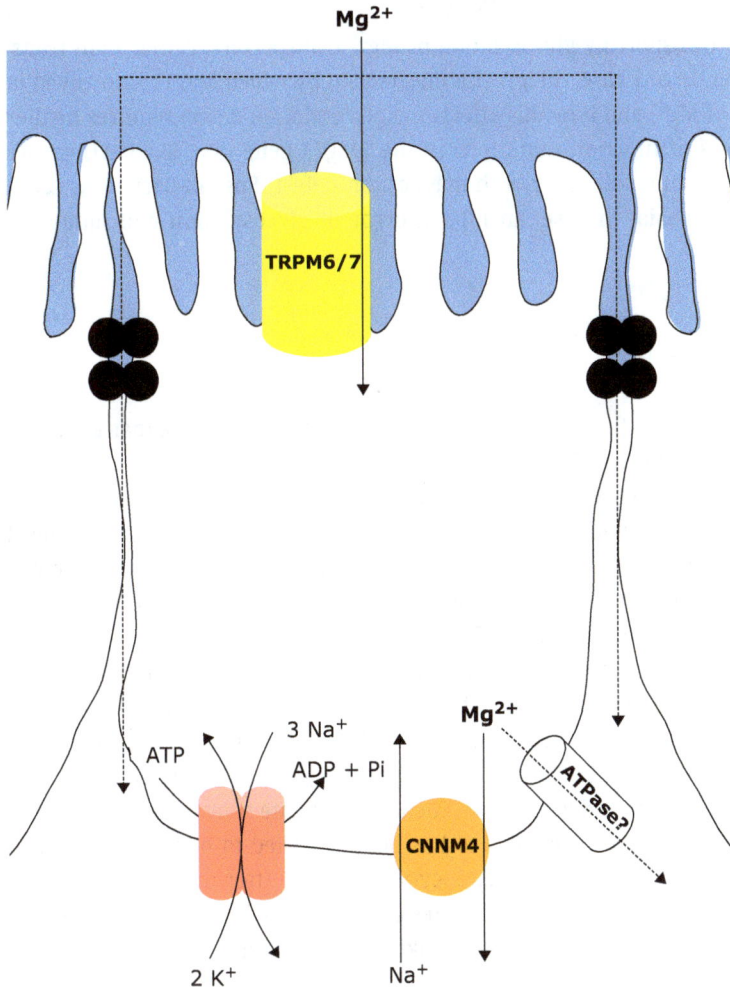

Figure 13.9: Mechanism of absorption of magnesium in enterocytes. See main text for references and further details.

The basolateral Mg^{2+} extrusion mechanism of the enterocyte has been suggested to occur via CNNM4, a Na^+/Mg^{2+}- antiporter, the driving force being established via the Na^+/K^+-ATPase. The occurrence of a Mg^{2+}-ATPase has also been postulated but not confirmed in humans [17, 70]. Parathyroid hormone (PTH), fibroblast growth factor-23 (FGF-23), luminal protons, pH-sensing channels and receptors, proton pump inhibitors like omeprazol, and the microbiome have been recently proposed as regulatory factors of small intestinal Mg^{2+} uptake [71].

Another Mg^{2+} transporter, MagT1, was initially described as a ubiquitously expressed Mg^{2+} channel involved in magnesium homeostasis; however, this protein was later identified as a part of the oligosaccharyltransferase complex, that functions in the N-glycosylation pathway in the endoplasmic reticulum. MagT1 has been proposed to regulate Mg^{2+} transport by glycosylation of Mg^{2+} transporters, rather than transporting Mg^{2+} itself. In any case, the precise mechanism by which MagT1 is involved in the homeostasis of Mg^{2+} and how this affects the glycosylation defect requires further investigation. Loss of function mutations in the MagT1 gene (aka SLC58A1) causes XMEN (X-linked immunodeficiency with magnesium defect, Epstein-Barr virus (EBV) infection, and neoplasia) disease, an inborn error of glycosylation and immunity [72, 73].

13.7.5 Iron (Figure 13.10)

Dietary iron exists in two forms, non-heme and heme iron (Chapter 7), that are taken up by independent pathways (Figure 13.10):

–   Non-heme iron is transported into intestinal epithelial cells by the divalent metal transporter 1 (DMT1), which is a proton-coupled transporter located on the apical membrane. The ferric form (Fe^{3+}) has to be reduced to the absorbable ferrous form (Fe^{2+}) by the duodenal brush border cytochrome b reductase 1 (DCYTB), as well as by dietary ascorbic acid [17, 74].
–   About 2/3 of total iron absorbed by the intestine is incorporated in heme; however, the mechanism of its transport into the enterocyte has not been clearly elucidated. Two proteins have been proposed to be involved in heme uptake: heme carrier protein 1 (HCP1/PCFT) and heme responsive gene-1 (HRG-1). HCP1 exhibits high affinity for folate and is likely to function as a folate transporter (Table 13.2). HRG-1 has high affinity for heme and has been reported to be involved in iron transport into the cytosol via the endocytic pathway. After uptake, heme is degraded by heme oxygenases and generates carbon monoxide (CO), biliverdin and ferrous iron (Fe^{2+}), which is metabolized like non-heme iron [74–76].

The iron chaperone poly C binding protein 2 (PCBP2) is considered to be responsible for delivering Fe^{2+} from the apical to the basolateral side of the enterocytes, due to its capacity to bind both to DMT1 and to ferroportin (FPN), the iron exporter at the basolateral membrane. Alternatively, PCBP2 can transfer Fe^{2+} to ferritin, where it is oxidized to Fe^{3+} by the H-subunit, for storage. Fe^{2+} cations exported out of the enterocyte by FPN are rapidly oxidized to Fe^{3+} by hephaestin, a multicopper ferroxidase, and

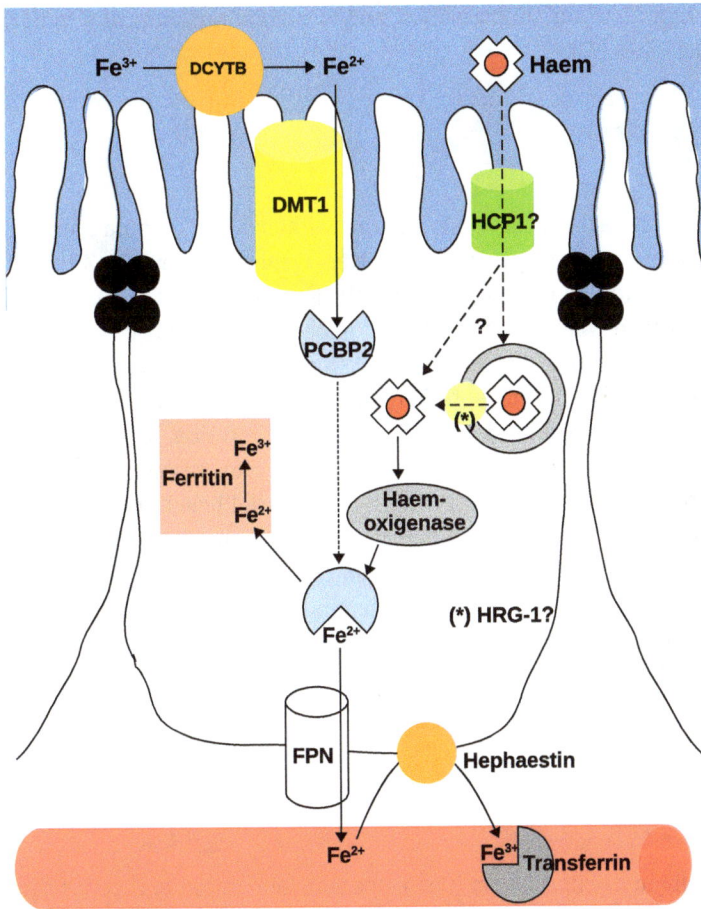

Figure 13.10: Mechanism of absorption of iron in enterocytes. See main text for references and further details.

then bound to transferrin for delivery to various tissues via circulation. FPN expression is regulated by hepcidin-mediated degradation. Hepcidin is a peptide hormone, synthesized and secreted by the liver, that binds directly to FPN and promotes its internalization and degradation. High iron levels stimulate hepcidin expression, thereby leading to a reduction of plasma iron; by contrast, hepcidin level is decreased under iron-deficient condition, thus maintaining FPN expression and iron delivery to the plasma [74].

Lactoferrin, an iron-binding glycoprotein present in milk, may act as an iron supplier to intestinal cells, although the mechanism of this action is still not clear. Hydrolysis of lactoferrin is minimal in infants, due to gastric pH, and, therefore, it probably has greater biological potential in infants than in adults. Lactoferrin has been reported to modulate immunity and inflammation, and to exert antioxidant, anti-tumor, and antimicrobial activities [77].

13.7.6 Zinc (Figure 13.11)

Zinc absorption occurs throughout the small intestine, although the jejunum seems to be the most absorptive region. Several zinc transporters (generally known as ZIPs and ZNTs) have been reported to be expressed in intestinal epithelial cells (Figure 13.11). ZIP4, located at the apical membrane, is considered to be the main transporter of dietary Zn^{2+} into the enterocyte. Zn^{2+} might also bind to cytoplasmic proteins such as metallothionein (MT), that acts as a metal buffer. The gene coding for MT is regulated by copper and zinc, thus establishing a link between the absorption, and hence the metabolism, of both metal ions. ZnT-1 is an efflux transporter expressed in the basolateral surface of enterocytes that transports Zn^{2+} into the portal blood. Other ZIP and ZNT proteins may contribute to trafficking of zinc, thus, ZIP-5 and ZIP-14 are localized on the basolateral surface of enterocytes and antagonize ZnT-1 and ZIP-4 functions by importing zinc from the blood circulation into the enterocytes. ZnT-2 and ZnT-4 to ZnT-7 are expressed in the structures of enterocytes, such as vesicles, secretory granules, endosomes, lysosomes, and Golgi complex where they facilitate zinc influx [74, 78, 79].

13.7.7 Copper (Figure 13.11)

Dietary copper, usually in its cupric form (Cu^{2+}), needs to be reduced to cuprous copper (Cu^+) before uptake across the apical membrane of intestinal epithelial cells at the upper small intestine. Enterocytes produce reductases such as steap2/3/4 to reduce cell surface Cu^{2+} to Cu^+ [80]. Cu^+ is supposedly transported across the apical membrane by copper transporter 1 (CTR1), however, the localization of this protein in intestinal cells is unclear, as some experimental evidences suggest that CTR1 is localized primarily at the basolateral membrane. Some authors have proposed that members of other transporter families, unidentified so far, might contribute to copper absorption. After uptake, Cu^+ is bound by the chaperone antioxidant-1 (ATOX1), which mediates its transfer to the copper-transporting Cu^+-ATPases ATP7A and ATP7B. Under low-copper conditions, ATP7A resides in the trans-Golgi network, but traffics to the basolateral membrane when intracellular copper concentration increases, then it pumps Cu^+ into portal circulation. This mechanism provides the primary route of copper efflux from enterocytes. ATP7B sequesters copper in vesicles whose role and final destination are unknown, although it is possible that this copper fraction may be released from the vesicles if needed [74, 81] (Figure 13.11). Divalent metal transporter 1 (DMT1) has been reported to transport Cu^{2+} or Cu^+ into the enterocyte; however, stud-

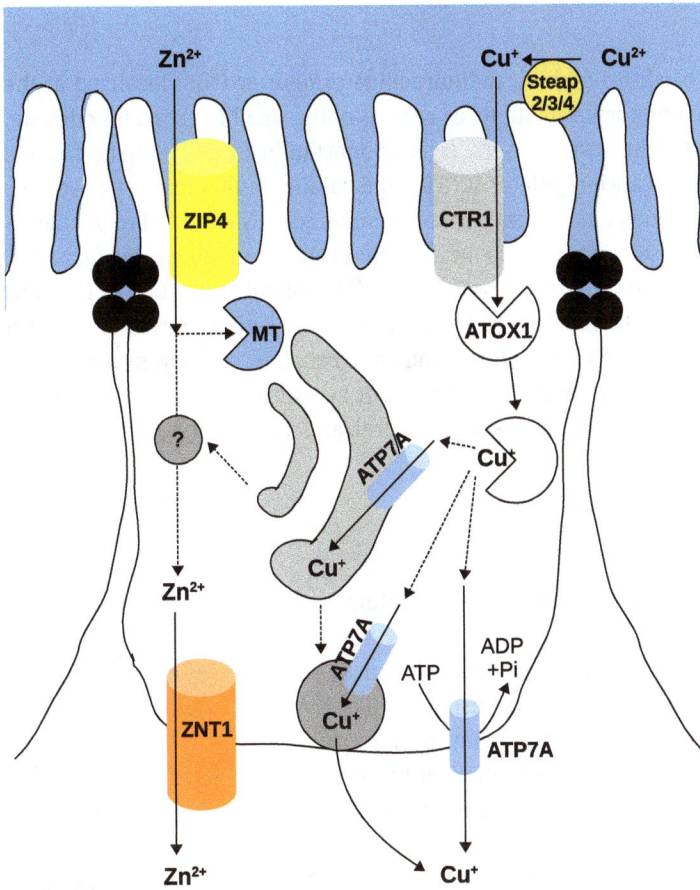

Figure 13.11: Mechanism of absorption of zinc and copper in enterocytes. See main text for references and further details.

ies performed in mice showed that DMT1 is not required for copper absorption under physiological conditions [82].

A recent study has shown that the intestinal mucin MUC2 presents two copper binding sites, a histidine-rich site that captures Cu^{2+} and a methionine-rich site that binds Cu^+. MUC2-bound Cu^{2+} can be reduced to Cu^+ by vitamin C or other dietary anti-oxidants, while bound Cu^+ is protected by MUC2 from oxidation and can be released for nutritional delivery to cells. This mechanism prevents copper toxicity, while permitting Cu^+ uptake into enterocytes [83].

13.7.8 Iodide

Iodide (I^-) anions are absorbed by a sodium iodide symporter (NIS) localized at the apical surface of the enterocytes in all three regions of the small intestine (duodenum, jejunum, and ileum). Other transporters, such as anoctamin 1 (ANO1), cystic fibrosis transmembrane conductance regulator (CFTR), and sodium multivitamin transporter (SMVT) may also be involved in I^- absorption. I^- is subsequently exported to the bloodstream through the basolateral surface of intestinal cells by unknown mechanisms, although chloride channel ClC-2 might be implicated. Intestinal absorption of I^- anions is remarkably efficient (>90%), whereas around 50% of the iodine content of ingested iodo-compounds is estimated to be reduced and absorbed. Interestingly, new evidence suggests that, along the gastrointestinal tract, I^- can be oxidized to hypoiodite (IO^-), a potent oxidant that acts as an antibacterial, antiviral, and antifungal agent [84].

13.7.9 Selenium

The mechanisms of intestinal absorption of selenium vary depending of the chemical form [85, 86]:

– Selenomethionine, the principal chemical form of selenium in most diets, is absorbed via intestinal methionine transporters like B^0AT1 (Table 13.1).
– Selenocysteine is much less abundant in plant proteins than selenomethionine and little is known about its absorption, although selenocystine, its oxidized form, inhibits cystine absorption and is taken up into cultured cells by intestinal transporters for dibasic and neutral amino acids, such as $b^{0,+}AT$ (Table 13.1).
– Selenite absorption varies in the range of 50% to 90%, being affected by diet constituents, whereas selenate absorption is nearly complete. The mechanism of selenite absorption is unknown, while selenate is transported by anion exchangers from the family of the SLC26 gene. Once absorbed, however, selenate must be reduced to selenite, and a significant amount of selenate is lost in the urine before it can be metabolized further.

Most of the absorbed selenium chemical species are used to produce hydrogen selenide (H_2Se) in the enterocytes. Selenomethionine can be converted to selenocysteine that will be further metabolized (along with that coming from the diet) to H_2Se by selenocysteine lyase. Alternatively, selenomethionine can be incorporated non-specifically into proteins, such as albumin and hemoglobin, replacing methionine. Selenite and synthetic selenium derivatives are also metabolized by different pathways to yield H_2Se. All H_2Se regardless of its origin is transported in the blood mainly linked to lipoproteins such as VLDL and LDL (Chapter 4, Section 4.14), as well as to other proteins (albumin and alpha-globulin) [86].

13.7.10 Manganese absorption

Intestinal absorption of Mn^{2+} (and Cd^{2+}) has been suggested to occur via the same pathway as used for iron (Section 13.7.5) because excess amounts of $Fe^{2+/3+}$ in the diet suppress intestinal Mn^{2+} absorption. Luminal Mn^{2+} is captured and taken up into the enterocytes from the apical side via DMT1 (and probably via Zn^{2+} transporter ZIP8). Mn^{2+} is probably effluxed into the portal vein across the basolateral border by ferroportin and binds to transferrin in the bloodstream. Other Zn^{2+} transporters, such as ZIP14 and ZIP10, might be involved in Mn^{2+} excretion from the enterocytes into the lumen. Genetic defects in ZIP8, ZIP14, and ZNT10 suggest that they do play essential roles in Mn metabolism in the body; however, the mechanisms underlying Mn^{2+} transport by these Zn^{2+} transporters remain to be elucidated [17, 87].

13.7.11 Other microelements

A brief account of the current knowledge on the absorption of other microelements such as molybdenum, chromium, fluorine, boron, and silicon, can be found in Chapter 7.

13.8 Absorption in the large intestine

The material (chyme) that enters the colon contains only a limited amount of macronutrients, mainly water, electrolytes such as sodium (Na^+), potassium (K^+), magnesium (Mg^{2+}), and chloride (Cl^-), and indigestible carbohydrates (i.e., fibers). The gastrointestinal microbiota is also involved in functions such as digestion and absorption of some nutrients and production of certain vitamins. The colonic bacteria transform indigestible carbohydrates, such as oligosaccharides, resistant starch, and soluble non-starch polysaccharides (Chapter 3), into gas (including H_2, CO_2, and methane) and short-chain fatty acids (SCFAs), mainly acetate, propionate, and butyrate. In the distal colon, proteolytic fermentation by the prevailing microbiota may also result in the production of SCFAs, together with branched-chain fatty acids [88].

The colonic epithelial surface is flat with crypts without villi, in contrast to the small intestine, and serves as a barrier against detrimental luminal substances and compounds derived from food and the microbiota. It is made of different specialized cells with some homology with those found in the small intestine (Section 13.1). Epithelial cells in the large intestine include colonocytes (see below), enteroendocrine cells (that release hormones mainly in response to microbial-derived compounds, mucus-secreting goblet cells (more abundant than in the small intestine) and Tuft cells. Paneth cells have not been detected in the large intestine, colonocytes being responsible for secreting antimicrobial peptides [9].

Colonocytes are absorptive polarized cells that are present mainly in the surface epithelium, that are responsible for net absorption of water and electrolytes. More specifically, the colon is a net absorber of Na^+ and Cl^- and a net secretor of K^+ and bicarbonate (HCO_3^-), and has an absorptive capacity of water up to 4 to 5 L per day [88]. The mechanisms involved in water and electrolytes movement require the participation of transporters similar to those found in enterocytes (Section 13.7) with some differences such as the presence of a H^+/K^+-ATPase in the apical membrane of colonocytes located in crypts. These mechanisms have been thoroughly reviewed elsewhere [89]. Water absorption allows the progressive dehydration of the luminal content along the large intestine, that becomes more solid in the rectal part.

Colonocytes display a high energy demand due to the rapid turnover of epithelial cells and to the Na^+/K^+-ATPase activity, which also regulates sodium concentration. ATP production can be obtained through the metabolism of compounds that enter colonocytes from both the apical and the basolateral sides. Luminal compounds supplied through the apical side do not include significant amounts of glucose but include amino acids derived from bacterial metabolism and endogenous sources [22]. Glutamine provided through the basolateral membrane by the arterial capillaries is a major fuel for colonocytes, being oxidized in the mitochondria to produce ATP. This process has been proposed to participate in the hypoxic conditions necessary for the maintenance of the colonic microbiota, dominated by obligate anaerobic bacteria [9].

SCFAs produced by colonic bacteria (mainly acetic, propionic, and butyric acids that dissociates into acetate, propionate and butyrate, respectively, and protons) are absorbed by colonocytes via passive diffusion or active transport. The latter can be mediated by H^+-dependent monocarboxylate transporters, such as MCT1 (Table 13.2 and Section 13.7.1), that transports SCFAs in an H^+-dependent, electroneutral manner, and the electrogenic sodium-dependent monocarboxylate transporter SMCT1 (Table 13.2). Acetate, propionate and, mainly, butyrate can be used as energy source by colonocytes, but they can also modulate the homeostasis and function of intestinal epithelial cell and act as mediators in several pathways, which include local, immune, endocrine effects, and microbiota-gut-brain communication. SCFAs seem to act through epigenetic mechanisms and via interaction with several receptors and tissues involved in the maintenance of glucose homeostasis, in pre-diabetes and in T2DM [90].

Some transporters involved in the absorption of amino acids and water-soluble vitamins in colonocytes are displayed in Tables 13.1 and 13.2, respectively.

References

[1] Neyraud E, Cabaret S, Brignot H, Chabanet C, Labouré H, Guichard E, et al. The basal free fatty acid concentration in human saliva is related to salivary lipolytic activity. Sci Rep 2017;7:5969. https://doi.org/10.1038/s41598-017-06418-2.

[2] Patton KT, Thibodeau GA, Hutton A. Anatomy and physiology adapted international edition e-book. Elsevier Health Sciences; 2019.

[3] Hong SN, Dunn JCY, Stelzner M, Martín MG. Concise Review: The Potential Use of Intestinal Stem Cells to Treat Patients with Intestinal Failure: Clinical Use of ISCs. Stem Cells Transl Med 2017;6:666–76. https://doi.org/10.5966/sctm.2016-0153.

[4] Clevers H. The Intestinal Crypt, A Prototype Stem Cell Compartment. Cell 2013;154:274–84. https://doi.org/10.1016/j.cell.2013.07.004.

[5] Animal organs. Digestive system. Small intestine. Atlas of plant and animal histology. n.d. https://mmegias.webs.uvigo.es/02-english/2-organos-a/imagenes-grandes/digestivo-delgado.php (accessed October 29, 2019).

[6] Pelaseyed T, Bergström JH, Gustafsson JK, Ermund A, Birchenough GM, Schütte A, et al. The mucus and s of the goblet cells and enterocytes provide the first defense line of the gastrointestinal tract and interact with the immune system. Immunol Rev 2014;260:8–20. https://doi.org/10.1111/imr.12182.

[7] Sun WW, Krystofiak ES, Leo-Macias A, Cui R, Sesso A, Weigert R, et al. Nanoarchitecture and dynamics of the mouse enteric glycocalyx examined by freeze-etching electron tomography and intravital microscopy. Commun Biol 2020;3:1–10. https://doi.org/10.1038/s42003-019-0735-5.

[8] Buttet M, Traynard V, Tran TTT, Besnard P, Poirier H, Niot I. From fatty-acid sensing to chylomicron synthesis: Role of intestinal lipid-binding proteins. Biochimie 2014;96:37–47. https://doi.org/10.1016/j.biochi.2013.08.011.

[9] Blachier F. Physiological and metabolic functions of the intestinal epithelium: From the small to the large intestine. Metab. Aliment. Compd. Intest. Microbiota Health. Cham: Springer International Publishing; 2023, p. 1–26. https://doi.org/10.1007/978-3-031-26322-4_1.

[10] Gomez-Martinez I. Fatty Acid Handling and Lineage Maturation by the Human Intestinal Epithelium. Ph.D. The University of North Carolina at Chapel Hill, 2023.

[11] McCauley HA, Matthis AL, Enriquez JR, Nichol JT, Sanchez JG, Stone WJ, et al. Enteroendocrine cells couple nutrient sensing to nutrient absorption by regulating ion transport. Nat Commun 2020;11:4791. https://doi.org/10.1038/s41467-020-18536-z.

[12] Schneeberger K, Roth S, Nieuwenhuis EES, Middendorp S. Intestinal epithelial cell polarity defects in disease: lessons from microvillus inclusion disease. Dis Model Mech 2018;11:dmm031088. https://doi.org/10.1242/dmm.031088.

[13] Lea T. Epithelial cell models; general introduction. In: Verhoeckx K, Cotter P, López-Expósito I, Kleiveland C, Lea T, Mackie A, et al., editors. Impact Food Bioact. Health, Cham: Springer International Publishing; 2015, p. 95–102. https://doi.org/10.1007/978-3-319-16104-4_9.

[14] De Santis S, Cavalcanti E, Mastronardi M, Jirillo E, Chieppa M. Nutritional Keys for Intestinal Barrier Modulation. Front Immunol 2015; 6. https://doi.org/10.3389/fimmu.2015.00612.

[15] Fung KYY, Fairn GD, Lee WL. Transcellular vesicular transport in epithelial and endothelial cells: Challenges and opportunities. Traffic 2018;19:5–18. https://doi.org/10.1111/tra.12533.

[16] Ferraris RP, Choe J, Patel CR. Intestinal Absorption of Fructose. Annu Rev Nutr 2018;38:41–67. https://doi.org/10.1146/annurev-nutr-082117-051707.

[17] Kiela PR, Ghishan FK. Physiology of Intestinal Absorption and Secretion. Best Pract Res Clin Gastroenterol 2016;30:145–59. https://doi.org/10.1016/j.bpg.2016.02.007.

[18] Koepsell H. Glucose transporters in the small intestine in health and disease. J Physiol 2020;472:1207–48. https://doi.org/10.1007/s00424-020-02439-5.

[19] Sun B, Chen H, Xue J, Li P, Fu X. The role of GLUT2 in glucose metabolism in multiple organs and tissues. Mol Biol Rep 2023. https://doi.org/10.1007/s11033-023-08535-w.

[20] Karasov WH. Integrative physiology of transcellular and paracellular intestinal absorption. J Exp Biol 2017;220:2495–501. https://doi.org/10.1242/jeb.144048.

[21] Jochems P, Garssen J, van Keulen A, Masereeuw R, Jeurink P. Evaluating Human Intestinal Cell Lines for Studying Dietary Protein Absorption. Nutrients 2018;10:322. https://doi.org/10.3390/nu10030322.

[22] Bröer S. Intestinal Amino Acid Transport and Metabolic Health. Annu Rev Nutr 2023;43:annurev-nutr -061121-094344. https://doi.org/10.1146/annurev-nutr-061121-094344.

[23] Wang B, Xie N, Li B. Influence of peptide characteristics on their stability, intestinal transport, and in vitro bioavailability: A review. J Food Biochem 2019;43:e12571. https://doi.org/10.1111/jfbc.12571.

[24] Healthy diet n.d. https://www.who.int/news-room/fact-sheets/detail/healthy-diet (accessed February 25, 2019).

[25] Wang TY, Liu M, Portincasa P, Wang DQ-H. New insights into the molecular mechanism of intestinal fatty acid absorption. Eur J Clin Invest 2013;43:1203–23. https://doi.org/10.1111/eci.12161.

[26] Samovski D, Jacome-Sosa M, Abumrad NA. Fatty Acid Transport and Signaling: Mechanisms and Physiological Implications. Annu Rev Physiol 2023;85:317–37. https://doi.org/10.1146/annurev-physiol-032122-030352.

[27] He Q, Chen Y, Wang Z, He H, Yu P. Cellular Uptake, Metabolism and Sensing of Long-Chain Fatty Acids. Front Biosci Landmark 2023;28:10. https://doi.org/10.31083/j.fbl2801010.

[28] Ko C-w, Qu J, Black DD, Tso P. Regulation of intestinal lipid metabolism: current concepts and relevance to disease. Nat Rev Gastroenterol Hepatol 2020;17:169–83. https://doi.org/10.1038/s41575-019-0250-7.

[29] Jay AG, Hamilton JA. The enigmatic membrane fatty acid transporter CD36: New insights into fatty acid binding and their effects on uptake of oxidized LDL. Prostaglandins Leukot Essent Fatty Acids 2018;138:64–70. https://doi.org/10.1016/j.plefa.2016.05.005.

[30] CD36 Gene – GeneCards | CD36 Protein | CD36 Antibody n.d. https://www.genecards.org/cgi-bin /carddisp.pl?gene=CD36 (accessed November 11, 2019).

[31] Li H, Herrmann T, Seeßle J, Liebisch G, Merle U, Stremmel W, et al. Role of fatty acid transport protein 4 in metabolic tissues: insights into obesity and fatty liver disease. Biosci Rep 2022;42: BSR20211854. https://doi.org/10.1042/BSR20211854.

[32] Siddiqi S, Mansbach CM. Dietary and biliary phosphatidylcholine activates PKCζ in rat intestine. J Lipid Res 2015;56:859–70. https://doi.org/10.1194/jlr.M056051.

[33] Carreiro AL, Buhman KK. Absorption of dietary fat and its metabolism in enterocytes. Mol. Nutr. Fats, Elsevier; 2019, p. 33–48. https://doi.org/10.1016/B978-0-12-811297-7.00003-2.

[34] Vergès B. Intestinal lipid absorption and transport in type 2 diabetes. Diabetologia 2022;65:1587–600. https://doi.org/10.1007/s00125-022-05765-8.

[35] Altmann SW. Niemann-Pick C1 Like 1 Protein Is Critical for Intestinal Cholesterol Absorption. Science 2004;303:1201–04. https://doi.org/10.1126/science.1093131.

[36] Garcia-Calvo M, Lisnock J, Bull HG, Hawes BE, Burnett DA, Braun MP, et al. The target of ezetimibe is Niemann-Pick C1-Like 1 (NPC1L1). Proc Natl Acad Sci U S A 2005;102:8132–37. https://doi.org/10.1073/ pnas.0500269102.

[37] Jia L, Betters JL, Yu L. Niemann-Pick C1-Like 1 (NPC1L1) Protein in Intestinal and Hepatic Cholesterol Transport. Annu Rev Physiol 2011;73:239–59. https://doi.org/10.1146/annurev-physiol-012110-142233.

[38] Morel E, Ghezzal S, Lucchi G, Truntzer C, Pais de Barros J-P, Simon-Plas F, et al. Cholesterol trafficking and raft-like membrane domain composition mediate scavenger receptor class B type 1-dependent lipid sensing in intestinal epithelial cells. Biochim Biophys Acta BBA – Mol Cell Biol Lipids 2018;1863:199–211. https://doi.org/10.1016/j.bbalip.2017.11.009.

[39] Nassir F, Wilson B, Han X, Gross RW, Abumrad NA. CD36 is important for fatty acid and cholesterol uptake by the proximal but not distal intestine. J Biol Chem 2007;282:19493–501.

[40] Reboul E. Vitamin E intestinal absorption: Regulation of membrane transport across the enterocyte. IUBMB Life 2019;71:416–23. https://doi.org/10.1002/iub.1955.

[41] Silva MC, Furlanetto TW. Intestinal absorption of vitamin D: a systematic review. Nutr Rev 2018;76:60–76. https://doi.org/10.1093/nutrit/nux034.

[42] Shearer MJ, Okano T. Key Pathways and Regulators of Vitamin K Function and Intermediary Metabolism. Annu Rev Nutr 2018;38:127–51. https://doi.org/10.1146/annurev-nutr-082117-051741.

[43] Antoine T, Le May C, Margier M, Halimi C, Nowicki M, Defoort C, et al. The Complex ABCG5/ABCG8 Regulates Vitamin D Absorption Rate and Contributes to its Efflux from the Intestine. Mol Nutr Food Res 2021;65:2100617. https://doi.org/10.1002/mnfr.202100617.

[44] Matsuo M, Ogata Y, Yamanashi Y, Takada T. ABCG5 and ABCG8 Are Involved in Vitamin K Transport. Nutrients 2023;15:998. https://doi.org/10.3390/nu15040998.

[45] Borel P, Desmarchelier C. Genetic Variations Associated with Vitamin A Status and Vitamin A Bioavailability. Nutrients 2017;9:246. https://doi.org/10.3390/nu9030246.

[46] Reboul E. Mechanisms of Carotenoid Intestinal Absorption: Where Do We Stand? Nutrients 2019;11:838. https://doi.org/10.3390/nu11040838.

[47] Dash S, Xiao C, Morgantini C, Lewis GF. New Insights into the Regulation of Chylomicron Production. Annu Rev Nutr 2015;35:265–94. https://doi.org/10.1146/annurev-nutr-071714-034338.

[48] Hui DY. Intestinal phospholipid and lysophospholipid metabolism in cardiometabolic disease. Curr Opin Lipidol 2016;27:507–12. https://doi.org/10.1097/MOL.0000000000000334.

[49] D'Aquila T, Hung Y-H, Carreiro A, Buhman KK. Recent discoveries on absorption of dietary fat: Presence, synthesis, and metabolism of cytoplasmic lipid droplets within enterocytes. Biochim Biophys Acta BBA – Mol Cell Biol Lipids 2016;1861:730–47. https://doi.org/10.1016/j.bbalip.2016.04.012.

[50] Borel P, Dangles O, Kopec RE. Fat-soluble vitamin and phytochemical metabolites: Production, gastrointestinal absorption, and health effects. Prog Lipid Res 2023;90:101220. https://doi.org/10.1016/j.plipres.2023.101220.

[51] Said HM, Nexo E. Intestinal absorption of water-soluble vitamins: Cellular and molecular mechanisms. Physiol. Gastrointest. Tract, Elsevier; 2018, p. 1201–48. https://doi.org/10.1016/B978-0-12-809954-4.00054-2.

[52] Nielsen MJ, Rasmussen MR, Andersen CBF, Nexø E, Moestrup SK. Vitamin B12 transport from food to the body's cells – a sophisticated, multistep pathway. Nat Rev Gastroenterol Hepatol 2012;9:345–54. https://doi.org/10.1038/nrgastro.2012.76.

[53] Eck P. Nutrigenomics of vitamin C absorption and transport. Curr Opin Food Sci 2018;20:100–04. https://doi.org/10.1016/j.cofs.2018.05.001.

[54] Koepsell H. Organic Cation Transporters in Health and Disease. Pharmacol Rev 2020;72:253–319. https://doi.org/10.1124/pr.118.015578.

[55] Hrubša M, Siatka T, Nejmanová I, Vopršalová M, Kujovská Krčmová L, Matoušová K, et al. Biological Properties of Vitamins of the B-Complex, Part 1: Vitamins B1, B2, B3, and B5. Nutrients 2022;14:484. https://doi.org/10.3390/nu14030484.

[56] Yamashiro T, Yasujima T, Said HM, Yuasa H. pH-dependent pyridoxine transport by SLC19A2 and SLC19A3: Implications for absorption in acidic microclimates. J Biol Chem 2020;295:16998–7008. https://doi.org/10.1074/jbc.RA120.013610.

[57] Barrett KE, Keely SJ. Electrolyte secretion and absorption in the small intestine and colon. In: Wang TC, Camilleri M, Lebwohl B, Lok AS, Sandborn WJ, Wang KK, et al., editors. Yamadas Textb. Gastroenterol. 1st edition. Wiley; 2022, p. 283–312. https://doi.org/10.1002/9781119600206.ch16.

[58] Das S, Jayaratne R, Barrett KE. The Role of Ion Transporters in the Pathophysiology of Infectious Diarrhea. Cell Mol Gastroenterol Hepatol 2018;6:33–45. https://doi.org/10.1016/j.jcmgh.2018.02.009.

[59] Barrett KE. Endogenous and exogenous control of gastrointestinal epithelial function: building on the legacy of Bayliss and Starling: 2015 Bayliss-Starling Prize Lecture. J Physiol 2017;595:423–32. https://doi.org/10.1113/JP272227.

[60] Goff JP. Invited review: Mineral absorption mechanisms, mineral interactions that affect acid-base and antioxidant status, and diet considerations to improve mineral status. J Dairy Sci 2018;101:2763–813. https://doi.org/10.3168/jds.2017-13112.

[61] Agarwal R, Afzalpurkar R, Fordtran JS. Pathophysiology of potassium absorption and secretion by the human intestine. Gastroenterology 1994;107:548–71. https://doi.org/10.1016/0016-5085(94)90184-8.

[62] Beggs MR, Bhullar H, Dimke H, Alexander RT. The contribution of regulated colonic calcium absorption to the maintenance of calcium homeostasis. J Steroid Biochem Mol Biol 2022;220:106098. https://doi.org/10.1016/j.jsbmb.2022.106098.

[63] Fleet JC. Vitamin D-Mediated Regulation of Intestinal Calcium Absorption. Nutrients 2022;14:3351. https://doi.org/10.3390/nu14163351.

[64] Diaz de Barboza G, Guizzardi S, Tolosa de Talamoni N. Molecular aspects of intestinal calcium absorption. World J Gastroenterol 2015;21:7142–54. https://doi.org/10.3748/wjg.v21.i23.7142.

[65] Saurette M, Alexander RT. Intestinal phosphate absorption: The paracellular pathway predominates? Exp Biol Med 2019;244:646–54. https://doi.org/10.1177/1535370219831220.

[66] Hernando N, Gagnon K, Lederer E. Phosphate Transport in Epithelial and Nonepithelial Tissue. Physiol Rev 2021;101:1–35. https://doi.org/10.1152/physrev.00008.2019.

[67] Centeno PP, Herberger A, Mun H-C, Tu C, Nemeth EF, Chang W, et al. Phosphate acts directly on the calcium-sensing receptor to stimulate parathyroid hormone secretion. Nat Commun 2019;10:4693. https://doi.org/10.1038/s41467-019-12399-9.

[68] Penido MGMG, Alon US. Phosphate homeostasis and its role in bone health. Pediatr Nephrol 2012;27:2039–48. https://doi.org/10.1007/s00467-012-2175-z.

[69] Luongo F, Pietropaolo G, Gautier M, Dhennin-Duthille I, Ouadid-Ahidouch H, Wolf FI, et al. TRPM6 is Essential for Magnesium Uptake and Epithelial Cell Function in the Colon. Nutrients 2018; 10. https://doi.org/10.3390/nu10060784.

[70] Schuchardt JP, Hahn A. Intestinal Absorption and Factors Influencing Bioavailability of Magnesium-An Update. Curr Nutr Food Sci 2017;13. https://doi.org/10.2174/1573401313666170427162740.

[71] Chamniansawat S, Suksridechacin N, Thongon N. Current opinion on the regulation of small intestinal magnesium absorption. World J Gastroenterol 2023;29:332–42. https://doi.org/10.3748/wjg.v29.i2.332.

[72] Ravell JC, Chauvin SD, He T, Lenardo M. An Update on XMEN Disease. J Clin Immunol 2020;40:671–81. https://doi.org/10.1007/s10875-020-00790-x.

[73] Schäffers OJM, Hoenderop JGJ, Bindels RJM, De Baaij JHF. The rise and fall of novel renal magnesium transporters. Am J Physiol-Ren Physiol 2018;314:F1027–33. https://doi.org/10.1152/ajprenal.00634.2017.

[74] Nishito Y, Kambe T. Absorption Mechanisms of Iron, Copper, and Zinc: An Overview. J Nutr Sci Vitaminol (Tokyo) 2018;64:1–7. https://doi.org/10.3177/jnsv.64.1.

[75] White C, Yuan X, Schmidt PJ, Bresciani E, Samuel TK, Campagna D, et al. HRG1 Is Essential for Heme Transport from the Phagolysosome of Macrophages during Erythrophagocytosis. Cell Metab 2013;17:261–70. https://doi.org/10.1016/j.cmet.2013.01.005.

[76] Zaugg J, Pujol Giménez J, Cabra RS, Hofstetter W, Hediger MA, Albrecht C. New Insights into the Physiology of Iron Transport: An Interdisciplinary Approach. CHIMIA 2022;76:996. https://doi.org/10.2533/chimia.2022.996.

[77] Conesa C, Bellés A, Grasa L, Sánchez L. The Role of Lactoferrin in Intestinal Health. Pharmaceutics 2023;15:1569. https://doi.org/10.3390/pharmaceutics15061569.

[78] Cousins RJ, Liuzzi JP. Trace metal absorption and transport. Physiol Gastrointest Tract. Elsevier; 2018, p. 1485–98. https://doi.org/10.1016/B978-0-12-809954-4.00061-X.

[79] Delbaere K, Roegiers I, Bron A, Durif C, van de Wiele T, Blanquet-Diot S, et al. The small intestine: dining table of host-microbiota meetings. FEMS Microbiol Rev 2023;47:fuad022. https://doi.org/10.1093/femsre/fuad022.

[80] Wang Z, Jin D, Zhou S, Dong N, Ji Y, An P, et al. Regulatory roles of copper metabolism and cuproptosis in human cancers. Front Oncol 2023;13. https://www.frontiersin.org/articles/10.3389/fonc.2023.1123420

[81] Pierson H, Yang H, Lutsenko S. Copper Transport and Disease: What Can We Learn from Organoids? Annu Rev Nutr 2019;39:75–94. https://doi.org/10.1146/annurev-nutr-082018-124242.

[82] Shawki A, Anthony SR, Nose Y, Engevik MA, Niespodzany EJ, Barrientos T, et al. Intestinal DMT1 is critical for iron absorption in the mouse but is not required for the absorption of copper or manganese. Am J Physiol Gastrointest Liver Physiol 2015;309:G635–47. https://doi.org/10.1152/ajpgi.00160.2015.

[83] Reznik N, Gallo AD, Rush KW, Javitt G, Fridmann-Sirkis Y, Ilani T, et al. Intestinal mucin is a chaperone of multivalent copper. Cell 2022;185:4206–15.e11. https://doi.org/10.1016/j.cell.2022.09.021.

[84] De la Vieja A, Santisteban P. Role of iodide metabolism in physiology and cancer. Endocr Relat Cancer 2018;25:R225–45. https://doi.org/10.1530/ERC-17-0515.

[85] Burk RF, Hill KE. Regulation of Selenium Metabolism and Transport. Annu Rev Nutr 2015;35:109–34. https://doi.org/10.1146/annurev-nutr-071714-034250.

[86] Ferreira RLU, Sena-Evangelista KCM, De Azevedo EP, Pinheiro FI, Cobucci RN, Pedrosa LFC. Selenium in Human Health and Gut Microflora: Bioavailability of Selenocompounds and Relationship With Diseases. Front Nutr 2021;8:685317. https://doi.org/10.3389/fnut.2021.685317.

[87] Fujishiro H, Kambe T. Manganese transport in mammals by zinc transporter family proteins, ZNT and ZIP. J Pharmacol Sci 2022;148:125–33. https://doi.org/10.1016/j.jphs.2021.10.011.

[88] Verbiest A, Jeppesen PB, Joly F, Vanuytsel T. The Role of a Colon-in-Continuity in Short Bowel Syndrome. Nutrients 2023;15:628. https://doi.org/10.3390/nu15030628.

[89] Negussie AB, Dell AC, Davis BA, Geibel JP. Colonic Fluid and Electrolyte Transport 2022: An Update. Cells 2022;11:1712. https://doi.org/10.3390/cells11101712.

[90] Portincasa P, Bonfrate L, Vacca M, De Angelis M, Farella I, Lanza E, et al. Gut Microbiota and Short Chain Fatty Acids: Implications in Glucose Homeostasis. Int J Mol Sci 2022;23:1105. https://doi.org/10.3390/ijms23031105.

14 Energy homeostasis and integration of metabolism

14.1 Introduction

Living organisms are extremely complex and organized physical systems. Maintaining such complexity and order requires the energy provided by the nutrients (that is, the different building blocks of biological structures) which also supply matter. Inside the cells, a complex network of chemical reactions makes sure that nutrients are used in such a way that the biological organization is built, maintained, and eventually transmitted to the following generation. In this process, nutrients are degraded to simpler molecules (waste products) that must be excreted, while some energy is constantly dissipated as heat into the external environment. The prevailing view is that the dissipation of energy and matter is necessary to fulfil the Second Law of Thermodynamics, that is, living organisms can decrease their entropy (increase their order) as long as they compensate for by increasing that of their surroundings. Therefore, from a thermodynamics point of view, living systems have to be open: they must exchange energy and matter with the environment [1].

Energy management is of paramount importance for life at any level, however, it is specially challenging in the case of multicellular organisms. In previous chapters, it has been described how nutrients enter the human body through the mouth, mainly as macromolecules, and then converted to smaller molecules that have to be absorbed and distributed to all cells. The average human adult has an estimated 37 trillion (3.72×10^{13}) cells [2], therefore, guaranteeing energy supply and waste removal to the basic units of our bodies is an enormous task. Moreover, because cells are organized into successive levels of increasing complexity (tissues, organs, systems, and organism), energy management must cope with a wide range of requirements and adapt to changing conditions caused by exogenous and/or endogenous factors.

The way the body uses and stores energy is usually known as "energy metabolism" and is influenced by many factors, such as the pattern of food intake, growth, stress, and metabolic rate (the amount of energy expended per unit of time). Energy metabolism involves mechanisms subjected to a coordinated regulation in which a series of biochemical messengers acting on different organs, the hormones, play a major role [3].

14.2 Energy homeostasis and energy balance

Homeostasis can be defined as "the tendency of an organism to maintain a stable internal state" [4] and energy homeostasis (EH), that is, maintaining the energy status, is considered to be critical for the survival of living organisms. In principle, EH aims

https://doi.org/10.1515/9783111111872-014

at achieving a balance between the energy intake and the energy expenditure, which means that energy stores are maintained, as living organisms also obey the First Law of Thermodynamics.

Energy intake = Energy utilization (biological work, energy store)

+ Energy output (heat dissipation, waste excretion)

For accurate calculations of energy balances, it must be considered that the energy content of body stores is not necessarily equivalent to its weight and composition. This idea can be illustrated with an example: if 100 g of glucose are absorbed and 100 g of triglycerides are oxidized, the body weight does not change, but the energy content decreases because glucose contains less energy per gram than triglycerides (see below) [3]. On the other hand, it has been proposed that during energy imbalance, not only changes in energy values have to be considered but also the rate at which these changes occur. This proposal introduces time as an important variable in the assessment of energy balance [5].

14.2.1 Energy utilization: the fate of the energy within the body

The flow of energy through the body is outlined in Figure 14.1. The total energy supplied by food is referred to as ingested energy. Incomplete digestion of food in the small intestine, in some cases accompanied by fermentation of unabsorbed carbohydrate in the colon, results in losses of energy (fecal energy and gaseous energy, the latter in the form of gases like hydrogen and methane produced by fermentation). Short-chain fatty acids are also formed in the process, some of which may be absorbed and used to obtain energy. The rest of the energy is internalized, while some losses occur through urine (mainly as urea), hair, nails, and cell turnover in the skin and intestinal epithelium. The energy that remains after accounting for these losses is known as metabolizable energy (ME). Some ME is used during digestion, absorption, and intermediary metabolism of food and is referred to as dietary-induced thermogenesis, or thermic effect of food, that varies with the type of food ingested. When the energy lost to microbial fermentation and obligatory thermogenesis are subtracted from ME, the result is the net metabolizable energy (NME). Some energy may also be lost as heat produced by metabolic processes associated with other forms of thermogenesis (non-obligatory thermogenesis), such as the effects of cold, hormones, certain drugs, bioactive compounds, and stimulants. The energy that remains after subtracting these heat losses from NME (referred to as net energy or NE) is used to support basal metabolism, physical activity, and other needs, like growth, pregnancy, and lactation [6].

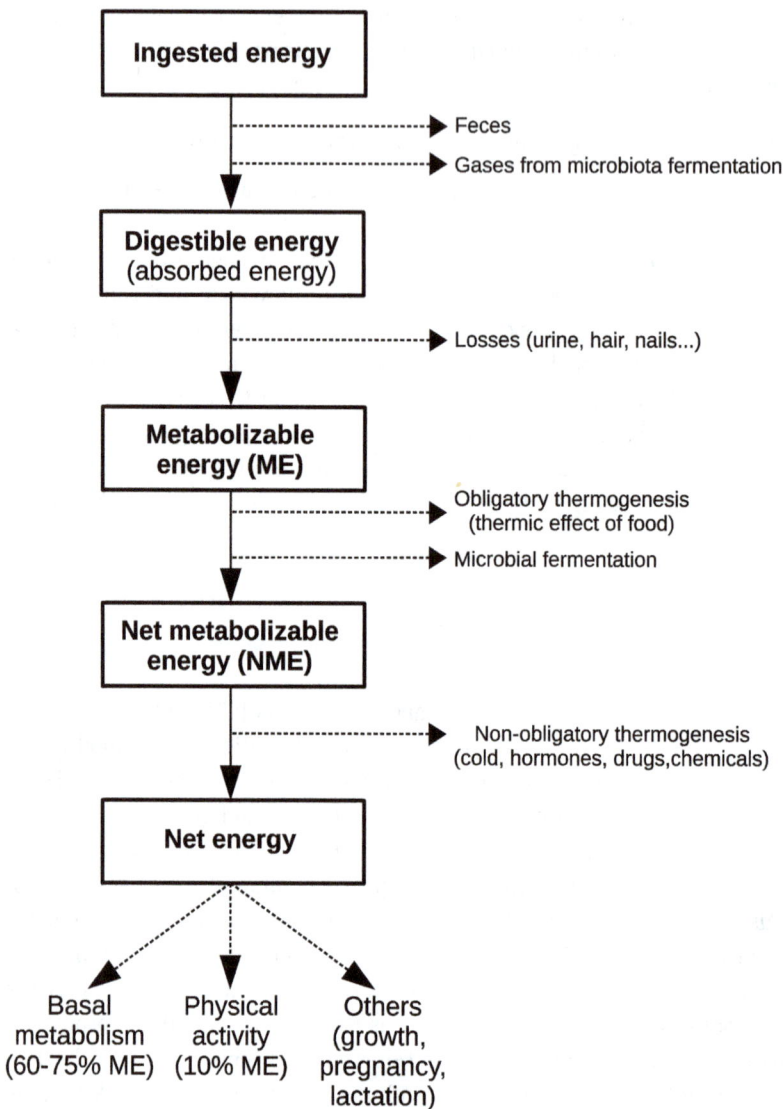

Figure 14.1: The flow of energy through the body. Adapted from references [6] and [7].

In adults, the energy used for basal metabolism, that is, the biochemical activities essential for life, accounts for approximately 60–75% of ME, while physical activity and food thermogenesis contribute with 15–30% and 10–15% ME, respectively [7, 8]. Basal metabolism is usually expressed as the basal metabolic rate (BMR), which is defined as the metabolic rate during rest while the person is awake and measured at least 12 h after the last meal and/or strenuous exercise. The air in the room should be comfortable with all sources of excitement removed. Another frequently used parameter is the resting energy expenditure (REE), which differs from the BMR in that its determination does not require fasting for 12 h [9]. Methods for the determination of BMR/ REE include:

- Direct calorimetry, which measures the heat quantity produced by the organism in hermetic cameras with insulating walls, where the subject is confined, and the heat produced and dissipated by radiation, convection, and evaporation is measured. This method is difficult to use in practice [9, 10].
- Indirect calorimetry, which estimates the quantity of total heat produced in the organism by measuring the volume of oxygen consumed and the volume of carbon dioxide exhaled. This widely used method is based on the assumption that energy is obtained by the complete oxidation of nutrients, whose stoichiometric equations in the case of fatty acids and sugars are:

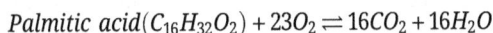

$$Palmitic\ acid(C_{16}H_{32}O_2) + 23O_2 \rightleftharpoons 16CO_2 + 16H_2O$$

$$Glucose(C_6H_{12}O_6) + 6O_2 \rightleftharpoons 6CO_2 + 6H_2O$$

The heat produced by these reactions in a given time can be calculated by Wier formula [11]:

$$Heat\ output\ (kilocalories) = 3.9 \times liters\ of\ O_2\ consumed + 1.1 \times liters\ of\ CO_2\ produced$$

This equation was adapted, also by Weir, to include protein utilization based on the idea that about 80% of nitrogen is eliminated through the kidneys as urea, and proteins contain approximately 16% of nitrogen by weight [11]:

$$Heat\ output\ (kilocalories) = 3.9 \times liters\ of\ O_2\ consumed + 1.1 \times liters\ of\ CO_2\ produced$$
$$- 2.17 \times g\ of\ urine\ nitrogen$$

The ratio of liters of CO_2 produced to liters of O_2 consumed at cellular level is referred to as the respiratory quotient (RQ), which is difficult to measure; therefore, the same ratio measured from exhaled air, the respiratory exchange ratio (RER), is normally used. RER is assumed to be equivalent to RQ under steady-state conditions (during rest). Other assumptions of the indirect calorimetry are that growth, interconversion of fuels, and anaerobic metabolism are not occurring [12].

Other methods for determining BMR include the doubly labeled water technique, heart rate monitoring, and bioelectrical impedance [9]. A number of equations that aim at predicting BMR are also available (Table 14.1).

The major factor determining basal metabolism is fat free mass (FFM), which shows a positive linear correlation with BMR [14]. Body weight, height, fat mass, and age have also been reported to influence BMR, although it is sometimes difficult to ascertain if they are independent factors, or they act through their effects on FFM. Several hormones and their functionality have also been described as contributors to energy expenditure. These include insulin and insulin sensitivity, leptin, and vitamin D

Table 14.1: Some equations used for the prediction of basal metabolic rate measured in kcal/day [13].

Authors	Equation
Harris and Benedict	66.47 + (13.75 × BW*) + (5.00 × height) – (6.76 × age)
Schofield: 18–30 years	[(0.063 × BW + 2.896) × 239]
FAO/WHO/UNU: 18–30 years	(15.3 × BW + 679)
Henry and Rees 21: 18–30 years	[(0.056 × BW + 2.800) × 239]
Mifflin-St. Jeor	(9.99 × BW) + (6.25 × height) – (4.92 × age) + 5

BW: body weight in kg. Height and age are expressed in cm and years, respectively.

status, in addition to thyroid hormones (Chapter 7, Section 7.3.4), that are known to be key regulators of BMR. Caloric restriction has been described as a factor that decreases overall energy expenditure, including resting and non-resting (physical activity and food thermogenesis) contributions [15].

14.2.2 Energy intake: the energy content of nutrients

The assessment of energy intake requires the determination of the energy content of foods. To this end, the Food and Agriculture Organization (FAO) of the United Nations (UN) has set the following guidelines [6]:
- The components of food that provide energy, mainly protein, fat, carbohydrate, and alcohol (ethanol), should be determined by appropriate analytical methods.
- The quantity of each individual component must be converted to food energy using a factor that expresses the amount of available energy per unit of weight.
- The food energies of all components must be added together to represent the nutritional energy value of the food for humans.

There are three main systems of energy conversion factors for foods in use by FAO. They are based on the concept of metabolizable energy (ME) as defined above:
- The Atwater general factor system, based on the heats of combustion of protein, fat, and carbohydrate, measured in a calorimeter, and corrected for losses in digestion, absorption, and urinary excretion of urea [16]. This system uses a single factor for each of the energy-yielding substrates expressed in kilojoules per gram (kJ/g) or kilocalories per gram (kcal/g):
  - 17 kJ/g (4.0 kcal/g) for protein
  - 37 kJ/g (9.0 kcal/g) for fat
  - 17 kJ/g (4.0 kcal/g) for carbohydrates (this value is determined by difference, and thus includes fiber)
  - 29 kJ/g (~7.0 kcal/g) for alcohol
- The extensive general factor system, derived by modifying, refining, and making additions to the Atwater general factor system. As an example, this system in-

cludes a division of total carbohydrate into available carbohydrate and fiber, thus, an energy factor for dietary fiber of 8.0 kJ/g (2.0 kcal/g) has been recommended. This factor is calculated by assuming that fiber is 70% fermentable and that some of the energy generated by its fermentation is lost as gas and some is incorporated into colonic bacteria and lost in the feces. General factors for organic acids (13 kJ/g) and polyols (10 kJ/g), as well as individual factors for specific polyols (glycerol, sorbitol . . .) and for different organic acids (acetic, propionic, butyric . . .), have also been calculated [17].

– The Atwater specific factor system, proposed by Merrill and Watt, emphasizes that there are ranges in the heats of combustion and in the coefficients of digestibility of different proteins, fats, and carbohydrates. This system derives different factors for proteins, fats, and carbohydrates, depending on the foods in which they are found, thus creating a set of tables that show a substantial variability in the energy factors applied to various foods. Energy conversion factors for protein range from 10.2 kJ/g for some vegetable proteins to 18.2 kJ/g for eggs; factors for fat vary from 35 kJ/g to 37.7 kJ/g, and those for total carbohydrate from 11.3 kJ/g in lemon and lime juices to 17.4 kJ/g in polished rice [18].

Although both the Atwater specific factor system and the extensive general factor system appear to be superior to the original Atwater general system, the latter has been widely used, in part because of its simplicity. Other systems based on the concept of net metabolizable energy (NME) are also available, but those based on ME are recommended [6]. It must be mentioned that other authorities, such as the United States Food and Drug Administration (FDA), have somewhat different methods to those used by FAO for calculating dietary energy, although they are also based on Atwater factors [19]. Another important consideration to make is that the actual energy content of some foods may differ from those calculated as described above due to differences in macronutrient digestibility and food structure. One of the best examples of this discrepancy is represented by nuts, such as almonds, walnuts, cashews, and pistachios [8].

14.2.3 Mechanisms of energy homeostasis

As stated above, living organisms have a tendency to maintain the balance between the energy intake and the energy expenditure, the so-called energy homeostasis. Today, a major task for scientists is to find out why homeostatic responses seem to be insufficient to cope with challenges (overfeeding, unbalanced diets, and lack of exercise) that are thought to be at the basis of the obesity epidemic. The very concept of energy homeostasis is a matter of debate and many aspects of its physiology remain to be established [4]. In any case, maintaining the energy balance involves a series of mechanisms that fulfil energy needs while adapting to changes in energy intake, thus, it must reduce intake when expenditure decreases and vice versa. In the case of animals, the presence of internal energy stores than can be mobilized when needed, or replenished when there is an excess of energy, introduces a factor that complicates the task of understanding the mechanisms that drive energy homeostasis. To make matters more complicated, metabolism can be modulated in response to environmental changes by different means, including epigenetic mechanisms (Section 14.8.5). Several physiological systems are known to be involved in energy homeostasis in humans [4]:

- Afferent hormones and vagal neurons, that link the peripheral system to the central nervous system.
- The hypothalamus, mainly the arcuate nucleus, the brain area where peripheral messages induce homeostatic responses that alter energy intake and energy expenditure.
- The brainstem, notably the nucleus tractus solitarius, that mediates the inhibition of intake triggered by gastrointestinal signals.
- The reward system.
- The cognitive frontal and cortical system, involved in decision-making, self-control, and executive function.
- The neuronal efferent pathway represented by the sympathetic nervous system, which not only modulates energy expenditure but may also influence the effects of all the previous factors on energy intake.

Many of these aspects are beyond the scope of this book, however, some of the underlying biochemical mechanisms involved in energy homeostasis are briefly described below.

14.3 The adipose tissue as an endocrine organ

Two major types of adipose tissues, white and brown, have been identified: brown adipose tissue (BAT) dissipates energy by producing heat (thermogenesis), whereas white adipose tissue (WAT) is specialized in the storage and release of fat. BAT is mainly composed of brown adipocytes, characterized by a high mitochondrial content and, consequently, a high capacity of lipid oxidation. White adipocytes, the major components of WAT, are cells with great energy storage ability, high pro-inflammatory profile, and capable of producing and secreting biomolecules related to energy homeostasis and inflammation modulation. Two other types of adipocytes, pink and beige, have also been described. Pink adipocytes are milk-producing epithelial cells derived from white adipocytes during lactation. They have a great potential for energy storage, albeit with a higher metabolic activity and lower ability to regulate inflammation compared to white adipocytes. Beige adipocytes combine the ability to produce heat with the highest metabolic activity among all four types of adipocytes [20]. Recent evidence suggest that, instead of regulating whole-body energy homeostasis, human BAT functions as: (i) a local thermogenic source and substrate sink to protect and ensure the adequate function of vital organs; (ii) an endocrine organ; and/or (iii) a biomarker of adipose tissue health (see reference [21] for an updated review on BAT).

An overwhelming body of evidence has shown that WAT, whose intracellular lipid metabolism is a key determinant of body weight and insulin sensitivity, is an important endocrine organ. WAT secretes numerous factors, mainly peptides and proteins (known as adipokines or adipocytokines), that play a role in various biological and physiological functions, among which feeding modulation, inflammatory and immune function, glucose and lipid metabolism, and blood pressure control. The effects of adipose tissue signaling range from central regulation of energy expenditure and satiety to the alteration of hormone secretion from the pancreas and other organs. Furthermore, reciprocal signals from other tissues communicate with adipose tissue and modulate adipokine production, thereby establishing a "crosstalk" among different organs [22–24].

Many adipocyte-derived products are physiologically important, however, three of them are known to exert a profound influence on systemic metabolic homeostasis: the peptide hormones leptin and adiponectin, as well as the so-called lipokines.

14.3.1 Leptin

Leptin is a polypeptide of 167 amino acids that is considered the main actor of an afferent path that links WAT and the hypothalamic areas involved in food intake and energy expenditure. Leptin carries the message that fat reserves are sufficient, thereby promoting a reduction in energy intake and increasing energy expenditure [25].

According to the classical model, in the arcuate nucleus of the hypothalamus (ARC) leptin stimulates the expression of the anorexigenic (appetite suppressing) peptides pro-opiomelanocortin (POMC) and cocaine and amphetamine regulated transcript (CART) by a specific group of neurons (the so-called POMC neurons). Leptin also inhibits the production and secretion of the orexigenic (appetite stimulating) peptides neuropeptide Y (NPY) and agouti-related peptide (AgRP) in a separate group of neurons (AgRP/NPY neurons) (Figure 14.2). Post-translational processing of POMC produces melanocortins, a series of peptides that include the adrenocorticotropic hormone (ACTH), and α-, β-, and γ-melanocyte-stimulating hormones (α-, β-, γ-MSHs). This system, known as the leptin/melanocortin pathway, is involved in energy homeostasis, sexual activity, exocrine secretion, as well as anti-inflammatory and immunomodulatory actions [26, 27]. It is worth mentioning that AgRP/NPY and POMC neurons function by projecting into other hypothalamic centers, such as the ventromedial nucleus, the paraventricular nucleus, the dorsomedial nucleus, and the lateral hypothalamus [28]. The classical view of the leptin/melanocortin pathway has been challenged by recent discoveries, thus, direct action of leptin on POMC neurons seems to regulate only glucose homeostasis, whereas the effects of leptin on food intake and body weight are apparently indirect. According to the new model, most POMC neurons do not express leptin receptors, being controlled mainly by AgRP/NPY and other gamma-aminobutyric acid (GABA)-releasing leptin-sensitive neurons. Other neurotransmitters released from different neurons in response to leptin may also be involved [29].

Leptin is also known to exert effects on other areas of the central nervous system, thereby modulating not only food intake and energy expenditure but also the perception of food palatability and spontaneous physical activity [4].

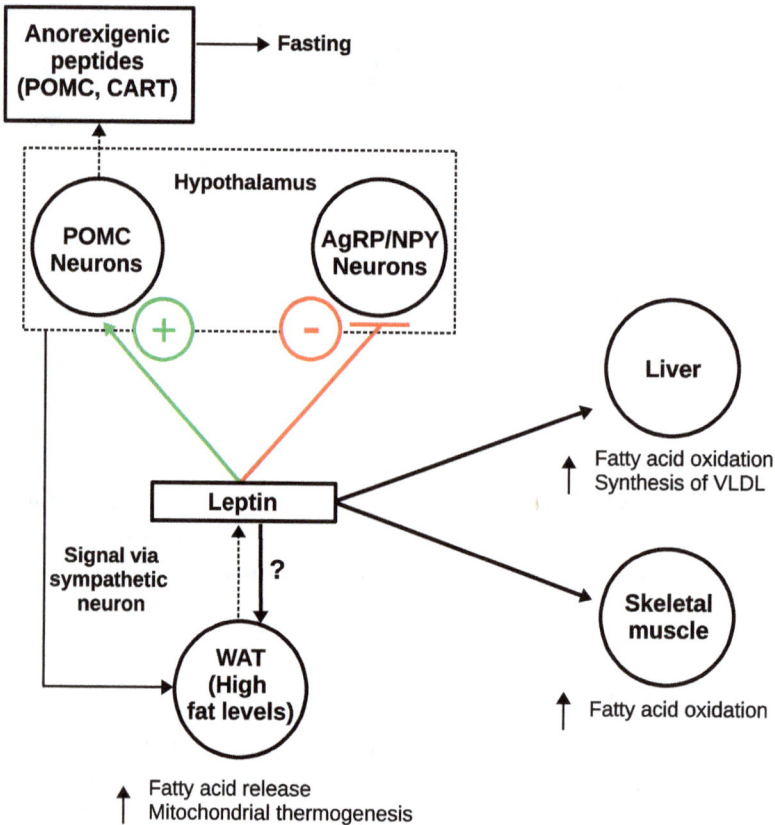

Figure 14.2: Simplified scheme of the effects exerted by leptin in the human body. WAT: white adipose tissue. Figure elaborated with information obtained from [25–27].

Leptin increases energy production from endogenous sources by mechanisms not completely understood, thus, it leads to increased fatty acid release from WAT, probably acting through the sympathetic nervous system, although a direct action on adipocytes (autocrine signaling) cannot be discarded. Other organs whose lipid and glucose metabolisms are affected by leptin include liver, skeletal muscles, pancreas, and kidney (Figure 14.2) [4, 23].

Circulating levels of leptin are proportional to fat mass and increase proportionally to weight gain, whereas they are reduced by fasting. Although it was initially thought that leptin administration might result in weight loss in obese individuals, this effect is only significant in those patients with low levels of this hormone. This suggests that body weight is a complex trait that is controlled by other factors in addition to leptin [4].

In summary, leptin exert pleiotropic functions, being reportedly involved in processes such as glucose metabolism, hematopoiesis, bone remodeling, neurogenesis and neuroprotection, interaction with the immune system, reproduction, angiogenesis, blood pressure, gastric emptying, glomerular filtration rate, and others. An updated review on the advances in the molecular mechanisms of leptin action and its role in pathological situations, including cancer and neurological disorders such as Parkinson's and Alzheimer's diseases, can be found elsewhere [30].

14.3.2 Adiponectin

Adiponectin is a polypeptide of 224 amino acids produced mainly by the adipose tissue, whose plasma concentration increases with fasting and shows an inverse correlation with fat mass. Its main functions can be categorized as anti-apoptotic, anti-inflammatory/anti-fibrotic, anti-lipotoxic, and insulin sensitizing. Adiponectin acts primarily on the adipose tissue (autocrine and paracrine action), heart, kidney, liver, and pancreas; however, the ubiquitous expression of its receptors suggests that this adipokine exerts effects in more tissues [31]. Some actions attributed to adiponectin are [23]:

- In adipocytes, it enhances lipid storage and glucose uptake. This effect prevents ectopic lipid accumulation.
- In liver, adiponectin decreases lipogenesis and gluconeogenesis, and increases fatty acid β-oxidation, thereby ameliorating the excess of lipid accumulation associated with hepatic dysfunction.
- Facilitates the metabolic effects of fibroblast growth factor 21 (FGF21), a peptide hormone secreted by adipose tissue, liver, and skeletal muscle. FGF21 restores normal concentration of glucose in blood (euglycemia), ameliorates hyperlipidemia, and decreases fat mass in animal models of obesity and type II diabetes mellitus.
- Stimulates glucose uptake in isolated skeletal muscle and cultured myocytes. A truncated globular form of adiponectin preserves insulin sensitivity by enhancing skeletal muscle fatty acid oxidation and decreasing intramuscular lipid accumulation. It is not clear whether the latter effects are exerted by the full-length version in vivo.
- Protects β-cells against the lipotoxic effects of overnutrition. In mice, adiponectin prevents obesity and the decrease in β cell mass induced by type 1 diabetes melli-

tus. Moreover, adiponectin stimulates insulin secretion in isolated mouse islets exposed to high glucose concentrations. Conversely, it is possible that insulin modulates adiponectin secretion.
–   Protects kidneys in several mouse models and clinical studies suggest that it is beneficial to kidney disease outcomes.
–   Regulates brain circuits involved in energy expenditure.

Adiponectin exerts many of its effects by activating AMP-activated protein kinase (AMPK), a protein present in most mammalian tissues that triggers a shift in cellular metabolism from biosynthesis toward energy generation [25].

Adiponectin is thought to be the major hormonal factor mediating the beneficial health effects of adipose tissue and numerous studies have shown an inverse correlation between its plasma concentration and diseases like type 2 diabetes, coronary artery disease, and myocardial infarction, visceral adiposity, chronic systemic inflammation, insulin resistance, metabolic syndrome, hepatic steatosis, and neurodegenerative diseases [31, 32].

14.3.3 Lipokines

Lipokines were initially described as lipid molecules with endocrine activity secreted by the adipose tissue, but later studies have shown that they can also be produced by other tissues. Lipokines include molecules such as palmitoleic acid, 12,13-dihydroxy-9Z-octadecenoic acid (12,13-diHOME), and fatty acid esters of hydroxy fatty acids (FAHFA) species (Figure 14.3). Other members of this group of biomolecules are lysophosphatidic acid, alkyl ether lipids, and oxidized lipid metabolites derived from polyunsaturated fatty acids. Lipokines can be directly obtained from diet or by endogenous synthesis in the adipose tissue by de novo lipogenesis and/or lipolysis in response to environmental and physiological cues, such as cold exposure, exercise, diets, and drugs. They are involved in regulating nutrient utilization, thermogenesis, and systemic metabolism by communicating with other metabolic organs. Lipokines can be produced by brown/beige adipose tissue and by WAT. In humans, serum concentration of palmitic-acid-9-hydroxy-stearic-acid (PAHSA), a specific FAHFA, is positively correlated with insulin sensitivity, however, mounting evidence suggests that PAHSA and the rest of lipokines play a pivotal role in the regulation of metabolism; actually, they are currently being proposed as treatment for disorders such as obesity, type 2 diabetes, and other chronic diseases [23, 33, 34].

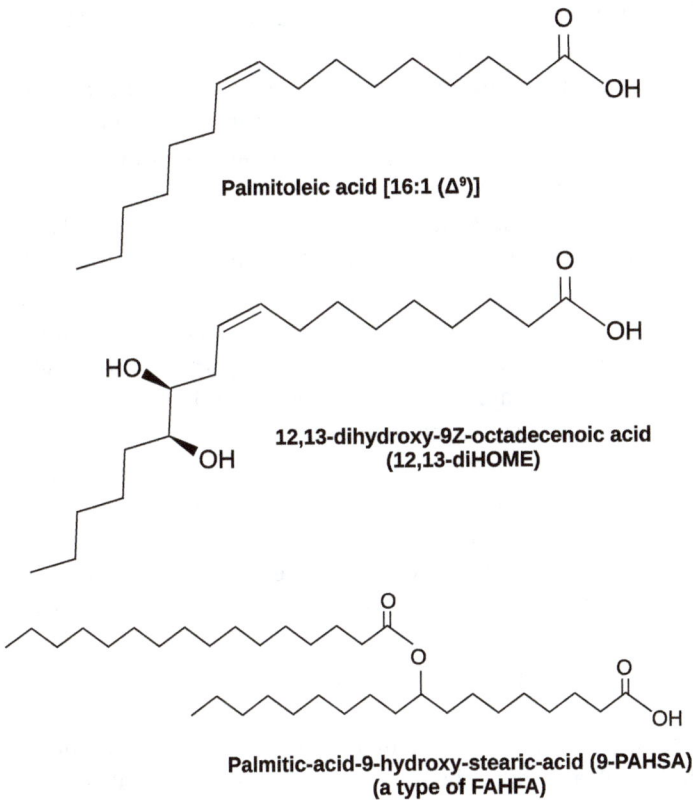

Palmitoleic acid [16:1 (Δ⁹)]

**12,13-dihydroxy-9Z-octadecenoic acid
(12,13-diHOME)**

**Palmitic-acid-9-hydroxy-stearic-acid (9-PAHSA)
(a type of FAHFA)**

Figure 14.3: Chemical formulae of some lipokines.
Formulae were elaborated using the structural information available at National Center for Biotechnology Information. PubChem Database (accessed on July 26, 2023):
Palmitoleic acid, CID = 445638, https://pubchem.ncbi.nlm.nih.gov/compound/Palmitoleic-acid
12,13-DiHOME, CID = 10236635, https://pubchem.ncbi.nlm.nih.gov/compound/12_13-DiHOME
9-Pahsa, CID = 72189985, https://pubchem.ncbi.nlm.nih.gov/compound/9-Pahsa

14.4 Pancreatic hormones

14.4.1 Insulin

Insulin is a peptide hormone of 51 amino acids secreted by the β-cells located in the pancreatic islets of Langerhans. It is initially synthesized as preproinsulin, a single-chain precursor that consists of four domains: the amino-terminal signal peptide, the B-chain, the C-peptide, and the carboxy-terminal A-chain. Proinsulin is produced from preproinsulin by cleaving off the signal peptide and, finally, insulin is produced within secretory vesicles by cleaving the C-peptide from proinsulin. Mature insulin consists of two independent peptide chains: the A-chain (21 residues) and the B-chain

(30 residues), which are connected by two disulfide bonds with an additional intra-chain disulfide bond within the A-chain [35].

Insulin promotes the synthesis of energy storage molecules and other processes characteristic of the so-called absorptive state, that is, the period of 3–4 h after a typical meal, during which nutrients are absorbed and the rate of energy input generally exceeds energy output. Insulin secretion is directly triggered by glucose and increased by the plasma concentrations of glucose, amino acids, the hormones glucose-dependent insulinotropic peptide (GIP) and glucagon-like polypeptide 1 (GLP-1), and by parasympathetic nerve activity. By contrast, sympathetic nerve activity and high plasma concentrations of adrenaline decrease insulin secretion. Insulin promotes energy storage by stimulating the synthesis of fatty acids and triglycerides in the liver and adipose tissue, glycogen in liver and skeletal muscle, and proteins in most tissues. At the same time, it inhibits the breakdown of proteins, triglycerides, and glycogen, and suppresses gluconeogenesis by the liver (Figure 14.4). Insulin acts on a variety of target tissues not only by influencing major aspects of energy metabolism but also by affecting the transport of nutrients across cell membranes (except in the liver and central nervous system). In this respect, it has been reported that insulin stimulates the uptake of amino acids in many cell types [3]; moreover, it has been reported that most, if not all, tissues of the body, including the brain, express insulin receptors (InsR) and are, consequently, insulin sensitive [36].

Along with leptin, insulin is the principal anorexigenic hormone, also reducing expression of NPY and AgRP, and increasing that of POMC in the hypothalamus (see above). In connection with this, studies performed in mice have shown that the effect of insulin of POMC neurons is more subtle than previously thought, thus, the proportion of these neurons activated by insulin depends on the regulation of the insulin receptor signaling. This regulation is accomplished by the phosphatase TCPTP, which is increased by fasting, degraded after feeding, and elevated in diet-induced obesity [41].

Insulin and leptin regulate body weight maintenance in a concerted way, although their signal transduction pathways are distinct. However, some experimental evidence suggest a convergence of leptin and insulin intracellular signaling, actually, it has been shown that insulin potentiates leptin-induced signaling in the central nervous system. Resistance to the central actions of leptin or insulin is linked to the emergence of obesity and diabetes mellitus [42].

Recent evidence suggests that insulin modulates cognition, memory, and mood, whereas central defects in insulin action may contribute to the development of Alzheimer's disease [36, 43].

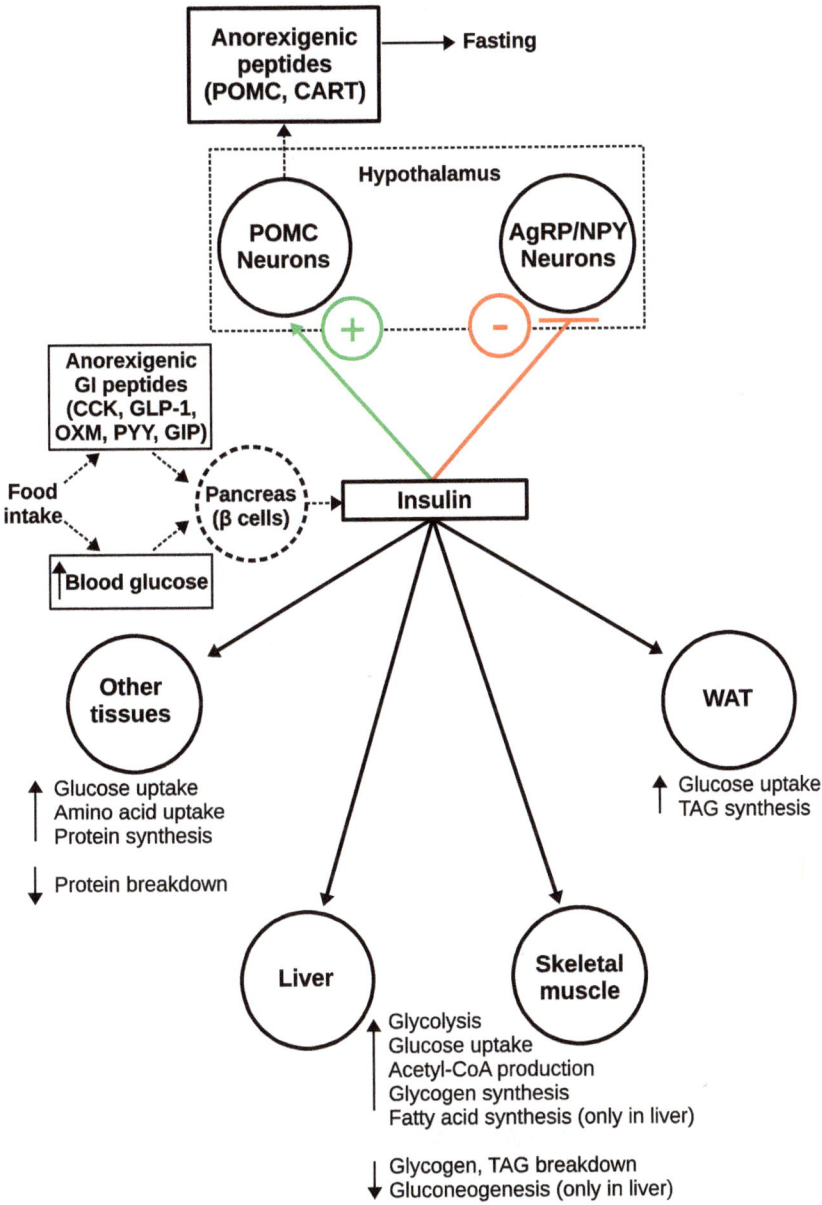

Figure 14.4: Simplified scheme of the effects exerted by insulin in the human body. WAT: white adipose tissue. Figure elaborated with information obtained from references [25, 37, 38].

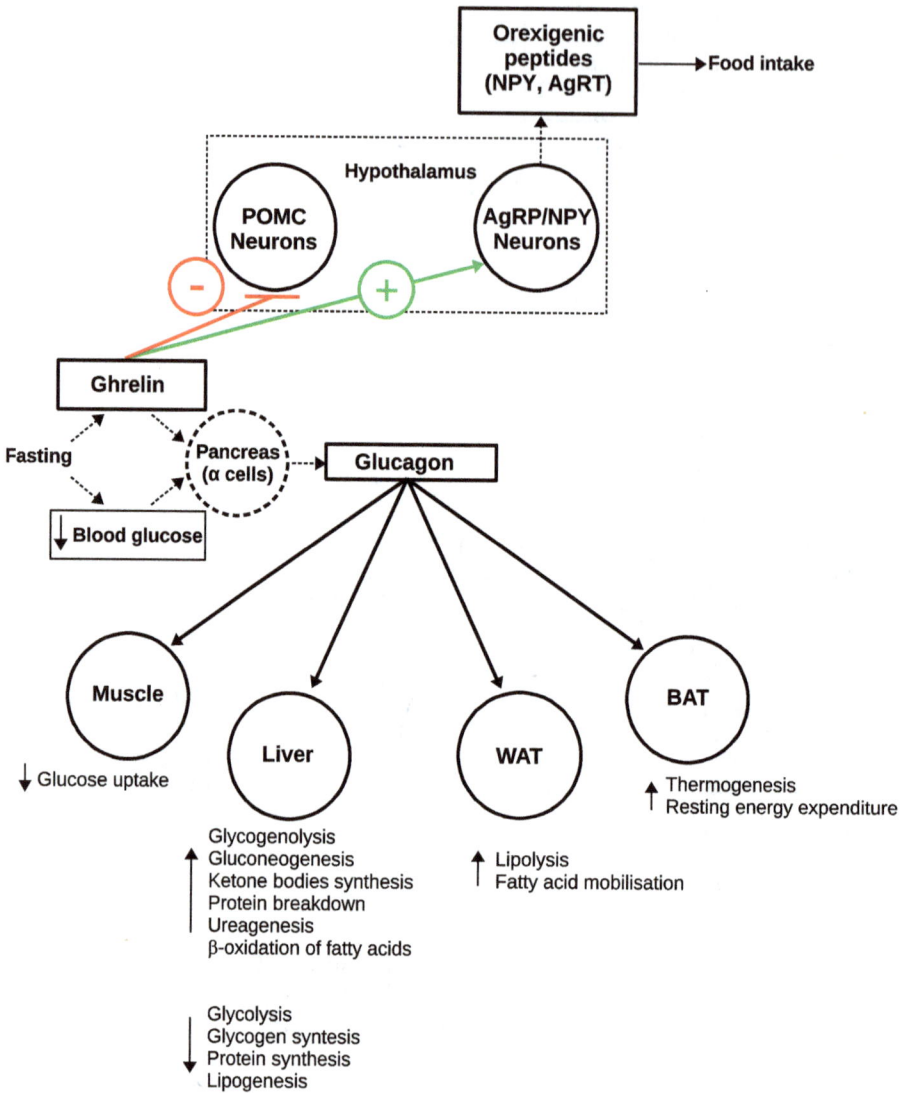

Figure 14.5: Simplified scheme of the effects exerted by glucagon and ghrelin in the human body. WAT: white adipose tissue; BAT: brown adipose tissue. The effects of glucagon on BAT require low insulin levels. Figure elaborated with information obtained from [25, 39, 40].

14.4.2 Glucagon

Glucagon is a peptide hormone of 29 amino acids produced by enzymatic cleavage of a proglucagon precursor. Processing of proglucagon is tissue specific and depends on the predominating prohormone convertase cleaving the prohormone into smaller entities. In pancreatic α-cells, proglucagon is cleaved to the fully processed 29-amino acid glucagon peptide (glucagon 33–61), in parallel with glicentin-related pancreatic polypeptide (GRPP) and a minor amount of the longer glucagon 1–61 peptide. In intestinal enteroendocrine cells, the prohormone cleavage results in equimolar amounts of glicentin, glucagon-like peptide 1 (GLP1 – Section 14.5) and GLP2 (although glucagon may also be secreted under certain circumstances). Gut-derived glicentin is further processed and released as oxyntomodulin and GRPP [39]. Glucagon is secreted in response to a number of factors [39, 44]:

- Inhibitors, that can be classified in:
  - Metabolic, that include hyperglycemia or glucose, carbohydrate meal, non-esterified fatty acids, isoleucine, and ketones
  - Chemicals such as ATP, gamma-aminobutyric acid (GABA), and Zn^{2+}
  - Hormones like pancreatic β-cells factors amylin, insulin, δ cells-derived somatostatin, the intestinal peptide GLP-1, leptin and secretin. It is unclear whether glucagon controls its own secretion in an autocrine manner.
- Stimulators:
  - Metabolic, that include hypoglycemia, protein meal, most amino acids, and some fatty acids
  - Physiological, such as stress, cold exposure, parasympathetic nerves, and sympathetic nerves
  - Hormones like adrenalin, cholecystokinin, gastric releasing peptide, glucose-dependent insulinotropic polypeptide (GIP), ghrelin, oxytocin, vasopressin, and vasoactive intestinal peptide

The multiple effects of glucagon in several target tissues are illustrated by the broad distribution of the glucagon receptor, that is mainly found in the liver but is also present in the kidneys, heart, adipocytes, lymphoblasts, spleen, brain, adrenal glands, retina and gastrointestinal tract [39].

Glucagon opposes the action of insulin by promoting processes during the time between meals when nutrients are not being absorbed, the so-called postabsorptive state. The overall effect of glucagon is the mobilization of energy stores and the synthesis of energy sources that can be used by tissues. In the liver, glucagon promotes glycogenolysis and gluconeogenesis (that is, glucose production), ketone bodies synthesis, and breakdown of proteins, while inhibiting both glycogen and protein syntheses. In adipose tissue, glucagon stimulates lipolysis and suppresses triglyceride synthesis. Excess glucagon in blood (hyperglucagonemia) has been related to obesity and liver fat content [3, 39].

The combined actions of insulin and glucagon control plasma glucose concentration through negative feedback (Figures 14.4 and 14.5), thus, an increase in plasma glucose concentration increases insulin secretion and decreases glucagon secretion from the pancreas, both of which cause a decrease in glucose levels. Conversely, a decrease in plasma glucose concentration is corrected by a decrease in insulin secretion and an increase in glucagon secretion. Moreover, glucagon and insulin control each other's secretion, with insulin inhibiting the secretion of glucagon from α-cells, and glucagon inhibiting the secretion of insulin from β-cells. Paradoxically, exogenous glucagon decreases food intake and promotes body weight loss in several species, including humans; however, the regulatory mechanisms controlling glucagon-induced satiety are poorly understood [3, 39].

14.5 Gut hormones

The gastrointestinal tract is the body's first point of contact with ingested nutrients and is a highly active organ from the metabolic point of view. At least 15 types of enteroendocrine cells are distributed along the mucosa of the entire GI tract. They produce and secrete more than 20 different hormones that act as autocrine, paracrine, and endocrine regulators of processes such as secretion, gut motility, and food intake and absorption. Among these hormones, ghrelin, cholecystokinin (CCK), peptide YY (PYY), and glucagon-like peptide-1 (GLP-1), have been identified as major regulators of food intake, as well as glucose and energy homeostasis [38, 40].

– Ghrelin is an orexigenic hormone of 28 amino acids in which the third residue (a serine) is modified by n-octanoic acid. It is produced mainly in the stomach and has unique functions in the regulation of energy homeostasis, including the abilities to communicate current peripheral nutrition status with the hypothalamus and perform energy compensation. Ghrelin stimulates the synthesis of the orexigenic peptides NPY and AgRP, and opposes the anorexigenic effects of leptin in POMC neurons (Figure 14.5). It also exerts actions in brain areas involved in reward, memory, and seeking behavior, and suppresses insulin secretion. Circulating ghrelin levels peak prior to nutrient ingestion and decrease rapidly after a meal, while its administration promotes a feeling of hunger and stimulates energy intake. Some studies suggest that ghrelin safeguards critical organs from metabolic stress and chronic metabolic inflammation (metaflammation) [4, 38, 45]. Ghrelin is the only metabolic signal known to potently activate the hypothalamic AgRP/NPY neurons [40].

– CCK is a peptide hormone secreted by the upper intestine (duodenum and jejunum), pancreas, and certain enteric and central neurons, in response to nutrient intake, in particular lipids and protein. It is released in the bloodstream 10 min after eating onset and peaks at 60 min, remaining elevated until 7 h. CCK is present in various forms (with 8, 22, 33, 39, and 58 amino acids) that inhibit energy intake and gastric secretion, while stimulating insulin secretion [4, 38]. CCK binds to the CCK1 receptor (CCK1R), that is widely expressed in the vagal afferents, brainstem, and hypothalamus [40].

– GLP-1 is released post-prandially by enteroendocrine cells located at the small intestine and the colon in response to nutrient sensing and through stimulation by bile acids. GLP-1 receptors are distributed on central and peripheral organs and tissues, including the hypothalamus, liver, skeletal, and muscle. It exerts anorexigenic effects, such as reduction of appetite and energy intake, delaying of gastric emptying, promotion of insulin secretion and β-cell proliferation, and suppression of glucagon secretion [38].

– Oxyntomodulin (OXM) is co-secreted with GLP-1 and binds to GLP-1 and glucagon receptors, as well as to unknown receptors in the hypothalamus. It decreases energy intake, delays gastric emptying, and causes glucose-dependent insulin secretion. Exogenous administration of OXM is associated with weight loss [38].

– PYY, and more specifically its endogenous isoform PYY_{3-36} (two amino-terminal tyrosine residues plus 34 amino acids), is released by the lower intestine (distal intestine, colon, and rectum), pancreas, and brainstem, and binds to various receptors in the hypothalamus, the vague, and the brainstem. It reduces appetite and energy intake, delays gastric emptying, and promotes insulin secretion. PYY_{3-36} levels in the blood rise after a meal and remain high for some hours [4, 25, 38].

– Glucose-dependent insulinotropic polypeptide (GIP) is secreted by enteroendocrine K-cells in response to dietary lipids and interacts with receptors at the pancreatic islet cells, hypothalamus, and adipose tissue. GIP stimulates insulin and glucagon secretion, exerts anti-apoptotic function in pancreatic beta cells, increases lipogenesis in white adipose tissue and reduces energy intake [38, 40].

Gut hormones usually act synergistically, thus, GLP-1 and PYY in combination have a more potent effect than either of the two hormones alone, whereas OXM, GIP, and CCK enhance the effects of GLP-1. Other effects exerted by gut hormones, as well as their underlying mechanisms, remain incompletely understood [38].

14.6 The liver

The liver is one of the metabolically most active and versatile organs. About 25% of its blood supply comes via the hepatic artery, the remaining 75% being delivered via the portal vein, which drains blood coming from the intestine. Consequently, the liver is the first organ to get in contact with absorbed nutrients, ingested toxins, and products from intestinal microorganisms. Its major functions are:

- Processing and storage of nutrients, guaranteeing an adequate nutrient supply for other organs during both the absorptive and postabsorptive state.
- Detoxification and excretion of xeno- and endobiotics into the bile.
- Participation in pathogen defense and immune functions and to trigger acute phase responses in inflammation.

–   Homeostatic functions such as maintenance of acid-base and glucose homeostasis, synthesis of most plasma proteins, triglyceride and cholesterol metabolism and transport, and hormone processing and secretion. Furthermore, bile acids (BAs) are synthesized in the liver, which aid in triglyceride digestion in the intestine and are increasingly recognized as important signaling molecules and coordinators of interorgan metabolism.

Seventy percent of the hepatic cell mass are made up by parenchymal cells, also called hepatocytes [46], that are responsible for the transformation of dietary nutrients into the fuels and precursors required by other tissues. The types and amounts of nutrients received by the liver vary with several factors, including the diet and the time between meals, whereas the demands of extrahepatic tissues varies among organs and with the level of activity and overall nutritional state [25].

In the absorptive phase, when plasma insulin levels increase, the liver synthesizes and stores glycogen from gluconeogenic precursors and from absorbed glucose, whereas in the postabsorptive state, when insulin levels are low and glucagon levels are high, glucose is mobilized from glycogen. After exhaustion of glycogen stores, gluconeogenesis (mainly from glucogenic amino acids, glycerol, and lactate) is stimulated through induction of enzymes of amino acid metabolism, gluconeogenesis, and the urea cycle. Therefore, the liver acts as a "glucostat," being the central organ of glucose homeostasis [46].

The liver is also a major organ for synthesis of triglycerides, cholesterol, and sphingolipids, and secretes very low-density lipoproteins. In the postabsorptive state, fatty acid oxidation provides energy for the liver, while ketogenesis and ketone bodies release provide energy for other organs, such as the brain. As far as the amino acid metabolism is concerned, the liver constantly renews its own proteins, and is also the site of biosynthesis of most plasma proteins. Alternatively, amino acids can be exported to other organs, used for the biosynthesis of nucleotides, hormones, and other nitrogenous compounds, or degraded to obtain intermediates of citric acid cycle and, eventually, energy. The liver can also metabolize amino acids that come from other tissues, especially alanine from muscles, that can be converted to pyruvate and then to glucose and exported back to the circulation. This glucose-alanine cycle contributes to maintain blood glucose levels between meals [25, 46].

14.6.1 Control of hepatic function

Hepatic energy metabolism is regulated by insulin, glucagon, and other metabolic hormones, which dynamically modulate gluconeogenesis, β oxidation, and lipogenesis in order to meet a systemic metabolic demand (Figures 14.2, 14.4, and 14.5). The mechanisms of hormonal action are complex and involve different levels of regulation, including modulation of enzymatic activities and control of gene expression through numerous transcription factors and coregulators. The liver also has close communications with extrahepatic tissues and organs, including adipose tissue, skeletal muscle, gastrointestinal tract, and brain. Dysregulation of metabolism in the liver promotes insulin resistance, diabetes, and nonalcoholic fatty liver diseases [47].

14.6.2 Metabolites of the liver affecting other tissues

In addition to its central role in metabolism and detoxification of xenobiotics, the liver serves as an endocrine organ by synthesizing and secreting signaling molecules, such as calcidiol (Section 6.4.2), angiotensinogen, hepcidin (Section 13.7.5), insulin-like growth factors 1 and 2, thrombopoietin, hepatokines, ketone bodies, and bile acids [48–50]:

- Angiotensinogen (AGT) is a prohormone and the unique substrate of the renin-angiotensin system, a hormonal cascade that regulates blood pressure and cardiac contractility, as well as renal sodium and water absorptions. AGT is converted into a series of angiotensin peptides through sequential cleavages, with angiotensin II being the major vasoconstrictive hormone.
- Insulin-like growth factors 1 and 2 (IGF-1 and IGF-2) are homologous peptide hormones that control growth, anabolic activities, metabolism, and aging. IGF-1 mediates the actions of growth hormone (see below) and binds to insulin receptors and hybrid IGF-1/insulin receptors; consequently, IGF-1, growth hormone, and insulin are hypothesized to constitute a regulated axis that inform cells about nutritional status [49].
- Thrombopoietin (TPO) controls platelet production.
- Hepatokines are proteins secreted by hepatocytes that influence metabolism through autocrine, paracrine, and endocrine signaling. They are the main mediators in the crosstalk between the liver, non-hepatic target tissues, and the brain. More than 500 hepatokines have been identified to date, including fetuin-A, fibroblast growth factor 21 (FGF21), activin E, Tsukushi, and glycoprotein nonmetastatic melanoma protein B (GPNMB).
- Ketone bodies and bile acids are involved in endocrine, paracrine, and autocrine effects already confirmed in humans. Bile acids might play an important role in modulating the microbiota–brain interaction.

The liver is also involved in the metabolism (activation and inactivation) of multiple endocrine hormones, including thyroid hormones, glucagon-like peptide-1, and steroid hormones, thereby contributing to the control of hormonal balance in the body [46, 49]. Moreover, novel liver endocrine functions have been unraveled in the last several years, such as endocrine regulation of pancreatic α-cells, adipose tissue, and insulin sensitivity [49].

14.7 Other hormones that affect energy metabolism

– Growth hormone (GH). The primary function of GH, a peptide hormone that is secreted by the anterior pituitary, is to promote postnatal longitudinal growth by inducing bone growth; however, in the last three decades GH has been recognized to play
an important role in health after the cessation of growth. Human studies have revealed important regulatory roles of GH in lipid, carbohydrate, nitrogen, and mineral
metabolism, as well as electrolyte balance. It increases lipolysis in adipocytes and
amino acid uptake and nitrogen retention in muscle, thereby decreasing body fat,
while increasing and maintaining muscle mass and strength. GH has effects on the
immune system, cardiovascular system, neurogenesis and the central nervous system,
and aging. The pleiotropic actions of GH explain the clinical picture of increased adiposity, reduced lean mass, and impaired physical and psychological function in GH
deficient adults [51, 52].

– Thyroid hormones (TH), mainly 3,3',5-triiodo-L-thyronine or T_3 (Chapter 7, Section 7.3.4), regulate metabolic processes essential for normal growth and development, as well as general metabolism in adults. TH influences key metabolic pathways
that control energy balance by regulating energy storage and expenditure through actions in the brain, white fat, brown fat, skeletal muscle, liver, and pancreas. TH status
correlates with body weight and energy expenditure, thus, excess thyroid hormone
(hyperthyroidism) promotes a hypermetabolic state characterized by increased resting energy expenditure, weight loss, reduced cholesterol levels, increased lipolysis,
and gluconeogenesis. Conversely, reduced TH levels (hypothyroidism) are associated
with reduced resting energy expenditure, weight gain, increased cholesterol levels, reduced lipolysis, and reduced gluconeogenesis [53]. A recent review on the biochemical
properties of TH can be found elsewhere [54].

– Glucocorticoids are steroid hormones produced at the adrenal cortex whose secretion is stimulated by the adrenocorticotropic hormone (ACTH) from the anterior pituitary, which in turn is stimulated by corticotropin releasing hormone (CRH) from the
hypothalamus. Cortisol is the primary glucocorticoid released from the adrenal cortex, being secreted in bursts with variable frequency along the day: higher in the
morning and lower at night. Stress, whether physical or emotional, is an important
stimulus for cortisol secretion. The presence of glucocorticoids is essential to mobilize
fuels in response to signals from other hormones, such as insulin and glucagon. The
primary actions of glucocorticoids are to maintain normal concentrations of enzymes
necessary both for the breakdown of proteins, fats, and glycogen, and for the conversion of amino acids to glucose in the liver; for this reason, they are necessary for survival during prolonged fasting. At doses that exceed normal physiological levels,
glucocorticoids inhibit inflammation and allergic reactions, however, they must be administered with caution because of their immunosuppressive actions [3]. It has been

recently proposed that glucocorticoid excess may impair hormonal appetite signaling and, consequently, eating control in the context of obesity [55].

– Adrenaline or epinephrine (IUPAC systematic name: 4-[(1R)-1-hydroxy-2-(methyla-mino)ethyl]benzene-1,2-diol) is a hormone derived from the amino acid tyrosine, which has profound effects on the cardiorespiratory system, thus, it increases heart rate, blood pressure and dilates respiratory passages. It is also a mediator of the fight-or-flight response and participates in the counter-regulatory response to hypoglyce-mia (low levels of glucose in blood) by promoting glycogenolysis in the liver and skel-etal muscle, gluconeogenesis in the liver, fatty acid mobilization in adipose tissue, and reduction of glucose uptake by tissues such as skeletal muscle. Adrenaline interacts with other hormones, thus, it promotes the secretion of glucagon and inhibits that of insulin, while its biosynthesis is strongly influenced by glucocorticoids [25, 56].

Many of the hormones described above are also involved in the regulation of blood glucose levels (Chapter 3, Section 3.9).

14.8 Obesity

14.8.1 Definition and measure of obesity

The World Health Organization (WHO) defines overweight and obesity as "abnormal or excessive fat accumulation that may impair health" [57]. Obesity is characterized not only by total fat mass but also by its distribution, including ectopic (i.e., in an ab-normal location) lipid accumulation, which can have a systemic influence on glucose/lipid metabolism and inflammation [34].

The body mass index (BMI) is a simple index of weight-for-height commonly used to classify overweigh and obesity in adults. It is calculated by dividing a person's weight in kilograms by the square of his height in meters (kg/m^2). Cut-off BMI values are [58]:
– Underweight < 18.5
– Normal weight 18.5–24.9
– Overweight 25–29.9
– Obesity class I 30–34.9
– Obesity class II 35–39.9
– Obesity class III ≥ 40

In the case of children and adolescents, age needs to be considered when defining overweight and obesity [57]:
– Under 5 years of age: overweight is weight-for-height greater than 2 standard de-viations above WHO Child Growth Standards median; and obesity is weight-for-

height greater than 3 standard deviations above the WHO Child Growth Standards median.

– Children aged between 5 and 19 years: overweight is BMI-for-age greater than 1 standard deviation above the WHO Growth Reference median; and obesity is greater than 2 standard deviations above the WHO Growth Reference median.

BMI has several limitations because is not a direct measure of adipose tissue mass, thus, on the one hand, it overestimates adiposity in patients with edema or in athletes with high muscle mass and, on the other, it underestimates adiposity in elderly patients with sarcopenia. Moreover, BMI does not reflect the impact of adiposity on the health of the patient, although this can be partially solved by incorporating waist circumference measurements, which provide additional information regarding fat distribution and cardiometabolic disease risk. In connection with this, it has been proposed that there are three separate fat depots within the body with divergent effects on the development of metabolic disease: (a) subcutaneous adipose tissue, that tends to confer less metabolic morbidity, (b) visceral adipose tissue, that is related to a higher risk of type 2 diabetes mellitus, and (c) ectopic (liver and muscle) fat, that is non-adaptive and associated with metabolic dysfunction. These depots are sexually dimorphic; on average, men have more visceral fat, while women have larger subcutaneous fat stores [59].

Nowadays, BMI is considered as a screening tool for obesity, but evaluation and diagnosis require a clinical evaluation to assess the risk and severity of weight-related complications. Clinical evaluation must include aspects such as physical examination, review of different organ systems (cardiovascular, skeletal, gastrointestinal . . .), and biochemical analyses (blood levels of glucose, glycated hemoglobin, total cholesterol, triglycerides, HDL cholesterol, LDL cholesterol, thyroid hormones . . .). Additional factors that must be considered in the evaluation of patients with obesity are the presence of other diseases, the psychological status, and the use of iatrogenic medication [58].

14.8.2 The prevalence of obesity

Recent data from the World Health Organization (WHO) suggest that more than half of the adult population worldwide is currently overweight or obese, and researchers project that the obesity prevalence will increase by 33% over the next two decades, impacting further on global health burden and cost [60]. Using data from 2016 the WHO has estimated that [57]:

– More than 1.9 billion adults aged 18 years and older were overweight. Of these over 650 million adults were obese.
– 39% of adults aged 18 years and over (39% of men and 40% of women) were overweight.

- Overall, about 13% of the world's adult population (11% of men and 15% of women) were obese.
- While just under 1% of children and adolescents aged 5–19 were obese in 1975, more 124 million children and adolescents (6% of girls and 8% of boys) were obese in 2016.
- The prevalence of overweight and obesity among children and adolescents aged 5–19 has risen from 4% in 1975 to just over 18% in 2016. The rise has occurred similarly among both boys and girls: in 2016 18% of girls and 19% of boys were overweight.
- The worldwide prevalence of obesity nearly tripled between 1975 and 2016.

Obesity is nowadays considered a pandemic, being linked to more deaths worldwide than underweight. The WHO estimates that, globally, there are more people who are obese than underweight, actually, this situation occurs in every region of the world except in parts of sub-Saharan Africa and Asia [57].

14.8.3 Causes of obesity

Although the fundamental cause of obesity and overweight is an imbalance between energy consumed and energy expended (Section 14.2), in reality, obesity is a complex pathology that results from the interaction of environmental and genetic factors:
- Environmental factors have been traditionally included in the so-called obesogenic environment (availability of inexpensive calorie dense food, technologies and structure of communities that reduce or replace physical activity, and inexpensive nonphysical entertainment); however, other factors have been recently suggested to contribute to obesity. These include microorganisms, viruses, increasing maternal age, fetal malnutrition, greater fecundity among people with higher adiposity, sleep debt, early exposure to endocrine-disrupting chemicals (EDC), and pharmaceutical iatrogenesis, among others [50, 59, 61].
- Genetic factors largely determine the susceptibility to gain weight, thus, twin and family studies report heritability values of 40%–75% for body mass index. In the past decades, the causal relationships between certain genes and obesity have been elucidated and several types of obesity have been described according to the number of genes involved [27, 62]:
  - Polygenic obesity, the most common clinical presentation, results from the cumulative effects of multiple genetic variants and their interactions with the environment. Genome-wide association studies (GWAS) of obesity have identified over 300 gene variants associated with obesity; however, GWAS evaluation also suggests that most of metabolic syndrome (Section 14.8.4) are explained by the environment rather than by genetics (85% vs. 15%) [59].

- Monogenic non-syndromic obesity is described as rare and severe early onset obesity associated with endocrine disorders. It is mainly due to mutations in genes of the leptin/melanocortin pathway (Section 14.3.1).
    - Monogenic syndromic obesity is associated with additional phenotypes (mental retardation, dysmorphic features, and organ-specific developmental abnormalities). Prader-Willi (PWS) and Bardet-Biedl (BBS) are the syndromes most frequently linked to obesity.
    - Oligogenic obesity is characterized by a variable severity of the disease, partly dependent on environmental factors and the absence of a specific phenotype. It is responsible for 2–3% of obesity in adults and children.

In humans eating behavior and appetite present homeostatic and hedonic aspects that are regulated by different mechanisms at the central nervous system. The physiological need to eat (homeostatic) is mainly controlled by the hypothalamus as previously described in this Chapter; on the contrary, hedonic feeding is mediated by mesolimbic reward circuits in response to food cues such as sight, smell, and taste. The hedonic aspect may override homeostatic regulation by increasing the desire to consume highly palatable foods even in situations of abundant energy stores. This situation illustrates the fact that the mechanisms that control the homeostatic and hedonic aspects of eating are deeply interconnected within our brains. The understanding of these interactions may have important implications for the treatment of obesity [55, 59, 63].

14.8.4 Obesity, health, and evolution

A general consensus exists about the association of overweight or obesity with several conditions and diseases [59, 64, 65]:
- Insulin resistance and type II diabetes
- Coronary heart diseases
- Cerebral vasculopathy and stroke
- Gallbladder lithiasis
- Dyslipidemia, that is, abnormal (normally, higher than normal) levels of lipids in the blood
- Osteoarthritis
- High blood pressure
- Metabolic syndrome (clustering of abdominal obesity, dyslipidemia, hyperglycemia, and hypertension)
- Asthma
- Liver (e.g., nonalcoholic fatty liver disease) and respiratory diseases
- Ovarian polycystosis
- Sleep apnea
- Reproductive problems

- Dementia
- Increased susceptibility to healthcare-associated (nosocomial) infections, wound infections, and pandemics, such as influenza and COVID-19
- Several types of cancer, such as meningioma, adenocarcinoma of esophagus, multiple myeloma, and cancers of thyroid, breast, liver, gallbladder, upper stomach, kidney, pancreas, endometrium, ovary, colon, and rectum [66].

A state of low-grade inflammation and immune dysfunction has been proposed to be the link between obesity and some of the pathologies listed above; on the other hand, the gut microbiome has also been found to play an important role in all the complications associated with an obese phenotype [59]. In this respect, a recent study performed in mice showed that an inflammatory mediator, prostaglandin E_2 (PGE$_2$), is induced in the hypothalamus by high-fat diet; moreover, PGE$_2$ can directly stimulate appetite-promoting neurons to exacerbate diet-induced obesity and fatty liver [67].

The Obesity Society (TOS) has stated that "obesity is not only an underpinning of major chronic diseases such as heart disease, cancer, stroke, and diabetes, to name a few, but can be a serious debilitating condition in its own right. TOS therefore advances an unequivocal position confirming obesity as a disease" [68]. In any case, the complexity of obesity and its health consequences is illustrated by the existence of the so-called obesity paradox (OP), that is, the existence of certain individuals with elevated BMI that show improved survival to cardiovascular diseases and cancer compared with normal-weight patients. OP has been explained by proposing that adiposity may not be necessarily unhealthy, but it depends on aspects such as regional fat distribution (Section 14.8.1), the adaptations of adipose tissue to excess caloric intake and the type of fat expansion [69, 70]. Conversely, a significant number of normal-weight individuals also show metabolic dysfunction, probably due to increased visceral fat; that is, they are "thin on the outside, fat on the inside" [59].

Obesity is not just limited to one group of people or to one range of age, but is prevalent in all ages and ethnic groups. This fact introduces two important questions from an evolutionary perspective: how has natural selection favored the spread of genes that increase the risk for an obese phenotype and how has such a susceptibility to obesity evolved? [65]. During the twentieth century, several ideas were put forward to answer these questions:

- The "thrifty genotype hypothesis" establishes that during human evolution, genes that enabled individuals to efficiently collect and process food and deposit fat during periods of food abundance would increase the probability to survive famine periods. These genes have become disadvantageous in modern societies, where food is easily available, resulting in the spread of obesity [71]. A critical point of this hypothesis is that many people with these thrifty genotypes do not develop the obese phenotype [65].
- In 1991, Lucas proposed the concept of "programming by early nutrition in man," a process "whereby a stimulus or insult at a critical period of development has

lasting or lifelong significance." The neonatal period was identified as one of these critical periods in preterm infants [72].

- Hales and Barker, also in 1991, proposed the "thrifty phenotype hypothesis" of the etiology of type II diabetes, according to which "poor nutrition in fetal and early infant life are detrimental to the development and function of the β-cells of the islets of Langerhans. Such defects of structure and function [. . .] predispose to the later development of Type 2 diabetes [. . .]." Indeed the complex interactions of the type and timing of nutritional defects in early life are suggested as underlying the pathogenesis of the variable abnormalities sometimes described as "Syndrome X" [73].

Nowadays, it is accepted that not only conditions of poor nutrition but also maternal fetal stress and exposure to toxic substances are capable of interfering with the programming of organs and tissues, playing a complex pathogenetic role. Barker's thrifty phenotype hypothesis has been expanded and is currently known as the Developmental Origins of Health and Disease (DOHaD) theory, defined by the DOHaD Society (https://dohadsoc.org) as: "a multidisciplinary field that examines how environmental factors acting during the phase of developmental plasticity interact with genotypic variation to change the capacity of the organism to cope with its environment in later life" [74]. According to Gluckman and colleagues "DOHaD should be viewed as a part of a broader biological mechanism of *plasticity* by which organisms, in response to cues such as nutrition or hormones, adapt their phenotype to environment. These responses may be divided into those for immediate benefit and those aimed at prediction of a future environment: disease occurs in the mismatch between predicted and realized future" [75]. The DOHaD theory aims at explaining the worldwide increase of chronic, degenerative, and inflammatory diseases such as obesity, diabetes, cardiovascular disease, neurodegenerative disease, and cancer, and their association with other factors such as environmental pollutants (obesogens, reviewed in [76]), gut microbiota, and specific nutrients [65].

The mechanisms that mediate the programming effects of diverse environmental insults, or how this memory is stored are unclear, but some have been postulated in the context of the DOHaD theory. These include: excessive exposure to glucocorticoids (GCs), dysregulation in the development of the hypothalamic–pituitary–adrenal axis, irreversible changes in organ structure, and alterations in gene expression through epigenetic mechanisms [77].

14.8.5 The role of epigenetics

The term "epigenetics" was introduced by Conrad Hal Waddington in 1942 and refers to the study of heritable changes in gene expression that do not involve changes to the underlying DNA sequence, that is, epigenetic change implies a variation of phenotype without a change in genotype. This is a natural phenomenon that can be influenced by factors such as age, the environment/lifestyle, and health status. These modifications not only underlie the manner in which cells differentiate, but they may also have damaging effects that result in diseases like cancer or obesity. The most important mechanisms that initiate and sustain epigenetic change include DNA methylation, histone modification, and noncoding RNAs (ncRNAs) (Figure 14.6) [65, 78].

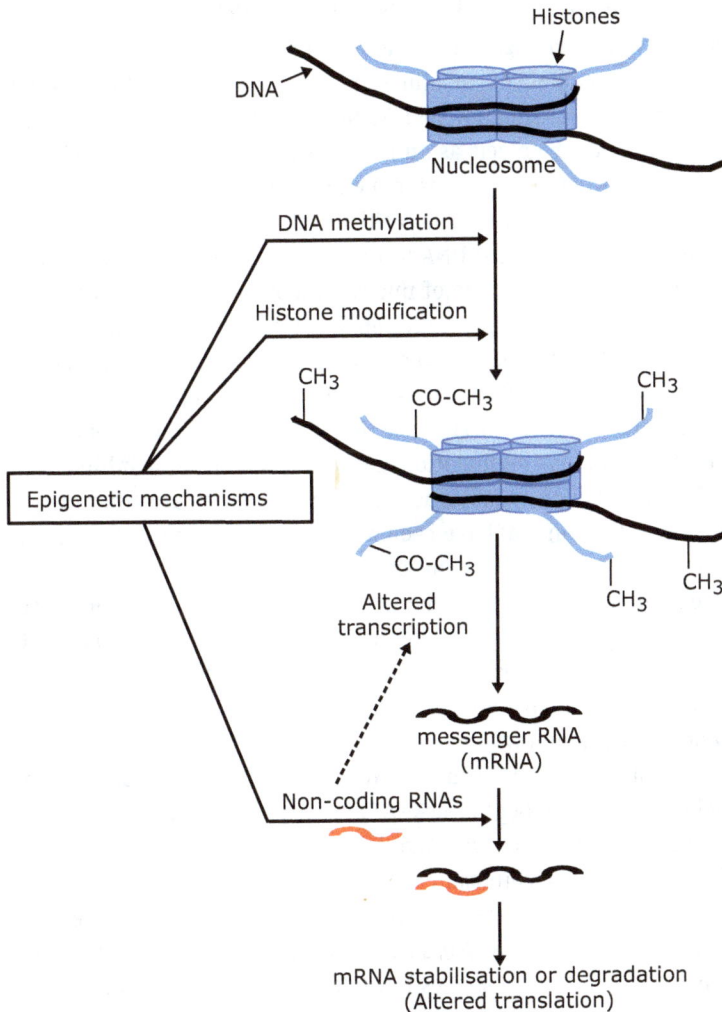

Figure 14.6: Overview of epigenetic mechanisms. Covalent modification of DNA and histones alter transcriptions of genes; noncoding RNAs may alter both transcription and translation.

- DNA methylation is the most widely studied epigenetic mechanism for the regulation of gene expression. It involves the covalent attachment of a methyl group to the 5' position of cytosine residues mediated by specialized enzymes, known as DNA methyltransferases. Methylation impedes the binding of transcription factors, thus resulting in repression of a particular locus. Changes in global patterns of DNA methylation is considered a hallmark of some diseases, particularly cancer; however, the association between global DNA methylation and obesity is in general inconsistent. In any case, studies in animal models and humans have demonstrated methylation changes in promoters of genes implicated in obesity, appetite control and metabolism, insulin signaling, immunity, inflammation, growth, and circadian clock regulation [65]. Methylated DNA can be oxidized into hydroxymethylated DNA, which acts as an intermediate for DNA demethylation, that has also been suggested to play a role in metabolic regulation [79].
- Histone modification. Histones are small positively charged proteins involved in the packaging of negatively charged DNA to form nucleosomes, the fundamental core unit of chromatin. The downsize of nucleosome stabilization is that it hinders transcription, replication, and DNA repair. It is therefore necessary to weaken the histone-DNA interaction for these cellular processes to occur. Modifications of N-terminal amino acids, such as acetylation, can reduce the positive charge on histone tails and hence destabilize the nucleosome. Other chemical alterations of histones, such as methylation, can either enhance or inhibit gene transcription. Phosphorylation, sumoylation, and ubiquitination are other mechanisms of histone modification that have been described. Enzymes involved in histone modification may have a role in obesity, thus, in response to high-fat diets and fasting, the medial hypothalamus changes the expression of neuropeptides involved in feeding, metabolism, and reproductive behaviors (see Section 14.3.1), which in turn regulate the expression of histone deacetylases. However, nowadays there is only a preliminary understanding about the dynamics of histone modification and its association with nutrient availability [65, 80].
- ncRNAs are functional RNA molecules that are transcribed from DNA but not translated into proteins. Two main groups of ncRNAs appear to be involved in epigenetic processes: short ncRNAs (<30 nucleotides) and the long ncRNAs (>200 nucleotides). In general, ncRNAs are involved in chromatin remodeling, as well as transcriptional and post-transcriptional regulation of gene expression. Among short ncRNAs there are three major classes: microRNAs (miRNAs), short interfering RNAs (siRNAs), and piwi-interacting RNAs (piRNAs). Several studies have demonstrated the involvement of both types of ncRNAs in body weight homeostasis and obesity comorbidities [81, 82].

Our genome cannot be changed to predispose the whole world to obesity in just a few decades, and epigenetic modifications provide a plausible mechanism by which the environment may have interacted with the genome to influence human health and disease. This view is currently being supported by results obtained from animal and human studies, actually, many genes are already known to be affected by environmental factors through epigenetic mechanisms. This epigenetic reprogramming has been suggested to affect critical pathways or regulatory processes, such as insulin signaling and secretion; adipocyte differentiation, transdifferentiation and function; mitochondrial function and redox regulation; lipid and glucose homeostasis; cytokine signaling and inflammation; cell cycle, apoptosis, and autophagy. The obesity spread can be explained, at least partially, by the evidence that environmental factors, such as lifestyle and nutrition, as well as pollutants and gut microbiota modifications, affect the epigenetic programming of parental gametes, the fetus, and the early postnatal development. Consequently, epigenetic marks induced *in utero* and in early life could determine a significant increase of obesity and other complex diseases, such as type II diabetes and cardiovascular disease, that can be transmitted transgenerationally [62, 65, 79]. Conversely, the beneficial effects of a balanced diet, physical activity, and even bariatric surgery might also be exerted through epigenetics mechanisms. This opens the possibility to explore alternative nutrition-based therapeutic approaches, and to develop tools for personalized diet to improve health and increase life expectancy [83].

14.8.6 Obesity and dietary patterns

Obesity is an extremely complex disease that is influenced by multiple factors; however, there is a consensus that dietary patterns play a key role in its development and prevalence. Consequently, dietary interventions are always an important part of the strategies designed to tackle obesity and other associated pathologies. A number of diets for weight loss and weight-loss management have been proposed, a list of some of these diets with brief descriptions and comments about them (based on information found in reference [84]) is shown below:

– Diets based on amount of food intake
   – Low-calorie diets. They involve consumption of 1,000–1,500 calories per day by typically restricting fats or carbohydrates (see below). Metabolic adaptations to decrease energy expenditure can lead to a plateau with this type of diet. It must be mentioned that calorie restriction (CE) (reduction of energy intake by ~10–40% while maintaining a balanced nutrient intake to prevent malnutrition) has been shown to have positive effects as an anticancer intervention in preclinical models [85]; moreover, several clinical and experimental evidence suggest that CE is able to delay aging and exert potent anti-inflammatory effects in different pathological conditions [86]. The mechanisms underlying the positive effects of CR on human health and longevity are the object of continuous research nowadays.
   – Very-low calorie diets provide <800 kcal a day. They are not recommended for routine weight management and should only be used in limited circumstances along with medical monitoring

- Very-low ketogenic diet. Consists of very-low-calorie (<700–800 kcal/day) and low-carbohydrate (<30–50 g/day) intake along with adequate protein consumption (equivalent to 0.8–1.2 g/day/kg of ideal body weight) for a short period, followed by a gradual switch to a low-calorie diet. It is recommended by some medical authorities in cases of severe obesity, sarcopenic obesity, and obesity associated with type II diabetes mellitus (T2DM), hypertriglyceridemia, and hypertension.
- Meal replacement. This diet is based on the substitution of soups, shakes, bars, and/or portion-controlled ready-made meals for "normal" food in one or more meals to reduce the daily calorie intake. They are not normally successful in the long term.
- Diets based on types of food eaten
  - Low-fat diets. They usually consists of diets with very low (≤ 10% of calories from fat) to moderate (≤30% of calories from fat and <7–10% from saturated fatty acids) fat content. Diets low in saturated fatty acids supplemented with good-quality fat and fibers and combined with total calorie restriction, are successful in order to achieve weight management and to prevent some types of cancer (colorectal and breast).
  - Low-carbohydrate diets encompass a range of carbohydrate intake from 50–130 g/day or 10%–45% of total energy from carbohydrates. With carbohydrate intake <10% of total energy (or <20–50 g/day), nutritional ketosis can occur (see below). In this situation, daily protein intake is usually 0.8–1.5 g/kg of ideal body weight to preserve lean body mass. Low-carbohydrate (low-carb) diets have been widely used not only for weight reduction, but also to manage T2DM
  - Ketogenic diets (KDs) are characterized by an extreme reduction in carbohydrate intake (<50 g/day) and a relative increase in the proportions of protein and fat. Ketogenic diet can suppress hunger during calorie restriction and may have some therapeutic effects on T2DM, polycystic ovary syndrome, and cardiovascular and neurological diseases. The standard KD is based on intake of ~80% fat, 10% protein, and <1% of carbohydrates by weight (95% of calories derived from fats), supplemented with a mix of vitamins and minerals to avoid malnutrition (there are variations with higher amounts of protein, accounting for 30–40% of the total caloric intake). The amino acid composition of these diets must be carefully considered to ensure the induction of a glucose-restricted physiological state. Standard KDs have shown promising results in cancer treatment using preclinical models [85].
  - High-protein diets. They usually refer to an increased protein intake to 30% of the total daily calories or 1–1.2 g/kg of the ideal body weight per day. They can provide a satiating effect that helps decrease energy intake and maintain successful weight loss; moreover, they increase the thermic effect of food, that is, the energy expenditure associated to food digestion and absorption (Section 14.2.1). Diets high in protein pose a potential risk to the kidneys due to their associated protein induced acid loads, such as the sulfuric acid produced from

oxidation of methionine and cysteine. They are also associated with increases in serum urea level and urinary calcium excretion, which might be related to a higher risk of kidney stone formation. It has also been suggested that long-term overactivation of mTORC1 (a eukaryotic protein complex that coordinates cellular growth and metabolism and responds to the availability of nutrients, such as amino acids) appears to be associated with adverse health outcomes [87].

- Mediterranean diet (See Chapter 15)
- Others
  - Paleolithic (Paleo) diet advises consumption of lean meat, fish, vegetables, fruits, and nuts while avoiding grains, dairy products, processed foods, and added sugar and salt. This is supposed to follow the nutritional patterns of humans who lived in the Paleolithic era (from more than 2 million years ago until about 10,000 years ago). This diet has been reported to have favorable effects on lipid profile, blood pressure, and circulating C-reactive protein concentrations, although the evidence is not yet conclusive. The Paleo diet can be high in saturated fats, which might increase the risk of cardiovascular disease.
  - Low glycemic index/glycemic load diet (See Section 3.7 and Table 3.3)
  - The Nordic diet is based on unprocessed whole grains, high-fiber vegetables, fish, low-fat dairy foods, lean meat of all types (beef, pork, lamb), beans and lentils, fruit, dense breads, tofu, and skinless poultry. It recommends more calories from plant foods (fewer from meat) and more foods from the sea, lakes, and the wild countryside. It has been reported that adherence to the Nordic diet significantly improved body weight.
  - Vegetarian/vegan diets. All dietary guidelines recommend diets based on foods of plant origin, that is, vegetables, fruits and whole grains (Section 1.3). Vegetarian diets can lower the risk of ischemic heart disease, T2DM, and cancer, reduce blood pressure, lipid profiles, and inflammatory biomarkers, while improving glycemic control and other cardiometabolic parameters. Strictly vegetarian diets exclude meat, fish, and poultry, but there are many variations like lactovegetarians and lacto-ovo-vegetarians diets. Veganism has been associated with adverse health outcomes, namely, nervous, skeletal, and immune system impairments, hematological disorders, as well as mental health problems due to potential deficits of micro (vitamins D and B12, calcium iron, zinc) and macronutrient (essential amino acids). These potential dangers are especially important in the case of vulnerable populations, such as children, adolescents, pregnant and breastfeeding women, and fetal outcomes in strict vegan mothers [88]. Consequently, strict vegetarian/vegan diets must be closely monitored by experts in nutrition.
  - Dietary approaches to stop hypertension (DASH) diet. This diet was originally developed to lower blood pressure without medication and includes many vegetables, fruits, and grains with an emphasis on whole grains. Low-fat or non-fat dairy foods, pulses, nuts, seeds, lean meats, poultry, and seafood are

also allowed. The diet limits sodium intake to 2,300 mg/day and can reduce the risk of cancer, cardiovascular diseases.

–   The portfolio diet is a vegan plan that emphasizes a "portfolio" of foods or food components that lower cholesterol. This diet recommends daily consumption of 2 g of plant sterols, 50 g of nuts, 10–25 g of soluble fibers from plant foods, and 50 g of soy protein; foods of animal origin are not allowed. The portfolio diet helps reduce LDL (Section 4.14) but its effect on weight loss is small.

–   Based on timing of meal consumption

  –   Intermittent fasting (IF) involves regular periods with no or very limited calorie intake. The three most widely used regimens are: alternate-day fasting, 5:2 intermittent fasting (or consuming 900–1,000 calories for 2 days each week), and daily time restricted feeding (fasting for 16–18 hours a day). IF not only reduces calorie intake, but also exerts effects on metabolic switching to reverse insulin resistance, strengthen the immune system, and enhance physical and cognitive functions. It has been suggested that IF could benefit patients with obesity and has effects comparable to daily calorie restriction. However, little is known about the long-term sustainability and health effects of this type of fasting. IF is a widely used approach to mimic the effects of fasting, that aims to cut caloric intake entirely for shorter periods, leading to depletion of circulating glucose and glycogen stores after ~24 h and consequent lipid mobilization and fatty acid oxidation in the liver. This process leads to many of the same adaptations observed with caloric restriction. Rather than a diet, fasting is a therapeutic intervention; however, it is challenging for patients to comply with and may lead to adverse events and malnutrition in diseased patients [85].

  –   Meal timing and the circadian rhythm have raised a novel issue in weight management; actually, timing of meals could have serious implications not only for weight management, but also for development of cardiovascular disease. The American Heart Association recommends distributing calories over a defined period of the day, consuming a greater share of the total calorie intake earlier in the day, and maintaining consistent overnight fasting periods.

Dietary intervention is a necessary tool not only in the prevention and treatment of obesity but also on other pathologies associated (Section 14.8.4). In this respect, a recent publication provides an overview of dietary interventions that may help in the treatment of several types of cancer and the possible molecular mechanisms involved [85].

References

[1]   Schrödinger E. What is life? The physical aspect of the living cell with, mind and matter & autobiographical sketches. Cambridge ; New York: Cambridge University Press; 1992.

[2]   Bianconi E, Piovesan A, Facchin F, Beraudi A, Casadei R, Frabetti F, et al. An estimation of the number of cells in the human body. Ann Hum Biol 2013;40:463–71.https://doi.org/10.3109/03014460.2013.807878.

[3]   Stanfield CL. Principles of human physiology. 6th edition. Boston: Pearson; 2017.

[4]   Chapelot D, Charlot K. Physiology of energy homeostasis: Models, actors, challenges and the glucoadipostatic loop. Metabolism 2019;92:11–25. https://doi.org/10.1016/j.metabol.2018.11.012.

[5]   Galgani J, Ravussin E. Energy metabolism, fuel selection and body weight regulation. Int J Obes 2008;32:S109–19.

[6]   Chapter 3: Calculation of the energy content of foods – energy conversion factors n.d. http://www.fao.org/3/Y5022E/y5022e04.htm#TopOfPage (accessed December 15, 2019).

[7]   Fonseca DC, Sala P, De azevedo muner ferreira B, Reis J, Torrinhas RS, Bendavid I, et al. Body weight control and energy expenditure. Clin Nutr Exp 2018;20:55–59.https://doi.org/10.1016/j.yclnex.2018.04.001.

[8]   Bo S, Fadda M, Fedele D, Pellegrini M, Ghigo E, Pellegrini N. A critical review on the role of food and nutrition in the energy balance. Nutrients 2020;12, 1161. https://doi.org/10.3390/nu12041161

[9]   Feher J. Energy balance and regulation of food intake. Quant Hum Physiol 2017;834–46. https://doi.org/10.1016/B978-0-12-800883-6.00082-3.

[10] Blasco Redondo R. Gasto energético en reposo; métodos de evaluación y aplicaciones. Nutr Hosp 2015;245–54. https://doi.org/10.3305/nh.2015.31.sup3.8772.

[11] Weir JBDB. New methods for calculating metabolic rate with special reference to protein metabolism. J Physiol 1949;109:1–9. https://doi.org/10.1113/jphysiol.1949.sp004363.

[12] Ahmad D, Joseph K, Halpin C. Nutrition and indirect calorimetry. In: Hoag JB, editor. Oncol Crit Care. InTech; 2016. https://doi.org/10.5772/64385.

[13] Krüger RL, Lopes AL, Gross JDS, Macedo RCO, Teixeira BC, Reischak-Oliveira Á. Validação de equações de predição da taxa metabólica basal em sujeitos eutróficos e obesos. Rev Bras Cineantropometria E Desempenho Hum 2014;17:73. https://doi.org/10.5007/1980-0037.2015v17n1p73.

[14] Cunningham JJ. A reanalysis of the factors influencing basal metabolic rate in normal adults. Am J Clin Nutr 1980;33:2372–74. https://doi.org/10.1093/ajcn/33.11.2372.

[15] Soares MJ, Müller MJ. Resting energy expenditure and body composition: Critical aspects for clinical nutrition. Eur J Clin Nutr 2018;72:1208–14. https://doi.org/10.1038/s41430-018-0220-0.

[16] Atwater W, Woods C. The chemical composition of American food materials. Office of Experiment Stations. US Dep Agric Bull 1896;11–41.

[17] Livesey G, Buss D, Coussement P, Edwards DG, Howlett J, Jonas DA, et al. Suitability of traditional energy values for novel foods and food ingredients. Food Control 2000;11:249–89.

[18] Merrill AL, Watt BK. Energy value of foods: Basis and derivation, Agriculture Handbook No. 74 Wash US Gov Print Off; 1973.

[19] Roberts SB, Flaherman V. Dietary energy. Adv Nutr 2022;13:2681–85. https://doi.org/10.1093/advances/nmac092.

[20] Corrêa H, Magalhaes. The impact of the adipose organ plasticity on inflammation and cancer progression. Cells 2019;8:662. https://doi.org/10.3390/cells8070662.

[21] Blondin DP. Human thermogenic adipose tissue. Curr Opin Genet Dev 2023;80:102054. https://doi.org/10.1016/j.gde.2023.102054.

[22] Vegiopoulos A, Rohm M, Herzig S. Adipose tissue: Between the extremes. EMBO J 2017;36:1999–2017. https://doi.org/10.15252/embj.201696206.

[23] Stern JH, Rutkowski JM, Scherer PE. Adiponectin, leptin, and fatty acids in the maintenance of metabolic homeostasis through adipose tissue crosstalk. Cell Metab 2016;23:770–84. https://doi.org/10.1016/j.cmet.2016.04.011.

[24] Recinella L, Orlando G, Ferrante C, Chiavaroli A, Brunetti L, Leone S. Adipokines: New potential therapeutic target for obesity and metabolic, rheumatic, and cardiovascular diseases. Front Physiol 2020;11.

[25] Nelson DL, Cox MM, Hoskins AA, Lehninger AL. Lehninger principles of biochemistry. 8th edition. New York, NY: Macmillan International, Higher Education; 2021.

[26] Wang W, Guo D-Y, Lin Y-J, Tao Y-X. Melanocortin regulation of inflammation. Front Endocrinol 2019;10:683. https://doi.org/10.3389/fendo.2019.00683.

[27] Huvenne H, Dubern B, Clément K, Poitou C. Rare genetic forms of obesity: Clinical approach and current treatments in 2016. Obes Facts 2016;9:158–73. https://doi.org/10.1159/000445061.

[28] Vohra MS, Benchoula K, Serpell CJ, Hwa WE. AgRP/NPY and POMC neurons in the arcuate nucleus and their potential role in treatment of obesity. Eur J Pharmacol 2022;915:174611. https://doi.org/10.1016/j.ejphar.2021.174611.

[29] Lavoie O, Michael NJ, Caron A. A critical update on the leptin-melanocortin system. J Neurochem 2023;165:467–86. https://doi.org/10.1111/jnc.15765.

[30] Casado ME, Collado-Pérez R, Frago LM, Barrios V. Recent advances in the knowledge of the mechanisms of leptin physiology and actions in neurological and metabolic pathologies. Int J Mol Sci 2023;24:1422. https://doi.org/10.3390/ijms24021422.

[31] Straub LG, Scherer PE. Metabolic messengers: Adiponectin. Nat Metab 2019;1:334–39. https://doi.org/10.1038/s42255-019-0041-z.

[32] Dezonne RS, Pereira CM, De moraes martins CJ, De abreu VG, Francischetti EA. Adiponectin, the adiponectin paradox, and Alzheimer's Disease: Is this association biologically plausible?. Metab Brain Dis 2023;38:109–21. https://doi.org/10.1007/s11011-022-01064-8.

[33] Hernández-Saavedra D, Stanford KI. The regulation of lipokines by environmental factors. Nutrients 2019;11. https://doi.org/10.3390/nu11102422.

[34] Tsuji T, Tseng Y-H. Adipose tissue-derived lipokines in metabolism. Curr Opin Genet Dev 2023;81:102089. https://doi.org/10.1016/j.gde.2023.102089.

[35] Ataie-Ashtiani S, Forbes B. A review of the biosynthesis and structural implications of insulin gene mutations linked to human disease. Cells 2023;12:1008. https://doi.org/10.3390/cells12071008.

[36] Chen W, Cai W, Hoover B, Kahn CR. Insulin action in the brain: Cell types, circuits, and diseases. Trends Neurosci 2022;45:384–400. https://doi.org/10.1016/j.tins.2022.03.001.

[37] Jais A, Brüning JC. Arcuate nucleus-dependent regulation of metabolism – pathways to obesity and diabetes mellitus. Endocr Rev 2022;43:314–28. https://doi.org/10.1210/endrev/bnab025.

[38] Mok JKW, Makaronidis JM, Batterham RL. The role of gut hormones in obesity. Curr Opin Endocr Metab Res 2019;4:4–13. https://doi.org/10.1016/j.coemr.2018.09.005.

[39] Hædersdal S, Andersen A, Knop FK, Vilsbøll T. Revisiting the role of glucagon in health, diabetes mellitus and other metabolic diseases. Nat Rev Endocrinol 2023;19:321–35. https://doi.org/10.1038/s41574-023-00817-4.

[40] Roh E, Choi KM. Hormonal gut–brain signaling for the treatment of obesity. Int J Mol Sci 2023;24:3384. https://doi.org/10.3390/ijms24043384.

[41] Dodd GT, Michael NJ, Lee-Young RS, Mangiafico SP, Pryor JT, Munder AC, et al. Insulin regulates POMC neuronal plasticity to control glucose metabolism. ELife 2018;7:e38704.https://doi.org/10.7554/eLife.38704.

[42] Thon M, Hosoi T, Ozawa K. Possible integrative actions of leptin and insulin signaling in the hypothalamus targeting energy homeostasis. Front Endocrinol 2016;7. https://doi.org/10.3389/fendo.2016.00138.

[43] Tokarz VL, MacDonald PE, Klip A. The cell biology of systemic insulin function. J Cell Biol 2018;217:2273–89. https://doi.org/10.1083/jcb.201802095.

[44] Wewer Albrechtsen NJ, Kuhre RE, Pedersen J, Knop FK, Holst JJ. The biology of glucagon and the consequences of hyperglucagonemia. Biomark Med 2016;10:1141–51. https://doi.org/10.2217/bmm-2016-0090.

[45] Yanagi S, Sato T, Kangawa K, Nakazato M. The homeostatic force of Ghrelin. Cell Metab 2018;27:786–804. https://doi.org/10.1016/j.cmet.2018.02.008.

[46] Häussinger D. Overview. In: Lammert E, Zeeb M, editors. Metab Hum Dis. Vienna: Springer Vienna; 2014. 173–79. https://doi.org/10.1007/978-3-7091-0715-7_27.

[47] Rui L. Energy metabolism in the liver. In: Terjung R, editor. Compr Physiol. Hoboken, NJ, USA: John Wiley & Sons, Inc.; 2014. 177–97. https://doi.org/10.1002/cphy.c130024.

[48] Cocciolillo S, Sebastiani G, Blostein M, Pantopoulos K. Chapter 18 – Liver hormones. In: Litwack G, editor. Horm Signal Biol Med. Academic Press; 2020. 425–44. https://doi.org/10.1016/B978-0-12-813814-4.00018-3.

[49] Rhyu J, Yu R. Newly discovered endocrine functions of the liver. World J Hepatol 2021;13:1611–28. https://doi.org/10.4254/wjh.v13.i11.1611.

[50] Garruti G, Baj J, Cignarelli A, Perrini S, Giorgino F. Hepatokines, bile acids and ketone bodies are novel Hormones regulating energy homeostasis. Front Endocrinol 2023;14.

[51] Lu M, Flanagan JU, Langley RJ, Hay MP, Perry JK. Targeting growth hormone function: Strategies and therapeutic applications. Signal Transduct Target Ther 2019;4:1–11. https://doi.org/10.1038/s41392-019-0036-y.

[52] Ho KK, O'Sullivan AJ, Burt MG. The physiology of growth hormone (GH) in adults: Translational journey to GH replacement therapy. J Endocrinol 2023;257. https://doi.org/10.1530/JOE-22-0197.

[53] Mullur R, Liu -Y-Y, Brent GA. Thyroid hormone regulation of metabolism. Physiol Rev 2014;94:355–82. https://doi.org/10.1152/physrev.00030.2013.

[54] Mugesh G, Giri D, Mondal S. Chemical biology of thyroid hormones. AsiaChem Mag 2023;3:136–45. https://doi.org/10.51167/acm00048.

[55] Kuckuck S, Van Der Valk ES, Scheurink AJW, Van Der Voorn B, Iyer AM, Visser JA, et al. Glucocorticoids, stress and eating: The mediating role of appetite-regulating hormones. Obes Rev 2023;24. https://doi.org/10.1111/obr.13539.

[56] Verberne AJM, Korim WS, Sabetghadam A, Llewellyn-Smith IJ. Adrenaline: Insights into its metabolic roles in hypoglycaemia and diabetes: Adrenaline and glucose homeostasis. Br J Pharmacol 2016;173:1425–37. https://doi.org/10.1111/bph.13458.

[57] Obesity and overweight n.d. https://www.who.int/news-room/fact-sheets/detail/obesity-and-overweight (accessed July 28, 2023).

[58] Timothy Garvey W. The diagnosis and evaluation of patients with obesity. Curr Opin Endocr Metab Res 2019;4:50–57. https://doi.org/10.1016/j.coemr.2018.10.001.

[59] Lustig RH, Collier D, Kassotis C, Roepke TA, Kim MJ, Blanc E, et al. Obesity I: Overview and molecular and biochemical mechanisms. Biochem Pharmacol 2022;199:115012.https://doi.org/10.1016/j.bcp.2022.115012.

[60] Sbraccia P, Kushner R. Editorial overview: Frontiers in obesity. Curr Opin Endocr Metab Res 2019;4:1–3. https://doi.org/10.1016/j.coemr.2018.12.002.

[61] McAllister EJ, Dhurandhar NV, Keith SW, Aronne LJ, Barger J, Baskin M, et al. Ten putative contributors to the obesity epidemic. Crit Rev Food Sci Nutr 2009;49:868–913.https://doi.org/10.1080/10408390903372599.

[62] Trang K, Grant SFA. Genetics and epigenetics in the obesity phenotyping scenario. Rev Endocr Metab Disord 2023; https://doi.org/10.1007/s11154-023-09804-6.

[63] Berthoud H-R, Münzberg H, Morrison CD. Blaming the brain for obesity: Integration of hedonic and homeostatic mechanisms. Gastroenterology 2017;152:1728–38. https://doi.org/10.1053/j.gastro.2016.12.050.

[64] De Lorenzo A, Gratteri S, Gualtieri P, Cammarano A, Bertucci P, Di Renzo L. Why primary obesity is a disease?. J Transl Med 2019;17:169. https://doi.org/10.1186/s12967-019-1919-y.

[65] Lopomo A, Burgio E, Migliore L. Epigenetics of obesity. Prog Mol Biol Transl Sci 2016;140:151–84. https://doi.org/10.1016/bs.pmbts.2016.02.002.

[66] Cancers Associated with Overweight and Obesity Infographic. Natl Cancer Inst n.d. https://www.cancer.gov/about-cancer/causes-prevention/risk/obesity/overweight-cancers-infographic (accessed December 17, 2019).

[67] Fang LZ, Linehan V, Licursi M, Alberto CO, Power JL, Parsons MP, et al. Prostaglandin E $_2$ activates melanin-concentrating hormone neurons to drive diet-induced obesity. Proc Natl Acad Sci 2023;120: e2302809120.https://doi.org/10.1073/pnas.2302809120.

[68] Jastreboff AM, Kotz CM, Kahan S, Kelly AS, Heymsfield SB. Obesity as a disease: The obesity society 2018 position statement. Obesity 2019;27:7–9. https://doi.org/10.1002/oby.22378.

[69] Antonopoulos AS, Tousoulis D. The molecular mechanisms of obesity paradox. Cardiovasc Res 2017;113:1074–86. https://doi.org/10.1093/cvr/cvx106.

[70] Lennon H, Sperrin M, Badrick E, Renehan AG. The obesity paradox in cancer: A review. Curr Oncol Rep 2016;18. https://doi.org/10.1007/s11912-016-0539-4.

[71] Neel JV. Diabetes mellitus: A "thrifty" genotype rendered detrimental by "progress"?. Am J Hum Genet 1962;14:353–62.

[72] Lucas A. Programming by early nutrition in man. Ciba Found Symp 1991;156:38–50.

[73] Hales CN, Barker DJP. Type 2 (non-insulin-dependent) diabetes mellitus: The thrifty phenotype hypothesis. Diabetologia 1992;35:595–601. https://doi.org/10.1007/BF00400248.

[74] Heindel JJ, Balbus J, Birnbaum L, Brune-Drisse MN, Grandjean P, Gray K, et al. Developmental origins of health and disease: Integrating environmental influences. Endocrinology 2015;156:3416–21. https://doi.org/10.1210/en.2015-1394.

[75] Gluckman PD, Hanson MA, Buklijas T. A conceptual framework for the developmental origins of health and disease. J Dev Orig Health Dis 2010;1:6–18. https://doi.org/10.1017/S2040174409990171.

[76] Heindel JJ, Howard S, Agay-Shay K, Arrebola JP, Audouze K, Babin PJ, et al. Obesity II: Establishing causal links between chemical exposures and obesity. Biochem Pharmacol 2022;199:115015. https://doi.org/10.1016/j.bcp.2022.115015.

[77] Mandy M, Nyirenda M. Developmental origins of health and disease: The relevance to developing nations. Int Health 2018;10:66–70. https://doi.org/10.1093/inthealth/ihy006.

[78] Epigenetics: Fundamentals, History, and Examples. What Epigenetics n.d. http://www.whatisepigenetics.com/fundamentals/ (accessed December 18, 2019).

[79] Cheng Z, Zheng L, Almeida FA. Epigenetic reprogramming in metabolic disorders: nutritional factors and beyond. J Nutr Biochem 2018;54:1–10. https://doi.org/10.1016/j.jnutbio.2017.10.004.

[80] Xu L, Yeung MHY, Yau MYC, Lui PPY, Wong C-M. Role of histone acetylation and methylation in obesity. Curr Pharmacol Rep 2019;5:196–203. https://doi.org/10.1007/s40495-019-00176-7.

[81] Izquierdo AG, Crujeiras AB. Obesity-related epigenetic changes after bariatric surgery. Front Endocrinol 2019;10. https://doi.org/10.3389/fendo.2019.00232.

[82] Non-Coding RNA. What Epigenetics n.d. http://www.whatisepigenetics.com/non-coding-rna/ (accessed December 18, 2019).

[83] Zhang Y, Kutateladze TG. Diet and the epigenome. Nat Commun 2018;9:1–3. https://doi.org/10.1038/s41467-018-05778-1.

[84] Kim JY. Optimal diet strategies for weight loss and weight loss maintenance. J Obes Metab Syndr 2021;30:20–31. https://doi.org/10.7570/jomes20065.

[85] Martínez-Garay C, Djouder N. Dietary interventions and precision nutrition in cancer therapy. Trends Mol Med 2023;29:489–511. https://doi.org/10.1016/j.molmed.2023.04.004.

[86] Procaccini C, De candia P, Russo C, De Rosa G, Lepore MT, Colamatteo A, et al. Caloric restriction for the immunometabolic control of human health. Cardiovasc Res 2023;cvad035. https://doi.org/10.1093/cvr/cvad035.

[87] Bröer S. Intestinal amino acid transport and metabolic health. Annu Rev Nutr 2023;43:annurev-nutr-061121-094344 https://doi.org/10.1146/annurev-nutr-061121-094344.

[88] Bali A, Naik R. The Impact of a Vegan Diet on Many Aspects of Health: The Overlooked Side of Veganism. Cureus 2023;15(2): e35148. doi:10.7759/cureus.35148.

15 The Mediterranean diet

Victoria Valls-Bellés

15.1 Origins of the Mediterranean diet

The Mediterranean diet has its origins in a piece of land considered unique in its type, the Mediterranean Basin, a place that historians call "the cradle of civilization," because it was within its geographical boundaries where the complete history of the ancient world developed [1].

The Mediterranean diet has been forged by the ancient civilizations of the Old World. In their passage through the Mediterranean Basin they managed to transmit a part of their knowledge and customs, interweaving the foundations of this dietary model [3]. However, such a lifestyle could not have occurred without the invaluable role of the region's microclimate. The humidity, the many hours of sun in the hot and dry summers, the winters not too extreme, and the generous rains during spring and autumn turn this landscape into fertile lands where practically everything can be cultivated [4].

The first foods from which all this framework was constructed came from the Levant (specifically from Lebanon, Israel, Palestine, Syria, Jordan, and Iraq), with the cultivation of cereals and legumes, and with the Phoenicians, Greeks, and Romans, that introduced the cultivation of olive trees (Figure 15.1), wheat fields, and vineyards. These last three cultivars are the founders of what is now known as the "Mediterranean trilogy," consisting of bread, oil, and wine [3].

Two dietary patterns could be distinguished during the classical era:

- The classic-Mediterranean model, deeply rooted in Ancient Rome and originally from Greece, was based on agriculture. They had a large selection of fruits and vegetables, as well as nuts and cheese, leaving in the background the consumption of animal products (little meat was consumed in general, having a predilection for the intake of fresh fish and seafood).
- The barbarian-continental model. Germanic peoples (more inclined to livestock breeding) lived mainly from hunting, fishing, and gathering of wild fruits; however, they also enjoyed small gardens where they could grow some vegetables and grains [5].

These two culinary cultures merged over the years until the agro-silvo-pastoral model was developed [6]. However, the Roman Empire never abandoned the great Mediterranean triad, which was gradually exported to the regions of Continental Europe by monastic orders.

Victoria Valls-Bellés, Unitat Predepartamental de Medicina (Area: Fisiologia), Universitat Jaume I, Castelló de la Plana, Spain

https://doi.org/10.1515/9783111111872-015

Later, the fall of Western Europe into Arab-Muslim hands left an indelible mark on the new Romanesque-Germanic nutritional model. The diet in Al-Andalus was very rich and brought a new world of flavors with the introduction of sugar cane, citrus, eggplant, spinach, almonds, pomegranates, and spices. On the other hand, they implemented a series of innovative agricultural techniques that marked a revolution in the world of farming. The exaltation of the diet as a fundamental pillar of health and human well-being is also attributable to them [7].

In addition, the Mediterranean Basin had the immense luck of having the most relevant routes and commercial enclaves, which allowed to import new and exotic food from the New World. With the discovery of America, there was a great exchange of foods, that included products such as potatoes, tomatoes, corn, peppers, chili, and different varieties of beans, coffee, and chocolate [1]. Other regions that were relevant in this aspect were Asia and Oceania (exporters of rice, rosemary, pepper, sesame, ginger, basil, cucumber, etc.).

It is somewhat anecdotal that such an ancient tradition has not been encompassed within the term "Mediterranean diet" until well into the twentieth century. During 1950s and 1960s there were numerous professionals who focused their interest in the nutritional study of Mediterranean lands.

The first scientific references to the Mediterranean diet were made by the epidemiologist Leland G. Allbaugh in his report "Crete: A case study of an underdeveloped area." He tried to analyze all the factors that could exert some influence on the health of the Greeks, in order to establish future intervention plans. To accomplish this task, a specific population of the island of Crete was studied. It was concluded that the levels of consumption of the different nutrients were extremely balanced and that these eating patterns were very well adapted both to their natural and economic resources and to their needs [8].

The actual person responsible for the existence of the term "Mediterranean diet" was American physiologist Ancel Keys, that would appear on the scene later on. He studied coronary heart disease and associated risk factors in his "Seven Countries Study" after World War II [9]. It was observed that the rural areas of southern Europe (corresponding to the Mediterranean Basin) and Japan had a lower incidence of heart disease, concluding that there should be something in the lifestyle of these lands which kept them away from these pathologies. This "something" was called "Mediterranean Way" or "Mediterranean Style," which would eventually be known as "Mediterranean diet" after the dissemination of the study [10].

Professor Grande Covián, when describing the elements that make up the lunch of a farmer who works from sun to sun, praises the intelligence of the Mediterranean diet for selecting these products: "A 'bota' sardine,[1] a piece of bacon, half a loaf of bread, some hanging tomatoes and a bit of wine." "Even NASA would not have

---

1 A type of salted pressed sardine, traditionally kept in wooden barrels, very popular in Spain

found foods that took up so little space and were so specific to prevent the fainting and dehydration of a man on a manual harvest day." Glucose, salts, proteins, and carbohydrates are wisely combined at lunch.

Trichopolou and collaborators were the first to define an index of adherence to the Mediterranean diet and thus compare the high or low adherence with the risk of mortality [11]. The Lyon Heart Study [12] concluded that the Mediterranean diet, rich in α-linolenic acid (ALA), was more efficient than other diets in preventing coronary processes. On the other hand, it has been observed that the incidence of cardiovascular diseases and cancer is much higher in Northern Europe than in Southern Europe, which has a direct relationship with the consumption of fruits and vegetables typical of the Mediterranean diet, that have a high content of compounds with antioxidant activity [13].

Additionally, it was observed that the Mediterranean diet was able to reduce total cholesterol, LDL, and triglycerides, increase HDL as well as enhances endothelial function and even showed an improvement in insulin resistance and a reduction in the metabolic syndrome [14].

On November 16, 2010, UNESCO included the Mediterranean diet in the representative list of the Intangible Cultural Heritage of Humanity [15, 16]. We are therefore faced with an ancestral cultural legacy that has evolved over millennia.

15.2 The concept of Mediterranean diet

Many people think they know with precision what the Mediterranean diet consists of, moreover, it seems logical to assume that what people in the Mediterranean regions usually consume should be part of it but, in fact, when it comes to defining such a diet only vague and intuitive notions come to us.

It would be a mistake to look for a brief and synthetic definition of what seems to be an ambiguous and extensive concept. The Mediterranean diet is not simply a nutritional guideline, a way of combining food to decorate a table or satiate hunger, but it constitutes something infinitely more intricate, developed over the centuries by our ancestors, that reflects the constant transfer of civilizations through the Mediterranean Basin.

The Mediterranean diet is a lifestyle that combines a healthy eating model (perfectly integrating native customs and traditions) with appropriate physical exercise habits.

The Mediterranean Diet Foundation describes the basic principles [12, 17]:
– The use of olive oil as the main addition fat. It is rich in monounsaturated fatty acids (MUFAs) and in vitamins E and K (Chapters 4 and 6), being the most used fat in the Mediterranean region.
– Eating foods of plant origin in abundance: fruits, vegetables, legumes, and nuts. Fruits and vegetables of different colors are recommended. In addition, it is advised that one of the vegetable rations must be eaten raw and the others can be

cooked if desired. This way, we ensure an adequate supply of vitamins, among other components with antioxidant activity.

– Foods from cereals, such as pasta, rice, and especially their whole grain products, like whole wheat bread, should be part of the daily diet. It is one of the main sources of energy for our body.

– The most adequate foods are minimally processed, fresh, and seasonal. It is important to take advantage of seasonal products since, especially in the case of fruits and vegetables, it allows us to consume them at their best, both for nutrient supply and also for aroma and flavor.

– Eating dairy products, mainly yogurts and cheeses, daily. Consumption of fermented milks (yogurt, etc.) is associated with a series of health benefits because they contain live microorganisms capable of improving the balance of the intestinal microbiota.

– Red meat should be consumed in moderation, preferably lean meats, and as part of dishes based on vegetables and cereals. Processed meats only in small quantities and as snack ingredients. Consumption of three or four eggs a week is a good alternative to meat and fish.

– Eating fish. The consumption of oily fishes is recommended at least once or twice a week. Their fats are rich in omega-3 (ω3 or n-3) polyunsaturated fatty acids (ω3-PUFAs) that have cardioprotective effects. The recommendations of the World Health Organization (WHO) issued in 2010 regarding the recommended consumption of omega-3 fats were: 0.250 g/day for adults (EPA + DHA[2]), 0.3 g/day for pregnant women (EPA + DHA). Maximum tolerable intakes: 2 g/day.

– Fresh fruit should be the usual dessert. Fruits are very nutritious foods, being a good alternative as a snack. Sweets and cakes must be eaten occasionally, as they have no nutritional interest.

– Water is the quintessential drink in the Mediterranean, being essential in our diet. An intake of 1 to 1.5 L/day is recommended. Wine should be taken in moderation and during meals. Wine is a traditional food in the Mediterranean diet that may have beneficial effects on health by consuming it in moderation and in the context of a balanced diet.

– Perform physical activity every day, as it is as important as eating properly. Staying physically active and performing a physical exercise adapted to our abilities on a daily basis is very important to maintain a good health.

Currently, all these postulates can be combined in what is known as the "Traditional Pyramid of the Mediterranean Diet" [17], that shows, in an illustrative and didactic way, which foods we should consume and in what proportion (Figure 15.2). Foods that are essential for the diet and must be consumed in abundance are located at the base

---

2 EPA: *all-cis*-5,8,11,14,17-(E)icosapentanoic acid. DHA: *all-cis*-4,7,10,13,16,19-Docosahexaenoic acid. See Chapter 4 (Section 4.2 and Table 4.2) for more information on fatty acids

of the pyramid, while products that must be eaten with moderation appear at higher (and narrower) levels. In addition, the composition and number of portions of the main meals are indicated [18].

15.3 Benefits of the Mediterranean diet

The Mediterranean diet provides us with many benefits against the most prevalent pathologies in our society. In recent years, some relevant studies have supplied numerous data that relate good health with nutritional patterns similar to those found in Southern Europe:

– **Reduction of cardiovascular disease risk**. High-fat diets, like those typical in North America, have been associated with the highest cholesterol levels and the highest mortality rates from acute myocardial infarction (AMI). This statement suggests certain inconsistency in the discourse of physiologists, since the Mediterranean diet has a daily fat intake of 30–35% (35% is acceptable as long as it is olive oil) of the total energy content. However, the factor that influences cardiovascular pathology is not the amount of fat consumed but rather the source it comes from, as well as its chemical composition (in the Mediterranean diet intake would be predominantly of vegetable fat, rich in MUFAs, while that in the North American diet it would be of animal origin and with a noticeable percentage of saturated fatty acids -SFAs-). In relation to this, Ancel Keys along with Francisco Grande Covián and Joseph T. Anderson studied the influence of the type of fatty acid (FA) consumed in the diet with cholesterol levels and formulated what is known as the equation of Keys, Grande, and Anderson. This equation allows to estimate the variation in blood cholesterol levels from the percentage of saturated or polyunsaturated fat consumed in the diet [19]. Moreover, it has been observed that a tomato extract (rich in lycopene) with extra virgin olive oil reduces lipid parameters and oxidative stress biomarkers involved in cardiovascular pathology [20].

The direct relationship between the excess of SFAs in the diet and the increase in cardiovascular risk has been supported by numerous studies [21–23], among them that of PREDIMED (Mediterranean Diet Prevention) [24]. It was concluded that both a diet supplemented with olive oil and a diet supplemented with nuts caused a reduction in cardiovascular risk. The study also highlighted a relative risk reduction of 30% among high-risk people who were free of disease at the start of the study. Another relevant fact that emerges from the trial is that the decrease in cardiovascular disease is more evident in cerebrovascular accidents. On the other hand, the consumption of oily fish at least once or twice a week is very healthy since these fishes have fats very similar to those of plant origin, which protect against heart disease. The Mediterranean diet is also rich in vegetables and fruits (it is advised to take 5 servings of fruit and vegetables daily), thus providing a high content of antioxidants and fiber, two factors that also prevent these diseases [25, 26].

– **Attainment of healthy aging.** Numerous studies suggest that the Mediterranean diet could be associated with a lower risk of mild cognitive impairment during the aging process, as well as during the transition stage between dementia and Alzheimer's disease [27, 28]. PREDIMED elaborated another study [29] in which a better cognitive performance was seen in those individuals with high initial cardiovascular risk who consumed a greater amount of foods rich in polyphenols (due to their antioxidant potential) [30, 31]. This contributes to reducing the risk of developing neurodegenerative diseases and cognitive impairment associated with age. When this type of diet was offered to a group of people, it was shown that they improved their memory and also verbal constructions. Foods that enhance memory and are part of the Mediterranean diet include nuts, which have ω3-PUFAs, and oily fishes that, additionally, contain many vitamins and minerals [32]. On the other hand, the Mediterranean diet reduces the likelihood of suffering from Parkinson's disease [33].

Olive oil was related to a better verbal memory, nuts with a better working memory, and wine with better results in the Mini-Mental test. It has also been observed that the polyphenols present in red wine could have a neuroprotective effect on Alzheimer's disease and Parkinson's [34]. According to the latest scientific evidence, extra virgin olive oil improves brain connectivity and reduces the permeability of the blood-brain barrier, which suggests that it is the extra virgin olive oil (EVOO) polyphenols that contribute to this effect [35].

The PREDIMED study confirms that a Mediterranean-style diet can be a useful tool in the management of metabolic syndrome characterized by high cholesterol levels, abdominal obesity, hypertension, and hyperglycemia.

– **Nutritional habits and cancer.** It has been previously mentioned that there may exist a relationship between nutritional habits and cancer [36, 37]. Thus, it has been observed that the polyphenols contributed by olive oil could protect us from damage and cellular stress associated with breast cancer, due to their antioxidant capacity [37]. Good adherence to the Mediterranean diet has also been shown to have a protective role against certain digestive neoplasms, especially against colorectal cancer [38].

– **Obesity reduction.** The Mediterranean diet can reduce the risk of overweight among children, a problem that is increasingly growing. A recent study performed by professionals of the Catalan Institute of Health (ICS) has shown that the promotion of the Mediterranean diet in children between 8 and 10 years old reduces the prevalence of overweight by 6.3 points [39]. Moreover, the recommendations to reduce the intake of unhealthy foods, such as many industrial foods, were successful and reduced the consumption of these foods by half among the children who took part in the study.

– **Prevention of diabetes.** The Mediterranean diet is rich in fruits, vegetables, and seasonal products, which makes it good for reducing diabetes. This is due to the low-fat diet that involves, along with whole grains, fish and the use of olive oil and herbs instead of butter and salt for cooking and to accompany various ingredients [40].

– **Reduction of bone fracture risk.** In the XIV National Congress of the Spanish Association for the Study of Menopause (AEEM) it was concluded that a greater adherence to the Mediterranean diet is related to a lower risk of hip fractures, due to the large calcium intake of this diet, among other reasons [41, 42].

Other pathologies in which the Mediterranean diet has been shown to have positive effects, can be pointed out, thus, it improves the proper functioning of the kidney and heart, reduces the consumption of preservatives and additives, etc.

In summary, adherence to the Mediterranean diet:

– Increases blood levels of high density lipoproteins (HDL), which are beneficial for health
– Decreases blood levels of total cholesterol and LDL
– Increases the body's antioxidant capacity
– Raises levels of vitamin C, E, beta-carotene, and polyphenols in the blood
– Lowers blood pressure levels because it has a low sodium content and is abundant in potassium and fiber
– Helps detoxify substances in the liver
– Reduces the risk of thrombosis, acting on the mechanisms of coagulation
– Protects the arteries, dilating them and stimulating the production of the enzyme nitric oxide synthase of the endothelium (inner layer of the arteries)
– Decreases inflammatory reactions
– Modifies the expression of genes and increases the immune capacity of defense

15.4 Adherence to the Mediterranean diet

Despite the growing evidence on the benefits of the Mediterranean diet (MD), recent data indicate that adherence to this eating pattern is declining in Mediterranean regions, particularly among children and adolescents. Cabrera et al. evaluated adherence to MD in a population of 24,067 children and adolescents using the KIDMED test; the results showed high adherence to MD in 10% of the population, while low adherence accounted for 21%, so there was a clear trend towards abandoning the Mediterranean lifestyle [43]. Globalization and the adoption of Western dietary patterns partly explain the low rate of adherence in adolescents residing in the Mediterranean region. García-Meseguer's group tried to characterize the eating habits of Spanish university students and assess the quality of their diet. 27.5% of the students showed a BMI above normal values and 75% claimed to have a sedentary life. The evaluation of the quality of the diet carried out by the healthy eating index received a low score, so that only 5.3% of the students had a high adherence to the MD [44]. In 2022, Obeid et al., carried out a data collection on adherence to MD, mainly carried out in the European Mediterranean basin, and adherence is still very moderate [45].

15.5 Current Mediterranean diet

Mediterranean diet foods are, as the name suggests, those foods that are grown, hunted, or caught in all regions around the Mediterranean Sea. The traditional Mediterranean diet, up to 1950s, would be a "natural diet" characterized by the consumption of local and seasonal products, of a rather high variety, and traditional techniques of cooking, preservation, and presentation.

While it is true that the Mediterranean diet is a nutritional pattern built over centuries, so is the fact that societies are constantly changing and developing. In an increasingly industrialized population, where fast food chains and foods with considerable amounts of additives are abundant, it is difficult to find healthy nutritional habits [46, 47]. Meals have stopped being an important moment of the day, now they are a small break from the constant stress in which we live. People have stopped having time to cook and the easy option offered by supermarkets directly threatens the lifestyle proposed by the Mediterranean diet.

The recommended nutritional goals for good nutritional habits (Table 15.1) have been replaced by a greater contribution of fat (from 30 to 40–42% of the recommended total energy intake), the increase being mainly in saturated fat. Proteins have also suffered a considerable increase, from 0.8 g of protein per kg of body weight we have moved to 2 g and even higher levels. This has decreased carbohydrates consumption from 55–60% to the current 40–45%, the majority of them being simple sugars, from pastries and soft drinks, that are associated with a series of pathologies, such as obesity, diabetes, high cholesterol levels, hypertension, etc., the so-called metabolic syndrome [48–51].

Table 15.1: Population nutrient intake goals for the prevention of diet-related chronic diseases.

Dietary factor	Goal (% of total energy, unless otherwise stated)
Total fat	15–30%
Saturated fatty acids	<10%
Polyunsaturated fatty acids (PUFAs)	6–10%
n-6 Polyunsaturated fatty acids (PUFAs)	5–8%
n-3 Polyunsaturated fatty acids (PUFAs)	1–2%
Trans fatty acids	<1%
Monounsaturated fatty acids (MUFAs)	By difference[a]
Total carbohydrate	55–75%[b]
Free sugars[c]	<10%
Protein	10–15%[d]
Cholesterol	<300 mg per day[e]
Sodium chloride (sodium)	<5 g per day (<2 g per day)
Fruits and vegetables	>400 g per day

Table 15.1 (continued)

Dietary factor	Goal (% of total energy, unless otherwise stated)
Total dietary fiber	From foods[f]
Non-starch polysaccharides (NSP)	From foods[f]

[a]This is calculated as: total fat – (saturated fatty acids + polyunsaturated fatty acids + trans fatty acids).
[b]The percentage of total energy available after taking into account that consumed as protein and fat, hence the wide range.
[c]The term "free sugars" refers to all monosaccharides and disaccharides added to foods by the manufacturer, cook, or consumer, plus sugars naturally present in honey, syrups, and fruit juices.
[d]The suggested range should be seen in the light of the Joint WHO/FAO/UNU Expert Consultation on Protein and Amino Acid Requirements in Human Nutrition, held in Geneva from 9 to 16 April 2002.
[e]Salt should be iodized appropriately. The need to adjust salt iodization, depending on observed sodium intake and surveillance of iodine status of the population, should be recognized.
[f]Wholegrain cereals, fruits, and vegetables are the preferred sources of non–starch polysaccharides (NSP). The recommended intake of fruits and vegetables and consumption of wholegrain foods is likely to provide > 20 g per day of NSP (>25 g per day of total dietary fiber).
Source: WHO, Joint and FAO Expert Consultation. "Diet, nutrition and the prevention of chronic diseases." *World Health Organ Tech Rep Ser* 916.i–viii (2003).

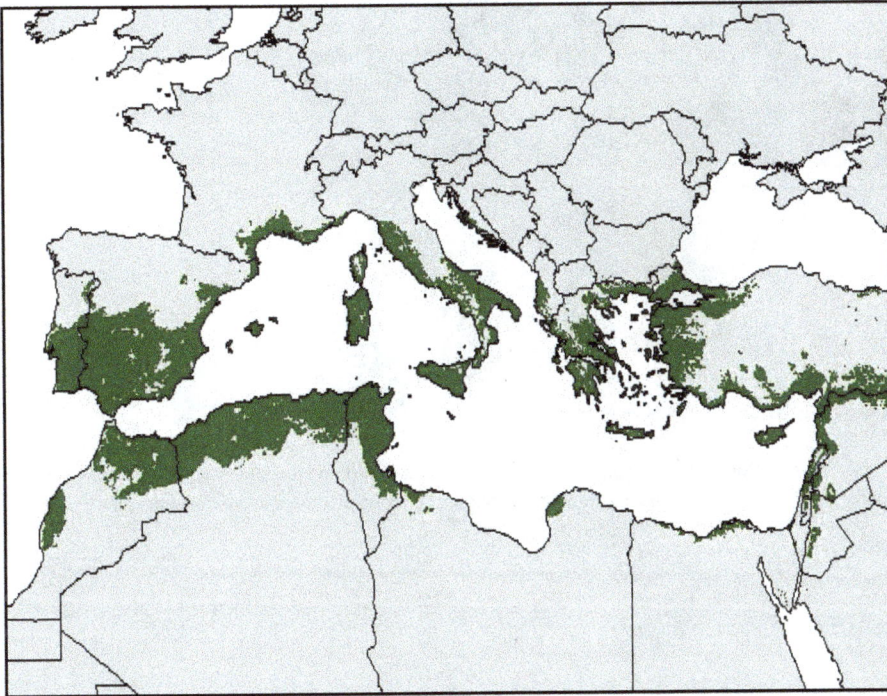

Figure 15.1: Potential distribution of the olive tree over the Mediterranean Basin.
Source [2]: Reproduced with permission.

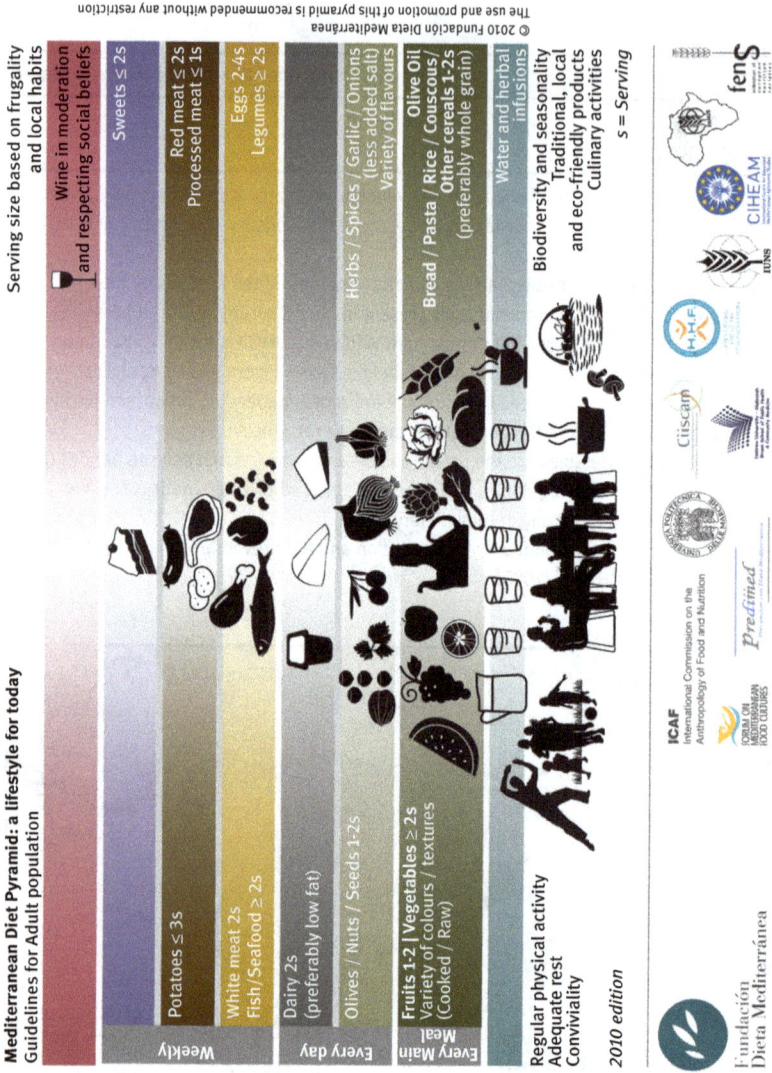

Figure 15.2: The Mediterranean diet pyramid. Reproduced with permission.

Source: The Mediterranean Diet Foundation, Barcelona, Spain, https://dietamediterranea.com/en/nutrition/.

The abandonment of these habits is evident when observing the general population. Obesity is the new epidemic of the twenty-first century and is promoted by highly manipulated products that we consume regularly without being aware of it. Moreover, metabolic diseases (diabetes, dyslipidemia, etc.) and cancer are the order of the day among the adult population [52]. An exercise in reflection and awareness about all this is important, otherwise, society will be destined to live with greater comorbidities and to experience a rather unsatisfactory old age [53–57].

Bibliography

[1]   Altomare R, Cacciabaudo F, Damiano G, Palumbo VD, Gioviale MC, Bellavia M, Tomasello G, Lo Monte AI. The mediterranean diet: A history of health. Iran J Public Health 2013;42(5):449–57.
[2]   Oteros J. Ph. D. Thesis. Universidad de Córdoba, Spain (2014). Thesis title. "Modelización del ciclo fenológico reproductor del olivo (Olea europaea L.)" DOI: 10.13140/RG.2.1.2690.8327. https://doi.org/10.13140/rg.2.1.2690.8327.
[3]   Anson R. Cultura gastrónomica del mediterráneo: Pasado, presente y futuro. Madrid: Lundwerg ediciones; 2015.
[4]   Capone R, El Bilali H, Elferchichi A, Lamaddalena NY, Lamberti L. Recursos naturales y alimentación en el mediterráneo. In: Alcaraz C, Cusí P, Crespin F, Fontan-Kiss F, López Francos A, Pérez A, Vandesande MG, editors. Terramed: la dieta mediterránea para un desarrollo regional sostenible, 3a ed. Madrid: INO reproducciones; 2012. 181–204.
[5]   Essid MY. La historia de la alimentación mediterránea. In: Alcaraz C, Cusí P, Crespin F, Fontan-Kiss F, López Francos A, Pérez A, Vandesande MG, editors. Terramed: la dieta mediterránea para un desarrollo regional sostenible. 3a ed. Madrid: INO reproducciones; 2012. 55–74.
[6]   Ferrari R, Rapezzi C. The Mediterranean diet: A cultural journey. Lancet 2011;377(9779):1730–31.
[7]   Salas-Salvadó J, Huetos-Solano MD, García-Lorda P, Bulló M. Diet and dietetics in al-Andalus. Br J Nutr 2006;96(S1):S100–4.
[8]   Allbaugh LG, Princeton N. Crete, a case study of an underdeveloped area. Am J Public Health 1953;43(1):928–29.
[9]   Menotti A, Puddu PE. How the seven countries study contributed to the definition and development of the mediterranean diet concept: A 50-year journey. Nutr Metab Cardiovasc Dis 2015;25(3):245–52.
[10] Keys A, Keys MCB. Sentirse Bien. La receta Mediterránea. Fundación Dieta Mediterránea, Ministerio de Agricultura, Pesca y Alimentación. Barcelona; 2006.
[11] Trichopoulou A, Kouris-Blazos A, Vassilakou T, Gnardellis C, Polychronopoulos E, Venizelos M, Lagiou P, Wahlqvist ML, Trichopoulos D. Diet and survival of elderly Greeks: A link to the past. Am J Clin Nutr 1995;61(6Suppl):1346S–1350S.
[12] De lorgeril M, Salen P, Rabaeus M. New and traditional foods in a modernized Mediterranean diet model. Eur J Clin Nutr 2019;Jul72(Suppl 1):47–54.
[13] Dg1 L, Astley SB. European research on the functional effects of dietary antioxidants – EUROFEDA. Mol Aspects Med 2002;23(1–3):1–38.
[14] Serra-Majem L, Roman B, Estruch R. Scientific evidence of interventions using the Mediterranean diet: A systematic review. Nutr Rev 2006;64(2 Pt 2):S27–47.
[15] González Turmo I. La dieta mediterránea como objeto patrimonial. Quaderns de la Mediterrània 2010;13:197.
[16] Dernini S. La erosión y el renacimiento de la dieta mediterránea: un recurso cultural sostenible. Quaderns de la Mediterrània 2011;16:253.

[17]   Fundación Dieta Mediterránea. Eur J Clin Nutr 2019;72(Suppl 1):47–54. 10.1038/s41430-018-0308-6.

[18]   Bach-Faig A, Berry EM, Lairon D, Reguant J, Trichopoulou A, Dernini S, Medina FX, Battino M, Belahsen R, Miranda G, Serra-Majem L. Mediterranean diet pyramid today. Public Health Nutr 2011;14(12A):2274–84.

[19]   Keys A, Anderson JT, Grande F. Serum cholesterol response to changes in the diet: IV. Particular saturated fatty acids in the diet. Metabolism 1965;14(7):776–87.

[20]   Álvarez M, Lopez-Jaen AB, Cavia-Saiz M, Muñiz P, Valls-Bellés V. Beneficial effects of olive oil enriched with lycopene on the plasma antioxidant and anti-inflammatory profile of hypercholesterolemic patients. Antioxidants 2023;12:x. https://doi.org/10.3990/xxxx.

[21]   Martínez-González MA, García-López M, Bes-Rastrollo M, Toledo E, Martínez-Lapiscina EH, Delgado-Rodriguez M, Vazquez Z, Benito S, Beunza JJ. Mediterranean diet and the incidence of cardiovascular disease: A Spanish cohort. Nutr Metab Cardiovasc Dis 2011;21(4):237–44.

[22]   Tektonidis TG, Åkesson A, Gigante B, Wolk A, Larsson SC. A Mediterranean diet and risk of myocardial infarction, heart failure and stroke: A population-based cohort study. Atherosclerosis 2015;243(1):93–98.

[23]   Turati F, Pelucchi C, Galeone C, Praud D, Tavani A, La Vecchia C. Mediterranean diet and non-fatal acute myocardial infarction: A case–control study from Italy. Public Health Nutr 2015;18(04):713–20.

[24]   Guasch-Ferré M, Hu FB, Martínez-González MA, Fitó M, Bulló M, Estruch R, Ros E, Corella D, Redondo J, Gómez-Gracia E, Fiol M, Lapetra J, Serra-Majem L, Muñoz MA, Pintó X, Lamuela-Raventós RM, Basora J, Buil-Cosiales P, Sorlí JV, Ruiz-Gutiérrez V, Martínez JA, Salas-Salvadó J. Olive oil intake and risk of cardiovascular disease and mortality in the predimed Study. BMC Med 2014;12:78.

[25]   Codoñer-Franch P, López-Jaén AB, Muñiz P, Sentandreu E, Valls Bellés V. Mandarin juice improves antioxidant status in hypercholesterolemic children. J Pediatr Gastroenterol Nutr 2008;47(3):349–55. ha sido uno de los artículos objeto del Year Book of Pediatrics del año 2009 (James A. Stockman III ed, ISBN: 978–1416057376)

[26]   Román GC, Jackson RE, Gadhia R, Román AN, Reis J. Mediterranean diet: The role of long-chain ω-3 fatty acids in fish; polyphenols in fruits, vegetables, cereals, coffee, tea, cacao and wine; probiotics and vitamins in prevention of stroke, age-related cognitive decline, and Alzheimer disease. Rev Neurol (Paris) 2019; pii: S0035-3787(19)30773-8, 10.1016/j.neurol.2019.08.005.

[27]   Qosa H, Mohamed LA, Batarseh YS, Alqahtani S, Ibrahim B, LeVine H, Keller JN, Kaddoumi A. Extra-virgin olive oil attenuates amyloid-β and tau pathologies in the brains of TgSwDI mice. J Nutr Biochem 2015;26(12):1479–90.

[28]   Petersson SD, Philippou E. Mediterranean diet, cognitive function, and dementia: A systematic review of the evidence. Adv Nutr 2016;7(5):889–904.

[29]   Valls-Pedret C, Lamuela-Raventós RM, Medina-Remón A, Quintana M, Corella D, Pintó X, et al. Polyphenol-rich foods in the Mediterranean diet are associated with better cognitive function in elderly subjects at high cardiovascular risk. J Alzheimers dis 2012;29(4):773–82.

[30]   Valls Bellés V. Presencia y actividad funcional de los polifenoles. Actualización en nutrición 2005. Evidencias en nutrición, Ed. Jesús R. Martínez y Carlos Iglesias 2005. 133–44. ISBN: 8460975886

[31]   Valls-Belles V, Torres C, Muñiz P, Beltran S, Martinez-Alvarez JR, Codoñer-Franch P. Effect of grape seed polyphenols before adriamycin toxicity in rat hepatocytes. European J Nutr 2006;10:1–7.

[32]   Di Renzo L, Gualtieri P, Romano L, Marrone G, Noce A, Pujia A, Perrone MA, Aiello V, Colica C, De Lorenzo A. Role of personalized nutrition in chronic-degenerative diseases. Nutrients 2019;11(8):pii: E1707 10.3390/nu11081707.

[33]   Cassani E, Barichella M, Ferri V, Pinelli G, Iorio L, Bolliri C, Caronni S, Faierman SA, Mottolese A, Pusani C, Monajemi F, Pasqua M, Lubisco A, Cereda E, Frazzitta G, Petroni ML, Pezzoli G. Dietary habits in Parkinson's disease: Adherence to Mediterranean diet. Parkinsonism Relat Disord 2017;42:40–46.

[34] Caruana M, Cauchi R, Vassallo N. Putative role of red wine polyphenols against brain pathology in Alzheimer's and Parkinson's disease. Front Nutr 2016;3:31.

[35] Kaddoumi A, Denney TS Jr, Deshpande D, Robinson JL, Beyers RJ, Redden DT, Praticò D, Tassos Kyriakides C, Lu B, Kirby AN, Beck DT. Extra-virgin olive oil enhances the blood-brain barrier function in mild cognitive impairment: A randomized controlled trial. Nutrients 2022;14(23):5102.

[36] American Institute for Cancer Research. World Cancer Research Fund, editores. Food, nutrition, physical activity and the prevention of cancer: a global perspective: a project of World Cancer Research Fund International. Washington, D.C: American Institute for Cancer Research; 2007. 517.

[37] Schwingshackl L, Hoffmann G. Adherence to Mediterranean diet and risk of cancer: An updated systematic review and meta-analysis of observational studies. Cancer Med 2015;4(12):1933–47.

[38] Braakhuis A, Campion P, Bishop K. Reducing breast cancer recurrence: The role of dietary polyphenolics. Nutrients 2016;8(9):547.

[39] Rotelli MT, Bocale D, De Fazio M, Ancona P, Scalera I, Memeo R, Travaglio E, Zbar AP, Altomare DF. IN-VITRO evidence for the protective properties of the main components of the Mediterranean diet against colorectal cancer: A systematic review. Surg Oncol 2015;24(3):145–52.

[40] Gómez SF, Casas Esteve R, Subirana I, Serra-Majem L, Fletas Torrent M, Homs C, Bawaked RA, Estrada L, Fíto M, Schröder H. Effect of a community-based childhood obesity intervention program on changes in anthropometric variables, incidence of obesity, and lifestyle choices in Spanish children aged 8 to 10 years. Eur J Pediatr 2018;177(10):1531–39.

[41] Liu X, Zheng Y, Guasch-Ferré M, Ruiz-Canela M, Toledo E, Clish C, Liang L, Razquin C, Corella D, Estruch R, Fito M, Gómez-Gracia E, Arós F, Ros E, Lapetra J, Fiol M, Serra-Majem L, Papandreou C, Martínez-González MA, Hu FB, Salas-Salvadó J. High plasma glutamate and low glutamine-to-glutamate ratio are associated with type 2 diabetes: Case-cohort study within the PREDIMED trial. Nutr Metab Cardiovasc Dis 2019;29(10):1040–49.

[42] Vignini A, Nanetti L, Raffaelli F, Sabbatinelli J, Salvolini E, Quagliarini V, Cester N, Mazzanti L. Effect of 1-y oral supplementation with vitaminized olive oil on platelets from healthy postmenopausal women. Nutrition 2017;42:92–98.

[43] Palomeras-Vilches A, Viñals-Mayolas E, Bou-Mias C, Jordà-Castro M, Agüero-Martínez M, Busquets-Barceló M, Pujol-Busquets G, Carrion C, Bosque-Prous M, Serra-Majem L, Bach-Faig A. Adherence to the mediterranean diet and bone fracture risk in middle-aged women: A case control study. Nutrients 2019;11(10):pii: E2508 10.3390/nu11102508.

[44] García Cabrera S, Herrera Fernández N, Rodríguez Hernández C, Nissensohn M, Román-Viñas B, Serra-Majem LKT. Prevalence of low adherence to the mediterranean diet in children and young; a systematic review. Nutr Hosp 2015;32(6):2390–99.

[45] García-Meseguer MJ, Cervera Burriel F, Vico García C, Serrano-Urrea R. Adherence to Mediterranean diet in a Spanish university population. Appetite 2014;78:156–64.

[46] Obeid CA, Gubbels JS, Jaalouk D, Kremers SPJ, Oenema A. Adherence to the Mediterranean diet among adults in Mediterranean countries: A systematic literature review. Eur J Nutr 2022;61 (7):3327–44.

[47] Laurentin A, Schnell M, Tovar J, Domínguez Z, Pérez B, López de blanco M. Transición alimentaria y nutricional. Entre la desnutrición y la obesidad. An Venez Nutr 2007;20(1):47–52.

[48] Monteiro CA, Moubarac JC, Cannon G, Ng SW, Popkin B. Ultra-processed products are becoming dominant in the global food system: Ultra-processed products: global dominance. Obes Rev 2013;14:21–28.

[49] FAO/WHO. Expert consultation on fats and fatty acids in human nutrition. interim summary of conclusions and dietary recommendations on total fat & fatty acids. 2008. Disponible en: http://www.who.int/nutrition/topics/FFA_summary_rec_conclusion.pdf

[50] EFSA. Opinion of the scientific panel on dietetic products, nutrition, and allergies on a request from the Commission related to labelling reference intake values for n−3 and n−6 polyunsaturated fatty acids. EFSA J 2009;1176:1–11.

[51] Aranceta J, Majem S, Senc L. Objetivos nutricionales para la población española. Consenso de la Sociedad Española de Nutrición Comunitaria 2011. Rev Esp Nutr Com 2011;17(4):178-199.

[52] World Health Organization. Food and agriculture organization of the united nations, editores. Vitamin and mineral requirements in human nutrition. 2nd edition. Geneva: Rome: World Health Organization; FAO; 2004. 341.

[53] Vandevijvere S, Chow CC, Hall KD, Umali E, Swinburn BA. Increased food energy supply as a major driver of the obesity epidemic: A global analysis. Bull World Health Organ 2015;93(7):446–56.

[54] Ruano C, Henriquez P, Bes-Rastrollo M, Ruiz-Canela M, Del burgo CL, Sánchez-Villegas A. Dietary fat intake and quality of life: The Sun project. Nutr J 2011;10(1):121.

[55] Sánchez-Villegas A, Verberne L, De Irala J, Ruíz-Canela M, Toledo E, Serra-Majem L, Martínez-González MA. Dietary fat intake and the risk of depression: The sun project. PLoS One 2011;6(1): e16268.

[56] García-Fernández E, Rico-Cabanas L, Rosgaard N, Estruch R, Bach-Faig A. Mediterranean diet and cardiodiabesity: A review. Nutrients 2014;6(9):3474–500.

[57] Choi E, Kim S-A, Joung H. Relationship between obesity and Korean and Mediterranean dietary patterns: A review of the literature. J Obes Metab Syndr 2019;28(1):30–39.

16 The foods of the future

Javier Vigara and José M. Vega

16.1 Demographic boom

Human population increases steadily on the planet: in 1950 it was below 3,000 million people, whereas in the year 2000 amounted to 6,100 million; in 2019, world population already exceeds 8,000 million and it is projected to reach 9,300 million inhabitants by 2050, although some estimate that we will reach 10,000 million by then. How can so many people be fed, if nowadays more than three million children under five die annually from causes related to malnutrition? It is clear that the demand for food grows every year and we have to prepare ourselves, in a scientific and rigorous way, to face the situation.

On the other hand, the available data unequivocally show the aging of the population. People over 65 years old accounted for 10% of the population in 2010, while the relative population of older people will be 16% by 2050. The elderly usually present certain metabolic problems; therefore, the diet has to be adjusted to their specific needs, which requires offering them a personalized diet that integrates new and imaginative foods. We have to produce better, higher quality food to respond to the demand of an increasingly large and aging population.

16.2 Food production and environmental impact

The bases of food production, par excellence, are agriculture, livestock, and fisheries, which are done intensively and are certainly aggressive with the environment.

16.2.1 Impact of agriculture

The extension of land dedicated to agricultural production occupies an area equivalent to 2,000 million of hectares, which represents almost the 50% of the land area not covered by ice on the planet, and the forecasts suggest that it would be necessary to increase the cultivation area for a sustained and sufficient production of food in an extension equivalent to that of Brazil, by 2050. Undoubtedly, this is not manageable

Javier Vigara, Departamento de Química "Carlos Vílchez Martín" Facultad de Experimentales, Universidad de Huelva, Huelva, Spain
José M. Vega, Departamento de Bioquímica Vegetal y Biología Molecular. Facultad de Química, Universidad de Sevilla, Sevilla, Spain

https://doi.org/10.1515/9783111111872-016

by the planet without the risk of seriously affecting biodiversity and the environment [1]. On the other hand, there are different factors that threaten agriculture, such as.

– CO_2, CH_4, and N_2O emission, with high warming effect and closely related with climate change.
– Soil degradation, a global problem associated with agriculture, which includes phenomena such as erosion, salinization, and nutrient losses.
– Water management, another important challenge, since agriculture uses 70% of all water from aquifers, streams, rivers, and lakes. About 40% of global increases in food production come from irrigated areas and, by 2050, it is estimated that irrigated crops will increase by 6% (Chapter 2).

Agriculture is one of the great contributors to global warming of the planet, since it emits more greenhouse gases than all cars, trucks, trains, and aeroplanes together. In addition, it is one of the main sources of water pollution, due to the use of fertilizers, and also accelerates the loss of biodiversity, as we have turned forests into farms, thus causing the extinction of many animals that inhabited those forests. It is clear that we must reflect on all this and consider new concepts in the production, storage, and processing of food [1].

The production of cereals, such as corn, wheat, and rice, has increased linearly in the last 50 years until reaching 1,036, 800, and 600 million tons, respectively, in 2018 and demand continues to grow. Given the evidence that the extension of land for cultivation cannot be increased, the main objective of agricultural activity is to increase production per hectare (ha) of crop. On the other hand, the evolution of the productivity of rice per hectare cultivated has increased up to 10 times in the last century, from 1.6 tones/ha in 1900 to 16 tones/ha in 2010, a barrier that has not been surpassed so far. The use of both, new agricultural techniques and genetically modified rice have been the keys to increasing productivity per hectare. It is therefore convenient to analyze these numbers and work in this direction for the immediate future, keeping the view that genetically modified foods are essentially required for the new agriculture revolution [2].

16.2.2 Impact of livestock

The production of animals for human food is reaching unsustainable limits for the planet, while demand continues to increase. According to FAO, the global demand for milk in the world increases by 2.5% per year and that of meat in general increases by 2.0% per year. Increased livestock activity undoubtedly has a negative impact on the environment, due to the high amount of methane and CO_2 that are expelled into the atmosphere and the direct contamination of water produced by livestock feces. Biogas from animal flatulences is composed of 60% methane and 40% CO_2. A cow expels annually between 60 and 120 kg of methane, depending on the species, a sheep 8 kg, a pig 1.5 kg, and a human 0.12 kg. Methane is a gas with an important greenhouse effect, which contributes significantly to global warming and climate change. It is expected that by 2050 the livestock impact will contribute a 30% to the increase in the emission of gases, especially CO_2 and methane. On the other hand, animal droppings contain significant amounts of nitrogen and phosphorus, which pollute waters and cause excessive growth of bacteria and microalgae. Given this situation, it is advisable not to increase the livestock population and look for alternative solutions, such as the genetic modification of animals in order to expel less nitrogen and phosphorus in their feces, an aspect that we have to study and assess.

The amount of land and resources required for animal husbandry is significantly higher than that used for plant production. Thus, an area of 39.9 m^2 is needed to produce a kilo of food from veal, 20.1 m^2 for pigs, 18.3 m^2 for chickens, 0.7 m^2 for fruit, 0.5 m^2 for potatoes, and 0.3 m^2 for vegetables.

16.2.3 Fishing impact

Fishing is another activity that is increasing dangerously producing the depletion of traditional fishing grounds. The numbers are quite eloquent, since we have gone from about 75 million tons of catches/year in 1985 to 90 in 2005. Current forecasts are that this figure cannot continue to increase because, otherwise, we would put the oceanic balance of the species at risk. Illegal fishing can represent up to 15% of annual catches and, in addition to economic damage, such practices threaten biodiversity and food security. As an alternative to the growing demand for fish, we are presented with aquaculture, whose production already exceeds that of catches. On the other hand, genetic modification of fish is poorly developed, but there are good and promising examples, such as the increase of size in salmons, obtained by over-expression of their own growth hormone.

16.3 Innovations in food production

The future of human food relies on a global plan based on five axes: (a) stopping the expansion of agriculture; (b) improving crop production per hectare; (c) using resources more effectively; (d) decreasing meat consumption; and (e) reducing waste. Forecasts indicate that sustainable food production could be doubled while relieving environmental aggressions in parallel [1]. Given this situation, work is already underway so that the future does not surprise us and food production can be not only effective in quantity and quality for humans, but also sustainable, so that the planet, our home, remains as clean and pleasant as possible. This involves working on different fronts, which are discussed below.

16.3.1 Urban agriculture

It is another concept of plant production, which consists of using urban spaces, buildings, or small areas of land near the city, that are exploited by several families, usually for their own livelihood [3]. Urban orchards can be located in vacant lots, inside the house or even in buildings. It is mainly an agriculture of fruit and vegetable production, which has given rise to different techniques, such as drip irrigation and hydroponic crops [4], which are those in which plants are held by an appropriate structure and their roots are immersed in running water that carries the necessary nutrients for growth.

Urban agriculture has important advantages, such as:
– It results in a significant reduction in the cost of the product, due to the low or zero cost of transportation (the product is consumed near where it is produced), the decrease in intermediaries involved in the sale of product, and the almost absence of storage and refrigeration costs.
– It can contribute not only to the fight against unemployment, but also to the well-being of citizens, due to the stress reduction caused by contact with nature and noise damping.
– Water contaminated with biological waste can be recycled and redirected to hydroponic crops, where they serve as a source of nutrients for plants, which produce purified water.

16.3.2 Aquaculture

Aquaculture is the cultivation of fish, crustaceans, and mollusks, both in salt water and in fish farms. In 2011, aquaculture production amounted to approximately 61.6 million tons. In 2016, production from catches and aquaculture was 90.9 and 80.0 million tons, respectively, which clearly indicates that marine catches stabilize and aquaculture continues to increase, so that forecasts for 2045 are 92 and 100 million tones, respectively. There is a long tradition of carp breeding in captivity, particularly in Europe and Asia. This species continues to prevail in aquaculture and represents most of fish production in fish farms, since it has the advantage of not being carnivorous and therefore it does not demand expensive protein-rich foods. Tilapia, the basis of domestic aquaculture production, occurs very frequently in Asia, especially China, the Philippines, and Thailand. Aquaculture provides half of the annual shrimp production. The main freshwater species produced by aquaculture in the period 1995–2015 were carp (31%), tilapia (17%), shrimp (15%), catfish (11%), salmon 7%, crustaceans (5%), and trout (2%). As far as marine species are concerned, around 8% of production was obtained by aquaculture.

16.4 Genetically modified foods

Genetic modification is a very powerful and controversial resource, which can be used to improve the quantity and quality of food produced by man. It is a process by which we introduce an extra gene into an organism, with the idea of providing it with new characteristics it did not have before. A genetically modified organism can be transgenic (if the extra gene comes from another species) or cisgenic (if the extra gene comes from its own species). Transgenesis is a non-risk-free process, where the main problem is that the researcher cannot determine the location where the foreign gene of interest is going to be integrated within the genome, that is, integration occurs randomly. Undoubtedly, the insertion of a DNA fragment into the genome of an organism can inactivate other genes vital to it, therefore experiments are often unsuccessful. Researchers have found a way to control transgenesis by the action of specific nucleases that act on the recipient DNA. These nucleases are: (a) zinc finger nucleases; (b) transcription activator-like effector nucleases (TALEN); and (c) the CRISPR-Cas system. Among these three methods, the third one allows the genetic modification of an organism with a high degree of precision, efficiency, and speed. The CRISPR-Cas9 technology for crop improvement to produce foods started on 2013. In the food sector, genetic modification has already arrived, not without controversy, to microorganisms, plants, and animals used by man for food production. Genetically modified foods can be of several types:

16.4.1 Knock-out foods

They come from organisms in which a gene that causes metabolic changes have been inactivated, or silenced, with ultimately beneficial effects. Accurate inactivation of a gene can be achieved by directed mutagenesis, interfering RNA (RNAi), or CRISPR-Cas9. This is the case of the *Flavr-Savr* tomato, which was the first genetically modified plant introduced in agriculture, back in 1996, in which the synthesis of the polygalacturonase enzyme was suppressed by means of the RNAi (gene silencing) technique. Modified tomatoes increased post-harvest life by 30 days before deteriorating. The CRISPR-Cas9 technique allows gene inactivation with precision and speed, thus the inactivation of the polyphenol oxidase gene has significantly increased the post-harvest half-life of mushrooms. Inactivation, also by this technique, of the *mstnba* gene, encoding the protein myostatin, an inhibitor of skeletal muscle growth, has allowed obtain carps with a muscle development superior to normal fish [5].

16.4.2 Transgenic foods

Transgenic agriculture uses genetically modified plants by recombinant DNA technology [6]. It uses all the resources of intensive agriculture, with additional substantial improvements based on the use of more resistant and productive plants to increase the yield per cultivated hectare. This agriculture was born in 1996 and has had a sustained increase of 3% of the cultivated land annually until reaching an area of approximately 198 million hectares in 2019. The progression experienced during recent years suggests that transgenic agriculture has not yet reached its limits.

The main transgenic crops used to date are aimed at obtaining:

- More resistant plants, which allow to increase the production per hectare of crop. This is the case of plants resistant to biotic stress, produced by pathogens, pests, fungi, bacteria, and virus. One of the greatest successes to date is *Bt-corn*, resistant to the caterpillar (larval stage) of European corn borer (*Ostrinia nubilalis*). Plants resistant to abiotic stress, such as herbicide resistant soybeans, salt-resistant rice, drought-resistant plants, and other adverse conditions, have also been obtained.
- More nutritious plants, such as golden rice, which has allowed the supply of vitamin A in countries where the deficiency of this vitamin was endemic, with serious adverse effects on the population, particularly women and children. Figure 16.1 shows the most used plant species to date in transgenic crops.

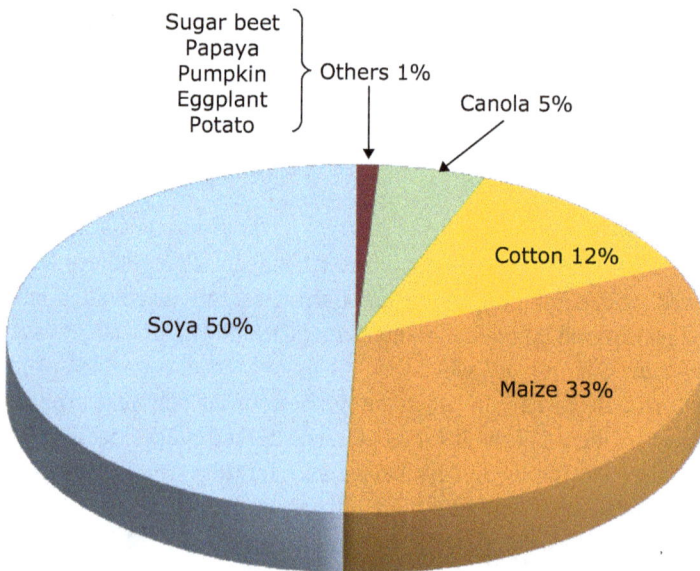

Figure 16.1: Transgenic species most used in agriculture. Source adapted from [7].

Quite often an agricultural crop faces several potentially dangerous situations, so that plants have to be modified with several resistance genes, thus obtaining multitransgenic organisms. The cultivation of corn plants with several simultaneous genetic modifications (stacked maize) is currently being imposed, so that, in the USA, a corn plant with four different genetic modifications is cultivated (Bt11, MIR162, TC1507, GA21). This plant is known as *Agrisure Viptera 3220* corn and is resistant to insects, herbicides, and drought, being more capable of facing several problems on the field. It was produced by conventional breeding crosses of several transgenic lines: the Bt11 corn is resistant to the European corn borer (see above); MIR162 is resistant to several types of insects and also capable of metabolizing mannose, by insertion of the enzyme mannose isomerase, which allows an easy selection of transformed plants; the TC1507 line is resistant to the larvae of the European corn borer and also to herbicides of the phosphinothricin type and ammonium glufosinate; and the GA21 line is resistant to glyphosate herbicide.

The genetic modification of mammals and birds has made great progress in obtaining: (a) more resistant animals to diseases and environmental conditions, such as mastitis resistant cows to produce more milk per animal; (b) production of higher quality milk, of low lactose and allergens, as well as with higher nutritional quality, rich in nutraceuticals, or in recombinant human proteins that help cure diseases; (c) production of higher quality meat, of lower fat content and higher proportion of ω-3 fatty acids; (d) and less polluting animals, such as a pig with salivary phytase, that excretes less phosphate into the environment.

The development of captive breeding of fish has allowed to obtain transgenic fish, with the idea of improving the quantity and quality of seafood. Fish are the main suppliers of polyunsaturated fatty acids (particularly ω-3) and iodine required by man, which make them a highly valued and necessary resource. Fish transgenesis has developed toward production of larger fishes, more resistant to diseases and environmental conditions, and more nutritious.

16.4.3 Cisgenic foods

Cisgenesis and transgenesis use the same recombinant DNA technology, but they differ in the nature of the transgene used. In the first case, foreign gene belongs to the same species of the organism to be transformed, while in the second, the genetic material belongs to another species. A cisgenic food is a genetically modified food, but it is not transgenic. This is the case of cows genetically modified with the gene encoding their own growth hormone to get more milk.

The greatest success in fish biotechnology lies in the size increase of the ocean salmon by overexpression of the gene encoding its own growth hormone. The modified salmon (North Atlantic, *Salmo salar*) has been obtained from two other varieties, and shows a size larger than normal, thus, after 2 years, the wild types weigh 2.0 kg while the cisgenic triples that weight. The transformation of different varieties of salmon, such as the Atlantic, Coho, and Chijnook, has also been successfully achieved [8]. *AquaAdvantage* transgenic salmon was authorized for human consumption in November 2015 and the market forecasts are very positive. Overexpression of growth hormone to increase size has also been used successfully in other fish such as carp (where it has been shown that this stimulus reduces mobility of the animal [9]), trout, and tilapia [10].

On the other hand, there are fish species that live in very cold waters (0–4 °C), because they produce and excrete in their blood proteins that lower the freezing temperature of the serum, thus protecting the animal. To date, two types of "antifreeze" proteins, AFPs and the glycoproteins AFGs, have been isolated and characterized. Most fishes of interest in aquaculture, such as salmon, trout, and tilapia do not naturally produce this type of protein and, therefore, they could not survive at temperatures between 0 and 4 °C. Other fishes such as rainbow trout (*Oncorhynchus mykiss*), common carp (*Cyprinus carpio*), catfish (*Ictalurus punctatus*), and hybrid tilapia (*Oreochromis hornorum* crossed with *O. aureus*) have been genetically modified, in this regard. The current situation is hopeful but cautious, because there is a risk of an uncontrolled interaction between natural fish and transgenic fish in oceanic waters.

16.4.4 Edited foods

The application of CRISPR/Cas9 to edit plant genome provide researchers with the ability to modulate crop-specific traits in a more precise and effective way. Gene edition through CRISPR-Cas9 has been successfully used to obtain rice rich in α-amylose and containing a resistant starch, which allows to increase its caloric power and also extends the hydrolysis time. This protects the consumer against diseases such as hypertension, diabetes, and colon cancer. Since the first edited crop by the CRISPR-Cas9 system in rice (2013), CRISPR-Cas9 has been used by many researchers to edit the genome of different plants, particularly rice, tomato, and oilseed rape. Improving the shape, size, color modification, palatability, biofortification of nutrient, fatty acids composition, eliminating anti-nutrients, and prolonging shelf life [11].

16.5 Alternative foods

16.5.1 Insects: source of quality protein

Eating insects, a common practice in many tropical countries, could be an interesting alternative to obtain quality proteins. Beetles and crickets provide excellent quality proteins and its production has no negative effects on the environment. To solve the emotional and palatability problems that insect consumption can produce in many people, insects can be processed and consumed in the form of dust or pastes. The proportion of protein and fat in crickets is similar to most meats and, for example, flour made from bed bug meal will be available. One hundred grams of insects contain 96 kcal, 72% protein, 16% lipids; 12% carbohydrates, 78.5 mg of calcium, 9.5 mg of iron, and 185.3 mg of phosphorus, and its production only generates 0.157 g of CO_2 of waste. A 100 g of meat contains 285 kcal, 52% protein, 48% fat, 0% carbohydrates, 30 mg of calcium, 9.1 mg of iron, and 10 mg of phosphorus, while its production generates 285 g of CO_2 and 2,193 liters of waste. Therefore, insect production emits fewer greenhouse gases and requires less land than pigs or cows [12]. Among the most consumed insects nowadays are beetles (31%); worms (18%); ants, bees, and wasps (15%); crickets, grasshoppers, and locusts (13%); dragonflies, flies, termites, and others (23%).

16.5.2 Milk analog

In general, many consumers are interested in decreasing their consumption of animal products, such as milk, because of health, environmental, and ethical reasons. Cow's milk is considered an essential food because of its nutritional value, providing macro (carbohydrate, quality protein, and fat) and micronutrients (calcium, phosphorous, and vitamins of B complex) in balanced proportions, however, low availability of certain minerals (iron) and vitamins (like C), with issues like milk allergy, lactose intolerance, and hypercholesterolemia prevalence, makes to some specific consumers to look for other alternatives, like plant based milk, dominated by soya bean milk, oat milk, almond milk, coconut milk, hemp milk, etc. [13].

Energetically and nutritionally, vegetable milks are poorest than cow's milk, but they are appreciated for their functionally active compounds, and the possibility to enriched them with probiotics or nutraceuticals [14].

16.6 Functional foods

Due to the gradual increase of certain diseases developed as a result of the lifestyle and eating habits we have the population began to realize the importance of diet to maintain and improve health. At the same time, the public authorities appreciated that a healthier population would alleviate the high health costs. The concept of functional food was developed in Japan during the 1980s and can be defined as that which, in addition to nourishing, is beneficial for health. Functional foods can be natural, such as breast milk or extra virgin olive oil; enriched, such as yogurt with probiotics, eggs with ω-3 fatty acids, juices, and other drinks with antioxidants; or genetically modified, as is the case with golden rice, rich in β-carotene. The maximal expression of functional foods will come with diets and personalized foods, as they can be used as specific supplements suitable for people with metabolic problems. Functional foods prevent chronic degenerative diseases, such as vascular accidents, hypertension, diabetes, cancer, etc. Nutraceuticals, which are described below, are key in the production of this type of food.

16.6.1 Nutraceuticals

The term "nutraceutical" is derived from "nutrition" and "pharmaceutical," and was determined in 1989 by Dr. Stephen DeFelice, president of the foundation for the innovation of medicine, in New Jersey (USA), at that time. DeFelice defined nutraceutical, as "The food, or part of a food, that provides medical or health benefits, including the prevention and/or treatment of diseases" [15]. Later, Shahidi described nutraceuticals as "Industrial food products made from natural foods and/or various non-food substances, which are consumed as pills, capsules, candies, cookies, refreshing and nutritious beverages, that are thought to be beneficial to health" [16]. They are active chemical or biological substances that can be found as natural components of food or added to them. Nutraceuticals have the ability to strengthen the healthy characteristics of a food, helping in the care and maintenance of health, either in the prevention of diseases and/or in the improvement of the physiological functions of the organism. Tomato is a functional food and lycopene (natural antioxidant especially abundant in tomato) is the nutraceutical, therefore, a nutraceutical converts a normal food into a functional one. Among the main types of nutraceuticals, we have:

– **Essential nutrients**. They can be amino acids, fatty acids, vitamins, and minerals. Man has to eat 8 of the 20 proteinogenic amino acids in the diet, since our body cannot synthesize them (Chapter 5). The FAO gives some recommendations about the average content of L-lysine and L-methionine that the ingested proteins must have to ensure a balanced nutrition. Conventional foods for humans are below these values, hence the need to enrich foods with these amino acids. Carotenoids, ω-3 and ω-6 polyunsaturated fatty acids, vitamins, and minerals are also essential nutrients.

Carotenoids can be divided into two fundamental groups: (a) carotenes, which are linear chain hydrocarbon carotenoids or with cyclations at one or both ends of the molecule and whose main exponent is β-carotene ($C_{40}H_{56}$); and (b) xanthophyll, which are oxygenated derivatives of carotenes. Oxygen atoms can be present as part of hydroxyl groups, as in lutein, or ketone groups, such as canthaxanthin, or a combination of both, as in astaxanthin, or other groups such as methoxyl, epoxide, carboxyl or acetate. Carotenoids are precursors of vitamin A, and excellent antioxidants (Chapters 4, 6 and 8), are especially abundant in microalgae.

Vitamins are essential substances for life and are needed in low quantity to develop their functions, which are usually very varied, as are their chemical structures. Its name comes from life (vita) and that the first one identified, vitamin B_1, had the chemical nature of an amine, hence its name of vita-amine, vitamin. They are classified, according to a criterion of solubility, in water-soluble vitamins, such as vitamin C and vitamins of the B complex. The other group of vitamins, A, D, E, K, are fat-soluble and mainly associated with vegetable fats. Its role in the body is to serve as antioxidants, co-enzymes, or hormones, among others (Chapter 6).

In industrialized countries, vitamin B_{12} is used as an additive to fortify food, due to its absence in plant-based foods and the tendency of the population to eat more foods of plant origin. Vitamin C (L-ascorbic acid) is an efficient antioxidant and cofactor in the biosynthesis of collagen, carnitine, and hormones, such as adrenaline. It is used as a food additive, under different formulas, E300 (free ascorbic acid), E301–E303 (sodium, calcium, and potassium ascorbate, respectively), E304 (esters with fatty acids), and E315 (erythorbic acid, which is a stereoisomer). Its main industrial use is in pharmacy (50%), food (25%), and drinks (15%).

– **Antioxidants**. In everyday life we are exposed to numerous and aggressive chemical oxidants, as a result of stress, environmental conditions, or harmful habits, such as smoking, drinking alcohol, or an inadequate diet. About 4% of the oxygen we consume daily degenerates toward the formation of reactive oxygen species, such as singlet oxygen, superoxide, and hydroxyl radicals and peroxides (ROS), which can irreversibly damage the biological tissue and contribute to the development of degenerative diseases, such as vascular type, cancer, diabetes, cataracts, or hypertension, among others (Chapter 8). The antioxidant nutraceuticals help to curb the effect of ROS and, therefore, to protect us against oxidative tissue degeneration. Among these foods we have those rich in phenolic compounds, anthocyanins, and flavonoids, such as grapes, red wine, cherries, pomegranates, blueberries, strawberries, blackberries, spinach, apples, and tea, among others. On the other hand, there are foods rich in carotenes (pro-vitamin A), such as carrots, mangoes, papayas, tomatoes, spinach, broccolis, celery, plums, asparagus, and nectarines; those rich in vitamin E, such as nuts, peanuts, almonds, paprika, and virgin olive oil; or those rich in vitamin C, citrus fruits, plums, guavas, strawberries, kiwis, persimmons, pineapple, and broccoli, among others. It should also be noted, among this type of foods, those containing selenium (eggs, meat, milk, chocolates) and phytosterols (nuts and peanuts).

– **Dietary fiber**. Dietary fibers are polysaccharides from plant foods and that we are unable to digest because we do not have the β-glycosidases enzymes, necessary for this (Chapter 3). Dietary fiber can be soluble or insoluble in water. Soluble fiber slows intestinal emptying and delays the absorption of glucose in the intestine. In addition, it serves as a fermentable substrate for the microbiota of the colon, facilitating mineral absorption, and producing vitamins of microbial origin. Foods rich in this type of fiber are oats, prickly pears, algae, and fruits. On the other hand, foods that contain insoluble dietary fiber stimulate salivation, chewing, and gastric secretions, therefore they facilitate digestion, increase the size of the bolus, and help to defecate because they stimulate intestinal peristaltic movements. Foods rich in insoluble fiber include whole grain vegetables, wheat bran, rice, and corn, as well as most fruits and vegetables.

16.6.2 Functional milk

Producing recombinant proteins in the milk implies having a functional food that can be either eaten directly or used as a starting raw material for the purification of the recombinant protein in question, thereby producing drugs. In order to get an animal to specifically produce a protein of interest in milk, it must be transformed with a transgene that integrates, at least, the structural gene corresponding to the recombinant protein to be obtained, fused with a promoter for its specific expression in mammary glands. The β-lactoglobulin promoter is the most frequently used (Figure 16.2).

Proteins began to be used as drugs in the 1920s, with insulin from pig pancreas as a pioneer. In the early 1980s, the production of recombinant human insulin by transgenic *E. coli* allowed all patients with diabetes to be treated with this protein. Several more proteins, including human growth hormone, are also produced by bacteria. However, this technique is limited because bacteria cannot synthesize protein complexes such as monoclonal antibodies, coagulation factors, and other proteins that have to undergo a post-translational modification process to be active. These processes include folding, proteolysis, association of subunits, glycosylation, among others, which can only be achieved in animal cells, cultured or in the animal itself.

Different transgenic species of animals can produce recombinant proteins, but so far only two systems are being implemented: the first is milk produced by transgenic ani-

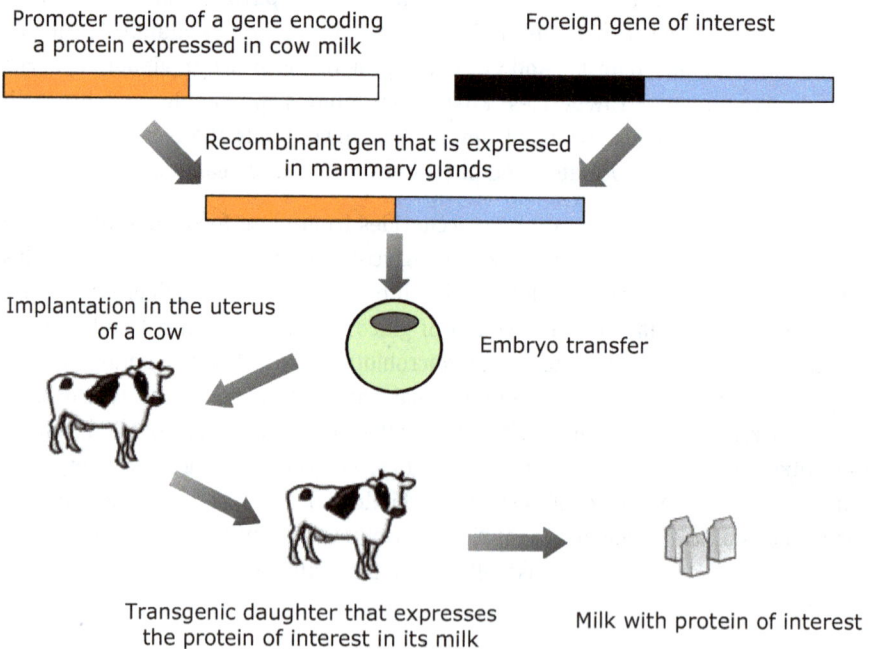

Figure 16.2: Production of recombinant proteins in milk. Source adapted from [17].

mals, which has allowed us to obtain interesting human proteins, such as coagulation factors [18]. Eggs of transgenic birds are second system that is currently being used. They have already allowed the successful obtaining of monoclonal antibodies, blood factors, hormones, growth factors, and erythropoietin [19]. Medicines for emphysema, hemophilia, rheumatoid arthritis, cancer or AIDS, among many other diseases, can be purified and prepared from milk and eggs. Large-scale production of human lactoferrin from the milk of cloned transgenic cows is particularly striking [20].

16.7 Therapeutic foods

Traditionally, vaccines consist in the injection into the person of attenuated viruses, or fractions of the virus or the microorganism causing the disease, so that a sufficient immune reaction that can protect us in the short term is induced. In this way, diseases of viral origin, such as smallpox, polio, etc., have been eradicated. The vaccine market is rising in economic terms but this type of prevention has its risks, since the attenuated microorganisms that are injected might cause unpredictable infections. On the other hand, vaccines must be stored in a refrigerator, they must be administered by specialized personnel and the syringes must be disposable to avoid contamination by other diseases. All these requirements make the immunization process more expensive and make it economically inaccessible to the most vulnerable countries to diseases. Every year millions of children in undeveloped countries do not have access to immunization programs.

Infectious diseases constitute a serious health problem for the world today and every year one third of deaths are due to this cause. Currently, around 2 million deaths in children under 5 years of age are due to diarrheal episodes and, according to the World Health Organization, 1.3 million of these deaths occur in Asian and African countries. The pathogens causing these events are rotavirus (25.4%), *Escherichia coli* (25.1%), *Shigella* (5.6%), *Campilobacter* (4.5%), and *Salmonella* (4.4%), among others. There are no effective vaccines available to stop this type of pathogens.

The possibility of using food for immunization against diseases is a reality in today's world. The idea is that foods can substitute for injections and significant progress has been made in this regard. Plants have been identified as a promising source of antigens for immunization of animals and humans, as they have sufficient capacity to elicit an adequate immune response. During the last decade, more than 100 recombinant proteins have been expressed in different plant tissues worldwide and initial estimations ensure the possibility of obtaining sufficient antigens to vaccinate millions of individuals from one hectare of cropland. The advances achieved in techniques of genetic modification of plants, as well as the development of protoplast suspension cultures have allowed to obtain a long list of species, such as potatoes, alfalfa, tobacco, corn, lettuce, tomatoes, spinach, rice, etc., which can be used for the production of recombinant antigens [21].

16.7.1 Therapeutic milk

Milk is the most promising vehicle to obtain substances with high added value, so far. The main companies in the biotechnology market have already launched the first experimental "pharmaceutical farms" in which herds of cows, sheep, and transgenic goats are producing milk for therapeutic use.

By combining nutritional and genetic interventions scientists obtain therapeutic milk, which is rich in specific components that have implications for health and disease treatment. In general, cow's milk is an attractive vehicle for large-scale production of biopharmaceuticals. Cows, goats, and sheep produce milk containing more than 60 different human proteins, including blood proteins, monoclonal antibodies, and vaccines. Transgenic technology also prevents all transmission of viral diseases associated with the use of human plasma.

In 1997, the first transgenic cow that produced milk enriched in α-lactoglobulin (rich in essential amino acids) was obtained for delivery to infants and elderly. In 1992, Tracy was the first sheep that produced therapeutic milk by expressing human α1-antitrypsin, under the control of the sheep promoter of β-lactoglobulin. α1-Antitrypsin appears in the milk at a concentration of 35 g/l and is used to treat several lung diseases. Although the first transgenic human protein obtained in milk was placed on the market in 2000, there are still improvements to make in the productive transgenic system, which is already used for large-scale production. Large-scale production of human recombinant proteins, such as growth hormone [22] and lactoferrin [20] are practical examples.

16.8 Probiotics, prebiotics, and symbiotics

Probiotics are foods that provide us with live and beneficial bacteria, so that they cross the digestive barrier and settle in our intestines, where they proliferate and displace pathogenic bacteria. The definition of probiotic used in the European Union is "live microorganisms that when administered in adequate amounts confer a health benefit on the host." The main human probiotics are lactic acid bacteria, as well as bifidobacteria [23].

Prebiotics are ingredients of plant origin that are not digested in humans and reach the colon where they serve as nutrients to the beneficial bacteria, *Lactobacillus* and *Bifidobacteria*, that proliferate preferentially over the others. These bacteria exert beneficial functions for the organism, as described in the previous section. Among the prebiotic substances are [24] (see also Chapter 3):

– Inulin and fructooligosaccharides (FOS), which are oligosaccharides that are fermented in the colon producing gases, organic acids, and short-chain fatty acids. Both substances facilitate a beneficial balance of intestinal microbiota, improve the bioavailability of calcium, regulate intestinal transit, and prevent colon cancer.

- Galactooligosaccharides (GOS), which are oligosaccharides of lactose-bound galactose. They are found naturally in breast milk along with others such as lacto-N-tetrose, lacto-N-fucopentose, and monosiasil-lacto-N-neohexose. They are not digested by human enzymes and facilitate the absorption of calcium and magnesium in the intestines, prevent colon cancer and inhibit the adhesion of pathogenic bacteria to intestinal epithelia.

- Lactulose and lactitol: lactulose is a semisynthetic disaccharide of galactose and fructose, which is obtained from lactose by isomerization in the presence of an aluminum salt. It is used as a medicine to combat constipation because it has an effect of stool softening, it also causes acidification of the colon and prevents the proliferation of germs. Lactitol is a polyalcohol that prevents the proliferation of proteolytic bacteria by increasing acidophilic bacteria.

- Polydextrose (PDX) is a glucose–derived polysaccharide that contains small amounts of sorbitol. It cannot be digested by the human body and is fermented by intestinal bacteria causing effects similar to the other prebiotics described.

- Isomaltose, stachyose, raffinose, soy oligosaccharide, pectin, and hemicellulose, which are carbohydrates present in vegetables, such as garlic, onions, leeks, asparagus, artichokes, tomatoes, bananas, and wheat. They are also recognized as prebiotics.

The most representative products on sale in this type of foods are some dairy preparations, beverages, baby foods, and cookies. In general, the average amount of prebiotics ingested in the diet usually ranges around 0.8 g/day, while experts recommend an ingestion of 2–6 g.

Symbiotic foods are a mixture of probiotics and prebiotics, such as *Lactobacillus* plus lactitol or *Bifidobacterium* plus FOS, and generate a beneficial effect on health, because they (a) regulate the composition of the intestinal microbiota; (b) mature and balance the immune system, thereby increasing resistance to infections; (c) prevent and help treat diarrhea caused by inflammation of the intestinal mucosa; (d) stabilize and improve diseases that affect the intestine, such as Crohn's disease and ulcerative colitis; (e) produce metabolic benefits, such as lowering serum cholesterol and triglyceride levels, as well as better glycemic control; and (f) improve the digestion and absorption of nutrients [25].

The beneficial effects that this type of food has for health are the following:

- Protection against pathogenic bacteria and viruses. Because they secrete antiviral, antibacterial, and antifungal substances, they effectively prevent against the invasion by pathogenic microorganisms. Currently, viruses are a very serious threat to human health, since more than half of the infections are caused by them. There are studies that show the help of probiotics in intestinal diarrhea caused by rotavirus, shortening the duration of the disease by 1–1.5 days. Probiotic bacteria can interact with the rotavirus and inactivate it. The effectiveness varies according to the strain in question and it has been documented that *L. rhamnosus* and *B. lactis* are the most efficient. In addition, strains of *L. rhamnosus* are currently used for the treatment of food in order to eliminate and/or prevent mycotoxins, particularly aflatoxins (produced by certain fungi in corn, peanuts, and some nuts). It seems that bacteria do not metabolize mycotoxins, but they rather adhere them to their outer surface and immobilize them, so they do not pass into the blood.

- They contribute to the maturation of the immune system. They form a physical barrier on the intestinal mucosa that induces the maturation of the immune system, thus controlling diarrhea caused by intestinal irritation and allergies in children.
- Promote mineral nutrition. They ferment food remains that generate an acidic environment in the intestine that facilitates mineral absorption and the production of vitamins for the epithelium.
- They prevent unwanted transgenic microorganisms from progressing into the microbiota. With the emergence of transgenic foods in the human diet, the presence of probiotics is no doubt more relevant, since they can displace any transgenic microorganism that has been able to reach our intestinal flora. The effect of transgenic microorganisms on microbiota has not been thoroughly studied yet, so it is very important to prevent the implantation of any of them that could cause unwanted side effects [23].

16.8.1 Genetic modification of probiotics

Probiotics acquire greater relevance, if possible, with the possibility of genetically modifying them, since they are not only suitable for displacing dangerous microorganisms in the flora, including some unfortunate transgenic, but also because they can be modified to increase their beneficial properties. The genetic modification of probiotics will undoubtedly be a new front to address the treatment and prevention of human diseases [26]. Progress has been made in the modification of lactic bacteria for the prevention and treatment of ulcerative colitis [27]. The use of genetically modified *Lactococcus lactis* to secrete cytokine interleukin 10, with anti-inflammatory effects, has been shown to be an effective treatment of irritable bowel, which opens the doors to treating diseases without effective remedies so far.

Among the next generation of probiotics, we must highlight:

- Genus *Bacteroide*, which account for 40% of the microflora and are involved in carbohydrate fermentation and short chain fatty acids production, in the modulation of the immune system and in the biotransformation of steroids and bile acids (thus affecting the cholesterol levels). Good candidates for experimentation are *Bacteroides xylanosolvens*, *B. acidifaciens*, and *B. uniformis*.

- Genus *Clostridium* is also very abundant in microflora. *Clostridium butiricum* has been used to fight *Helicobacter pylori* infections and also reduces cholesterol levels.
- Genus *Lactobacillus*, such as *L. casei* and *L. delbruekii*, produce biogenic amines like histamines and tyramines, which have hypotensive effects.
- *Faeclibacterium prausnitzii* strains, which accounts for 5% of colon microbiota have anti-inflammatory properties. On the other hand, *Akkermansia muciniphila* can restore the functions of the intestinal barrier and reduce inflammation. This organism is significantly reduced in obese people.
- Transgenic probiotics that stimulate the production of tumor suppressor proteins can help control colorectal cancer.

16.9 Cellular agriculture

The highly negative impact of food production on the environment, particularly those produced on animal bases, requires a high degree of innovation in order to keep the sustainability of our planet, with high degree of biodiversity. Cellular agriculture includes the use of single-cell organisms, cell cultures, and bioreactors for the industrial production of food and food ingredients, instead of traditional agriculture or animal husbandry. Its development is motivated by the need to develop a sustainable food production system, addressing food and water safety, environmental footprint and animal welfare [28]. Among these activities we find the production of single-cell proteins (SCP), cultured meat, and artificial milk.

16.9.1 Single cell protein

The so-called single-cell protein (SCP) comes from unicellular microorganisms, such as algae, fungi, yeasts, and bacteria, which are used as a protein supplement for human and animal feed. Since ancient times, microorganisms have been used directly or in precision fermentation processes as a food supplement, particularly in times of urgent need, as in the great European wars, when yeasts of the *Saccharomyces cerevisiae* type were used as a source of protein in countries involved in the conflict. The increase in the protein deficit we face is a constant concern for man and microorganisms are being produced for use in animal and human food since 1996 [29].

16.9.1.1 Microalgae: source of protein quality and nutraceuticals
Microalgae have a protein concentration between 40% and 60%, a level high enough to be viable for their use in animal feed. With respect to the quality of these proteins, the content of essential amino acids (L-lysine, L-methionine, L-threonine, L-tryptophan,

L-leucine, L-isoleucine, L-valine, and L-phenylalanine) is very diverse, depending on the type of microalgae and the culture conditions used to produce them. However, on average, microalgae proteins show higher contents of essential amino acids than do conventional foods for humans, such as cereals. Nevertheless, microorganisms have an important problem as a direct food: its high content of nucleic acids, which ranges between 3% and 8% (w/w) in the case of microalgae. This imposes limitations to their consumption or forces to separate nucleic acids from the corresponding biomass, which would undoubtedly make the product more expensive. Consumption of an excess of nucleic acids in humans raises the uric acid content in the blood and therefore the risk of suffering gout or kidney stones, among other problems. The nucleic content of a food must be reduced to 2% of its dry weight to make it suitable for human use [30].

According to the data previously reported, microalgae (eukaryotes and cyanobacteria) appear to be very suitable microorganisms for human and animal feed for several reasons: (a) they can be grown throughout the year, with a productivity of 50 tones of dry weight per hectare and per year and minimum nutrient requirements; (b) the space used for its production does not compete with traditional food production in agriculture; (c) the microalgae protein has a quality comparable to that of vegetable protein; and (d) its nucleic acid content is the lowest among the different microorganisms studied.

The annual production of microalgae is more than 10,000 tones of dry weight, distributed among different land areas. This biomass is collected by various techniques, dried and used directly in the form of powder, tablets, capsules or tablets. The most cultivated species for this purpose are the eukaryotic microalgae *Haematococcus pluvialis* (33%), *Chlorella vulgaris* (22%), and *Dunaliella salina* (13%), as well as the filamentous cyanobacterium *Spirulina platensis* (also called *Artrospira platensis*) (32%). The microalgae market is expected to multiply by five in 2025, given its industrial possibilities for the human and animal food industry (around 60% of the biomass produced), and also for the production of biodiesel (~ 40%).

But microalgae also have their drawbacks: (a) biomass production currently has a high cost, so studies are underway to reduce the price of the process; (b) they have a cell wall, which represents 10% of the dry weight of the biomass, which is not digestible by human, therefore, the biomass should be treated to break the wall, so that the proteins are accessible to digestive enzymes. In this context, a variety of methods can disrupt the cell wall: chemical (organic solvents or acids), enzymatic (cellulases), and physical and mechanical (bead milling, high-pressure homogenization) can be used, but these treatments can be omitted if the SCP is used as feed for cattle, as they have cellulose-degrading symbiotic bacteria [31]; (c) the nucleic acid content is excessive for the current food standards; and (d) possibility of heavy metal contamination, which tend to accumulate in the cellular interior, if the medium is contaminated. In fact, among the defense mechanisms that microalgae have for their defense/adaptation to contaminated media is the accumulation of metals in their cellular vacuoles, where they cannot exert their toxic activity to the algae; however, this makes the biomass of these algae not recommended for human and/or animal use.

16.9.1.2 Mycoprotein: meat substitute

Mycoprotein is a type of SCP derived from the microfungi *Fusarium venenatum*, which is combined with egg white, wheat protein or other ingredients, and used as a substitute for meat in vegetarian diets due to its low calories content and texture analogous to that of meat. It is also characterized by its high protein and fiber content and low cholesterol and nucleic acid content (Table 16.1). It has an important content of essential amino acids, resembling that of microalgae proteins.

Mycoprotein has important benefits for the consumer:

- Being a preparation rich in protein and fiber, while low in calories, is an excellent food for weight loss. It has an effect of increasing satiety and decreasing the production of ghrelin, the hormone that encourages eating (Chapter 14). The fiber it contains produces improvements in intestinal transit and digestive health.
- Due to its high content of essential amino acids, it is a very complete protein that provides substances necessary for the body. It is very suitable for vegetarian diets, which have the problem of being deficient in essential amino acids.
- Lowers cholesterol. Studies of animal protein replacement by mycoproteins have been done, and results in a 9% decrease in "bad cholesterol," that is, LDL-cholesterol, while increasing HDL-cholesterol ("good cholesterol") by 12%.
- Regulates the level of blood glucose, reaching a decrease of 36% compared to the control in clinical studies.
- They also have the positive effect of reducing animal slaughter by 5% and greenhouse gases associated with food production by 25%.

Mycoproteins can be used to improve the nutritional value and functional features of food items, in addition to being an excellent source of protein-rich nutrients in and of itself. Other species of fungi used for SCP production are *Aspergillus flavus*, *A. niger*, *A. ochraceus*, *A. oryzae*, *Cladosporium cladosporioides*, *Monascus ruber*, *Penicillium citrinum*, and *Trichoderma viride* [31]. On the other hand, mycoproteins have produced allergic reactions in particularly sensitive people, and also have the disadvantage that their potential allergens are poorly understood. However, these allergic reactions are not very numerous, but it is important to continue research in this area.

Table 16.1: Nutritional composition of mycoprotein per 100 grams of fresh weight.

Nutrient	Quantity
Energy (Kcal)	85.0
Protein (g)	11.0
Fat (g)	2.9
Saturated fatty acids	0.7
Monounsaturated fatty acids	0.5
Polyunsaturated fatty acids	1.8
Dietary fiber (g)	6.0
Soluble sugars	0.5
Sodium (mg)	5.0
Cholesterol (g)	0.0
Iron (mg)	0.5
Zinc (mg)	9.0
Selenium (mg)	0.02

Source [35].

16.9.2 Cultured meat

There are at least three reasons to look for alternative sources of meat for human feeding. The first is related to the increase in the pace of global meat consumption; the second is the increasingly documented concern of the environmental impact that meat production has on the planet; and the third has to do with social concerns about the conditions of breeding, transport, and slaughter of animals, which demand hygiene and minimum acceptable conditions.

An innovative technique in meat production is cultured meat, which consists in the cultivation of muscle tissue in the laboratory, based on stem or embryonic cells, in nutrient-rich media, using grids or plates of a biodegradable material with hamburger shape. Muscle cells can also be cultured in a bioreactor and the resulting mass is vacuum packed or sent to a processing plant. Vitamins and minerals, as well as flavoring agents can be added to the meat during the growing process [33]. Currently, the two main stem cells considered most suitable for culturing meat are embryonic stem cells or satellite cells. The production of cultured meat involves the isolation of animal stem cells and the proliferation and differentiation of these cells in a suitable cell culture medium. During the growth process scaffolding materials such as collagen-like gel polymers are often added to form a support network for tissue development [34]. The biggest technical advantage of cultured meat is that it is done under very controlled and highly operable conditions. In addition, cultured meat has the great advantage of its low environmental impact which is significant in traditional meat production based on beef, lamb, pork, and chicken, as shown in Table 16.2.

Table 16.2: Environmental impact of meat production.

Environmental parameter	Beef	Pork	Poultry (%)	In vitro
Energy use	100	59.5	40.5	55.7
Land use	100	36.8	33.9	1.0
Water use	100	35.7	22.1	4.3
GEG[a] emission	100	21.8	23.2	5.0

[a]GEG: Greenhouse effect gases. Source [32]

The implementation of an alternative meat production allows the possibility of offering new products to the market, whose acceptability must be proven. Consumers see important benefits for the environment and food security in cultured meat, which contrast with a concern about social changes that would occur in rural areas [36]. Cultured meat is technically possible today and its small-scale production seems particularly promising from a social acceptance point of view. However, its economic possibilities emerge as a major obstacle [37]. Actually, the "Believer Meat Company," operating in Rehovot (Israel) produces 500 Kg meat/day which gives 5,000 hamburgers of 100 g each, at a cost of 1.5 USD. A project for a new factory in Wilson (North Carolina, US) is growing to produce 10 million Kg meat/year. A forecast for meat culture market for 2040 is expected to grow up to 500 million USD.

Although cultured meat is carried out in laboratories with high safety measures, possible health problems are related to the use of animal serum, during the production process. Thus, among other institutions, the European Medicines Agency has indicated that the bovine serum used must be free of viruses such as bovine viral diarrhea virus, bluetongue virus or rabies virus, among others [38].

16.9.3 Lab-grown milk

Global cow milk production, in 2021, was 544.1 million of Tons, which suppose an increase of about 10% in the last five years, and the demand is still increasing. The cows in the world are actually about 1,000 million and it is not possible to increase them due to the negative impact on the environment. In order to increase 20%, the milk produced per cow, there have obtained transgenic cow using the gen of the proper hormone growth. In addition, cows producing the enzyme lysostaphin in the milk make animal resistant to mastitis, which also increases the amount of milk produced by a cow during the year. Date from the US indicates that they obtained a production of milk, per cow during 2020, of 10,705 Kg, which suppose an increase of 12% during the last ten years.

Cellular agriculture allows us to produce animal proteins through microbial precision fermentation [39]. The mechanism to produce artificial milk is based on the use of transgenic yeast, which was modified by including in the genome eight genes coding for the main milk proteins, four *LACAS* genes for caseins (α, β, Y, κ), two *LGB* genes for α, and β lactoglobulins, respectively, one *LALBA* gene for lactalbumin, and one *LAF* gene for lactoferrin. The modified yeast is growth in a bioreactor, under optimal fermentation conditions with glucose in a liquid medium. After a few days the yeast are harvested, and the milk proteins extracted. These proteins are suspended in an amount of water, keeping the adequate proportion and we obtained a dairy food. Furthermore, the addition of a carbohydrate, which could be fructose, vegetable fats, and some minerals and vitamins, we obtain fresh milk ready to drink or use to produce dairy products.

Artificial milk supposes a new source of cow fresh milk, without the requirement of cows. In addition, this milk can be produced without lactose, cholesterol, hormones, and antibiotics, which suppose an advance to satisfy the milk demand, and a significant contribution to the sustainability of the planet. Global precision fermentation market is expected to grow from 1.6 billion of USD in 2022 up to 36.3 billion USD in 2030 [40].

16.10 Conclusions

The future will be conditioned by a requirement of more and more quality food, which allows the integration of personalized diets, according to the needs of each person. In this context, the predominance of functional and therapeutic foods, as well as probiotics, prebiotics, and symbiotics, will be a constant requirement for the food industry.

The genetic modification of living beings for the production of more resistant and productive plants, animals and microorganisms of high quality will be a very preferred activity in the world, as well as aquaculture, with the objective of meeting the demand for fish, very valuable food for its supply of ω-3 fatty acids and iodine to the human diet. Alternative food source, such as insects and plants milk, will be present in our diet, as a source of quality protein and essential nutrients.

Cellular agriculture and precision fermentation with microorganisms are activities with a great future to produce fresh and highly nutritive foods and nutraceuticals for animals. All of this must occur with a rigorous respect for the environment, to make Earth, our home, a liveable, and pleasant planet.

Bibliography

[1]   Foley JA, Steinmetz G, Richardson J. A five-step plan to feed the world. Nat Geog 2014;225:26–59.
[2]   Zhang F, Chen X, Vitousek P. Un ensayo agrícola a gran escala. Investigación y Ciencia 2013;11:80–83.

[3] Despommier D. Agricultura vertical. Investigación y Ciencia. Septiembre 2010:74–81.

[4] Smeets K, Ruytinx J, Van Belleghem F, Semane B, Lin D, Vangronsveld J, Cuypers A. Critical evaluation and statistical validation of a hydroponic culture system for Arabidopsis thaliana. Plant Physiol Biochem 2008;46:212–18.

[5] Zhong Z, Niu P, Wang M, Huang G, Shuhao X, Sun Y, Xiaona X, Hou Y, Sun X, Yan Y, Wang H. Targeted disruption of sp7 and myostatin with CRISPR-Cas9 results in severe bone defects and more muscular cells in common carp. Sci Rep 2016;6:22953. 10.1038/srep22953.

[6] Freedman DH. Cultivos transgénicos: sigue el debate. Investigación y Ciencia 2013;446:72–79.

[7] ISAAA. The International Service for the Acquisition in Agri-biotech Applications 2018. www.isaaa.org.

[8] Friesen EN, Higgs DA, Devlin RH. Flesh nutritional content of growth hormone transgenic and non-transgenic coho salmon compared to various species of farmed and wild salmon. Aquaculture 2015;437:318–26.

[9] Li DL, Fu CZ, Hu W, Zhong S, Wang YP, Zhu ZY. Rapid growth cost in "all-fish" growth hormone gene transgenic carp: Reduced critical swimming speed. Chin Sci Bull 2007;52:1501–06.

[10] Alimudding K, Faridah N, Yoshizaki G, Nuryati S, Setiawati M. Growth, survival, and body composition of transgenic common carp Cyprinus carpio 3rd generation expressing tilapia growth hormone cDNA. HAYATY J Biosci 2016;23:150–54.

[11] Camerlengo F, Frittelli A, Pagliarello R. CRISPR towards a sustainable agriculture. Encyclopedia 2022;2:538–58. https://doi.org/10.3390/encyclopedia2010036.

[12] Van Huis A, Van Itterbeeck J, Klunder H, Mertens E, Halloran A, Muir G, Vantomme P. Edible insects: Future prospects for food and feed security. Food Agric Organ United Nations; 2013. ISBN:978-92-5-107595-1.

[13] Paul AA, Kumar S, Kumar V, Sharma R. Milk Analog: Plant bases alternatives to conventional milk, production, potential and health concern. Crit Rev Food Sci Nutr 2020;60:3005–23. https://doi.org/10.1080/10408398.2019.1674243.

[14] Reyes-jurado F, SotoReyes N, Dávila-Rodríguez M, Lorenzo-Leal AC, Jiménez-Mungia MT, Mani-López-Bello E, López-Malo A. Plant-based milk alternatives; Types, processes, benefits characteristics. Food Res Int 2021. 10.1080/87559129.2021.1952421.

[15] DeFelice SL. The nutraceutical revolution: Its impact on food industry R&D. Trends Food Sci 1995;6:59–61.

[16] Shahidi F. Nutraceuticals and functional foods in health promotion and disease prevention. Acta Hortic 2005;680:13–24.

[17] Vigara J, Vega JM. Alimentos del futuro. Impacto de los transgénicos. Editorial Universidad de Sevilla. Sevilla. España 2016. ISBN:978-84-472-1839-4.

[18] Shepelev MV, Kalinichenko SV, Deykin AV, Korobko IV. Production of recombinant proteins in the milk of transgenic animals: Current state and prospects. Acta Nat 2018;10:40–47.

[19] Kwon MS, Koo BC, Kim D, Nam YH, Cui X-S, Kim NH, Kim T. Generation of transgenic chickens expressing the human erythropoietin (hEPO) gene in an oviduct-specific manner: Production of transgenic chicken eggs containing human erythropoietin in egg whites. PLoS One 2018;13(5): e0194721.

[20] Wang M, Sun Z, Yu T, Ding F, Li L, Wang X, Fu M, Wang H, Huang J, Li N, Dai Y. Large-scale production of recombinant human lactoferrin from high-expression, marker-free transgenic cloned cows. Sci Rep 2017;7:10733. 10.1038/s41598-017-11462-z.

[21] Concha C, Cañas R, Macuer J, Torres MJ, Herrada AA, Jamett F, Ibáñez C. Disease prevention: An opportunity to expand edible plant-based vaccines. Vaccines 2017;5:14. 10.3390/vaccines5020014.

[22] Jiang BC, Yu DB, Wang LJ, Dong FL, Kaleri HA, Wang XG, Ally N, Li J, Liu HL. Doxycycline-regulated growth hormone gene expression system for swine. Genet Mol Res 2012;11(3):2946–57.

[23] Ng SC, Hart AL, Kamm MA, Atagg AJ, Knight SC. Mechanisms of action of probiotics: Recent advances. Inflamm Bowel Dis 2009;15:300–10.

[24] Slavin J. Fibre and prebiotics: Mechanism and health benefits. Nutrients 2013;5:1417–35. 10.3390/nu5041417.

[25] Olagnero G, Abad A, Bendersky S, Genevois C, Granzella L, Montonati M. Alimentos funcionales: Fibra, prebióticos, probióticos y simbióticos. Diaeta 2007;25:20–33.

[26] Chua KJ, Kwok WC, Aggarwal N, Sun T, Chang MW. Designer probiotics for the prevention and treatment of human diseases. Curr Opin Chem Biol 2017;40:8–16.

[27] Moreno de leblanc A, Del Carmen S, Chatel J-M, Miyoshi A, Azevedo V, Langella P, Bermúdez-Humarán LG, Leblanc JG. Current Review of genetically modified lactic acid bacteria for the prevention and treatment of colitis using murine models. Hindawi Publishing Corporation Gastroenterology Research and Practice. Volume 2015, article ID 146972; 2015. http://dx.doi.org/10.1155/2015/146972.

[28] Ercili-Cura D, Barth D. Cellular agriculture: Lab grown foods. Editorial American Chemical Society; 2021.

[29] Ritala A, Häkkinen ST, Toivari M, Wiebe MG. Single cell protein state-of-the-art, industrial landscape and patents. Front Microbiol 2017;8:1–18.

[30] Forján E, Vílchez C, Vega JM. Biotecnología de microalgas. Servicio Publicaciones Universidad de Huelva. Huelva; 2014. ISBN: 978-84-617-2314-0.

[31] Bajic B, Vučurovic D, Vasic D, Mučibabic RJ, Dodic S. Biotechnological production of sustainable microbial proteins from agro-industrial residues and by-products. Foods 2023;12:107.

[32] Tuomisto HL, Teixeira de mattos MJ. Environmental impacts of cultured meat production. Environ Sci Technol 2011;45(14):6117–23. 10.1021/es200130u.

[33] Post MJ. Cultured meat from stem cells: Challenges and prospects. Meat Sci 2012;92:297–301.

[34] Kumar A, Sood A, Han SS. Technological and structural aspects of scaffold manufacturing for cultured meat: Recent advances, challenges, and opportunities. Crit Rev Food Sci Nutr 2023;63:585–612.

[35] Finnigan TJA, Wall BT, Wilde PJ, Stephens FB, Taylor SL, Freedman MR. Mycoprotein: The future of nutritious nonmeat protein, a symposium review. Curr Dev Nutr 2019;3:nzz021.

[36] Verbeke W, Marcu A, Rutsaert P, Gaspar R, Seibt B, Fletcher D, Barnett J. Would you eat cultured meat? Consumers' reactions and attitude formation in Belgium, Portugal and the United Kingdom. Meat Sci 2015;102:49–58.

[37] Van der weele C, Tramper J. Cultured meat: Every village its own factory. Trends Biotechnol 2014;32:294–96.

[38] Hadi J, Brightwell G. Safety of alternative proteins: Technological, environmental and regulatory aspects of cultured meat, plant-based meat, insect protein and single-cell protein. Foods 2021;10:1226.

[39] Broad GM, Tomas OZ, Dillard C, Bowman D, Le Roy B. Framing the futures of animal-free dairy: Using focus groups to explore early-adopter perceptions of the precision fermentation process. Front Nutr 2022;9:997632. 10.3389/fnut.2022.997632.

[40] Upadhyay S. Precision fermentation, growing scale of lab-grown milk and future of dairy farmers. RethinkX, SCIRO, Forbes: Good Food Institute; 2022. http://www.shirish.net.in/2022/08/precision-fermentation-growing-scale-of.html.

Appendix 1: Dietary reference values

Dietary reference values (DRVs) are science-based nutrient reference values for healthy populations. It is an umbrella term for a set of nutrient reference values. *DRVs vary by life stage and gender.* They have various purposes such as assessing the nutritional quality of diets of individuals or groups, designing diets (e.g., school meals), creating nutrition guidelines, dietary counseling, setting reference values for food labeling, and for the development of nutrition and food policies. *DRVs should not be viewed as recommendations for individuals.* Rather, DRVs are scientific references for professionals who use them when setting nutrient goals for populations or recommendations for individuals.

In its scientific opinions, the European Food Safety Agency (EFSA) used four types of DRVs:

- Average requirement (AR)
- Population reference intake (PRI)
- Adequate intake (AI)
- Reference intake (RI) range for macronutrients

(Sometimes the tolerable upper intake level (UL) is included: the maximum amount of a nutrient that can be consumed safely over a long period of time.)

The *AR* is the level of (nutrient) intake estimated to satisfy the physiological requirement or metabolic demand, as defined by the specified criterion for adequacy of that nutrient, in half of the people in a population group, given a normal distribution of requirement. Synonym: Estimated AR (USA/Canada) is defined as the average daily nutrient intake level that is estimated to meet the requirements of half of the healthy individuals in a particular life stage and gender group.

The *PRI* is the level of (nutrient) intake that is adequate for virtually all people in a population group. On the assumption that the individual requirements for a nutrient are normally distributed within a population and the interindividual variation is known, the PRI is calculated on the basis of the AR plus twice its standard deviation. This will meet the requirements of 97.5% of the individuals in the population.

PRI (UE) corresponds with the recommended dietary allowance (USA/Canada). A similar concept is the reference nutrient intakes (UK).

AI is the value estimated when a PRI cannot be established because an AR cannot be determined. An AI is the average observed or experimentally determined approximations or estimates of nutrient intake by a population group (or groups) of apparently healthy people that is assumed to be adequate. The practical implication of an AI is similar to that of a PRI, which describes the level of intake that is considered adequate for health reasons. The terminological distinction relates to the different way in which these values are derived and to the resultant difference in the "firmness" of the value.

https://doi.org/10.1515/9783111111872-017

The *RI range* is used for energy-yielding macronutrients. It is expressed as the proportion (%) of energy derived from that macronutrient. RIs represent ranges of intakes that are adequate for maintaining health and are associated with a low risk of selected chronic diseases.

The *tolerable UL* is the maximum chronic daily intake of a nutrient (from all sources) judged to be unlikely to pose a risk of adverse health effects to humans.

Appendix 2: Reference intakes (RI) for total fat and carbohydrates and adequate intakes (AIs) for fatty acids, dietary fiber, and water

Age group (years)	Total C.H. (E%)[a]	Dietary fiber (g/day)[b]	Total fat (E%)[a]	SFA	LA (E%)[a]	ALA (E%)[a]	EPA+ DHA (E%)[a]	DHA (E%)[a]	TFA	Age group (years)	Water (L/day)[b,c] Male	Female
7–11 (months)[d]			40[b]	ALAP	4	0.5		100	ALAP	6–12 months		0.8–1.0
1	45–60	10	35–40	ALAP	4	0.5		100	ALAP	1		1.1–1.2
2–3	45–60	10	35–40	ALAP	4	0.5	250		ALAP	2–3		1.3
4–6	45–60	14	20–35	ALAP	4	0.5	250		ALAP	4–8		1.6
7–10	45–60	16	20–35	ALAP	4	0.5	250		ALAP	9–13	2.1	1.9
11–14	45–60	19	20–35	ALAP	4	0.5	250		ALAP	14–17	2.5	2.0
15–17	45–60	21	20–35	ALAP	4	0.5	250		ALAP			
≥18	45–60	25	20–35	ALAP	4	0.5	250		ALAP	≥18	2.5	2.0
Pregnancy												
			20–35	ALAP	4	0.5	250	+100 −200[e]	ALAP			2.3
Lactation												
			20–35	ALAP	4	0.5	250	+100 −200[e]	ALAP			2.7

ALA, α-linolenic acid; ALAP, as low as possible; DHA, docosahexaenoic acid; E%, percentage of energy intake; EPA, eicosapentaenoic acid; F, female; L, liter; LA, linoleic acid; M, male; SFA, saturated fatty acids; TFA, transfatty acids.
[a]RI, reference intake range.
[b]AI, adequate intake.
[c]Includes water from beverages of all types, including drinking and mineral water, and from food moisture.
[d]The second half of the first year of life (from the beginning of the seventh month to the first birthday).
[e]In addition to combined intakes of EPA and DHA of 250 mg/day.

https://doi.org/10.1515/9783111111872-018

Appendix 3: Summary of average requirement (AR) for energy expressed in kcal/day for different groups of population

Infants

Age (months)	AR (kcal/day)		AR (kcal/kg of body mass per day)	
	Boys	Girls	Boys	Girls
7	636	573	76	76
8	661	599	77	76
9	688	625	77	76
10	725	656	79	77
11	742	673	79	77

Children and adolescents

Age (years)	AR (kcal/day)	
	Boys	Girls
1	550–777	503–712
2	727–1,028	669–946
3	830–1,174	775–1,096
4	888–1,615	826–1,502
5	942–1,712	877–1,594
6	996–1,811	928–1,687
7	1,059–1,925	984–1,790
8	1,126–2,046	1,045–1,899
9	1,191–2,165	1,107–2,013
10	1,196–2,416	1,125–2,273
11	1,264–2,554	1,181–2,385
12	1,345–2,717	1,240–2,505
13	1,444–2,916	1,299–2,624
14	1,555–3,142	1,346–2,719

https://doi.org/10.1515/9783111111872-019

(continued)

Age (years)	AR (kcal/day)	
	Boys	**Girls**
15	1,670–3,374	1,379–2,786
16	1,761–3,556	1,398–2,824
17	1,819–3,675	1,409–2,846

Adults

Age (years)	AR (kcal/day)
Men	
18–29	2,338–3,340
30–39	2,264–3,235
40–49	2,234–3,192
50–59	2,204–3,149
60–69	2,017–2,882
70–79	1,984–2,834
Women	
18–29	1,878–2,683
30–39	1,813–2,590
40–49	1,798–2,569
50–59	1,783–2,547
60–69	1,628–2,326
70–79	1,614–2,305
Pregnant women[1]	
First semester	+70
Second semester	+260
Third semester	+500
Lactating women[1]	
0–6 months postpartum	+500

[1]In addition to AR for nonpregnant women.

Notes

Values for children, adolescents, and adults were calculated to take account of different levels of physical activity (hence, the intervals) and are based on an assumed healthy body mass index of 22 kg/m^2.

Energy values can be converted to kilojoules per day (kJ/day) by multiplying by 4.184.

References and more detailed information

DRV Finder, available at: https://multimedia.efsa.europa.eu/drvs/index.htm

EFSA (European Food Safety Authority), 2017. Dietary reference values for nutrients summary report. EFSA support publication 2017;14 (12):e15121. 98 pp. 10.2903/sp.efsa.2017.e15121

EFSA Panel on Dietetic Products. Nutrition and Allergies (NDA); Scientific opinion on dietary reference values for energy. EFSA J 2013;11(1):R3005. 112 pp. 10.2903/j.efsa.2013.3005.

Institute of Medicine (US) Food and Nutrition Board. Dietary Reference Intakes: A Risk Assessment Model for Establishing Upper Intake Levels for Nutrients. Washington (DC): National Academies Press (US); 1998. What are Dietary Reference Intakes? Available from: https://www.ncbi.nlm.nih.gov/books/NBK45182/

Appendix 4: Alcohol as a nutrient

The International Union of Pure and Applied Chemistry (IUPAC) defines alcohols as "compounds in which a hydroxy group, –OH, is attached to a saturated carbon atom R_3COH" [1]. Any alcohol can be toxic if ingested in large enough quantities. In general, isopropanol, methanol, and ethylene glycol (the latter being a compound with two hydroxy groups linked to adjacent carbon atoms) have been traditionally referred to as "toxic alcohols" [2]. In normal speech, alcohol refers to ethanol (or ethyl alcohol: CH_3-CH_2OH), a molecule produced by many microorganisms via the so-called alcoholic fermentation of sugars. This process involves the decarboxylation (with concomitant release of CO_2) and reduction of pyruvate, the end product of glycolysis (Section 3.8.1):

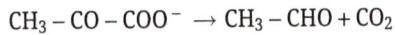

$$CH_3 - CO - COO^- \rightarrow CH_3 - CHO + CO_2$$
$$\textit{pyruvate decarboxylase}$$

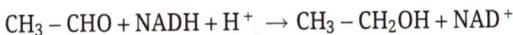

$$CH_3 - CHO + NADH + H^+ \rightarrow CH_3 - CH_2OH + NAD^+$$
$$\textit{alcohol dehydrogenase (ADH)}$$

The main ethanol-producing microorganism in food industry is the yeast *Saccharomyces cerevisiae*, a eukaryotic single-celled microorganism that belongs to the fungus kingdom. *S. cerevisiae* (and other yeasts) can produce ethanol from sugars contained in fruit juices, as well as cereal worts and flours. This is an essential process in the production of alcoholic beverages and bread, which have been part of the human diet for thousands of years. Alcoholic beverages such as sake, wine, beer, and whisky have been incorporated into the eating habits and, therefore, into the culture and social customs of many civilizations [3].

Part of the ingested ethanol does not enter the systemic circulation because it is oxidized in the stomach (the so-called first-pass metabolism), and the rest does not require digestion and is directly absorbed through the gastric and intestinal mucosae by passive diffusion down its concentration gradient. The rate of absorption primarily depends on the alcohol concentration of the beverage consumed and the presence of food in the stomach, which reduces absorption by retarding gastric emptying. Ethanol concentration in blood is determined by the balance between the rates of absorption and elimination (see below), while its distribution from the blood into all tissues is proportional to the relative water content of the latter because ethanol is extremely soluble in water and poorly soluble in lipids. An equilibrium is normally established between the ethanol concentration of the plasma and that of a given tissue, except in the liver, whose concentration is significantly higher than that measured from the peripheral blood. This is due to the fact that ethanol is directly transported from the stomach and small intestine via the portal vein [4, 5].

Liver eliminates more than 90% of ingested ethanol by oxidation, a reaction mostly catalyzed by alcohol dehydrogenases (ADHs), a family of proteins that use NAD^+ as cofactor and yield acetaldehyde and NADH. Other systems that can oxidize ethanol are the

https://doi.org/10.1515/9783111111872-020

microsomal ethanol oxidizing system (MEOS), which involves several members of the cytochrome P450 family, and the peroxisomal catalase, that uses H_2O_2 [6].

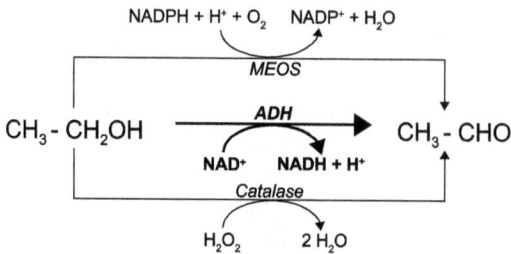

In a subsequent reaction, acetaldehyde is converted to acetate by aldehyde dehydrogenase (ALDH):

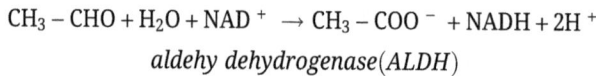

$$CH_3 - CHO + H_2O + NAD^+ \rightarrow CH_3 - COO^- + NADH + 2H^+$$
$$aldehy\ dehydrogenase(ALDH)$$

Acetate can be transported to peripheral tissues where it is activated to acetyl-CoA. This compound is also produced in the catabolism of carbohydrates, fatty acids, and proteins (Chapters 3–5), and is a central metabolite that has several destinations (oxidation to CO_2, or synthesis of fatty acids, ketone bodies, and cholesterol) depending on the energy state and the nutritional and hormonal conditions. Overall, the rate of alcohol elimination depends on a series of factors such as body size, percentage of body fat, liver mass, sex, age, race, food consumption, biological rhythms, physical activity, alcoholism, drugs, and even genetic factors [4, 6]. Ethanol has a caloric value (7 kcal/g) that is only lower than that of fats (9 kcal/g, Chapter 14); consequently, its consumption has to be considered when calculating dietary energy balances.[1]

Apart from its caloric content, ethanol is a sedative and mild anesthetic that produces a sense of well-being, relaxation, disinhibition, and euphoria. These effects are believed to be exerted by the activation of the pleasure or reward centers in the brain. It also triggers physiological changes such as flushing, sweating, tachycardia, and increases in blood pressure and urine production, the latter due to the osmotic effect of alcohol and inhibition of secretion of the antidiuretic hormone [5]. Excessive consumption can lead to a state of intoxication, which can be sporadic (acute poisoning causing damage to the central nervous system that may even lead to death) or as a result of a sustained intake over time (chronic poisonings). The latter can manifest after years and affect both the nervous system and the liver [3]. Many of the negative effects produced by alcohol in the body are attributed to the toxicity of acetaldehyde [4, 6].

---

1 Several calculators of the caloric content of different alcoholic beverages are available in the Internet. One of them can be found at: https://www.drinkaware.co.uk/understand-your-drinking/unit-calculator.

According to the World Health Organization: "The harmful use of alcohol is one of the leading risk factors for population health worldwide and has a direct impact on many health-related targets of the Sustainable Development Goals (SDGs), including those for maternal and child health, infectious diseases (HIV, viral hepatitis, tuberculosis), noncommunicable diseases and mental health, injuries and poisonings" [7].

Alcohol is used to enhance social interactions and its moderate consumption might have some positive effects; thus, healthy people who drink alcohol in moderation (two drinks/day for men and one drink/day for women[2]) have been reported to live longer and experience less cardiovascular disease than nonconsumers [8]. In any case, although it is clear that excessive alcohol drinking is harmful, there is a great controversy about its benefits when consumed in moderation. For some authors there is no safe level of consumption, whereas others defend that, as long as there is a positive assessment by a health professional, "low to moderate wine consumption with the meals (in a non binge-drinking pattern) has more health benefits than risks and can be part of a health lifestyle" [9].

References and further reading

[1] Nič M, Jirát J, Košata B, Jenkins A, McNaught A, editors. Alcohols IUPAC Compendium of Chemical Terminology. 2.1.0, Research Triangle Park. NC: IUPAC; 2009. https://doi.org/10.1351/goldbook. A00204

[2] Alcohol Toxicity: Practice Essentials, Pathophysiology, Epidemiology 2019. https://emedicine.med scape.com/article/812411-overview (accessed May 27, 2020).

[3] Fundación Alimentación Saludable: Su alimentación > El alcohol n.d. https://www.alimentacionsalud able.es/alim_alcohol.htm (accessed May 27, 2020).

[4] Cederbaum AI. Alcohol metabolism. Clin Liver Dis 2012;16:667–85. https://doi.org/10.1016/j.cld.2012. 08.002

[5] Paton A. Alcohol in the body. BMJ 2005;330:85–7. https://doi.org/10.1136/bmj.330.7482.85.

[6] Jiang Y, Zhang T, Kusumanchi P, Han S, Yang Z, Liangpunsakul S. Alcohol metabolizing enzymes, microsomal ethanol oxidizing system, Cytochrome P450 2E1, catalase, and aldehyde dehydrogenase in alcohol-associated liver disease. Biomedicines 2020;8. https://doi.org/10.3390/ biomedicines8030050.

[7] World Health Organization. World Health Organization, World Health Organization, Management of Substance Abuse Team. Global status report on alcohol and health. 2018.

[8] Pownall H, Rosales C, Gillard B, Gotto A. Alcohol: A nutrient with multiple salutary effects. Nutrients 2015;7:1992–2000. https://doi.org/10.3390/nu7031992.

[9] Fradera U, Stein-Hammer C. From scientific evidence to media and policy: Wine – part of a balanced diet or a health risk? Bio Web Conf 2019;15:04004. https://doi.org/10.1051/bioconf/20191504004.

---

2 A typical drink is 125 mL of wine with a 12% (v/v) content of ethanol. This contains 15 mL (12 g) of ethanol, which corresponds to 1.5 units of alcohol, generally defined as 10 mL (8 g) of pure ethanol (https://www.drinkaware.co.uk/facts/alcoholic-drinks-and-units/what-is-an-alcohol-unit).

Index

https://doi.org/10.1515/9783111111872-021